T0199922

Regardfully Yours
Selected Correspondence of
Ferdinand von Mueller

Life and Letters of Ferdinand von Mueller

PETER LANG
Bern · Berlin · Bruxelles · Frankfurt am Main · New York · Oxford · Wien

Regardfully Yours

Selected Correspondence of Ferdinand von Mueller

Volume III: 1876–1896

edited by
R. W. Home, A. M. Lucas, Sara Maroske,
D. M. Sinkora, J. H. Voigt and Monika Wells

PETER LANG
Bern · Berlin · Bruxelles · Frankfurt am Main · New York · Oxford · Wien

Bibliographic information published by Die Deutsche Bibliothek
Die Deutsche Bibliothek lists this publication in the Deutsche Nationalbibliografie; detailed bibliographic data is available on the Internet at ‹http://dnb.ddb.de›.

British Library and Library of Congress Cataloguing-in-Publication Data:
A catalogue record for this book is available from *The British Library*, Great Britain, and from *The Library of Congress*, USA

Cover design by Philipp Kirchner, Peter Lang AG

ISBN 3-906757-10-2
US-ISBN 0-8204-7559-9

© Peter Lang AG, International Academic Publishers, Bern 2006
Hochfeldstrasse 32, Postfach 746, CH-3000 Bern 9, Switzerland
info@peterlang.com, www.peterlang.com, www.peterlang.net

All rights reserved.
All parts of this publication are protected by copyright.
Any utilisation outside the strict limits of the copyright law, without the permission of the publisher, is forbidden and liable to prosecution.
This applies in particular to reproductions, translations, microfilming, and storage and processing in electronic retrieval systems.

Printed in Germany

An Herrn Baron

Sir Ferd. von Mueller K.C.M.G.

Sehr geehrter Herr Baron,

Hochgeschätzter Landsmann und Freund!

Ihre Landsleute in Melbourne können die Feier Ihres siebzigsten Geburtstages nicht vorübergehen lassen, ohne die Gefühle der Hochachtung und Verehrung, von denen sie für Sie beseelt sind, Ihnen kund zu geben.

Wenn auch Verdienste um die Wissenschaft nur von Berufsgenossen völlig gewürdigt werden können, so haben doch Ihre von reichem Erfolg begleiteten Forschungen auf neu erschlossenen Länder- und Wissens-Gebieten Ihren Namen weit hinaus über die eigentlichen Fachkreise getragen und ihn unter jene eingereiht, welche die Nation, als ihr zum Ruhm und zur Zierde gereichend, für sich in Anspruch nimmt. Fürsten, Universitäten und andere gelehrte Körperschaften haben die Bedeutung Ihres Schaffens erkannt und Sie mit Auszeichnungen überhäuft. Die australischen Colonieen aber sind Ihnen ganz besonderen Dank schuldig für Ihre rastlose und hingebende Thätigkeit in der Erforschung der heimischen Flora wie in der Einführung geeigneter Culturpflanzen aus anderen Erdtheilen, wodurch Sie einen wesentlichen und segensvollen Einfluss auf die Entwicklung dieser zukunftsreichen Gemeinwesen gewonnen haben.

Dass Sie neben unermüdlichen wissenschaftlichen Arbeiten Zeit und Musse für die Förderung der schönen Künste und zahlreicher Werke der Barmherzigkeit gefunden haben, darf nicht mit Stillschweigen übergangen werden. Für Ihre unerschütterliche Anhänglichkeit aber an unser gemeinsames Vaterland und die deutsche Sache sagen wir Ihnen unseren aufrichtigen und tiefempfundenen Dank.

Am heutigen Festtage wünschen wir von ganzem Herzen, dass die Vorsehung Ihnen noch eine lange Reihe durch Gesundheit, Freundschaft und wissenschaftliche Erfolge verschönte Jahre gewähren möge: der Menschheit zum Nutzen, dem deutschen Namen zur Ehre und Ihren australischen Landsleuten zu freudiger Genugthuung.

Der Deutsche Verein von Victoria,

J. G. Luchmann, Präsident — D. Diercks, Secretair

Melbourne 1ten Juli 1895

Frontispiece. Illuminated manuscript commemorating Mueller's 70th birthday (30 June), presented to him by the Deutsche Verein, Melbourne, on 1 July 1895. Courtesy of the Library, Royal Botanic Gardens Melbourne.

Table of Contents

Introduction

Ferdinand von Mueller was fifty years old at the commencement of the period covered by this third and final volume of his selected correspondence. In the twenty-eight years since he had arrived in Australia as an enthusiastic young botanist, recently graduated from university and determined to make his mark in science, he had achieved an astonishing amount. His most cherished ambitions had, however, turned to ashes. Several attempts to marry and to sire children who would carry on his name had come to nothing, and as he grew older he led an increasingly lonely existence, despite the high social standing he enjoyed as a result of the hereditary barony and other honours that had come his way. Authorship of the comprehensive flora of Australia that he had dreamed of writing had been given to George Bentham instead, while Mueller had been pressed into assisting Bentham by allowing him access to his herbarium collections. Worst of all, in May 1873 the directorship of the Melbourne Botanic Garden, to the development of which Mueller had devoted most of his energies for the previous sixteen years, had been taken from him amidst widespread criticism of his administration. To be sure, thus freed, as those responsible for his losing the position proclaimed, from the 'duties of minor importance' with which he had previously been encumbered, he was able thereafter to devote his time more exclusively to his scientific work, with the result that his already impressive rate of scientific publication increased significantly during the last twenty years of his life.[1] Yet to Mueller, the loss was a body-blow, both deeply humiliating and disruptive, he said, of the extensive programme of applied botanical research he had been pursuing in the Garden side-by-side with that in his rapidly expanding herbarium of dried plants. Too proud to venture ever again into the Garden that had

1 See graph in Churchill et al. (1978), p. 2. In the period 1856-75, M's publication rate averaged around 29 titles per year, whereas in the period 1876-95 the rate was around 45 titles per year.

been taken from him, he would henceforth have to confine himself largely to herbarium-based research.

As a result of that research, however, and the publications that flowed from it in an unceasing stream, his scientific standing continued to grow and honours continued to be showered upon him. By the time he died on 10 October 1896, he may have had more publications to his credit than any scientist had ever had. His public standing in Victoria also grew, perhaps partly as a result of his no longer being a centre of controversy but also no doubt due to a widening recognition of his scientific achievements. The various civil awards he had received were international acknowledgement of those achievements, and at the same time they themselves added to his social status locally: indeed, as Vance Palmer noted, Mueller became 'a great functionary – the second citizen in Victoria, while governors came and went'.[2] His funeral was a major public occasion. Afterwards two separate public appeals were conducted to memorialize him, one leading to the construction of a large monument on his grave, the other endowing the Mueller Medal awarded by the Australasian Association for the Advancement of Science (later ANZAAS) for 'the most important contribution, or series of contributions, to natural knowledge, published originally within His Majesty's dominions within a period of not more than five, nor less than one year, of the date of the award, preference always given to work having special reference to Australasia'.[3]

Government Botanist

When he was ousted from the Garden, Mueller felt betrayed by the leaders of British science, especially Joseph Hooker and Charles Darwin. For years afterwards he complained over and over in his letters that, if only the British scientific community had supported

2 Palmer (1940), p. 98.

3 See pp. 759-68 below and *Report of the ninth meeting of the Australasian Association for the Advancement of Science, Hobart, 1902*, p. xlvii.

him better in his hour of need, in the way it had rallied behind Hooker when his position was threatened, he could have fought off his enemies. Perhaps Mueller was right in this but it seems unlikely, since it is clear that those enemies had been working towards his dismissal for years and were determined to see him go, once the political pendulum swung in their favour. Later, as Mueller manoeuvered to recapture the Garden, he was dismayed by reports that Hooker and others had expressed the view that he was better off without it. Even when made in private conversation, such comments, Mueller cried, would be used against him. For several years, in fact, Mueller remained hopeful that the Garden might be restored to him and lobbied vigorously for this whenever an opportunity presented itself. His best chance came when a parliamentary board of inquiry was set up in 1877 to review his situation. Even though the board was favourably disposed towards him, however, and fully accepted his argument that 'his proper place is in the midst of his living plants which he needs for daily study', it did not propose putting him in charge of the Garden again. With Mueller's replacement, William Guilfoyle, already beginning to transform the Garden into the place of wondrous beauty it was to become under his administration, Mueller's time there had clearly passed. Instead, the board recommended that several acres of ground be allocated to him, close to the city and the existing botanic garden, for a new 'Scientific Botanical Garden' – the suggested name implied a recognition that the existing Garden could no longer in any meaningful sense be regarded as scientific – in which he could pursue his research.[4] No definite recommendation was made as to a site but several possibilities were noted, including the land immediately surrounding Mueller's existing Herbarium building.

The board's proposal would have done nothing to restore Mueller's wounded pride but it would undoubtedly have facilitated his research in applied botany, especially if the further recommendations had also been adopted that an area of State Forest be put at his disposal for test plantings, and that he be given additional staff, a

4 L. Smith to G. Berry, 11 July 1877; pp. 745-8 below.

larger building to house the Herbarium, and a new laboratory to replace the one he had had in the Garden but had lost. Whatever hopes the report aroused were, however, quickly dashed. He was offered ten acres of ground in the Domain, in the vicinity of the Herbarium, but rejected this as inadequate.[5] Funds were also allocated for extending the Botanical Museum, the building that housed the Herbarium, but these were never spent and did not reappear in the following year's budget. None of the board's other recommendations were taken up, and the report was simply filed away. Not until many years after Mueller's death did the Victorian Government again move to provide better facilities for its Government Botanist, or better accommodation for the vast herbarium Mueller had assembled in its name (and even then it required the prompting of a large private benefaction). In Mueller's lifetime, the only addition to the land or buildings under his control was a galvanized iron extension to the original Herbarium building that was erected in 1883 to house Wilhelm Sonder's vast collection after this was purchased from Sonder's widow.

In addition to making recommendations about resources, the 1877 board of inquiry addressed another matter on which Mueller held passionate views that had arisen as a result of his being dismissed from the Botanic Garden. This concerned his standing as Government Botanist vis-à-vis Guilfoyle's as curator of the Garden. Not surprisingly, many botanists and officials around the world did not understand that there were two positions involved, and from time to time botanical inquiries were directed to Guilfoyle that Mueller, who scornfully dismissed Guilfoyle as a mere gardener without any scientific standing or expertise, thought should have gone to him. Mueller learned of a case of this happening at the time the board was sitting – the guilty party on this occasion being Kew Gardens where people should certainly have been better informed – and this no doubt prompted him to bring the problem to the board's attention. The board fully supported Mueller's point of view, recommending that 'all communications bearing upon the science of Botany should be addressed to the Government Botanist,

5 M to J. von Haast, 25 December 1879.

and that any not so addressed, should be transferred to him to be dealt with'. The sting, of course, was in the final clause. Mueller, however, did not wait for instructions to be issued but took action himself, whenever the opportunity presented itself, to set his correspondents right on the matter.[6]

Though ousted from the Botanic Garden, Mueller remained, as Victoria's Government Botanist, head of a (very small) government agency. From the date of its separation from the Garden, this reverted to the Chief Secretary's Department while the Garden remained in the Department of Agriculture (at that time a subdepartment within the Lands Department). For a time, Mueller was subjected to constant harassment by his government masters in an apparent effort to persuade him to resign. This eventually ceased, however, and thereafter he was left in relative peace. Even though several of the recommendations of the 1877 board of inquiry were ignored, the terms in which the board was set up, 'to inquire into the present position of Dr. Mueller, in relation to his professional duties, with the view to advise, what alteration, if any is necessary, to afford him reasonable facilities for the due discharge of his scientific labours', suggest a much more friendly attitude towards him than had previously prevailed. So, too, does the fact that he was offered ground for planting and funds were allocated to extend the Museum building. It was Mueller's political naiveté, not government neglect, that led to the money for the building not being spent. The sum allocated was only half what was required. Whereas any normal bureaucrat faced with this situation would have made sure the funds allocated were committed and then sought to have the remaining amount allocated in the following year, Mueller opted not to spend any of what was granted to him and then to request, instead, twice as much in the following year! It is scarcely surprising that the strategy failed, or that his later attempt at least to retrieve the £900 that had gone back into government coffers was likewise turned down.[7]

6 M to J. Hooker, 7 July 1877.

7 M to J. Patterson, 23 April 1880; also M to C. Pearson, 15 August 1878 [Collected Correspondence].

One might similarly query Mueller's decision to reject the ground for planting that he was offered. He did so, he said, 'because I planted annually at the rate of 30 acres and they would give me almost no means of production at all', which suggests that he was unwilling to do anything less than return to the scale of operations on which he had been able to work when he had control of the Garden. A more sensible strategy might once again have been to accept what he was offered and then to work later towards having the allocation increased.

A further opportunity to have the long-planned extension to the Museum erected arose in 1891 but once again Mueller allowed the opportunity to pass, acceding all too easily to the suggestion that construction be delayed until the following year – even though, as he explained, this would entail putting the herbarium collections at risk because it meant that fumigating them with carbon bisulphide would have to be deferred.[8] Soon afterwards, the Victorian economy collapsed and all hope of having the extension built within Mueller's lifetime disappeared.

These are not the only examples of Mueller's naiveté – some would say incompetence – as a bureaucrat. A further remarkable instance may be seen in his comment to the then Chief Secretary, Bryan O'Loghlen, in 1879, two years after the parliamentary board of inquiry into his position had reported to O'Loghlen's predecessor, that he had never seen the board's report nor sought access to it, 'being assured, that what the Committee recommended would be reasonable and just'.[9] If true, he was surely being extraordinarily naive; but in any case, why should he not have sought access to the report, since there was no reason why it should have been kept from him? Again, Mueller regularly spent a large fraction of his own salary in propping up the budget of his department, in the unrealistic expectation that when the powers-that-be saw him doing so, they would somehow be persuaded to take up the shortfall. Conversely, he regarded himself as personally liable for any over-

8 M to T. Wilson, 12 March 1891.
9 M to B. O'Loghlen, 18 May 1879.

spending by his department[10] – a view that could only be maintained, even for a moment, because his department's budget was so small. Equally unrealistically, he proposed that, in calculating the pension entitlements of retiring employees, time spent on his personal payroll should be included in the computation.[11]

Nevertheless, Mueller was treated sympathetically by his chiefs throughout the 1880s. Ministers such as Graham Berry and Alfred Deakin to whom he was answerable during this period clearly recognised his scientific standing and valued the work he did, and were prepared to tolerate his occasional idiosyncrasies. Having the funds allocated to purchase Sonder's herbarium was a major coup for him, and he was also allowed to manipulate his budget to keep a couple of junior assistants on the payroll, even when there were no established positions for them in his department. In addition, significant sums were made available almost every year to support Mueller's publishing programme. Some of his works, such as the successive editions of *Select extra-tropical plants*, sold well, and in these cases the government soon got its money back. In other cases, however, such as the monographic works that involved the preparation of large numbers of expensive lithographic plates, there was no prospect of recouping the money from sales. These works thus represented a substantial investment by the Victorian Government in basic science.

Mueller's book, *Select extra-tropical plants, readily eligible for industrial culture or naturalisation*, was first published in book form in 1876 under a slightly different title, *Select plants readily eligible for industrial culture or naturalisation in Victoria*, that more accurately indicated its original intention.[12] The demand for successive editions of the work and their publication by the Victorian Government attest to the value placed by his fellow-colonists on this aspect of Mueller's work. The book distilled a vast amount of information on a subject close to the heart of every colonist, namely

10 M to G. Berry, 28 June 1883.

11 M to G. Berry, 7 February 1884.

12 B76.13.03. The work grew out of separate catalogues of timber trees and other kinds of plants compiled by M some years earlier; see B71.13.01, B72.13.02 and B74.13.06.

how best to extract value from an environment ripe for exploitation but seemingly deficient in native plants of economic value. Mueller was the acknowledged expert to whom people turned for advice and he shared his knowledge unstintingly, both directly and by means of his book. This, he recognised, was what he was principally employed to do, the chief basis for whatever political support he enjoyed. Moreover, to help in this way in the development of the colony was for him a moral obligation that he felt he owed to the British Queen whose loyal subject he had become when he was naturalized.

From Mueller's correspondence we learn that he not only wrote about the acclimatization of plants, he remained directly involved in transferring plants around the globe. In particular, he continued to supply on request large quantities of seed of his favourite species of *Eucalyptus* and *Acacia* for planting in other parts of the world, and developed a new enthusiasm for exporting species of saltbush from central Australia for planting in desert environments elsewhere, such as in southern California and Algeria. He also continued to advocate experimental plantings in Victoria of economically important species such as tea and cinchona, and to provide advice and assistance to landholders willing to become involved.[13]

For the most part, preparing works for publication, while an onerous task, accounted for only a small fraction of Mueller's daily workload. Indeed, he appears to have done most of what he called his 'literary work' – that is, his writing for publication – as well as most of his letter-writing outside ordinary office hours. An exception was his dichotomous *Key to the system of Victorian plants*, which proved far more difficult to write than he (or presumably those who pressed him into doing it) had envisaged. It seems that the *Key*, in which he tried to balance the artificiality of the dichotomous approach with the subtleties of the natural system of classification, became, for several years in the mid-1880s, Mueller's chief preoccupation. He came under considerable pressure to finish the work, with questions even being asked in Parliament about its progress. It is a constant refrain in his letters from that period

13 See, for example, G. Robinson to M, 16 August 1877.

that work on the *Key* had put him well behind with all his other work.

A major responsibility that clearly usually took up much more of both his time and that of his assistants was to provide a plant identification service for the Victorian public, especially the colony's farmers. This would have been mostly routine work of which few records survive beyond Mueller's occasional references to it in his correspondence, and no doubt much of it was done by Mueller's assistants under his general supervision, not by him personally. For the most part, the surviving record relates to less humdrum cases that he dealt with himself, most notably instances in which specimens sent for identification proved to represent new taxa – the record survives in these cases because it was Mueller's practice to add such specimens to his herbarium and to file the relevant correspondence with them – or cases where poisoning was suspected, whether of stock or people, in which he was asked to identify the source. Less straightforward tasks of this kind came to him not just from around Victoria but also from the other Australian colonies and New Zealand: as noted in the introduction to Volume 2,[14] Mueller functioned in this way as a national resource, long before the Australian federation came into being in 1901. When the staff at the Herbarium was drastically cut back in the 1890s, the plant identification service became an oppressive burden for Mueller personally, the more so as he seems to have felt a deep commitment to helping the colony's rural community through the economic crisis of those years.[15]

It was not only the general public who looked to Mueller for assistance and advice. So, too, as the letters published in this volume reveal, did professional botanists in both Australia and New Zealand. Mueller identified plants for them, encouraged their publishing ambitions and urged them to follow his views on systematics. He expected them in return to acknowledge his standing as the primary reference point for the study of Australian plants. He became outraged if other Australian workers tried to establish their own, in-

14 *Selected correspondence*, vol. 2, p. 14.
15 M to A. Engler, 31 January 1893 [Collected Correspondence].

dependent lines of communication with leading overseas botanists. If Australian materials were to be sent to overseas experts for analysis, he maintained, they should be channelled through him so as to avoid the confusion that would arise from duplicated sendings.

The herbarium collections themselves as they continued to grow generated more and more work, simply to maintain them, while the cramped conditions in which they were stored made it increasingly difficult to pursue research efficiently, or even to keep track of materials properly. Probably only Mueller's extraordinary memory enabled him to continue functioning effectively. The richness of the collections could not be disputed, however, and Mueller was understandably hurt when they were overlooked by Alphonse de Candolle in a survey of the world's leading herbaria. His indignant letter of complaint to de Candolle,[16] in which he spelled out in considerable detail what he had assembled, left no doubt that under his administration the Melbourne Herbarium had become one of the world's great collections, and pre-eminent in its holdings of Australian plants, even before Sonder's vast collection, rich in Australian types, was added to it.

An activity to which Mueller continued to devote a great deal of energy until his final years was preparing displays for the Victorian stands at the seemingly endless round of intercolonial and international exhibitions held in Australia and elsewhere around the world, throughout the period in question. The rationale for these exhibitions was always to boost trade, each territory's exhibit being designed to show off its produce – whether raw materials or manufactured goods – and its potential for development to best advantage.[17] Mueller understood this perfectly and designed his contributions to Victoria's displays to show off a range of the colony's plant-based products or potential products. He did so with considerable flair, even though the products exhibited varied little over the years, and his exhibits continued to be awarded trophies and commendations. Typically, his display would include a range of dressed timber specimens prepared from Australian trees,

16 M to A. de Candolle, 4 August 1880.

17 See Greenhalgh (1988).

distillation products – especially eucalyptus oil – from Australian plants, and the set of plaster models he had had prepared at an early stage of fruits grown in Victorian orchards. An innovation of which he was particularly proud was to have timber specimens prepared in the shape of books which could then be displayed as if they formed a library.[18] For Australian audiences, he also included in his displays educational materials intended to enhance understanding of the Australian flora. At the Adelaide Jubilee International Exhibition in 1887, for example, he included in his display one of the so-called 'educational collections' of dried specimens of Victorian plants that he had assembled some years earlier for distribution to Victorian educational institutions,[19] albums of dried ferns and grasses, and a 'photograph view of trees', all of which received certificates of commendation.[20] During the period of this volume, two major international exhibitions were held in Melbourne, in 1880-81 and 1888, and these generated even more work for Mueller because he was appointed to several of the juries set up to oversee and assess the different exhibits.

As Government Botanist, Mueller was also called on from time to time for official advice on botanical matters including, until the separate position of Plant Pathologist was created within the Department of Agriculture in 1890, the diseases of plants. As the vineyards of Europe were devastated by infestations of Phylloxera in the 1870s, Mueller passed on warnings to his chiefs. Later, when it became clear that the warnings had come too late and that Victoria's vines were already catastrophically infested, he provided advice on means of combating the scourge, becoming an early advocate of what proved to be the best long-term solution, the wholesale re-planting of vineyards with root stock of Phylloxera-resistant American species raised from 'well sifted seeds' – to prevent the inadvertent importation of Phylloxera eggs if cuttings were imported – on to which European varieties could be grafted above

18 M to J. Hooker, 1 October 1882.

19 See Maroske (1995).

20 RB MSS M200, Library, Royal Botanic Gardens Melbourne.

ground level.[21] He resisted suggestions, however, that he should go to Europe in 1881 to represent the Australian colonies at an international conference on the Phylloxera problem that was to be held at Bordeaux, proposing that Joseph Hooker be asked to represent the colonies instead.[22]

The letters published in this volume show other instances of Mueller's being called on for expert advice. He was, for example, a member of the commission of inquiry established in 1878 in response to concerns about the supply of suitable barks for the tanning industry as readily available sources were worked out,[23] while in 1881 his advice was sought about suitable plantings around the Yan Yean reservoir that supplied Melbourne's water, after concerns arose about the quality of the supply.[24] He provided advice to local government authorities as well, for example to those along the west coast of Victoria about plantings of Marram Grass (*Ammophila arenaria*) and other species that successfully stabilized the sand dunes that were rapidly over-running large areas of farmland.[25]

Even before his appointment in 1871-2 to a government-appointed Royal Commission on Foreign Industries and Forests, Mueller had become an outspoken advocate of measures to protect Victoria's forests, which he saw were being devastated under the impact of mining and closer agricultural settlement. In the years that followed, he seized every opportunity to urge upon the Victorian Government the desirability of creating a Forests Department to conserve and manage what remained. One such occasion came in 1887 when he was called as an expert witness before the long-running Royal Commission on Vegetable Products, but his advice was ignored. The Commission focused instead on promoting dairy farming, a policy that quickly (albeit inadvertently) led to the destruction of vast areas of the colony's high-rainfall for-

21 M to J. Casey, 17 May 1873 [vol. 2, pp. 664-6]; M to T. Wilson, 16 June 1881, and M to J. Pescott, 17 August 1895 [both Collected Correspondence].

22 M to J. Hooker, 22 July 1881; also M to S. Wilson, 16 June 1881 [Collected Correspondence].

23 M to J. Forrest, 14 February 1878; see also B78.14.01.

24 M to G. Langridge, 21 March 1881.

25 Heathcote & Maroske (1996).

ests.[26] Mueller was also an early advocate of the setting aside of reserves to protect areas of botanical significance such as Wilson's Promontory and the valleys in East Gippsland that were home to the cabbage palm groves that he so greatly admired. He supported the efforts of his friend Friedrich Krichauff that led to the creation of a Forests Department in South Australia in 1875, the first such department to be formed in any of the Australian colonies, and also similar efforts being made by others in New Zealand. Mueller himself was engaged by the Western Australian Government to prepare a report on the forest resources of that colony.[27] Not until 1888, however, did the Victorian Government make its first move to protect the colony's forests, when George Perrin was appointed as the colony's first forests officer. When Perrin needed taxonomic advice, Mueller provided it.[28]

In the early 1880s, new regulations were introduced governing the Victorian public service that had significant implications for Mueller. They were administered by the Public Service Board, a powerful new body established by Act of Parliament in 1881 with the aim of doing away with the nepotism and jobbery that had previously determined too many appointments within the service. Among other things, the new regulations fixed a standard retiring age of 60 for Victoria's public servants (later raised to 65). Mueller, with his life still focused wholly on his science and unfettered access to his herbarium being crucial to the science that he did, had no desire to retire; moreover, he claimed that he could no longer afford to do so after having for so many years poured so much from his own resources into the work of his department. At first both Mueller and his then Minister, the Chief Secretary Graham Berry, appear to have treated the regulation governing retirement lightly and to have assumed that it could be easily over-ridden. Mueller sent a brief note to Berry on 27 June 1885, only three days before his 60th birthday, requesting that he be allowed to continue in his position, and Berry then formally recommended to the Governor-

26 Gillbank (1993); Frost & Harvey (1997).

27 B79.13.10.

28 See G. Perrin to M, 10 September 1892.

in-Council a week later – that is, after Mueller's official retirement date had already passed – that Mueller 'continue to perform his duties, until it shall be further ordered'. Both men clearly assumed that that would be all that was needed for Mueller to be allowed to retain his position indefinitely. They had not reckoned on the Public Service Board which promptly flexed its muscles, declaring that in its opinion, 'the retention in the service of any officer after he has attained the age of sixty years must be determined not upon individual claims but solely upon public grounds'. Berry, humbled, had to write again, setting out specific grounds as to why it was in the public interest to retain Mueller's services, before the Public Service Board would agree to his recommendation. Even then, the Board refused to allow the open-ended extension of Mueller's appointment that Berry had suggested, but agreed only to extend it for twelve months.[29] Thereafter, Mueller had to seek a further extension of his appointment each year and later, each six months. He thus lost all security of tenure and became vulnerable in a way he had not previously been to the whims of the Government.

As he grew older, Mueller's ambitions for his department clearly declined, and in his later years it seems that all he wanted was to be left undisturbed so that he could get on quietly with his work. Opportunities to re-visit Europe that in earlier and more confident times he might have seized, he now allowed to pass by, despite the urgings of his friends[30] and his own often-stated desire to make the journey. Most striking of all is his response in 1890 to a proposal that his department should be enlarged by transferring to it the Government's recently appointed plant pathologist, Daniel McAlpine. All that was required was for Mueller to make a formal request for an assistant in vegetable pathology and the Minister, Alfred Deakin, anxious to establish McAlpine's position on a satisfactory basis, would have obliged. There was every prospect, McAlpine assured Mueller, that he would bring with him a laboratory, a museum and a specialist library. With Deakin's support already secured, here was a chance to restore the fortunes of Mueller's

29 See M to G. Berry, 27 June 1885 and the notes thereto.
30 E. Hilgard to M, 16 October 1893.

department. Mueller, however, seems to have viewed the proposal as a threat rather than an opportunity and his response was wholly negative, dwelling on perceived difficulties where none existed and the insufficiency of the resources of his department to support McAlpine's work when new resources were already promised. McAlpine's position should be established, Mueller said, not within the Government Botanist's Department but within the Department of Agriculture. He pleaded with Deakin 'not to enlarge my establishment and burden additional obligations on it' at this 'very late time' in his life; all he wanted was to 'quietly continue and finish that particular work, which I have laid out for myself during the remainder of my earthly career, while strength still may last me'.[31] Deakin, tolerant as ever of Mueller's foibles, did not insist, and McAlpine was placed in the Department of Agriculture instead.

When the Victorian economy collapsed in the early 1890s, the Government was forced to cut costs savagely, wherever it could. Given Mueller's lack of tenure, he now looked an easy target, and in January 1892 the decision was taken to abolish the office of the Government Botanist. Mueller, resigned to his fate, began to make arrangements for the break-up of his department. His prime concern was naturally the fate of the great herbarium he had assembled. He recommended that this should be transferred to the control of the trustees of the Public Library and Museum in preference to several other possibilities that had evidently been canvassed, and offered his services thereafter on an honorary basis as its custodian.[32] Mueller was also concerned for his staff and sought to protect his chief assistant, Georg Luehmann, in particular.[33] His political masters had reckoned without the local press, however, most of which, including Mueller's old tormentor, the *Argus*, came out fighting vigorously on his behalf, to such good effect that, before the month was out, the Government retreated.[34] While Mueller's

31 D. McAlpine to M, 18 July 1890; M to A. Deakin, 19 July 1890.

32 M to T. Wilson, 16 January 1892.

33 M to A. McLean, 17 January 1892.

34 M to T. Wilson, 23 January 1892 and 31 January 1892 [both Collected Correspondence]. See also *Argus*, 23 and 25 January 1892; *Daily telegraph*, 26 January 1892; and *Evening standard*, 1 February 1892.

department was still abolished, his position was maintained, albeit with a working budget reduced to a derisory £400, only one quarter of the already pitifully small allocation he had had before. This was sufficient to keep Luehmann and a junior, James Tovey, on his payroll, but alternative arrangements had to be made for his other assistants, James Minchin, Charles French Jr and Gerhard Renner. Renner soon moved to a vacant position in another department, but with the evident connivance of senior officers within the Chief Secretary's Department, Mueller managed to keep Minchin and French with him for many months, paid for out of supplementary allocations while they waited, so it was said, for suitable vacancies to arise elsewhere.[35] Eventually, however, in January 1895, the Public Service Board became aware that the two were still working for Mueller even though they had been officially declared to be 'in excess', and immediately transferred them to the Department of Agriculture despite Mueller's protests that they were needed at the Herbarium.[36] Minchin was transferred back to the Herbarium three months later but French's position was lost for good. In the new financial year that commenced in July that year, Mueller's branch lost its separate budget line within the Chief Secretary's Department.[37] Thereafter, for the remainder of Mueller's life and indeed for many years beyond it, the office of the Victorian Government Botanist limped along at the level to which it had now shrunk, able to do little more than conserve the great herbarium Mueller had assembled and maintain the plant identification service.[38]

35 M to T. Wilson, 23 March 1892; M to T. Wilson, 25 August 1892; M to A. McLean, 26 September 1892 [all Collected Correspondence].

36 M to F. Reddin, 28 January 1895.

37 C. Topp to M, 28 June 1895.

38 The period 1905-19, when Alfred Ewart was Government Botanist, is a partial exception to this generalization, for Ewart maintained a very active research programme, much of it taxonomic work on aspects of the Australian flora. However, Ewart was also professor of botany at the University of Melbourne and it could be argued that much of his research output should be attributed to his university appointment. When his position at the University became full-time and he resigned as Government Botanist, research at the Herbarium effectively ceased. See Cohn (forthcoming, 2005).

'Literary work'

When Mueller was dismissed from the Botanic Garden, those responsible claimed that their intention was to free him to pursue his scientific researches without interruption from 'duties of minor importance'. Not for a moment did Mueller accept this argument. Nevertheless, in the years following his dismissal, botany was his chief consolation and new scientific publications poured from his pen at an astonishing rate.

By far the largest number were devoted to formal descriptions of newly identified plant taxa. In most cases, the specimens on which these descriptions were based no longer derived from Mueller's own fieldwork, as had been the case in earlier years. Now, they had been collected by others – whether members of exploring expeditions, professional collectors employed by Mueller, fellow botanists who sought his opinion on taxonomic matters, or private correspondents who sent materials to him for identification. The letters published below richly document the wide range of sources upon which Mueller drew. For some years, his principal outlet for publishing descriptions of new taxa remained his *Fragmenta phytographiae australiae*, new issues of which continued to appear regularly until the early 1880s. This long-running publication ceased, however, with the appearance of the 94th part in December 1882. Mueller clearly intended the series to continue, since the issue in question was identified as the first part of a new volume, Volume 12, and even as late as 1886 he was apparently still hoping to complete this.[39] However, the launching of new Melbourne-based scientific journals in which he could publish his descriptions – the short-lived *Southern science record* (1880-85) and *Australasian scientific magazine* (1885) and the much more long-lived (and indeed still running) *Victorian naturalist* (commenced 1884) – meant that it was no longer so important to him to keep the *Fragmenta* going, and in the event he allowed it to lapse.

At the commencement of our period, George Bentham's monumental *Flora australiensis* was not yet finished and Mueller was

39 M to W. Thiselton-Dyer, 17 February 1886.

committed to lending him, in order, the remaining sections of his ever-growing herbarium of Australian plants. There was a long gap between the appearance of Volume 6, in 1873, and the publication of the seventh and final volume in 1878, caused partly by the disruptions to Mueller's life consequent upon his dismissal from the Botanic Garden that for a time made it difficult for him to assemble the materials Bentham needed. The seven volumes covered only phanerogams and ferns and Bentham made it clear that he had no intention of extending the work to take in the cryptogams as well. Mueller was, however, keen to see them included. Recognising that he did not have sufficient expertise to take them on himself, he tried to recruit an international team of specialists to do the job under his co-ordination. Various experts compiled lists of the different species of Australian cryptogams with locations where species had been collected and Mueller published these as supplements to Volume 11 of his *Fragmenta*, but this was as far as the project went.

Another vexing question was the preparation of a supplementary volume to the *Flora*. Both Bentham and Mueller assumed from the outset that a supplement would be required that incorporated new species that had later been discovered and new information on geographical distribution, as well as ordinary corrections. With this in mind, Mueller sent Bentham corrigenda to the proofs of the different volumes as he received them, even though he knew the corrections could not reach Bentham in time for them to be incorporated in the relevant volume. Bentham, Mueller assumed, would want to publish all such corrections in the proposed supplement or supplements. Bentham quickly decided, however, not to publish supplements to earlier volumes in later volumes in the series, but instead to accumulate all the supplementary material in a separate volume to be published at the end of the project. Then, as the seventh volume was being printed, he asked Mueller to take over responsibility for the supplementary one.[40] During the next few years, Mueller from time to time reported progress on the supplement, but his enthusiasm waned as he took on other projects

40 G. Bentham to M, 27 June 1877.

that to some extent could be seen as providing an alternative to it. In particular, his *Systematic census of Australian plants* (1882, 2nd edition 1889)[41] included references to the descriptions of species published since the appearance of the relevant volume in Bentham's series, enabling them to be tracked down readily enough, while his great monograph on the eucalypts, *Eucalyptographia* (1879-84), directly addressed what Bentham himself had regarded as one of the weakest sections of his own work, on which he had urged Mueller to undertake a more comprehensive study.[42]

The question of a supplementary volume was revived in the 1890s when Queensland's Government Botanist, F. M. Bailey, announced without consulting Mueller that he intended to publish such a volume. Mueller was outraged since it seemed that all Bailey proposed to do was re-publish the descriptions of new Australian taxa that had been published since Bentham's work appeared, most of which were his! In some of the letters selected for the present volume, we see Mueller protesting vigorously against what he saw as an attempted theft of his intellectual property, and rallying other leading botanists in support of his right to issue the supplementary volume himself. Bailey, after some posturing that he would only withdraw if Mueller agreed to follow Bentham's taxonomic order slavishly and not introduce the modifications he had been advocating for several years, was forced to yield.[43] Mueller, it seems, was prompted to resume work on the supplement but it remained unpublished at the time of his death. Though the executors of his estate announced that they would be publishing it, they never in fact did so.

Bailey's anxiety about Mueller's taxonomic innovations arose principally from Mueller's persistently advocating doing away altogether with the Monochlamydeae, a very mixed order of plants in the classificatory system of Jussieu and de Candolle that was regarded widely and not just by Mueller as unsatisfactory, and re-

41 B82.13.16, B89.13.12.

42 B79.13.11, B80.13.14, B82.13.17, B83.13.07, B84.13.19. See also G. Bentham to M, 18 October 1865 [*Selected correspondence*, vol. 2, pp. 326-8].

43 W. Thiselton-Dyer to M, 20 May 1893; M to F. Bailey, 31 May 1893; and F. Bailey to M, 3 June 1893. See also Clements (1998) and Lucas (2003).

distributing the constituent elements elsewhere within the sequence of orders. Mueller is seen promoting this change in a number of the letters published below, but unfortunately he never published a fully-argued rationale for adopting his view. Instead, he relied on a brief statement of his position in the introduction to his *Systematic census of Australian plants* – a work in which he then carried his proposal into effect in the order in which he treated the different groups of plants – and otherwise on his correspondence, to persuade individuals of the merits of his views. Twenty years earlier, he had adopted the same approach in relation to Darwin's theory of the origin of species, in that case publishing a brief statement of his views in the introduction to his account of the flora of the Chatham Islands but otherwise again relying on his correspondence. It is hardly surprising that Mueller's opinions had little impact in either case, because in science ideas become widely accepted not through private exchanges of views among a few individuals but as a result of convincing arguments in support of them being published and thereby rendered open to public scrutiny and debate. Mueller seems never to have understood this or, if he did, to have been prepared to expose himself to challenge in this way. As a result, his high standing as a botanist rested solely on his massive contributions to the delineation of the Australian flora and not at all on any contributions of his to botanical theory.

One matter that Mueller did argue in public as well as in his correspondence was the absolute precedence that he thought should be given to priority in the naming of species. This led him, in his *Systematic census of Australian plants*, in a significant number of cases to abandon generally accepted names in favour of names which, while published first, had since been neglected. For this he was heavily criticized by Bentham and others at Kew who were willing under certain circumstances to abandon strict priority, and who dismissed labouring over priorities as useless philological work.[44] Mueller, however, was unrepentant and his views eventually prevailed in the international code of nomenclature.[45]

44 See G. Bentham to M, 25 April 1883.

45 Stevens (1991); Stevens (1997).

The appearance in 1879 of the first Decades of Mueller's *Eucalyptographia* represented a new style of publication so far as he was concerned, an authoritative monograph covering a major genus of Australian plants, in which each species discussed was illustrated by a splendid lithograph prepared by one of the excellent Melbourne-based botanical artists to whom Mueller extended his patronage. The preparation of the work was accompanied by an upsurge of related correspondence. During the 1880s and early 1890s the work was followed by illustrated volumes of the Acacias and cognate genera, the Myoporineae, the Salsolaceae and – until the collapse of Mueller's budget in 1892 halted publication with only one Decade in print – the Candolleaceae. These later series, however, lacked the extended discussion of each species that was a feature of the work on the Eucalypts. Mueller's correspondence also reveals him encouraging others to undertake definitive studies of aspects of the Australian flora over which he ceded authority. These included the English mycologist Mordecai Cooke, who at Mueller's urging undertook a study of the Australian fungi, and the Swedish botanist Otto Nordstedt whose never-completed monograph on the Australian Characeae Mueller subsidized out of the resources of his department until the collapse of his budget made this no longer possible.

From his correspondence, we learn that Mueller envisaged other monographic publications as well, committing himself to contributing the sections on some of his favourite families of plants to both of the great international taxonomic compendia then in progress, the continuation of de Candolle's famous *Prodromus systematis naturalis regni vegetabilis* being edited by Alphonse de Candolle and his son Casimir, and the ambitious project recently launched by the Berlin botanists Adolf Engler and Karl Prantl, *Die natürlichen Pflanzenfamilien*, publication of which was to continue well into the twentieth century.[46] Sadly, none of Mueller's promised contributions to these series eventuated. Mueller blamed his workload as Government Botanist, especially the pressure on him to

46 M to A. de Candolle, 24 October 1883, 15 May 1886 and 7 May 1887 [all Collected Correspondence]; M to A. Engler, 13 July 1885.

complete his *Key to the system of Victorian plants*, which as already mentioned turned out to be a much more difficult and time-consuming task than he had envisaged. Later, he took offence at Engler's failure to protect him from being criticized by one of the junior Berlin botanists, and withdrew from Engler's project in protest.[47]

Undoubtedly the most successful of all Mueller's publications was his handbook for would-be acclimatizers of plants, *Select extra-tropical plants readily eligible for industrial culture or naturalization*, which went through nine editions in English between 1876 and 1895 (including editions published in India and America), and which was also published in German and (some years after Mueller's death) Portuguese translation, and in a French-language adaptation. The book drew on Mueller's experience over many years both in introducing exotic plants to Australia and in distributing seeds of Australian plants that he judged to have economic potential to other parts of the world. Much of it, however, was a distillation of his extensive reading (and also, unacknowledged, that of his assistant, Georg Luehmann)[48] on economic botany. The history of the work, which had its origin in reports prepared for the Acclimatisation Society of Victoria on plants suitable for cultivation in the colony[49] but whose scope quickly expanded to embrace the world's extra-tropical regions more generally, may likewise be traced in the correspondence published below. Thus we see Mueller indefatigably casting his net ever more widely in search of additional information to include in future editions, negotiating the publication of those new editions and promoting the distribution of copies, trying without success to prompt a New Zealand edition of the work, and equally unsuccessfully encouraging one of his Portuguese correspondents to undertake a translation into his native language and arrange for its publication.

47 M to A. Engler, 22 February 1888.

48 There is an interleaved copy of the 1876 edition of the book at the Royal Botanic Gardens Melbourne, in which all the additions noted for the next edition are in Luehmann's handwriting, with corrections only by M.

49 B71.13.01, B72.13.02.

Another publication that reflected his commitment to the use of plants for economic purposes was his translation of the treatise by the German pharmacist Georg Christian Wittstein on the organic constituents of plants.[50] The translation was initially prepared by Ludwig Rummel, a chemist whom Mueller employed in his department for several years, but was then extensively edited by Mueller before being published at his expense.[51] For years afterwards he complained in his letters that the project had been a financial disaster from his point of view, with few copies ever being sold – a scarcely surprising result when the Australasian colonies between them constituted such a small potential market for a work of this kind, and the Melbourne publisher had no established distribution system in other parts of the English-speaking world where more satisfactory sales might have been achieved.

Mueller's correspondence also reveals his continuing commitment to educational activities of various kinds. Though he vigorously resisted all suggestions from those who wanted to abolish his government appointment that a position might be found for him at the University of Melbourne, he did serve as an examiner there and at both the other universities in Australia at that time, in Adelaide and Sydney, and also at the School of Mines in Ballarat and the Pharmacy College and the Veterinary College in Melbourne. A Ballarat examination paper reproduced here illustrates his approach to the task.[52] The preparation of his *Key to the system of Victorian plants* as a guide to budding naturalists has already been noted. In addition, Mueller devoted considerable effort to composing what he called his 'Victorian school flora', a work that he envisaged being used as a text in the colony's schools that would enable students to identify the plants around them. A text that introduced children to the world of plants by focusing on the local flora would have been a major innovation and a welcome change from the imported texts focusing on British plants that were all that had been available until then. As Mueller initially planned

50 B78.13.06.

51 See M to J. von Haast, 28 July 1881.

52 Examination paper, 23 August 1882.

the work, however, it amounted to a set of formal descriptions of the colony's indigenous plants. When the then Chief Secretary, John MacPherson, saw some sample pages, he judged it, with some justification, as unsuitable as a school text, and stopped publication.[53] Mueller later had the material he had prepared issued with a new title page, without the 'textbook' label, as *The native plants of Victoria*.[54] Meanwhile, he prepared a new school text on less formal lines that was published in 1877 as his *Introduction to botanic teachings at the schools of Victoria*.[55]

Exploration

As Mueller grew older, he spent less time botanizing in the field and became more and more a herbarium-based scientist. His last major period of field-work was in 1877, when he spent several weeks travelling through the settled districts of the south-west of Western Australia after being commissioned by that colony's government to survey the forest resources of the region.[56] By then his days as an explorer of regions previously unknown to Europeans were long over. From his correspondence, however, we learn that he continued to be actively involved in promoting the exploration of inland Australia. In the years covered by the present volume, the exploration of New Guinea and Antarctica also became major preoccupations, with Mueller himself at one stage contemplating going to New Guinea.[57] At first, he promoted expeditions on a personal basis, lobbying governments and wealthy individuals such as the South Australian pastoralist Sir Thomas Elder to fund new ventures into parts of central and northern Australia that had not yet been adequately explored. Often he contributed significant sums himself to the funding of such expeditions. Later, he used the

53 See M to W. Odgers, 12 August 1876, and the notes thereto.
54 B79.13.08.
55 B77.13.07.
56 See B79.13.10.
57 See M to E. Ramsay, 22 August 1878.

Geographical (later Royal Geographical) Society of Australasia, over the Victorian Branch of which he presided from its foundation in 1883 until his death, as a vehicle for promoting new expeditions. In return for his support, he expected expeditions to botanize systematically as they went and to send their plant collections to him for classification and description. In this way, he was associated with virtually every significant exploring expedition in Australia and New Guinea, and also many lesser ones, during the final three decades of his life, and was responsible for publishing the botanical results of most of them. His involvement is extensively documented in the letters published here.

Also documented is Mueller's continuing close relationship, at least until well into the 1880s, with the great German geographical publishing house of Justus Perthes, publisher of the famous geographical journal, *Petermanns geographische Mittheilungen*. Mueller corresponded regularly with the founder of the journal, August Petermann[58] and his successor Ernst Behm, keeping them abreast of the latest news on the exploration of inland Australia and furnishing them with the most up-to-date data to be incorporated in the publisher's maps. Chiefly as a result of Mueller's activities, these maps were during these years generally recognised as the best maps of Australia available anywhere.

Explorers grateful for Mueller's support bestowed his name on various mountains, rivers, creeks and glaciers in Australia and elsewhere. Offering donors the possibility of having their names bestowed on one or other of the geographical features that was discovered was one of his own favourite stratagems in seeking funds for new expeditions. For such name-granting rights to retain their credibility as an inducement for donors, however, some guarantee was essential that the names proposed would be adopted and maintained. Part of the attraction to Mueller of his connection with the Justus Perthes organization was doubtless the fact that having his preferred names published on its maps went a long way towards ensuring that the names would stick. Mueller went further than this, however, urging without success the adoption internationally

58 See Voigt (1996).

of a priority rule for geographical names, similar to that which already operated with regard to the naming of plants.[59]

To the end of his days, Mueller remained fascinated by the fate of Ludwig Leichhardt, who with his companions and all the animals they took with them had disappeared without trace in 1848 while attempting to cross the continent from Moreton Bay (Brisbane) to Perth. The previous volume of this series documented Mueller's close links with the Ladies' Leichhardt Search Committee, formed at his instigation in 1864 to raise funds for an expedition to follow up supposed traces of Leichhardt that had been found in central Queensland, and also a later plan for Mueller himself to lead an expedition east from the settled areas of Western Australia to investigate another report. Stories of white men (or children of white men) living with Aboriginal tribes in remote parts of the continent where Europeans were not known to have penetrated, or of unidentified parties of white men perishing in different places in the interior, or of other supposed traces of the explorers, continued to appear from time to time, and Mueller took a close interest in all of them. Any that seemed linked at all plausibly with Leichhardt, he sought to have followed up.[60] Even in the 1890s, when a new expedition under the leadership of David Lindsay was being organized to explore the north-west of the continent with funding from Sir Thomas Elder, Mueller arranged for Lindsay's instructions to include investigating a story attributed to Aborigines from far out in the Great Sandy Desert of a party of white men having perished, years earlier, in that area. Unfortunately, Lindsay's expedition broke up in disarray without doing this, leaving Leichhardt's fate as great a mystery as ever.[61]

New Guinea and other parts of Melanesia were a focus of much anxiety in Australia in the 1870s and 1880s as rival powers to the

59 See M to A. Petermann, 24 October 1876. M submitted a paper, 'On the rules of priority of geographic names', to the Royal Geographical Society in 1876, but it was never published; see M to Royal Geographic Society, November 1876 [Collected Correspondence].

60 See, for example, M to H. Parkes, 21 May 1881.

61 D. Lindsay to M, 27 April 1892. For a late-in-life summing-up of M's views on Leichhardt, see M to A. Macdonald, 25 September 1893 [Collected Correspondence].

British moved into the region. Alarmed by German commercial activity in New Guinea, Queensland in 1883 annexed the south-eastern part of the island; and though this action was immediately repudiated by the British Government, a British protectorate was declared over the south-east coast in the following year, prompting the Germans ten days later to declare a protectorate over the north-east coast and the nearby Bismarck Archipelago. The subsequent setting up of a colonial administrative apparatus provided the necessary underpinning for the further exploration of the territory, which inland from the coast remained largely unknown to Europeans. This opportunity was seized upon by the newly-founded Geographical Society of Australasia, which promptly began sponsoring expeditions. Mueller as President of the Victorian Branch of the Society played a central role in setting the strategy for this, as well as in lobbying the various colonial governments for the necessary funding. Several of the letters published here are concerned with his activities in this regard.

From Mueller's correspondence, we can also trace the leading role that he played in a concerted Australian campaign during the 1880s and early 1890s to promote the scientific exploration and economic exploitation of Antarctica. Though he did not live long enough to witness the wave of exploring expeditions that invaded the great southern continent from the last years of the nineteenth century onwards, he did preside over the celebrations when the Geographical Society feted the Norwegian whaling skipper Leonard Kristensen and his crew when they returned to Melbourne after making the first landing on the Antarctic mainland on 24 January 1895. Mueller and his Victorian colleagues succeeded in bringing Antarctica on to the international stage and in establishing that the exploration of the region should be primarily (though not in their view exclusively) for purposes of scientific research rather than the mere traversing of territory. They also set an agenda for this research that came to be widely accepted as the first major expeditions were mounted.[62] Throughout the campaign to get an

62 See Home et al. (1992).

expedition under way, Mueller was everywhere, his social and scientific standing once again a major asset in attracting attention and funding to the cause.

Personal

In earlier years, occasional tensions had emerged in Mueller's correspondence with Joseph Hooker and George Bentham at Kew. As all three men grew older, the relations between them, as expressed in their letters, noticeably softened, even when they disagreed. From their Kew-centred perspective, Hooker and Bentham continued to disapprove of Mueller's habit of publishing descriptions of new species in publications to which they did not always have convenient access, as well as of his insistence on adhering rigorously to priority in the naming of species and his tendency to lump together species that most people regarded as separate. Nevertheless, as Bentham finally signed off on *Flora australiensis*, he was full of praise for Mueller's contribution to the work, and though he later penned strong criticisms of aspects of Mueller's *Systematic census of Australian plants*, he seems then to have thought better of sending them to Mueller.[63] Hooker played a significant role in Mueller's being knighted KCMG in 1879, at the conclusion of the *Flora australiensis* project, and was a frequent source of good advice that Mueller, to his cost, did not always follow. Hooker's later letters, especially following his retirement from the directorship at Kew in 1885, are much more relaxed and chatty than they had been earlier. He was obviously very grateful to Mueller for his efforts to find a suitable position for his son Brian after he went to Australia.[64] Meanwhile the more formal aspects of Mueller's link with Kew that had most potential for generating tension were taken over by Hooker's son-in-law and successor as Director, William Thiselton-Dyer.[65]

63 G. Bentham to M, 25 April 1883.

64 S. Hooker to M, 5 January 1894; M to J. Hooker, 5 August 1894; J. Hooker to M, 17 November 1895 [Collected Correspondence]; J. Hooker to M, 2 March 1896.

65 See, for example, W. Thiselton-Dyer to M, 20 May 1893.

The letters in this volume include several that document honours and awards being bestowed upon Mueller. Other letters make it clear that in some cases he actively solicited the award in question.[66] He was deeply perplexed by Bentham's disdain for all such awards.[67] So far as Mueller was concerned, such marks of recognition were not merely sops to his vanity – though vain he certainly was – they gave him the assurance he needed, isolated as he was from the major centres of scientific activity, that his work was well-regarded by his scientific peers. Mueller well knew that unless his work was held in high esteem, these awards would never have come his way, no matter how often he asked for them. No other scientist working in Australia received one of the medals awarded by the Royal Society of London, or was elected to either the Académie des Sciences in Paris or the American Academy of Arts and Sciences in Boston, for another half-century and more after Mueller.

Many of the honours bestowed on Mueller, though not the major scientific ones just mentioned, were in return for his vast sendings of botanical, zoological and ethnographical specimens to museums and botanic gardens in other parts of the world. In some cases, including the barony bestowed on him by the King of Württemberg, it is clear that part of Mueller's motivation in sending this material was his hope of receiving an honour. It is also clear, however, that this was only ever part of his motivation, and that his sendings, the cost of which were a major drain on his private finances, were also prompted by a simple desire to advance science. Throughout the period covered by this volume – that is, long after he received his barony – Mueller continued to send significant numbers of zoological specimens to the natural history museum in Stuttgart, which he had settled on, years earlier, as the prime recipient of his zoological materials. From Stuttgart, duplicate specimens were sent on to other institutions that Mueller had designated such as the university in his home town, Rostock. Mueller remained anxious to see his material being put to good use and

66 e.g. M to P. Sclater, December 1887; also A. Milne-Edwards to M, 9 February 1894, replying to M to A. Milne-Edwards, 1 January 1894 [Collected Correspondence].

67 M to J. Hooker, 10 October 1879.

described scientifically, not just added to a museum's collection. As the Australian natural history museums (especially the Australian Museum in Sydney) expanded their research activities in the 1880s, Mueller urged Ferdinand von Krauss in Stuttgart to be prompt in describing the materials he was sending him, lest he be forestalled by Australian researchers.[68]

Mueller's pleasure at the honours that came his way did not derive solely, however, from the recognition of his work that they represented. He was also a terrible snob who delighted in the enhanced social standing that came with them. His obsequiousness towards his social superiors and especially towards royalty grates badly on modern ears.[69] Even in his own day, his attitude was very different from that of colleagues such as Bentham and Hooker, and was also seriously at odds with prevailing views in colonial Australia.

Mueller's letters also show how he in turn sought to dispense scientific patronage by arranging for those he wished to reward to be elected to various metropolitan scientific societies – most notably the Royal Society, the Royal Geographical Society or the Linnean Society – or to have awards bestowed upon them.[70] From his nominations to the Royal Society, where the elections were highly competitive, we can see how out of touch Mueller was with the institutions of metropolitan science. A high proportion of those whose nominations he organized were never elected, even after Joseph Hooker gently coached him in what needed to be done if a nomination were to have any hope of success.[71] An even more striking illustration of Mueller's lack of familiarity with the conventions involved is his proposing to William Thomson, Lord Kelvin, during the latter's presidency of the Royal Society that the undistinguished German physicist K.H. Knoblauch be either awarded one of the Society's medals or elected a Foreign Mem-

68 See Introduction to *Selected correspondence*, vol. 2, pp. 32-3, and M to F. von Krauss, 21 April 1885.

69 See, for example, M to J. Hooker, 27 May 1879, and M to E. Rowan, 3 March 1896.

70 See Lucas (1988).

71 J. Hooker to M, 7 April 1885.

ber.[72] The proposal shows that Mueller could not have had any familiarity with either Knoblauch's work or the basis on which the Society made these awards, but was under the mistaken impression that the office Knoblauch held as president of one of the major German scientific societies both implied that he must have a distinguished scientific record and in itself provided a sufficient basis for such an award.

Always somewhat hypochondriacal, Mueller as he grew older worried more and more about his health. It is a constant refrain in his letters that he did not expect to live much longer. He remained convinced that he had inherited a weakness in his chest that sooner rather than later would carry him off, as it had both his parents and also his sisters Iwanne and Bertha; and indeed from time to time he seems to have suffered severe bronchial attacks, the most serious of which persisted for several months during 1884 and led him for a time to move in with his doctor and friend Alexander Buettner for closer medical supervision, and later to leave Melbourne for some weeks in search of a healthier climate. Mueller's reluctance ever to go out in public without his scarf wound around his neck to ward off possible chills became a source of general amusement, but he had good grounds for being careful. Mueller also suffered painfully from time to time from piles. He had twice undergone surgery for this in 1859, and in 1878 he had to submit to the surgeon's knife again.[73] No doubt he suffered considerable discomfort at various other times, without its being severe enough to result in surgical intervention.

Mueller's brother-in-law Eduard Wehl died in February 1876, leaving Mueller's sole surviving sister, Clara, to bring up their twelve surviving children alone at their property, 'Ehrenbreitstein', near Millicent, SA. Though Mueller seldom visited Clara thereafter, or his dead sister Bertha's husband George Doughty and their two children living in nearby Mount Gambier, surviving letters show that the family remained in regular contact and that Mueller took a kindly, avuncular interest in the well-being of his nieces and nephews.

72 M to Lord Kelvin, 23 April 1893.

73 M to R. Owen, 30 October 1878.

He encouraged them to take an interest in plants and supported his nieces Louise and Marie Wehl when they showed a talent for flower-painting. He was thrown into deep mourning by the deaths of his two brothers-in-law and later of Bertha's son, also named George.[74] As the children grew up, he sometimes tried using his influence to help find suitable positions for them; however, he discouraged Louise Wehl from moving to Melbourne to become his housekeeper.[75] The children were clearly fond of Mueller and also proud of what he had achieved, and they and Clara did their best, after his death, to safeguard his memory.[76]

Mueller's sentimental attachment to his family extended to his memories of his upbringing in Germany. Though he remained one of Queen Victoria's most loyal subjects, he also felt a deep devotion to his German homeland. He sent funds to Rostock, where he had been born, for the maintenance of his father's grave, and dreamed of revisiting the town and again seeing the Mönchentor where he and his family had lived.[77] He maintained close links with his German scientific correspondents and also with the German community in Australia. To members of both groups, he always wrote in German, even when – as, for example, in the case of Eugene Hilgard who had lived in America for many years – they wrote to him in English.[78] To colleagues in Germany, he reminisced wistfully from time to time about his early life. Working at a distance, as he did, from the centre of world science, Mueller relied heavily on his correspondence to keep in touch with developments in his field. Though his links with the British scientific community and especially with Kew Gardens were very important to him in this regard, he had many more links with German science. As a unified Germany came to be seen as a potential rival to British imperial power, unfamiliar tensions appeared in the previously friendly relations between British and German immigrants in Australia, and Mueller himself may have been a victim – certainly

74 M to H. Deane, 25 November 1893.

75 M to L. Wehl, 26 February 1885.

76 See Introduction to *Selected correspondence*, vol. 1, pp. 41-3.

77 M to H. Massmann, 3 July 1876.

78 See E. Hilgard to M, 16 October 1893 and 3 December 1895.

he was inclined to blame the attacks on him partly on his being a foreigner (that is, non-British).[79] He was evidently troubled by the resulting potential for divided loyalties and when opportunities presented themselves, he did his best to strengthen the bonds between German and British Australians.[80]

From Mueller's letters we can see that he moved regularly in upper circles of Melbourne society. He was personally acquainted with every Governor of Victoria from Latrobe onwards and, when the need arose, could write personally to Governors and government leaders in the other Australasian colonies or even provide letters of introduction for prominent Australians such as the wealthy South Australian pastoralist Thomas (later Sir Thomas) Elder when they visited England.[81] His formal public lectures were major set-piece occasions with the Governor in the chair, but he also frequently lectured less formally but still invariably at great length to church and other groups. Mueller clearly enjoyed being seen, festooned with medals, at Government House functions and on other formal occasions. If there was dancing, he threw himself into it with gusto.[82] He seems to have had a coterie of upper-class friends, many of them medicos, with whom he dined or exchanged visits from time to time.[83] No doubt they shared his very conservative political opinions that caused him to feel alienated from the democratic spirit of the society around him.[84] He also clearly enjoyed the company of several of his fellow council members in the Victorian Branch of the Royal Geographical Society of Australasia, like-minded enthusiasts all for geographical exploration.[85] He loved music and actively supported Melbourne's musical life, being for several years Vice-President of the Melbourne Liedertafel, a supporter of efforts to establish a local opera company, and offer-

79 See, for example, M to J. Hooker, 7 September 1872 (*Selected correspondence*, vol. 2, p. 647).

80 See M to W. Thiselton-Dyer, 7 February 1896.

81 M to H. Parkes, 1 November 1887; M to Lord Russell, 5 October 1877.

82 M to W. Thiselton-Dyer, 7 September 1883.

83 M to H. Wehl, 6 July 1886.

84 M to F. von Krauss, 15 March 1881; M to E. Ramsay, 13 May 1891.

85 M to A. Macdonald, 23 December 1893; M to J. Shillinglaw, 30 July 1894.

ing patronage and support in other ways to a number of local musicians including the young child prodigy Percy Grainger.[86]

When Mueller went to Australia, he had significant financial resources at his disposal that he had inherited from his parents – enough, he later told William Thiselton-Dyer, effectively to cover all his expenses during his first years in the country.[87] These funds he was able to supplement by working for part of the time as a pharmacist in Adelaide and perhaps by sending botanical collections that he assembled to Europe for sale. Then from 1853 until his death, except for a brief period in the mid-1850s he enjoyed a substantial salary as Victoria's Government Botanist. Throughout that period, however, he spent his own funds freely to prop up the work of his department, for example by purchasing additional books and journals and employing additional assistants and collectors beyond those covered by his departmental budget. As a result, by the time he died, his fortune was exhausted and he was living in cramped and straitened circumstances, barely able any longer to keep up with the obligations that pressed upon him as a result of his position in society. After his death, when probate was declared, his estate amounted to only a few hundred pounds. Though he seems to have been able to survive on less sleep than most people, he was also worn out from years of over-work. Photographs of him taken in the 1890s invariably show him looking run down and exhausted. Yet he continued working sixteen-hour days until the end, a sad and rather lonely figure who, as plant physiology steadily displaced taxonomy at the leading edge of botanical research, was even beginning to be left behind by the science to which he had devoted his life. Mueller achieved a tremendous amount during his lifetime and that achievement was widely recognised. The high reputation that he enjoyed at the time of his death, as Australia's leading scientist of the nineteenth century, was thoroughly deserved. One thing he seems never to have achieved, however, was lasting happiness.

86 M to J. Hooker, 22 November 1881; R. Grainger to M, 23 August 1894.
87 M to W. Thiselton-Dyer, 9 February 1896.

The Correspondence

In the present volume we publish a further 326 letters and associated documents to or from Mueller, together with a further six letters not written by or to him but in which he figures large. The letters to or from Mueller have been selected from a total of nearly 6000 letters from the period covered by this volume that have so far been located through a world-wide search. In other words, some 5.4% of the letters located have been selected, a slightly higher percentage than that achieved in previous volumes. As stated in the earlier volumes, we believe that what we have found probably represents less than five per cent of what once existed. Once again, many of the letters that have not been selected because of space limitations are of great interest. In referring in footnotes to letters that have not been selected, we give their location as 'Collected Correspondence', by which we mean the comprehensive edition of all the letters that we have located that is currently being prepared for electronic publication. Letters referred to in footnotes without this label have been published in one of the three volumes of selected correspondence.

In making the selection for this volume, we have followed the same principles as in the earlier volumes. Letters have been chosen on the basis of their scientific importance and their centrality to our understanding of Mueller's life and work, and we have also tried to give a sense of the wide range of Mueller's correspondents and of the subjects on which he corresponded with them. As in Volume 2, with only a few exceptions letters, no matter how interesting, that come from contemporary published sources have been systematically excluded. On the other hand, the fact that letters have subsequently been published by historians has not been held to exclude them from the selection, for that would have ruled out some of the most interesting and important letters of all.

The letters that have been selected are published in the standard format used in earlier volumes. As before, letters have been published in the language in which they were written, with an accompanying English translation for those not originally written in English. Where appropriate, notes provide further details concerning the original manuscripts, as well as contextual and other relevant information. Biographical information about individuals referred to in the letters is not provided in the notes but in Appendix A, the Biographical Register. However, in order to conserve space, entries provided in the Biographical Registers in the earlier volumes are not repeated here. Hence the absence of an entry for a person in the Biographical Register in this volume may mean either that we have been unable to discover any information about the person beyond what is revealed in the letter in which the name appears, or that the person has an entry in one of the earlier volumes. (These entries will be consolidated into a searchable list in the electronic publication.)

Once one gets used to it, Mueller's handwriting is generally clear, though some of his letter forms are somewhat idiosyncratic and require getting used to. It seems that not everybody had the patience to do this, for it is noticeable that there are typewritten transcriptions of increasing numbers of Mueller's letters to his Ministers in the files from the 1880s and 1890s at the Public Record Office of Victoria – presumably prepared for the Minister's convenience as typewriters became more common in government offices. The handwriting of some of Mueller's correspondents, most notably that of Joseph Hooker, is much harder to read. As Mueller grew older, his writing got larger, presumably to compensate for deteriorating eyesight. Occasional grammatical mistakes also began appearing when he was writing in German: we have noted these only when necessary to clarify the sense.

The following table, that parallels one given in Volume 2, gives more detailed information about both the letters that have survived and the selection that has been made, in the cases of those to or from whom the largest numbers of letters have survived from the period covered by this volume. (The numbers set out relate only to this period.) Not recorded in the table, however, are the numerous letters to or from Mueller's successive ministerial chiefs, or the much larger number to or from the departmental functionaries in the Victorian Public Service with whom he had constantly to deal.

Name	Letters from Mueller		Letters to Mueller		Total	
	Known	Selected	Known	Selected	Known	Selected
O. Tepper	284	5	19	0	303	5 (1.7%)
W. Thiselton-Dyer	214	17	1	1	215	18 (8.4%)
J. Hooker	157	16	44	5	201	21 (10.4%)
M. Holtze	198[88]	1	3	0	201	1 (0.5%)
A. Macdonald	142	2	2	0	144	2 (1.4%)
R. Tate	135	5	8	0	143	5 (3.5%)
J. Agardh	104	9	23	1	127	10 (7.9%)
E. P. Ramsay	118	7	8	0	126	7 (5.6%)
F. Bailey	16	1	70	3	86	4 (4.7%)
F. von Krauss	58	5	6	1	64	6 (9.4%)
J. Maiden	12	0	46	0	58	0
W. Woolls	8	0	49	1	57	1 (1.8%)
W. Gill	49	3	4	0	53	3 (5.7%)
G. Bentham	27	3	9	2	36	5 (13.9%)
O. Nordstedt	28	1	8	0	36	1 (2.8%)
A. de Candolle	32	2	2	0	34	2 (5.9%)
A. Milne-Edwards	20	0	8	1	28	1 (3.6%)
A. Engler	28	3	0	0	28	3 (10.7%)
L. Dejardin	25	1	0	0	25	1 (4.0%)
J. Stirling	17	2	7	0	24	2 (8.3%)
J. von Haast	24	2	0	0	24	2 (8.3%)
E. von Regel	21	2	3	0	24	2 (8.3%)

As in the previous volume, the importance of Mueller's correspondence with the leading figures at Kew Gardens is immediately apparent. While his correspondence with George Bentham was clearly tapering off as the work for *Flora australiensis* came to an end and Bentham's health failed, his correspondence with Joseph Hooker continued long beyond Hooker's retirement, ending only with Mueller's death, while the continuing significance for him of the connection with Kew is documented by the growth of his correspondence with William Thiselton-Dyer who succeeded Hooker as Director. Several other Kew botanists are also represented in the letters published below, even though not enough correspondence with Mueller has survived for them to appear in the table.

88 The number of letters to Holtze may be over-stated. Many of the surviving items are lists of plants that may not originally have been separate documents.

The table also reveals, however, the strength during Mueller's later years of his links with other Australian naturalists, with names such as Otto Tepper, Maurice Holtze, Ralph Tate and E. P. Ramsay featuring high on the list (though Holtze's count may be higher than it ought to be). As with Mueller's overseas correspondents, the vagaries of fortune have determined which letters have survived. For example, many of Joseph Hooker's letters to Mueller survived while Thiselton-Dyer's did not because, fortuitously, Charles Topp had separated them out from the bulk of Mueller's papers before the papers were lost. More generally, the table reflects what was noted in earlier volumes, namely that as a result of the bulk of Mueller's papers being lost, there are many fewer incoming than outgoing letters among those that have survived. There are, however, exceptions to this generalization such as F. M. Bailey and J. H. Maiden who, as government officials, kept file copies of their outgoing mail that have survived, whereas their incoming mail has not fared so well. Others who held official positions – E. P. Ramsay, for example – appear, like Mueller himself, not systematically to have kept copies of the letters they wrote.

The table, and also the selection of letters published in this volume, vividly brings out the breadth of Mueller's correspondence with naturalists around the world. Most of those to whom he wrote were of course botanists, but Mueller also corresponded regularly with zoologists such as Ferdinand von Krauss in Stuttgart, Alphonse Milne-Edwards in Paris, and Richard Owen in London. While his links with the Geographical Institute at Gotha became weaker following the death of August Petermann in 1878, his continuing interest in matters geographical appears in the large number of letters he wrote to Alexander Macdonald, the hard-working Secretary of the Victorian Branch of the Royal Geographical Society of Australasia over which he presided.

The table makes plain the subjective nature of the selection we have made of letters to include in this volume. A much higher proportion of the correspondence we have found with Bentham, Joseph Hooker and the great German systematizer Adolf Engler has been included, on account of what we judge to be its inherent interest, than of Mueller's more routine exchanges with Macdonald,

Tepper and Holtze. Most strikingly of all, none of the 58 letters to or from Joseph Maiden that we have located has been included, being all comparatively short exchanges on routine business matters. Among correspondents represented by smaller numbers of surviving letters, strikingly high percentages have been selected in some cases – for example, Asa Gray (4 of the 9 letters found selected, or 44.4%), Jean Müller (4 of 21, or 19.0%), Georgina King (2 of 13, or 15.4%) and Eugene Hilgard (2 of 16, or 12.5%). Between them, the letters selected shed fascinating light on the final two decades of the extraordinary career of an extraordinary man, as well as, more generally, on the cultural and intellectual history of Australia at a crucial stage of its development and on the history of the field sciences in the final years of the nineteenth century.

Acknowledgements

A project of the kind that has given rise to this edition of Ferdinand von Mueller's correspondence inevitably depends for its success on the co-operation of many individuals and institutions. We record here our thanks to all those who have helped us bring the work to fruition. In particular, we once again gratefully acknowledge the support and encouragement we have received from the institutions where we have been employed and at which the project's offices have been based, namely the University of Melbourne and its Department of History and Philosophy of Science, King's College London, the Royal Botanic Gardens Melbourne, and the Historisches Institut of the University of Stuttgart.

Sources of the funding that underpinned earlier stages of the project were acknowledged in Volumes 1 and 2. Additional funding to support the preparation of the present volume was provided by the Government of the Federal Republic of Germany, the Baker Foundation, the Royal Botanic Gardens, Melbourne, the Friends of the Royal Botanic Gardens, Melbourne, and anonymous donors in Melbourne and London.

For the subsidy that has made publication of this volume possible, we are grateful to King's College London and to Dr Günter Klatt, proprietor of the Einhorn-Rats-Apotheke in Husum, the pharmacy at which Ferdinand Mueller received his training in the 1840s.

For continued encouragement, moral support and wise advice during the preparation of the present volume, we thank Dr Philip Moors, Director of Melbourne's Royal Botanic Gardens, the late Dr Sophie C. Ducker, formerly Reader in Botany at the University of Melbourne, and an informal editorial advisory group comprising Ms Helen M. Cohn, Dr T. A. Darragh and Professor J. H. Ross.

Helen Cohn has been a bounteous source of information, especially on bibliographical matters and on the history of the Melbourne Herbarium, and has also taken primary responsibility for preparing the Bibliography, while Jim Ross has been a fount of wisdom on technical botanical questions. Various other members of staff at the Royal Botanic Gardens Melbourne have also provided advice on botanical matters, most notably Dr Tom May.

Tom Darragh has freely shared with us his vast knowledge of the social and scientific history of nineteenth-century Victoria and especially its German community, and has spent many hours searching the collections of the Public Record Office of Victoria for Mueller-related material. Since 2001 he has also taken primary responsibility within the project for transcribing and translating letters and other documents written in German, and most of the translations from German that appear in this volume are due to him.

In London, Paula Lucas has continued as research assistant, allowing the project to benefit from her detailed acquaintance with the archives of nineteenth-century scientific London.

Assistance in translating documents from other languages into English has been provided by Nancy Wallace and Associate Professor Roger Scott, who helped with materials in Latin, Dr Stephen Kolsky (Italian), and Professor Manuel Thomaz (Portuguese). The State Library of Victoria allowed us long-term use of its largest and most comprehensive Langenscheidt German-English dictionary.

Peter Cavanagh and Bob Patey – both, sadly, now deceased – contributed countless hours to proof-reading, which in their hands was an art-form. Herbert Meyer and Hans Schroeder assisted with proof-reading of German-language materials. Lawrence Cohn has also assisted with proof-reading and layout. In addition, he took on the enormous task of compiling the Biographical Register. Marjorie Home extracted details of author's inscriptions from Mueller's large surviving collection of offprints, pamphlets and books.

A vast number of librarians and archivists around the world have facilitated our search for letters to or from Mueller and for additional information about him and his work. Many of those whose help was acknowledged in Volumes 1 and 2 have likewise assisted with the material for Volume 3: we particularly acknowledge the continued help of Walter Struve, who has pursued numerous queries for us at the State Library of Victoria; Lise Summers, who has done likewise at the State Records Office of Western Australia; Jill Thurlow, assistant librarian at the Royal Botanic Gardens Melbourne; Kate Pickard, Archivist, Royal Botanic Gardens, Kew; and Susan Snell, Archivist, Natural History Museum, London. We also acknowledge here the help provided by Dr Ulrich Sieber, Deputy Director of the Universitätsbibliothek Stuttgart, and his staff; staff of the Württembergische Landesbibliothek, Stuttgart, and of the Landesarchiv Schleswig-Holstein, Schleswig; Michele Losse, Assistant Archivist, Royal Botanic Gardens, Kew; Carol Gokce, Natural History Librarian, Natural History Museum, London; and Gina Douglas, Archivist, Linnean Society, London. We are grateful, too, to the following who, in addition to those acknowledged in Volumes 1 and 2 or beyond what was acknowledged there, have shared with us their specialized knowledge of various subjects bearing on Mueller's career: Richard Aitken (19th-century Victorian horticulture), Anthea Bundock (*Australian dictionary of biography* biographical files), Linden Gillbank (botany in Melbourne), Darrell Lewis (Leichhardt traces, Mueller publications in provincial newspapers), Graeme Powell (Mueller-related materials in the National Library of Australia).

For granting permission to publish material in this volume that is in their possession or over which they hold copyright, we are grateful to the following: Alexander Turnbull Library, Wellington; American Academy of Arts and Sciences, Library of the Boston Athenaeum, Boston, Massachusetts; Bancroft Library, University of California, Berkeley; Archives, Academy of Sciences, St Petersburg; Archives du Conservatoire botanique de Genève; Auckland Institute and Museum, Auckland; Barr Smith Library, University of Adelaide; Battye Library, Perth; Benedictine Community, New Norcia, Western Australia; Berlin-Brandenburgische Akademie der Wissenschaften, Berlin; Biblioteca Nazionale Centrale, Florence; Bibliothèque Centrale du Muséum National d'Histoire Naturelle, Paris; Botanisk Centralbibliotek, Copenhagen; Field Naturalists' Club of Victoria; Gray Herbarium Archives and Library, Harvard University; Hocken Library, University of Otago; Imperial College Archives, London; Justus Perthes Verlag, Gotha; Keeper of Public Records, Victoria; Kungl. Vetenskapsakademien, Stockholm; Laboratoire de Cryptogamie, Muséum National d'Histoire Naturelle, Paris; Linnean Society, London; Magyar Természettüdományi Múseum, Budapest; Mecklenburgisches Landeshauptarchiv Schwerin; Melbourne Necropolis, Springvale; Mitchell Library, State Library of New South Wales, Sydney; Mortlock Library, State Library of South Australia, Adelaide; National Archives of Canada, Ottawa; National Archives, Washington; National Herbarium of New South Wales, Sydney; Natural History Museum, London; National Library of Australia, Canberra; Orto Botanico, Università degli Studi di Padova, Padua; Queensland Herbarium, Brisbane; Royal Australian Historical Society, Sydney; Royal Botanic Gardens, Edinburgh; Royal Botanic Gardens, Kew; Royal Botanic Gardens, Melbourne; Royal Geographical Society, London; Royal Society of London; Royal Society of New South Wales, Sydney; Schleswig-Holsteinische Landesbibliothek, Kiel; School of Mines and Industries, Ballarat; Staatliches Museum für Naturkunde, Stuttgart; Staatsbibliothek zu Berlin – Preussischer Kulturbesitz; State Library of Victoria, Melbourne; State Record Office of Western Australia, Perth; State Records of South Australia, Adelaide; Universidade de Coimbra, Coimbra; Universitätsarchiv, Vienna; Universitetsbibliotek, Lund; University of Melbourne Archives; University of Tasmania Archives, Hobart.

Illustrations

Frontispiece. Illuminated manuscript commemorating Mueller's 70th birthday (30 June), presented to him by the Deutsche Verein, Melbourne, on 1 July 1895. Courtesy of the Library, Royal Botanic Gardens Melbourne.

Figure 1. Eucalyptus globulus, illustrated by Emil Todt, published in *Eucalyptographia* (1879) (B79.11.13). Reprinted courtesy of the Library, Royal Botanic Gardens Melbourne. p. 88

Figure 2. Ernest Giles, photographed by The Temple of Light, Adelaide, contained in Mueller's photograph album. Courtesy of the Library, Royal Botanic Gardens Melbourne. p. 90

Figure 3. Alphonse de Candolle. Published in *Archives des sciences, physiques et naturelles* (1893), 3s, vol. 30. Reprinted courtesy of the Library, Royal Botanic Gardens Melbourne. p. 92

Figure 4. Eucalyptus howittiana, illustrated by Emil Todt and published in *Eucalyptographia* (1879) (B79.11.13). Reprinted courtesy of the Library, Royal Botanic Gardens Melbourne. p. 105

Figure 5. John Forrest, photographed by Johnstone O'Shannessy & Co., Melbourne, contained in Mueller's photograph album. Courtesy of the Library, Royal Botanic Gardens Melbourne. p. 115

Figure 6. The first page of Mueller's promotion to Knight Commander of St Michael and St George, signed at the top by Queen Victoria. Courtesy of the Library, Royal Botanic Gardens Melbourne. p. 144

Figure 7. Asa Gray. Published in the *Bulletin of the Torrey Botanical Club* (1888), vol. 11, frontispiece. Reprinted courtesy of the Library, Royal Botanic Gardens Melbourne. p. 181

Figure 8. The old Melbourne Herbarium, showing the extension erected to house the Sonderian collection, photographed about 1890. The two men on the steps are probably Charles French Jnr and Gerhard Renner, M's assistants. Courtesy of the Library, Royal Botanic Gardens Melbourne. p. 316

Figure 19. Mueller in the 1890s wearing some of his medals. Lithograph printed by Charles Troedel. Mueller sent inscribed copies of this portrait to several people. Courtesy of the Library, Royal Botanic Gardens Melbourne. p. 724

Figure 20. The Governor-General, Lord Hopetoun, unveiling the monument over Mueller's grave, 26 November 1901. Mueller was buried at St Kilda Cemetery, Melbourne, on 13 October 1896. Courtesy of the Library, Royal Botanic Gardens Melbourne. p. 767

Editorial Conventions

Transcriptions are made up of the following elements:

Item number entry. Each item has a unique number based on the date of the item, e.g. 77.06.27a = second item of date 27 June 1877. Incomplete elements of the date are represented by '00' e.g. 80.09.00 = item of date September 1880. Item numbers are used as citations in the appendices, e.g. L83.12.15 = item number 83.12.15. The 'L' prefix distinguishes item numbers from bibliography numbers e.g. B85.13.06. (For an explanation of the coding system used for identifying Mueller's publications, see Appendix B, Volume 1.)

Correspondent entry. Provides information on the individual(s) to whom the item is addressed or from whom it comes, e.g. *From Joseph Hooker*, or gives a descriptive title, e.g. *Examination Paper.*

Location entry. The archival source of the item, e.g. RB MSS M1, Library, Royal Botanic Gardens Melbourne. Where the item is known only in a published form, details of the publication are given in a footnote.

Text. Letters are transcribed into a standardized format consisting of address and date, salutation, letter paragraphs, valediction, addressee and postscripts. Marginal notes are interpolated in the body of the letter following the author's marks or, in the absence of such marks, treated as postscripts and placed at the end of the letter.

Spelling and punctuation follow the MS. Corrections such as deletions and overwritings are usually noted only where they have a number of possible readings. Editorial comments on the text of the MS are given in footnotes in italic, e.g. *illegible.* Underlined text is transcribed in italic.

Square brackets are used to perform a variety of editorial functions but mean *uncertain reading* unless otherwise specified in a footnote such as '*editorial addition*' or '*illegible* – obscured by binding tape'.

Items in languages other than English are followed by translations into English. The translations are intended to assist readers to gain an understanding of the contents. We have not standardized the translating style of the various editors. In items that are partly printed and partly handwritten, a non-standard font is used for the printed text. Physical information about the MS may be provided in a footnote at the end of the correspondent entry, e.g. MS black edged.

Footnotes. References to sources used by the editors are given in the author (date) format in footnotes, e.g. Agardh (1889). Where there is more than one author of the same surname, as many initials are provided as are needed to distinguish them. Full citations are provided in the Bibliography. References to newspaper articles and parliamentary papers are given in full in the footnotes. Biographical details on correspondents and on individuals mentioned in letters are given in Appendix A, or in the corresponding appendices in earlier volumes.

Measurements

Quantitative measurements that appear in the letters are reproduced as written. The following conversion table may be applied:

inch (in)	1 in	=	2.54 cm
foot (ft)	1 ft	=	30.5 cm
yard (yd)	1 yd	=	0.914 m
mile	1 mile	=	1.61 km
ounce (oz)	1 oz	=	28.3 gm
pound (lb)	1 lb	=	454 gm
ton	1 ton	=	1.02 tonne

Botanical names

Publication of a name in this edition is not intended to confer any formal status on a taxon under the rules of botanical nomenclature. We have also not attempted to identify botanical synonymy, nor to identify plants that are given manuscript names only in the text.

Attention is drawn to apparent misspellings but not to recognized orthographic variants. For information on orthographic variants, readers should refer to standard works such as APNI.

Identifications are suggested for plants that are alluded to in the text but not named. For taxa that Mueller suggests as new, readers can find publication details in Muir (1979) (our Appendix C, Volume 1). Publication details of Mueller's taxa not listed in Muir (1979) are given in footnotes. In cases where no published record for a name of an Australian taxon has been found, a footnote is added stating 'Not in APNI'.

Abbreviations

The following abbreviations have been used in the notes:

ADB	= *Australian dictionary of biography* (1966-90) vols. 1-12
APNI	= Chapman (1991)
B	= item in Mueller's bibliography (see Appendix B, Volume 1)
c.	= circa
DND	= *Dictionary of national biography* (1917-90) vols. 1-22
IK [Index Kewensis]	= Jackson (1893-5) and supplements
KEW	= Herbarium, Royal Botanic Gardens, Kew, London
L	= item in Mueller's correspondence
M	= Mueller
MEL	= National Herbarium of Victoria, Royal Botanic Gardens Melbourne
MS	= manuscript
n.d.	= no date
pers. comm.	= personal communication
PRO	= The National Archive, London (formerly the Public Record Office, London)
PROV	= Public Record Office, Victoria
RBG, Kew	= Royal Botanic Gardens, Kew, London
SMH	= *Sydney morning herald*
TL2	= Stafleu & Cowan (1976-88), Stafleu & Mennega (1992-2000)

Calendar of Selected Mueller Letters and Other Items, 1876-1897

1. 76.03.29 Circular
2. 76.06.10 to Eduard Fenzl
3. 76.06.10a to Ferdinand von Krauss
4. 76.06.14 to John MacPherson
5. 76.07.03 to Magnus Massmann
6. 76.07.10 to Edward Ramsay
7. 76.08.12 to William Odgers
8. 76.09.23 to Joseph Hooker
9. 76 10.24 to August Petermann
10. 76.12.25 to William Branwhite Clarke
11. 76.12.28 to Count Louis Torelli
12. 77.06.02 from Ernest Giles
13. 77.06.05 to Alphonse de Candolle
14. 77.06.27 to William Odgers
15. 77.06.27a from George Bentham
16. 77.07.07 to Joseph Hooker
17. 77.08.07 to Graham Berry
18. 77.08.16 from G. Robinson
19. 77.09.05 to Edwin Jephcott
20. 77.10.05 to Lord Russell
21. 77.10.06 from the Adelaide German Club
22. 77.11.20 to Harry Ord
23. 78.02.14 to John Forrest
24. 78.05.12 to William Thiselton-Dyer
25. 78.06.04 to Edward Ramsay
26. 78.08.22 to Edward Ramsay
27. 78.09.18 from Richard Kippist
28. 78.10.26 from Alexander Forrest
29. 78.10.30 to Richard Owen
30. 78.11.17 to George Bowen
31. 79.01.13 to Ernst Behm
32. 79.03.20 to William Munro
33. 79.03.31 to Edward Bage

34. 79.05.18 to Bryan O'Loghlen
35. 79.05.24 from Queen Victoria
36. 79.05.27 to Joseph Hooker
37. 79.06.08 to Thomas Huxley
38. 79.06.08a to George Bentham
39. 79.06.15 to James Agnew
40. 79.06.20 from John Thurston
41. 79.07.24 from Ann Timbrell
42. 79.08.04 to George Day
43. 79.08.16 to Joseph Hooker
44. 79.09.09 to Friedrich Wöhler
45. 79.09.11 to [John Forrest]
46. 79.09.20 from Charles Moore
47. 79.10.06 to Otto Tepper
48. 79.10.10 to Joseph Hooker
49. 79.11.05 from Frederick Bailey
50. 79.11.22 to Ralph Tate
51. 79.12.25 to Julius von Haast
52. 79.12.31 to George Grey
53. 80.01.01 to Oliver Jones
54. 80.01.19 from Robert Fitzgerald
55. 80.02.21 to Asa Gray
56. 80.04.23 to James Patterson
57. 80.04.27 from Kate Taylor
58. 80.05.16 to Asa Gray
59. 80.06.21 to Joseph Hooker
60. 80.07.08 to Louise Wehl
61. 80.08.04 to Alphonse de Candolle
62. 80.09.00 from R. W. Moran
63. 81.01.12 to Jean Müller
64. 81.01.20 from Hermann Kempe
65. 81.02.01 to Richard Owen
66. 81.03.00 to the Linnean Society, London
67. 81.03.15 to Ferdinand von Krauss
68. 81.03.21 to George Langridge
69. 81.05.21 to Henry Parkes
70. 81.06.06 from William Woolls
71. 81.06.10 from Joseph Hooker

72. 81.07.22 to Joseph Hooker
73. 81.07.26 from D. Jones
74. 81.07.28 to Julius von Haast
75. 81.08.09 to Annie Walker
76. 81.08.24 from Umberto I, King of Italy
77. 81.09.02 from Alfred Selwyn
78. 81.09.10 to William Thiselton-Dyer
79. 81.10.08 to Joseph Hooker
80. 81.10.27 to Ralph Tate
81. 81.11.10 to Annie Krefft
82. 81.11.22 to Joseph Hooker
83. 81.12.08 from Benedetto Scortechini
84. 82.01.07 to Maurice Holtze
85. 82.02.10 to Paul Ascherson
86. 82.02.13 from Kalakaua, King of Hawaii
87. 82.02.13a to William Mitten
88. 82.02.16 to Louis Smith
89. 82.03.00 from Annie McCann
90. 82.03.06 from Eugene Fitzalan
91. 82.03.28 to William Thiselton-Dyer
92. 82.03.29 to Jean Müller
93. 82.04.24 from Andrew Berry
94. 82.05.02 to Bryan O'Loghlen
95. 82.05.09 to Edward Ramsay
96. 82.06.05 to Jacob Agardh
97. 82.06.13 from Frederick Gladman
98. 82.06.17 from W. Anthony Persieh
99. 82.07.24 to Joseph Hooker
100. 82.08.16 to William Thiselton-Dyer
101. 82.08.22 to Robert Pohlman
102. 82.08.23 Examination Paper
103. 82.09.19 from Paul Foelsche
104. 82.10.01 to Joseph Hooker
105. 82.10.02 from John Thurston
106. 82.10.16 from Sidney Johnson
107. 82.11.02 to James Stirling
108. 82.11.08 to George Bentham
109. 82.11.22 to Lajos Haynald

110. 82.12.25 to Archibald Liversidge
111. 83.01.02 to Frederick Broome
112. 83.02.21 from Charles Fawcett
113. 83.02.26 to Malcolm Fraser
114. 83.02.28 to Messrs Watson & Scull
115. 83.04.25 from George Bentham
116. 83.05.07 to Thomas Wilson
117. 83.05.11 to Alfred Deakin
118. 83.05.15 to James Service
119. 83.05.23 to William Brodribb
120. 83.05.25 to Ebenezer Thomas
121. 83.06.28 to Graham Berry
122. 83.08.06 from Edwin Daintrey
123. 83.08.25 to Asa Gray
124. 83.08.28 from William Dobson
125. 83.09.07 to William Thiselton-Dyer
126. 83.10.20 to Ralph Tate
127. 83.10.24 to Graham Berry
128. 83.11.05 from Sarah Brooks
129. 83.11.26 from William Allitt
130. 83.12.00 to George Doughty Jr
131. 83.12.22 from William Fisher
132. 83.12.30 to George Brown
133. 83.12.31 to George Bentham
134. 84.01.17 from Joseph Hooker
135. 84.02.07 to Graham Berry
136. 84.02.14 to Thomas Wilson
137. 84.03.14 from Joseph Hooker
138. 84.03.15 to Ralph Tate
139. 84.03.30 to Robert Watson
140. 84.04.08 to Graham Berry
141. 84.05.03 to [Bernhard Perthes]
142. 84.06.17 Last Will and Testament
143. 84.07.14 from Robert Johnston
144. 84.08.16 to William McCrea
145. 84.08.27 to Thomas Wilson
146. 84.09.06 to Thomas Wilson
147. 84.09.08 to Joseph Hooker

148. 84.10.03 to Georgiana von Hochstetter
149. 85.01.23 to Franz Stephani
150. 85.02.26 to Louise Wehl
151. 85.04.01 to Heinrich Fayenz
152. 85.04.07 from Joseph Hooker
153. 85.04.15 from Gerhard Renner
154. 85.04.21 to Ferdinand von Krauss
155. 85.05.15 to Otto Tepper
156. 85.05.18 to Adolpho Möller
157. 85.05.25 to William Thiselton-Dyer
158. 85.05.30 to Joseph Hooker
159. 85.06.08 to Ferdinand von Krauss
160. 85.06.10 to Otto Tepper
161. 85.06.24 to Alfred Moloney
162. 85.06.26 from Ferdinand Wehl
163. 85.06.27 to Graham Berry
164. 85.06.29 to William Thiselton-Dyer
165. 85.07.11 to Johan Lange
166. 85.07.13 to Adolf Engler
167. 85.09.10 from N. Samwell
168. 85.09.21 to Eduard von Regel
169. 85.10.10 to Edward Strickland
170. 85.10.15 to David Lindsay
171. 85.11.03 to John Buchanan
172. 85.11.28 to William Thiselton-Dyer
173. 85.12.12 to Mary Kennedy
174. 85.12.14 to Alexander Macdonald
175. 85.12.21 from Karl Kirchhoff
176. 86.01.18 from Julian Thomas
177. 86.02.17 to William Thiselton-Dyer
178. 86.03.05 to Jean Müller
179. 86.04.14 to Paolo Dattari
180. 86.05.22 to Francis Barnard
181. 86.05.23 to Alexander Magarey
182. 86.06.20 to William Carruthers
183. 86.07.06 to Henrietta Wehl
184. 86.08.21 to Alfred Deakin
185. 86.10.04 from Frederick Hagenauer

186. 86.11.03 to Thomas Whitelegge
187. 87.01.10 to William Thiselton-Dyer
188. 87.01.25 to Henry Ranford
189. 87.02.11 to Edward Ramsay
190. 87.03.02 to Charles Moore
191. 87.03.20 to John Brooking
192. 87.03.30 to William Thiselton-Dyer
193. 87.04.28 to Henry Moors
194. 87.05.11 from James Dickinson
195. 87.05.17 to Henry Ridley
196. 87.05.21 to Charles Riley
197. 87.06.11 to Ferdinand von Krauss
198. 87.09.16 to William Thiselton-Dyer
199. 87.10.26 to Henry Forbes
200. 87.10.26a to Rudolph Virchow
201. 87.11.01 to Henry Parkes
202. 87.11.14 from William Macleay
203. 87.11.24 from William Tietkens
204. 87.12.00 to [Philip Sclater]
205. 87.12.16 to Rudolf von Fischer-Benzon
206. 88.01.26 to Asa Gray
207. 88.02.22 to Adolf Engler
208. 88.03.19 to William Lawes
209. 88.04.18 to Paul Maistre
210. 88.05.24 to Samuel Davenport
211. 88.08.01 to Thomas Wilson
212. 88.09.18 to Walter Gill
213. 88.11.02 from Jacob Agardh
214. 89.01.01 to George Stokes
215. 89.01.27 to Isaac Balfour
216. 89.01.28 to Malcolm Fraser
217. 89.02.05 to Alfred Deakin
218. 89.03.17 to Ralph Tate
219. 89.05.09 from William Armstrong
220. 89.05.12 from Ferdinand von Krauss
221. 89.05.29 to Eduard von Regel
222. 89.07.12 to Friedrich Franz III, Grand-Duke of Mecklenburg-Schwerin

223. 89.07.26 from James Thomson
224. 89.08.10 to Arthur Lucas
225. 89.09.01 from Mordecai Cooke
226. 89.09.30 from Jemima Irvine
227. 89.12.07 to Thomas Wilson
228. 90.01.28 to William Thiselton-Dyer
229. 90.03.24 from Hans Gundersen
230. 90.04.03 to Francis Barnard
231. 90.07.02 from Thomas Elder
232. 90.07.18 from Daniel McAlpine
233. 90.07.19 to Alfred Deakin
234. 90.08.01 to Jacob Agardh
235. 90.08.27 from John MacKay
236. 90.08.28 from Friedrich Schmitz
237. 90.08.30 from Wilhelm Bäuerlen
238. 90.11.11 to Edward Ramsay
239. 90.11.14 to Léon Dejardin
240. 91.01.21 to Joseph Hooker
241. 91.01.24 to Jane Gray
242. 91.03.12 to Thomas Wilson
243. 91.05.13 to Edward Ramsay
244. 91.06.22 to Thomas Wilson
245. 91.07.00 to Walter Gill
246. 91.07.01 to [William Thiselton-Dyer]
247. 91.07.27 to James Stirling
248. 91.08.14 to Jacob Agardh
249. 91.10.04 from Richard Helms
250. 91.12.25 to Otto Tepper
251. 92.01.12 to Nicholas Holtze
252. 92.01.16 to Thomas Wilson
253. 92.01.17 to Allan McLean
254. 92.02.05 from George Murray
255. 92.02.20 to Otto Tepper
256. 92.03.15 to Otto Nordstedt
257. 92.04.27 from David Lindsay
258. 92.05.16 to Thomas Sisley
259. 92.08.05 from Frederick Bailey
260. 92.08.06 to [Adolf Engler]

261. 92.08.19 to William Thiselton-Dyer
262. 92.09.03 to Josiah Cooke and Charles Jackson
263. 92.09.10 from George Perrin
264. 92.12.10 from William Potter
265. 93.00.00 to Jacob Agardh
266. 93.01.14 to Veit Wittrock
267. 93.02.24 to Paul Ascherson
268. 93.04.15 to Friedrich Krichauff
269. 93.04.23 to Lord Kelvin
270. 93.05.20 from William Thiselton-Dyer
271. 93.05.31 to Frederick Bailey
272. 93.06.03 from Frederick Bailey
273. 93.06.20 to Pier Saccardo
274. 93.08.20 to Annie Walker
275. 93.08.21 to Franz Stephani
276. 93.10.15 to Henry Deane
277. 93.10.16 from Eugene Hilgard
278. 93.11.03 to William Thiselton-Dyer
279. 93.11.25 to Henry Deane
280. 93.12.19 to Joseph Hooker
281. 93.12.23 to Alexander Macdonald
282. 93.12.25 to Thomas Wilson
283. 94.01.05 from Sophie Hooker
284. 94.02.09 from Alphonse Milne-Edwards
285. 94.04.19 to Walter Gill
286. 94.06.08 from Thomas Kirk
287. 94.07.30 to John Shillinglaw
288. 94.08.05 to Joseph Hooker
289. 94.08.23 from Rose Grainger
290. 94.09.04 to Charles Musson
291. 94.10.16 from Georgina King
292. 94 10.29 from John Brooks
293. 94 11.11 to Jacob Agardh
294. 94.11.17 to Benjamin Robinson
295. 95.01.05 from Lancelot Lindley-Cowen
296. 95.01.28 to Francis Reddin
297. 95.03.04 to Jean Müller
298. 95.06.02 to Veit Wittrock

299. 95.06.18 to Jacob Agardh
300. 95.06.28 from [Charles Topp]
301. 95.07.01 from the Deutsche Verein von Victoria
302. 95.08.01 to Charles French
303. 95.08.29 to Clara Ryan
304. 95.09.03 to Veit Wittrock
305. 95.09.26 to Thomas Cheeseman
306. 95.10.01 from May Wise
307. 95.10.05 to Jacob Agardh
308. 95.10.25 to Jessie Hussey
309. 95.12.01 to John Keltie
310. 95.12.03 from Eugene Hilgard
311. 96.01.17 to Baldwin Spencer
312. 96.02.04 to Jacob Agardh
313. 96.02.09 to [William Thiselton-Dyer]
314. 96.03.02 from Joseph Hooker
315. 96.03.03 to Ellis Rowan
316. 96.03.11 from Ignatz Urban
317. 96.03.31 to Casimir de Candolle
318. 96.04.12 to Walter Rothschild
319. 96.04.27 to Ada Armytage
320. 96.06.14 to John Shillinglaw
321. 96.07.02 to Louise Matthiessen
322. 96.07.14 to Georgina King
323. 96.07.18 to Jacob Agardh
324. 96.07.22 from Robert Etheridge, Jr
325. 96.08.16 to Thomas Hart
326. 96.09.08 to Édouard Bornet

Letters about Mueller (see p. 745)

327. 77.07.11 Louis Smith to Graham Berry
328. 90.01.22 James Walker to Mary Walker
329. 94.05.23 Kurt Lampert to Gustav von Silcher
330. 95.06.24 Charles Naudin to the Académie des Sciences, Paris
331. 97.00.00 Grave Monument Appeal
332. 97.11.00 Baron von Mueller National Memorial Fund

Selected Correspondence, 1876-1896

Circular[1]

76.03.29

K76/3065, unit 876, VPRS 3991/P inward registered correspondence, VA 475 Chief Secretary's Department, Public Record Office, Victoria.[2]

For the completion of an Universal work on Australian indigenous plants (on which mainly under the auspices of the Victorian Government the Undersigned has been engaged in this continent for more than 28 *years*) it is desirable to obtain additional collections of plants[3] in a pressed and dried state, particularly from districts far inland or recently settled. It is an important aim by these means to trace out the exact geographic limits of the many thousand species, which constitute the original Vegetation of Australia, in order also that all observations on their respective utilitarian value, whether for pastoral cultural medical or industrial purposes, may become recognized and applicable to the widest extent. Moreover it is necessary to study still further the degrees of variability, to which all kinds of plants are more or less subject, with a final view of circumscribing the exact characteristics of each species. It is to be impressed on those, who may feel interested in the promotion of such researches, not to exclude from local collections any plants, merely because they appear frequent or insignificant,[4] The process of drying plants for permanent collections is simple and easy in the extreme, it needs hardly any explanation, beyond perhaps the remark, that the parcels of paper, containing any recently gathered plants, after a few hours pressure should be divided into thin

1 See M to J. MacPherson, 29 March 1876 [Collected Correspondence].

2 For published versions of this text see B76.04.02 and B76.05.04.

3 On 10 April 1876 MacPherson submitted a 'Schedule of Transfers submitted by the Chief Secretary to the Governor in Council for Approval' requesting a transfer of £50 from 'Publication of work on Australian plants' to 'Collecting botanical specimens, clerical assistance &c'. The reason given was 'Insufficiency of the vote'. His request was approved by the Governor in Council on the same day (K76/3574, unit 873, VPRS 3991/P, PROV).

4 (for none of these objects of nature are called forth without a divine design for distinct purposes, which we may well endeavour to ascertain). *deleted.*

76.03.29 sets, and be spread out on a dry or warm place, to facilitate and to spread the exsiccation, and to lessen also the requirements of shifting plants from paper which became moist, into dry paper. Small plants should be gathered with their roots, and all not merely in flower but in fruit also, as indeed from the latter generally the main-characteristics are derived. Water weeds, rushes, sedges, mosses, lichens, fungi (and on the seacoast also algae), even if ever so small, should not be passed in collecting. Transits are best effected early after the preparation of the specimens in small parcels closely packed by successive mails. Whoever wishes to become scientifically acquainted with the native plants of his vicinity or of localities otherwise accessible to him, can obtain the specific names, if a duplicate set is retained, in which the specimens are numbered correspondingly to those of the transmitted set. An intimate knowledge of the indigenous vegetation, while it largely indicates climatic and geologic circumstances, tends also to afford an insight not only into the natural vegetative resources of any tract of a country, but also into much of the pastoral or cultural capabilites of the respective localities. Researches of these kinds become furthermore the sources of educational works, and unfold to well trained and intelligent minds pure recreative and healthful pleasures, inexpensively everywhere within reach.

[5]Ferd. von Mueller,
M & Ph.D., FRS.

5 Baron *deleted.*

To Eduard Fenzl[1]

76.06.10

Eduard Fenzl Nachlass, Archiv der Universität Wien, Vienna.

Melbourne
10/6/76.

Nehmen Sie meinen besten Dank hin, edler Herr, für die Freundlichkeit, mit welcher Sie mich über Ihre schöne Pomaderris obcordata belehrten. Ich erbitte nun wieder Auskunft von Ihrer Güte.

In Endlichers Atakta ist eine Anamirta Baueriana abgebildet, aber der text geht nicht so weit, um etwas über die Entdeckungsstelle zu erfahren. Im Allgemeinen sind Bauer's Pflanzen, die unter Ihres grossen Freundes Endlicher Bearbeitung kamen, entweder von dem Festlande Australiens oder von der Norfolk-Insel. Recht leicht wäre es möglich, dass Bauer seine Pflanze an der Carpentaria-Bucht fand; indessen scheint niemand sonst eine Art dieses genus dort oder sonstwo in Australien gesehen zu haben. Die Tropenküsten sind bei uns freilich immer noch nicht vollständig untersucht, selbst nachdem ich mehrere hundert Arten von denselben als supplementar zu Bentham's u meinem Werke beschrieben habe. Der Nachweis des Vorhandenseins von Anamirta in Australien nach den Original-vorlagen von Bauer wäre immerhin wünschenswerth. Das Vorkommen dieses Genus bei uns würde auch nicht befremden, da es sich ja wie Stephania vom Festland Indiens nach Ceylon und dem Archipelagus erstreckt. Dann entstände eine zweite Frage. Ist A. Baueriana wirklich von A. Cocculus (oder besser und gerechter gesagt A. paniculata) verschieden. Mein indisches Material ist nicht sehr vollständig, aber so weit wie ich urtheilen kann, sind keine Unterschiede da.

Nun, ehrenwerther Freund, schonen Sie sich im Herbst des Lebens. Ich werde längst begraben sein, ehe ich Ihr venerables Alter erreiche, in dem Sie noch solche Jugendfrische für die Wissenschaft bewahren.

Verehrungsvoll der Ihre
Ferd. von Mueller

1 MS black edged – M's brother-in-law, Eduard Wehl, died on 11 February 1876.

76.06.10

Melbourne
10 June 1876.

Accept my best thanks, noble Sir, for the kindness, with which you informed me about the lovely *Pomaderris obcordata*.[2] Now I ask once more information from your kindness.

In Endlicher's *Atakta*[3] an *Anamirta baueriana* is illustrated, but the text does not extend far enough to find out something about the locality, where it was discovered. Generally speaking, Bauer's plants, which were worked on by your great friend Endlicher, came either from the Australian mainland or from Norfolk Island.[4] It is quite possible, that Bauer found his plants at the Gulf of Carpentaria; however, nobody else appears to have seen a species of this genus there or anywhere else in Australia. Admittedly our tropical coasts here are not yet fully explored, even after I have described several hundred species from there as a supplement to Bentham's and my work.[5] Proof of the presence of *Anamirta* in Australia from Bauer's original specimens would certainly be desirable. The occurrence of this genus here would not surprise as, like *Stephania*, it extends from the Indian mainland to Ceylon and the Archipelago. That leads to a second question: is *A. baueriana* really different from *A. cocculus* (or better and more justly *A. paniculata*)? My Indian material is not very complete, but as far as I can judge, there are no differences.

Now, worthy friend, take care of yourself in the autumn of your life. I shall be buried long before I reach your venerable age, in which you still maintain such youthful enthusiasm for science.

Respectfully your
Ferd. von Mueller.

2 See M to E. Fenzl, 28 October 1875 [Collected Correspondence].
3 Endlicher (1833).
4 Ferdinand Bauer accompanied Matthew Flinders on his circumnavigation of Australia, 1801-3.
5 In a letter to G. Bentham, 14 April 1876 [Collected Correspondence], M claimed to have described in his *Fragmenta* over 500 species additional to those described in the first six volumes of *Flora australiensis*.

To Ferdinand von Krauss[1]

76.06.10a

Staatliches Museum für Naturkunde, Stuttgart.

Privat

10/6/76

Endlich einmal wieder ein Lebenszeichen, edler Freund. Ich habe eben mit dem Dampfschiff Northumberland eine Kiste an Prof Dr Eichler in Kiel zu senden, u. darin ein Kästchen, welches für Sie 2 schöne Exempl. von Sphenodon punctatum[2] aus dem nördlichsten Theil Neuseeland. Dies Thier ist ausserordentlich selten u schwer zu erlangen, da es nur auf den kleinen fast unlandbaren und unersteigbaren Inselfelsen vorkommt.

Auch ein paar Gorgonien ex. sind aus Polynesien beigefügt. Endlich ist auch wieder eine Fischsammlung begonnen. Die Stockung der Sendungen entstand theils aus dem gänzlichen Mangel an Räumlichkeit u Einrichtung für selbst meine Amtsarbeiten in den letzten 3 Jahren, theils bei den zerrütteten Zuständen, in dem in *dieser* Colonie irgend eine Anstalt für *eine Weile* verfallen mag. Denken Sie nur an die Zeit der Commune jüngst in der hoch civilisirten französ. Hauptstadt. Im Staatsdienst müssen wir uns *hier* den Verhältnissen fügen, gerade wie im Militärdienst, während der letztere doch sichernde Privilegien hat.

Von der Anstellung eines Oberrichters zu der eines Telegraphboten, von der Anstellung eines Museum-Trustee zu dem eines[3] Hospitalsmagd, von der Anstellung des Befehlshabers einer geographischen Expedition zu der eines Gartenburschen hängt hier fast Alles von *politischem* Einfluss ab, u diesem wird *Alles* geopfert,

1 MS black edged – M's brother-in-law, Eduard Wehl, died on 11 February 1876. MS annotation by Krauss: 'an Eichler d. 19. Sept geschr. Rep. 26. Oct. 76. Dank für Schriften Dromaeius, Ceratodus, Ornithorhynchus, Thylacinus, Sarcophilus, Dingo' [wrote to Eichler 19 September. Replied 26 October 1876. Thanks for papers Dromaeius, Ceratodus, Ornithorhynchus, Thylacinus, Sarcophilus, Dingo].

2 MS annotation interlined by Krauss: '20th erhalten' [received 20th].

3 einer?

76.06.10a damit die Machthaber für eine Zeit das Land zu ihrem Vortheil ausbeuten können. Natürlich giebt es auch darin löbliche Ausnahmen. Mein jetziger Minister ist ein Mann von *Bildung* u gutem Character, so dass ich besseren Zeiten trauen darf, u damit mehr auch für *Sie* werde wirken können.

Stets der Ihre
Ferd. von Mueller

Private

10/6/76

At last once again a sign of life, noble friend. I have just sent off a box to Prof. Dr. Eichler in Kiel by the steamer *Northumberland*, and in it a small box with 2 splendid specimens of *Sphenodon punctatum* from the northernmost part of New Zealand for you. The animal is extraordinarily rare and difficult to get, since it occurs only on the small almost unlandable and unscalable rock islands.

Also a couple of *Gorgonia* specimens from Polynesia are included. At last a collection of fish is also begun again. The stoppage of the consignments originated partly from the complete lack of space and accommodation for even my official work in the last 3 years, partly by the unsettled conditions in which in *this* Colony any kind of institution may decline for *a while*. Just consider the period of the Commune recently in the highly civilised French capital. In the state service we must accommodate ourselves to the circumstances, just as in the military service, whereas the latter do have protecting privileges.

From the appointment of a chief judge to that of a telegraph messenger, from the appointment of a Museum trustee to that of a hospital maid, from the appointment of the commander of a geographical expedition to that of a garden boy, almost all here depends on *political* influence, and *everything* is sacrificed to this, so that the rulers can exploit the country to their advantage for a time. Naturally there are also honourable exceptions to this. My present minister[4] is a man of *education* and good character, so that I may trust in better times and therewith also will be able to produce more for *you*.

Always your
Ferd. von Mueller

4 John MacPherson.

To John MacPherson

76.06.14

K76/6174, unit 930, VPRS 3991/P inward registered correspondence, VA 475 Chief Secretary's Department, Public Record Office, Victoria.

Melbourne, 14/6/76.

The honorable John Macpherson,
Chief Secretary.

Sir,

In compliance with your request I have the honor of transmitting to you herewith a copy of four of the volumes, issued by our Acclimation-Society, in which treatises on industrial plants are contained.[1] If my suggestion of their reprint in a copious edition[2] for wide distribution here met with your approbation, then it might be advisable to have the whole rearranged into one alphabetic series, for which purpose the needful number could be put to each species, on returning to me for this purpose the now transmitted copy. I beg to add, that the Accl. Society had only a limited number of copies printed, and that none of these are obtainable now any more.

I have the honor to be, Sir,
your obedient servant
Ferd. von Mueller.[3]

1 B71.13.01, B72.13.02, B74.13.06, B75.13.11.

2 B76.13.03.

3 On 4 July 1876, Sir S. Wilson wrote to MacPherson noting that these papers by M were out of print, and urging 'that a few thousand copies be reprinted for sale or distribution' (K76/7042, unit 930, VPRS 3991/P, PROV). M subsequently recommended that, the Government Printing Office being overwhelmed with work, a tender he had obtained from McCarron, Bird & Co. for printing the volume be accepted; see M to J. MacPherson, 21 September 1876 [Collected Correspondence]. MacPherson approved, and later allocated additional funding to permit a total of 3,000 copies to be printed.

76.07.03 *To Magnus Massmann*[1]

Melbourne
3/7/76

Es war recht gütig von Ihnen, edler Herr, dass Sie dem Wunsche des gütigen Herrn Bürgermeisters Passow nachkamen, und auf eine so befriedigende Weise für die Instandhaltung der Grabstätte meines verehrten Vaters sorgten. Es war mir nicht bekannt geworden, dass der ehrenwerthe Hr Passow auch schon aus dem Leben geschieden sei. Wollen Sie gütig seiner Familie meine Condolenz ausdrücken. Sollten Sie, geehrter Herr Doctor, es für anwendbar halten, dass eine grössere Summe für die Erhaltung des Grabes meines Vaters belegt werde auf dortiger Sparcasse, so bitte ich Sie, mich darüber zu benachrichtigen. Der Familie des Herrn Bürgermeisters Passow schuldige ich auch noch den Ausspruch des Dankes, für die Aufmerksamkeit, welche der Verstorbene in mitten anderer Angelegenheit so sorgfältig auf meine verwandte.

Gern hätte ich spät im Leben das Grab des Vaters und das Vaterhaus (das Mönchenthor) noch einmal wieder gesehen, und vielleicht ist mir diese Freude noch einst vergönnt, wodurch mir auch die Ehre zu Theil würde Ihnen persönlich meine Aufmerksamkeit beweisen zu können. Darf ich Sie bitten das beigefügte Photogram als ein anspruchsloses Souvenir annehmen zu wollen.

Mit Ehrerbietung
der Ihre
Ferd. von Mueller

Herrn Prof. Dr Roeper bitte ich meinen Gruss auch gütig entbieten zu wollen.

1 Letter not found. The text used here is from a photograph of the MS published in Kobert (1911), p. 33 (fig. 9); also reprinted in Jahn & Schmidt (1996), appendix.

Melbourne, 3 July 1876. 76.07.03

It was very kind of you, noble Sir, to have complied with the wishes of the kind Mayor Passow, and cared for the upkeep of the grave of my venerated father in such a satisfactory manner. I had not heard that the esteemed Mr Passow had died. Will you kindly express my condolence to his family. Please notify me, esteemed Doctor, if you should consider it useful, that a larger sum of money be deposited in a savings bank account there for the upkeep of my father's grave. I also owe it to the family of the late Mayor Passow to express my gratitude for the attention which, in the midst of other concerns, the deceased accorded mine with so much care.

I should like very much to have seen once more late in life my father's grave and home (the Mönchenthor), and perhaps this great joy will be granted to me yet, whereby I should also have the honour to pay you my respects in person. May I ask you to accept the enclosed photogram as a modest souvenir.

With respect your
Ferd. von Mueller.

Please be so kind as to convey my regards to Professor Roeper.

To Edward Ramsay 76.07.10

ML MSS. 562, Letters to E. P. Ramsay 1862-91, Mitchell Library, State Library of New South Wales, Sydney.

10/7/76

Herewith, dear Mr Ramsay, *fresh* seeds of the best Cinchona, the red one, so rich in Quinin. The plant ought to become important for your sheltered fernglens.

Did you or your brother ever see a true Bamboo in the N.[1] Of course I do not mean the Phragmites, the common cosmopolitan species, which dies down every year

1 Ramsay had spent some time in the 1860s in the Clarence River and Richmond River districts of northern NSW, and in working a sugar plantation near Maryborough, Qld. In the latter venture he was accompanied by his younger brother Percy (b. 1848).

76.07.10 We want all these Gramineae timely for vol. VII.[2] Is still any one of your horticultural establishment[3] in the North? Remember that whoever is there has a *rare chance*, the best yet in Australia, to distinguish himself by collecting[4] carefully all the *jungle* plants in flower – & fruit – specimens, particularly those of trees. I feel somewhat disappointed, that after several years of writing not a single one of my N. Correspondents has managed to send me a bundle of real good or rare specimens, when the finder would earn lasting merit from very little toil.

Regardfully
Ferd von Mueller

76.08.12 *To William Odgers*

J76/8599, unit 883, VPRS 3991/P inward registered correspondence, VA 475 Chief Secretary's Department, Public Record Office, Victoria.

Melbourne,
12/8/76.

W. H. Odgers Esqr,
Undersecretary.

Sir

I have the honor to enquire, whether the hon. the Chief Secretary will be pleased to arrange with the hon. the Minister of Mines, that through Mr M'Lellans Department a small map may be issued, which could be placed against the title-page of my School-Flora. For this purpose the Mines Department has facilities, which I do

2 Of *Flora australiensis*; i.e. Bentham (1878).
3 The Ramsay family's Dobroyd plant and seed nursery, Ashfield, NSW.
4 Marginal insertion by M: 'there may be any day competition from abroad'.

not enjoy in my establishment, and Mr M'Lellan is sure to be willing to aid me thus far. 76.08.12

The map would require to be slightly smaller than the one transmitted herewith;[1] it would need to be complete up to date; it should be correct in detail (Mt Buller for instance being placed wrongly on the accompanying map;) but no necessity exists, to have the names and limits of the counties on it, nor need the map to be colored

I have the honor to be,
Sir, your obedient
Ferd. von Mueller.[2]

1 Map not found.

2 On 15 August 1876, the Chief Secretary, J. MacPherson, minuted: 'Perhaps the Mines dept will inform me if facilities exist there for the performance of such work and if it will be undertaken without inconvenience I should have thought the Lands dept the most appropriate for such work'. The Minister of Mines, W. McLellan, replied on 28 August: 'Enquiries have been made in this matter, and it is considered that the map should be engraved on copper. The Lands Department employ the only officers who can undertake such work, and it is suggested that the engraving might be done at the Lands Office, and that the maps might be subsequently printed by contract.'

On 22 November, on a page of Rough Proof that had been sent to him, MacPherson wrote to the Government Printer, J. Ferres: 'As a work for the use of our State School children such a work as this represents would be of little or no value – & from that point of view it is needless to proceed with it' (K76/13910, unit 883, VPRS 3991, PROV). From the proof, it is evident that the proposed work was very different in character from the educational text M published a year later, namely B77.13.07. The work here proposed was never published. M subsequently used the material that had been typeset in a different work, namely B79.13.08.

On 30 November Odgers noted: 'The C.S. [Chief Secretary] has stopped the printing of the book'.

76.09.23 *To Joseph Hooker*[1]

RBG Kew, Kew correspondence, Australia, Mueller, 1871-81, f. 183.

23/9/76.

Let me express in first instance, dear Dr Hooker, my best gratulation to your new matrimonial union with the distinguished Lady, who is to grace your home.[2] I wished only that a similar happiness would brighten my future, but it will not likely be, unless I can regain my Department, my House and position, and prosperity

I trust your Lady will soon be styled Lady Hooker. Pray express also to her my reverence, my remembrance of the distinguished editor of Natural History Library.[3] My Card herewith.

In the undeserved oppression which I have still to endure the unexpected gift of one of the three portions of Mr J. [St] Mill's collection, has been a great moral support to me, and I consider it as one of the triumphs of my life. Two of the daily journals of Melbourne have noticed in becoming terms the honor thus far conferred by your generosity and Miss Taylor's friendliness on this Colony.[4] I have endeavoured to render this country honored and respected abroad in my branch of knowledge, but have gone myself thereby to ruin.

The young Doctor, Mr Baker's[5] disciple, can express to him and you how sad, homeless and cheerless he found me. If however an earnest appeal from *Europe* was made, I might come as a Phoenix

1 MS black edged – M's brother-in-law, Eduard Wehl, died on 11 February 1876. MS annotation by Hooker: 'An[swere]d. Dec 14'. See J. Hooker to M, 15 December 1876 [Collected Correspondence].

2 Hooker married Hyacinth (née Symonds), widow of Sir William Jardine, on 22 August 1876.

3 i.e. W. Jardine.

4 While living in France in his youth, John Stuart Mill 'took up botany as an amusement … under the influence doubtless of George Bentham ... and was always an enthusiastic collector, although not a scientific botanist' (DNB). Bentham (1997) has scattered references to collecting with Mill. A portion of Mill's collection is still held at MEL.

5 J. G. Baker.

out of the Ashes; here no one carries the influence of a Hooker, Darwin, Paget, Bentham, Cobbold, Owen &c. 76.09.23

With my best wishes for your happiness,
Ferd. von Mueller.

To August Petermann[1] 76.10.24

Briefsammlung, Archiv, Justus Perthes Verlag, Gotha, Germany.

24/10/76.

Diesmal sende ich Ihnen, lieber Herr Professor, wieder allerlei über Explorationen. Die Notiz von der Queensland Nordwest Exp.[2] bedarf einiger Erörterung. Der Name Diamantina ist substituirt für den Fluss, den mir M'Kinley widmete. Hier sollte Priorität entscheiden, u. ich hoffe, dass Sie Ihren Landsmann seines besten geograph. Monumentes nicht beraubt sehen wollen. Dies wäre eine gute Gelegenheit einmal die Prioritätsfrage in der Geographie zu besprechen, zumal da man den Namen des Mt Hotham, den ich schon 1854 gab, jetzt überall zu unterdrücken sucht, was doch auch des Gouverneurs Hotham Andenken unterdrückt, obwohl er hier starb und man ihn ehren sollte, wie ich es wünschte. Sie haben ja selbst die Entwicklung der Geographie Australiens verfolgt, und können also leicht ausfinden, in welchem Jahre der Name Diamantina entstand, und ob er einem *Nebenflusse* oder *Hauptflusse* zukommt. Sie sind ja doch die grösste practische Autorität der Jetztzeit in diesen Fragen; so verlässt sich ein deutscher Gelehrter auf Sie besonders und ruft Ihre Gerechtigkeit an.

Stets Ihr
Ferd. von Mueller.

Dieser Brief ist *nicht* für Veröffentlichung geeignet

1 MS annotation by Petermann: 'Erh. 31. Dez. 1877.' [Received 31 December 1877.] MS black edged – M's brother-in-law, Eduard Wehl, died on 11 February 1876. For a published version of this letter see Voigt (1996) pp. 131-2.

2 Expedition.

76.10.24

24 October 1876.

This time, dear Professor, I am sending you again a miscellany about exploration. The notice about the Queensland North-west expedition[3] requires some comment. The name *Diamantina*[4] has been substituted for the river dedicated to me by McKinlay.[5] Priority should decide here, and I hope that you do not want to see your compatriot robbed of his best geographical monument. This would be a good opportunity to discuss the question of priority in geography, especially as they now seek everywhere to suppress the name Mount Hotham, which I gave already in 1854, which in turn suppresses the memory of Governor Hotham, even though he died here and should be honoured as I wished him to be.[6] You yourself have followed the development of Australian geography and could easily find out, in which year the name Diamantina was first used, and whether it was attached to a *tributary* or a *major river*. After all, you are the greatest practising authority of the present time in these questions; thus a German scientist relies particularly on you and appeals to your justice.

Always your
Ferd. von Mueller.

This letter is *not* suitable for publication.

3 An expedition led by William Hodgkinson to explore the country west of the Diamantina River to determine the extent of pastoral land in the region.

4 The Diamantina River, about 900 km long, with its source in western Qld, combines during the rainy season with the Georgina River to form the Warburton River, which flows into Lake Eyre, SA.

5 McKinlay in 1862 had named 'Mueller's Creek' a 'magnificent stream' that he had followed for a time while searching for Burke and Wills; see M to J. McKinlay, 7 June 1863 [*Selected correspondence*, vol. 2, pp. 198-9]. This subsequently proved to be a stretch of the same river that had been named Diamantina by William Landsborough, leader of another of that year's Burke and Wills relief expeditions, in honour of Lady Diamantina Bowen, wife of Sir George Bowen, Governor of Qld; and Landsborough's name prevailed despite the priority of McKinlay's naming.

6 See M's November 1876 MS 'On the rules of priority of geographic names', submitted to the Royal Geographical Society but not published [Collected Correspondence]. The mountain named 'Mt Hotham' by M is now known as Mt Feathertop, while the name Hotham has been bestowed on a nearby peak. The compass bearings M recorded for the peaks he named could not be reconciled with later surveys, due, it is now recognised, to magnetic interference from the basalt rocks of the area introducing errors into M's compass readings. See Gillbank (1992).

To William Branwhite Clarke 76.12.25

ML MSS.3608 Clarke papers, Mitchell Library, State Library of New South Wales, Sydney.

Christmasday
25/12/76.

Private

Let me thank you for your friendly letter,[1] rev. and enlightened friend, and for the sending of the learned essays, among which the splendid array of facts, explained in that "on the effects of forests on climate" particularly interested me.[2] I have to thank you for the generous manner, in which you alluded to myself in your writing; this is a great support to me, while my once celebrated Department remains almost annihilated, and while almost all my preaching on the importance of forests culture and forest-conservancy was to deaf ears here. It is a sad spectacle in your colony to see even a Legislator publicly write against the conservation and maintenance of woods; and this seemingly one, who lost most of his heards[3] in treeless runs, and who barely escaped with his life in a subsequent flood.[4] It matters little, my venerable friend, whether

1 Letter not found.

2 Clarke (1876). An offprint of Clarke's article, inscribed 'With the Author's regards', is at the Royal Botanic Gardens Melbourne.

3 herds?

4 Clarke read his paper before the Royal Society of NSW on 1 November 1876. It was initially published in the *SMH* on the following Saturday, 4 November. On 8 November an editorial in the *SMH* declared: 'The Rev. W.B. Clarke has done well once more to call attention to this subject, as Baron von Mueller has done in Melbourne. We wish that the Government and the public would give it the consideration which it deserves.'

On 13 November the *SMH* published a letter from L. Fane de Salis, MLC, defending the widespread ring-barking of eucalypts by graziers – 'an improvement known by us to give value to worthless scrub land'. Clarke's reply was published on 14 November, followed by another editorial on 16 November and a further letter from de Salis on 20 November, referring to Clarke's 'doubtful commercial figures, as well as those figures by which he may prove to us, on the authority of Baron von Mueller, that man can *create* water'.

A further, long letter from de Salis was published in the *SMH* on 15 December, vigorously criticizing Clarke and also 'German professors' who 'adept enough

76.12.25 Mr Salis gives to a downtrodden man an other stab in his ruin. As I find now late in life, that I followed a *wrong plan of life*, it is immaterial how much I am persecuted. But apart of myself, I cannot see why Mr S. takes the "German Professors" to task merely. Have not French, Scandinavian and American writers said as much as my countrymen have on the forest question and in the same train not to speak of your own English Mr Marsh?[5] England with its coals, its combustible turf, has less felt the want of forests, as moreover its insular position in the cold-temperate zone keeps its climate moist, and as it[6] numerous harbours allow ready import of timber, *so long as that import may last* & be possible.

Even in the Gilbert-Group,[7] from whence I have just received through the Rev. Mr Whitmee the very few plants existing there, even on these minute specks of corals in the wide ocean the rain falls often not for very many month, because besides some Coconuts hardly any trees exist there.

As for the value of timber after ringing, it must be considered, that our hard woods yield not well to the saw when indurated

in their own speciality, are somewhat given to "inconclusive" theories on other topics':

> Humbly and blindly acknowledging God's wisdom, we yet know that, at a late period of the creation, he placed man here as a conquering fighting animal. German professors might even quote Scripture to the effect that conquering man ought to extirpate other races of men that differ from him in language or in ideas – I won't. But I know that it is man's instinctive right and function to extirpate all interfering insects, beasts, trees, vegetables, fungi – everything opposed to man's progress – and that, acting in direct accordance with this truth – however German professors may slander the present climate of these countries – our forefathers in England, Ireland, Scotland, France, Germany, Belgium, Holland did right in cutting down the forests – AND HISTORY PROVES MY ASSERTION.

The *SMH* of 18 December carried a brief note from Clarke, declining to reply to de Salis's 'extraordinary' communication: 'its style of argumentation and its tone render that impossible'.

5 Clarke in his essay refers to George P. Marsh (1864) *Man and nature, or physical geography as modified by human action*. However, Marsh was American, not English.

6 its?

7 Gilbert Islands, Pacific Ocean.

after a protracted period; hence such wood cut after years can only be used for fuel, and that inferior to fresher woods; but how much waste all this involves on public lands. Surely everyone knows, that trees bring up from lower strata the *mineral* elements of nutrition particularly the phosphates of lime & potash. See, how the foliage of European forests is sought for manure! There are many grasses, which will live under trees, many mentioned in my new edition of the "industrial plants".[8] The naturalisation of such grasses and (herbs) should be encouraged. 76.12.25

Possibly your rivers may contain as much water as before, though the Cedarbrushes are cut down; but where are the incomparable Red Cedars to come from for your nepotes? Besides there is a vast difference between water merely running down rivers & there being lost, between water soaking gradually down slopes, sending out vapor on the way, *renewing rains*; surely it must be apparent to anyone's understanding, that even if the Rivers are still as high, that the fall of rain must have diminished, as in the forest regions formerly only a small *portion* of the water run down at all to the brooks & streams. If those, who remove the trees, *did really sow Luzern, Sanfoin,*[9] *dense nutritious perennial grasses*, then the diminished rain might still be retained, but to my mind there is an enormous difference between foliage surface of evergreen trees and the grasses, parched from Christmas til Easter, when we wish most to prevent the heating of the soil & the escape of water.

I admire the youthful freshness of your mind. May providence leave that blessing to you for many years to come yet in best bodily health.

With best salutation to the new year.
Ferd. von Mueller

Cereals would give *some* compensation for forest trees, but the stubble is a poor concern after harvest also, so far as climate is concerned in hot & dry countries. Perhaps you have occasion to promulgate in some of your ingenious writings these views of mine.

8 B76.13.03.
9 Sainfoin (i.e. *Onobrychis viciifolia*)?

76.12.25 Every tree, however miserable helps to shade & cool a certain space of ground.

The two main-principles, which I have all along laid down for Australian forest-culture at the *present* time are
1, local forest boards,
2, Revoting of the revenue derived from forests for the maintenance and augmentation of the woods.

It would be unreasonable to blame so excellent a friend as yourself for any hostile remarks of Mr Salis, especially as you so well defended me

76.12.28 *To Count Louis Torelli*[1]

Melbourne 28 décembre 1876

Monsieur LE COMTE,

Votre communication m'honore beaucoup, ainsi que la confiance, que vous me montrez en me consultant touchant les plantations d'Eucalyptus qui doivent être entreprises dans les marécages fievreux des environs de Rome, sous les auspices éclairés du gouvernement italien. Des essais ont prouvé, que l'Eucalyptus globulus n'est pas toujours le mieux adapté aux plantations dans les terres marécageuses; de plus cette espèce est plus sujette aux gélées, que beaucoup d'autres. J'ai en conséquence conseillé dans plusieurs cas, que l'Eucalyptus rostrata fût choisi pour les régions marécageuses, pour lesquelles il est le plus propre.

1 Letter not found. For the text used here see B78.13.07.

Pour l'Italie centrale il serait aussi préférable, comme il supporte un plus grand degré de froid, que l'Eucalyptus globulus. Le bois est aussi d'une valeur beaucoup plus considérable pour l'ebénisterie, pour les traverses de chemin de fer, ainsi que pour les constructions souterraines. 76.12.28

Pour ces usages seuls l'Eucalyptus rostrata devrait être cultivé en Italie comme un des bois de construction les plus durables de tout le globe!

Par l'obligeance du chevalier Marinucci, le digne représentant de l'Italie dans notre colonie, j'ai l'honneur de vous faire parvenir une quantité de graines fraîches suffisante pour faire venir des milliers d'arbres; et après avoir consulté le chevalier, j'y ai ajouté les graines de plusieurs autres Eucalyptus, afin que l'on pût établir à Rome une déduction concernant la vigueur des autres sortes d'Eucalyptus dans cet endroit.

Si l'Eucalyptus amygdalina, comme je l'espère, supporte le climat de l'Italie, vous obtenez alors une espèce plus puissante, que n'est l'Eucalyptus globulus, parceque celui-là contient quatre fois autant d'huile volatile (Eucalyptöl).

Je prendrai toujours le plus vif intérêt aux progrès de la culture de l'Eucalyptus dans votre pays, et serai heureux d'y aider; ainsi donc, cher Monsieur, veuillez mettre mes services à votre disposition en tout temps sans la moindre hésitation.

Daignez agréer, Mons. le comte, l'assurance de ma haute considération.

Firm. VON MUELLER

A. M. le comte Louis Torelli,
Sénateur d'Italie
Rome.

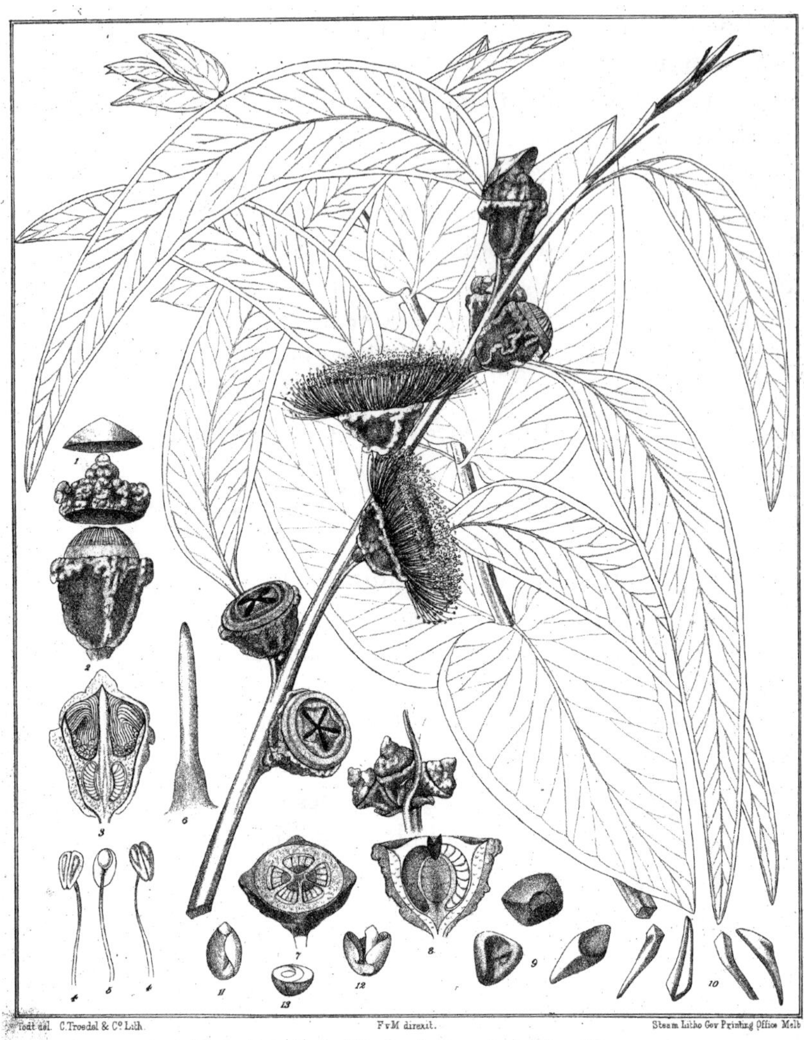

Figure 1. Eucalyptus globulus, illustrated by Emil Todt, published in *Eucalyptographia* (1879) (B79.11.13). Reprinted courtesy of the Library, Royal Botanic Gardens Melbourne.

Melbourne 28 December 1876 76.12.28

Dear COUNT,

Your communication[2] greatly honours me, and likewise the confidence you show in me by consulting me regarding the plantings of Eucalyptus that ought to be undertaken in the fever-ridden marshes near Rome, under the enlightened auspices of the Italian government.[3] Trials have proved that *Eucalyptus globulus* is not always the most suitable for planting in marshy ground; moreover, this species is more liable to frosts than many others. I have as a result advised in several cases that *Eucalyptus rostrata* be chosen for marshy regions, for which it is the most appropriate.

For central Italy, too, it would be preferable since it withstands a greater degree of cold than *Eucalyptus globulus*. The wood is also much more valuable for cabinet work, for railway sleepers, and for underground constructions.

For these reasons alone, *Eucalyptus rostrata* ought to be cultivated in Italy as one of the most durable construction timbers in all the world!

Through the obligingness of the chevalier Marinucci, Italy's worthy representative in our colony, I have the honor of sending you a quantity of fresh seeds sufficient to give rise to thousands of trees; and after consulting the chevalier, I have added seeds of several other Eucalyptus, so that you can reach some conclusion in Rome about the vigour of other kinds of Eucalyptus there.

If *Eucalyptus amygdalina* tolerates the Italian climate as I hope, you will have a more potent species than *Eucalyptus globulus*, because it contains four times as much volatile oil (Eucalyptol).

I shall always take the liveliest interest in the progress of Eucalyptus culture in your country, and shall be happy to help in that regard; therefore, dear Sir, take advantage of my services at any time without the least hesitation.

With the greatest respect, dear Count, I am
(signed) VON MUELLER

To Count Louis Torelli
Senator of Italy
Rome.

2 Letter not found.

3 For further details of M's part in introducing *Eucalyptus* to the marshes near Rome, see Barclay-Lloyd (1994). See also M to L. Haynald, 22 November 1882.

Figure 2. Ernest Giles, photographed by The Temple of Light, Adelaide, contained in Mueller's photograph album. Courtesy of the Library, Royal Botanic Gardens Melbourne.

From Ernest Giles

77.06.02

Cons 5000, SDUR/G4/325C, unregistered letters from settlers, Lands and Survey Office, State Records Office of Western Australia, Perth.[1]

A memorial to be registered of a paper writing of which the following is a copy
Duplicate[2]

Melbourne
2nd June 1877

To Baron Ferdinand von Mueller CMG
Melbourne

I hereby agree to transfer and by this document do transfer all my right title and interest in and to a grant of 2000 acres of Land granted to me in Western Australia by the Secretary of State for the Colonies and which His Excellency Governor Robinson's letter attached to the original of this declares I am entitled to[3] in consideration of having already received the sum of Eight Hundred pounds sterling (£800.0.0) being the price agreed to be paid to me by Baron von Mueller for the Land in question.

(signed) Ernest Giles[4]

C. A. [Graves]
Witness

1 The cover sheet has: 'Ernest Giles Esq | & | Baron F. von Mueller | Agreement for Sale | Received a memorial of the written agreement for registration at 12.45 p.m. on 10th July 1877 and registered the same, Book 7 No 2511 | Lawrence S. Eliot'.

2 A duplicate of this letter is annotated: '*2511* | Received of Chas G. H. Cooper Law Clerk Perth at 12.45 p.m. on 10th July 1877 and registered in Book 7 No 2511 – | Lawrence S. Eliot | Registrar of Deeds.' (book 7, no. 2511, Titles Office, Perth). The title that was issued to Giles is attached to this copy.

3 William C. Robinson's letter is dated 'Perth 20 Jan 1877' and includes: 'I have the great pleasure to inform you that I have received a Despatch from Lord Carnarvon authorising me to make you a moderate grant of Land in acknowledgement of the service which by your Explorations you have rendered to Western Australia.' Carnarvon did not specify how much land but 'from the tenor of his despatch' and after consideration, 2000 acres had been decided on (SDUR/G4/325E, State Records Office of WA).

4 See also E. Giles to J. Forrest, 2 June 1877, in which Giles accepted Forrest's offer to identify a suitable piece of land for him, which should then be transferred to M

Figure 3. Alphonse de Candolle. Published in *Archives des sciences, physiques et naturelles* (1893), 3s, vol. 30. Reprinted courtesy of the Library, Royal Botanic Gardens Melbourne.

To Alphonse de Candolle 77.06.05
Conservatoire et Jardin botanique, Geneva.

Melbourne
5/6/77.

Being aware, dear and honored Sir, that you have sought the aid of phytographers in the elucidation of the plants for your series of monographie's, I beg to ask, whether you would entrust the elaboration of the xerocarpic myrtaceae to my care. Altho only about 30 species have been added to the Australian xerocarpic Myrtaceae since under my aid Benthams elaboration for the third volume of the flora Australiensis did appear,[1] yet numerous additional observations on the geographic range of the species, on their characteristic diagnosis &c are to be recorded, irrespective of many errors, which crept into the Flora Australiensis. To finish the Myrtaceae as a whole, would of course require a special worker for the chylocarpic genera, which perhaps your excellent son[2] would find interesting, more particularly as Berg's[3] disquisition of the Brazilian Myrtaceae needs a careful revision. Should you think a contribution of mine acceptable, then pray tell me, how much time you would grant me. There are several small orders or tribes which I could manage also, such as Philydreae &c.[4]

Regardfully your
Ferd. von Mueller.

I do not yet exactly know the plan, adopted for the monograms.

(SDUR/G4/325B, State Records of WA). A year earlier, M had unsuccessfully sought a free grant of land for himself on Sturt's Creek, WA, in consideration for his having been a member of the North Australian Exploring Expedition, 1855-6. He continued to seek this grant, even while negotiating about Giles's grant. See M to M. Fraser, 7 August 1876 and 1 August 1877 [both Collected Correspondence]. See also M to J. Forrest, 14 February 1878.

1 Bentham (1866), pp. 185-261.
2 Casimir de Candolle.
3 Berg (1857-9).
4 M renewed his offer in M to A. de Candolle, 24 October 1878 [Collected Correspondence].

77.06.27 *To William Odgers*[1]

L77/7293, unit 952, VPRS 3991/P inward registered correspondence, VA 475 Chief Secretary's Department, Public Record Office, Victoria.

Melbourne 27 June 1877

W. H. Odgers Esqr
Undersecretary.

Sir

In compliance with the request of the hon. the Chief Secretary,[2] I have the honor to transmit herewith four ounces of fresh seeds of Eucalyptus globulus, for experimental culture in the Punjab. As this species does not endure a tropic clime so well as the Red Gum tree (E. rostrata) I have added 4 ounces also of that species.[3]

I have the honor to be,
Sir,
your obedient
Ferd. von Mueller.

1 On 16 May 1877 G. Batten, Officiating Secretary to the Department of Revenue, Agriculture and Commerce, Government of India, wrote to the Colonial Secretary, Victoria, from Simla asking for '4 ounces of good seed of the *Eucalyptus Globulus* for experimental cultivation in the Kangra district of the Punjab' to be sent to the departmental office in Calcutta (L77/7108, unit 952, VPRS 3991/P, PROV). A file annotation records that the Chief Secretary, G. Berry, wrote to Batten on 7 July 1877, reporting that the seeds supplied by M had been sent.

2 The file (see note 1) indicates that a letter was sent to M on 26 June 1877. Letter not found.

3 On 3 October 1877 G. Lane, Secretary to the Tender Board, wrote to Odgers saying a case of seeds for the Central Forest Department, India, had arrived from the Under Secretary's Office without the 'usual requisition or other authority'. The seeds had been forwarded, but he awaited the paperwork (M77/10982, unit 1027, VPRS 3991/P, PROV).

From George Bentham

77.06.27a

RB MSS M3, Library, Royal Botanic Gardens Melbourne.

25, WILTON PLACE. S.W.
London
June 27 /77

My dear Sir

I send by this post four sheets of the new vol. of Flora Australiensis[1] the printing goes on but slowly though the copy is all in the printer's hands to the end of Cyperaceae – the delay however will I hope enable me to have the Gramineae ready in time – though they take me longer than I expected The Order has been thrown into such confusion by Nees[2] & by Steudel[3] that notwithstanding the great help I get from General Munro I have great difficulty in identifying species and in fixing some kind of limits to genera

Your ferns are I believe in town and will I hope be at Kew in a day or two. I shall very soon send off a box of Gramineae – but as yet I have only finished the Paniceae and Andropogoneae

Many thanks for the completion of vol VIII of your Fragmenta[4] – the want of the Index was a great impediment to consulting it. When I have published this volume of the Flora I hope you will prepare a methodical supplement with all your additions. This will give you the opportunity of giving the characters of the genera as you understand them where your views differ from mine I trust you will at the same time give a sketch of the physical features and characteristic floras of the different parts of Australia which no-one is in a position for doing as well as yourself. With regard to

1 Bentham (1878).
2 Presumably Nees von Esenbeck (1841).
3 Steudel (1855).
4 In an earlier letter Bentham wrote: 'I have your Fragmenta complete except the title page and index to vol VIII which I have never had – have you a copy to spare? in order that I may bind the volume' (G. Bentham to M, 30 April 1877 [Collected Correspondence]).

77.06.27a the general geographical distribution of the Australian Flora there is but very little to add or alter from what Hooker gave in his preface to the Tasmanian Flora[5] which makes me the less regret that I have no time to work it out notwithstanding the attention I have long paid to the subject

Yours very sincerely
George Bentham

Baron F. von Mueller[6]

5 J. Hooker (1855-60).

6 This is Bentham's last letter in the collection held in the Library, Royal Botanic Gardens Melbourne. The final sheet in the folder is an unidentified notice recording Bentham's death in 1884:

DEATH OF MR. GEORGE BENTHAM.

LONDON, SEPT. 19.

The death is announced, at an advanced age, of Mr. George Bentham, the distinguished botanist, and the author, in conjunction with Baron von Mueller, of *Flora Australiensis*.

To Joseph Hooker 77.07.07

RBG Kew, Kew correspondence, Australia, Mueller, 1871-81, ff. 197-8.

7/7/77.

A few days ago, dear Dr Hooker, the attached print appeared in the Argus.[1] I feel sure, that by some oversight or inadvertence this communication did not go to me as Gov. Botanist, because I am certain, that you wish my official position here respected, particularly since it is so much impaired. A honored colleague like you will at a moment receive,[2] particularly as a professional Gentleman, that if even on occasions like this I am passed, nothing will remain of my position. My other Colleagues in large numbers still continue there[3] sendings of seeds & other horticultural concerns to me, as you will see by the enclosed ticket from Dr King,[4] who sent as usual seeds by last mail Even suppose, that the Gardener at the bot Garden had applied for the Prosopis seeds, which I do not

1 Attached to the letter is a clipping from the *Argus*, 23 June 1877, that begins:

'THE MESQUIT BEAN.

We have been informed by Mr. W. R. Guilfoyle, director [director *is heavily scored through*] of the Botanic gardens, that he has received important information respecting the above plant, of which a full description was lately given in this journal. We subjoin copies of the letters sent to Mr. Guilfoyle, from which it will be seen that grave doubts exist as to whether the plant is not, after all, highly detrimental, and even poisonous, instead of being the desirable acquisition previous reports pronounced it. A very large number of persons having been induced by our account to obtain seeds from Mr. Guilfoyle for propagation, the warning given by the high authorities quoted will be worth their attention: –'

The article then reproduces letters from W. Thiselton-Dyer to W. Guilfoyle, 4 May 1877, and from R. Thomson [of the Cinchona Plantations, Jamaica] to W. Thiselton-Dyer, 6 April 1877, concerning experimental feeding of the pods of *Prosopis pubescens* to a horse, which died.

2 perceive?

3 their?

4 Ticket not found. Presumably George King, Director, Royal Botanic Garden, Calcutta.

77.07.07 think, you would surely allow me as Gov. Botanist to be the channel of any *professional* information, such as the Subdirectors letter conveys. Excuse me, what might Mr Dyers[5] feeling be, supposed the case was reversed. As an occasion arises, to plead for your confraternal consideration in reference to the Prosopis, perhaps you will allow me a few special remarks thereon. I do not for a moment compare the value of the Prosopis-pod, irrespective of any poison-principles, to that of the Ceratonia. The Spaniards at their conquest more than 300 years ago took the Ceratonia at once to South America, but I doubt whether they would have set the same value on the Prosopis, which as yet is not cultivated in the countries at the mediterranean Sea. Had the Ceratonia been a native of Peru or Chili, the Conquistadors would have doubtless brought the plant at once home with them. Nevertheless I thought it advisable to accept Prosopis dulcis among the "select plants" of my little work, altho I have mentioned P. pubescens merely as a hedge plant.[6] Mr Thomson's idea, that the germinating of the seeds in the stomach of the horses causes their death, is surely based on some erroneous conceptions. Even if after crushing or mastication any germs were developed, which I consider unlikely, they would rather be in a state to serve for assimilation into chyme & chyle. I deem it far more likely, that the enormous quantity of catecho-tannic acid, known to exist in the pericarp of Prosopis, acts as a powerful adstringent[7] on the coat of the stomach, dries up the mucous secretion and gastric use,[8] prevents the secretion of bile, may possibly even contract the pylorus, and most likely set up enteritis and peritonitis. I have given my views with freedom, as you may feel induced to investigate this subject further and set at rest, through physicians in Jamaica, whether it is the tannin or perhaps a distinct poison-principle (such as now known in Lathyrus, Laburnum Swainsona, some Lotus, Gastrolobium &c), which causes the mischief.

5 W. Thiselton-Dyer.
6 In B76.13.03, *Prosopis dulcis* was mentioned as 'yielding the sweetish Algoraba-pods for cattle-fodder, and utilized even in some instances for human food'.
7 astringent?
8 juice?

I am making just now with the aid of a few friends a powerful effort to resuscitate my Department under the new Ministry, and have for this purpose been twice examined in the course of a few days before a parliamentary Committee.[9] Your kind issue of the Xanthorrhoea plate[10] is a support to me in my "struggle for existence" and had your communication on the Prosopis gone to me it would have been a professional support in my undeserved trials also. 77.07.07

Regardfully your
Ferd von Mueller.

The Committee is likely to recommend at least that I get office-buildings again, laboratory, Director title and some votes

Dr Kings sending of seeds by last mail was as usual spontaneous just as in the years of my Directorship.

The direction of this communication to the Garden has caused here in professional circles much surprise.[11]

9 On 23 June 1877 the Chief Secretary, G. Berry, established a Board comprising L. Smith, J. Bosisto and M. King, to 'inquire into the present position of Dr. Mueller, in relation to his professional duties, with the view to advise, what alteration, if any, is necessary, to afford him reasonable facilities for the due discharge of his scientific labours'. The committee reported on 11 July 1877; see pp. 745-8 below.

10 J. Hooker (1877). He wrote: '*X. minor* was sent to Kew some years ago by Baron von Müller from the rich collections of the Melbourne Botanic Gardens, of which he was the director, and it flowered in February this year.'

11 The final two post-scripts are placed adjacent to the reference to Dr King sending seeds to M.

77.08.07 *To Graham Berry*[1]

N78/2312, unit 1018, VPRS 3991/P inward registered correspondence, VA 475 Chief Secretary's Department, Public Record Office, Victoria.

Copy

Melbourne, 7/8/77.

Sir

In compliance with the request conveyed in your letter of 26th July,[2] I have the honor to submit the following remarks on Dr C. Brown's publication on the forest-schools in Europe,[3] in reference to their applicability to our own colony.

His ideas arose from the circumstance, that the needful extension of the botanic garden of Edinburgh, by adding an arboretum, might serve to make simultaneously arrangements for training scientific foresters. This plan is simple and comparatively inexpensive, and could in time advantageously be adopted in Victoria. For the study of surveying, engineering, natural history including that of plants, chemistry, climatology, and cultural operations, so far as requisite for the scientific education of forest managers, various facilities exist already in Victoria, sufficient (in my opinion) for the requirements of initiating here a highly advanced system of forestry. For adopting here at once a management such as that of Germany, France, or Scandinavia, an expenditure would be needed entirely out of proportion to the area and population of our colony as compared to that of European countries; – thus in Bavaria about 2500 forest officials are employed, at which rate 7500 employees would be needed for the three times larger area of Victoria.

1 MS written by G. Luehmann, who also signed the letter for M. M's original letter went astray and a copy was sent some months later. See M to W. Odgers, 5 March 1878 [Collected Correspondence].

2 See W. Odgers to M, 26 July 1877 [Collected Correspondence].

3 J. C. Brown (1877).

Foreseeing that the Government and Legislature of this colony 77.08.07
would not likely for many years sanction an expenditure for forest management on an European scale, I suggested in my lecture on "Forest Culture", delivered at the Melbourne Technologic Museum on 22nd June 1871,[4] and indeed in oral communications and occasional official notes for years before, that the two first cardinal measures for protecting, utilizing and enriching our forests should consist:

1, in the appointment of local forest-boards, under some general supervision by a central board;

2, in placing at the disposal of such local forest-boards the whole, or at least a portion of the revenue emanating from the forest land under the control of each.

By these means a circumspect and inexpensive system of forest management would be created, out of which in time, according to the general and special requirements of our colony, ample measures would arise.

On the first of these suggestions of mine – which claimed originality – was at once acted by ministerial dicision;[5] but the second measure, without which the local forest-boards were to a great extent powerless, received if I am rightly informed, ministerial sanction only this year.

In my lecture, the print of which was widely circulated at the time and which deserves reproduction anew, it was further pointed out, that the arrangements adopted for the maintenance of forests in colder countries, where alone as yet scientific forestry is largely developed, were not fully applicable to a territory in a hotter clime like that of Victoria. Contrarily it was shown, that the far greater dangers of forest conflagration had to be coped with, that the trees of all extra-tropic countries and not merely those of the cold zone could be reared in our clime, and that the trees in our native forests and their products and educts were very different from those of the countries, in which forest culture has become scientifically controlled. All this required some special modifications of forest-

4 B71.13.03.

5 decision?

77.08.07 laws for our colony, though some of the fundamental arrangements for forestry would everywhere remain the same.

I took then also already the opportunity to enumerate in published form[6] the many hundreds of trees available as strictly utilitarian for our forests, in comparison to the much more limited number, which would endure the North- or Middle-European clime, where again indeed mainly attention is bestowed on the very few kinds of trees indigenous there. It is perhaps also but just to observe, that ever since I became Director of the Botanic Garden in 1857, great efforts were made to introduce foreign forest-trees and to distribute them widely over our country, in order to become thus in numerous places the nuclei of forest-culture. It was besides one of my principal aims to surround myself with the greatest number of different trees, which with my means and appliances I could maintain at the botanic garden for display and information to the public, and for the various daily observations of my own. In this direction is evidently also Dr. Brown's aim; and thus the question arises, whether the Government would entrust the formation of an arboretum to me, perhaps on the rises from Prince's bridge to the private ground of Government house, where all the existing trees were reared by me, and where moreover the buildings now used as alms houses could be converted without much sacrifice of time and expense into offices for the re-organisation of my department.[7] It will be evident, that for each of the three principal collections of a scientific arboretum, namely for the geographic, the systematic and the technologic ones, a comparatively ample area would be needed, hence such trees as would live on the flat and partially gravelly sandy or saline ground of Albert Park and perhaps also Fawkner Park, might be planted there, by which means those extensive areas would be turned to the purpose, for which more than 30 years ago His Excellency Governor Latrobe and his advisers had destined them with other parks. This, as regards Albert Park, deserves all the more consideration, as it has continued to be the

6 B71.13.01.

7 M continued to press this proposal unsuccessfully for some time; see M to B. O'Loghlen, 18 May 1879.

most abandoned of all the metropolitan reserves, though it is surrounded by the thrifty and populous towns of Emeraldhill,[8] Sandridge,[9] St. Kilda, Prahran and South Yarra. By the means, which I respectfully suggest, the way for forest-culture would also aidingly be paved, no enormous outlay would be incurred, and a measure be adopted, which would gain great and deserved popularity. 77.08.07

In fluctuating health and advancing age (after 30 years uninterrupted Australian service and nearly 10 years previous fieldwork) I might perhaps even feel strong enough to attend to the requirements of a strictly departmental model-forest, for which the reservation of a large timbered area on the new Gippsland railway[10] would be recommendable, as best accessible and as likely affording some natural boundaries by streams from the higher backranges. The various local forest-boards might receive aid in this manner, and the public mind would be gradually led to perceive the advantage of a systematic and scientific supervision of our forest-resources.

In reference to the technologic products of our forests I beg to point out, that I considered it my duty as director of the botanic garden to prepare gradually from the trees and other forest-plants, whether indigenous or foreign, under my control, a series of tars, acids, oils, potash, dye-principles, tannin, varnishes, wood-alcohol and other substances, which operations however were totally interrupted and discontinued by the destruction of my laboratory and the removal of all my apparatus, laboratory instruments and appliances, together with nearly the whole of the modest votes for my former directorial purposes.

In the United States of North America the Government promotes forest-interests by the extensive distribution of tree seeds and by the abundant issue of of[11] printed volumes bearing on utilitarian cultures; this is effected from a central office at Washington, and as far as I am aware, forest-maintenance is left there in a great

8 Now South Melbourne.

9 Now Port Melbourne.

10 The railway line from Melbourne to Sale in Gippsland was opened in stages between June 1877 and April 1878.

11 *word repeated.*

77.08.07 measure to the foresight and resources of private land proprietors; so also in Europe many extensive forests are private property. Nevertheless I am impressed with the necessity, that in a largely unsettled community like that of a comparatively young colony, an extensive area of forest-land should remain unalienated from the Crown, not merely because we have here not yet discovered any workable coal-field or accessible turf-pits for fuel, but also because our native forests are devoid of trees fit for flooring boards, masts, spars, light furniture and many other kinds of timber, which in generations to come are not likely any more obtainable from the more or less decreasing forest areas of Europe and America, on which parts of the globe we now entirely rely for the enormous daily need of such kinds of timber.

I may also be permitted to point out, that forests ought to be created in our colony, where over wide areas, such as the north-western, only scrub exists. Scientific arboreta with concomitant other important plants, all of which can be arranged with scenic taste, if formed at or near the city, would lead visitors to comprehend the importance of well organized forest-measures, and would become the means of extensive instruction without large outlay being involved.

Moreover I have always held the view, that each country should provide finally within its own geographic limits every kind of timber, of which the local growth can be effected, so far as the climate allows.

I would further recommend, that a more copious and vigorous introduction of trees be adopted than I could effect with the slender means at my command while at the botanic garden, although I did distribute a few hundred thousand, many of these now bearing seeds available for forest culture. The kinds of trees not at all introduced yet into our colony are very numerous still, as will be perceived by reference to my "Select Plants",[12] in which work also the conditions necessary for the growth of each species are briefly indicated. Supplemental notes for a new edition are written.

12 B76.13.03.

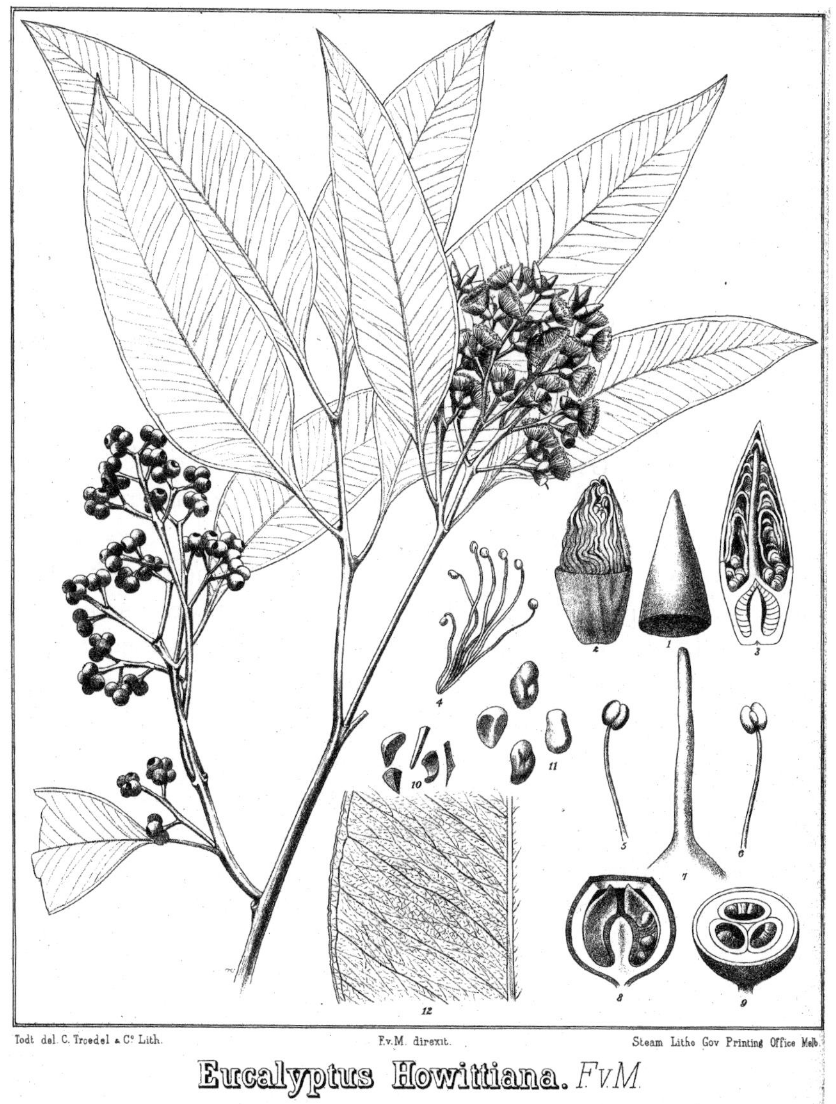

Todt del. C. Troedel & C° Lith.
F.v.M. direxit.
Steam Litho Gov Printing Office Melb.
Eucalyptus Howittiana. FvM

Figure 4. Eucalyptus howittiana, illustrated by Emil Todt and published in *Eucalyptographia* (1879) (B79.11.13). Reprinted courtesy of the Library, Royal Botanic Gardens Melbourne.

77.08.07 The long contemplated and now forthcoming descriptive Eucalyptus Atlas[13] approved of by you, Sir, will tend also to develop our forest resources. (Proof plates appended herewith). Many of such trees could be reared on now unsaleable or on permanently reserved crownlands, such as sandy heaths, sea shores, railway lines, in the snowy mountains, rocky ranges etc., with the prospect of a very remunerative revenue in the course of time, and much enhanced value of the planted ground.

Whatever plans may finally be adopted by the Government for any special forest-schools of ours, the measures respectfully indicated by me cannot fail to pave preliminarily the way for rational future forest instruction, just as what I had commenced at the botanic garden was in accordance with what Monsieur de Vilmorin carried out on his estate at Barres,[14] for rendering known and testing trees in the interest of forest-science.

It may however be worthy of consideration, whether a professional forester, specially trained at one of the great forest schools of Europe, should be attached to one or more of the local forest-boards, to become their principal functionary.

In conclusion it is hardly necessary to assure you, that I shall cheerfully explain my views on Australian forestry at greater length if desired, and that I would gladly promote by my departmental advice and professional knowledge, so far as it could be justly expected from me in my position as Government Botanist, any measures, which you and your honorable colleagues may wish to bestow on so important a national subject as that now under consideration.[15]

I have the honor to be
Sir
Your most obedient servant
(signed) Ferd. von Mueller.

13 The first decades of M's *Eucalyptographia* were published in 1879 (B79.13.11).

14 Vilmorin (1864).

15 See also M to W. Odgers, 5 March 1878 [Collected Correspondence].

From G. Robinson

77.08.16

M77/9339, unit 961, VPRS 3991/P inward registered correspondence, VA 475 Chief Secretary's Department, Public Record Office, Victoria.

Hillsley
Berwick[1]
Aug 16th 1877.

Dear Baron v Mueller

In reply to yours of Monday 13 ins[2] requesting to learn what are my ideas & experiences in relation to the growth of the Cinchonas in this locality.

In the first place I must mention that the plants you kindly let me have, the half of them I took to a selection that I had in the Dandenong State forest, these survived the first summer but they were neglected as I had so much to do that they became choked with weeds and so I lost them, the rest, (six) I kept at home here near Berwick I planted out in different places in the garden. I lost two of them through the excessive heat in Summer and a third one was injured by a cow getting in the garden, this one never recovered, so that I have only three left, one of these being sheltered from the hot winds has done very well, it is now seven feet 3 inches high and it first came into flower last April and it has continued flowering at intervals through the winter and it seems now, (August) to be still coming on with fresh flower buds, the flowers do not come out all at once but singly, as shown on the specimen enclosed. I do not know if this is its natural manner of flowering or not, or the result of winter flowering. The frost that we have had this winter has not touched or had any effect the tree itself or flowers, although it did two or three times during very severe frosts in past years. I do not think I shall get any seed from the winter flowers but I expect to do so from those that are to come out yet. From this tree at 2 feet from the ground I cut off a small peice of the Cinchona bark this I enclose for you as a small specimen of

1 Vic.
2 Letter not found.

77.08.16 Colonial grown bark, of course I do not expect it will be rich in Quinin as I think this is not the proper time of the year to strip. The second tree that is also sheltered from the Hot Winds is growing vigorously, but suffered from the late frosts a little. The other tree has been exposed to all weathers from the first, it does not seem to thrive very well that I attribute to my own fault, as I have not attended to it at all in any way. The frosts do not appear to affect it or the hot winds either. From my experience I am quite satisfied that the Cinchona will thrive here almost as well as any other tree that is with ordinary care and attention. I am certain that if my plants had been at all delicate that I should have lost them for I let them take their chance so as to give them a good trial[.][3] I should like to get another supply of Cinchonas, so as to form a small plantation. Should I succeed in saving seed from my Cinchona do you think I could I get them propagated at the houses in the Bot Garden, Melbourne, for I feel certain that if I tried here I would lose them. Wether the cultivation of the Cinchona will be profitable or not I cannot say, the only way is to form a small plantation and then to give them the same attention, as would be given to vines, olives or Hops or such like plants. My experience only shows that with scarcely any attention the Cinchonas thrive very fairly, with care and attention I see no reason why they should not grow equally as well [as][4] any other cultivated plant.

I have jotted down these notes in a hurry as I am afraid I may miss the mail.

If I can be of service to you in any way please let me know

I am yours respectfully
G. A. Robinson
Shire Secretary & Surveyor

Would you please tell me the Bot name of the Cinchona[5]

3 *editorial addition.*

4 *editorial addition.*

5 MS is with a plant label written by M: 'Cinchona Calisaya of Peru & Bolivia flowered since April 1877 at Berwick (Robinson)'. See also M to W. Odgers, 19 August 1877 [Collected Correspondence], in which M reported Robinson's success: 'I write this note ... because the Cinchona culture is totally disrecommended in official documents, which found their way into the public press to my disadvantage'.

To Edwin Jephcott 77.09.05

Private hands.

5/9/77.

The Grevillea, sent by you, dear Mr Jephcott, is Grevillea ericifolia of Robert Brown. The Lobelia-like plant is Isotoma axillaris of Lindley, and – as you observed – it is more powerful even than L. inflata, indeed very poisonous even in a small dose. The work on the "plants of Victoria" is still in manuscript, as my Department is almost ruined, and I have no means of getting it printed and the woodcuts done.[1]

I examined twice the Mount Coskiusko[2] range, in 1854 from the Maneroo-side,[3] and 2 years ago from Mr Finlay's place.[4] I had however to go up from the South East side of the Range, and never examined the north west portion, which is nearer to Walwa.[5]

If I took horses straight on to the *North part*, from Mr Watsons could one climb the range there on foot? and could I get any help in the district, so that I could camp out for a night with one or two companions?

I have not prospered after 30 years toil in Australia, and find that these journeys are too costly for my sole means. Among the granit hills is also much yet to be discovered as regards plants there, as I examined only the one near Mr Watsons and lost the *spring* vegetation. Try with your young son to make a good & full collect-

1 B79.13.08?

2 Mt Kosciusko, NSW.

3 Monaro, NSW. M's spelling was common at that time. For M's response to this first trip see B55.11.01.

4 M visited the upper Murray River region and the neighbouring mountains in January 1874; see M to J. Francis, 27 January 1874 and M to G. Bentham, on the same date [both Collected Correspondence].

5 Vic.

77.09.05 ion now of all the spring-plants there, including those of the rivers & *lagoons* & swamps!

Amongst the *minutest* plants would be most novelty.

Regardfully
Ferd. von Mueller

I found several plants, *new* to science on the granit mountains near you, but did not get the flowers

77.10.05 *To Lord Russell*

Royal Geographical Society, Archives, London, RGS correspondence 1871-80, Mueller, Sir Ferdinand.

On bord of the "Siam"[1]
5/10/77

Allow me, my Lord, to introduce to you the generous supporter of the geographic enterprises of Colonel Warburton and of the two last expeditions of Mr Giles. The hon. Th. Elder is on a visit to his native country, and I feel sure, he will regard it an honor to be introduced to the Royal Geographic Society by your Lordship,[2] while the great British forum of geographic research will doubtless recognize the generosity of this excellent man, who rendered known by his private means most extensive tracts of the Australian Continent, and who is likely to promote still further geographic mapping after his return to his Australian Home.

Regardfully
your Ferd von Mueller

1 M was en route from Melbourne to WA.
2 Lord Russell was Foreign Secretary of the Royal Geographical Society.

From the Adelaide German Club 77.10.06

RB MSS M200a.41, Library, Royal Botanic Gardens Melbourne.

Herrn Baron Ferdinand v. Müller
C.M.G. und Inhaber verschiedener anderer Orden.
Doctor der Medizin und der Philosophie, Regierungs-Beamter für die Colonie Victoria, Mitglied zahlreicher wissenschaftlicher Vereine u.s.w.

Einige zwanzig Jahre sind verflossen seit Sie, Herr Baron nach mehrjährigem Aufenthalte Süd-Australien verliessen, wo Sie den Grund legten zu den ausgezeichneten botanischen Werken, zumal über die Flora Australiens, welche seitdem auf dem ganzen Erdenrunde als klassisch betrachtet werden.

Trotz der kurzen Nachricht beschloss der Adelaider "Deutsche Club" Ihnen wenigstens durch seinen Vorstand diese Adresse zu überreichen, mit dem Bedauern, dass Sie Adelaide bei dieser Gelegenheit nicht wenigstens einen kurzen Besuch machen, um sich an den wesentlichen Fortschritten der damals so kleinen Colonie zu freuen und die Bekanntschaft so mancher alten Colonisten zu erneuern, welche Sie in treuem Angedanken behalten und Ihre wissenschaftlichen Triumphe mitgefeiert haben.

Die Mitglieder des Clubs wünschen Ihnen eine glückliche Reise und Heimkehr und hoffen dass Sie nach dieser Erholung als ächter, unermüdlicher Jünger der Wissenschaft Ihre Forschungen noch lange fortsetzen können.

Der Vorstand des "Deutschen Clubs["][1]
F. Krichauff Präsident
F. Basedow, Vice-Präsident
Oscar Wehrstedt, Secretair.
C. Wittig, Cassirer

Adelaide, den 6ten October 1877.

1 *editorial addition.*

77.10.06 To Baron Ferdinand von Mueller, C.M.G. and holder of various other Orders, MD & PhD, Government Botanist of the colony of Victoria, Member of numerous scientific societies, etc.

More than 20 years have passed since you, Baron, after several years' residence in South Australia, left our colony, where you laid the foundation for the excellent botanic works, especially in reference to the Flora of Australia, which throughout the whole globe are considered as classical.

Notwithstanding but short notice it was resolved by the German Club of Adelaide, to present at least through its Committee this address, and to express our regret that on this occasion you cannot even pay a short visit to Adelaide, with a view to witnessing the substantial progress of what was at your time so small a colony, and also to renewing the acquaintance of many of the old colonists, who kept you in true memory, and who rejoiced also in your scientific triumphs.

The members of the club wish you a happy voyage and return, and hope that after this recreation you as a genuine and indefatigable disciple of science may yet continue your researches for a long period.

The Committee of the German Club: –
F. Krichauff, President
F. Basedow, Vice-President
Oskar Wehrsted, Secretary
C. Wittig, Treasurer.

Adelaide, 6th October 1877.[2]

2 For a published English version of this letter see *Herald* (Perth), 20 October 1877, p. 3, where it was introduced: 'Address to Baron von Mueller | On the arrival of the P. & O. s.s. Siam at Glenelg, South Australia, from Melbourne, the following complimentary address was presented by the Adelaide German Club'.

To Harry Ord[1]

77.11.20

Colonial Secretary's Office, letters received, Acc. 36, vol. 852, ff. 119-20, State Records Office of Western Australia, Perth.

Perth, 20 Nov. 1877.

To his Excellency General Sir Harry Ord, C.B., K.C.M.G., Governor of West Australia, &c.

Sir.

I have the honor to solicit your Excellency, to be admitted for naturalisation in West-Australia. My oath of allegiance to her Britannic Majesty has been rendered already 30 years ago in South-Australia and 20 years since in Victoria, and I seek now to obtain the same privilege in the colony governed by your Excellency. I was born in Rostock on the Baltic shores in 1825, on the 30. June. I studied at the University of Kiel, am a Doctor of Philosophy and also of Medicine, hold the appointment of Gov. Botanist of Victoria uninterruptedly since 1852,[2] was created a Baron (with hereditary rank) of the Kingdom of Wuerttemberg in 1871, and was honored by her Majesty with the Companionship of the order of St Michael and George, being included among those first appointed when that royal distinction was extended to the colonies. I have traversed a portion of the northern territory of West-Australia as Naturalist of the expedition, sent out by his Grace the Duke of Newcastle under Mr Aug. Gregory, C.M.G., in 1855 and 1856.[3] West Australia was revisited by me in 1867 for the phytologic examination of the Stirling's Ranges and the surrounding country; it is my intention

1 MS annotation: 'Ans[were]d 26/11/77'. See R. Goldsworthy to M, 26 November 1877 [Collected Correspondence], in which M was informed that the Lieutenant Governor regretted 'that under the conditions of the Naturalisation Act 35 Victoria No. 2 he is unable to entertain your application in as much as it does not appear that you have come to reside in the Colony with the intention of settling therein'.

2 M's appointment as Government Botanist dated from 26 January 1853.

3 North Australian Exploring Expedition, 1855-6; M was botanist, J. R. Elsey was the naturalist.

77.11.20 to continue my phytographic researches in your territory by periodic visits, having shared largely in the investigation of the West-Australian plants since 1847, and this to a great extent on my private expense. I intend moreover to invest in land and other property in your colony, and shall always take a vivid interest in the development of its resources.

I have the honor to be,
your Excellency's
obedient and humble serv.
Ferd. von Mueller.

My full name is Ferdinand James Henry von Mueller.

78.02.14 *To John Forrest*

M8/900K, Land and Survey Office, unregistered letters from settlers, Cons 5000, SDUR, State Records Office of Western Australia, Perth.

14/2/78.

Private

Amidst the Valentine joys of others, dear Mr Forrest, this is a day of sad disappointment to me. I went to Mr Fraser[1] this morning, he having just arrived, and to my sorrow he will not help me to obtain an other grant, though he said, that if it was done in his absense, he of course would acquiesce. I felt hurt, that I am to be regarded as an ordinary selector, that my right of final approval is not recognized & more that while I in good faith trusted to an arrangement being made, by which I could at least regain my £800

1 Malcolm Fraser, Surveyor-General of WA.

Figure 5. John Forrest, photographed by Johnstone O'Shannessy & Co., Melbourne, contained in Mueller's photograph album. Courtesy of the Library, Royal Botanic Gardens Melbourne.

78.02.14 at any time, I now am in danger to loose the last I had in the world, the savings of several years, and this late in life, while in fluctuating health & while my official position is not for a moment secure. My own impression is, my dear friend, that I could not raise more than £200 cash from the land, for which (or I understood) you *temporarily* applied, because 400 acres are the maximum of fair land of the whole 2000, the rest being hilly Jarrah country. You call it a valuable estate. If so the difficulty would be solved at once, because then it could be sold for what I gave, and I should be content to loose the interest of the money, for some of it for six years, while in the meanwhile the sum might have been doubled & tripled by investment in Sheep.

You will kindly remember, that whenever I came to W.A. I told you at once, that whatever application was made for any land, it must be such, as admitted of my realizing at any time the full purchase sum. It was on the strenght of your own letter to Giles, that you could get at any time 8/ (like for your own selection) that I advanced Giles the rest of the purchase money £400 cash, he having had £350 (without interest) before.[2] But like in every thing else of monetary affairs since my early orphanage I seem also in this instance doomed to disappointment, unless you acted at once on my telegram of to day[3] & solicited from his Excellency, that he allowed an other application to be made. I told Mr Fraser, that I could not accept the land selected & wished to sell the right of selection. My widow sister[4] is in-distress, I got into debt through my journey to W.A., where I certainly hoped to have my saying in the selection. I have in two months to raise £250 to pay the printer for the issue of my augmented edition of Wittsteins chemistry of plants;[5] and my expenses here have been great during my absense. I fear my journey across to your colony, has done me here great injury, because the people say here "I had land as extensive as Europe there, and could raise from it £10000 or £20000 annually

2 See E. Giles to M, 2 June 1877.

3 Telegram not found.

4 Clara Wehl.

5 Wittstein (1878).

revenue not two £ as offered me! &c, you can understand that the idea of my prosperity will hinder all resuscitation of my Department. 78.02.14

The land in question can doubtless be *made* into a valuable estate by spending £10 an acre on clearing the timber, fencing, building, and by ploughing the small share of its areable land. On all this I wrote from Geograph Bay at once[.][6] I told your surveyor *not* to *survey* it for me, and he said, that he had not solely come for that, because other land had to be surveyed; why then now force this on me? especially as I am in difficulties. Such land should be taken up by a working man, who with a number of young strong children can do fencing, ploughing &c himself. Not even natural boundaries exist, the river with its fine flats is gone

Surely there is a wide difference between a grant from Earl Carnarvon & a solicitor taking up a small plot somewhere, "Nulla regula sine exceptione".[7] – I do not think that I have deserved from W.A to be met in any other but the kindest & most generous spirits, and I am harrassed already enough here. I had *such* a *faith* in W.A., I thought that there *for once* in life I might meet with some generosity & help, to allow me to build up some little fortune. Perhaps it is yet so. For one thing I am sorry, but cannot help it now, to have given *you* so much trouble in this land affair, but had it not been your own spontaneous offer, I never would have thought of such a thing as purchasing the grant. I feel you may be displeased, my dear Mr Forrest, after all exertions for me & your goodness to me; but see, I am not young & strong & prosperous like you & this last ruinous transaction spoiles again my hopes for domestic happiness!

I have sent some drawings by last mail to Mr Frasers adress, & some more will come by this mail. There will be no hurry for the 2 reports,[8] because the lithographer will have work to do for months

6 *editorial addition.*
7 No rule without an exception.
8 B79.13.10?

78.02.14 and I have to be now in the field for several weeks on the Tanners bark commission.[9]

Regardfully
Ferd von Mueller

Best compliments to your generous Lady, to whom I have forwarded Lohengrin for kind acceptance.

You ought as *acting Surveyor General* be able to settle this land difficulty & cancel the grant.

I cannot express to you, my dear Forrest, how sad and forlorn I feel.[10]

9 M was a member of the Board established by the Victorian Government on 7 January 1878 to inquire into the supply of wattle bark for tanning; for the Board's report, see B78.14.01.

10 On 2 March 1878, Forrest forwarded M's letter to the Colonial Secretary, R. Goldsworthy. On 18 March, Giles wrote to the Governor of WA, H. Ord, supporting M's request, as did Forrest on 4 April to his chief, M. Fraser. On 2 May, despite Fraser's opposing the application on the grounds that it would create a dangerous precedent, the Governor granted Giles permission to re-select the 2,000 acres to which he was entitled. Six months were allowed in which to make the selection.

On 7 July, M wrote to M. Fraser [Collected Correspondence] seeking an extension of time in which to make the new selection; he was given until the end of the year.

On 8 November 1878, M wrote (with a duplicate letter on 26 November), applying for a block of land between Perth township and the coast; see M to H. Ord, 8 and 26 November 1878 [both Collected Correspondence]. Unknown to M, however, the land in question had been set aside as Commonage; see R. Goldsworthy to M, 26 November 1878 [Collected Correspondence]. M therefore sought a further extension of time; see M to R. Goldsworthy, 26 December 1878 [Collected Correspondence].

On 16 January 1879, Goldsworthy asked Forrest where matters stood, so that he could reply to M's request. Forrest replied: 'I do not think he need be answered. I have sold his right & the application will be made in a day or two & I do not anticipate any further difficulty' (Acc. 527, no. 345/8, State Record Office of WA). Since there is no further mention of the matter in the files, it would appear that the sale must have gone ahead, enabling M to retrieve his money.

To William Thiselton-Dyer 78.05.12

RBG Kew, Kew correspondence, Australia, Mueller, 1871-81, ff. 211-14.

12/5/78.

Strictly *private*[1]

Let me thank you for your kind letter, dear Mr Dyer, of 18 febr.[2] It is not surprising, that a Dipteraceous plant is found in N. Guinea; indeed I hope to demonstrate the occurrence of the order some day in Australia yet, in the N.E. part, where even Paradise-birds occur. I had a visit from Dr Beccari; what a *humiliation* of the "Princeps of Australian Botany" (as you considerately call me)[3] to be obliged to tell the *Director of the* botanic Garden of Florence, that his colleague here was a Nurseryman,[4] and that I could not set my foot into my own garden since 5 years with any self-respect.[5] Pray look on my report of 1869[6] & the *Gardenplan*; nearly every walk is changed, nearly every tree moved or destroyed to remove all traces of my work and to convert a scientific Garden into a Cremorne.[7] There was even the Vict. reg. *first* through me in Australia,[8] the first Glass-houses in Austr, islands in the lake, Geyserfountain, Bamboos, Palms, and *many thousand species* of plants, systematically arranged, others geographically, others industrially. All this with 1/4 of the means available after my time. Look to the 600 000 pl largely Himalaian & Californ Pines which I distributed. Now how is it, that altho' I am praised as a Botanist,

1 Each page of the letter is headed *Private*; subsequent headings are not transcribed here.

2 Letter not found.

3 Source not identified.

4 William Guilfoyle.

5 See also M to J. Hooker, 15 July 1873 [Collected Correspondence].

6 B69.07.01.

7 Cremorne Gardens was a commercially-operated amusement park or 'pleasure gardens' near Battersea Bridge, London. The name was also given to amusement parks established in Melbourne and Sydney in the 1850s.

8 See Maroske (1992).

78.05.12 *Kew* has never given me any support in my struggles as a *Horticulturist*. This is an enigma to me and my friends here and abroad.

One single letter from Kew, such as I could have shown to Ministers here might have saved me from *ruin*! See, my dear Dyer, you have a young Lady to grace your home, but I am forbidden to build up a household, because when I was driven for the sake of a common Gardener [&] (who usurpates the Director-title & boasts to be the Colleague of Hooker, but who is Gazetted only *Curator* and should not be acknowledged by any professional University-man otherwise), well –, when I was deprived of House and home and staff & votes and the very *living* plants, which I as well as Hooker & yourself want daily as much as our Museum plants, for research, – I had then in this *expensive* country to maintain the wreck of a once illustrious department mainly out of my salary; indeed I would be infinitely better off with £300 & my house & former votes in the bot Garden than with £800 out of it. So my future is *blighted*, for I was forsaken as a Director even by Kew.

Two days ago I waited with the Explorer D'Albertis on the Governor,[9] when his Exc. distinctly asserted, that Sir Joseph told him that he wished to be placed like myself! This very statement was made by Sir Geo. Bowen to my former Minister, the hon. John M'Pherson, and altho' this expression may have been made hastily yet Dr Hooker was fully informed, that I was ruined as a Head of my Department & that without the bot Garden I could *not* do my work as Gov Botanist.

These young colonies do not care about my sorting Museum specimens; they want me to elaborate for them means for new industries & cultures; but even my laboratory was pulled down & my apparatus taken away, & the greater part of my library is stored away since 5 years!

Now, could not a single letter have been written to me – or could not even now a letter be written, *not to Sir George Bowen nor to the Ministry*, as it would cause only additional annoyance and trouble & perhaps professional *disgrace* has Kew not so much

9 Sir George Bowen.

consideration for me as the leading Gov. establishment of England 78.05.12
and the leading institution for bot science, *to point out in a letter to me*, which I could present to Members of the Ministry, that the place of the Gov Botanist is *in* the bot Garden & not out of it, and the place of a Landscape Gardener (whatever that may mean that unlogic word), is on the Parks & Reserves, that I want *living* plants more than dry plants, that a traveller & naturalist knows the treatment of living plants better than *any* gardener, and that I am placed in ruinous & false position without buildings & but very scanty votes, and thus prevented to do justice to my work & duties, and *stopped* in most, even though I have not a single servant to save what little I have to carry on the wreck, while enormous sums are squandered on my former place in senseless and destructive changes or on plebejan[10] cultures, for which latter ample scope is on all the parks & other Gardens of the Government round Melbourne.

What a change to me since 1857 when Sir William gave me such glorious support also in the horticultural branch. I will do Sir Joseph justice that he not willingly wished to injure me, when he made the uncautious remark to Sir G. Bowen, but surely it would be impossible for any Gov Botanist, whether of England or any country, to do his work without the bot. Garden & all that pertains to it. The *deep injury* which that remark made at Kew (if it was made) inflicted on me, nothing will make good again & it has hindered me even to get married! Now can this not be remedied? Cannot a fair statement be made, setting forth in *a letter to me, what* a Governm. Botanist requires.

This is perhaps the last letter, which I may ever write to Kew, my dear Mr Dyer, for *grief* has brought on an almost permanent insomnia, and it is a wonder I become not already insane; for not only have I to do my daily duty as Gov. Botanist under extreme obstacles but have to spend all that I have and have to waste an enormous time in the mere "struggle for existence",[11] while I have *lost* all hold on the country, and may any day be called on to hand over to the Nurseryman (a cousin of *Mr Casey* the Paris Exhib

10 plebian?
11 The title of ch. 3 of Darwin (1859).

78.05.12 Commissioner) also my library & my Museum, if the *only room* built since 1857 for my dried plants can be called such!

As perhaps you will never get a letter from me again, you will as a *professional* man allow me to claim your brotherhood as an University man also, by pointing out an other deep & lasting injury inflicted on me, i.e., by *depriving me of the authorship of the Australian flora.* Bentham's work is everywhere quoted as his sole and own; I have myself done so, simply the diagnoses are written by him; but the *real main work,* in travels, prior publications & accumulating & examining the vast material the main work of my best years of life done by me since 1847 uninterruptedly is *far* greater than Benthams.

Now, why was that not left for me? Would it not have been far more just to leave *me* that work, when Kew was teaming[12] with unexamined treasures from tropical Africa & other parts of the Globe. Or why did the venerable Bentham not concentrate his great talent on the "genera",[13] which by this time would then have been ready. No doubt, you will say, I merely out of *vanity* wished to reserve the Austral Flora to myself; it is not so; but if you had heard the endless derogatory remarks *made here* to injure me concerning the "Flora Australiensis or rather Australiana" you would have *sympathy* with me. I lost that *stronghold* also, which the *coauthorship* of the flora would have afforded me; and altho' I have with manly straightforwardness kept all my promises as a mere "assistant", yet I deeply regretted ever since the first volume was issued, that I entered into a compact, which has had a damaging influence on my position here.

Be it enough! and excuse the bitterness of this letter, but my whole future is *spoiled.* Should you send any return for the seeds from West Australia, please send such direct to me not to the bot Garden. I am striving to get my own creation back again; failing that to get a new piece of ground & new *buildings* & votes. But, *how cruel* even then to ask me to commence again at the age of 53,

12 teeming?

13 Bentham & Hooker (1862-83).

when it will take til the end of the century, to build anew what I left in 1873. 78.05.12

Regardfully your
Ferd von Mueller

Is there a single friendly sentence about *me* except in the preface, in the 7 volumes of [B][14]

I hope this appeal to the P. R. S.,[15] who sits in Newtons chair, from a man of science in distress, will not be in vain.

To Edward Ramsay 78.06.04

ML MSS.562, Letters to E. P. Ramsay 1862-91, Mitchell Library, State Library of New South Wales, Sydney.

4/6/78.

I believe, dear Mr Ramsay, that we are both equally sorry, to have mixed ourself up with Mr Goldie.[1] Our united means would have been sufficient to send a collector independent of him. He *never* writes, and I have no faith whatever in him. As regards any letters of mine to him, I cannot now remember, that I said anything about your collector; if I did, I certainly said nothing that could hurt him, as his own plants had to come also to me, he being merely a collector himself, and that a very illiterate one too; of course we know his vanity and quarrelsome disposition, and the former was exempliced by what never occurred in the worlds history before, that he named a river after himself!

14 See Lucas (2003).

15 Joseph Hooker completed his five-year term as President of the Royal Society in 1878. See M to J. Hooker, 18 May 1878 [Collected Correspondence].

1 Andrew Goldie.

78.06.04 I enclose a drawing of a leaf of the aralioceous plant with white square spots, as left me by d'Albertis. Perhaps that plant did survive. His great Arad with slit leaves must have been a Raphidophora Perhaps it would be best, if *you* kindly asked Mr Ingham J.P at Port Moresby to pick up any plants (including Algae) for me, as you are befriended with that Gentleman – The enclosed circular[2] applies as well to New Guinea as to Australia. Is there no chance of my for once enjoying the privilege of a voyage in one of her Majestys Ships, as Commodore Goodenough desired so much, and as Commodore Hoskins is willing also to arrange obligingly. But the Naval Station being *there* not here, I lost the two brillant[3] opportunities of joining Capt Moresby! My God! what chances lost! and I did *not even* know of the chances. Pray, my kind & generous friend, keep watch for me, when an other opportunity arises. I feel so inexpressibly unhappy here, that it will be a charity to me to take me away at least for a time. If the Library & Museum collections were all my own, I would quit the colony, but for Gods sake, mention this to no one, as an expression to that effect would prevent me to succeed in rebuilding my Department and regaining my position.

I am sorry, that unintentionally I should have been the cause of some troubles of yours with Goldie.

I have again looked for the few specimens of Rushes; but in the overcrowded only Museum room, to which has not been added for 21 years! (since 1857) things are stored away into all corners & recesses that it is not possible to keep order in the daily augmenting collections. If your trustee would only mention the names of the Rushes, I could then easily pick out of the Normal collection the specimens in a far better state than the specimens sent to me.

With regardful remembrance your
Ferd. von Mueller.

2 Possibly Circular, 29 March 1876.

3 brilliant?

To Edward Ramsay 78.08.22

ML MSS.562, Letters to E. P. Ramsay 1862-91, Mitchell Library, State Library of New South Wales, Sydney.

22/8/78.

I was aware, dear Mr Ramsay, that I missed my chance for going to New Guinea in the War-Ship. But even, had I been in time, I could not have gone, because I underwent an operation for haemorr-oides,[1] & as my health had suffered so much in "the struggle for existence"[2] I suffered very greatly & was kept on the bed for two weeks and shall not be able for several days yet, to move about

Moreover the rainy season will just have set in by the time I came there or perhaps soon. My estimates are also not yet passed, and if I do not watch the still trifling votes and counteract the constant traducing misrepresentations concerning my administr-ative ability, I might drift into the debt of ruin so it will be best, to wait to the commencement of the next dry season.

I have interchanged letters with Baron Micklucho-Maclay[3] since he is in Sydney[4]

Many thanks for your kindness to offer me shelter under your roof. I am glad you can manage to procure some Diatomaceae. I am still excessively weak, I have not slept for 12 nights

Your letter came only *this* day

Regardfully
Ferd. von Mueller[5]

1 M had undergone two operations in 1859 for a similar complaint; see Darragh (2001), p. 393.
2 The title of ch. 3 of Darwin (1859).
3 N. Miklouho-Maclay.
4 Letters not found.
5 The file contains an envelope addressed by M to 'Edw. Ramsay Esqr FLS, CMZS., &c &c Director of the Museum *Sydney*'.

78.09.18 ## *From Richard Kippist*[1]

RB MSS M1, Library, Royal Botanic Gardens Melbourne.

Linnean Society,
Burlington House, London, W.
Septr. 18th, *1878.*

Dear Baron von Mueller,

In accordance with your request I send by way of *loans*, (for your own information, but as at this time of year, it has not been possible to obtain the formal permission of the Council, *please not to publish* the fact) one or two small fragments from the *type-specimen* in Smith's Herbarium, of the *Eucalyptus emarginata*[2] described by him in our Transactions.[3] I also send a sketch from his specimen which will give you a pretty accurate idea of the proportions of the leaves, peduncles, &c., but as it is the first time for many years that, with my failing sight, I have tried to draw a plant, and as I have, moreover, been hard pressed for time, I am afraid it is very far from being as good as I could wish. I have copied the inscription on the sheet, and you will observe that it confirms the statement in the Transactions, as to the seeds having been obtained from *Port* Jackson.[4] There is no mention of King George's Sound,[5] or of Menzies, and I cannot make out on what grounds Mr Bentham assumes that the specimen in the Banksian Herbarium which accords precisely with Smith's,[6] is specifically identical with the much more *coriaceous*-leaved *floribunda* of Hügel, with its narrow falcate, acuminate, nearly *concolorous* leaves, and with a differently

1 MS found with a specimen of *Eucalyptus marginata* (MEL 1611903).
2 marginata?
3 Smith (1802).
4 NSW.
5 WA.
6 MS footnote by Kippist: 'Our Library having been closed for a few weeks, I have been enabled to compare the two'.

formed calyptra, much longer, as it seems to me, in proportion to the calyx tube.[7] Of this plant which I have only seen from the *West* Coast, I send you, for comparison, a leaf and flowers which need not be returned. They are unquestionably from K. George's Sound; not from Hügel, but Mr Bentham seems to have adopted my identification 78.09.18

I don't imagine that any Herbarium exists (Dr Trimen never heard of such a colln, any more than myself[)][8] in connexion with Donn's Hortus Cantabr:,[9] and if there were, as the work is a mere *List*, Sir James Smith having been the first to *describe* the species, his name must, of course, stand.

It appears to me that the East coast plant is, in addition to the difference in texture and the paler under surface, further characterised by the prominence of the reticuled venation on the *upper* surface.

Thus much for the *Eucalyptus*, of which you will now be in a position to judge for yourself.

In answer to your enquiry with respect to your Subscription, I may say that it has always been very promptly and regularly paid by Messrs Dulau, and as they are really very methodical people, and very regular in claiming what is due in return to the Australian Fellows whose subscriptions are paid by or through them, I should be very sorry that the arrangement should be disturbed. By our regulations, Fellows abroad, who have not compounded, are required to name a London Agent. The plan of sending Annual Subscrns by Bill or P.O. order often gives a deal of trouble, to us, and prevents the Fellow from receiving his publicns so early as he otherwise would.

7 Bentham (1866), p. 209: 'The species was originally described by Smith from specimens raised at Kew from seeds brought by Menzies from King George's Sound.' M discussed the taxonomic confusion surrounding *E. marginata* in *Eucalyptographia*, Decade 7 (B80.13.14), citing but not endorsing Kippist's view that the specimen Smith described came from Port Jackson.

8 *editorial addition.* (Dr Trimen ... myself [)] *is a marginal note by Kippist.*

9 Donn (1800), p. 61, refers to *Eucalyptus marginata* but provides no description or diagnosis.

78.09.18 I was truly grieved to hear of the death of M. Thozet. I handed over the printed notice of him enclosed in your last to Dr. Trimen, who will probably insert it in the next no. of the Journal of Botany.[10]

I had much pleasure in receiving the Certificate in favour of ..... a new member (I am writing away from home, spending a week by the sea-side, so cannot recall his name) I cannot but regret, however, that you continually send informal Certificates, *signed only by yourself*. Our bye-laws recognize only *personal* knowledge, "either of the Candidate or of his works." Our Fellows here, therefore, naturally object to sign the Certif[ica]te of one of whom they know nothing especially as we have now so many Australian Fellows, many of them recomd by yourself, who would doubtless gladly second your recommendn, and would be gratified by being appealed to, and be all the more likely, if reminded of their privilege, to bestir themselves in introducing new Fellows. I may remind you too, that we now rarely see either Sir Joseph Hooker or Mr Bentham, who used formerly to endorse your signature On the present occasion, however, Mr. Carruthers & Dr Trimen have kindly signed the Certificate, tho' the promised pamphlets have not yet arrived.

With kindest regards, believe me,
Yours faithfully
Rich. Kippist

Baron von Mueller
F.R.S., L.S. &c.

10 Trimen was the editor of the *Journal of botany*: an obituary notice of Thozet appeared under 'Botanical news' in the October issue (vol. 16 [1878], p. 320).

From Alexander Forrest 78.10.26

Briefsammlung, Archiv, Justus Perthes Verlag, Gotha, Germany.

Perth
26th Oct 78

My dear Baron

I leave here in a month for the Grey River.[1] I am then to explore the whole of the watershed on the North Coast[2] as far as Port Darwin. This will be a very interesting trip & valuable results no doubt will accrue from this Expt[3]

I am sorry you have not the means of paying a Botanist. I have sent to Melbourne for a practical miner who knows something of Geology[.][4] I shall write to you before I leave giving you more detail of the proposed Expt perhaps you might be able to give me some information of the Northern Territory as you were on Mr Gregory's Expt.[5] were the natives at all troublesome & other particulars. I send you a whole lot of plants collected on the Cane & Ashburton Rivers[6] which no doubt will be interesting to you

About your land I am afraid the Govt. will not approve of your application as they no doubt will lay it out in Suburban lots[7] As soon as I know for certain I will telegraph to you. Mr Whitefield offered me from 6/ to 7/- per acre. I have not much faith in the land about Perth it is so very poor however you may rest assured we will do our best for you My sister in law is writing to you,[8] she has been several times to Mr Goldsworthy about your lands. You must however either select or sell before the end of the year.

1 De Grey River, WA.
2 i.e. of WA.
3 Expedition.
4 *editorial addition.*
5 North Australian Exploring Expedition, 1855-6.
6 WA.
7 See M to J. Forrest, 14 February 1878, and the notes thereto.
8 Margaret Forrest. Letter not found.

78.10.26 You say that you want land for your *widow*. So you are going to join the happy state. I wish you every happiness. I was sorry to hear that you had been laid up. Saw Mr Brown[9] who has gone on to Champion's Bay.

I remain
My Dear Baron
Yours sincerely
Alex Forrest

78.10.30 *To Richard Owen*

Natural History Museum, London, Archives, DF200/14, Letters 1878 L-Z, f. 379.

30/10/78

These lines are written to you, dear Professor Owen, from my sick-bed, on which I have been thrown after an operation for varicous veins of the rectum.[1] You will therefore kindly excuse my writing with graphit-pencil; but I did not wish to loose a monthly mail to bring under your notice an offer for sale of a splendid collection of Coleoptera formed by a Gardener, who for many years worked under me in the bot Garden.[2]

It is on my own suggestion, that Mr French now offers these collections to the British Museum, it being before his intention to sell it to some other less important institution To your glorious Museum it is of course an object to secure particularly such coll-

9 Maitland Brown.

1 i.e. haemorrhoids. M had had a similar operation two months earlier; see M to E. Ramsay, 22 August 1878.

2 See C. French to M, 28 October 1878 [Collected Correspondence].

ections, as contain many novelties, so as to afford material for adding to the celebrated records emanating from the British Museum. 78.10.30

I consider the price *cheap*, considering the *high expenses* for travelling in any part of Australia!

Should you and the trustees be able to secure this almost unique collection, then Mr French must of course *insure* the sending and be responsible for its safe arrival[3]

Trusting, venerable Sir, that you are in the enjoyment of perfect health & happiness, and that a long serene evening after an eventful life of discoveries will be before you, I remain your

Ferd. von Mueller.

Pray Remember me Kindly to Sir Charles Nicholson & Sir Henry Barkly

3 The collection comprised about 700 species of Longicorn beetles in 890 specimens, for which French wanted £150, and about 3,500 specimens in other families, which French priced at £200, in each case without the cabinets containing the collections (particulars by Charles French, Natural History Museum, London, Archives, DF200/14, Letters 1878 L-Z, f. 379c).

Owen, in a letter to A. Günther, Keeper of Zoology, 23 December 1878, commented: 'I have, in acknowledging receipt of enclosed, told von Mueller [letter not found] that his letter with Mr French's statement have been refered to you with whom rests the initiative of recommendation. You will perhaps therefore, if you think the proposition worth entertaining, communicate to Dr Frd. von Mueller, (Melbourne Victoria) any suggestion tending to facilitate your wishes on the subject.' (Natural History Museum, Archives, DF200/14, Letters 1878 L-Z, f. 379a.)

No evidence of purchase can be found in the records of zoological accessions to the Natural History Museum, London.

78.11.17 *To George Bowen*

M78/57, unit 31, VPRS 1096, VA 466 Governor's inward correspondence, Public Record Office, Victoria.[1]

Melbourne 17. Nov. 1878.

To his Excellency Sir G. Bowen,
G.C.M.G., Governor of Victoria.

Sir.

I have the honor to acknowledge the receipt of a copy of a despatch received by your Excellency from the Right hon. the Minister[2] of State for the colonies, accompanied by copies of letters from Sir Jos. Hooker and the honorable Sir[3] R. G. W. Herbert, respecting the completion of the first series of volumes of the Australian Flora,[4] on which Mr G. Bentham has been engaged under my constant cooperation for the last 16 years.[5] It is most gratifying to observe, that the Right Hon. Sir Mich. Hicks-Beach appreciates so highly the services thus rendered to the British and Colonial Empire of her Majesty, by which efforts the foundation work on

1 There is a copy of this letter at CO 447/30, Order of St. Michael and St. George, 1878, vol. I, Despatches, warrants, letters &c, Victoria no. 488, enclosure to despatch no. 217 of 22 November 1878, Public Record Office, London.

2 Minister *deleted and replaced with* Secretary *presumably by H. Pitt the Governor's Private Secretary.*

3 the honorable Sir *deleted and replaced with* Mr *presumably by H. Pitt the Governor's Private Secretary.*

4 Bentham (1863-78).

5 See G. Bowen to M, 7 October 1878 [Collected Correspondence], enclosing a copy of a circular dated 19 August 1878, sent by the Secretary of State, M. Hicks Beach, to the Governors of Britain's seven Australasian colonies (unit 32A, VPRS 1087, dispatches of the Secretary of State to the Governor, vol. 32, part 2, July-December 1878, PROV). The circular includes copies of the letters from J. Hooker to R. Meade, Under Secretary at the Colonial Office, 24 July 1878, and R. Herbert to J. Hooker, 9 August 1878, to which M refers. (A copy of the circular is held at the PRO London, CO 854/19, Colonial Office, Circular Despatches 1878, 9309/78 Australia, ff. 176-7.)

the plants of one of the five great divisions of the globe is recently brought to a close. Every one must share the admiration, expressed by the distinguished Director of the Royal bot. Garden of Kew for Mr Benthams exertions, to elucidate the plants of the whole Australian continent, and this all the more as he at his venerable age has continued with youthful enthusiasm to work not merely with my aid on the plants of the greatest of the Queens Dominions, but has been engaged all the time also with Sir Joseph Hooker on an equally extensive work, namely the plant-genera of the whole globe.[6] Both works involved the almost daily tracing of microscopic minutiae and an amount of detail-studies, which must have deprived him, whom we all recognize as the greatest phytographer of the day, of all repose even at the evening of a long and labourious life,[7] and I trust, that this great man, while he is still spared us, will receive for his brilliant services the fullest imperial recognition. In justice to the colony, of which your Excellency is the vice-regal ruler, I may be permitted to observe more fully, what share Victoria has had in the issue of this certainly unique work, after Sir Jos. Hooker referred already in generous terms to my own efforts for rendering known the vegetable treasures of a continent nearly as large as Europe, inhabitable in all its zones, and rich in an endemic vegetation of much industrial value.[8] 78.11.17

6 Bentham & Hooker (1862-83).

7 J. Hooker, in his note of appreciation of the honor of CMG conferred on G. Bentham, wrote to C. Cox, 27 May 1879: 'he is in his 80th year I believe & works at Kew daily! Summer and winter!' (PRO London, CO 447/34, Order of St Michael and St George, vol. 2, 1879, Misc. Offices no. 8687).

8 In his letter to R. Meade, Hooker described how M, 'with singular generosity', 'transmitted to this country his immense Australian Herbarium, the examination of which was indispensable to the proper elaboration of the work. He, furthermore, allowed duplicate specimens to be taken from his collections for preservation at Kew as the authentic types upon which the descriptions published in the "Flora" had been based. I trust that the completion of this, the most important of the series of Colonial Floras projected by my late father, will be recognized as evidence of the value of Baron von Mueller's services, and of the botanical Establishments of Melbourne and Kew in furthering the development of the inexhaustible vegetable resources of our Colonial empire.'

78.11.17 In frankly setting forth my own engagements for the Flora of Australia, it is not done with any desire to earn any praise or reward whatever, but merely to show, that I endeavoured to acquit myself honorably of a task, which I commenced when I arrived 31 years ago (in 1847) in Australia, after the needful prior University education for the specialities of such work. Since then uninterruptedly engaged in Australia, I have extended the lines of my personal field-observations to between 27,000 and 28,000 miles, involving much of the toils, privations and dangers of early geographic exploration but affording me also unusual experience; the material thus accumulated by myself and supplemented by amateur-collectors or departmental emissaries exceeds one hundred thousand specimens from all parts of the Australian continent as far as hitherto mapped, embracing about 2/3 of all the localities recorded in the Flora Australiana. The whole of these enormous collections, nearly all examined by myself, were transmitted since 1862 gradually on loan to the illustrious main-author of the Flora, after my own successive observations on the species had to a great extent been recorded in the ten volumes of the "fragmenta phytographiae Australiae" and in some minor descriptive works, while the utilitarian value of many of the Australian plants was traced in my chemical laboratory, tested by my experimental cultures in the bot Garden of Melbourne, recorded in distinct publications and brought gradually at the successive great exhibitions before the commercial and technological world.

The first literary foundation of the Flora of Australia was laid (in 1810) by Robert Brown, the celebrated Naturalist of Flinders Expedition in his "prodromus,"[9] comprising several thousand plants either of his own collections or those of others, chiefly of Sir Joseph Banks. Valuable scattered memoirs of many authors, now mostly numbering with the death,[10] have followed, Mr Bentham's own early contributions dating as far back as 1837, all scattered except the researches of a number of men of science of several

9 R. Brown (1810). Brown accompanied Matthew Flinders on his voyage of exploration, 1801-2, in which he charted much of the south coast of Australia.

10 dead?

nations on the plants of West-Australia, collected by Prof. Lehmann[11] into two volumes, and except also two splendid and illustrated volumes by Sir Jos. Hooker on the plants of Tasmania.[12] 78.11.17

It was a rare facility, enjoyed by Mr Bentham, to be able to compare the original material chiefly among the great treasures at Kew, which the enlightened statesmanship of a great nation caused to be accumulated and there to be turned to unrivalled scientific account. In facilities like these I could only have participated by one or more visits to England, had it fallen to my sole share to elaborate connectedly the Flora of the fifth continent, a vast territory in the sole possession of Britain. It is now my intention, as announced by Mr Bentham in the preface to the 7th volume,[13] – if providence grants me life and health and if my researches should continue to obtain the needful fair and intelligent support –, to issue supplemental volumes for the description of those plants, discovered gradually since the Flora was issued;[14] for the first volumes particularly extensive additions have been obtained through the geographic disclosures of vast regions of the interior during late years. It will furthermore be needful to edit separate volumes on mosses, lichens, fungi and algae,[15] though of these the latter are to a great extent already elaborated by the late Professor Harvey in his five illustrated volumes, the result of his two years travels along the Australian coast.[16]

If I have at some perhaps undue length entered into this exposition, the Right honorable the Secretary of State will kindly consider, that I have made the study of plants of all Australia an object of life, that I have sacrificed for it nearly all that is dear to us in the world, and that it is with some pardonable pride when I own to have lived through the greater portion of the century of main-discovery of natural history and to have helped to unfold largely

11 Lehmann (1844-7).
12 J. Hooker (1855-60).
13 Bentham (1878), p. v.
14 No supplements were published.
15 No cryptogamic volumes were published.
16 Harvey (1858-63). For Harvey's expedition see Ducker (1988).

78.11.17 the vegetable objects, which nature with prodigal richness has strewed over the grandest possession of the British crown.[17]

I have the honor to be,
your Excellencys very obedient
Ferd. von Mueller

79.01.13 *To Ernst Behm*

Briefsammlung, Archiv, Justus Perthes Verlag, Gotha, Germany.[1]

13/1/79.

Es liegt mir ob, sehr geehrter Herr Doctor, Ihre gütige Zuschrift, welche mit letzter Post anlangte, anzuerkennen, u Ihnen und den Herren Ihrer Anstalt erneuet meine Versicherung auszudrücken, dass das traurige und unerwartete Hischeiden[2] des berühmten

17 The copy held in London is accompanied by a minute paper bearing, among others, the following minutes: 'There is nothing I presume to be done on this. Dr Mueller expresses a hope that Mr Bentham "will receive for his excellent services the fullest Imperial recognition" – Dr Von Muellers own name has been more than once brought under notice in connection with a KCMG. [WD]' and 'Sir G Bowen makes no recommendation. I think they both deserve to be noted for consideration – Dr Mueller for KCMG & Mr Bentham for CMG. [JB] 14/1.' and 'Note them for consideration, [for] the next Birthday Gazette. RGWH Jan 18.'

M was Gazetted as KCMG in the Supplement to the London Gazette of Friday 23 May (24 May 1879), p. 3597, and Bentham as CMG on p. 3598. The Governor of Victoria was notified by telegram, 23 May 1979: 'Von Mueller appointed K.C.M.G. – Bentham C.M.G.– & Colonel Scratchley C.M.G.' In the draft letter (17 May 1879, CO 447/34 (M8452/79)) to Queen Victoria recommending various appointments to the Order, the reference to M is: 'Dr Ferdinand von Mueller, C.M.G. He has spent his life in most successful and valuable investigations of the natural history of Australia.'

For Bentham for CMG, the description reads: 'For Services – together with Dr. von Mueller – in collecting and illustrating the Flora of Australia.'

1 For a published version of this letter see Voigt (1996) pp. 134-5.

2 Hinscheiden?

Freundes Petermann in Bezug auf dasjenige, welches ich für Sie dort zu thun vermag, keinen behindernden Einfluss haben wird. 79.01.13

Ob aber überhaupt viel von meiner Seite geschehen kann, um die geographischen Nachrichten zu fördern, bleibt ungewiss. Freilich ging ich mit dem Gedanken um, Mr Giles wieder auszusenden, aber nun hat Sir Thomas Elder eben Anstalten getroffen, um Mr Jesse Young, den Begleiter Giles's in seiner 3ten Reise mit der Führung einer neuen Dromedar Expedition zu betrauen, was hindernd in den Weg von Giles tritt

Mr Goldie ist nun auch nach 2 oder 3 jähriger Abwesenheit nach Sydney zurückgekehrt; ob seine letzten Resultate von geographischem Belang sind, weiss ich noch nicht; aber ethnologisch ist es interessant, nun durch ihn zu wissen, wie die Fabel über die geschwänzten Menschen entstanden sei.

Einige der in den Bergen Neu Guineas lebenden Stämme scheinen sich als Zierrath einen nachgeahmten künstlichen Schwanz anzulegen, was mit den mannigfachen andere Ideen dieser Naturkinder ganz im Einklang steht. Eine Wirbel-Verlängerung in einem Caudal Anhang bei der menschlichen Species gehört freilich auch ja zur Unmöglichkeit.

Ich danke Ihnen für Ihre Aufmerksamkeit, mir die näheren Umstände des Familien Lebens unseres armen Petermann, die zu seinem unzeitigen Ende beitrugen, geschildert zu haben. In unsern Zeitungen war die Schuld zum Theil den anonymen Briefen der ersten Gemahlin zugeschrieben; wollen wir es mit dem Mantel der christlichen Liebe zudecken!

Vom ärztlichen Standpunkt erscheint es, als ob das überarbeitete Gehirn des grossen Geographen, am leichtesten durch eine längere wissenschaftliche Reise zur Ruhe und ins Gleichgewicht gekommen sein würde.

Die geographischen Mittheilungen hätten sicher unter keine bessere neue Leitung als die Ihrige kommen können als eine würdige Folge von Petermann.

Verehrungsvoll der Ihre
Ferd. von Mueller

13/1/79

It is incumbent upon me, very respected Doctor, to acknowledge your kind letter that arrived with the last mail[3] and to express anew[4] to you and the gentlemen of your institution my assurance that the sad and unexpected death of the celebrated friend, Petermann, will have no hindering influence in respect to those things which I am able to do for you here.

But whether much from my side can happen at all to promote geographical information remains uncertain. Of course I turn over the idea to send out Mr Giles again, but now Sir Thomas Elder has just made arrangements to entrust Mr Jesse Young, the companion of Giles on his 3rd journey, with the leadership of his new expedition with dromedaries, which will place obstacles in the way of Giles.

Mr Goldie[5] has now also returned to Sydney after 2 or 3 years' absence; whether his last results are of geographic importance I do not know yet, but ethnographically it is of interest now to know, through him, how the fable about the men with tails arose.

Some of the tribes living in the mountains of New Guinea seem to attach an imitation artificial tail as decoration, which stands in complete harmony with the manifold other ideas of these children of nature. A vertebral extension in a caudal appendage in the case of the human species of course belongs indeed to the impossible.

I thank you for your attention in having described to me the particulars of family life of our poor Petermann that contributed to his untimely end. In our newspapers the blame was in part attributed to the anonymous letters of the first wife; let us cover it up with the the cloak of Christian love![6]

From the medical standpoint, it seems as if the overworked brain of the great geographer would have most easily come to rest and into equilibrium by a longer scientific journey.

The *Geographische Mittheilungen* could not have come under a better new management than yours as a worthy follower of Petermann

Respectfully your
Ferd. von Mueller

3 Letter not found.
4 The earlier condolence that M apparently sent has not been found but was probably included in his letter of 14 December 1878, an extract from which was later published in *Petermann's geographische Mittheilungen* (B79.13.09).
5 Andrew Goldie.
6 Petermann committed suicide on 25 September 1878.

To William Munro 79.03.20

RBG Kew, filed with Herbarium specimen Gramineae, Phragmites australis.[1]

20/3/79

Allow me, dear General Munro, to consult you as the great investigator of Gramineae, in reference to the Australian species of Phragmites. Mr Bentham has admitted in the flora Australiensis only one species;[2] but it seems, that the tropical Phragmites, which occurs not only in North Queensland, but which I have also recently received from the Nickol-Bay district of West Australia, does not die down annually, like the British common Phragmites, which latter seems however clearly identical with our southern Australian species. The difference between what I suppose is Phragm. Roxburghii and Phragm. communis is thus very striking, and probably it is supported by other characteristics, such as the limited material here at my disposal does not readily allow me to trace out but which you will long since have studied fully. It seems to me, also, that P. Roxburghii is a larger plant than P. communis.

I have a few other grasses from several parts of Australia, which were discovered since the 7th vol. of the Flora Australiensis appeared; of these I will gladly send you specimens, if you should desire it, by which means the Australian material obtained from my museum-collections will be kept complete with you.

With reverence your
Ferd. von Mueller, M.D.

1 MS found with a specimen sheet marked: 'No 64 Arundo Phragmites River Yarra, Melbourne 7/4/53'.

2 *Phragmites communis*, Bentham (1878), p. 636.

79.03.31 ## *To Edward Bage*

Private hands.

31/3/79.

The high Eucalypts, dear Mr Bage, which I measured between Fernshaw and Marysville[1] were all near the macadamized road, though some deep down in the vallies. I have my manuscript notes stored away, & cannot give you without much loss of time the precise data, but the highest of the trees came up to about 400 feet and many to 350. Still higher measurements are on record, but for their absolute accuracy I cannot vouch, as for them we have to rely on others. You must however have observed yourself that at the top of the highest trees often a single slender terminal branch may straggle out 20-30 feet long, which may break off any time and thus change the measurement of an individual tree. But the largest trees known are between Berwick & the Upper Yarra;[2] these I never came to myself. I intend to institute more extensive measurements this autumn, if I can get time. The tree, here always the highest is Eucalyptus amygdalina of a variety called "*regnans*".

Regardfully your
Ferd. von Mueller

I saw kind Dr & Mrs Lewellin a few days ago, and they are happily quite well again.

1 Vic.
2 Vic.

To Bryan O'Loghlen

79.05.18

P79/4536, unit 153, VPRS 3991/P inward registered correspondence, VA 475 Chief Secretary's Department, Public Record Office, Victoria.

Melbourne
18/5/79.

To the honorable Sir Bryan O'Loghlen, Bt,
Acting Chief Secretary.

Sir

I have the honor to solicit, now while the estimates for the new finance-year are under the consideration of the Government again, that you and your honorable Colleagues will be pleased to give to the Report of the Committee of Dr L. Smith, Mr Bosisto and the late Mr King, MLA,[1] your favorable consideration for the resuscitation of my Department in its proper scope. The report of the above named Gentlemen, who were appointed by the hon Graham Berry, to advise him respecting the means, required by me for the proper functions of Gov. Botanist, was submitted long ago, but I have not sought to obtain an insight into this document, being assured, that what the Committee recommended would be reasonable and just. I believe however, that I shall be in consonance with the general views of the Committee on my case, if I beg of you, to cede to me the buildings, now occupied as temporary almhouses or immigrants homes at the City bridge, it being doubtless the intention of the Government to remove early the paupers from the close vicinity of Government House, on sanitary considerations.[2] I would further ask, in probable conformity of the Committee's recommendation, that all the ground in Gov. House reserve, not enclosed for the private use of His Excellency the Governor, be allotted to my control, from the boathouses up to the St Kilda Road and towards the Barracks and the vicinity of the Observatory. It needs

1 See L. Smith to G. Berry, 11 July 1877, pp. 745-8 below.

2 M had previously made similar requests to have these buildings allocated to him; see M to J. MacPherson, 26 June 1876 [Collected Correspondence] and M to G. Berry, 7 August 1877.

79.05.18 not my explanation, that in a clime like ours a few acres would be utterly insufficient for my purposes, as a full collection of different trees from all parts of the globe, so far as hardy here, would occupy alone a very extensive area, not to speak of the numerous other kinds of plants, which as Gov. Botanist I should have in a *living* state under my daily observation. Indeed I may be allowed to remark, that in every other country the Gov. Botanist is provided with a botanic Garden; so it is with Sir Jos. Hooker at Kew & everywhere else. The administration of the new grounds, asked by me, *need not involve a heavy annual expense, nor would my planting scheme be solely scientific but ornamental as well.* In the ground asked for, almost every tree has been raised by me, so that I would return to a creation of my own, utterly neglected since; and by removing from me the undeserved and deeply saddening *humiliation*, which also *impedes my work* in most directions, my hopes of life by your kindness would become once more brightened.

I have the honor to be,
Sir,
Your obed. serv.
Ferd. von Mueller[3]

3 On Wednesday 4 June 1879 L. Smith wrote to B. O'Loghlen, Acting Chief Secretary: 'Dr Mueller having written to me stating that you have granted him an interview on friday but having omitted the hour will you please inform me the time you have appointed. I will then bring up with me as Chairman the copy of the Report.' (P79/5117, unit 1086, VPRS 3991/P, PROV).

On 7 June 1879 O'Loghlen minuted: 'Will the Commissioner of Public Works be good enough to direct some leading officer of his department to report on the buildings now occupied as the Immigrants Home on the *Eastern* side of the St Kilda Road with a view to their suitability for a Botanical Museum and the Government Botanists Residence'.

A week later, on 13 June, Charles Barrett, Chief Assistant Architect of the Public Works Office, reported:

> I have the honor to report that this establishment consists chiefly of six substantial brick buildings in a fair state of repair fuller particulars of which are given on the other side
>
> The Hospital could be converted into a residence for the Government Botanist at a cost of about £250.0.0. The conversion of the Dormitories, Mess Room &c into Botanical Museums including repairs to Fencing and drainage is estimated to cost an additional £225.0.0.
>
> As I have not been furnished with particulars of the number of buildings it is

From Queen Victoria[1]

79.05.24

RB MSS M202, Library, Royal Botanic Gardens Melbourne.[2]

Victoria R. I.[3]

AUSPICIUM MELIORIS AEVI[4]

Victoria by the Grace of God of the United Kingdom of Great Britain and Ireland Queen, Defender of the Faith, Empress of India, Sovereign and Chief of the Most Distinguished Order of

> proposed to occupy as a Museum, or of the nature and extent of the Fittings, Furniture &c required I am unable to estimate these latter but believe they would form a serious item in the cost of the work.

The letter includes a sketch of the buildings and a brief description of each, with dimensions. Barrett's report was forwarded to the Departments of the Chief Secretary and Public Works.

On 18 June 1879, the day after the Chief Secretary, G. Berry, resumed office upon returning from a visit to London, M wrote seeking an interview with him prior to the Estimates being finally settled; see M to W. Odgers, 18 June 1879 [Collected Correspondence].

L. Smith wrote to Berry on 12 July 1879: 'As Chairman of the Commission re Baron Mueller I am requested by my colleague Mr Bosisto & by Mr Dow to ask you to receive us as a Deputation to place the Baron's case before you, will you please name day & hour'. On 14 July W. Odgers, Under Secretary instructed: 'Ack[nowledge] & say that C.S. is fully acquainted with all the particulars of the case and doubts if any good purpose w[oul]d be served by the interview but if the Gent[leme]n as Members of Par[liamen]t particularly desire to see him Mr B. will be happy to receive them at noon on Thursday' [i.e. 17 July] (P79/6404, unit 1086, VPRS 3991/P, PROV).

In the event, no action was taken on M's proposal.

1 MS bears the seal and the richly coloured insignia of the Order of St Michael and St George at the top of the letter.

2 A draft of this Royal Warrant, authorised by George, Duke of Cambridge and Charles Cox, Grand Master and Chancellor of the Order, is held at PRO (CO 447/33, Order of St. Michael and St. George, 1879, vol. I, Despatches, warrants &c.).

3 Regina Imperatrix [Queen Empress]. The Queen has signed the document at the top.

4 'Token of a better age', the motto of the Order of St Michael and St George embossed on the seal.

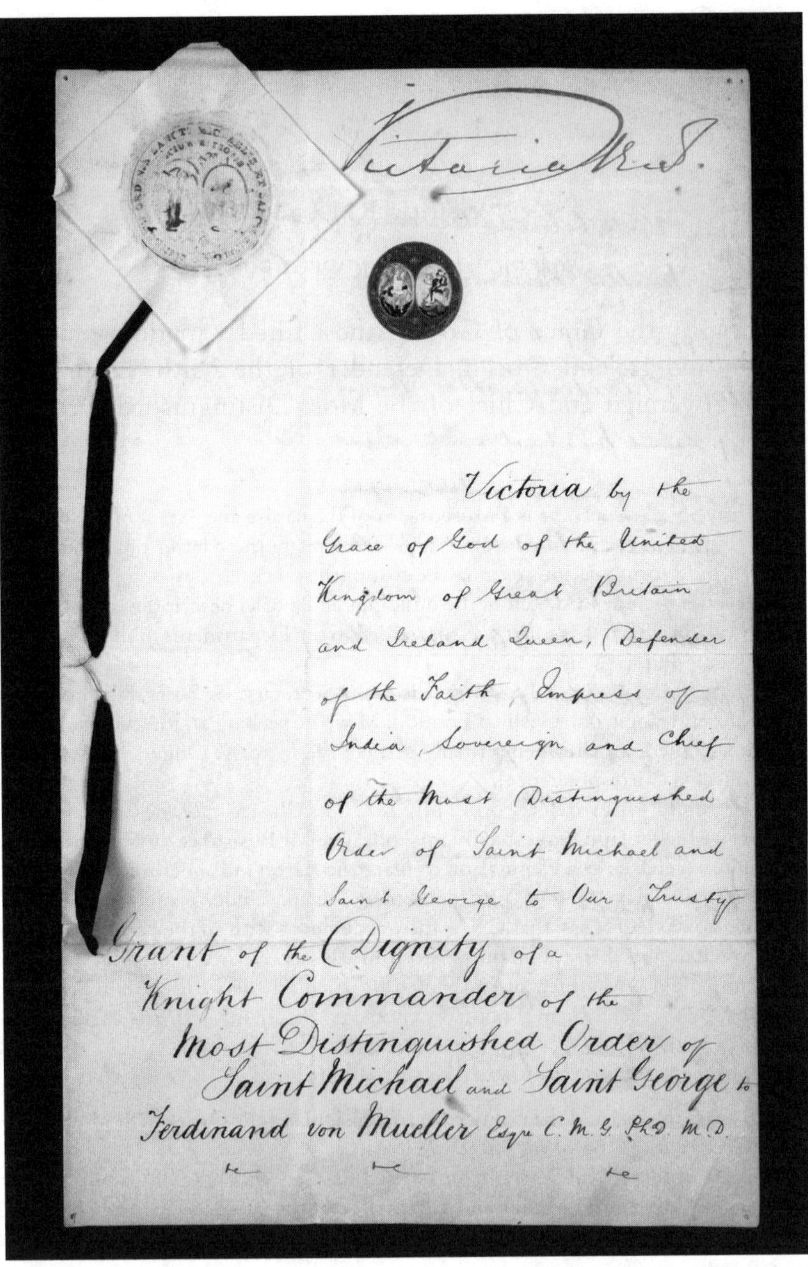

Victoria R.I.

Victoria by the Grace of God of the United Kingdom of Great Britain and Ireland Queen, Defender of the Faith, Empress of India, Sovereign and Chief of the Most Distinguished Order of Saint Michael and Saint George to Our Trusty

Grant of the Dignity of a Knight Commander of the Most Distinguished Order of Saint Michael and Saint George to Ferdinand von Mueller Esqre C.M.G. PhD. M.D.

&c &c &c

Figure 6. The first page of Mueller's promotion to Knight Commander of St Michael and St George, signed at the top by Queen Victoria. Courtesy of the Library, Royal Botanic Gardens Melbourne.

Saint Michael and Saint George to Our Trusty and Well-beloved Ferdinand von Mueller, Esquire, Companion of Our said Most Distinguished Order of Saint Michael and Saint George, Doctor of Philosophy Doctor of Medicine Government Botanist for Our Colony of Victoria: Greeting. 79.05.24

Whereas We have thought fit to nominate and appoint you to be a Member of the Second Class or Knights Commanders of Our Most Distinguished Order of Saint Michael and Saint George, We do by these Presents grant unto you the Dignity of a Knight Commander of Our said Most Distinguished Order. And We do hereby authorize you to have, hold and enjoy the said Dignity as a Member of the Second Class or Knights Commanders of Our said Most Distinguished Order, together with all and singular Privileges thereunto belonging or appertaining.[5]

Given at Our Court at Balmoral under Our Sign Manual and the Seal of the said Order this Twenty fourth Day of May 1879 in the Forty second Year of Our Reign.[6]

By the Sovereign's Command

George
Grand Master

By the Sovereigns Command

C. Cox[7]

Grant of the Dignity of a Knight Commander of the Most Distinguished Order of Saint Michael and Saint George to
Ferdinand von Mueller Esqre C.M.G. Ph.D. M.D.
&c &c &c

5 The draft held in London differs in minor details from the final version transcribed here.

6 See also R. Herbert to M, 30 June 1879 [Collected Correspondence], in which the Queen's Secretary forwarded the Warrants and Insignia of M's new rank as Knight Commander of the Order, and asked that he return the Insignia he had worn as a Companion of the Order.

7 The last two lines are in different hands.

79.05.27 *To Joseph Hooker*[1]

RBG Kew, Kew correspondence, Australia, Mueller, 1871-81, ff. 229-30.

27/5/79.

Since Queen's Birthday, dear Sir Joseph, I have been in a whirl of joyous excitement through the utterly unexpected honor conferred on me by our gracious Sovereign.[2] If I had received a telegram, that some unknown stranger had at his death left me a fortune, I could not have been more surprised, than when H Exc.[3] the Marquis of Normanby in a most gracious note informed me of the high distinction bestowed on me at the throne of Britain.[4]

Though I have worked in her Majesty's Australian territory for one third of a century uninterruptedly, I never gave it any thought, that such a mark of royal grace was in store for me, more particularly, since the breaking up of my Department seemed to debar me from any further real advance in any direction. Hence my astonishment was all the greater, and I must think, that this bestowal of a high British Order is rather intended as a mark of recognition of Australian science generally (I being the senior now in Australia) than as any reward deserved for my own work. Numerous have been the felicitations, among them a telegram from H.E. General Sir Harry Ord, the Governor of W.A.[5] My countrymen are particularly elated, as this is the first distinction of its kind (since Count Strzelecky's time, who not was strictly a German[6]) bestowed on one originally belonging to the German Nation. I just learn, that in celebration of this glorious event my countrymen here intend to give me a Banquet.

1 MS annotation by Hooker: 'An[swere]d July 27 /79'. Letter not found.

2 See Queen Victoria to M, 24 May 1879, appointing M as Knight Commander of the Order of St Michael and St George.

3 His Excellency.

4 Note not found.

5 Telegram not found.

6 The explorer P. de Strzelecki was born a citizen of Prussia, but of Polish parentage. He was appointed KCMG in 1869.

I am completely unconscious, to whom particularly I owe my promotion in the order, being certainly in Australia the senior with the former Attorney General of N.S. Wales;[7] but I cannot but think, that among the generous friends, who have advocated my cause, you have taken with Sir Henry Barkly and Sir Charles Nicholson the lead. So, my dear Sir Joseph, let me thank you for your disinterested goodness, and rejoyce with me, that such an effusion of favor from the fountain head of honor must greatly tend to resuscitate my Department and therewith to brighten my future. 79.05.27

Only in one way my joy is dimmed, at seeing the great and venerable Bentham placed as CMG on the honor list.[8] When yesterday waiting personally on H. Exc. I at once alluded to this: but the Marquis replied, that unless an exception was made in favor of a Governor, everyone had to go through his grade in the Order; but I anticipate, that Mr Benthams promotion will be next anniversary of her Majesty's Birthday, if indeed not earlier.[9]

Now I would ask you a question. Would Mr B.[10] be pleased and would you approve of it, if I made an application to the different Australian Governments, more particularly that of this colony, N.S W., QL. & S.A. to obtain for our venerable friend a monetary recognition; and if so, should I ask for £250 from each Government or should it be more? I am not acquainted with Mr B. wordly affairs, & though he inhered[11] Jeremiah Bentham's fortune,[12] he may have since spent it in his researches. Let me be guided by your views on this subject, and I will do my utmost for him, as his claims on the gratitude of all Australia are immense and highly just.[13]

7 Sir William Manning?

8 See minutes filed with M to G. Bowen, 17 November 1878 (PRO, London, CO 447/30, Order of St. Michael and St. George, 1878, vol. 1, Despatches, warrants, letters &c, Victoria no. 488, Enclosure to despatch no. 217 of 22 November 1878), where the officials in the Colonial Office decided that Bentham should be recommended for CMG and M for KCMG. See also Lucas (2003).

9 Bentham was never promoted.

10 George Bentham.

11 inherited?

12 Bentham's uncle, the social philosopher Jeremy Bentham.

13 *This paragraph is marked with a line in the margin and annotated by Hooker*: 'Declined. J H'.

79.05.27 I write this for an extra mail (pr. Northumberland) and have not much time left for other correspondence. So allow me to conclude with grateful remembrance your

Ferd. von Mueller

I am of course under the deepest obligation to the Right Honorable Sir Michael Hicks Beach but can only formally express through the Governor my gratitude when Patent and insignia arrive.[14]

79.06.08 *To Thomas Huxley*

Huxley papers, vol. XXIII, pp. 109-10, Imperial College Archives, London.

Melbourne
8/6/79

Allow me, dear Professor Huxley, to solicit your powerful support for the election of Prof McCoy, FGS, into the Royal Society next November.[1] As a fellow worker in the same specialities you will know how worthy a candidate I propose, and doubtless the Professor would long since have earned this reward in science also, as lately that of the Murchison Medal,[2] had he allowed his modesty to step forward as a Candidate. It is solely at my action that he now seeks the honor of FRS.

14 See M to M. Hicks Beach, 29 September 1879 [Collected Correspondence].

1 MS annotation: 'elected 3.6.1880'. Under its Statutes, the Royal Society of London held a single election of new Fellows each year, not at its Anniversary Meeting in November as M here supposes, but on the first Thursday in June. McCoy was elected at the first attempt on 3 June 1880. See *Record of the Royal Society* (1940), pp. 302, 488.

2 McCoy was awarded the Murchison Medal of the Geological Society in 1879.

I have written simultaneously to Sir Jos. Hooker, Sir John Lubbock, Prof Williamson,[3] Prof Stokes and Mr Carruthers[4] to see whether they unitedly would secure Prof McCoys election, in which effort you will doubtless gladly share. 79.06.08

Let me hope, my honored Sir, that your health is quite firm again, so that you can proceed on your luminous path of investigations uninteruptedly.

Regardfully your
Ferd. von Mueller

Though *all* the illustrious men on the Council of the RS are known to me by their works, I have refrained from adressing those, whose researches are in different directions to my own poor work. Nevertheless all would likely respond to a kind call of yours to honour Prof McCoy.

To George Bentham 79.06.08a

RBG Kew, Kew correspondence, Australia, Mueller, 1871-81, f. 232.

Allow me, dear Mr Bentham, to offer you my best felicitation to your membership of the Order of St Michael & St George,[1] though I wished that I at once could congratulate you to[2] the long earned British Knighthood, which is sure early to follow after this first

3 Presumably A. W. Williamson, who was Foreign Secretary of the Royal Society at this time.

4 See M to J. Hooker and to W. Carruthers, both 8 June 1879 [Collected Correspondence]. Letters to Lubbock, Williamson and Stokes not found.

1 Bentham was created a Companion of the Order of St Michael and St George (CMG) on 24 May 1879, gazetted in the same honours list that contained the knighthood for M by promotion from CMG to KCMG.

2 on?

79.06.08a stage[3] of royal approval of your illustrious researches through half a century & more. Of all men of science *you*, the most labourious, deserved the *highest* reward for your disinterested furtherance of one of the most important of all sciences.

What a pleasure & delight must it be to you, to be conscious, that hardly a square mile in[4] the whole globe, so far as inhabitable, does exist, where plants do not speak & will speak for all ages of you. If we came to count up the species named by you, I believe they would exceed even those of L's & D C's[5] and Hookers! Well may you be proud of that; no earthly other glory will compare to such.

Let me hope, that your health amidst your great labours will remain firm, and that you autumn of life will be long & serene.

Regardfully always your
Ferd. von Mueller

3 Three lines of the letter, to this point, are each marked in the margin with a cross. Bentham, on seeing the Gazette announcement 'wrote to remonstrate' (RBG Kew, Bentham's diary, vol. 19, 24 May 1879). Bentham believed that he had no claim to the award as the announcement described him as 'George Bentham Esqr of Victoria author of the Flora Australiensis'. His letter, 24 May 1879, to R. Herbert, Secretary to the Order, continued: 'I have not nor have I ever had any connection with any of the Colonies and therefore can have no claim or right to be placed on the Order of St Michael and St George and would only be considered as an intruder there and cannot conceive how the mistake has originated – probably the honor may be intended for someone else. At any rate I hope that the error may be corrected and the nomination cancelled' (PRO London, CO 447/35. Order of St Michael & St George, Original correspondence 1879, Individuals; B, no. 8686). Herbert replied on 29 May that 'the mistake in the Gazette arose from the fact that the Services recognised by the Queen were rendered principally in connection with that Colony. The value of those services have been repeatedly brought before H. M. Govt. by the Gov. of Victoria as well as by Sir Jos. Hooker; & there can be no question that their recognition in the manner that H. M. has been pleased to adopt will be very satisfactory to all Australians interested in the Natural History of the Country' (Draft reply with the copy of the letter from Bentham). Bentham replied on 30 May: 'Although I may be still of [the] opinion that personally I am not qualified for the distinction yet I accept it thankfully as an honorable acknowledgement of the value of science' (G. Bentham to R. Herbert, PRO London, CO 447/35, Order of St Michael & St George, Original correspondence 1879, Individuals; B. no. 8789).

4 on *over* in, *or* in *over* on.

5 Linnaeus; De Candolle.

To James Agnew 79.06.15

RSA/B.13(14), Archives, Royal Society of Tasmania, University of Tasmania, Hobart.

15/6/79.

Your kind letter, dear Dr. Agnew, remained for some or many days unanswered, as I was surrounded by pressing duties for some time past. Excuse under the circumstances the delay of my acknowledgment. The fossils, transmitted by Mr Wintle, are very interesting & particularly so the leguminous fragments representing two species. It will take some time to search (up to date) through the palaeontologic literature to determine the affinity of these plants, and it would be well if the zealous observer and clever generalizer, who found these vegetable remnants, could procure more specimens, so that a full diagnosis of the respective species could be elaborated. As there seems to be no urgency about settling the specific position of these vegetable relics, I will wait to learn, whether a[ddi]tional[1] specimens are procurable

About the fossil shell I will obtain Professor MCoys most valuable opinion. The whole will of course be returned to your Museum.

Let me remain, dear Dr Agnew, with regardful remembrance your

Ferd. von Mueller.

I do not like to sacrifice the very few fragments by dissection

1 *editorial addition* – MS damaged.

79.06.20 ## *From John Thurston*[1]

RB MSS M76, Library, Royal Botanic Gardens Melbourne.

[Fiji][2] 20th June 1879

My dear Sir.

I enclose you a Circular from the Italian Government relative to the diseases &c of the Citron family Also an extract from a newspaper and one from a book I lately read.

I am quite alarmed at the destruction caused here by one of the "Coccidae" – It has only appeared here of late years, but it is ruining every orange and lime tree in the place.

It attacks numerous other plants and I have even found it in the far bush on ferns – The natives have no name for it, and it was not here when I came to the country in 1865.

Most likely this coccus is the one referred to in the printed extract enclosed[3] and we may have imported it as there described.

Some years ago I remember that in Mauritius they suffered from a parasite to which the local name of "pou blanc" was given but a "pou rouge"[4] was introduced which ate up its white brother and at same time did no harm.

If with your great knowledge of plant life you could give me any advice on this point I should be greatly obliged.

It appears to me that the use of "Kerosene or linseed oil" even if effective could never be carried out to any extent and I am anxious to know whether the insect has a "natural enemy" in Australia that could be imported.

I am preparing a set of Ferns which I hope to send you shortly.

Believe me
My dear Sir
Very faithfully yours
Jno B Thurston.

1 MS annotation by M: 'Answ 7/8/79.' Letter not found.

2 *editorial addition.*

3 Enclosure not found.

4 white louse; red louse.

From Ann Timbrell

79.07.24

RBG Kew, Miscellaneous reports 7.7, Victoria, Miscellaneous 1861-1916, f. 345.[1]

Silk Farm.
Plenty Road.
Collingwood.[2]
Thursday. July 24. 1879

My dear Baron: –

I am sincerely obliged to you for your kindness; and am delighted to hear of your accession to the title of K.C.M.G. but I am afraid that you have over-estimated my small silk samples from the Osage Orange.

I now enclose another curiosity. Some months ago, I noticed a paragraph in one of the newspapers, about a M. Brunet having succeeded in obtaining "*Silk from the bark of a mulberry tree*": so, last night, I tried an experiment, with what result, you can now judge from the accompanying paper.[3]

It is only interesting as a novelty in Sericultural Science; and, please to remember, it is only my first effort.

1 M forwarded this letter, together with the samples associated with it, to J. Hooker, see M to J. Hooker, 27 July 1879 [Collected Correspondence].

2 Melbourne.

3 A small piece of paper (about 9 × 11 cm.) is glued to the front of the folio. Two samples of fibre, one off-white, the other pinkish-purple, are attached by thread on one side of the paper, and below them Timbrell has written: 'Respectfully presented as a novelty in sericultural science to Baron Ferd. von Muëller. K.C.M.G. &c &c'.

Below the off-white sample she has written: 'natural: as extracted from mulberry bark' and below the purple one: 'dyed: with magenta'.

Opposite these she has written: 'Mrs Timbrell's *first* attempt at procuring a substitute for Silk, from *the fibre of mulberry bark*. The experiment was made on the night of Wednesday, July 23. 1879: at her Silk Farm, Plenty Road, Collingwood near Melbourne. Probably, the next operation may be an improvement on the above result.'

79.07.24 I am going to try again: and, will send you some Silkworm eggs very soon.

Allowe me to remain
Yours sincerely
Ann Timbrell[4]
[...][5]

79.08.04 *To George Day*

Private hands.[1]

4/8/79

On enquiry I find, dear Mr Day, that the vote for my Department will not likely come on for some weeks; hence it would be unwise to take any action at present regarding any engagements, involving lengthened expense; and the question arises, whether it would be better to carry out in first instance your original plans for the Mallee-Country during the next few months. I gave my former collectors £3 pr week, out of which they had to bear their travelling expenses, but I supplied collecting material extra, and paid also the freight for the collections.

The Mallee country is botanically so much exhausted, that I should not feel justified to spend departmental means on collections from thence; altho' commercially a collecting-tour, such as suggested by Mr Luehmann, may possibly turn out quite payable, if the seed-season should prove favorable this year. Rest assured,

4 The signature is in a much larger and freer hand than the text of the letter and the notes, in which the writing is small and almost Copperplate and may be that of an amanuensis.

5 *illegible* – Obscured by binding.

1 The letter is accompanied by an envelope addressed in M's hand to 'George Day Esqr Little Bendigo near Ballarat' and dated by M '13/3 1879', with a Melbourne postmark dated 14 March 1879.

that no one would like more than myself to turn your energy & enthusiasm into a channel profitable to my favorite science. But my hands are tied at the present. 79.08.04

Regardfully your
Ferd. von Mueller

Your legislative friend must *not* raise any debate on my vote, but must act in concert with Dr L.L. Smith, Mr Bosisto, Mr Dow & Mr Blackett concerning any attempts of resuscitating my Department, not by a discussion in Parliament, but by friendly remonstrances in private with the ministry.[2]

To Joseph Hooker[1] 79.08.16

RBG Kew, Kew correspondence, Australia, Mueller, 1871-81, ff. 241-2.

16/8/79.

Let me express to you my best thanks, dear Sir Joseph, for your generous felicitation to the *unexpected* high distinction, which the grace of her Majesty conferred on me.[2] A congratulation from *you* is particularly gratifying. That Mr Bentham did not *at once* attain to the same degree, made me quite sad at the time, after his 60 years working on plants systematically. When I think of it, what a lenght of time!, almost as much already as your fathers of

2 Smith and Bosisto were members of a Parliamentary Board of Inquiry into M's position that had provided strong support for M; see L. Smith to G. Berry, 11 July 1877, pp. 745-8 below. J.L. Dow and C.R. Blackett were also members of the Victorian Parliament.

1 MS annotation by Hooker, obscured by binding: 'an[swere]d [Oct] 28/79.' Letter not found.

2 KCMG. See Queen Victoria to M, 24 May 1879. Hooker's letter of congratulation has not been found.

79.08.16 unperishable memory. When I studied in Kiel in 1846, I went during the autumn vacation to Sylt & Föhr on the West coast of Schleswig to examine more closely its vegetation, gathering Juncus pygmaeus (which you have now also in your Students Flora)[3] and many other rare plants. When I came back, the first word, Professor Nolte said, you have missed Bentham visit![4] So it was, he then came to Kiel also on his continental tour to gather every then possible information on the Labiatae for the 12th vol. of D.C.[5] – and now, he is still as fresh and labourious mentally as then, when I was a youngster. –

It is really very good of you, that you have spent some of your *precious time* to see my W.A. Forest-report[6] through the press, which thus obtains additional value, which ought to be acknowledged publicly; this will be done in the decades of the Eucalyptography, of which the 4th is nearly ready, though through delays of the lithographer not even the first decade is actually out.

At last a *forest bill* is to be introduced here,[7] but our best forests are by this time quite demolished. I preached here to deaf ears on this like on so many other subjects.[8] So, some years ago, *I induced* Mr Krichauff, a member of the Adelaide legislature, and an University friend of mine, to bring in *there* a Forest-bill, which he did and out of which arose at once then in Adelaide a proper Forest-

3 Hooker (1878).

4 M misremembered which of his absences from Kiel that year had coincided with Bentham's visit. Bentham visited the Kiel University Botanic Garden not during the autumn vacation but on 1 July 1846. He commented in his diary: 'It is small and has but few funds to keep it up. Prof Nolte is however very zealous. He is chiefly occupied with the flora of the duchy of Schleswig Holstein' (RBG Kew, Archives, Bentham, G., Diary vol. 2 [1845-7]). M was most likely at Heiligenhafen on the Baltic coast of Schleswig-Holstein at the time.

5 Bentham (1848).

6 B79.13.10.

7 A Forest Bill was introduced into the Victorian Legislative Assembly in 1879 but did not proceed. A similar bill was introduced in October 1881 but again did not proceed. See Moulds (1991), p. 14.

8 See B71.13.03.

Department,[9] for which latterly Mr J. E. Brown has done such excellent work. 79.08.16

I will take advantage of every opportunity to examine Salicorniae in a fresh state, but the garden here is no longer devoted to science-culture nor accessible to me, and the Saltbush-desert commences only 150 miles from here. In the same way I am *daily hampered* for forest-investigations, for which I want the rich collection of *living trees*, established by me in the bot Garden, including numerous species of Eucalypts, many of which however since destroyed by ignorance and in senseless changes. I shall *not* be able to do anything even for the Sydney or Melbourne great Exhibitions;[10] as not even my laboratory has been restored to me, & I am provided with no means of any other kinds to share in the Exhibitions.

With every feelings of grateful regards
your
Ferd. von Mueller

Have you noticed, that in the genus Poranthera, species occur with *opposite* leaves (e.g. P. microphylla); perhaps you like to take notice of it in the "*genera*".[11] What a *splendid work*; I hope your & Mr Bentham's united labors will bring it in a few years to conclusion. I can furnish you with some addenda, if you like, for any supplemental notes in the last volume.

9 For a brief account of Krichauff's successive initiatives to encourage forestry in SA, see J.E. Brown (1881), pp. viii-ix. See also M to J. Hector, 10 August 1874 [*Selected correspondence*, vol. 2, pp. 710-11].

10 International Exhibition, Sydney, 1879-80; International Exhibition, Melbourne, 1880-1.

11 Bentham & Hooker (1862-83).

79.09.09 *To Friedrich Wöhler*

Archiv-Sammlung Wachs, no. 407, Berlin-Brandenburgische Akademie der Wissenschaften, Berlin.

9/9/79.

Mit dieser Monats-Post erlaube ich mir, edler Herr, Ihnen Mr John Forrest's neue Vermessungs-Karte des Nickol-Bai-Districts vorzulegen, auf welcher dieser berühmte Geograph Ihnen eine Berghöhe gewidmet hat, und zwar auf meinen besonders ausgesprochenen Wunsch. Allerdings ist dies nur eine unbedeutende Gabe aus dem fernen Australlande; es bleibt aber doch immerhin eine dauernde, so lange die gegenwärtige Schöpfungs Epoche bestehen wird.

Nehmen Sie also, hochverehrter Herr Geheimrath, diesen kleinen Tribut freundlichst hin.

Mit Ehrerbietung
der Ihre
Ferd. von Mueller.

Herrn Geheimen Medicinalrath
Prof Dr Wöhler
Göttingen

9/9/79.

With this month's mail I venture, noble Sir, to submit to you Mr John Forrest's new survey map of the Nickol Bay district,[1] on which this celebrated geographer has dedicated a mountain height to you,[2] and in fact at my particular expressed request. Admittedly this is only an insignificant gift from the far south land; but all the same it does remain a lasting one as long as the present creation epoch will exist.

Therefore accept most kindly, highly respected Privy Counsellor, this small tribute.

With respect your
Ferd. von Mueller

Privy Medicinal Counsellor
Professor Doctor Wöhler
Göttingen

1 Map not found.
2 Mount Wohler, WA.

To [John Forrest][1]

79.09.11

Acc 527. 345/with item 12, State Records Office of Western Australia, Perth.

11/9/79.

By the P. & O. Steamer[2] I have just received the W.A. forest-report,[3] dear friend, my booksellers (Dulau & Co in London) having *bought* five copies for me, for which they had to pay to Reeve's firm £3.8, besides cost of transmission; three of these copies have gone to Italy, France & Prussia. The price is enormously high, especially as you said your Council had voted £300 for the issue of this publication and as I furnished manuscript and drawings *gratis*! With exception of the shading in of some details of the anatomic plates the work is *done well* under Sir Joseph Hookers direction. This document ought to be sent to the Sydney Exhibition by your colony. I suppose, that you have plenty of copies supplied to your Government, if actually £300 was voted for this Report. If so, pray ask Mr Fraser, to send me at least a few,[4] as I am too poor to purchase many, now particularly when for *eleven* years I am to pay off at the Building Society for my Cottage (for Office work only, not private dwelling) –

The Report ought to be a great help to the development of colonisation in W.A.[5]

Regardfully your
Ferd. von Mueller.

1 MS annotation: 'This letter is written to Mr J. Forrest'.

2 Peninsula and Orient Steam Navigation Company.

3 B79.13.10.

4 The file includes a telegram from M. Fraser to J. Forrest, 24 November 1879: 'You may send baron von Mueller six copies of his report for private use & I will decide how many shall be sent to the Victorian Govt. when I come back' (Acc 527, 345/10); and another from Fraser to Forrest, 28 November 1879: 'You can send fifty copies of Muellers report on our forests to Colonial Secretarys Office & request Mr Eliot to forward them to Chief Secretary Victoria asking him to let the Baron have twenty five & further to enquire whether they wish any more copies' (Acc 527, 345/9).

5 MS annotation: 'The Baron expresses himself very modestly'.

79.09.11 If the *main bulk* of the copies is gone to Swan River, then I would like to write a dedicatory page to Sir Joseph Hooker, which (I suppose) could be printed in Perth & inserted yet in many of the copies. It would be an acknowledgment to Hooker.[6]

79.09.20 *From Charles Moore*[1]

RB MSS M50, Library, Royal Botanic Gardens Melbourne.[2]

Sydney 20 Sepr
1879

My dear Baron

I have been & am so much engaged, & besides so much depressed in spirits, that my correspondence has been wholly neglected. You of all others I wished to write to, but could not till now. I cannot enter into the subject of my late dear brothers death, but thank you most sincerely for your kind & most sympathious letter.[3] Since that sad news reached me, I feel as if all object in life had left me, but enough of this.

I have now nearly finished getting in the specimens of woods for our Exhibition,[4] and will forward to you in a few days such of

6 This proposal appears not to have gone ahead. MS annotation: 'The Report was written for this Governments [an unknown amount of text missing]'.

1 MS black edged – Moore's brother David, curator of the Glasnevin Botanic Garden, Dublin, died on 9 June 1879. MS annotation by M: 'Answ 6/9/79'. M presumably meant 6/10/79. Letter not found.

2 MS found with a specimen of *Eucalyptus largiflorens* (MEL 1610556), collected at the Bogan River, NSW, in 1879.

3 Letter not found.

4 International Exhibition, Sydney, 1879-80.

the specimens of the leaves and flowers as I cannot determine at first sight. I fancy from the names attached to Fawcetts[5] specimens he must have been in communication with you; but still more inspection will be necessary & I very much regret that the material which I have to send you for naming is not better. I fear that many of them will be quite undeterminable, but I feel assured that with your usual kindness you will do all you can for me.

Enclosed is a specimen of an Eucalypt, which has just reached me. It is called the red flowering Box from the Bogan.[6] the bark on the upper or smaller branches is of a very dark colour and does not peal off like most of the species. the leaves have a silvery tinge, and the branches droop even to the ground like the myrtle or willow. It is usually found near swamps or creeks.

Trusting this will find you well & that we shall see you at the Exhibition.

I remain
most faithfully yours
Charles Moore

Baron Von Mueller K.C.MG
&c &c &c &c

P.S. Would you like to have any of our wood specimens? if so I will save them for you.

5 Charles Fawcett.
6 Bogan River, NSW.

79.10.06 *To Otto Tepper*[1]

RB MSS M198, Library, Royal Botanic Gardens Melbourne.

6/10/79.

Es macht mir Freude, lieber Herr Tepper, Ihnen die eben geschickten Pflanzen benennen zu können.

Da Sie in Ihrer Halb-Insel ein so natürlich begrenztes Gebiet haben, so könnten Sie solches für eine abgesonderte Special-Flora sehr leicht wählen, u durch Hülfe von Beiträgen anderer Sammler u weitere eigne Reisen leicht eine pflanzengeographisch interessante Enumeration der dortigen Arten liefern.

Es wäre ein hübscher Anhaltspunkt für andere Arbeiten in ähnlichen Gebieten. Die eingeführten Arten sollten aber am Ende der Liste für sich stehen, um das einheimische Pflanzenbild nicht zu entstellen. Eine solche Liste würde für viele südaustral. Schulen nützlich werden.

Ich werde zur Zeit (deo volente) bereit sein die Liste zu revidiren.

Ich würde die Namen für Sie auch bereitwillig systematisch ordnen

Alle Euphrasia Arten die nicht *gelb* blühen sind bei Ihnen nur Formen von E. Brownii Es sind mehrere sehr seltne Pflanzen von der Nähe des Port Lincoln bekannt; solche möchten sich vielleicht noch auf Ihrer Halbinsel auch finden lassen. –

Haben Sie noch einige Exemplare des blattlosen Prasophyllum zur Untersuchung abzulassen? Ein halbes Tausend wirklich verschiedener Pflanzen Arten werden sich in Ihrer Halbinsel schon finden lassen

Stets der Ihre
Ferd. von Mueller.

Besten Dank für die Manna Aphida. Können Sie mir einige andere Exemplare senden, um die Gattung u Art festzustellen?

1 MS annotation by Tepper: 'Reply 26.10.79'. Letter not found.

6/10/79. 79.10.06

I will be happy, dear Mr Tepper, to name for you the plants just sent.

Since you have in your peninsula[2] such a naturally defined region, you could very easily choose it for a separate special flora, and through the help of contributions of other collectors and your own further travels easily produce a phytogeographically interesting enumeration of the plants there.

It would be a nice reference base for other workers in similar regions. The introduced species should, however, be at the end of the list by themselves in order not to distort the image of the indigenous plants. Such a list would be useful for many South Australian schools.

I will be prepared at the time (God willing) to revise the list.

I would also be willing to arrange the names for you in systematic order.

All *Euphrasia* species that are not *yellow* flowering are in your case only forms of *E. brownii*. There are several very rare plants known from the vicinity of Port Lincoln;[3] such may also perhaps still be found on your peninsula.

Have you still some specimens of the leafless *Prasophyllum* to forward for examination?[4] Half a thousand species of truly different plants are found already in your peninsula.

Always your
Ferd. von Mueller

Best thanks for the Manna aphids. Could you send me some other specimens to determine the genus and species?[5]

2 Yorke Peninsula, SA.
3 SA.
4 See also M to O. Tepper, 11 October 1879 [Collected Correspondence].
5 M subsequently sent Tepper's specimen to Richard Owen for identification; see M to O. Tepper, 18 February 1880, M to R. Owen, 2 March 1880 and M to O. Tepper, 3 March 1880 [all Collected Correspondence].

79.10.10 *To Joseph Hooker*

RBG Kew, Kew correspondence, Australia, Mueller, 1871-81, ff. 246-7.

10/10/79.

It was with much gratification, dear Sir Joseph, that I received this day through the Governor your two very kind letters.[1]

The views, taken by our venerable Bentham, on honors from the throne are extreme; – and as he has *rendered important services* to *the Australian Colonies*, through his elucidation of their Floras, he comes quite within the reach of an order, confined to the mediterranean & colonial possessions of Britain.[2]

In many instances men of science must make some sacrifices of their own feelings for the sake of their social status, family connections, official position, & where a distinction has a value even for a Lady, she must also be considered.

When you see Sir Henry Barkly, pray remember me kindly to him, and thank him for his continued goodness towards me. I wished he was here again! Assuredly the British Government will still further utilize his great administrative abilities.[3]

Sir Michael Hicks-Beach must be a particularly kind and enlightened man. I hope that some day it will be in my power, to evince to the right Honorable Baronet actively my gratitude for the consideration, which he has shown me.[4]

1 Letters not found.

2 At that period, the Order of St Michael and St George was so confined.

3 Barkly, who was retired on a pension in 1877, was nominated as one of the 'Commissioners on the defence of British possessions and commerce abroad' on 8 December 1879. He did not receive a further appointment as a colonial Governor (DNB).

4 See M to M. Hicks Beach, 29 September 1879 [Collected Correspondence].

Let me thank you also for devoting again a plate to one of my plants. It almost brings *tears* into my eyes! The foliage of this 6445 will bring *Guichenotia* to mind, G. ledifolia in its small state being not dissimilar to Lasiopet. Baueri[.][5] I gathered seeds of it in 1867 in W.A. My "native plants"[6] will not be free from shortcomings, but a foundation for a more perfect work will thereby be gained. 79.10.10

Perhaps you will remember, dear Sir Joseph, that I strongly urged to you & Mr Bentham, when you commenced the genera,[7] to include the Monochlamydeae among the higher developed plants,[8] to which you mainly objected, because it would be better to keep to D.C., as the prodromus[9] was everywhere in use. Still, I foresee, that in a new edition you will in due time amalgamate the Apetalae with Calyci- & Thalamiflorae. But a few instances of difficulty occur, e.g. with Casuarinae.

It is kind of you to support Profess. M'Coy at the Council of the R.S.,[10] and I will act on your advice concerning the certificate.[11]

With regardful
remembrance
Ferd. von Mueller

5 *editorial addition.* J. Hooker (1879), t. 6445, illustrates *Lasiopetalum baueri* 'raised from seeds sent ... by Baron Sir Ferdinand Mueller'.

6 B79.13.08.

7 Bentham & Hooker (1862-83).

8 See for example M to G. Bentham, 24 January 1862 [*Selected correspondence*, vol. 2, pp. 124-7]; M to W. Hooker, 29 October 1862 [Collected Correspondence].

9 A.P. de Candolle (1823-73).

10 Royal Society. See M to T. Huxley, 8 June 1879, and M to J. Hooker, 8 June 1879 [Collected Correspondence].

11 See Royal Society nomination certificate for Frederick McCoy [Collected Correspondence].

79.11.05 *From Frederick Bailey*[1]

RB MSS M5, Library, Royal Botanic Gardens Melbourne.[2]

Robert St Brisbane
November 5th 1879

Dear Baron Mueller

On the 11th of last August[3] I sent you a shoot of a supposed poison bush which I had gathered at Maroochie it was numbered 13 of the few things then sent although I could at the time neither obtain flowers or fruit I still thought you might be able to detect it from perhaps some little peculiarity of the foliage[.][4] I have this day received some flower specimens and I now send you them It is now nearly 2 years since I received two fruit which I have mislaid but it was something like this – [sketch] have a peach-like pubescence From the flowers now to hand I find it belongs to Proteaceae and to me seems a connecting link between the genera Helicia *Lour* and Macadamia F. v. Muell. for you will observe it has an irregular perianth. very revolute segments anthers on short broad filaments, connective of medium length but very obtuse free unequal hypogynous glands The fruit had not the hard putamen of Macadamia[.] I have no specimens of the genus Helicia *Lour* and in fact never saw any – The man who brought me the specimens and who also showed me the plant growing is the one who say[5] only about half a nut nearly killed him in fact he was so ill after eating it that he had to get off his horse and *lie* down for some time before he could get home – Is not this a strange story about a proteace which

1 MS annotation by M: 'Answ 14/11/79'. Letter not found.

2 MS found with a specimen of *Helicia youngiana* (MEL 93791).

3 Letter not found.

4 *editorial addition*. All [.] in the following text have this meaning.

5 said?

I have always considered safe and wholesome[.] I have read up the 79.11.05
Flora species of Helicia[6] but it agrees with none[.] I would have waited and got the fruit again and than[7] have described the plant only thought you might have a form of the genus Helicia to associate it with. For forgive me but I feel some how that you scarcely care for any things I forward unless they happen to be what you ask for – I am most grateful for all your kindness to me. My collect[ing] correspondence &c &c is for the most part done and has been done for the love of the science at my own expense – Even now as keeper of the Herbarium for which I receive a paltry pittance – I am not found in a single book[.] I have to trust to my own poor library which thanks to *Father Woods* contains some good works beside I must acknowledge by[8] indebtedness to your writings[.] I always consider that your extensive knowledge of the flora of Australia should be taken advantage of by other workers so as to keep down as much as possible the multiplicity of names and for that reason I would much rather not name myself – By the bye I have had the other day sent to me although in a young state plants of Xanthium strumarium Linn did you know of its being in Australia notice[9] it in the Fragmenta[?][10] I think I'm right but it was loose fruit and young plants that were sent to me and they were said to be poisoning cattle

Your obediently
F M Bailey

6 i.e. those described in *Flora australiensis*, vol. 5, i.e. Bentham (1870).
7 then?
8 my?
9 or notice?
10 *editorial addition.*

79.11.22 *To Ralph Tate*

Barr Smith Library, University of Adelaide.

22/11/79.

In the 'Flora",[1] dear Prof. Tate, the localities are most scantily given, and thus also those of Isoetopsis graminifolia. It *is* however recorded from the Murray River in S. Austr,[2] where I found it already in 1848, *when* I gave this little plant the manuscript-name Rhizocephalum angustifolium, recognizing it as a new genus. But a set of my manuscripts, soon subsequently forwarded for Europe, was lost in a wreck of a Ship at the Cape of Good Hope and before, after some delays, the new manuscripts arrived in Germany, (to be published in the Linnaea) Turczaninow, who received the same plant from W.A. published it in Moscou (1851).[3] So of course my earlier appellation was lost. Bentham, though with access to extensive manuscript notes on the range of very many species, (these notes being deposited at Kew) preferred to record only localities, from which he had actually seen specimens. I have since recorded a vast number of "habitat" in the fragmenta, and will do so still more fully in the new edition of the Flora, which – deo volente[4] – I hope yet to bring out, with all the additional species (about 700[5] hitherto published) and all the corrections.

I am not the least surprised at the gross mistake made there[6] in naming the Scaevola spinescens as Lycium australe; altho' I have sent named specimens of the former often to Adelaide, and altho' any one who knew the meaning of the word Scaevola should rec-

1 Bentham (1863-78).
2 See Bentham (1866), p. 556.
3 Turczaninow (1851), p. 175. See also Lucas (1995).
4 God willing.
5 800 *deleted.*
6 i.e. by Richard Schomburgk, director of the Adelaide Botanic Garden.

ognize any species of that genus on a glance, when in flower.[7] 79.11.22
Entre nous I may mention, that from the garden there[8] even for years the common Yorkshire fog (Holcus lanatus) was sent out as the gigantic Tussockgrass of the Falkland-Islands (Dactylis caespitosa); – and in a same manner paraded in public prints from the garden there for years the old Plinian & Dioscoridean Andropogon Haleppensis as the Coapim of W. trop. Africa (Panicum specabile)![9] If you will kindly spare any extraspecimens of any kind of plant from new localities for my collection, you will get due credit for them in my works.[10] Well known plants of S.A., found by me thirty years, ago are in some instances left as S.A. unrecorded in the Flora, either by oversight in London, or for the reasons which I explained.

In 1860 I proved the Ranunculus sessiliflorus of RBr[11] as identical with R. parviflorus L. – I had it from many South Australian stations, as it is not uncommon in your Colony. Your Brachycome from the cliffs of the Bight[12] must be quite distinct from B. graminea, as your additional more developed specimen now proves. I wished much, that ripe fruit of it could be procured, as it may be a thick leaved form of a tender-leaved species, coast-plants usually becoming succulent.

Should the species (on obtaining *ripe* fruit) prove new, I will be happy to name it after you.

Regardfully your
Ferd. von Mueller.

7 I am not ... flower *has been crossed through with a single line.*

8 The sheets bearing the text that follows are filed separately from the first pages of the letter, but the continuity of the text shows that they are part of the same letter.

9 i.e. *P. spectabile.*

10 even for years ... my works *has been crossed through.*

11 Robert Brown.

12 i.e. Great Australian Bight, WA and SA.

79.12.25 *To Julius von Haast*

MS papers 37, folder 211, Alexander Turnbull Library, Wellington.

Privat Weihnacht 1879

Ihr so freundlich gemeintes Schreiben, theurer Herr Landsmann, hat mich in meiner trüben u. niedergeschlagenen Stimmung nur noch trauriger gemacht, u Sie müssen mir erlauben, *Ihnen*, dem wissenschaftlichen Freunde, nun noch einmal brieflich vorzulegen, was ich mündlich auseinandersetzte, wie ich Sie und Ihre gütige Gemahlin hier zu treffen das Glück hatte, denn selbst *Sie* haben sich in Bezug auf meine jetzige Stellung gänzlich missleiten lassen. Mag es mir denn gestattet sein noch einmal und ohne Übertreibung zu wiederholen, dass ich social, finanziell u häuslich fast ganz *ruinirt* bin, und *literarisch* nicht mehr halb das zu leisten vermag, seit man mir den Garten (das einzige welches ich im Leben hatte) so grausam nahm! Wer ist es denn gewesen, Bester, der Sie so gänzlich missleitete? Es muss doch jemand gewesen sein, der dabei interessirt ist, mich niederzuhalten und den Fortschritt meiner Forschungen zu hemmen! Meine Rohmaterialien u meine Beobachtungs Materialien sind *hauptsächlich* im Garten u. unter den lebenden Pflanzen! Was sollte Hooker denn machen, wäre es Ayrton gelungen ihn aus Kew zu stossen? Das Bot. Magaz. müsste doch gleich eingehen u er könnte nicht mehr die glorreichen Rapporte liefern deren wir gerade jetzt den letzten vor uns haben. Wie sollte *er* ausserhalb Kew seiner Stellung als Regierungsbotaniker gerecht werden? in England. Wie sollte er seinen Untersuchungen selbst im Herbarium practischen Werth geben? Nein! Besser ich muss mich Ihnen gegenüber frei aussprechen, denn sonst *arbeiteten* Sie meinen Feinden ja erst in die Hände, u *das* können Sie doch nicht wollen. Der Platz eines Regierungs Botanikers ist *im* botanisch Garten u nicht ausserhalb desselben. Er braucht, namentlich für die Industrien neuer Colonien, seine *lebenden* Pflanzen viel mehr wie die getrocknet[1] Herbarien. *Im*

1 getrockneten?

Garten sind die mir nöthigen Betriebs-*Gelder*, der von mir zugelernte *Stab*, die nöthigen Gebäude. Der Eindringling, der mit der Hülfe seines Vetters (des damaligen Ministers) mich spät im Leben in unverdientes Unglück stürzte, hat sogar mein Laboratorium niederreissen lassen damit ich doch ja nicht darin fortarbeiten solle und meine in Jahren zusammengebrachten Apparate des Laboratoriums sind mir auch entzogen. Ich habe daher *nichts* in der Philadelphia Ausstellung gehabt, nichts in der letzten Pariser, nichts in Wien, nichts in Sydney u werde auch nun in Melbourne nichts haben! Wie kann ich *ohne* ein Departement etwas dem Lande nützendes herstellen? als etwa eine Sammlung trockner Pflanzen, die von fast gar keinem Belang wäre! O Gott! wie haben sich die Zeiten für mich unglücklich geändert. Als ganz junger Mann war ich bereits Commissionär für die erste Pariser Ausstellung, wohin ich unter Anderm das Eucalyptus-Öl sandte, dass jetzt *tonnenweise* ausgeführt wird. Seit ich aus meiner Schöpfung getrieben u an die Strasse geworfen wurde, bin ich selbstverständlich aus allen Commissionen ausgeschlossen, u ein ganz ungebildeter Gärtner, der einfach den Sydney *Zier* Garten nachahmt (weil er nie etwas anderes im Leben sah) ist der *College* von Hooker (*nicht ich*), und dieser unwürdige Mensch kann sich mit *meinen* Schätzen nicht nur täglich vor dem Publicum brüsten mit der Hülfe der von mir jahrelang eingeschulten Gehülfen aus *meinen* seit 25 Jahren gezogenen Pflanzen Sachen für die Ausstellungen senden, wie Sie in Sydney wahrgenommen haben. 79.12.25

Soviel Ehrgefühl u professioneller Stolz sind mir noch geblieben, dass ich Melbourne während der Monate der Ausstellung verlassen werde, denn als *abgesetzter Gartendirector* möchte ich mich doch den grossen Gelehrten aus andern Ländern nicht präsentiren u zum Lachstock werden! Nein! *Das* kann ich nicht, zumal da man mir absichtlich alle Culturmittel verweigert, damit ich ja nicht in Concurrenz mit dem Sydney nurseryman trete. Freilich hat mir das jetztige Ministerium, das mir nicht unfreundlich gesinnt ist, zehn (sage 10) Acker angeboten von den 2000 Acker Park u Gartenland des Landsdepartements um Melbourne. Da ich aber nach der Rate von 30 Ackern jährlich pflanzte u man mir fast gar keine Betriebsmittel geben würde, so habe ich dies abgelehnt!

79.12.25 Was möchte wohl *Ihr* Gefühl sein, edler Freund, wenn der Vetter eines dortigen Ministers Ihnen Ihr Museum plötzlich nähme u unter solchen empörenden Beleidigungen, dass Sie es *nie* wieder betreten könnten; wenn ein solcher Eindringling mit Ihrem Stab fortarbeitete ohne selbst Kenntnisse mitzubringen, u Sie Ihrer Amts Wohnung entzogen würden, u kümmerlich aus dem *Salair* u einigen kleinen Mitteln Ihre Stellung in der wissenschaftlichen Welt aufrecht zu erhalten wäre!, während *Ihre* früheren Mittel für das Museum aus Nepotismus ver 3- oder 4-facht würden!

Ich bin gänzlich verarmt, kann nicht einmal einen Sammler mehr halten, u so ist selbst von den "*Fragmenten*" seit fast 2 Jahren nichts mehr mehr erschienen. Hooker kann dies selbstverständlich nur *lieb* sein, denn er wünschte es *nicht*, dass jemand an Pflanzen in brittischen Colonien unabhängig arbeitete. Der Eucalyptus-Atlas, das einzige meiner letztjährigen Thätigkeit, beruht hauptsächlich auf die Beobachtungen früherer Jahre, u selbst dafür ich[2] mir meine reiche Sammlung von etwa 60 Arten Euc. entzogen, die ich im Garten cultivirte, u die nun grossentheils gewöhnlichem Rasen, wie Sie solche auf jedem "Bowling Green u Cricket Ground" sehen Platz gemacht haben!

Es ist in der ganzen neuern Geschichte der Wissenschaft kein ähnlicher Fall, dass ein Mann in meiner Stellung *unverschuldet* in den Staub getreten wurde. Aber ich bin Deutscher! Meine Correspondenz u die literarischen Arbeiten, soweit solche kürzlich noch fortgesetzt werden können, verrichte ich in einer kleinen "Cottage", die ich durch eine Baugesellschaft gekauft, u woran ich noch jahrelang zu zahlen habe! Wittstein wurde *schon im Garten* übersetzt, u es kostete mir die Ersparnisse zweier Jahre, den Druck zu bestreiten. Selbstverständlich ist fast nichts vom Werke verkaufbar gewesen.

Nach Sydney bin ich nicht gewillt zu gehen. Was nützt es mir, *ohne Garten* mich da zu zeigen? Etwaige neue Erfahrungen könnte ich ja doch practisch nicht verwerthen, u es würde für mich nur *peinlich* sein, mich aus allem practischen Mitwirken ausgestossen

2 ist?

zu sehen. Verzeihen Sie, dass ich so freimüthig spreche, aber wenn 79.12.25
selbst Sie, der *wissenschaftliche Landsmann*, die Worte meiner Feinde aufnehmen freilich unbewusst, so ist um so weniger an den Aufbau eines ehrenhaften Departements nicht länger zu denken, u ich habe doch noch immer "gehofft gegen Hoffnung". – *Im* Garten hätte ich selbst *literarisch* doppelt soviel leisten können, u. jetzt muss es grossentheils auf Kosten derjenigen Privatmittel geschehen, aus welchen ich so gern einen häuslichen Heerd u Familien Glück begründet hätte, aber selbst *das* gönnt man mir hier nicht. Bei mir, Freund, ist es nicht Gefühlssache bloss, sondern *Ruin!* u Sie haben sich täuschen lassen, wenn Sie jetzt schreiben "ich hätte meine ehrenvolle Stellung *nicht* verloren!"

So werden wir uns wohl im Leben nie wiedersehen. Genehmigen Sie u Ihre Gemahlin die Anerkennung Ihrer Güte vieler früherer Jahre, u. mag das neue Jahr Ihnen u Ihrer schönen Familie recht viel Freude bringen

Stets Ihr Ferd von Mueller

Austausche von Samen u Pflanzen für selbst Industrie Culturen kann ich auch nicht mehr unterhalten.

Im Garten hatte ich selbst durch den Stab in Correspondenzen u literarischen Arbeiten Vorhülfe welche ich seither entbehren musste.

Auf Ihre hoffnungsvollen Söhne müssen Sie recht stolz sein!

Dass Ihre liebe Gemahlin ganz wiedererstarkte, muss Sie recht beglücken. Dr Martin ist nun auch schon dahingeschieden! u seine Witwe beinahe unversorgt.

Christmas
1879

Private

Your so kindly meant letter, dear fellowcountryman, has only made me even sadder in my troubled and depressed mood, and you must permit me to put before *you*, the scientific friend, again in a letter what I explained verbally, when I had the good fortune to meet you and your good wife here, since even *you* have let yourself be completely misled in respect to my present position.

79.12.25 May I then be permitted to repeat once again and without exaggeration, that I am almost completely *ruined* socially, financially and domestically, and *from a literary aspect* no longer able to produce half as much since they took the garden from me so cruelly (the only thing which I had in life). Who has it been then, best of friends, who has misled you so completely? It must have been someone who is interested in keeping me down and retarding the progress of my research!

My raw materials and my observation materials are *mainly* in the garden and among the living plants! What would Hooker have done, if Ayrton had succeeded in pushing him out of Kew?[3] The *Botanical Magazine* would have had to cease immediately and he could no longer have produced the glorious reports, the last of which we just now have before us. How should *he*, outside Kew, have justified his position as Government Botanist in England? How should he have given his investigations practical value even in the herbarium? No! Better I must express myself towards you freely, lest you *worked* in fact really into the hands of my enemies and *that* you do not want. The place of a government botanist is *in* the botanical garden and not outside it. He needs, especially for the industries of new colonies, his *living* plants much more than the dry herbarium. *In* the garden are the operational funds necessary for me, the *staff* trained by me, the necessary buildings. The intruder,[4] who with the help of his cousin (the minister of the time[5]) overthrew me late in life into undeserved misfortune, has even had my laboratory torn down, so that I certainly shall not continue to work in it and my laboratory apparatus brought together over the years is also taken from me! Therefore I have had *nothing* in the Philadelphia Exhibition, nothing in the last Paris, nothing in Vienna, nothing in Sydney and will also now have nothing in Melbourne![6] How can I *without* a department produce anything useful for the country? than perhaps a collection of dried plants, which would be of almost no importance at all! O God! how the times have changed unhappily for me. When quite a young man I was already Commissioner for the first Paris Exhibition,[7] to which I among others sent eucalyptus oil, that now is exported *by the barrel*. Since I was driven out of my creation and thrown on the street, I am of course excluded from all commissions and a quite uneducated gardener, who simply copies the Sydney *flower* garden (be-

3 See MacLeod (1974).

4 William Guilfoyle.

5 J. J. Casey.

6 Centennial Exposition, Philadelphia, 1876; Exposition Universelle, Paris, 1878; Weltausstellung [International Exhibition], Vienna, 1873; International Exhibition, Sydney, 1879-80; International Exhibition, Melbourne, 1880-1.

7 Exposition Universelle, Paris, 1855.

cause he never saw anything else in his life) is the *colleague* of Hooker (*not I*), and this unworthy person can not only daily give himself airs before the public with *my* treasures, with the help of the assistants schooled for years by me, he can send things for the exhibitions from *my* plants raised for 25 years, as you have noticed in Sydney. 79.12.25

So much of my sense of honour and professional pride still remain that I will leave Melbourne during the months of the Exhibition, since as *deposed garden director* I would certainly not like to be presented to the great scholars from other countries and be a laughing stock! No! *That* I cannot, especially since they deliberately refused me all means of cultivation, so that I cannot enter into competition with the Sydney nurseryman. Admittedly the present ministry, which is not unfriendly disposed to me, has offered me ten (I say 10) acres of the 2000 acre park and garden land of the Lands Department around Melbourne, but because I planted annually at the rate of 30 acres and they would give me almost no means of production at all, I have declined this!

What would *your* feelings be, noble friend, if the cousin of a minister there suddenly took your museum from you and under such outrageous insults, that you could *never* again set foot in it; if such an intruder continued working with your staff without bringing knowledge with him, and you were deprived of your official dwelling, and were to support your position in the scientific world wretchedly from the *salary* and some small means!, whereas *your* former means for the museum were trebled or quadrupled from nepotism!

I am completely reduced to poverty, cannot even keep a collector any longer, and so even the "*Fragmenta*" has not appeared for almost 2 years. This can of course only *please* Hooker, since he did *not* wish that anyone worked independently on plants in British colonies. The *Eucalyptus* Atlas, the only activity of my last years, is based mainly on the observations of earlier years, and even for that I am deprived of my rich collection of about 60 species of euc. that I cultivated in the garden, and that now for the most part have made way for common lawn, such as you see on every bowling green and cricket ground.

There is no similar case in the whole recent history of science, that a man in my position was *undeservedly* trodden into the dust. But I am a German! My correspondence and literary work, as far as such can still briefly be continued, I will carry out in a small "cottage", that I purchased through a building society, and for which I still have to pay for years! Wittstein was translated *already in the garden* and it cost me the savings of two years to pay for the printing.[8] Of course almost nothing from the work has been saleable.

8 B78.13.06.

79.12.25 I do not intend to go to Sydney.[9] What good is it for me to show myself there *without the garden*? I certainly could not make practical use of possible new experiences, and it would only be *embarrassing* for me, to see myself removed from all practical collaboration. Excuse me that I speak so frankly, but if even you, *the scientific countryman*, take up the words of my enemies, admittedly unknowingly, then it is so much less to think any longer of the building of an honourable department, and I have certainly still "hoped against hope". *In* the garden I would have been able to accomplish even double as much *from a literary aspect*, and now it must happen for the most part at the expense of that private means from which I would so gladly have founded a household and family happiness, but even *that* they begrudged me here. In my case, friend, it is not merely a matter of feeling, but *ruin*! and you have let yourself be deceived, when you now write "I would *not* have lost my honourable position!"

So we will probably never see one another again in life. May you and your wife accept the acknowledgement of your kindness of many previous years, and may the new year bring you and your splendid family very much joy.

Always your
Ferd. von Mueller

I can also no longer undertake exchange of seeds and plants for even industrial cultivation.

In the garden I even had preliminary help by the staff in correspondence and literary work which since then I have had to do without.

You must be very proud of your promising sons!

You must be very pleased that your dear wife has grown quite strong again. Dr Martin[10] has now departed this world too! and his widow nearly unprovided for.

9 Haast and his wife visited the International Exhibition, Sydney, in September 1879, returning to NZ in November.

10 Probably Lawrence Joseph Martin, MD (Melbourne), LRCS (Edinburgh), whose death at Cannes, France, on 19 January 1879 was noted in the *Australian medical journal*, April 1879, p. 204.

To George Grey

Grey papers, GL M50(14) GLC, Auckland Public Library.

Newyears eve 1879

It will afford me much pleasure, dear Sir George, to procure for you the Decades of the Eucalypts Atlas, which here appeared,[1] and I will write to the West Austr. Government about the portion of the work, done for the colony there, but published in London.[2]

The Report on the Wattel-barks[3] shall be forwarded simultaneously, also seeds of the best Tanners Wattels.

I am much beholden to you for sending me the charming book "his island home",[4] which makes me all the more deeply regret, that my anticipation to spend with you a few weeks exploring in New Zealand could not be realized 20 years ago, when you so generously invited me. I had *then* just undertaken to form a large Department for the Vict. Government, which for a dozen years rendered this colony so far illustrious, but which through envey and nepotism led finally to my *undeserved* ruin, so that now late in life it is with deep regret of mine that I choose Victoria as a home.

Had I then responded to your friendly call, it would have fallen to my share to make known nearly the whole of the *alpine* flora of your grand island. But I did not like to infere[5] with Sir Jos. Hooker, to whom more particularly N.Z. belonged as a phytographic territory.

You did me the honor to say, dear Sir George, that you had actually *read* through the whole of my "select plants". An enlarged edition is just issued by a [...] Government in the Northern hemisphere.[6] I still think, that the reissue of the work (now enlarged) in N.Z. and its extensive distribution among the colonists, would open

1 The first four decades of M's *Eucalyptographia* were published in 1879 (B79.13.11).
2 B79.13.10.
3 B78.14.01.
4 Grey (1879).
5 interfere?
6 Presumably the Indian edition, B80.13.07.

79.12.31 mines of vegetable wealth now lying concealed. The book is still quite unique and N.Z. is quite welcome to reproduce it.[7]

With deep regards your
Ferd. von Mueller

80.01.01 *To Oliver Jones*

Acc. 2461A/2, Battye Library, Perth.

Newyears day
1880.

Let me offer to you, dear Corporal Jones, my best felicitation on the new year, may this new one be to you & your family one replete with happiness. I have not heard of you for a long while; perhaps you have been far away. Is there anything I can send you; and how have the various seeds grown, which I from time to time forwarded? When lately I looked over my W.A. accounts, I found that a few £ more were in [the][1] bank than my notes showed. Did you perhaps never cash my little cheque, or is otherwise this still due to you if so I will forward a post office order for it.

Regardfully your
Ferd von Mueller.

I often think with pleasurable recollection of our joint travels,[2] and should like it very much to be once more out with you. Who now occupies Dirk Hartog's Island?[3]

7 The book was not published in NZ. For discussions about the proposal see M to J. Hector, 10 August 1874 (*Selected correspondence*, vol. 2, p. 710); and G. Cooper to M, 6 October 1874; M to G. Cooper, 3 November 1874 and notes thereto; and M to G. Grey, 12 July 1877 [all Collected Correspondence].

1 *editorial addition*. – Paper damaged.
2 Jones evidently accompanied M during part of his visit to WA in 1877, most likely as he traversed the country between Champion Bay (Geraldton) and Shark Bay.
3 WA. M visited the area in late 1877.

From Robert Fitzgerald[1]

80.01.19

RB MSS M9, Library, Royal Botanic Gardens Melbourne.[2]

Surveyor Generals
Office Sydney
19 Jan 80

Dear Baron

I send you a specimen of a plaint said to have poisoned a number of sheep at Burke.[3] If without the flowers you can make out what it is I should be obliged for the name and your opinion of it as a poisoner – I also send you a drawing[4] of this plant I found at Hill End which I have no doubt from the places I found it in is a native. Cheranthera is not to be found in the Flora[5] so I could not examine as to your conjecture. The anthers were as shown in every flower and could not have been united unless in the hub which I did not examine, thinking I could readily find it in some genus close to Isotoma Please return this drawing as it is one of the illustration[s] of my private copy of the flora, the genera of which, (where possible) I have been similarly representing

I remain Dear Baron
Yours truly
Robt D Fitzgerald

Hill End is about sixty miles *north west of Bathurst*[6]

1 MS annotation by M: 'Answ 20/1/80 FvM'. Letter not found.
2 MS found with a specimen of *Phyllanthus* sp. (MEL 1595507).
3 i.e. Bourke, NSW.
4 Drawing not found; M presumably returned it to Fitzgerald as requested.
5 Contrary to what Fitzgerald asserts, *Cheiranthera* appears in Bentham (1863), pp. 127-8.
6 NSW.

80.02.21 *To Asa Gray*

Gray Herbarium Archives, Harvard University, Cambridge, Massachusetts.

21/2/80

Private

The February-mail from North-America, dear Dr Gray, brought me your kind sending of the biography of Prof Joseph Henry,[1] which I read with all the more interest, as (through the Smithsonian Institute) I came myself into communication with that illustrious man for many years.[2] It is a beautiful piece of writing again, this of yours, – though – may I venture to say so – Ritchie, Cannstadt,[3] Faraday and Wheatstone might claim a place in the series of those, who brought electric telegraphy about. Thinking of the Smithsonian Institute, may I ask, whether any desirable volumes are missing in the series of my own poor publications, kept there. At antipodal distance and under *great adversities* of later years, I may not have managed, to get volume after volume regularly across to you, and some may have been lost on the way. – But now to an other subject.

Some years ago,[4] when you alluded to the fallacious remarks of Governor Bowen, propounded in conversation with you, regarding my miserably changed position, you made in your usual generosity the remark to me: "rest assured, if I cannot help you, I shall at all events not harm you". When now I read your kind review of my work on the Eucalypts,[5] you allude feelingly to the difficulties, which beset my path of research even in a work of this kind, but you – my honored friend – neutralize completely any good effect, which a word from *you* might have done me by the phrase "let him consider, how much valuable time he saves for true botanical work by his riddance from the multifarious cares, which garden-superintendence involves". I hope earnestly that this review by you will not

1 Gray (1880).

2 See J. Henry to M, 25 October 1869 [*Selected correspondence*, vol. 2, p. 522], and M to J. Henry, 31 January 1870 [Collected Correspondence].

3 i.e. P. L. Schilling von Cannstadt.

4 Letter not found.

5 Gray (1879).

Figure 7. Asa Gray. Published in the *Bulletin of the Torrey Botanical Club* (1888), vol. 11, frontispiece. Reprinted courtesy of the Library, Royal Botanic Gardens Melbourne.

80.02.21 fall into the hands of my adversaries here, as it would *smash* the strenous[6] efforts of myself and a few scientific friends, to resuscitate at least to a small extent my formerly illustrious & so highly *useful* department.[7] How far a general principle, such as you lay down in the above sentence, may apply to the Garden-administration of any University, I should not venture to pronounce. Even in such, *local circumstances* must greatly affect such a general proposition, and I should be loath to give on it a public opinion regarding *any* institution, which I had never visited & on which I could not judge fairly at a distance, lest I might inflict – however unwillingly & however unprovoked – an injury. To the position of *mine* or *any Gov. Botanist*, the proposition or principle, which you lay down, – my generous Sir, – is *utterly inapplicable*, as I practically find out in the daily execution of my work! – Though J. Hooker once, in disgust of Ayrton's conduct,[8] actually wanted to relinquish the Kew Garden, I feel sure he is glad to have thought better of it, and I certainly, when he told me of his intention, advised Sir Joseph, to remain *Generalissimus*.[9] But, however that may be, Melbourne should not be judged even by Kew, as Sir Joseph would at all events probably have no actual hostile intrusion on his position, had he merely kept the Museum, though I cannot see, how his Museum work can be severed from the horticultural work & vice versa, without causing impediments, intrusion, misunderstandings &c. in all directions. *Here* my being driven out of house & home, away from *my* thousand of kinds of living plants, from the staff trained by me, from my laboratory & seed magazine and indeed all I had, except the *small Govern.* bot. Library and the only herbarium room, here this senseless and cruel measure, dictated by envey & nepotism, has had a *most disastrous effect*! It took away from me even the means of keeping a collector in the field for the continuation of the fragmenta; it stopped my introduction of plants for forests & fields & pastures, it led to the pulling down of my Laboratory & the withdrawal of

6 strenuous?

7 See L. Smith to G. Berry, 11 July 1877, pp. 745-8 below, and also M to B. O'Loghlen, 18 May 1879.

8 See MacLeod (1974).

9 See M. to J. Hooker, 6 November 1873 [Collected Correspondence].

my apparatus, it tied my hands so that I could not do a single thing for the Philadelphia Exhibition[10] nor any other Exhibition since, though as far back as 1855 I brought out the Eucalyptus-oil for the first French Exhibition,[11] an article, the export of which up to this time (and steadily increasing, represents about £15000 in the Export-trade of Melbourne already!) the same may be said of the export of ferntrees, Eucalyptus seeds etc started by me. Not even as much as an office-room was left me, my library had to be stored away for years and I have only lately got it set up again, having bought a small cottage through a Building Society. Doubtless you (& perhaps Hooker & others) will say, that I enjoyed a splendid Salary, & there ought to be no difficulty to provide for all my work out of that. Now it is nominally, not actually so. – Melbourne is an excessively costly place to live in; my income moreover is consumed to keep the wreck of my once illustrious Department afloat, though I never touch a card or billard-ball, give no parties, live the simplest of life (forlorn life too!), never visit races, but do not like to be *left quite behind* in the race with my compeers. Imagine Hooker *out* of Kew!! Though I have a high feeling for professional honor, what I say is not sentimental, but stern reality, and if you, Professor Gray, only for one hour visited my place here, you would deeply sympathize with me, and would do for me, what Paget, Holland, Carpenter, Huxley, Tyndall, Darwin & Bentham did for Hooker! – Let me give you an instance only *two days ago* I placed here as a member of the local branch of the British Medical Association Dr Kings new Calcutta febrifuge before the members as a cheap, but highly efficient therapeutic remedy in hospital practice & for indigents, – not crystalline nor quite purified, but a mixture a[12] Quinine, Cinchonin, Cinchonidine & Quinidine & perhaps also the amorphous alkaloid, the Quinine however prevailing.[13] Why

10 Centennial Exhibition, Philadelphia, 1876.

11 Exposition Universelle, Paris, 1855.

12 of?

13 The *Australian medical journal* includes a report of the meeting of the Victorian Branch of the British Medical Association on 20 February 1880 at which 'Baron von Mueller submitted a sample of mixed cinchona alkoloids, and presented a copy of his translation of Dr. Wittstein's work on the Constituents of Plants' (new series, vol. 2 [1880], p. 144).

80.02.21 do I mention this? – If I had not been ejected *without any fault of mine* out of my institution, I would have had certainly Quinine from my own trees in the Melbourne Exhibition this year,[14] whereas I shall have there nothing of *any* kind, because my hands are kept tied, and all working material, to give *vitality* to my research, is taken from me. Ignorance destroyed my Cinchonas and ever so much else, that we could grow in one or the other region of our colony to advantage, just as you could grow in the South of California in warm sheltered forest glens the Peru-Bark trees. I feel sure, you will not argue with me from so far a distance on what my position here ought to be, but perhaps you may say, let others grow the Cinchonae! But, my dear Sir, I am to be the *leader here of vegetable new industries*, and I must *observe*, and that daily & hourly, to help the colonists *practically*, and if I cannot do that and make annually a good show, (look only to Hookers Kew Reports!) then I am *bound* to go finally to ruin. The daily question in this community is already: What is he doing? What is the good of him? &c. What do the colonists care about my "*Flora*" as a whole? For *that* they would never for any period maintain my position, and as I sunk *all I had* in my researches (the printing of Wittstein[15] cost me – exampli gratia – £220 out of my private purse and only about one tenth of the copies were sold), I should be sent adrift as Gov. Botanist *without a pension*, should be forced to give up the Library & Herbarium also & spend the last days of my life in obscurity & poverty, getting only once for ever under the Civil Service law here a compensation for loss of Office. Count de Castelnau (with whom Dr. Weddell was out) wanted me also always according to his unfortunate talk here to "be relieved of drudgery", but when one morning he saw it suddenly announced, that I had to leave my creation & even the House, which I built in 17 years, he wrote at once a doleful lament,[16] though he helped unwittingly to undermind my position. However, he also numbers now with the death, & I was one of the pallbearers of his Coffin *this* month.[17]

14 International Exhibition, Melbourne, 1880-1.

15 B78.13.06.

16 Not identified.

17 Castelnau died on 4 February 1880.

You must pardon, dear Dr Gray, when I inflict this long letter of sadness on you. I fancy, it is my last! I have a presentiment that I do not live till the end of the year! I am since a long time under constant insomnia, and if I ever sleep a little, I am in dreams again in my little paradise, in my house, among my plants, among *my gardeners*, who only wanted one hours daily attention of mine as a rule, and did the drudgery for me, as they are now doing for the Sydney Nurseryman,[18] who is now the *Directorial Colleage* of Hooker, Regel, Eichler &c &c. If I spent a few additional hours daily in the garden, it refreshed my mind, invigorated by a thorough oxygenation of my blood my physical strenght, *originated daily new observations*, not only phytographical (for which the money grubbing Australian Communities do *not* care), but industrial tests also &c. I am forced to leave Melbourne at the time of the Exhibition, – for as a discarded Director and as a head of a Department without a Department, I cannot present myself to the illustrious strangers, while an unscrupulous intruder praises himself up daily with *my treasures*, on which the sweat of my brow adheres since the last quarter of a century. I have said as much to the excellent representative of your states in a letter to him in Sydney, Dr. Cox.[19] But I fear I may not live so long, and then you may have still after all that even misled *you* a few kind words for me in any little necrolog, you may possibly deem me worthy of. My God! what could I have done to advance science in Australia and give it a practical and useful bearing in new colonies, had I been left only with slender means in my creation, or had only half the sums, mostly squandered away, since I left (and largely increased) been at *my* command. 80.02.21

And now, good bye, dear Dr. Gray, and may providence watch over you also in future, as it protected you on the brink of an abyss lately, and let no clouds disturb your bright career, such as obscured mine.

Ferd. von Mueller.

I did *double* as much phytographic work in the bot Garden with all the facilities there, than since I am out of it.

18 William Guilfoyle.

19 Letter not found. The American commissioners for the Melbourne International Exhibition did not include anyone named Cox, and neither was this the name of the American consul in either Melbourne or Sydney.

80.04.23 *To James Patterson*[1]

S80/4126, unit 1154, VPRS 3991/P inward registered correspondence, VA 475 Chief Secretary's Department, Public Record Office, Victoria.

Melbourne, 23rd April 1880

Copy

Sir

In compliance with instructions received from the office of the Honorable the Chief Secretary,[2] I have the honor to submit my views of the requirements of my Department as regards buildings during the financial year 1880-1881. I may be allowed to remark that during the last seven years no additional provision for buildings has been made for my departmental service, though I have not even office-rooms nor laboratory buildings ever since. Four years ago the sum of £900 was voted by Parliament, through the action of the Honorable J. Macpherson (then Chief Secretary) for the commencement of new building-accommodation in my Department, which sum I allowed to lapse, with a desire of seeing it doubled in the next financial year. May I therefore hope that I may at least now recover the £900, saved then by the lapse of that sum to the revenue. I leave the consideration of this however entirely to your judgement, as the additional claims on the Treasury during the Exhibition-year[3] may render it expedient to postpone the grant of building-votes for my departmental purposes for an other year.[4]

I have the honor to be
Sir
Your obedient servant
Ferd. von Mueller

The Honorable the Commissioner of Public Works

1 MS written by Georg Luehmann and signed by M.
2 Instructions not found.
3 International Exhibition, Melbourne, 1880-1.
4 MS annotation: 'Adhere to works'.

From Kate Taylor

80.04.27

RB MSS M1, Library, Royal Botanic Gardens Melbourne.[1]

Albany[2] April 27 [1880][3]

My dear Baron Von Mueller

Thank you very much for your letter & kind present of Bulbs, I have planted them in my garden & they are growing beautifully. My dear Mother is now able to walk into the garden with my help, to see their progress each day, which is a great pleasure to us both My brother Campbell has just come from his Station on the "Thomas river"[4] he brought me a few seed of a very curious Eucalyptus which I send you. My brother says they are very strange looking trees, with their branches hanging down all round to the ground, like a "weeping Willow". Our wild flowers are just begining to blossom if there are any you would like me to dry for you please let me know, but I can *only* get those that grow within a short riding distance from town or if you would like a few young plants that I could get, I would send them with pleasure again thanking you for your kindness, I remain yours

very truly
Kate L. Taylor

P.S. I gave your letter to Mr Spencer[5] I think he is collecting & will write soon

K T

1 MS found with the Type of *Eucalyptus sepulcralis* (MEL 1612463), described by M in B82.13.17.
2 WA.
3 *editorial addition.* Year established on the basis of MS annotation by M: 'Answ 6/5/80. F.v.M.' Letter not found.
4 WA.
5 Not identified.

80.05.16 *To Asa Gray*

Gray Herbarium Archives, Harvard University, Cambridge, Massachusetts.

Whitsunday 16/5/80.

Since writing my first letter for this mail to you,[1] revered friend, I have received the volume of the two lectures,[2] which you so considerately presented to me. These discourses are worthy of an Asa Gray; the thoughts & the language are equally powerful in these discourses, and the views of the science of the day are blended with deep religious persuasion.

Much will we and the next generations yet have to learn, before the history of the bodily creation will be revealed to us, though the origin and nature of the godlike vital force will be concealed for ever to mortal eyes! To discuss the great questions of these essays of yours with so great a man as yourself is not within the scope of a letter, even if a younger and far less authorative observer dared to express himself frankly on the subject. But this I might venture to say, that the so called *protoplasm is to my view not an uniform substance* through the great empires of living beings, and that the present creation does *not entirely comprise forms of higher perfection*, but many of less development than those of the past, both animal and vegetable. Nor do I think it possible, that the present plants & animals in their marvellous diversity and copiousness could have sprung from one or few types, and I feel satisfied, that no mechanical efforts of nature would *give us in ages back again* the Dodo, the Moas, the *extinct* plants of St Helena and other byegone organism, even of the simpler types.

It is good for mankind, for our earthly happiness, that we cannot penetrate to the greatness of godlike power by any human investigations. But a religion, built on observations on the beauty & wisdom displayed in nature, preached in churches of her own

1 M to A. Gray, 13 May 1880 [Collected Correspondence].

2 Gray (1880a). The copy in the library at the Royal Botanic Gardens Melbourne is inscribed: 'Sir Ferd. Müller with kind regards of the Author'.

or in the freegods world, would greatly elevate the spirit of those, who cannot cling to christian revelations. 80.05.16

Homo sapiens remains after the study of thousands of years an unalterable species; and I feel convinced, that in the same manner other true species move within defined absolute limits, but our observations are as yet far too scanty to circumscribe their *real* specific boundaries; that will be the work of coming centuries. Herbert's early observations on the *fertility of hybrids*[3] when extended will give us many a new insight also into the value of specific forms, now often kept very apart. I rejoyce to understand from your discourses, that you do not deem the mere idea of selection sufficient to account for the development of higher organized from lower creatures. How could any one from a medical point alone! To my mind we must grasp the question of the creation of organized species from considerations of the *whole creation of the universe*. Could the eye of the mere housefly with its thousands of lenses & optic adjuncts gradually originate by evolution? And even if so, which I do *not* believe, is not our world of organisms, wonderfully varied as it is, a mere speck in the universe, "with worlds without ends"? Must not the grandest planets with their sun originate from the same godly power, which called forth the wondrous optic apparatus of insects, neither the one nor the other having changed in the least since science began to record its observations! I fancy, that it is *god's own breath* as well in us human beings as in the simplest of organism, which gives vitality. Under the ordinance of *such* a ruler we may rest secure, that we are watched & may anticipate a happy futurity, of which Religion gives us an earthly forethought. And why should we poor mortals try to narrow Gods creative power on this mere atom of the world down to a primordial germ, without support and evidence. Can we not in religious belief concede to the supreme power the might of calling forth distinctly the organic species? The primordial germ, if such existed, must be the most marvellous of earthly wonders anyhow, to be capable to develop in hundreds of thousands of species, easily recognized & classified even by human understanding.

3 Herbert (1837). Gray referred to Herbert's work in the first of his two lectures, pp. 43-4.

80.05.16 It must be a great consolation to you, dear & honored friend, that after the toils of enquiries through a long life you can still like Brewster[4] and so many others of the wise, cling to the comforts and assurances of religion.

Ever yours
Ferd. von Mueller

80.06.21 *To Joseph Hooker*

RBG Kew, Kew correspondence, Australia, Mueller, 1871-81, f. 277.

21/6/80

Private

It has just been announced, that H.R.H. the Prince of Wales will honor the International[1] with his Visit in November. Should you, dear Sir Joseph, happen to meet the Prince, will you then kindly say a few words of my *Directorial Career* here, which you can measure by what I did for Kew *Garden* even with my small directorial means. If you do *that* for me, it will be a *set-off* against the misrepresentations about my Garden Management, which are sure to be made against me in the most unscrupulous manner during the Exhibition. I *may* be able to defend myself in my communications with the foreign Commissioners, but I may not have the opportunity to speak to H R H, now as I can see the Prince no longer in my Garden.

I only hope that my reason will not give way in this mental torture, which I feel more bitter than ever now in my helplessness at the approach of the Exhibition! – A little support from *you* in time 7 years ago *might have saved me*, and may even now mitigate my distress and protect my *professional honor*.

Always your
Ferd. von Mueller.

4 Sir David Brewster.

1 International Exhibition, Melbourne, 1880-1.

If I possibly can manage it, I will *fly* out of Melbourne during the Exhibition, to save myself against the mortifying humiliation, to which I am so faultless subjected by nepotism & envey and misguiding journalism. 80.06.21

To Louise Wehl 80.07.08

RB MSS M100a, Library, Royal Botanic Gardens Melbourne.

8/7/1880.

Just, dear niece, a few lines at once in answer to your letter, for if it is deferred, it may be long before I could respond, as really the correspondence is apt to heap up from day to day & be left at last unnoticed. My eyesight is not so good as it was formerly & the endless writing tries me much.

Regarding a Governess, I do not think that any educated person with any fitness for teaching would go to your place for £25, even if you could afford her all the comfort, which she would expect. Only yesterday I tried to help a young Lady to a place, as Governess, whose parents I have known since 20 years, but the Salary is £80. If however a really good person can be found for £25, and it will not tell on your mother's resources to have an additional member in the household to support, you can let me know.

I feel sorry, that you did not commence your correspondence in the native language of your parents. Accept my best thanks for your felicitation to my birth-day; perhaps it is the last for me to celebrate, as my health is sadly impaired Pray, thank Marie for her pretty painting[1]

Sir Thom. Elder named the Racehorse after me;[2] so I may go this time to the Races, where I never was before all my life.

With my best wishes for you,

Ferd. von Mueller

Miss Louise Wehl

1 Louise Wehl's sister Marie was a talented flower painter; see May, Maroske & Sinkora (1995).

2 Dwyer (1996).

80.08.04 *To Alphonse de Candolle*

Conservatoire et Jardin botanique, Geneva.

4/8/80

Cher M. De Candolle.

Il est de mon devoir d'exprimer mes remerciments pour l'attention, que vous avez eue en m'envoyant votre "Phytographie"; ce qui est un autre volume, que vous avez ajouté aux importantes publications, qui ont rendu votre nom illustre. La raison pour laquelle dans les "native plants" j'ai donné les nomes des auteurs et leur oeuvres en les citant complétement c'est parce que ce livre est principalement fait pour les écoles de Victoria, ou les précepteurs ne comprendraient pas les abreviations de ce genre.

Dans l'allusion, que vous faites à la "Flora Australiensis" je vois, que ma collaboration est entièrement ignorée. Vous devez donc penser, que j'ai eu une part insignifiante dans l'élaboration, et je m'en contenterais, si j'etais un individu privé. Mais comme mes moyens propres ainsi que ceux de mon Departement furent devoués a cet objet, il serait très injuste pour le Gouvernement de Victoria, si je consentais a ce, qu'on crut, que la Flora Australiensis (melius Australiana) fut l'ouvrage de M. Bentham seulement. Cet homme illustre écrivit tout le texte, mais *deux-tiers* du material (echantillons preéxaminées) furent fournis par moi même. Mes collections Australiennes sont deux ou trois fois plus grandes, que celles de Kew. J'ai envoyé de 1862-1877 à Kew comme un emprunt mon herbier normal d'environs 70 grandes caisses, remplies de 90,000 échantillons complétement arrangées, qui par moi même priorément ont été examines, et que j'ai recuillis depuis 1847 avec l'aide d'amis et d'hommes payés; pour obtenir ces collections j'ai voyagé à cheval et à pied 28,000 milles anglaises! Enfin 30,000 échantillons ont été procurés et graduellement ajoutés par moi, depuis que les volumes de la Flora Australiensis parurent successivement, ce qui fait, qu'environ 650 espèces ont été ajoutés dans les "fragmenta", parmi lesquelles etaient plusieurs genres et quelques orders naturels. M. Dyer dans son recent discours à l'Institut colonial a reconnu ma part dans l'ouvrage.

Dans votre liste des grands herbiers je vois, que celui de Melbourne n'y est pas. Il est le plus grand dans l'hémisphère australe (et pour l'Australie largement unique?), contenant des plantes de toutes les parties du globe, lesquelles j'ai obtenues moi même depuis 1840 ou en faisant des achats et des échanges. Je calcule, que le nombre d'échantillons (qui ne sont pas doubles) sont environ 300,000, desquels il y en a 120,000 Australiens et 180,000 extra-australiens. Il n'est pas facile de déterminer, combien d'espèces cette vaste collection contient réellement. Le nombre des espèces australiennes et[1] d'environ 10,500, i.e. 80.08.04

Dicotyledoneae	7000)	considerées très-conservativement.
Monocotyledoneae	1600)	
Acotyledoneae	1900)	

Les Algae (par example) ont été augmentées dejà depuis Harvey de 600 à environ 800 principalement par Sonder d'après mes contributions.

La collection privé, que J. Drummond avait faite pour lui même, passa après sa mort en ma possession. On peut voir dans les "fragmenta", que les trois quarts des collecteurs botaniques de l'Australie ont envoyé leur "specimens" a moi, quoique j'ai permis successivement, qu'on en garda les duplicates à Kew. J'ai aussi reçu de Sir Jos. Hooker de nombreux exemplaires de R. Brown et A. Cunningham, et j'ai obtenu une collection complète du Dr Milligan, ainsi que beaucoup des plantes de Sieber, la plupart de celles de Preiss et toutes celles de Leichhardt. Dans mon Muséum les plantes de Australie, celles de Nouvelle Zélande et celles de la Polynésie sont apart; mais toutes les autres ont été inserées et arrangées (selon le système Candolleen) pour former une seule collection compacte. Mon Museum contient des plantes de Areschoug (Scandinavia et Algae), Aucher-Eloy, Ayres (Mauritius), Sir H. Barkly (Jamaique & Essequibo), Beccari (Sumatra), Boivin (Réunion), Bourgeau, Chickering (Amerique du Nord), Cosson, Cuming (Philippines),

1 est.

80.08.04 Durieu (Algerie), D'Albertis (Nouvelle Guinée), Eaton (Amerique boreale), Traill-Green (N. Amer.), Ecklon (toute sa collection d'Allemagne, ainsi que la plupart de cette du Sud de l'Afrique), Falconer (Indes), Fortune (Chine), Franqueville, Glaziou (par Hooker), Griffith (Indes), Hance (Hong-Kong & Chine), Haast (Nouvelle Zélande), Hauck (Algues) Harvey (Algues), Van Heurck, Heer (Suisse), Hillebrand (Filices de Hawaia), Hohenacker, Hooker (collections diverses et extensives), Hooker & Thomson (Indes), Hepp (lichenes), Kurz (grand collection Indienne), Kotschy, Labrador (Latrobe) Lehmann (Melastomées &c.), Lénormand (Algues), Leighton (Lichenes) Lindheimer (Texas par Asa Gray), Maillard, Meller (Madagascar), Lindberg (Musci), MacOwen, M'Ken (Natal), Stuart-Mill (Grande Bretagne, seulement trois collections, une des quelles m'a été donnée par sa niéce suadente Hookero), Miquel (Inde Hollandaise), Nonneprediger (Brezil), Moritzi (Venezuela), J. Mueller (Lichenes), Pancher (Nouvelle Caledonie), Parlatore (Italie), Parry (Amerique du Nord), Philippi (Chili), Regel (Russie), Riedel (Brezil par Regel), Rochel (Banat), Sartwell (Amerique boreale), Samson (Chine), Salzmann (Bahia), Scheffer (Myrsinées, Palmae &c.), Shuttleworth, Schimper (Abyssinie, par Hochstetter), Short (Amerique du Nord), Sonder (particulierement une très-grande collection des Monocotylédones), Steetz (toute sa très-grande collection privée), Thuemen (Fungi), Teyesman (Java &c.), Thwaites (Ceylon), Todaro (Sicilie), Travers (Nouvelle Zeelande et l'isles de Chatham), Vieillard (Nouvelle Caledonie par Lénormand), Wallich, Welwitsch (ouest de l'Afrique par le Comte Ficalho), Wight (Indes), Zeyher (Afrique australe), &c &c &c

Je pourrais ajouter plusieurs autres nomes de Botanistes et de Collecteurs, mais ce que je dis peut suffire a vous montrer, que mon herbier est extensif. En effet, je calcule, que mon Musée contient 40000 espèces *bien-fondées*, sans compter celles de l'Australie, de la Nouvelle Zelande et de l'Oceanie. Comme vous savez la plus grande partie des plantes Australiennes est endemique. Cela donne un grand total de 50000 espèces approximativement.

Vous aurez peut être l'occasion d'utilizer ceci avec d'autre notes pour une contribution pour quelque supplement à votre precieuse "phytographie", ou bien vous pourriez le trouver d'un interêt suffisant pour le publier dans un des nombreux journeaux, qui sont a votre disposition. 80.08.04

Avec tous les égards et respects je reste votre

Ferd. de Mueller.

4.8.1880

Dear Mr De Candolle.

I must express my thanks for your attention in sending me your "Phytography";[2] another volume you have added to the important publications that have made your name famous. The reason why I have, in the "Native Plants",[3] given the names of authors and their works with full citations is that this book is principally intended for the schools of Victoria, where the teachers would not understand abbreviations of this kind.

I see that in your allusion to the "Flora Australiensis", my collaboration is completely ignored. You therefore must think that I had an insignificant part in its preparation, and I should be content with that, if I were a private individual. But as both my own means and those of my Department were devoted to this work, it would be most unjust to the Government of Victoria, were I to allow it to be believed that the Flora Australiensis (better Australiana) was the work of Mr Bentham alone. This illustrious man wrote all the text, but *two-thirds* of the material (pre-examined specimens) were furnished by me. My Australian collections are two or three times larger than those of Kew. From 1862 to 1877 I sent to Kew on loan my standard herbarium of about 70 large crates, filled with 90,000 completely arranged specimens previously examined by me, which I have gathered since 1847 with the help of friends and paid men; to obtain these collections I travelled on horseback and on foot 28,000 English miles! Finally, 30,000 specimens have been procured and gradually added by me since the volumes of the Flora Australiensis successively appeared, which means that about 650 species have been added in the "Fragmenta", among them several genera and some natural orders. Mr Dyer in his recent speech at the Colonial Institute recognized my part in the work.[4]

In your list of large herbaria, I see that Melbourne's is not included. It is the biggest in the southern hemisphere (and largely unique for Australia?), containing plants from all parts of the globe that I have obtained since 1840

2 A. L. P. P. de Candolle (1880).

3 B79.13.08.

4 Thiselton-Dyer (1880).

80.08.04 myself or through purchases or exchanges. I calculate that the number of specimens (which are not duplicates) is about 300,000, of which 120,000 are Australian and 180,000 extra-Australian. It is not easy to determine how many species this vast collection really contains. The number of Australian species [is] about 10,500, i.e.

Dicotyledoneae	7000)	considered very conservatively.
Monocotyledoneae	1600)	
Acotyledoneae	1900)	

The Algae (for example) have already been increased since Harvey[5] from 600 to about 800 principally by Sonder from my contributions.

The private collection that Drummond made for himself passed after his death into my possession. One can see in the "Fragmenta" that three-quarters of the botanical collectors of Australia have sent their "specimens" to me, though I have successively allowed the duplicates to be kept at Kew. I received from Sir Jos. Hooker numerous specimens of R. Brown and A. Cunningham, and I obtained a complete collection from Dr Milligan, likewise many of Sieber's plants, the majority of Preiss's, and all Leichhardt's. In my Museum the plants of Australia, those of New Zealand and those of Polynesia are apart; but all the rest have been inserted and arranged (according to the Candollean system) to form a single compact collection. My Museum contains the plants of Areschoug (Scandinavia and Algae), Aucher-Eloy, Ayres (Mauritius), Sir H. Barkly (Jamaica & Essequibo), Beccari (Sumatra), Boivin (Réunion), Bourgeau, Chickering (North America), Cosson, Cuming (Philippines), Durieu (Algeria), D'Albertis (New Guinea), Eaton (north America), Traill-Green (N. Amer.), Ecklon (his whole German collection, along with most of that of South Africa), Falconer (India), Fortune (China), Franqueville, Glaziou (through Hooker), Griffith (India), Hance (Hong-Kong & China), Haast (New Zealand), Hauck (Algae), Harvey (Algae), Van Heurck, Heer (Switzerland), Hillebrand (filices of Hawaii), Hohenacker, Hooker (diverse and extensive collections), Hooker & Thomson (India), Hepp (lichens), Kurz (large Indian collection), Kotschy, Labrador (Latrobe)[6] Lehmann (Melastomeae &c.), Lénormand (Algae), Leighton (Lichens), Lindheimer (Texas through Asa Gray), Maillard, Meller (Madagascar), Lindberg (Musci), MacOwen,[7] M'Ken (Natal), Stuart-Mill[8] (Great Britain, only three coll-

5 Harvey (1858-63).

6 Labrador (Latrobe) *interlined above* Lehmann (Melastomées &c.) *M presumably meant* Latrobe (Labrador).

7 P. MacOwan.

8 i.e. J. S. Mill.

ections, one of which was given to me by his niece by Hooker persuaded), Miquel (Dutch Indies), Nonneprediger (Brazil), Moritzi (Venezuela), J. Mueller (Lichens), Pancher (New Caledonia), Parlatore (Italy), Parry (North America), Philippi (Chile), Regel (Russia), Riedel (Brazil through Regel), Rochel (Banat), Sartwell (north America), Samson (China), Salzmann (Bahia), Scheffer (Myrsinaceae, Palmae &c.), Shuttleworth, Schimper (Abyssinia, through Hochstetter), Short (North America), Sonder (particularly a very large collection of Monocotyledons), Steetz (the whole of his very large private collection), Thuemen (Fungi), Teyesman[9] (Java &c.), Thwaites (Ceylon), Todaro (Sicily), Travers (New Zealand and the Chatham Islands), Vieillard (New Caledonia through Lenormand), Wallich, Welwitsch (west Africa through Count Ficalho), Wight (India), Zeyher (south Africa), &c &c &c. 80.08.04

I could add several other names of Botanists and Collectors, but what I say can suffice to show you that my herbarium is extensive. Indeed, I calculate that my Museum contains 40,000 *well-founded* species, without counting those of Australia, New Zealand and Oceania. As you know, the greatest part of the Australian plants is endemic. That gives a grand total of 50,000 species approximately.

Perhaps you will have occasion to use this with other notes for a contribution for some supplement to your valuable "phytography", or you could even find it of sufficient interest to publish it in one of the numerous journals that are at your disposal.

With all regards and respect, I remain yours

Ferd. von Mueller.

9 J. E. Teijsmann.

80.09.00 *From R. W. Moran*

RB MSS M1, Library, Royal Botanic Gardens Melbourne.[1]

[September 1880][2]
N. W. Police Barracks
Cloncurry River
via Townsville[3]

Sir,

I have the honor to enclose a specimen of a plant that grows on the banks of the rivers in this district, the blacks call it "*Loungha*", it is used by them in killing fish in the waterholes, the[4] cut it up fine, and throw it into the water, soon after the fish come to the top dead. – As you may not have seen the plant before I thought you might be interested in the matter. –

I have the honor to be
Sir
yr obt Sevt
R. W. Moran

Baron Von Mueller
Melbourne

1 MS found with a specimen of *Tephrosia* sp. (MEL 2047624).

2 *editorial addition*. Letter dated on the basis of MS annotation by M: 'Answ. 3/10/80. F.v.M.' Letter not found.

3 Qld.

4 they?

To Jean Müller 81.01.12

Conservatoire et Jardin botanique, Geneva.

12/1/81

Die Post bringt mir eben, hochgeehrter Herr Professor, Ihren freundlichen Brief vom 26. Nov., u. danke ich Ihnen für die Benennung der letzten Flechten. Die schnelle Durcharbeitung solcher Materialien ermuthigt recht sehr zum Weitersammeln, u. hoffe ich Ihnen noch mancherlei Schönes aus Australien zuwenden zu können.

War es denn Ihr *einziger* Sohn, den Ihnen das herbe Geschick entriss? Nach Ihrer trauervollen Bemerkung, nur noch Muth u Trost in der Flechten Kunde zu finden, muss ich es fast fürchten.

Ich selbst stehe jetzt spät im Leben ganz vereinsamt da, ohne Familie, ohne irgend Vermögen, ohne eine auch nur annähernd würdigen Stellung. So ist auch *mein* Muth gesunken u. ich habe alles Interesse am Leben verloren. Das gänzliche u wie es scheint *absichtliche* Übergehen A.D.C's in seiner Phytographie über meinen Antheil an der Flora Australiensis (besser Australiana); sein gänzliches Schweigen über meine enormen Herbarium, (etwa 300.000 Exemplare, vielleicht 60.000 oder 70.000 *gute* Arten umfassend,) schmerzt mich mit andern Zurücksetzungen tief. Ich bin nun einmal von Natur so grenzenlos grossmüthig, dass ich auch meine Beobachtungen u Sammlungen so danklos Bentham aufopferte. Von meinen Fragment. spricht D.C. garnicht. Ich sehe es jetzt ein, dass ich Bentham hätte *ohne* meine Sammlungen das in Europa befindliche Material austral Pflanzen ausarbeiten lassen sollen; in welche zahllose Fehler wäre er gefallen ohne meine Formenreihen, wie viele hundert Arten hätte B. gar nicht gehabt u welche Unvollständigkeit der Standorte wäre dann die Folge gewesen, während ich mit einer vollendeten Flora Austr. selbst hätte vortreten können.

Stets Ihr
F. v. M

81.01.12 Krempelhubers Liste meiner Lichenen werden Sie in den Abhandlungen der zool. bot. Gesellschaft von Wien gesehen haben. Es sind für Continental Austr. freilich nur wenig über 100 Arten, da ich *nie* in flechtenreichen Gegenden lebte. Ich hoffe noch mancherlei aus Nord-Ost Australien zu bekommen; leider hat Bailey, den ich erst überhaupt zum Sammeln in *irgend* einer Abtheilung der Pflanzen heranzog, (Durch meine Correspondenz) zu meiner Verwunderung mir seine Lichenes entzogen u solche Dr Stirton zugewandt.

Würden Sie mir freundlich Addenda (einfach Namen u Standorte) zu Krempelhuber's Abhandlung geben? für eine vollständige Flechtenliste Australien's (*nicht* Neu-Seelands) So erinnere ich, dass Sie eine Flechte specifisch als leucophthalmas beschrieben haben, aber in meiner ganz engen u temporären Behausung ist der Druck verlegt. Ich glaube, es war eine Flechte vom Richmond-River. Nach Krempelhuber möchten Sie fast glauben, dass Mad. Hodgkinson u die andern an ihn selbst geschickt. Einige von K. neuen Arten scheinen mir unhaltbar zu sein. Das *Ganze* kam von mir, wurde auf meinen Antrieb u manchmal auch auf meine Kosten gesammelt; u nicht *eine einzige* Art hat K mit meinem Namen verknüpft, so dass mein Name ausser in einer schon 185[1][1] an Hampe gesandten Art gar nicht mit der Lichenologie verknüpft ist. Nun ist der arme Hampe denn ja auch dahin! –

Ich habe hier gerade einen Index der Moose Austr. unter der Presse von ihm, wohl seine *letzte* Arbeit.

Wird Nylander sein Werk nicht vollenden? ich habe nur 2 Theile. Ist je mehr erschienen?

Werden *Sie* uns eine Lichenography universalis geben können?

12/1/81

The post just brings me, highly esteemed Professor, your friendly letter of 26 November[2] and I thank you for the naming of the last lichens. The rapid working through of such material encourages further collecting very much and I

1 *final digit illegible*. – Established by the date of Hampe's publication; see note 8.
2 Letter not found.

hope to be able to let you have additional various kinds of nice things from Australia. 81.01.12

Was it then your *only* son, whom harsh fate snatched away from you? After your sad remarks about finding courage and consolation only in the study of lichens, I must almost fear it.

I myself stand now late in life quite lonely, without family, without any fortune, without even an approximately worthy position. So *my* spirit has also sunk and I have lost all interest in life. The complete and as it seems *deliberate* omission by ADC in his *Phytographie*[3] of my share in the *Flora australiensis* (better *Australiana*); his complete silence about my enormous herbarium (about 300,000 specimens, comprising perhaps 60,000 or 70,000 *good* species) hurts me deeply with other slights. I am just so boundlessly generous by nature, that I even sacrificed my observations and collections so thanklessly to Bentham. Of my *Fragmenta*, D.C. says nothing at all. Now I see that I should have let Bentham work up the Australian plant material held in Europe *without* my collections; into what innumerable errors he would have fallen without my series of forms, how many hundreds of species would B not have had at all and what incompleteness of localities would then have been the consequence, whilst I would have been able to produce a complete Flora Austral. myself.

Always your
Ferd. von Mueller

You will have seen Krempelhuber's list of my lichens in the *Abhandlungen der zoologisch-botanischen Gesellschaft von Wien*.[4] There are for continental Australia of course only a few over 100 species, because I *never* lived in lichen-rich regions. I still hope to get many kinds from northeast Australia; unfortunately Bailey, whom I, on the whole, first interested in collecting in *any kind* of division of plants (through my correspondence), to my surprise removed his lichens from me and let Dr Stirton[5] have them.

Would you kindly give me addenda (simply names and localities) to Krempelhuber's paper for a complete list of lichens of Australia (*not* of New Zealand)? Thus I remember that you have described a lichen specifically as

3 A. L. P. P. de Candolle (1880). M's complaint is surely justified. His role in the preparation of Bentham's *Flora australiensis* was ignored by Candolle in his account of the work even though M's name appears on the title page; the Melbourne herbarium was not included in Candolle's list of public herbaria; and in the list of 'authors and collectors' M was noted only for having sent 'numerous specimens' to various European collections. See M to A. de Candolle, 4 August 1880.

4 Krempelhuber (1881).

5 Stirton (1881).

81.01.12 *leucophthalma*, but in my quite confined and temporary lodging the paper is mislaid.[6] I think it was a lichen from the Richmond River.[7] According to Krempelhuber you might almost believe that Mrs Hodgkinson and the others sent them to him themselves. Some of K.'s new species seem to me to be untenable. The *lot* came from me, was collected on my initiative and sometimes also at my expense; and K. has not attached my name to *a single* species, so that my name is not connected at all with lichenology except in a species sent to Hampe as early as 185[1].[8] Now the poor Hampe is also deceased!

I have here just now an index of the mosses of Australia by him in press, probably his *last* work.[9]

Will Nylander not finish his work? I have only 2 parts.[10] Has more ever appeared?

Will *you* be able to give us a universal lichenography?[11]

81.01.20 *From Hermann Kempe*[1]

RB MSS M69, Library, Royal Botanic Gardens Melbourne.[2]

Hermannsburg Jan. 20. 81.

Hochgeehrter Herr Baron!

Sehr wenig ist es, was ich Ihnen diesmal zuzuschicken im Stande bin. Die Hitze ist dies Jahr so grausig u. lange anhaltend, dass selbst die Eingebornen nicht gerne am Tage umhergehen. Mehrere Wochen nach einander haben wir bis jetzt noch alle Tage Heisswind

6 Species not identified.
7 NSW.
8 *final digit illegible*. – M is presumably referring to *Stricta muelleri*, named by Hampe (1852) but subsequently renamed *Heterodea muelleri* by Nylander (1868).
9 M published Hampe's list as one of the appendices in B81.13.12.
10 Probably Nylander (1858-69), which was complete in two volumes.
11 Müller never published such a work.

1 MS annotation by M: 'Beantw 28/2/81.' [Answered 28.2.1881.] Letter not found.
2 MS found with a specimen of *Eucalyptus papuana* (MEL 705185).

gehabt, wobei das Therm. bis zu 55° R. in der Sonne stieg. In Folge davon sind auch alle unsre Pflanzen u. Bäumchen, die wir mit vieler Mühe noch lebend erhalten hatten, total verdorrt, nicht aus Mangel an Feuchtigkeit im Boden, sondern allein von der Sonnenhitze. Ich schicke Ihnen ausser Samen *zu No 341* noch Blüthen von einem Eucalyptus No 8. welchen die nativs "ilumba" nennen. Dieser Baum erreicht unter günstigen Umständen eine Höhe von 150 Fuss. 81.01.20

Rinde glatt u. schneeweiss bis in die kleinen jährigen Zweige. Die Zweige hängen gleich Ranken lang herunter u. geben dem Baume ein ziehrliches Ansehen. Die Krone ist sehr dicht. Das Holz ist weich, jedoch wie es scheint, nicht so dauerhaft, als anderes Eucalyptenholz. Hier wächst dieser Baum nur auf u. in der Nähe von Bergen, doch habe ich ihn auch schon auf sandigen Ebenen gesehen, z. B. an dem Hugh river.

Samen werde ich nachschicken, sobald derselbe reif ist.

Anbei sende ich auch etwas Samen von Eucalyptus No 1. Es scheint mir, dass alle die Eucal. welche hier in u. an den Creeks wachsen, eine Sorte sind, wenigstens kann ich in Blüthen & Samen keinen Unterschied erkennen. Auch folgt hierbei Samen zu No 379. Beefwood u. einige Beeren Canthium latifolium.

Ich dachte schon letzte Post einen Brief von Ihnen zu erhalten, es wird doch nicht ein oder mehrere Packete verloren gegangen sein. Ich schreibe Ihnen hiermit, an welchen Tagen ich Packete mit Pflanzen u. Sämereien an Sie abgeschickt habe u. wie viele.

Sept. 9. 2 Packete.
Okt. 8. 1 do.
Nov. 4. 3 do.
Dec. 1. 4 do.
do. 31. 1 do.
Jan. 25. 1 do.

Hoffend dass Sie alles wohl erhalten haben,

mit herzlichem Gruss
Ihr
H Kempe.

81.01.20 Hermannsburg,[3] 20 January 1881.

Highly esteemed Baron!

It is but very little, what I am able to send you this time. The heat this year is so terrible and so persistent, that even the natives do not like to go about during the day. For several consecutive weeks we have had hot winds daily, with the temperature rising to 55° R.[4] in the sun. As a result all our plants and young trees, which we had kept alive only with great difficulty, are completely withered, not from a lack of moisture in the soil, but purely from the heat of the sun.

Apart from seed *of No 341* I am also sending you flowers of *Eucalyptus* No 8, called 'ilumba' by the natives. Under favourable conditions this tree reaches a height of 150 feet.

Bark smooth and snow-white right down to the small annual twigs. The branches are pendulous like creepers and give the tree a dainty appearance. The crown is very dense. The wood is soft, but, it seems, not as durable as the timber of other eucalypts. Here the tree grows only on or near mountains, but I have seen it on sandy plains, for instance at the Hugh River.[5] Seed shall follow, as soon as it is ripe.

Enclosed I also send some seed of *Eucalyptus* No 1. It seems to me that all the eucalypts growing in and along the creeks here, are the same kind; at least I cannot see any recognisable difference in flowers or seeds. Herewith also seed of No 379, Beefwood, and some berries of *Canthium latifolium*.

I had thought I might receive a letter from you already with the last mail, I hope one or more of my parcels have not got lost. I give you here the dates on which I mailed parcels with plants and seeds to you, and how many.

9 September [1880][6]	2 parcels.
8 October	1 do.
4 November	3 do.
1 December	4 do.
31 do.	1 do.
25 January 1881	1 do.

In the hope that you received everything in good order,

with sincere regards
Your
H Kempe.

3 Lutheran mission near Alice Springs, central Australia.
4 i.e. Réaumur (= 69° C).
5 Central Australia.
6 *editorial addition.*

To Richard Owen[1] 81.02.01

Natural History Museum, London, General Library, Owen correspondence, vol. XIX, ff. 386-7.

1/2/81

Allow me, dear Prof. Owen, to send to you for kind identification specimens of the insect, which produces the Melitose (or Eucalyptus Manna) on Euc: *viminalis*. Though much controversy about the origin of this saccharine crumblike substance has been going on for very many years, I have only this season been able by the help of an Assistant of mine[2] to confirm, that it is the small insect, now forwarded, which secretes the *Melitose* in a liquid state assimilating and elaborating it from the sap of that particular Eucalypt.[3] I doubt, whether any where in Australia the generic & the particular specific position of this little creature can be ascertained beyond doubt;[4] so I proceed to you, as you are at the fountainhead of zoologic Knowledge in your grand establishment. The Melitose drops in the summer so copiously from *Euc. viminalis*, that it is of some interest to set questions about the [...]rial[5] source of this substance at rest.[6]

Let me hope, honored and venerable friend, that this year may again be to you one of unimpaired health and scientific triumphs, and that providence will grant you yet many other years for your luminous career & worldly happiness.

Regardfuly your
Ferd. von Mueller

1 MS annotations by Owen: 'R'ecd. from Prof. v. Mueller – a new-years gift – R. O. (1881)' and '(Valued Autograph of the Professor)'.

2 C. A. Groener?

3 Almost a year earlier M had sent Owen specimens of an insect that O. Tepper had identified as producing sugary secretions on other species of *Eucalyptus* near St Vincent's Gulf, SA; see M to R. Owen, 2 March 1880 [Collected Correspondence].

4 MS annotation: 'char. of this insect is known.'

5 *illegible* – obscured by binding.

6 M extensively discusses the various sources of manna in Australia in the entry for *E. viminalis* in B84.13.19, but makes no mention of information from Owen. See also M to W. Thisleton-Dyer, 29 January 1884 [Collected Correspondence].

81.03.00 *To the Linnean Society, London*

Linnean Society, London, Certificates of Fellows, Foreign Members and Asociates, 1877-82.[1]

We the undersigned beg to propose for election:

John Forrest Esqr., Gold-medalist and F.RGS, Chevalier of the Order of the Italian Crown, Hon. Member of the Royal Italian Geographic Society, &c to the honor of being elected a fellow of the Linnean Society of London, In consideration of his efforts for the elucidation of Australian plants during his extensive geographical journeys.

Ferd. von Mueller, M.D., F.L.S.
William Woolls Ph.D., F.L.S.
Robert Fitzgerald, F.L.S.[2]

March 1881

81.03.15 *To Ferdinand von Krauss*[1]

Staatliches Museum für Naturkunde, Stuttgart.

15/3/81

Wie entsetzlich das Geschick des armen Czars, edler Freund; ich kann noch gar nicht in meinem Gedankengange realisiren, dass so Schauervolles möglich sei! Möge doch jeder streben, der Religion förderlich zu sein, damit die ganze Menschheit durch die Lehren der Kirche veredelt werde! Dann werden solche Abscheulichkeiten, wie das Petersburger Attentat, auch nicht wiederkehren und Schandflecke in der Geschichte künftiger Jahrhunderte sein! – Dies grässliche Ereigniss muss der Welt die Augen öffnen, um die Unterdrückung

1 The MS is in M's hand; the other supporters have signed individually.

2 Forrest was proposed on 20 May 1881, and elected on 16 June.

1 MS annotation by Krauss: 'in Cannes bis 20 Mai die Königin deshalb ignorirt Rep. 15. Oct.' [In Cannes until 20 May, therefore the Queen ignored Reply 15 October.] Letter not found.

der Socialisten, Communisten, Nihilisten und wie die Banden der Atheisten sonst heissen mögen, herbeizuführen. Mich hat die Catastrophe in der russischen Kaiserstadt besonders schmerzlich berührt, denn der arme Czar ist auch gegen mich überaus gütig und grossmüthig gewesen, so sehr, dass ich zu bitten wagen,[2] soweit es die Etikette gestattet, wenn Ihre hehre Königin das nächste mal Ihr Museum besucht auch ein Wort der Condolenz von den Antipoden für mich ausdrücken zu wollen über das Unerhörte; – denn in *allen* Theilen der Erde u so ja auch hier ist der Ausdruck der Entrüstung über das Unerhörte ein lauter, u das Mitleid mit der Schwester des Kaisers muss ja auch auch bei Ihnen den tiefsten Nachklang finden. 81.03.15

Mit Ehrerbietung
der Ihre
Ferd. von Mueller

15/3/81

How shocking the fate of the poor Tsar, noble friend;[3] I can by no means yet realise in my train of thought that such a dreadful thing were possible! May each indeed strive to promote religion, so that the whole of mankind be improved by the teachings of the church! Then such abominations as the Petersburg outrage will not be repeated and be a blemish in the history of future centuries! – This shocking event must open the eyes of the world to bring about the suppression of socialists, communists, nihilists and whatever the gangs of atheists may otherwise be called. The catastrophe in the Russian imperial city has painfully affected me particularly, since the poor Tsar has also been extremely kind and generous to me, so much so, that I dare to ask you, as far as etiquette permits, when your august queen[4] visits your museum the next time to express also a word of condolence from the antipodes for me over the outrage; – because in *all* parts of the world and so even here the expression of indignation about the unheard of thing is a loud one, and the sympathy for the sister of the Emperor must certainly also find the deepest echo with you.

With respect
your
Ferd. von Mueller

2 wage?

3 Tsar Alexander II of Russia was assassinated in St Petersburg on 1 March 1881 Old Style (i.e. 13 March 1881 New Style).

4 Queen Olga of Württemburg was the sister of Tsar Alexander II.

81.03.21 ## *To George Langridge*

No. 1881/235, unit 2, VPRS 8609/P28 VA 744 Board of Land and Works, Public Record Office, Victoria.[1]

Melbourne,
21 March 1881.

To the honorable G. D. Langridge, M.L.A.,
Minister of public works

Sir,

In compliance with your request I had the honor to accompany Mr Le Cren, the Secretary, to the Yan-Yean[2] on the 19th instant, with a view of advising, whether by cultural operations around the Lake and on any of its drainage-areas the Yan-Yean water could be kept freer of such impurities, as very recently were found to deteriorate more or less the water, supplied to the city and its vicinity. In inspecting the Reservoirs at Preston and at the place, where the open aqueduct is connected with the main-pipes, and in proceeding for further inspection to the Yan-Yean Lake and its surroundings, we found the surface of the water in several spots both in the reservoirs and Lake covered by some green scum, drifted together by the wind or current. This film, which is of algic growth, consisting mainly of short spiral coils of a species of Spirillum, allied to the European S. volutans in the group of Oscillariae, and to a slight degree also visible as floating in the deeper parts of the Lake, did not occur in particular copiousness (so I am informed) anywhere in the Yan-Yean during former years; still the water of the Lake itself was at the time of our visit not to any appreciable degree turbid or badly odorous. Whether to the development of this scum and its subsequent decomposition, (especially when as debris blown

1 The file includes two handwritten copies of a summary of M's report headed 'Baron von Mueller's report condensed'. On one of them, prepared for the Minister of Public Works, George Langridge, M's recommendations regarding plantings on Bear's Flat and Dry Creek, and on Whittlesea Flat, are marked with a large '#' and initialled 'GDL'. A copy of the letter, together with the summary, is at the Library, Royal Botanic Gardens Melbourne (RB MSS M101).

2 Yan Yean reservoir, Vic., then the source of Melbourne's water.

together) the unpleasant odor and taste, noticed lately in the supply-water of Melbourne, is to any extent attributable, can only be ascertained by further extended observations. Under any circumstance it seems desirable to adopt some simple and inexpensive contrivance to skim off this scum, when largely drifted together and especially when copiously interspersed (as we found it in one of the reservoirs) with decaying larvae of small insects, before in stormy weather the particles of the film may become dispersed again and be drawn off in the supply-pipes, with the possibility of undergoing decay-fermentation on the long way of the water to the city, especially during hot weather. But whatever the direct cause or causes of the deterioration of the Yan-Yean may be, there can be no doubt, that the increasing influx of decaying particles of vegetable growth and of organic egesta into the Lake should be checked by all available means; but here the difficulty is encountered, that more than half of the Land, surrounding the Lake, is held by private proprietors since the early period of our colony, and as this land is utilized for grazing purposes and only a very narrow strip of Governments Ground separates the pastures from the brink of the Yan-Yean, the drainage from the nearest grazing ground into the Lake is likely to become more and more contaminated, indeed even at present sheep-droppings being found washed down to the very edge of the Lake. Moreover the flats, hitherto used as town-common at Whittlesea and as drover's resting-places near Jack's Creek, send their swampy waters with much of impurities also into the Lake after very heavy showers, while operations for obtaining by Government's licences timber from the slopes of the Plenty-Ranges, even where facing the Yan-Yean, are apt to tear up the soil within the drainage area and to contribute additional impurities to the feeders of the Yan Yean, irrespective of the contaminations arising on many farms along the River Plenty above the Lake, pasture-animals of all sorts approaching there without restriction any portion of the very stream, which has been chosen to supply the city and suburbs with water for domestic purposes. I am aware, that turbid flood-water is not admitted into the Yan-Yean Lake, but turned off at the byewash; yet the water even after settling cannot possibly be of the limpid purity, in which it arose at the springs and

81.03.21 in which it meandered as rivulets and brooks originally into the Plenty-River, when now it has to pass through so much farm-land. All this, to be contended with, has been pointed out by the officers of the Water-Department; and I merely allude to these difficulties here once more in connection to planting-measures, on which my opinion is especially desired, for the amelioration of the Yan-Yean Water by such means.

If any rearings of plants on the banks below high-water-mark were to be effected, it could be only with such, as will submit to occasional and even lengthened submersion without decay; mainly silicious-coated rushes and sedges would suffer this; but even these, when withering in the annual decay of their stems or when dried up by the lowness of the Lake, may become obnoxious; indeed as a rule the growth of miscellaneous plants between variable high- and low-water mark should be avoided as apt to give rise to putrid fermentation. Actual water-plants again ought to be kept out of Lakes and Reservoirs intended for drinking and cooking water, as small fragments of such aquatic vegetation – ever torn off – would be sure to enter the supply-pipes. One of the best plants, for holding its ground on the banks, would here be the hollow-rush (Heleocharis sphacelata), which occurs as indigenous on a few favorable places on the margin of the Lake and could readily be transferred, if that be found desirable. But I doubt very much, whether the extensive rearing of any rushy plants would much impede the ingress of impurities into the Lake. I would rather recommend, that in first instance a narrow belt of vigorous vegetation be established across the two northern vallies, vize Bear's flat and Back- or Dry-Creek, through which much of the drainage, flowing directly into the Lake, has to pass. The larger and stronger kinds of perennial grasses, fit for humid soil, might readily be sown on a strip of ploughed ground across the above mentioned depressions and on other portions of the surroundings of the Yan-Yean, while the Silt-grass, (Paspalum distichum) might be established in more frequently inundated ground, plants for the purpose being available at the banks of the Yarra, established there from the botanic garden long ago. In percolating through such a dense grassy vegetation, the drainage from the pastures would be freed to a considerable

extent of adherent impurities either mechanically or by absorpt- 81.03.21
ion and chemical changes; but a still more powerful barrier might be set to the flowing of any contaminating matter into the Yan-Yean, if a belt of the tall Donax-reed and of the several kinds of Bamboos, early introduced into the colony, was likewise planted over the ground of the principal drainage near the Lake, whereby moreover the lacustrine landscape would become more embellished; and as the Yan-Yean will always be here a favorite place for rural pleasure-parties, its scenic ornamentation might be still further enhanced with every advantage to the Lake itself by the ready raising of masses of our tall Gippsland Fan-Palm, of Date-Palms and other hardy species of that noble order of plants. Whether means could be devised readily, to filter any water of the Plenty-River through sand, before it enters the Lake, as is done with the water of the River Elbe for the supply of the City of Hamburg, is beyond the scope of this special report.

In reference to your request, to be informed, what kinds of trees ought to be chosen for plantations on the Whittlesea-flat, I would submit, that the choice should be largely of such kinds, as would render their culture finally one of permanent remunerativeness in a commercial point of view, unless it is merely desired, to plant quick-growing trees, such as Eucalypts, Willows, Poplars &c for the absorption of of[3] humidity. The European Walnut-tree and the North-American congeners as well as the best of Hickories would remain lasting for the yield of their nuts, the Cork-Oak for its bark, the Valonia-Oak for its tan-acorns, the Sugar-Maple for its sap. According to the wetter or drier ground of the reserve many kinds of Pines could be chosen, especially such as can be tapped for turpentine, as the Scotch Fir, the Pinaster, the Haleppo-Pine[4] and from the United states the Southern Swamp-Pine, the Pitch-Pine, the Frankincence-Pine, the Lambertian Fir, besides several from the higher mountain-regions of India. Some of the best Nut-Pines could also be grown to advantage, especially the American sorts. In the wettest places the gigantic Virginian Swamp-Cypress

3 *Word repeated.*

4 Aleppo pine (*Pinus halepensis*).

81.03.21 could be reared. All these kinds of trees were brought experimentally to Victoria already many years ago. Under any circumstances it would be advisable, to have a large interspersion of Conifers into the intended tree-plantation, in as much as the decaying leaves of most deciduous trees, when occasionally flooded into the Lake, might impair by their decomposition the purity of the water, whereas the tardily falling foliage of Pines on account of its antiseptic oil will not impart any unpleasant odor or taste to water. As the drainage of Whittlesea-flat finds now and then an outlet to the Yan-Yean, it would not be recommendable, to cause at any future time a heavy traffic on it by the rearing of trees solely for timber; but so favorable an area might partly be devoted without any very heavy expenditure to a general Arboretum for the local test-growth of a great variety of trees and perhaps even of a host of utilitarian shrubs, and might thus become a nucleus for forest-culture in the Yan-Yean region for the ultimate benefit of the extensive lands, there permanently reserved by the Government.

In conclusion I may be permitted to express a hope, that the metropolitan and suburban communities will cooperate with the Government more generally in providing tanks for the storage of rain-water, than has hitherto been done. For altho' the magnificent Yan-Yean works yield under ordinary circumstances a more copious supply of water of fair quality, than many other cities and towns elsewhere enjoy, yet no water of any aqueducts of great length and supplied from a vast drainage area can possibly compete in purity with rain-water, collected from well-kept roofs. If therefore each household would only provide for itself a tank (or in high-lying grounds perhaps a cistern) the purest of drinking and cooking water would become everywhere independently available, the demands on the Yan-Yean for general purposes in days of excessive heat would become lessened, and an enormous addition to the water-supply of the metropolis and adjoining cities would be gained from thousands of roofs thus for now unutilized.

I have the honor to be, Sir,
your obedient servant
Ferd. von Mueller.

To Henry Parkes 81.05.21

A925 Parkes correspondence, vol. 55, pp. 247-52, Mitchell Library, State Library of New South Wales, Sydney.

Melbourne
21/5/81.

Allow me, Sir Henry, to adress to you a few lines more in the interest of the proposed and likely last search after traces of Dr Leichhardt's party. The Governments of South-Australia and Victoria *have* promised kindly to provide £300 each for this new enterprise, if the Governments of New South Wales and Queensland will do the same. As now the *realisation* of this new object in the cause of humanity and geography also will depend on the cooperation of your colony and Queensland, I would beg of you to give this proposition of looking once more for the lost explorers your generous consideration and support. If even the tale of Mr Skuthorpe should prove correct, of which we have no evidence yet, it would account only very partially of the fate of the eight people, which formed the missing party.[1] The area, over which the search is to be conducted has become more and more ennarrowed, so that it is almost sure, that the new expedition would bring final tidings. The cool season would be the best to commence operations & Mr Giles is ready to take the field again. The expedition will necessarily also be of advantage to *all* the Australian colonies for extended pastoral settlement, and the subsequent traffic augmented into the interior must find its way to the coast, a long line of which is commanded by New South Wales also for commercial purposes. Besides new spaces are required for the surplus herds and flocks of the settled districts, so that the colonists of your territory

1 In late 1880 a cattleman, J. R. Skuthorpe, claimed to have found the journals of Leichhardt and a member of Leichhardt's party, Adolf Classen, together with other relics. However, he failed to produce them when asked to do so.

81.05.21 will also share in the practical benefits, which are sure to arise from further geographic exploits.

Very regardfully
your
Ferd. von Mueller.

The honorable
Sir Henry Parkes
K.C.M.G., &c

81.06.06 *From William Woolls*

RB MSS M46, Library, Royal Botanic Gardens Melbourne.[1]

Richmond[2]
June 6/81

My dear Baron,

I sent you last week Miss Scott's[3] drawing of the *Dampiera* (which is certainly indigenous near Sydney), & also a copy of the remaining portion of my article on Eucalypts.[4] It has been well recd, & I think it will be reprinted in the S. Mail.[5]

My old pupil Mr C. Brown from Melbourne called to see me. He spoke very kindly of you. I also had a chat about you with another old pupil Mr James Fairfax. He seemed pleased with the

1 MS found with a specimen of *Eucalyptus siderophloia* (MEL 231683).
2 NSW.
3 Harriet Scott.
4 Woolls (1880).
5 Gilbert's bibliography of Woolls's publications does not include any such reprinting in the *Sydney mail*; see Gilbert (1985).

book you sent him. He thinks of going to England to place his sons at the University of Oxford.[6] 81.06.06

As I had not time to run over to Cabramatta[7] to see the Iron Bark Box, I wrote to a friend to get some specimens. I forward to you what he has sent, but the trees will not flower for three months, or more. As I said before, the Iron Bark Box differs from *E. siderophloia* principally in the bark. The leaves are very similar, but generally, not so large or thick. The two marks however, of distinction are the operculum (which is much shorter than that of *E. Sid.*), & the shape of the fruit. We must wait for the anthers, unless you can decide from the buds previously sent. My friend says that this tree grows generally in company with *E. siderophloia* & *E. hemiphloia*. The true *E. sid.* has always a long operculum.

I send you what my friend collected. There is a great difficulty about the seedlings, but I hope some day to look for myself. Mr Shepherd says that the saplings had more of the appearance of Box, than of Iron Bark. This tree occurs also near Liverpool[8]

Yours very sincerely
W. Woolls

Mrs Forde[9] got one of Sand's[10] prizes for drawings. Miss Scott did not send anything, as she was working for Turner & Henderson.[11]

6 James Fairfax's second and third sons, Geoffrey and James Oswald Fairfax, entered Balliol College, Oxford in 1881.

7 NSW.

8 NSW.

9 Helena née Scott.

10 i.e. John Sands, printers, Sydney.

11 Another firm of printers in Sydney.

81.06.10 *From Joseph Hooker*

RB MSS M3, Library, Royal Botanic Gardens Melbourne.

Royal Kew Gardens[1]

June 10 /81.

Dear Von Mueller

I enclose Oliver's note[2] about the Acacia, our Brunonian specimens of *deltoidea* are miserable – we are now overhauling a whole heap of Brown's unarranged plants to complete our set, & are putting aside the first set of duplicates for you. Brown's collections were *enormous*, it is marvellous what he did in that little ship! – incredible. – After the Brit. Mus. had sent us the set intended for us by Bennett, it sent a further load, & it is this latter we are now sorting & selecting from previous to distribution.[3]

The Cycad will be most acceptable to Dyer[4] – who is perfecting our Cycad Herb & living collection indefatigably – We should be thankful for flat-dried fronds, old & young, of any Cycads & especially of this new one. Dyer is forming a Herbarium collection on large paper to which contributions will be most acceptable – the size is 2 x 1 3/4 ft.

Our collection of drawings is also very complete, & of the photographs also

Very sincerely yr
Jos D Hooker

1 Embossed letterhead.

2 Note not found.

3 M acknowledged Hooker's news about Brown's specimens in M to J. Hooker, 29 July 1881 [Collected Correspondence], and then acknowledged receipt of five cases of Brown's plants in M to J. Hooker, 22 and 29 May 1882 [both Collected Correspondence].

4 W. Thiselton-Dyer.

To Joseph Hooker 81.07.22

RBG Kew, Miscellaneous reports 1.52, Phylloxera, Bordeaux Congress 1881, ff. 56-7.

22/7/81.

You will have received a communication from the Agent General of Victoria, dear Sir Joseph, after a telegram from here, soliciting you to become the representative of Victoria and indeed also of N.S. Wales and South Australia at the Congress of Bordeaux early in September, when the Phylloxera-questions are further to be discussed.

A telegram from the Right. Hon the Secretary of State for the Colonies, invited us to send a delegate. New South Wales proposed, that you should name a Representative at home; but I went a step farther and recommended that *yourself* should be asked to be the delegate of these Colonies. As the Congress takes place at the period when usually you make a tour to the Continent, as moreover it takes you so little time to go to Bordeaux from Kew, but above all as you as our "Premier" in an Official position are interested in this Phylloxera-question for the sake of *all* the British Colonies, it was but fair to offer the mission to *you* in first instance, and in the event of your being pressingly prevented, to ask you to nominate the Delegate perhaps Mr Dyer[1] if he can be spared for the purpose. Of course all expenses incurred by you, will be paid by the three colonies jointly.

Some Reports & correspondence will be transmitted by the Cuzco-mail[2] to you through the Agent General, among these a letter of mine,[3] which explains, why – beyond a private visit to the Geelong district to see the ravages of the Phylloxera vastatrix

1 W. Thiselton-Dyer, whom Hooker did indeed nominate; see M to J. Hooker, 29 July 1881 [Collected Correspondence].

2 *Cuzco* was one of the mail steamers operated on the England-Australia run by the Orient Steam Navigation Company.

3 Letter not found.

81.07.22 personally [–][4] I became in no way identified with this, now to Australian Viticulture most serious question. This exclusion of mine from all cultural-concerns is one of the many ill consequences of my loosing the bot. Garden, and has made me powerless to do anything in matters of this sort or in any other cultural pursuits.

Your despatch of 1876[5] came too late to prevent the ingress of the Phylloxera, which found its way into Victoria (Geelong) insiduously soon after I left the bot. Garden in 1873. On this occasion I would now only add, that altho' it was taked[6] about by some, that I should go to the Congress and by others that the Secretary of our Agricultural Department should go, I stated that after I had not even my own large collection of Vines seen for the last eight years, and had been excluded from *all* cultural operations, I had no courage now to share in a task, of this kind, unless the Government here insisted on my going; *you* can also understand, that it would "break my heart" to meet the Directors of Kew, of Montpellier and all the other Garden-Directors, a colleague of whom I was til 1873, unless I was myself Director again at least of some one garden here; hence it is not likely that my intention of 1869 when my Garden was first invaded, to visit Europe, the homestead of my boyhood, my paternal house (and then also the great horticultural institutions of Europe) once more, will ever be realized!, not to speak of destroyed hopes of life and overthrow of my career in *most* respects from all of which I might have been saved by one single powerful appeal from England!

Regardfully
your
Ferd von Mueller

4 *editorial addition* – obscured by binding.

5 J. Hooker to Under-Secretary of State, Foreign Office, 9 February 1876, published in the Melbourne *Argus*, 8 May 1876. See also M to T. Wilson, 16 June 1881 [Collected Correspondence].

6 talked?

From D. Jones[1] 81.07.26

RB MSS M1, Library, Royal Botanic Gardens Melbourne.[2]

Port Mackay
Queensland
July 26, 1881

Dear Baron Mueller,

I have gathered a few plants on Flat Top Island, off Port Mackay, within the tropics, which I shall esteem it a great favour if you will name. The beans saved the lives of a shipwrecked crew a few years ago.

Very truly yours
D Jones

Baron Mueller

(Please address Post Office, Port Mackay.

To Julius von Haast 81.07.28

MS papers 37, folder 211, no. 610, Alexander Turnbull Library, Wellington.

28/7/81

Durch Ihren Einführungs-Brief, edler Freund, bin ich mit Mr Ivey bekannt geworden, obwohl er erst gerade vor seiner Abreise mich in meiner Abgeschiedenheit aufsuchte, da ich nicht einmal wusste, dass er hier sei. Er war von einem jungen Chemiker begleitet, desssen Namen mir in dem kurzen Gespräch nicht bekannt wurde, für den ich aber einen Abdruck *meiner* Übersetzung von

1 MS annotation by M: 'Answ 5/8/81'. Letter not found.
2 MS found with a specimen of *Canavalia* sp. (MEL 726785).

81.07.28 Wittsteins Chemie der Pflanzen beifüge, bittend ihm solche zu zustellen. Ich sage ausdrückliche *meine* Übersetzung, da ein deutscher Apotheker hier öffentlich behauptete, es sei nicht meine sondern die seinige. Die Sache verhält sich so. Dieser Mann brachte vor etwa ein Dutzend Jahren einen Einführungsbrief von einem mir befreundeten deutschen Gelehrten, u da sein Photographie ihn nicht ernähren wollte, u ich damals für den 5tn Band der Flora Australiens sehr beschäftigt war[,][1] engagirte ich ihn, auf meine Privatkosten eine *vorläufige* Übersetzung des Werkes zu machen. Ich hatte das Buch von meinem vieljährigen Freunde Wittstein erhalten; das Übersetzungsrecht hatte er u. der Verleger sich vorbehalten u *mir* für die engl. Sprache anvertraut. Wie ich die Vor-Übersetzung machen liess, sagte ich dem Manne, dass ich es in der Vorrede anerkennen würde aber aus obigen Grunde Wittstein u. Beck gegenüber es keine gemeinschaftliche Arbeit sein könne, zumal da ich 2 Capitel oder Abtheilungen ganz umarbeiten müsse, u. ich selbst die Cladde *Reihe für Reihe* zu revidiren hätte. Dies war vor 12 Jahren. Das *revidirte* Manuscript blieb bis 1878 liegen, da mitlerweise[2] meine unglückliche Verstossung aus dem bot Garten eintrat, eine Massregel über die ich noch jetzt jeden Tag in Zorn entbrenne. Der Angriff *jetzt* gegen mich geschah, weil ich den Gehülfen nicht länger mit einem bedeutenden Gehalt beschäftigen konnte! Die Mittel liessen es nicht länger im Departement zu.

Doch genug davon, aber dies könnten Sie vielleicht gewogentlich dem chemischen Gehülfen des Hr Ivey auseinandersetzen zur Steuer der Wahrheit.

Über die Neuseeländische Ausgabe meiner Select plants habe ich absichtlich nicht mit Mr Ivey gesprochen, ich will nicht aufdringlich erscheinen, und über ihre freundliche Fürsprache möchte ich nicht hingehen. Der Verlust ist ja nicht meiner, sondern für Neu Seeland; es war gewiss grossmüthig, nichts für das Recht einer neuen Auflage eines *solchen* Werkes, welches das Resultat vieljährigen Nachlesens u zahlreicher *Original* Beobachtungen ist, *gratis* anzubieten. Es werden bald eine deutsche, französische u portu-

1 *editorial addition.*

2 mittlerweile?

giesische und vielleicht auch eine italienische Übersetzung erscheinen, da das Buch in der Literatur der Welt einzig dasteht. 81.07.28

Ich nahm die Gelegenheit wahr, Mr Ivey u seinem Begleiter, eine klare Auseinandersetzung meiner Verhältnisse zu geben, zumal da er hier früher im Agricultur Departement wirkte, u dessen Dirigent mir noch täglich entgegensteht u viel zu meiner unverdienten Zurücksetzung vor 8 Jahren beitrug.

Sie ehrend der Ihre
Ferd. von Mueller.

Der Druck der Übersetzung von Wittsteins Werk geschah auf meine Privatkosten; ich habe in dieser Sache ungefähr £200 zugesetzt, da kaum £50 Exemplare verkauft wurden.

Ich habe selbst nebenher 2 Jahre Chemie unter Pfaff u Himley in Kiel studiert

28/7/81

Through your letter of introduction, noble friend, I became acquainted with Mr Ivey, although he only called on me in my seclusion just before his departure, since I did not even know that he was here. He was accompanied by a young chemist,[3] whose name did not become known to me in the short conversation, but for whom I include a copy of *my* translation of Wittstein's chemistry of plants,[4] requesting that it be delivered to him. I say expressly *my* translation, because a German pharmacist[5] here publicly maintained that it was not mine, but his. The matter stands as follows. About a dozen years ago this man brought me a letter of introduction from one of my German scholarly friends, and since his photography could not support him and at that time I was very occupied with the 5th volume of the *Flora australiensis*, I engaged him at my private cost to make a *preliminary* translation of the work. I had received the book from Wittstein, my friend of many years. He and the publisher had reserved the right of translation and entrusted *me* with that for the English language. As I had the

3 Not identified.
4 B78.13.06.
5 Ludwig Rummel; see B78.13.06, p. ix.

81.07.28 preliminary translation made, I said to the man that I would acknowledge it in the preface, but for the above reason towards Wittstein and Beck[6] it could not be a joint work, especially since I had to completely revise the 2nd chapter or section, and I would have to revise the rough draft *line by line* myself. This was 12 years ago. The *revised* manuscript was left until 1878, since meanwhile my unfortunate expulsion from the Botanic Garden occurred, a measure over which I still now burn in anger every day. The attack *now* against me happened because I could no longer employ the assistant with a significant salary![7] The funds no longer permitted it in the department.

But enough of this; however, you could perhaps explain this favourably to the chemistry assistant of Mr Ivey for the sake of the truth.

I have deliberately not spoken to Mr Ivey about the New Zealand edition of my *Select plants,* I do not want to appear obtrusive and I would not like to pass over your friendly recommendation.[8] The loss is certainly not mine, but for New Zealand; it was certainly magnanimous to offer *to charge* nothing for the right of a new edition of *such* a work, which is the result of many years of gleaning and numerous *original* observations. There will soon be published a German,[9] French[10] and Portuguese[11] and perhaps also an Italian[12] translation, because the book is unmatched in the literature of the world.

I was aware of the opportunity to give Mr Ivey and his companion a clear explanation of my circumstances, especially since he worked here previously in the Agriculture Department, and whose director[13] stood in my way even daily and contributed to my undeserved disdainful treatment 8 years ago.

Respectfully your
Ferd. von Mueller

The printing of the translation of Wittstein's work occurred at my private expense; I have sacrificed about £200 in this matter, because scarcely £50 worth of copies were sold.

I myself studied incidentally 2 years chemistry under Pfaff and Himley[14] in Kiel.

6 C. H. Beck was Wittstein's publisher.
7 Ludwig Rummel was dispensed with in 1880.
8 See M to G. Grey, 31 December 1879, and the notes thereto.
9 B83.13.06.
10 B87.14.06.
11 The expected Portuguese edition was not published but there was a posthumous edition, B05.13.01.
12 Not published.
13 A. R. Wallis.
14 i.e. K. Himly.

To Annie Walker

81.08.09

Am 27/1, Mitchell Library, State Library of New South Wales, Sydney.

9/8/81

I would advise you, dear Miss Walker, to send a copy of your illustrated work[1] to Mr Mullen, the Bookseller of Collins Street, Melbourne, and to get his aid for the sale of the work. As he keeps also the most fashionable circulating Library, Carriages of the leading Ladies of Melbourne are at his door daily. Thus your work will come under the notice of the very people, who would like to have it on their drawing room-table.

Mr Robertson's is another great book-business here, of which a branch exists in Sydney. I strongly advise you, to communicate with them; it is [in][2] their money-interests to push work like yours, and Mr Robertson has also a branch-business in London, all the divisions of the firm advertising regularly. I live out of Melbourne,[3] see seldom any one, have no leisure for mercantile pursuits, but am willing to contribute some notes to the text of your work, and if it is of any advantage to you, you may use my name as a referee in the advertisement of your publication.

Regardfully your
Ferd. von Mueller.

1 Walker (1887). Walker published the work herself, having it printed by the Sydney printers Turner & Henderson. The Royal Botanic Gardens Melbourne has a copy inscribed by Walker to M.

2 *editorial addition* – word left out by M.

3 In South Yarra, some 2 km. from the edge of the downtown area.

81.08.24 *From Umberto I, King of Italy*[1]

RB MSS M200, Library, Royal Botanic Gardens Melbourne.

S. M. UMBERTO I
per grazia di Dio e per volontà della Nazione
Re d'Italia
GRAN MAESTRO DELL'ORDINE DELLA CORONA D'ITALIA

Ha firmato il seguente decreto:

Sulla proposta del Nostro Ministro Segretario di Stato per gli affari Esteri:

Abbiamo nominato e nominiamo il Barone Ferdinando Von Mueller, Cavaliere dell'Ordine della Corona d'Italia e di quello dei SS. Maurizio e Lazzaro, ad Uffiziale dell'Ordine della Corona d'Italia con facoltà di fregiarsi delle insegne per tale Equestre grado stabilite.

Il Cancelliere dell'Ordine è incaricato dell'esecuzione del presente Decreto, che sarà registrato alla Cancelleria dell'Ordine medesimo.

Dato a Monza addi 22 Iuglio 1881

Firmato Umberto = Controssegnato: Mancini = Visto: C. Correnti

IL CANCELLIERE DELL'ORDINE DELLA CORONA D'ITALIA

dichiara che in esecuzione delle soprascritte venerate Regie disposizioni il predetto Signor Barone Cavaliere Ferdinando Von Mueller venne inscritto nel Ruolo dei Uffiziali (Esteri) al No. 655 e ne spedisce il presente documento al Decorato.

Roma 24 Agosto 1881

Il Cancelliere dell'Ordine Cesar Correnti
Il Capo del Personale V. Crema

1 MS is a certificate.

H. M. UMBERTO I
by the grace of God and the will of the Nation
King of Italy
GRAND MASTER OF THE ORDER OF THE CROWN OF ITALY

81.08.24

Has signed the following decree:

On the proposal of Our Minister the Secretary of State for Foreign Affairs:

We have elected and appoint Baron Ferdinand Von Mueller, Cavaliere of the Order of the Crown of Italy and of that of Saints Maurice and Lazarus, as Officer of the Order of the Crown of Italy with authority to wear the established insignia for such rank of Knighthood.

The Chancellor of the Order is charged with the execution of the present Decree, which will be registered at the Chancellery of that Order.

Given at Monza on 22 July 1881

Signed Umberto = Countersigned: Mancini = Endorsed: C. Correnti

THE CHANCELLOR OF THE ORDER OF THE CROWN OF ITALY
declares that in execution of the above venerated Royal ordinance the aforesaid Baron Cavaliere Ferdinand Von Mueller has been inscribed in the Roll of Officers (Foreign) at No. 655 and the present document sent to the person honoured.

Rome 24 August 1881

Chancellor of the Order Cesar Correnti
Chief of Staff V. Crema

81.09.02 *From Alfred Selwyn*[1]

Record group 45, vol. 78, p. 347, National Archives of Canada, Ottawa.

2 September [1881][2]

Baron Ferd von Mueller
Melbourne
Australia

My dear Baron von Mueller

Accept my sincere thanks for the copy of your very useful book on extra Tropical plants[3] The copy you have so kindly sent me I would like to keep in my own library both on account of its intrinsic value and as a remembrance of an old friend. But I would also wish to have a copy for our dept. […][4] Either for purchase or in Exchange for any of the publications of the Canadian Geol Survey which I hope you receive regularly. I have been very busy all summer removing our Museum & headquarters from Montreal, where they have been for nearly forty years, to this the political Capital of the Dominion. In some respects the change will I hope prove beneficial but it involves a great amount of labour and anxiety as well as disruption to my work. Your old aid Mr [Hoffmann] now presides in our laboratory & is doing good work tho' much of his time is occupied in giving information [to the public].

[…][5] a […] of labour which makes but [little] […] and interferes a good deal with special work. I never hear now from my old Colleagues Newbery & Ulrich,[6] are they still connected with the

1 Throughout Selwyn's letter-copy book, the ink has faded badly, making some words uncertain or illegible; illegible words are marked by […].

2 *editorial addition.* Date based on when the Canadian Geological Survey moved its headquarters.

3 Presumably B81.13.10.

4 *About eight words illegible.*

5 *Several words illegible.*

6 James Newbery, Georg Ulrich.

Melbourne Museum? I send our reports regularly to the Royal Society of Victoria but receive nothing in exchange. I would be glad to have a set of the Society Transactions for our library & any other Victorian scientific publications. With much kind regards believe me always sincerely 81.09.02

Yours

Alfred R. C. Selwyn

Remember me kindly to Prof McCoy & other [friends]

To William Thiselton-Dyer 81.09.10

RBG Kew, London, Kew correspondence, Australia, Mueller, 1871-81, ff. 318, 320-4.[1]

10/9/81.

This day, dear Mr Dyer, I got your letter from 20. July, regarding the genus Vahea.[2] Altho' I do not adhere to all the rules of Alph. de Candolle's Code,[3] yet I do think, that it is a just law to admit a correct plate without description as authority and concede to it accordingly also priority. Assuredly Lamarck went to some trouble to establish his Vahea,[4] and as Poiret, Bojer, Endlicher and with great care also Alph. de Candolle[5] took up the genus, it would not

1 There is no f. 319, but the letter is complete – the folios have been misnumbered.

2 Letter not found. In B80.13.07 and B81.13.10 M had introduced the names *Vahea florida* and *V. owariensis* in preference to Bentham's *Landolphia florida* and *L. owariensis*, arguing that 'the genus Vahea was fully established by Lamarck so early as 1791'.

3 So far as is known, M did not have a copy of the original version of the Code [i.e. A. L. P. P. de Candolle (1867)], but he did have copies of the German and English translations, both of which were published in 1868.

4 Lamarck & Poiret (1791-1823), vol. 2, part 5:2, pl. 292.

5 Poiret (1810-17), vol. 5, p. 409; Bojer (1837), p. 207; Endlicher (1836-40), pp. 586, 1397; A. L. P. P. de Candolle (1844), p. 327.

81.09.10 be equitable to deprive Lamarck of his right; indeed Palisot went to no more trouble in this respect as an investigator than his illustrious countryman, and ought by rights have been able to recognize the identity of his Landolphia-genus with that of Vahea.[6] My own opinion is, that if we once set aside a priority, we might then at once abandon all adherence to it. It is no doubt a point of soreness, to all those workers who see after much toil and study of theirs afterwards by others from better material and easier opportunities their genera abandoned, and as Mr Bentham through facilities which cannot & will not reoccur to *any* mortal, takes possession of *thousands* of species of which other observers thought they might be proud, he surely may allow in this instance the law to take its regular course. The restitution of Vahea is effected already in 3 books, namely in my enlarged English Edition of Wittstein's organic constituents of plants p. 258 & 268 (anno 1878),[7] in the Calcutta-edition of my select extratropical plants p. 344 & 386 (anno 1880)[8] and in the Sydney edition of the same work 349 & 393 (anno 1881);[9] nor could I see my way clear to return to Landolphia in the German edition now under the press. A. de Candolle in his Codex goes even so far, as to allow a printed label *preceeding authority*, and so it is with Wallich's list[10] &c. Vahea is taken up by [1]7 authors at least!, so far as I can see in my imperfect library, among them Blume[11] and Miquel![12] therefore: "suum cuique".[13] – As I happened to have entered on the theme of literary rights, I may single out just one glaring case among many of

6 Palisot de Beauvois (1806, p. 54) erected *Landolphia* in describing *L. owariensis*.

7 B78.13.06.

8 B80.13.07.

9 B81.13.10. In both these editions, and in later ones, M included the sentence: 'The genus Vahea was fully established by Lamarck as early as 1791.' Muir (1979) gives B80.03.07 as the creation by M of the new combinations *Vahea florida* and *V. owariensis*; in both cases M treated the names within *Landolphia* as synonyms.

10 Wallich (1828).

11 Probably Blume (1849-51), vol. 1, pp. 150-1.

12 Miquel (1855-[59]), vol. 2, p. 394.

13 'to each his own'. See B82.09.01, in which M discusses the priority of *Vahea* in some detail, quoting from A. de Candolle to M, 27 July 1882 [Collected Correspondence].

great injustice. The two Forsters established fairly by words & engraving the Composite genus Laxmannia.[14] Brown incorrectly joins this St. Helena tree to the Bidens-herbs, widely different as they generically are;[15] but when finding out his remarkable mistake he does not give back to the Forsters what he took from them, but renames their Laxmannia anew (Petrobium),[16] the excuse being, that the several published species of his genus Laxmannia ought not to receive a new generic name, there being several against one. But where is the justice. At R BR. time however the priority was not held so *sacred*, as in later years; hence though modern authors have maintained Petrobium, it is sure to fall hereafter to the ground forever, there being not even the excuse of inadvisability to change a name of a well known plant, as this endemic tree of a small island is nearly extinct & (I believe) never was anywhere under cultivation and existed but in a few herbaria. Contrast this with the readiness of our having accustomed ourselves to Hevea against the genus Siphonia, a name with which we were familiar not only in phytographic but also medical & technological works for 80 years.[17] So if in your important treatise[18] you maintain Landolphia, then Siphonia might also stand again, altho certainly Aublet gave some letterpress with his plate. See with what scrupulous rigor in *some* instances (but not in others) the first right in nomenclature is maintained! Centipeda,[19] a strange name of no meaning out of zoology, took readily the place of Myriogyne,[20] by which name we all knew for many years a widely distributed plant. Still I approve of that, as in conformity with the laws now unalterable thus far. Even Floscopa became adopted against rules of logic as well grammar, and I appealed in vain for Floriscopa, wishing in every way to

14 Forster & Forster (1776), p. 93, pl. XLVII.
15 R. Brown (1810), pp. 285-6.
16 R. Brown (1817), p. 113.
17 *Hevea* was erected by Aublet (1775), vol. 3, p. 871, pl. 335; *Siphonia* by Richard in Schreber (1789-91), vol. 1, p. 656.
18 Not identified.
19 See Loureiro (1790), p. 492. M owned the 1793 edition of Loureiro's work.
20 See Lessing (1831), p. 219.

81.09.10 leave Loureiro (except the mutilation of the word) his own.[21] So it is with Calostrophus, of which Labillardière explained himself correctly the Etymology.[22] Now, not to adopt such reasonable changes, against which Mr Jackson fights so violantly, is mere *pedantery*. The same I would say of natural orders. Whether we say for instance Jasmineae or Jasminaceae, the order belongs to A. L. de Jussieu, and even B. de Jussieus Jasmina should in such case not be overlooked in fairness, though he put one wrong genus into it. In the "Papuan plants"[23] I have at p. 38-40 given those Linnean botanic species-names, which in simple binary nomenclature were used before him; & had L. imagined, that subsequently his L. was to be put to them, he would certainly have protested. – Passingly may I say, that it was not Willdenow, who first put authorities to species-names (as Ascherson said in one of his publications) but G. Forster in his florulae, to do justice to Solander.[24] I write in the new number of the fragmenta[25] Commerçonia as the mere sound of the ç in french gives in this as in hundreds of other cases no right to mutilate a name and make a dedication unmeaning. CommerSonia conveys the idea that the dedication was intended for an English man or perhaps Scandinavian, certainly not for a real Frenchman. Forster at all events dedicates it to Commerçon, who however may have been of English descent.[26] Loeffling wrote rightly Buettneria, L. wrongly afterwards Buttneria.[27] Bougainvillaea is by B. & H. rightly admitted (III, 7)[28] against the original Buginvillaea of Comm-

21 See Loureiro (1790), p. 192; B82.13.13, p. 122. *Index Kewensis* treats M's 'correction' to *Floriscopa* as a typographical error. M persisted with the correction in B89.13.12, p. 206.

22 M's *Calostrophus* (B73.08.01, p. 86) is considered an orthgraphic variant of Calorophus (Labillardière [1804-6], vol. 2. p. 78, t. 228).

23 B76.06.01.

24 J. G. A. Forster (1786), pp. 5-8.

25 B81.08.03, p. 114.

26 Forster & Forster (1776), p. 43, t. 22. Modern French authorities such as the *Dictionnaire de biographie française* give the name as Commerson.

27 Loefling (1758), p. 313, had Byttneria; Schreber's edition of Linnaeus (Schreber [1789-91], vol. 1, p. 145) used *Buttneria*.

28 Bentham & Hooker (1862-83), vol. 3, part 1, p. 7.

erçon,[29] on Moquin's proper remonstrance.[30] Bélis must stand, 81.09.10
whatever Salisbury's other shortcomings may have been; [31] for it as little liable to be confounded with Bel-lis (against Bé-lis) as Brassia (of the rejector of Belis[32]) with Brassica or Bassia. If the adding in *all* instances of the year of publication to a systematic name could be enforced in actually descriptive works, as suggested by me some years ago in the Botanische Zeitung,[33] the priority would become clearly apparent, no matter if the name of a species had to be changed generically. In a systema naturae (or as now somewhat vagely is said in Biology and not so well) we can neither have generic names quite the same in Zoology & Botany. Therefore I prefer Sturmia to Liparis;[34] but where nomina propria[35] give the genus as in Kraussia I would suggest, that the Zoologists & Botanists should agree to some acknowledged slight change in the word, the one writting Kraussea the others Kraussia.[36] I will bring this long epistle on some of *my* views on nomenclature to a close by referring to Stylidium. Why should Loureiro not keep his genus? Candollea would then be reestablished (gone in Dilleniaceae).[37] If once agreed on, the Stylidiums would become everywhere wonder-

29 See Jussieu (1789), p. 91.

30 i.e. Moquin-Tandon. His 'remonstrance' has not been identified.

31 Salisbury (1807), p. 345. See also Bentham & Hooker (1862-83), vol. 3, part 1, pp. 435-6.

32 i.e. R. Brown. Bentham & Hooker (1862-83), vol. 3, part 1, pp. 435-6, rejected *Belis* Salisb. in favour of *Cunninghamia* R. Br., noting that Salisbury's name was rejected by Brown and Richards as sounding too similar to *Bellis*, Linn.

33 B79.08.01.

34 Botanically *Liparis* (= *Sturmia*) is an orchid; zoologically it has been used for both a genus of moth and of fish.

35 'proper names'.

36 Botanically *Kraussia* was used twice, in Compositae and Rubiaceae; zoologically it is an arthropod genus.

37 Louriero (1790), p. 220, erected *Stylidium* as the name of a genus in the Cornaceae; Swartz's genus *Stylidium* in the Stylidiaceae was applied by Willdenow (1805), vol. 1, part 1, p. 146. It was this latter genus that had become associated with *Stylidium* by usage. M was arguing for a return to using *Stylidium* to refer to the genus in Cornaceae, thus allowing Labillardière's 1805 name *Candollea* to be used for the genus in Stylideaceae (which would become Candolleaceae). *Candollea* had also been used (by Labillardière!) for the name of a genus in the family Dilleniaceae.

81.09.10 fully quick Candolleas, without my wishing for a moment to have my name attached to the species, which would be arrogance.[38] As I am on the subject of genera, let me once more express a hope, that Sir Joseph & Mr Bentham will be able to see in conformity with the Cotyledonar plants[39] also the Acotyledoneae all done. We then would have *all plant genera* and all names (synonyms &c) together! Besides we want much the genera of cryptogams worked up. You have for Mosses in England Mitten, for fungs Berkeley & Cooke, for Lichens Babington; – and for algs surely someone would fuse Harveys, Agardhs & Kuetzing labours into a connected form. It would be a great pity, if B & H left their work incomplete, while Endlicher brought his also in this respect up to the requirements of the time.[40]

You have strangely misunderstood what I intended to convey in the Calcutta- & Sydney-Edition of the "select plants" regarding Vahea Owariensis. I gave clearly West Africa as its native country *only*, being aware since 40 years that Palisot was in no other part of Africa. But I did not express myself accurately in the following sentense, which should run: this climber *and* several other Vaheas *yield* the West African & Madagascar Caoutchouc" I will alter it thus in the new edition, so that no longer any ambiguity can exist about this phrase.[41] I *do* trust, that in your important work you will not make me guilty of having transferred Vahea Owariensis to Madagascar! But you may have found out other &

38 M's name has become attached to the species in synonymy: he transferred the Swartzian *Stylidium* species to *Candollea* in B82.13.16, where he also renamed *Marlea vitiensis*, the only species of Cornaceae recorded in Australia, as *Stylidium vitiense*. M had earlier commented on name priorities in B73.04.02, p. 41.

39 Bentham & Hooker (1862-83).

40 Endlicher (1836-40).

41 The original sentence in B80.13.07, p. 344 and B81.13.10, p. 349 reads: 'This climber, with several other Vaheas, yields the West African and Madagascan caoutchouc'. In B84.13.22, pp. 387-8, the corresponding sentence reads: 'This climber, with several other Vaheas, yields the West African, and others the Madagascar caoutchouc.' In later editions the sentence was changed again: 'This climber, with several other Vaheas, yields the West African caoutchouc; others furnish the Madagascar-sort, particularly V. gummifera (Lamarck), now cultivated also in India' (B85.13.26, p. 393; B88.13.02, p. 432; B91.13.10, p. 503).

real errors, for the frankly pointing out of which I shall be obliged, just as I feel it almost obligatory to explain to my correspondents where I find them wrong. Besides from your rich store of knowledge you may be able to give addenda to the "select plants", some appearing already from my own pen in the German Edition. Do you care to have it, when it comes out?[42] The book in any of its editions ought to facilitate reference to extratropical plants of utility in a very handy manner and I think you will find it more reliable and *within its scope* more complete than any other literary production of any language on this subject. Of course, it could for each plant be much extended, if thereby the price was not so very much increased and its spread thus also impaired. But at Kew, where Mexican plants are elaborated now, you might ascertain for me perhaps, whether it is Quercus magnoliifolia, (e.g. urticifolia not urticaefolia, the real root of the word being from the Ablative pluralis)[43] which yields Silk-Cocoons in Mexico, also what Oaks yield there the best timber & tan-bark, what Pines the best wood, what grasses & shrubs the best pastorage & fodder. Hitherto I have in vain for years written for information on these important subjects to several parts of Mexico.[44] The Uplands of tropical Africa must also be rich in trees, shrubs, herbs & grasses, which should find a place in my "select plants" for the benefit of Victoria and other countries. What is Gaertner's genus Athecia?[45] It is omitted in B & H genera,[46] but will perhaps be referred to as unrecogniz- 81.09.10

42 B83.13.06. A card bearing M's coat-of-arms and printed: 'Baron Ferd. von Mueller K.C.M.G. M & Ph.D. F.R.S.', is pasted on to the back of the title page in the copy at Kew library. There is no inscription.

43 M is presumably drawing attention to his use of the form *magnoliifolia*, rather than *magnoliaefolia* which was used by Nees, the original author of the name.

44 Letters not found; but see M to L. Haynald, 22 November 1882.

45 Gaertner (1787-91), vol. 1, p. 141.

46 Bentham & Hooker (1862-83). MS annotation: 'Mr Bentham gives up Athecia as insoluble. Its structure restricts it to narrow range [of] Orders: as *Goodeniaceae*, *Santalaceae*, – a few Corneae? – possibly Rubiaceae –? The *Trientalis* I strongly suspect would be a *Crassula* of the "Tuberisae" Section *[nr.] C. Saxifraga & C. Septas* DO [Daniel Oliver].' See also J. Hooker to M, 1 May 1882 [Collected Correspondence], where Athecia is identified as *Royena glabra* from the original specimen sent on loan from Tübingen.

81.09.10 able at the end of that great work; possibly it is rubiaceous. What is the second Trientalis, mentioned as S. African in the first edition of L. sp. pl.?[47] This may perhaps yet be found out in Leyden.

Regardfully your
Ferd. von Mueller

Why should there be an exception made in quoting the first edition of L. sp. pl., the very book which above all others deserved right of priority!!

R. Brown however, great as he was, showed in regards of rights towards Labillardiere, Loureiro, Andrews, Cavanilles, Salisbury &c himself occasionally very little minded, as pungently pointed out by Lindley.[48]

I for one should not object to make Goodenia henceforth Goodenoughia; we have already Goodenoviaceae & Dioscorideae

Good generic authorities before L. such as Tourneforts[49] should in justice not be confined to oblivion.

What a pity that the Apetaleae could not have been transferred to the higher orders in the genera as I recommended strongly at the commencement of the work. Surely it will be done in [a future ed.][50]

I think, that the note on the sweet fruit of Vahea Owariensis I got from Welwitsch.[51] I will see, but it is not easy to find a book in the small cottage, where even my bedroom is overcrowded with books, documents & plants.

47 Linnaeus (1753), p. 344.

48 Lindley reference not identified.

49 M is probably referring to Tournefort (1700), a copy of which is in the library, Royal Botanic Gardens Melbourne (though with nothing to prove beyond doubt that it belonged to M).

50 For M's recommendation see M to G. Bentham, 24 January 1862 [*Selected correspondence*, vol. 2, p. 127]. There was no second edition of Bentham & Hooker (1862-83).

51 This information probably came from Collins (1872), p. 27, quoting Welwitsch '(in MS to Collins)': 'The fruit is about the shape of a middle-sized orange, containing ... a sweet, rather acidulous pulp'. M's description reads: '*V. Owariensis* produces edible fruits as large as middle-sized oranges with sweet and slightly acid pulp.' (B81.13.10, p. 349, and later editions).

To Joseph Hooker 81.10.08

RBG Kew, Kew correspondence, Australia, Mueller, 1871-81, ff. 331-2.

8/10/81

The circumstance, dear Sir Joseph, that neither the Gardeners Chronicle nor any communication from Kew alludes to so sad an event as the loss of Bentham, awakens my hope, that the great systematist is spared us yet, and that Caruel's information, derived from Planchon, as very implicitly and with very emphatic expressions of sorrow communicated by letter to me,[1] is based on some mistake. Let me hope, that he will long live to be your collaborator and to enlarge our favorite science![2] –

Poor Dr Cooke writes to me by last post, concerning his severe illness.[3] I have advised him[4] to take a *seavoyage to a healthy part of Central-America*, so as to give his mind perfect rest on the ocean, to enjoy the bracing fresh sea-air with its ozonous refreshing effect and to delight in rambles after fungs in the jungles of the tropics. It is so easy now, to reach any of the Central American countries, where a *special mycologist never yet explored*, that such a tour for some month would not only build up again his mental working power, but give him an opportunity of doing an immensity of research in a branch of science there, but scantily as yet carried on merely mechanically as a byework by collectors. What work could an experinced mycologist like Dr Cooke do during a few month in the virgin forests of America! –

Remember the vast material, gathered by Dr Ernst merely in Caracas. You are a Physician yourself, and will agree, that only by

1 Letter not found, but see T. Caruel to M, 21 November 1881 [Collected Correspondence].

2 See also M to J. Hooker, 19 September 1881 [Collected Correspondence]. M was correct, Bentham had not died.

3 Letter not found. Cooke appears to have suffered a series of slight strokes leading to a temporary paralysis of his arm; see English (1987), p. 200.

4 Letter not found.

81.10.08 a plan like this so valuable a life as that of your Mycologic Officer can be much prolonged, relapses being after any early resumption of arduous study-work almost inavoidable. Of Dr Cooke's private affairs I know nothing, as it was only a year or two ago, when Berkeley at his venerable age referred me to Cooke. Like most real son's of science (unless they had property by inheritage) Cooke is probably poor & may have a family to support. If financial obstacles stand in the way I will pay £10 or £20 towards his expenses, for which help he might send some of his spare specimens obtained in Central America. Others of his scientific friends would doubtless do the same, and so a few hundred £ might be got together for him.

Central America could be cheaper & quicker reached, than any other great unexplored mycologic area, and he could operate safer there than anywhere (with equal prospects) in the East.

If the enormously rich brother in law of Dr Harvey[5] had given that poor friend of ours a modest independence, so that as a *private man* he could have lived at the Cape of Good Hope, or in any other warm part of the globe, quietly engaged in his darling studies, I feel sure his life would have been spared for many years.

The same advise, which I give concerning Cooke, I proffert to my friend Professor Keferstein, the young promising Zoologist of Goettingen, a series of years ago, but it came *too late*.

Regardfully your
Ferd. von Mueller.

I suppose you met our delegate, my friend Dr Ruddall at your Medical Congress[6]

5 Presumably Thomas Harvey Todhunter.

6 James Rudall was a founding member (1879) of the Victorian Branch of the British Medical Association and represented it at the Seventh International Medical Congress, London, August 1881, during a tour of Europe and America. Rudall did not meet Hooker at the Congress; see M to J. Hooker, 31 December 1881 [Collected Correspondence].

To Ralph Tate 81.10.27

Barr Smith Library, University of Adelaide.

27/10/81

The umbelliferous annual, sent by you, dear Prof Tate, is Bupleurum tenuissimum L., a plant of wide distribution in Europe (inclusive Britain & Ireland), in N. Africa & W. Asia. It is evidently immigrated.

As regards the rights of Quinetia being considered a native, I beg of you to reconsider the case fully.[1] If you found that little annual near any place, where the K.G.Sound[2] Steamers are disembarking their goods, you may be sure that it is a hospitant. If away from any port or landing place, it will have to be considered, by what likely agency[3] it became disseminated there. The following reasons are against its being a native.

1, Quinetia is gregarious in S. W. Australia, not near you.

2, If indigenous, it ought to have spread extensively like other annual composites.

3, In five years close searches near Adelaide I never saw it.

4, It is common at KGS.,[4] and thus any few seeds of it may with embollage have become spread near Adelaide.

5, As early as 1847 I found in the vallies under Mt Lofty already Browns unioloides[5] quite spontaneous, evidently disseminated without other foreign grasses near it.

6, We know now several hundred immigrated plants in Australia, and every year adds to them, especially by the increasing traffic.

7, When botanizing in the little-inhabited places between Swan River & "Geographe"-Bay[6] 4 years ago, I found a Portuguese

1 See also M to R. Tate, 12 October and 7 November 1881 [both Collected Correspondence].

2 King George Sound, WA.

3 As regards ... agency *is marked off by short lines at the beginning and end of the section.*

4 King George Sound.

5 Presumably *Cyperus unioloides*, described in R. Brown (1810), p. 216.

6 WA.

81.10.27 Galium there, that any one would have taken for a native. Indeed as the genus Galium *is* indigenous, I believed to have discovered a new species, until a critical examination of the plant undeceived me.

In the case of Quinetia, it is the wonder, not that it is now found, but that it did not come across long ago from KGS. Many plants will yet come across from thence with the steamers, and Quinetia is one which most easily would spread.

The value of your index[7] is enhanced by the rigorous exclusion of immigrated plants; if once that exclusiveness is broken, the pure original aspect of the S. Austr. flora becomes disturbed. Hence I induced also Dr Woolls to banish his immigrated plants to a place at the end of his list.[8] One of the values of my early searches will be to show, what is native, what not.

I look forward with eagerness to sendings from your Kang. Isl.[9] friends, and want particularly for my monography *all* Eucalypts from thence, especially as I am not yet quite clear about E. cneorifolia, under which name Bentham mixed two very distinct species. I *know*, what both are, but do not know what intermediate stages the one of Kangaroo Island may show to E. oleosa. *Fruit*-specimens (branchlets) can – at all events – be got at any season of any of our myrtaceous plants.

I am sorry to hear of the death of my former assistant.[10] He left me in good health, on his own accord, to establish his seed-business there. I raised him over the head of all other employees, more than doubled in the course of time his salary, gave him an opportunity by his being with me in my office to add daily to his knowledge, and have shown him numerous favors, which he could never reciprocate.

Regardfully your
Ferd. von Mueller.

7 Probably Tate (1880).

8 Woolls (1880a). There is a marginal query by M on this page: 'Where is Golden Grove?'. See also M to R. Tate, 7 November 1881 [Collected Correspondence]. The *Quinetia* in question had been found at Golden Grove, SA, which M thought was near the coast. When he learned that it was in fact some distance inland, he accepted Tate's view that the plant was indigenous.

9 Kangaroo Island, SA.

10 Ernst Heyne.

Please thank the council for the kindness of according space to me in the new volume of your R.S. I find however at this moment myself hard pressed to finish off some arrear work in the Department here; and it would be some weeks before I could send you any manuscripts; kindly tell me what would be the latest "termin" to furnish my contribution. If it cannot be a lengthy one, I will send at least some brief notes in time.[11] 81.10.27

Regardfully your
Ferd. von Mueller.

To Annie Krefft 81.11.10

A263, Mitchell Library, State Library of New South Wales, Sydney.

Private.

10/11/81.

It is with much sympathy, dear Madame, that I hear of your distress and difficulties. It is very sorrowful, to loose such a genial consort as Mr Krefft was, but still sadder that you with your young family should be unprovided.[1] Unfortunately in Melbourne are no institutions, – so far as I am aware – to relieve sufferings like yours, and whether the English Society for literary men in distress has also a widow-fund, I am not aware. I thought Mr Krefft had insured his life, and that a sum of £1000 was due to him on resignation. As death ended his career, that sum would morally be due to

11 The deadline set by Tate was 18 November; see M to R. Tate, 13 November 1881 [Collected Correspondence]. M did send, in time, as promised, a short paper in which he described two new species; see B82.01.02.

1 Gerard Krefft died on 19 February 1881. Shortly before this his estate was sequestered with liabilities of £1131 (*ADB*).

81.11.10 you, though perhaps not legally, but I feel quite sure, that the N.S.W. Government would recognize all fair claims of yours, and the Gentleman, after whom your lamented husband named the wonderful fish Ceratodus,[2] which alone raises to him a permanent monument in Zoology, would doubtless aid you with his advise and influence to obtain for you some recognition & help from the authorities in Sydney.

I am myself quite without property, beyond my library and a very small cottage, in which I do my office-work; and since I left the bot Garden I am so out of all Society, that I do not know anyone, to whom I could appeal in a case of yours. I sunk all, I had, in my travels, my researches and my books, and am in fluctuating health after 34 years incessant work in Australia. The friends of the family of the late Mr Blandowski applied also to me from Breslau on their behalf, and all I could do, was to advise them to ask through the German Consulate your Government, to purchase his drawings and manuscripts, in the elaboration of which more than 25 years ago your husband gave gr[eat][3] help. If a subscription list was opened on your behalf, I would willingly contribute a little out of my slender means, but do not know any one here, to whom I could submit it for reasons explained. Dr Bennett[4] was always a generous supporter of Mr Krefft, and I feel sure he would be willing to give you his counsels. The photogram of Mr Krefft[5] I shall treasure much as a sad souvenir.

With sympathy, dear Madam,
yours
Ferd. von Mueller

2 Krefft named *Ceratodus forsteri* after its discoverer, the NSW politician William Forster, MLA (*SMH*, 18 January 1870).

3 *editorial addition* – Text obscured by binding tape.

4 George Bennett.

5 Photogram not found.

To Joseph Hooker 81.11.22

RBG Kew, Kew correspondence, Australia, Mueller, 1871-81, ff. 341-2.[1]

[22 November 1881][2]

I have a chronologic list of the genera of all Australian plants under the press[3] but shall not likely be able to send it before february It includes also the cryptogames, so the list will comprise nearly 2000 Genera. We all look forward with deep interest to your & Benthams finishing volume of the genera;[4] had either of you not done anything else, it would raise a grand Monument for you at all times, as your genera *must* be for ever on the table of all Botanists! –

I have not yet got to the sorting of my duplicates for Kew &c, but the Exhibition[5] has driven everything back, so much so that I have not even brought out a single part of the Eucalyptography in 1881.

Your position at the Brit Associat in the geographic Section was quite a proud one, & your essay or adress excellent.[6]

Last evening I attended as acting President (or Senior Vice Pres) a splendid Concert of the Liedertafel, at which our Conductors magnificent Cantata to Longfellows Orion was for the first time produced.[7] On my request a special *full* copy of the Music will be made out for presentation to Mr Longfellow, of whose poetry I am a great admirer, and as Dr Asa Gray is a friend of his, I shall

1 The folios may belong to separate letters. Folio 342 begins at 'I have instructed Mr Webb to send you a gin-case ...'. Both folios are in M's hand.

2 *editorial addition.* The Liedertafel concert to which M refers was held on 21 November 1881.

3 B82.13.08.

4 Bentham & Hooker (1862-83).

5 International Exhibition, Melbourne, 1880-1.

6 J. Hooker (1882).

7 Julius Siede's cantata to Longfellow's 'Occultation of Orion' was never published. However, a full score and also a vocal score are preserved in the Liedertafel Collection, Grainger Museum, University of Melbourne.

81.11.22 forward this music to Kew,[8] so that he may present it to his illustrious countryman, this course taking away any air of intrusiveness.

I have been President for several years of the Liedertafel

Dr Rudall is president now [–] we gave him credentials to the British Medical Congress from the Med. Association here, of which I am an ordinary member[9]

I have instructed Mr Webb to send you a gin-case full of Andersonias packed horizontally only with their natural moisture between layers of Cladonia retipora (common at KGS),[10] to nail down the lid[11] and put the box in a cool place into the hold of the vessel as ordinary good. I believe, the plants would endure the 6 week imprisonment quite well, but it would then require great care to pot them & establish them in a cold frame. –

If any Ship, which brings *frozen meat* from Australia took a few plants,[12] they would arrive probably well without trouble. I will see, what can be done in that way.

I have written to all my correspondents, who have Cycadeae within fair reach, on behalf of Mr Dyer,[13] for his important work, & you may rest assured, that whatever is attainable for me shall pass on to him, but it is a matter of time & expense.

8 In retirement, Asa Gray spent a year in Europe, 1880-1. See also M to A. Gray, 27 March 1882 [Collected Correspondence].

9 Rudall represented the Victorian Branch of the British Medical Association during a tour of America and Europe in 1881.

10 King George Sound, WA.

11 only ... lid *is marked with a line in the margin.*

12 If any ... plants *is marked with a line in the margin.*

13 W. Thiselton-Dyer.

From Benedetto Scortechini[1]

81.12.08

RB MSS M1, Library, Royal Botanic Gardens Melbourne.

Roma[2] 8th December 1881

Dear Baron

I have been tossed over 300 miles west of my usual residence. Till the end of the expiring year I shall knock about this place. The trips which I planned for Wilson's Peak Fraser's Island, Bundberg[3] have been blown to pieces by adverse winds. My Wilson's Peak trip on horseback was superseded by a trip to Warwick by rail. For the last few months I have not been master of my movements, however, after Chrismas I hope to have more time for myself.

Last Saturday week I was home, where I got your brief letter.[4] Perhaps further correspondence from you will wait there my return. In answer to your last in hand I beg to say that the large oval fruit is that of Endiandra Muellerii your fruit; in the note which I have appended to said plant[5] I mention my sending you this fruit. From the Flora I see that it is undescribed.[6] It is worth describing it in your Fragmenta. It seems to be very large for its congeners. – The best specimen I could gather of Clematis No 175 has been sent to you. It was growing at the top of a stony mound isolated in the midst of a plain. It was scarse there and out of flower with no fruit. I took particular notice of the leaves which I found, every one of them, simple. As the plant withers, there is no change[7] of getting better specimens till next spring. I will look out. – I plucked all the flowers of Cupania No 126, they were few, so when I went again to look for fruit, there was none. – of Myrtus No 169 and Leptospermum No 154 will gather fruit, I hope, next month. I know only one little nest of Loranthus Bidwillii growing on a pine at the

1 MS annotation by M: 'Endiandra? with large fruit'.
2 Qld.
3 Bundaberg, Qld?
4 Letter not found.
5 Note not found.
6 In Bentham (1870), p. 302, Bentham notes of *E. Muelleri*: 'Fruit not seen'.
7 chance?

81.12.08 mouth of Nerang Creek. If the flowers are out at my next visit to that district, the berries will be there I hope.

Since I wrote to you last I have seen the Revd. Tenison-Woods and mentioned to him your kind offer of proposing him a fellow of the Royal Society.[8] He would be glad of the honor. He told me he was going to write to you on the subject. I expect he has done it already.[9] You may kindly take steps for his proposal. His scientific labours deserve some honorable recognition. Yesterday he started for Bathurst N. S. Wales. Between scientific and ecclesiastic work he has been labouring hard in the past few months.

I am terribly disappointed with this place in a botanical point of view. Few desert plants make their appearance here. There[10] only one Eucalyptus, which I have not identified as yet, the *Acacia pendula*, and another brigalow; sometimes I meet Acacia Farnesiana very stunted in growth. Two heremophilas,[11] Mitchelli and longifolia, gejera parviflora,[12] and Apophyllum anomalum constitute mostly the whole scrub and forest vegetation. Boerhavia diffusa, Tribulus terrestris, and several Chenopodiaceae monopolize the small growth. To-morrow I will be out exploring, with little success, I expect. In a hundred miles of this country there are not many more varieties than in one single acre.

I do not know whether you may wish to have some particular observation made on some plants, which may be growing about here; if so let me know it at once. You may address your letter to Roma, from where I shall start towards the end of the month, so an early letter from you will find me here, any other will get me at the Logan.[13]

With my best regards I remain
Your humble Servant
B Scortechini

8 J. Tenison Woods was an unsuccessful candidate for election to the Royal Society of London in 1883, with M as his principal sponsor, and again in 1884 and then each year from 1886 to 1888.

9 Letter not found.

10 is *omitted*?

11 Eremophilas?

12 *Geijera parviflora*?

13 Logan River, Qld.

To Maurice Holtze 82.01.07

GRG 19/391, Public Library Museum and Art Gallery, State Records of South Australia, Adelaide.

7/1/82.

Vor längerer Zeit, lieber Herr Holtze, sandten Sie mir mit andern Pflanzen unter N. 73 ein sehr seltnes Gewächs, welches seit Capt Flinders' 1802 im Golf von Carpentaria war, nicht wieder gefunden wurde, – und so nur aus R Brown's Sammlung bekannt war. Ich hatte die Pflanze nie gesehen, bis *Sie* mir selbige sandten. Es ist Pleurocarpaea denticulata. Den neuen Standort, von Ihnen dargelegt, werde ich jetzt veröffentlichen.

Wären *reife Same* und mehr getrocknete Exemplare zu bekommen? Unter welchen Verhältnissen kommt das Gewächs da vor. Ist es krautig oder Strauch-artig? Was ist die Höhe.

Mit bestem Glückwunsch zum neuen Jahre der Ihre

Ferd. von Mueller.

Kann ich Ihnen etwas senden?

Gewiss fahren Sie fort zu sammeln, es muss dort noch manches zu entdecken sein.

7/1/82.

Quite a while ago, dear Mr Holtze, you sent me, as No. 73, a very rare plant together with other specimens. It hasn't been found again since Capt Flinders was in the Gulf of Carpentaria in 1802 and thus it was only known from Mr Brown's collection. I had never seen the plant until *you* sent it to me. Its name is *Pleurocarpaea denticulata*. I am going to publish the new location, as disclosed by you.[1] Would it be possible to obtain *ripe seeds* and more dried

1 See also M. Holtze to M, 21 May 1882 [Collected Correspondence]. M appears not to have proceeded with the proposed publication. Bentham had erected *Pleurocarpaea* in Bentham (1866), pp. 460-1, on the basis of a single species, *P. denticulata*, collected by R. Brown on islands in the Gulf of Carpentaria, but never formally described by him.

82.01.07 specimens? Under which circumstances does the plant grow there? Is it herb-like or shrub-like? What is its height?

With best wishes for the new year yours

Ferd. von Mueller

Can I send you something?

Undoubtedly you will continue collecting, surely there still must be something to be discovered.

82.02.10 *To Paul Ascherson*

Staatsbibliothek zu Berlin, Preussischer Kulturbesitz, Berlin.

10/2/82

Hierbei sende ich Ihnen, hochgeehrter Herr Professor, einen Abdruck von Mr Tepper's Abhandlung über die Keimung und Entwicklung von Cymodocea (oder wohl nun besser Amphilobis) antarctica)[1]. Sie werden vermuthlich auch von Herrn Tepper selbst diese Druckschrift erhalten haben, und es mag Sie wundern, dass Ihrer wissenschaftlichen Erklärungen der Organe, welche Mr Tepper von dieser Pflanze beschreibt, keiner Erwähnung geschehen. Dies erfordert eine Auseinander Setzung der Gründe.

Wie Ihr Brief mit den Bemerkungen über die Cymodocea an mich gelangte, war ich als einer der Commissäre u Preisrichter (für Chemie, Forstproducte und alle andern Pflanzenstoffe) so sehr für die Weltaustellung in Anspruch genommen, dass alles nicht officiell Dringende in meinem Departement liegen bleiben musste; ja ich konnte nicht einmal Zeit gewinnen den Eucalyptus-Atlas fortzusetzen. So häufte sich eine enorme Correspondenz in meinem Departement auf, und auch Ihr Brief wurde damals zur Seite gelegt, da ich erwartete, dass Herr Tepper mir seine Abhandlung

1 Presumably Amphibolis? *The superfluous parenthesis is M's error.*

und Zeichnungen vorlegen würde, ehe solche in den Druck gelangten, zumal da er durch *mich* überhaupt erst in die Pflanzenkunde gezogen wurde u. zwar erst seit 2 Jahren, denn seine früheren wissenschaftlichen Beschäftigungen waren in der Entomologie u Geologie. Jetzt plötzlich erscheint seine Abhandlung u. nicht nur ich bin ausgeschlossen, sondern auch Sie, was noch viel mehr zu bedauern ist. Selbstverständlich hätte er der Amphibolis gar keine specielle Aufmersamkeit gewidmet, hätte ich ihn nicht auf den Umstand aufmerksam gemacht, dass wir über die Befruchtungs Organe noch sehr im Dunkelen schwebten. 82.02.10

Er *wohnte* an einem Theil der Küste, wo diese keineswegs sehr weit verbreitete Pflanze vorkommt, u. hatte so bessere Gelegenheit solche zu beobachten, als ich sie je gehabt. In Gerechtigkeit zu Tepper muss ich bemerken, dass er mir einiges Material in Alcohol schickte, auch einige *kurze* Notizen, aber, da alles dies noch *unvollständig* war, blieb auch dies bis nach der Ausstellung liegen, u ich habe mich aus dem Rückständigen vielerlei Art noch jetzt nicht ganz herausgearbeitet. Die Correspondenz, welche ich eigenhändig in meinem Departement führe, beläuft jährlich sich auf zwischen 2000-3000 Briefe; und nun kann ich unter den noch nicht absolvirten Papieren Ihren Brief über Amphibolis nicht finden! Seit ich nach dem Nordamerikanischen System (êntre nous) in einer so sinnlosen, grausamen und unverdienten Weise aus dem Garten gestossen wurde, der meine Schöpfung war, um einem ganz unwissenschaftlichen Vetter einer Frau Ministerin Platz zu machen, der nie einen andern Staats-Garten als den von Sydney sah, – bin ich selbst nachdem ich das für meine Arbeiten in 17 Jahren gebaute Haus räumen musste, in einer so kleinen u. engen Wohnung, dass leicht etwas verlegt wird, und so ist es leider mit Ihrem Briefe gegangen. Sie werden mir verzeihen, wenn ich bemerke, dass ich eine Menge Werk im ersten Versuch nicht zu entziffern vermochte, sonst hätte ich den Brief gleich an Tepper abgesandt.

Wie die Sache jetzt steht, möchte ich Sie bitten, Ihre Erklärung der von Tepper gegebenen Vorgänge über die Reproduction von Amphibolis in einer deutschen bot. Zeitung zu veröffentlichen, u vielleicht *einfach* bemerken, dass ich seine Aufmerksamkeit auf diese Pflanze lenkte.

82.02.10 Haben Sie ein Exemplar von Prof Eichler's Syllabus seiner syst. Anordnung der Pflanzen abzulassen, u würde Sie so freundlich sein darin zu notiren, in welchen Punkten Sie von seinen Ansichten abweichen. Ich sende Ihnen hiermit mein eignes Verzeichniss der Pflanzen-Gattungen Australiens.

Sie ehrend der Ihre
Ferd. von Mueller

Herr Tepper ist jetzt in eine Innland Localität übergesiedelt.

10/2/82

Enclosed I am sending you, highly respected Professor, an offprint of Mr Tepper's paper[2] on the germination and development of *Cymodocea* (or I suppose now better *Amphibolis) antarctica*. Presumably you will also have received this publication from Mr Tepper himself,[3] and it may surprise you that no mention occurred of your scientific explanations of the organs, that Mr Tepper describes from this plant. This requires a discussion of the reasons.

When your letter with the remarks about the *Cymodocea* reached me,[4] my time was so much taken up as one of the commissioners and judges (for chemistry, forest products and all other plant materials) for the International Exhibition,[5] that everything not officially urgent in my department had to be left; indeed I could not even get time to continue the Eucalyptus Atlas. So an enormous correspondence accumulated in my department and your letter was also put aside at the time, because I expected that Mr Tepper would show me his paper and drawings before they reached the press, especially since he was after all first drawn to the study of plants through *me* and in fact only in the last 2 years, because his previous scientific activities were in entomology and geology. Now suddenly his paper appears and not only am I excluded but also you, which is even more to be regretted. Of course he would not have devoted any special attention at all to the *Amphibolis*, had I not made him aware of the fact that we were groping very much in the dark about the fertilization organs.

2 See Tepper (1882).
3 From Ascherson's letter to O. Tepper, in Ascherson (1882), we learn that this was indeed the case.
4 P. Ascherson to M, undated [Collected Correspondence].
5 International Exhibition, Melbourne, 1880-1.

He *lived* on that part of the coast where this by no means very widely distributed plant occurs and had much better opportunity to observe it than I ever had. In justice to Tepper I have to remark that he sent me some material in alcohol, even some *short* notes, but because all this was still *incomplete*, this was also left until after the Exhibition and I have not yet quite worked my way out of the arrears of many different kinds. The correspondence that I carry on in my own hand in my department amounts to between 2000-3000 letters annually; and now I cannot find your letter about *Amphibolis* among the pages not yet completed! Since I was pushed out of the gardens, which was my creation, according to the North American system (entre nous) in such a senseless, cruel and undeserved manner to make a place for a quite unscientific cousin of the wife of a minister, who never saw another state garden than that of Sydney – after I had to move out of the house built for my work in 17 years, I am in such a small and cramped dwelling, that something is easily mislaid and so it has happened with your letter unfortunately. You will excuse me if I remark that I was not able to decipher much of the work at the first attempt,[6] otherwise I would have sent the letter straight to Tepper. 82.02.10

As the matter now stands, I would like to ask you to publish your explanation of the details given to Tepper about the reproduction of *Amphibolis* in a German botanical journal and perhaps *simply* remark that I drew his attention to this plant.[7]

Have you a copy of Professor Eichler's syllabus of his systematic arrangement of plants[8] to give away and would you be so kind to note in it in which points you differ from his views? I am sending you herewith my own list of the plant genera of Australia.[9]

Respectfully your
Ferd. von Mueller

Mr Tepper has now moved to an inland locality.[10]

6 Ascherson's handwriting is indeed hard to read.
7 Ascherson's explanation, including due acknowledgement of M's role, was presented by Tepper to the Royal Society of SA on 13 June 1882 and subsequently published in the Society's *Transactions*; see note 2 above and O. Tepper to M, 18 June 1882 [Collected Correspondence].
8 Eichler (1875-8).
9 B82.13.08.
10 Tepper moved from Ardrossan, on the western shore of St Vincent's Gulf, to Clarendon in the Mount Lofty Ranges, SA, in August 1881.

82.02.13 *From Kalakaua, King of Hawaii*[1]

RB MSS M200a.50, Library, Royal Botanic Gardens Melbourne.

KALAKAUA,
KING OF THE HAWAIIAN ISLANDS.

To all who shall see these Presents–Greeting.

Know ye, That

We have Appointed and Commissioned, and by these presents We Appoint and Commission

Baron Ferdinand Von Müeller
to be
Knight Companion

of Our Royal Order of KALAKAUA, to exercise and enjoy all the Rights, Pre-eminences and Privileges to the same of right appertaining, and to wear the Insignia as by Decree created.

In Testimony Whereof, We have caused these Letters to be made Patent and the Seal of Our Kingdom to be hereunto affixed.

Given Under Our Hand, at Our Palace, in Honolulu, this thirteenth day of February in the Year of Our Lord One Thousand Eight Hundred and Eighty two.

Kalakaua Rex.

BY THE KING,
The Chancellor of the Royal Order of Kalakaua,
Jno O. Dominis

1 MS is a certificate. See also W. Allen to M, 13 February 1882 [Collected Correspondence].

To William Mitten 82.02.13a

RBG Kew, Letters to W. Mitten, 1848-1905, ff. 210-13.

13/2/82.

You[1] kind letter of the 28. Dec.,[2] dear Mr Mitten, together with the large and important list of mosses reached me a few days ago by last mail. I am particularly beholden to you for constructing this extensive list, which obtains additional value through the new species recorded in it. As remarked in my letter of end of January,[3] I had sent into print[4] the list of mosses, which I had added to Hampe's[5] from the Flora Tasmanica,[6] from Jaeger and Sauerbeck[7] and from msc. notes from you & C. Mueller; but I had kept the fungs and mosses in type, as Dr Cooke intended also to forward some more additions by the end of the year, and I was thus enabled to *append* to my compilation timely the additional species contained in your list.[8] I thought this the better course in justice to yourself, as you use in many cases genera, different to those adopted by J. & S.,[9] and as you follow also an other systematic arrangement. My list (strictly Australian) was nearly as large as yours, as Tasmania alone gave a great addition from Sir J. Hooker's work. When I requested our lamented friend Dr Hampe to enumerate *merely the species of Continental Australia*, I aimed not at a full enumeration of the whole cryptogams of Australia for the XI vol.

1 Your?
2 Letter not found.
3 Letter not found; but see M to W. Mitten, 7 November 1881 [Collected Correspondence]. Years earlier M had urged Mitten to prepare an index of Australian mosses; see M to W. Mitten, 2 September 1876 [Collected Correspondence].
4 B81.13.12, pp. 107-15.
5 B81.13.12, pp. 45-52.
6 J. Hooker (1855-60).
7 Jaeger & Sauerbeck (1870-9).
8 For Mitten's list see B81.13.12, pp. 114-15.
9 Jaeger & Sauerbeck.

82.02.13a of my fragmenta; but the extra duties for the international Exhibition[10] here rendered it impossible to finish, as intended, the vol. in 1880; and this gave me time, to complete the different indices to date, only the venerable Dr Gottsche having sent no addenda, having as I am aware, but little to add from material within *his* reach. Hence I accept all the more gratefully *your* offer of additions also of Hepaticae.[11]

My total sum of Australian Cryptogams (ferns excluded) is now 3417, as *printed off*, a large share of these having come from me or from collections formed by correspondents on my request. No doubt, future research will add much in all branches of Acotyledoneae, but it will be a matter of time. That you work up novelties so quickly is a great encouragement, and as Dr C. Mueller seems so much engaged with the editorship of "die natur", I shall send in future all mosses to you, if you will name them kindly. That Dr Hampe should not have possessed the "Flora Tasmanica" seems astonishing to me, especially when I think, that I presented my own private copy of that costly work to Prof Lindberg of Helsingfors, after his lamentations in this respect; and I may add, that I did not earn the *slightest* gratitude from him![12] –

That you determined the Fissidens from the Victoria-River,[13] is particularly pleasing, as I found in all the journeys in 1855 & 1856 from North West Australia to the ranges of the tropical East coast (3-4000 miles) til we reached the eastern slopes of the Ranges only 3 Mosses in North Australia; but it must be remembered, that we were mostly away from the narrow belt of coast jungles. In Central Australia over wide stretches, equal in extent to half Europe, no mosses of any kind occur, so far as known, though a few fungs and lichens. But in the crevices of the very limited number of mountains possibly some may yet be latent. Do any Mosses

10 International Exhibition, Melbourne, 1880-1.

11 M did not include a specific list of additional names for 'Musci Hepatici' in his *Additamenta* (see note 8 above).

12 See M to J. Hooker, 2 December 1868 and M to O. Lindberg, 2 February and 6 September 1869, and for examples of anxiety about their arrival, M to O. Lindberg, 19 April 1871 and 14 July 1873 [all Collected Correspondence].

13 NT. *Fissidens Victorialis*, described by Mitten in B83.05.01.

occur in the Sahara-region? and What does Central South-Africa offer in this respect. The Lake-Regions ought to be rich in them. 82.02.13a

I have sent you by this post the Census of the plant-genera of Australia.[14]

Your list of Austral. & Polynesian Mosses will be submitted with a few introductory words from me to the R.S. of Victoria at the first session (after the Dec-febr. vacation) in March.[15] I am beholden to you, for the generous spirit, in which you met me and cooperated with me, and if you will tell me, how I can show my gratitude, it shall be done so far as I can. Perhaps you have lots of duplicates of Cryptogams for disposal, and I could purchase them readily for my Museum.

So soon as the XI vol is bound, the alphabetic index being now in type, I will forward a copy. It may however be some weeks before the vol. is actually out.

Is there really no chance of the issue of a cryptogamic vol. with B. & H. genera?[16]

Regardfully your
Ferd. von Mueller

Please correct any errors in the bryologic portion of the Census of genera, & let me know. Where are the Riccieae from Drummond described?[17]

14 B82.13.08.

15 In the event, the March meeting of the Society was the Annual Meeting and devoted to business matters. Mitten's account of Australian mosses was presented at the next meeting, on 20 April 1882 (Mitten [1883]). His material on Polynesian mosses was communicated by M to the Linnean Society of NSW and was published in the Society's *Proceedings* with an introduction by M; see Mitten (1883a).

16 i.e. in Bentham & Hooker (1862-83), a universal enumeration of phanerogamic genera.

17 Some of Drummond's specimens were acquired by M, and this may have prompted this inquiry. Mitten's answer has not been found, but Taylor (1846), pp. 414-7 describes ten species of *Riccia* collected by Drummond.

82.02.16 *To Louis Smith*[1]

W82/1723, unit 1418, VPRS 3991/P inward registered correspondence, VA 475 Chief Secretary's Department, Public Record Office, Victoria.

Melbourne
16/2/82.

To the honorable Dr L. L. Smith, M.P.
&c &c

In reply to your question, dear Dr Smith, I beg to observe, that the Sonderian Collection of dried plants contains specimens from all parts of the globe, including even numerous species from the least accessible parts of tropical South America, India and other hot regions of the globe.[2] Indeed it is one of the very richest ever formed by a private Gentleman, and its historic value consists in the exceedingly large number of *autographic specimens* connected with published works. this authentic material reaching back to the earlier part of this century, when Dr Sonder commenced his interchanges with aged botanists. Numerically the collection comprises very many thousand of *species*, and each of them is represented by a series of specimens indicative of the geographic range and forms of varieties, thus the whole forms a huge mass of specimens, and would constitute a magnificent supplement to what I have gathered myself since 42 years.

Among the gems of the collection is the unique set of Algae (Seaweeds), on which sorts of plants Dr Sonder was one of the three greatest workers of this age. Indeed as a whole the collection is so valuable, that any other *colony even near us* would gladly secure it for their botanic Museums, such authenticated collections being of incalculable value for all times for reference. As instances, how much collections of great bot. authors are sought, I may remark, that some years ago Dr Meissners collection (then at Basel

1 See also M to L. Smith, 6 February 1882 [Collected Correspondence].

2 See also M to T. Wilson, 26 December 1881 [Collected Correspondence], in which M seeks an interview with the Chief Secretary to urge on him the desirability of acquiring Sonder's collection.

and offered to *me* by himself in first instance) was purchased for £*2000* by an American Merchant and presented to the City of New York.[3] When the great *Lindley* was on his last sickbed, he also was anxious that I should secure his highly important collections for Victoria,[4] after Sir Joseph Hooker had secured the *Orchids* (dried specimens) for which alone he paid *five hundred* £! The collection was subsequently bought for a London Institution.[5] Therefore no difficulty exists of disposing of the Sonderian collections, which the British Museum is eager to get, but which Mrs Dr Sonder in accordance with the wish of her late husband prefers seeing pass into my hands. If you deemed it desirable, the Sonderian collection could be put into the Exhibition-building, and I could go occasionally to see to its proper keeping, for which one youth would suffice; hence but very little annual expense incurred for maintenance. 82.02.16

I would however like during the probably only *few years* of my remaining life, to keep my own collection as a distinct department, as its removal to the Exhibition-Building could not possibly promote my work, and would take away that quietude, which I enjoyed at the bot. Garden for study and so far enjoy still. My collection is *as accessible to the public*, where it is now, as the Observatory is to visitors, but of course like the astronomic instruments, so my bot. collections, are only of real value to men of science or for professional and literary purposes. So is it with the large Herbarium of the Royal bot Garden of Kew.

In answer to your question about the form of the collection, I may observe, that Dr Sonder kept them as usual in parcels, covered by pasteboard. There are many hundreds of such parcels. They may either be kept in metal-cases or put on shelves covered by doors. The price would not exceed £*900*, *delivered here*, as the tin-lined packing cases, freight, insurance and agency expenses would not exceed £*100*, and perhaps be less. The transit and the payment would of course be effected through the Agent General in London.

3 There is no record of M's being offered Meisner's herbarium. It was purchased by John J. Crooke and presented to Columbia University, New York, and is now at the New York Botanical Garden (see TL2).

4 See J. Lindley to M, 29 August 1865 [Collected Correspondence].

5 Most of Lindley's herbarium went to Cambridge University.

82.02.16 Allow me to add, that I feel persuaded of not a single member of the Legislative assembly objecting to the acquisition of such *unique treasures of permanent value* by the Colony Victoria, if the honorable members of the Ministry would place the sum of £900 (as a not recurring item) among the miscellanea kindly on the estimates.[6]

With regardful
remembrance
your
Ferd. von Mueller.

82.03.00 *From Annie McCann*[1]

RB MSS M1, Library, Royal Botanic Gardens Melbourne.

Rock Alpine House
Snowy Creek
Beechworth[2]

Baron Ferd. von Mueller, K.C.M.G.

Dear Sir,

Words are but weak vehicles to convey all the thanks I would fain offer to you for your valuable work on "Botanic Teachings"[3] which you have so kindly sent *the children* –

6 MS annotation by the Premier, B. O'Loghlen, on 20 February 1882: 'The Government have decided to place on the Estimates a sum sufficient to make the purchase say not exceeding £900'. The Under Secretary, T. Wilson, forwarded the news to M who replied on 22 February: 'While thanking the Government for this act of enlightened generosity I beg to inform the hon. the Chief Secretary, that I have written to Mrs Dr Sonder by the Liguria, apprizing her that it is the intention of the Government to place £900 on the estimates for the purchase, packing, freight and insurance of these collections, and requesting her to hold them ready for shipment through the Agent General of Victoria in London by the time, when she will learn, that the money has actually been voted.'

1 MS annotation by M: 'Answ 29/3/82'. Letter not found.

2 Vic.

3 B77.13.07.

Believe me it shall long be a prized *souvenir* as well as a warm 82.03.00
incentive in promoting that study of which I am such an ardent worshipper –

I considered myself indeed highly favoured by your esteemed correspondence and by being advised of the names of so many of our local plants.

I trust, however, that I do not occasion you too great an excess of trouble. I send by this mail a small packet containing some more *fruits* of the "Persoonia" as requested. When I first saw this tree, I thought it was a native Olive, it seemed so prolific in fruit. It is but rarely met with here.

No 1, I found floating on the surface of a lagoon, it looked so fresh and delicate a green that I was tempted to risk a wetting to get to it – but when that was achieved I was in raptures, and for the moment thinking that it was some beautiful *Algae*, from the strong, peculiar smell of the Sea, which it emitted. My beloved Sea! which I have not seen for many a year! – Is it a water-weed? I am desirous of knowing the names of these shrubs and the creeping plant No 2. And of the enclosed Ferns, both seem to grow from the same root.

And now dear Sir, with all my compliments – and all the gratitude I can thus convey –

I beg to remain
your most obdt servt
Annie W. D. McCann

P.S.
Snowy Creek is a tributary of the Mitta Mitta running due North – and takes its rise in some of the Bogong Spurs.

Are the fruit of the "*Persoonia*" dangerous? – Children are often curious about such things and eat them freely when they meet them.

Please say if the Mosses sent in last collection were of a different species to those sent previously?

82.03.06 *From Eugene Fitzalan*[1]

RB MSS M8, Library, Royal Botanic Gardens Melbourne.

Willow Vale Bowen[2] March 6th 1882

The Baron von Mueller.

Dear Sir

When I received your telegram[3] I suspected that the specimens I sent you must have got damaged and was afraid I might be too late as the season for cones was nearly over, and our wet season had fairly set in it was raining heavily when I got your telegram on Wednesday and rained all thursday & Friday on Saturday it cleared a little and on Sunday morning I resolved to act on a suggestion you once threw out with regard to Mount Elliott viz – To make a dash for it – so I started on Sunday morning and camped among the Cycas late that night the distance being about 50 miles = In the morning to my dismay I found that since my last trip the Black's had been there gathering their usual Harvest of the Cycas fruit and had burned the Country for miles and that the Cycas leaves were all hanging down dead & brown many having fallen off altogether and this made matching the cones far more difficult as it was mostly by the leaves I was guided; and only the cones on the tallest trees had escaped scorching and to get some of these without breaking them was difficult and there was hardly a nut to be found as the Blacks do their harvesting most effectually – – – I rode miles trying to find a patch that had escaped the fire but in vain – it had been burned before the rain, and had not been burned the year before so it swept every thing before it – I got however 4 cones 2 of Normanbyana and one each of the two other species and I believe I have matched them correctly, I have dried them and

1 MS annotation by M: 'Gum of Cycas Answ 22/3/82'. Letter not found.

2 Qld.

3 Telegram not found.

sprinkled them with Kerosene and carefully sowed them up in 82.03.06
mosquito net and put nothing in the box with them except a few well dried wood shavings which do not readily heat or decay and I think this time they will reach you safely[.][4] I send some young rachis with the ovules and some full grown fruit on the rachis by which you will see, that where the ovules do not all swell into fruit their place and number are clearly shown, on the rachis thus Normanbyana will sometimes have only one full grown and one abortive nut, but I do not think it ever shows 3 or 4, and media may have only 2, 3 or 4 nuts but it will always show the places of at least 6 and sometimes 7 or 8 = of no 2 which you have kindly named after me[5] I send good rachis with fruit ripe and rachis with the ovules only formed off two different trees – and you can see no trace of more than 4 fruit on each = it is not until you get well up among the mountains that the two species which bushmen designate "Curly" from the peculiar appearance which the curved frond gives them is found no 2 appearing even more so than no 1 from the second upward curve of the frond  In this one the frond is not near so long in proportion to size of stem – as either of the others and the stem of several that I noticed lately appear to be more massive at the butt and thicker at the Top than either of the others, But this may be accidental, The stems that I have shipped away to Goldie & others might have a few no 1 amongst them but not likely no 2 – They would mostly be good specimens of media (no 3) as for the sake of carriage I got them as low down the ranges as possible and then it was 40 miles and very difficult to get at with teams

The Normanby ranges commence about 40 miles from Bowen in a southerly direction; on the highest portion of them 60 miles from Bowen the Normanby diggings are situated a reefing field of considerable extent which has been abandoned owing to the hard-

4 *editorial addition.* All [.] in the following text have this meaning.

5 No cycad named by M after Fitzalan has been identified. In B82.03.01, M named *Cycas Kennedyana*, collected in the Normanby Ranges by Fitzalan, in honour of Sir Arthur Kennedy, the Governor of Qld.

82.03.06 ness of the sinking and most of the reefs though rich were narrow – and though some of them gave 2 oz & more to the ton miners said it did not pay and it is now deserted; these ranges are about 30 miles long by about 10 to 15 wide E & W – and the whole of this space is covered with Cycas in countless thousands I may even say millions the stems of them were used for building Humpies & for slabbing shallow shafts on the Diggings being more plentiful than any timber

The ranges are all Grey Granite except just on the diggings where they strike a hard blue Rock at 30 to 40 feet though the surface even there shows Granite Boulders –

I know of no Zamia nearer here than the neighbourhood of Peak Downs which I believe is their most northern boundary they extend from there south into New South Wales in Isolated patches – being most plentiful I believe near Springsure[.] I never saw but one species (Spiralis) with fruit of an irregular oblong square – I have frequently had the fruit sent me from Springsure;[6]

I mention this because when camped at a station about 60 miles west from Bowen one of the young gentlemen showed me what I took to be a Zamia nut made into a match box; he said it was a Cycas and came from somewhere near the Barcoo (SW) it was oval shaped like a Cycas nut but more than twice the size being about 2 inches long by about 1 1/4 in wide with the texture of Zamia seed rather than Cycas – and beatifully[7] polished He declined to part with it –

With regard to Macrozamia Denisonia[8] or M. Hopeii or Katakidozamia I know nothing except what I saw on the Daintree and it appeared to be identical with M Denisonia that I had seen just previously in the Botanic Garden Sydney – and this must have been Mr Hills Catakidozamia as there was no other large Cycad either on the Daintree or Mossman or any of the Rivers in Trinity Bay and I saw a good deal of the Country much more than any

6 The sentence ends as shown.

7 beautifully?

8 *Macrozamia denisonii*?

other Collector[.] I found these macrozamias or whatever they 82.03.06
are from the size of ones fist up to 12 or 14 feet high and a foot thick[.] I found one with a cone of fruit but as it was only half grown I did not take it – I brought away a lot of young ones – I enclose sample of leaf of small plant – – I would like to have about £50 to spend on collecting between Cardwell and Cooktown[.] I could do more now for £50 than any one could have done 4 or 5 years ago for three times that amount as there is now more or less of settlement on all the Rivers and I would know exactly how to get about and where to go – and any thing that I could not get at the time of my visit I would point out to some resident to get for me at the proper time – But though I would like such a trip I could not make it pay – and it is hard to get about new country without money –

Our resident photographer is at present in England perfecting himself in the business but will be back in May – when it will be possible to get Cycas views on reasonable terms –

yours faithfully
Eugene Fitzalan

82.03.28 *To William Thiselton-Dyer*

RBG Kew, Kew correspondence, Australia, Mueller, 1882-90, f. 9.

28/3/82.

I have two more species of Cycas-Amenta ♂ for you at the Photographers, dear Mr Dyer; but they had decayed on the long way, and will make bad illustrations. You shall have these photograms by next post, also one of ♂ & ♀ Macrozamia Fraseri. I will try to forward a ♂ of the latter ipsa natura[1] by *this* post, which Mr Forest kindly sent on my request.[2] So you see, that your forthcoming opus is not forgotten.

Both last evening late & the evening before I had as Chairman in the one case and as Patron in the other, to deliver speeches in Social (*not Socialists!*) meetings; so, – that has taken much of the time away, wanted for the mail, & I must therefore be brief.

Perhaps I get away to Kew by *this* mail the 8th Euc. Decade[3] & the 11th vol. of the fragmenta; but I am not sure about the *binding* being done in time.

Regardfully your
Ferd von Mueller.

In the 11th vol over 3500 evascular Acotyledoneae are recorded.[4]

1 its natural self.
2 Letter not found.
3 B82.13.17.
4 In the separately paginated *Supplementum* to vol. 11 of the *Fragmenta* M published 'Indices Plantarum acotyledonarum complectens', comprising contributions by specialists on the various groups, and some additional comments and listings by himself.

To Jean Müller

Conservatoire et Jardin botanique, Geneva.

82.03.29

29/3/82

Nur wenige Worte, gelehrter Freund, um meine Unzufriedenheit zu rechtfertigen in Bezug auf den Sammler Karsten. Wenn ich sage, dass sein Grossonkel, der Prediger Karsten, der noch in Schwerin als General Superintendent der Kirche lebt, mich taufte, u. bei meines theuern Vater's Beerdigung die Grabrede hielt, so werden Sie wohl glauben, dass ich für diesen Abkömmling der Familie K. die besten Gefühle hatte, zumal auch Prof. Roeper aus meiner Vaterstadt Rostock den Mann sehr empfahl, freilich nur in allgemeinen Ausdrücken. Da der junge K Seemann gewesen, u. so an Thätigkeit und Klettern u. einfache Kost gewöhnt war, u gar keine Arbeit finden konnte, (warum, ist mir jetzt ziemlich klar,) so war ich thöricht genug, ihn für ein Jahr als Pflanzensammler nach N. Ost Australien zu senden. Für die damit verbundenen Ausgaben von fast *1000* (nicht 100) Thaler! hat er etwa 100 (nicht 1000) Pflanzen Arten gesandt in miserabelen Exemplaren. Die wenigen Flechten waren dabei das Beste, u. darauf hatte ich ihn besonders aufmerksam gemacht. Bedenken Sie, dass ja ein Schwarzer-Eingeborner ebenso gut u. viel besser einige Flechten u. Moose hätte von von[1] Baumstämmen abkratzen können im Urwald, ja dass 100 anstatt 10 Arten hätten gefunden werden können. Ob der Mensch K. monatelang absolut auf der Bärenhaut gelegen, ob er nach Gold anstatt nach Pflanzen gesucht, ob er gedankenlos hin u. her gelaufen ohne es der Mühe werth zu halten, sich nach einer Pflanze zu bücken, das weiss ich nicht; das aber weiss ich, dass er nicht krank gewesen, und keine schwere Unfälle erlitten hat. Also werde ich den Namen Karstens aus meinen Schriften bannen.

Ebenso traurige ja noch schlimmere Erfahrungen mache ich jetzt mit einem anderen jungen Deutschen, namens Wilhelmi; ich bin aber so vorsichtig gewesen, ihn nicht regulär zu engagieren, son-

1 *Word repeated.*

82.03.29 dern nur sein Fahrt nach N. Ost Austr zu bezahlen, und eine Summe Geldes vorzuschiessen. Nachdem das letztere in Liederlichkeit (so höre ich nun) dort verbraucht war, schicke ich noch einmal Geld nach. Das Resultat ist *nil*, wenigstens bisher nach mehreren Monaten, u das mitgegebene viele Papier, die Pappen &c &c. sollen in Beschlag genommen sein, um Schulden zu bezahlen. – Nun genug davon, aber es ist begreiflich, dass ich sehr *erbittert* bin. –

Mit nächster Post hoffe ich den 11ten Band der Fragm. senden zu können; er ist jetzt in den Händen des Buchbinders. Ich konnte im letzten Nachtrag noch die von Ihnen zuletzt verzeichneten Pflanzen aufnehmen. Da Bailey fort u fort an Dr. Stirton Flechten sendet, so wäre es sehr wünschenswerth, ja immer *recht bald* das von mir etwa Einlaufende zu absolviren, was Sie ja auch früher immer so gütig gethan haben. Verehren Sie mir doch auch gelegentlich Ihr Photogram. Mit bestem Grusse

Ferd. von Mueller

29/3/82

Just a few words, learned friend, to justify my dissatisfaction in reference to the collector Karsten. When I say that his great uncle, the preacher Karsten, who still lives in Schwerin as General Superintendent of the church, baptised me and gave the funeral oration at my father's burial, you will surely believe me when I say that I had the best feelings for this descendent of the Karsten family, especially as Professor Roeper from my home town of Rostock also recommended the man very much, admittedly only in general terms. Because the young Karsten had been a seaman and so was used to activity and climbing and simple food and could find no work at all (why is now rather clear to me), I was foolish enough to send him to northeast Australia for a year as a plant collector. For the expenditure associated with it of almost *1000* (not 100) Thaler! he has sent about 100 (not 1000) plant species in miserable specimens. The few lichens among them were the best, and I had particularly drawn his attention to them. Consider that even a Black Aboriginal could have scraped off some lichens and mosses just as well and much better from tree trunks in the forest, indeed that 100 instead of 10 species could have been found. Whether this person K. lounged about absolutely for months or whether he went after gold instead of after plants, whether he ran back and forth unthinkingly without considering it worthwhile to stoop to pick up plants, I do not know, but I do know that he had not been ill and had suffered no severe accident. Thus I will banish the name of Karsten from my writings.[2] 82.03.29

I had just as sad – indeed even worse – experiences with another young German by name of Wilhelmi;[3] but I was careful not to engage him on a regular basis but only to pay his journey to north-east Australia and advance a sum of money. After the latter was spent there in dissoluteness (so I now hear), I sent some more money. The result is *nil*, at least until now, after several months, and the great deal of paper, the cardboard &c &c provided are to be seized to pay debts. – Now enough of that, but it is understandable that I am very *bitter*. –

I hope to be able to send the 11th volume of the *Fragmenta* with the next mail; it is now in the hands of the bookbinder. I still was able to include in the last supplement the plants last recorded by you.[4] Because Bailey continually sends lichens to Dr Stirton,[5] it would be very desirable always to finish *very soon* the things arriving from me, which indeed you also have formerly always so kindly done. Do also give me your photograph some time.

With best greetings
Ferd. von Mueller.

2 In an earlier letter to Müller, 22 October 1881 [Collected Correspondence], M made the same complaints about Karsten, and asked Müller, if he had not yet published the new *Sticta* proposed to be named after Karsten, to use a different species name. However, the request came too late; Müller described *Sticta karstenii* in *Flora*, vol. 64, p. 505 (1881).

3 Harry Wilhelmi (1859-85), son of M's former assistant Carl Wilhelmi, arrived in Victoria in September 1881. He died in Qld.

4 B81.13.12, p. 115.

5 See M to J. Müller, 12 March 1882 [Collected Correspondence]: 'Mr Bailey seeks to outflank me in lichens, and therefore has put himself in communication with Dr. Stirton in Glasgow independently of me'. See also M to J. Hooker, 24 July 1882.

82.04.24 *From Andrew Berry*[1]

Letterpress copy book 1881-2, p. 253, School of Mines and Industries, Ballarat, Victoria.

[Ballarat][2] April 24th 1882.

Sir,

By direction of the Vice-President I am to inform you that the Council of this Institution are desirous of establishing a series of popular science lectures to be delivered at, and in connection with, The School during the coming winter months[3]

2. And respectfully to convey to you the cordial invitation of the Council to assist them in carrying out the object in view by the delivery of one or more of the series

3. I am to ask you to be so good, as to inform me if you can oblige by accepting the invitation, and if so the probable date of delivery. – All expenses will of course be defrayed by The School[4]

I have the honor to be
Sir,
Your most obedient Servant
Andw Berry
Acting Registrar

Baron von Müeller, K.C.M.G.
&c. &c. &c.
Melbourne

1 For reply see M to A. Berry, 26 April 1882 [Collected Correspondence].

2 *editorial addition.* Vic., formerly Ballaarat.

3 At a special Council meeting on 19 April 1882, M was included in a list of speakers to be invited to lecture at the School (School of Mines Ballaarat, minute book, School of Mines and Industries Ballarat Limited).

4 M gave a lecture entitled 'General Observations on the Flora of Australia' at the School on 13 September 1882 (*School of Mines Ballarat, annual report 1882* (1883), Ballarat, p. 27). See B82.13.07.

To Bryan O'Loghlen 82.05.02

W82/3955, unit 1373, VPRS 3991/P inward registered correspondence, VA 475 Chief Secretary's Department, Public Record Office, Victoria.[1]

Melbourne, 2 May 1882.

The honorable the Chief Secretary.

Sir.

I have the honor to acknowledge the receipt of your communication of yesterday[2] accompanied by a sample of poison-herb, which proved detrimental to some pigs at Caulfield. With a view of ascertaining particulars of the case and inspecting the locality, I proceeded to Mr T. W. Nicholson's estate, where the poisoning case occurred. The plant, to which the mischief has to be adscribed is a tall umbelliferous herb, well known botanically since the last fourty years as Didiscus pilosus; but altho' this plant is sparsely distributed from Port Phillip through some of the inland-regions to near the Gulf of Carpentaria[3] northward and to Swan-River[4] westward, and also extends to the north of Tasmania, – no record exists, – so far as I am aware –, of any injury to pastoral animals ever having arisen from this herb; notwithstanding that the plant abounded about 30 years ago on the now bare sands between Melbourne and

1 This letter also published in the *Argus*, 4 May 1882, and in the *Australasian*, 13 May 1882, p. 603.

2 Letter not found. M's letter was in response to a letter from T. W. Nicholson to the Premier, B. O'Loghlen, although according to the *Australasian* it was the Chief Secretary, J. Grant, who contacted M. Nicholson wrote on 26 April:

> I have the Honor to inform you that I herewith forward to you a "Plant" same as is now growing at my private residence at Caulfield for your information
>
> There was 4 Plants (same as one accompaning) pulled up by one of my son's on Monday morning and was casually thrown into the Piggs styes at about 10 am and on Tuesday morning the 25th the 16 sixteen Piggs was found Dead and only one has survived and that one was a sow much older = The 16 Piggs Poisoned by the accompaning Plant was about 6 month old[.] I also beg to state for your information that about 3 month ago a Gentleman lost 3 cows and 100 Piggs in one night from the effect of eating the same plant.

3 Qld.

4 WA.

82.05.02 Sandridge,[5] where stock then used to browse, it actually passing at that time under the treacherous name of "native Parsnip". – As this Didiscus is fortunately biennial only, it would become suppressed by preventing it from flowering. – On enquiry I heard at Mr. Nicholson's place, that a number of half-grown pigs, to which the Didiscus-plant was given, – (as additional green-feed, under the idea, that it was similar to the Carrot-plant) – had died in a day or a day and a half, the animals readily devouring the luxuriant foliage. The symptoms produced were vomiting and purging, which set in some hours afterwards; but, as far as I can learn, no paralysis was produced For ascertaining any after-death appearances no dissections were made by Mr Nicholson's people. The poison seems to have acted so virulently, that a single plant, though only partly devoured sufficed to cause the death of two pigs. As in this instance we have to do with an absolutely new kind of poison-plant, which moreover has in various parts of Australia and even in new Caledonia several congeners of doubtless equally deleterious property, it will be important to find out, what precise chemical principle operates so powerfully and so unexpectedly in this innocent-looking herb. With this view I have placed myself in communication with W. Johnson Esq., the Government analytic Chemist, who with the facilities at his command has kindly undertaken, to search for and isolate the poison-principle, as I suspect, that it will turn out to be analogous to Coniin and Cicutin among the Umbelliferae and therefore a volatile Alkaloid. It is also my intention with the cooperation of a leading Veterinary Surgeon to investigate the physiologic action of this new vegetable poison on animals, to trace the post-mortem appearances and thus to devise antidotes; and more, – as providence does not call forth organisms of any kind without some beneficent purpose, we may also now obtain a clue, how this new vegetable principle could become a powerful therapeutic auxiliary perhaps in the hand of physicians, to alleviate or even subdue human maladies.

I have the honour to be, Sir,
your obedient servant
Ferd. von Mueller.

5 The old name for Port Melbourne.

To Edward Ramsay

82.05.09

ML MSS.562, Letters to E. P. Ramsay 1862-91, Mitchell Library, State Library of New South Wales, Sydney.

9/5/82.

I feel, dear Mr Ramsay, that I ought not to send merely a formal receipt for the splendid enumeration of the Crustacea by Mr Haswell,[1] but should specially express my thanks to the trustees of the Sydney-Museum for the attention shown me in presenting this volume; and more, every acknowledgement is due to the trustees for the measures taken by them, to see through the able Mr Haswell and through your own excellent supervision such large material of Crustacea brought together and elucidated. In former years I have responded to the calls of zoologic friends in Europe, – some friends since my University time – to furnish them with specimens of Australian Crustacea, but if I should acquire additional material now, I will forward it for elaboration to your grand institution.

Regardfully your
Ferd. von Mueller

1 Haswell (1882).

82.06.05 *To Jacob Agardh*

Handskr. Avdl., Universitetsbibliotek, Lund, Sweden.

5/6/82.

Sie werden, edler Herr, wie ich und soviel Andere den Tod unseres genialen Freundes Dr Sonder betrauert haben. Nicht nur verliere ich durch Ihn meinen *ersten* bot. Freund im Leben, mit dem ich schon in meinen Knaben-Jahren correspondirte, sondern auch den fleissigen Helfer für den algologischen Theil der Flora Australiens. Nun möchte ich vorfragen, ob Sie am Abende, u. ich hoffe recht langem und heiterem Abende Ihres bewegten Lebens es der Mühe Werth erachten, weiter an Algen Australiens zu arbeiten, die ja nur selten noch etwas ganz Neues liefern, obgleich Manches für die Feststellung der Gattungen u. Arten zu beobachten übrig bleibt.

Aus Ihrem letzten schönen Bande ersehe ich freilich, dass Sie Ihre Aufmerksamkeit ganz besonders auf die Morphologie der Algen gelenkt haben, und in diese Arbeiten auf einem Felde, desses Grossmeister Sie sind, sollte niemand störend eingreifen. Vielleicht ist es aber doch für Sie gelegentlich eine Erholung, Algen auf Gattung u. Art zu bestimmen, u. gewiss würde ich Niemand lieber als Ihnen meine neuen Sendungen zuwenden. Es liegen auch noch einige Packete Algen bei der Witwe des Herrn Dr Sonder, der noch 4 Tage vor seinem plötzlichen Tode an der Bearbeitung des kürzlich vorher von mir eingesandten Materials beschäftigt war. Falls Sie also Vergnügen daran finden u Musse dafür haben, für die specielle Algologie Australiens weiter zu wirken, so bedürfte es nur eines Briefes an die Frau Doctorin Sonder in Hamburg (Papenstrasse) und das Ganze des noch unbearbeiteten Materials würde an Sie sofort abgesandt werden. Sie könnten über den ganzen Dubletten Vorrath verfügen wie Sie wünschen, ausser dass ich gern 1 Exemplar benannt von jeder Art u jedem Standort zurück hätte. Vielleicht könnten Sie auch eine Kleinigkeit an Sir J. Hooker in Kew u Conde Ficalho in Lissabon abgeben, bei denen ich recht verschuldet bin.

Sie ehrend der Ihre
Ferd von Mueller

5/6/82. 82.06.05

You, noble Sir, will have mourned as I and so many others the death of our brilliant friend Dr Sonder.[1] Not only do I lose through him my *first* botanical friend in life, with whom I corresponded as early as my boyhood years, but also the hard-working helper for the algological part of the flora of Australia. Now I would like to enquire whether you in the evening, and I hope very long and happy evening of your active life, consider it worth the trouble to work further on the algae of Australia, which indeed only rarely produce something quite new, although much remains to observe for the diagnosis of genera and species.

Admittedly I gather from your last fine volume[2] that you have directed your attention quite particularly to the morphology of the algae, and in this work, on a field whose grand master you are, no one should interfere. However, perhaps it is for you at times a relaxation to determine algae to genus and species, and certainly I would prefer to let no one have my new consignments other than you. Also there are still some packets of algae at the widow of Dr Sonder's place, who was occupied even 4 days before his sudden death on the working up of the material sent by me shortly before. Thus in case you find pleasure in and have time for further working on the special algology of Australia, it only needs a letter to Mrs Sonder in Hamburg (Papenstrasse) and the whole of the still unworked material would be sent to you immediately. You could have at your disposal the whole of the duplicate stock as you wish, except that I would gladly have back one named specimen of each species and each locality. Perhaps you could also give away a trifle to Sir J. Hooker in Kew and Conde Ficalho in Lisbon, to whom I am very much indebted.

Respectfully your
Ferd von Mueller

1 Sonder died on 21 November 1881.
2 Agardh (1879)?

82.06.13 *From Frederick Gladman*[1]

X82/5675, unit 1327, VPRS 3991/P inward registered correspondence, VA 475 Chief Secretary's Department, Public Record Office, Victoria.

Training Institution
Education Department,
Melbourne, June 13th *1882*

Baron von Mueller,

You are probably aware that the study of Elementary Botany has been introduced into our Institution for training teachers. – I noticed that you, in your enthusiasm for botanical science, have prepared fascicles of pressed & dried native plants, which are available at various mechanics' institutions &c.[2]

If such a collection could be obtained for use by our students, it would be of great value, & I write to ask whether you can & will kindly help us in the matter.[3] – I can promise that the specimens will be highly valued, & well used.

With great respect,
I remain, dear Baron von Mueller
Yours faithfully
Fredk. Jno. Gladman
Supt of Training

To Baron Ferdd von Mueller,
K.C.M.G. – &c. &c.

1 MS annotation by M: 'Answ 14/6/82'. Letter not found.

2 See Maroske (1995).

3 See M to T. Wilson, 20 June 1882 [Collected Correspondence], in which M asks whether there are any sets still available, and T. Wilson to M, 24 June 1882 [Collected Correspondence], in which Wilson reports that a set has been supplied for the Training Institution.

From W. Anthony Persieh[1] 82.06.17

RB MSS M1, Library, Royal Botanic Gardens Melbourne.[2]

Cooktown Jun. 17 82.

Herr Baron!

Sie erhalten heute pr Coreer den Rest meiner gesammelten Fungis für diese Season da die Regenzeit bereits seit einem Monat aufgehört hat u jetzt nicht mehr vorkommen Ich finde die Pilze sehr intressant ihrer mannigfaltiegen Formen u farben u wünsche die oberflächiege classifierung durch Ihre gefl Mittheilungen zu lernen um mit besseren Erfolg nächste Season dieselben zu sammeln Ich bin während der Regenzeit 6 Wochen fast täglich 5 M. nach dem Shrub gegangen wo dieselben fast ausschliesslich an den zergangen Bäumen zu finden waren, da die Umgebungen von Cootown[3] meistens zu sandig sind um Pilze zu sammeln Ich bitte Ihre gefl Mittheilungen was werthvolle oder selten u neue spc sind besonders die 2 spc in Spiritus gesandt ob solche neu sind u erbitte den botanischen Namen derselben bin darum oft hier befragt geworden.

Die Flora ist augenblicklich sehr unbedeutend ausgenommen Acacien die sich jetzt entwicklen Die früher gesandten 2 spc Acacia in Blüthe u Frucht No 2 u 5 erbitte gefl spc Namen

Ich hoffe innigst durch Ihre Güte mit den Fernrohr baldigst beehrt zu werden da in einigen Wochen die Prospekting Parthey in den höheren Wald Regionen abzugehen gedenken u ich gerne diese Gelegenheit benutzen möchte da einzelnd niemand hinsichtlich der Eingebornen diese Gegenden passiren kann u in den halbdunkeln Jungeln die hohen Bäume mit Schlingpflanzen verwachsen mit dem nackten Auge sehr wenig zu entdecken ist u besonders

1 MS annotation by M: 'Beantw [Answered] 27/6/82'. Letter not found.

2 MS found with a specimen of *Gomphandra australiana* (MEL 2240582), Endeavour River, Qld, Persieh, 1882.

3 i.e. Cooktown.

82.06.17 klein parasitische Gewächse über sehen werden Sie werden hoffentlich die übersandte neue Lagerstomia Blatt u Frucht erhalten haben u verspreche so viel als es mir möglich ist die aufgegebene neue spc Ihnen das Material zu liefern u erbitte was sonst neu oder selten sein mag in den letzten Sendungen gefl baldigst zu bemerken

Erbitte gutigst wen Sie schreiben Herr Baron Genus u spc Name des übersandten dunkelgrünen glanzender Blatt u rother Fruchtboden (Capsel) so wie 362 Blüthe u Frucht u Blatt gesandt Genus u spc Namen

Erbitte gefl einige Fragmentas sehr nützlich oft für mich als Leitfaden

Hoffentlich Ihre gefl baldige Mittheilungen entgegen sehend

Hochachtungsvoll ergebenst
W A Persieh

Die beifolgende Blätter u Frucht ist ein Baum mittler grösse die Zweige 8-10 F. lang niederhangend vom Stam ähnlich wie Trauerweide S. babilonica dunkelgrau glatte Baumrinde bisher diese spc nicht gesehen solte es neu sei bitte baldigst Ihre Mittheilungen habe gegenwärtig nur 2 Bäume gefunden

derselbe

Cooktown,[4] June 17 82.

Baron, Sir,

Today you will receive per courier the rest of the fungi, which I collected this season. The rainy season has ended already a month ago, and nothing more is found now. I find the fungi very interesting in their manifold forms and colours, and wish to learn their basic classification through your respected communications, so that I can collect them more successfully next season. I went for 5 miles to the shrub country almost daily for 6 weeks during the wet season, where I found them almost exclusively on rotting trees because the surround-

4 Qld.

ings of Cooktown are generally too sandy to collect fungi. I request your respected communications on what is valuable, or rare, and new species. Particularly the 2 specimens sent in spirit; if they are new, I request their botanical name. Have frequently been asked about it here. 82.06.17

The flora is at present very insignificant, except acacias, which are developing now. Respectfully request species name of the two previously sent *Acacia* species in flower and in fruit, No 2 and 5. I sincerely hope to be honoured soon with the binoculars through your generosity because the prospecting party to the higher forest area intends to depart in a few weeks and I would like to use this opportunity, because nobody can pass through these regions alone on account of the aborigines. With the naked eye very little can be discovered in the semi-dark jungles, with tall trees interwoven with climbing and trailing plants, and small parasitic plants in particular are overlooked. You will, I hope, have received the leaf and fruit of the new *Lagerstroemia*. I promise as far as is possible to me to procure material of the new species sent and request kindly to be notified soon of anything else perhaps new or rare in the recent transmissions.

Request respectfully, when you write, dear Baron, the genus and species name of the dark green shiny leaf and red fruit base (capsule sent), as well as genus and species name of No. 362, flower and fruit sent.

Request respectfully some *Fragmentas*, often very useful to me as guidance.

Hoping expectantly for your early respected communication.

Very respectfully devoted
W. A. Persieh

The enclosed leaves and fruit, from a medium size tree, branches 8-10 feet long, pendulous from the trunk similar to the Weeping Willow *Salix babylonica*, dark grey, smooth bark. Not seen this species before. Should it be new, please notify soon. Have found only 2 flowers of it to the present.

The same.

82.07.24 *To Joseph Hooker*

RBG Kew, Miscellaneous reports 7.2, Queensland Botanic Gardens etc., 1873-1919, ff. 37-40.

24/7/82.

My best acknowledgement is due to you, dear Sir Joseph, for an act of consideration of yours, of which I became only aware within the last days. It concerns a collection of mosses and some other evascular Acotyledoneae, sent to Kew from Melbourne, and collected in the Cape Otway Ranges[1] by a Mr Lucas.[2] I appreciate your action in the matter greatly, and Mr Lucas has now brought these plants to me, so that they can be elaborated in connection with allied species. Of course you saw, that scattering the material of one place without concert, would lead to complications, double working, increased synonymy & so forth. Besides, what I value far more, it was a tribute to my professional standing in a special public Department.

On similar matters I felt somewhat discontent with Mr Bailey, with whom I however maintain friendly relations, because unknown to me he entered into arrangements about the naming and description (if new) of his Evasculares, when he would have had quite as much credit and as much attention paid to his collections, had the[3] gone with my own, whereas now clashing interests have arisen, as more than one specialist independently of others works on similar and perhaps identical material, and thus valuable time in fleet life of men of science is thrown away. I felt all the more hurt, especially as it was made here to appear in public print, that *only he* gave attention to cryptogamic plants, as Mr Bailey is in reality my disciple. Through my animating him to collect and especially the Evasculares, he commenced forming collections when

1 Vic.

2 Possibly the bushman and carpenter Robert Lucas who testified to the Royal Commission on Vegetable Products (1887) that he had been in the 'Colac Forest' for about 33 years.

3 they?

he came to Queensland about a dozen years ago. I knew him as a young Gardener as far back as *1847* in his fathers nursery in Adelaide, where I first saw Clianthus Dampierii,[4] Doryanthes excelsa and other remarkable Austr. plants in flower. He then for 6 years in S.A. never collected, nor for a series of years afterwards during his stay in N.Z, when he had prior to Dr Hector, Dr Haast, Mr Travers, Prof Kirk and Mr Buchanan a splendid opportunity to distinguish himself, especially as he could have the rare benefit of your superb work for study,[5] the first that ever shed any extensive light except your antartica[6] on the Evasculares of the S. Hemisphere. In Brisbane he was so near (within a few miles) of the jungles, that I was eager to get plants from him, especially Cryptogams, which during my few days stay there in 1855 & 1857 I could not get, and for which others had no eye. Through collecting, only then initiated, Mr Bailey got an insight into the Austr. flora & Bentham's work & also my publications helped him on. But more, I have often sat at late night-hours to name plants for him like for dozends[7] of other Amateur-collectors and fear I have impaired my eyesight by having done this late night work for correspondents during the last 30 years & more in Australia. He has a good eye for plants, he is assiduous, and I have aroused in him a spirit for scientific distinction & I help him on still. So only lately I described under our united authority a Bauhinia,[8] closely allied to B. Malabarica, which he thought might be B. glaucescens (a Central American spec)[.][9] I do not think he discovered over a dozen actually new Phanerogams, and of these the half I named after him.[10] But he has given us many new localities and sometimes better mat- 82.07.24

4 *Clianthus dampieri*?
5 J. Hooker (1853-5).
6 J. Hooker (1844-7).
7 dozens?
8 *Bauhinia gilesii* (B82.07.04, p. 151).
9 *editorial addition* – text partly obscured by binding.
10 For example, *Dendrobium baileyi* (B74.04.01, p. 173), *Bolbophyllum baileyi* (B75.02.01, p. 5), *Indigofera baileyi* (B75.05.05, p. 43), *Leptorrhynchus baileyi* (B77.02.03, p. 101), *Eucalyptus baileyana* (B78.11.04, p. 37).

82.07.24 erial and he is obliging. But as he has no knowledge by personal observations of the European Flora, he is apt to fall into mistakes; thus not long ago he named the cosmopolitan Salsola Kali as Kochia brevifolia, sent Zieria obcordata as a new Boronia, and Angophora as a new Eucalyptus &c. I do not mention this in an unfriendly spirit, but to point out, that it is not safe to take his data without seeing his specimens, though of course after my so much naming for him (only this month Bosistoa) he is often quite correct. The latin diagnoses, which now & then he has published, were written for him by Father Woods, which however remains unacknowledged by him publicly, so that words like pannicula occur. The Rev. Tenison-Woods however is an excellent zoologist and a hard worker among invertebrata, so much so, that on my own accord I have brought his candidature for next year at the R.S.[11] forward, got the signatures of all the Austr FRS.[12] whom I asked, and left matters in Sir Henry Barkly's hands, so that you, & Mr Dyer,[13] Mr Oliver & Mr Baker could support him kindly also.[14] The Curator of the bot Garden of Melbourne[15] has suddenly gone to North Queensland, to the surprise of many people here, as it is just in the middle of the planting season, and for *Landscape-gardening* he was imported here from Sydney. He seems to have a morbid vanity to pass as a scientific man; thus lately publishing an article on Duboisia, of which he knows nothing – copying what Dr Bancroft and myself rendered known of course under suppression of my name[16] as he did about Fiji plants with Dr Seemans.[17] Whether he in any way intends to interfere with my working on the Austral

11 Royal Society.

12 Fellow of the Royal Society.

13 W. Thiselton-Dyer.

14 Though Tenison Woods was a candidate for election to the Royal Society of London for several years, he was never elected.

15 William Guilfoyle.

16 Guilfoyle (1882). M had written about *Duboisia* (Pituri) in B77.02.02 and B78.07.02. Guilfoyle, however, cites only Bancroft (1877), a copy of which, inscribed by Bancroft to him, is at the Royal Botanic Gardens Melbourne.

17 Seemann (1865-73) p. 10, notes the publication of Guilfoyle (1869), and in a footnote remarks: 'Much of the information in this sketch is taken verbatim, and without acknowledgment, from the various publications that issued from my pen.'

Flora as regards North Queensland[s] I do not know; but I may mention, that I failed last year after much expenditure with the collector, introduced to me by Prof Roeper of Rostock;[18] in renewing this year my effort, to make a finishing mainstroke for the flora of Australia I subsidise a zoolog. Collector, Persieh, also pay for what I get from an other countryman of mine,[19] for all of whom I besides name *every* plant, they keeping sets, so that lots of names can be picked up by others also locally at Cooktown and elsewhere by these means. possibly such names appearing in public reports without the slightest allusion to their origin. Now, to conclude these rambling statements, Sir Joseph, I may add, that I have placed also some months ago £50 at the disposal of Mr Fitzalan, who for his nursery at Port Denison[20] wanted Orchids, Palms, Cordylines &c from North Queensland, so that passingly he is to gather Museum specimens also. Some novelty doubtless will come forward, but probably not very much, as at the whole vegetation is similar to that of Rockinghams Bay[21] also on the Rivers further north. 82.07.24

I have for the Census of Austral plant-species[22] given the supplemental collections which accumulated since the volumes of Bentham's Flora were issued some fuller attention. Thus I have in toto 30 additional species (well marked) of Acacia, and yet but very few other the 2[9]3 are abolished by me from augmented material (A. obscura &c), but cleared up some errors. I dare say you all at Kew will find this Census acceptable as facilitating references.

Let me remain with regardful remembrance your

Ferd von Mueller.

I almost forgot to express my pleasure of Kew sending dried plants to the Brisbane Museum. I had only just written to Mr Bailey, that he could have many duplicates of N. Zealand plants but I am often

18 See M to J. Müller, 29 March 1882, for an account of the problems with the collector Karsten.
19 Not identified.
20 Qld.
21 Qld.
22 B82.13.16.

82.07.24 out of reckoning in my time, & the making up of collections takes hours and hours of time for which there is nothing adequate to show, while at the approachig evening of my life I have to be more and more parsimonious in the disposal of my time. Mr Mitten writes,[23] that all his Mosses (therefore also what ever I sent) will finally go to Kew, and so I suppose it will be with Dr Cooke's fungs.

The practical benefit of Bailey's collecting has been that a Gov. position fairly supported was created for him.

82.08.16 *To William Thiselton-Dyer*

RBG Kew, Kew correspondence, Australia, Mueller, 1882-90, ff. 24-5.

16/8/82.

Only a few hurried lines this time, dear Mr Dyer, just before the close of the mail. The *best*, the dearest, the noblest of my friends in this part of the world passed suddenly away; – a loss, which nothing in the world can replace![1] There was scarcely an eye without tears in the large congregation, when we sung the mourning hymn in the church before the coffin was removed from the Altar. I had the privilege, often as only guest, to spend after church in the evening the remaining hours of the Sunday with this genial man, and cannot realize that we shall see his kindly countenance no more, nor here[2] again his friendly voice. He was one of the very few here, who had a comprehension for a professional position, like mine, who disinterestedly on principle stood up for my Depart-

23 Letter not found.

1 Rev. John Stobbs. See also M to R. Tate, 20 August 1882 [Collected Correspondence], where M describes Stobbs as 'one of the few genuine friends I ever had since my early orphanage in life'.

2 hear?

ment; who had the courage of his opinion even though offended on Sir James M'Culloch's dinner-table for my sake, who never changed towards me, and gave me the benefit of his great influence in undeserved adversities – 82.08.16

I had occasion to examine at last specimens of Cycas Seemanni, and shall follow this enquiry up, as possibly there may be a second Cycas in Fiji. The specimens are not yet dry; but you shall have – of course – some of them by one of the next mail. But my daily work in the Department is so difficult on account of all want of convenience in the little private cottage here, that I often have not even the time to look for a specimen under tables or chairs even in the bedroom when the mail-day comes. Some stems of Macrozamia Denisonii are procured, and as this is an outlying species of the genus I would certainly advise to sacrifice one of the stems for anatomic dissections, as you are sure to wish contrasting the anatomic structure of Cycas, Zamia &c more fully then[3] hitherto done. The stems are quite fresh, and ought to grow, as they have been lifted with special care. If you not require them at Kew, they will be useful for interchanges I will try to get them on board next week.

Pray give my regards to Sir Joseph and let me remain your friend

Ferd. von Mueller

I shall have something to say to morrow night about poor Mr Stobbs as Patron of the young mens Christian association of the Church.

The slip is from the Melbourne "Age" of this day.[4]

The Rev. Mr Stobbs arrived two years after my Department was smashed, otherwise I would be in the garden still.

3 than?

4 There was no article about Stobbs in either the *Age* or the *Argus* on 16 August 1882. Both carried obituaries on 11 August, and the *Argus* also had an extensive report on 14 August of his funeral, held on 12 August, at which M was a pall-bearer, and of memorial services held on Sunday 13 August. M's elegy was not reported.

82.08.22 ## *To Robert Pohlman*[1]

Letter book 4, p. 41, Melbourne General Cemetery.

Melbourne
22/8/82

To the Chairman of the trustees[2] of the General Cemetery

Sir.

Since some time I have been anxious to bring under the notice of the trustees of the General Cemetery a subject, which in my opinion needs consideration, but which in all probability was never brought before yourself and your honored Colleagues.

On all occasions, when I was among those, who paid the last homage to a departed friend, by following his mortal remains to the place of interment, I was feeling, that the solemnity of the moment, when the coffin is sunk into the grave, was impaired by the utter want of a becoming dress of the workmen, who aid in sinking the Coffin into its place. The dress of the men I noticed often torn, neither clean nor exhibiting a vestige of mourning. Indeed it seemed, that the men appear in exactly the clothing, which they wear, while engaged in the labor of sinking the hole. Now, doubtless, the poor men are not in the position out of slender wages, to provide a fitting dress, while appearing in front of the mourners; but I am sure, the trustees would not willingly see anything done under their direction, which might hurt the feelings of those, who spend always freely for Coffin-decorant and emblems of Mourning, all of which as a rule is beautiful and often gorgeous; but in what miserable contrast to all this is the dress of the workmen, who sink the magnificent casing, the floral wreaths &c into the grave? I am convinced, that each family, to whom the sorrowful occasion arises, of conveying one of those, dear to them, to a last resting place, would willingly pay the few shillings extra, which

1 MS annotation by A. Purchas: 'Recd 29.8.82 AP.'

2 Robert Pohlman (Chairman), Charles McCarthy, David Ogilvy, Robert Smith, Frederick Cooper, Isaac Hart.

would cover their share of the expense, were it incurred, to provide a change of dress to the grave-labourers, while the Coffin is in their hands, especially as a simple Calico Gown of black color and black Calico trousers would merely be needed for the short while of receiving the Coffin. I felt painfully this short-coming for a long time past, but never so much as on the occasion of the funeral of Sir Redmond Barry the hon. Rob. Ramsay and the Rev. John Stobbs.[3] Under these circumstances I hope, that the trustees will excuse me in the interests of the dignified position held by them, and for which they sacrifice in a philanthropic manner so much of their time, if the shortcoming, which I beg to point out, is brought under their consideration, more particularly as it affects all members of the community at various times in sacred moments. 82.08.22

I am, Sir, with much obedience your
Ferd von Mueller.[4]

Examination Paper[1] 82.08.23

Uncatalogued archive box, School of Mines and Industries Ballarat Limited Museum, Victoria.

1, Set forth the main-differences between Dicotyledonous and Monocotyledonous plants and contrast therewith again the marks of distinction of Acotyledonous Plants.

3 M had attended Stobbs' funeral a few days before he wrote this letter; see M to W. Thiselton-Dyer, 16 August 1882. Ramsay had died the previous May, and Barry eighteen months before that.

4 See also M to A. Purchas, 23 September 1882 [Collected Correspondence].

1 MS is in M's hand and signed and dated by him. Annotated on the back: 'M.S. Question papers. July Term Exam papers'. The register of letters inwards, School of Mines and Industries, Ballarat, records the receipt of a letter from M 're examinations', dated 23 August 1882; letter not found. See also A. Berry to M, 22 August 1882 [Collected Correspondence].

82.08.23 2, Give the principal characteristics of the Order of Malvaceae.
3, Enumerate the most important utilitarian plants of the Order of Cruciferae.
4, Explain the anatomic structure of Eucalyptus wood.
5, State what plants of the Order of Ranunculaceae occur in the vicinity of Ballarat.

Ferd von Mueller
23/8/82.

82.09.19 *From Paul Foelsche*[1]

RB MSS M61, Library, Royal Botanic Gardens Melbourne.[2]

Port Darwin 19th Sept. 1882

Geehrter Herr Baron

Mit dieser Post schicke ich Ihnen noch einige Arten Acacia, kann aber von einigen weder Blüthen noch Früchte finden; was von diesen Exemplaren mit langem schmalen Blatt kommt 10 Meilen von hier vor, kann aber keine Blüthen und Früchte davon finden.

Die zwei nervige Sorte, wo von Sie Früchte zu haben wünsche, habe ich so eben 3 Meilen von hier mit reifen Früchten gefunden und Ihnen zu geschickt.

In Selbigen Pakete finden Sie auch ein Exemplar eines Strauches mit Blüthe und reifen Saamen, der Strauch wächts[3] bis 5 Fuss

1 MS annotation by M: 'Beantw [Answered] 21/10/82 FvM'. Letter not found.

2 MS found with a collection of *Eucalyptus foelscheana* (MEL 1608047-1608050), and an accompanying note by Foelsche: 'Rinde und reife Früchte der klein Blättrigen Euc wo von ich for einigen Monaten Blüthen und Knospen geschickt habe.' [Bark and ripe fruit of the small-leaved Eucalypt, of which I sent flowers and buds some months ago.]; annotated by M: 'Eucalyptus Port Darwin 1882 Paul Foelsche'. Foelsche's passages describing the two Eucalypts have been marked off along the margin, probably by M.

3 wächst – *Foelsche consistently transposes the final two letters of this word.*

hoch und ist am Adelaide Fluss gefunden. Auch finden Sie im Selbigen Paket eine Pflanze die in Sümpfen wächts, an den Wurzeln wachsen kleine Knollen die die Eingebornen essen. 82.09.19

Ich dachte dass ich Ihnen über die Euc. die ich für grandifolia hielte und die Sie mir zu Ehren Foelscheana genannt haben, Beschreibung gegeben hätte. Die junge Pflanze war 15 bis 18 Zoll hoch bekommt einen Bündel Knospen so gross wie eine Faust oben am Zipfel welche sich in grosse weisse Blumen öffnen, wenn ich mich nicht irre habe ich Ihnen mit den Blättern grüne Früchte geschickt. Die voll gewachsene Bäume, so weit ich sie bemerkt habe, werden bis 20 Fuss hoch, sehen aber krächzellich aus und werden sellten über 9 Zoll dick, und haben eine dunkel graue rauhe Rinde und die Blätter sind ungefähr halb so gross wie die ich Ihnen geschickt habe. Der Baum wächts in mageren sandigen Boden und das Holz ist blos zu Brennmaterialen nützlich; Blumen und reife Früchte werde ich, wenn die rechte Zeit kommt schon besorgen.

Die Euc. wo Sie mir eine Probe davon zurück geschickt haben um reife Früchte davon zu bekommen und was Sie denken der ältere Zustand von Euc. Foelscheana sein mag, ist eine ganz andere Art, d. h. wenn es die Sorte ist wovon ich Ihnen vor einigen Monaten Blüthen (weiss gelb) und Knospen geschickt habe, und wo von ich Ihnen jetzt in einem andern Pakete reifen Saamen schicke, so wie ein Stück Rinden eines voll gewachsenen Baumes. Der Baum wird bis 60 Fuss hoch und bis 2 Fuss von Duchmesser; in jungen Bäumen ist die Rinde am oberen Halbe des Baumes weiss gelb, das Holz ist hart und zähe. In manchen Bäumen fliesst aus der Rinde ein rothes Harz welches von den Eingebornen zur Heilung alter Wunden benutzt wird. Der Baum wächts in steinigem Boden und kommt häufig in der Umgegend von Port Darwin vor und auch inland wo er als Telegraphen Pfähle benutzt wird. In Selbigen Paket finden Sie auch eine Pflanze die an trocknen Zweigen und Bäumen wächts und Ihnen wahrscheinlich bekannt sein wird, da sie aber grade in voller Blüthe ist schicke ich Ihnen ein Exemplar.

Die Ipomaea, welche Knollen wie der süssen Kartoffel hat sind jetzt alle tod, ich werde aber aufpassen und Pflanzen mit Blüthen besorgen.

82.09.19 Bambus Blüthen kann ich noch immer nicht bekommen und glaube ich die Eingebornen wissen selbst nicht wann die Blüthe Zeite ist, sie werden aber keine Ruhe haben bis sie Blüthen finden.

Mr Little ist jetzt an Powell Creek, 600 Meilen inland an der Telegraphen linien und wird in ungefähr 2 Monaten zurück kommen, wenn ich Ihnen wahrscheinlich näheres über den Nan Keen gum tree mittheilen kann. Ich interessire mich sehr in Australischen Explorations, und habe Alle Reisebeschreibungen die in Druck sind gelesen konnte aber nie die unseres Landmanns Dr Leichardt's auftreiben, und da ich weiss dass Sie einer seiner besten Freunde waren mogen Sie vielleicht seine Reisebeschreibungen besitzen, sollte dies der Fall sein möchte ich Sie bitten mir Selbige lesen zu lassen ich werde sie sicher wieder zurück schicken.

Herzlich grüssend schliesst
Achtungsvoll
Ihr
Paul Foelsche

Port Darwin,[4] 19 September 1882.

Dear Baron,

With this mail I am sending you several species of *Acacia*, but can find neither flowers nor fruit of some of them. Those of the specimens with the long narrow leaves come from 10 miles from here, but I cannot find any flower and fruit of it.

I have just found the kind with the two veins, of which you desired to have fruit, 3 miles from here with ripe fruit, and have sent it to you.

In the same parcel you will also find a specimen of a shrub with flower and ripe seed, the shrub grows up to 5 feet high and was found at the Adelaide River.[5] You will also find in the same parcel a plant that grows in swamps; little tubers grow from the roots[6] that are eaten by the natives.

I thought I had given you a description of the Eucalypt, which I suppposed to be *E. grandifolia*, and which you named *E. foelscheana*[7] in my honour. The

4 NT.
5 NT. MS annotation interlined by M: 'Osbeckia Australiana Naudin'.
6 MS annotation interlined by M: 'Heliocharis sphacelata' (= *Eleocharis).*
7 *Eucalyptus foelscheana*, B 82.11.01, p. 56.

young plant was 15 to 18 inches high, with a bunch of buds the size of a fist at the apex, which open into large white flowers; if I am not mistaken I sent you green fruit with the leaves. The fully mature trees, as far as I have noticed, grow to 20 feet high, but look scraggy and rarely get more than 9 inches thick; they have a dark grey, rough bark and the leaves are about half the size of those I sent to you. The tree grows in poor sandy soil and the timber is only good for fuel. I shall obtain flowers and ripe fruit at the appropriate time. 82.09.19

The Eucalypt of which you sent me a sample in order to get ripe fruit, and which you believe may be the more mature state of *Eucalyptus foelscheana*, is a completely different species, that is, if it is the kind of which I sent you flowers (whitish yellow) and buds a few months ago, and of which I now send you ripe seed in another parcel as well as a piece of bark from a fully mature tree. The tree grows to 60 feet tall and up to 2 feet in diameter. In young trees the bark of the upper half of the tree is whitish yellow, the timber is hard and tough. In some trees a red resin flows from the bark, which is used by the Aborigines for the healing of old wounds. The tree grows in stony ground and frequently in the surrroundings of Port Darwin, also inland where it is used for telegraph poles.

In the same parcel you will also find a plant, which grows on dry branches and trees and is probably known to you. But as it is just now in full bloom I send you a specimen.

The *Ipomaea*, which have tubers like the sweet potato, are all dead now, but I shall watch them and send plants in flower.

I still cannot get any Bamboo flowers and believe the Aborigines themselves do not know at what time they flower,[8] but they will not rest until they find flowers.

Mr Little is now at Powell Creek,[9] 600 miles inland on the telegraph line, and will return in about 2 months, when I shall probably be able to give you more information on the Nankeen gum tree.

I am very interested in Australian exploration and have read old travelogues available in print, but have never been able to find that of our compatriot Dr Leichhardt. As I know, that you were one of his best friends, you may perhaps own a description of his travels. Should this be the case I should like to ask you to let me read it, I shall certainly send it back again.

I close with sincere regards,
respectfully your
Paul Foelsche.

8 Bamboos flower only rarely and spasmodically.
9 NT.

82.10.01 *To Joseph Hooker*

RBG Kew, Kew correspondence, Australia, Mueller, 1882-90, ff. 32-5.

1/10/82.

Whenever during my bid[1] of remaining life, dear Sir Joseph, I shall see Ottelia ovalifolia growing here, I shall pleasingly think of having been the first to introduce it through Princely Kew into Horticulture.[2] It will be of interest to see, whether it will ripen under glass its seeds readily. That the Experiment with the Livistona did not succeed,[3] is not surprising, the distance of shipping is too far. Stems, quite similarly treated, have grown here in Gardens quite well and made at once a magnificent show, but then it took only 2 or 3 weeks, to bring the stems from Illawarra[4] to Port Phillip.[5] Negative results however teach also something. More important still is the experiment with the Heath-plants from King Georges Sound,[6] as it involves a widely applicable principle, and if we succeed in the plan, suggested by me, a new era will commence in the transit of growing plants. Success, if any, will depend on three or four points: 1, the right season for shipping; 2, previous lifting or rather cutting loose at the spot in the season before; 3, Nicely *balancing humidity*; 4, proper choice (lichens, mosses) of *antiseptic* surrounding in packing. We must not be discouraged by any failures at first, but seek profit from the lessons thus thought[7]

1 bit?

2 See M to J. Hooker, 15 March 1882 [Collected Correspondence]. Seeds received at Kew on 27 April 1972, 'packed in clay, in bottle & in paper', were reported as having 'germinated and flowered 1882' (Kew inwards book 1878-83 [Kewensia], p. 380, record 152a).

3 A trunk of *Livistona australis*, eight feet long, arrived at Kew in 1880 and was described as 'live' in the accession record (Kew inwards book 1878-83 [Kewensia], p. 260, entry no. 384, 29 September 1880). No *Livistona* receipts are recorded for 1881.

4 NSW.

5 Vic.

6 WA.

7 taught?

I almost thought the woods would be extraspecimens there;[8] 82.10.01
but I *did not wish to pass* Kew, and you will do me the justice to affirm, that I never was guilty of that in the more than 30 years of my intercourse with Kew. I am glad, that the timber went to Prof. Sergeant,[9] who really does take a lead in America as regards Xylology. The specimens, as *named* with *exactitude*, were valuable, for it is not so easy, to make sure of names botanically, when one goes even to much expense in getting timber specimens from *mixed* natural forests. Some of the kinds of trees moreover will soon be extinct. Was I really the *first* to design woodbooks? I never heard of the method, til I adopted it in 1862 for the second London Exhibition, when the series from here, went, I think, by my request to Kew. We had none in the Exhibition here last year, but woodbook in several patterns *were* exhibited in Philadelphia;[10] but that was a dozen years after my having a series of them made. I hope you approve of the cheaper and simpler plan, which my sample of Acac. Melanoxyl. demonstrates. As regards an 8th vol. of the Flor. Austr., of course there would be *no sense* to adopt an other arrangement than Bentham's, as otherwise confusion would arise, notwithstanding an index. It is otherwise with any *distinct* work, such as the species-census of Australian plants,[11] for which any method may be adopted without inconvenience. I wrote my views on systematic sequence in a long letter to your excellent son in law recently.[12]

I stand not alone in attempting to make changes for the better in D.C.'s or Juss. system,[13] as Thunberg[14] ventured to do for Linne's.

8 See M to J. Hooker, 22 May 1882 [Collected Correspondence], for a list of timber specimens retained at Kew.

9 C. S. Sargent.

10 Centennial Exposition, Philadelphia, 1876.

11 B82.13.16.

12 No letter to Thiselton-Dyer written between January and October 1882 that fits this description has been found, but M to W. Thiselton-Dyer, 30 March 1883 [Collected Correspondence], implies that Thiselton-Dyer had made some comment in the letter to which M was then responding.

13 A. P. De Candolle (1816-21); Jussieu (1789).

14 Linnaeus (1791).

82.10.01 I will not speak of Caruel's notions on Phanerogams,[15] but there are now (after Grisebach) two men who have *seen much*, namely Eichler[16] and J. Mueller,[17] who have adopted systems of their own, which they are quite free to do, but which only those are *forced* to adopt, who may be dependent on them. Real good innnovations, such as abound in your & Bentham's genera,[18] as well force their way to permanency, by their intrinsic merit; but therewith is not said, that we have as yet arrived at the *best system*. Eichler's is not giving the best consociations of orders in many cases. J. Mueller's, only published in outline, is objectionable, because with Fries he places the Monopetaleae at the upper end of the system, the idea being (you having perhaps not seen his system), that by inserting the Monochlamydeae into Thalamiflorae & Calyciflorae, he got some *supposed* low orders, de[m]anding to stand beneath Monopetaleae in which (curious enough) a corolla is always developed or nearly alway. Now, – this is not logic reasoning; because with D.C. we must regard apocarpous development as the most potent evolution of pistillary organs. Moreover Euphorbiaceae are really petaliferous plants, which are subject to deficiencies in the corolla; but this very fact gives us the true place for Urticeae & so forth.

I never called my arrangement a system of my own; though in an overgenerous spirit my friend Woolls calls it so;[19] it merely wants to put away the Monochlamydeae, (as Thunberg abolished the Monoecia, Dioecia and Polygamia, so as to get together more closely [–:] Gramineae, Cyperaceae &c).

Jussieu & De Candolle have evolved the natural system so well, that all the improvement, to which it is susceptible, consists in finding the true places of the Monochlamydeae, and in doing that I have perhaps not been far from the mark, though I had not the benefit of knowing J. Mueller's details. I shall say something *dignified* and *just* on the same subject in the preface to the Census[20] at

15 Caruel (1881).
16 Eichler (1875-8).
17 Relevant reference not identified.
18 Bentham & Hooker (1862-83).
19 Woolls (1880a).
20 B82.13.16.

the end of this year, when the Monocotyledoneae will also be out (I hope) being unfortunately unable to stop printing, which is under private contract, so that I shall probably be deprived of the advantage to consult your genera for Monocotyledoneae.[21] As I am on this subject, let me still say, without wishing to be prolix, that I originally intended the census of the Austr. gen. and also of the species of plants for my Museum use. That it got printed, is not only an advantage to my Departm., but also to other institutions. It is *not* an elaboration of late years. I suggested it in 1862, when you commenced the genera. How awkward the Monochlamydeae are as stumbling blocks, I find out by any junior assistant in the Museum. Even lately an excellent Botanist of European Experience sent me, as belonging to *one* genus several Polycarpaeas & Gomph[renas], and you found yourself, in dealing with Santalaceae and Proteacea & Loranthaceae, that you had to place the latter into Monochlamydeae, where few would look for them among beginners. 82.10.01

Always very regardfully your
Ferd von Mueller.

Last evening I had to preside and to make the presentation of a gift of the Liedertafel to Mr Moroney.[22]

It remains a singular fact, that the very genus with which D.C. syst. commences (Clematis) *is* in most spec. apetalous!

21 Published in April 1883 in part 2 of vol. 3 of Bentham & Hooker (1862-83).

22 Moroney, a bass singer, was leaving for England, and M gave him letters of introduction; see M to J. Hooker, 1 October 1882 (second letter) [Collected Correspondence].

82.10.02 *From John Thurston*

RB MSS M76, Library, Royal Botanic Gardens Melbourne.

Suva Fiji
2[d] October 1882

My dear Sir,

I have to thank you very much for your several kind letters[1] and interesting enclosures all of which reached me safely and would have been acknowledged before but for an event which I think I may predict will not happen again.

We have just removed the seat of Government and all the Staff of offices from Levuka in the island of Ovalau to the Port of Suva in Viti Levu. The worry and difficulty, and the dislocation of all our habits of routine are things I cannot describe to you.

Directly that I have settled down in my new abode I will make attempts to secure the male amentum of the Cycas. I observe they grow in my neighbourhood but whether the oval or round seeded variety I have not yet had time to ascertain. One that I saw the other day was certainly 35 feet high – but I intend measuring it.

The large bananas – "Soanka" in the Fijian such as I sent you some time since per s.s. 'Suva' are of three or four varieties. The one ordinarily seen is erect, with bright red fruit about 3 inches long. The very large varieties are only found in the depth of the unfrequented forests

Of these there are two sorts – the one I sent you – and another in which the fruit is nearly as round and as large as a coconut. I am planting each sort in my garden to observe their peculiarities more closely –

Your suggestion to send a good sized Cycas to Kew shall be carried into effect, but I fear that if it arrived in England at the beginning or middle of winter it would perhaps die –

1 Letters not found.

I am in frequent correspondence with Mr Thiselton Dyer, in fact almost every Mail, and never lose an opportunity of sending him plants which I think may be of interest and as a rule my cases reach England in splendid condition 82.10.02

Mr Weber[2] has quite recovered from the effects of his rough treatment in Samoa – I believe that he has left for Germany but am not sure – He was not a pleasing young gentleman. I gave him free passage all over the group in Government vessels, store room for his boxes & collection – free quarters at several places which he visited, and did everything I could to help a young Botanist who I sensed was not overburdened with money But he never showed me a single thing that he collected, nor said 'thank you' for the trouble I had taken about him In short he seemed to consider himself injured because I was not disposed to pay all his expenses while in Fiji

There are two Casuarinas in Fiji 'Equisetifolia' and 'nodosa'[3] I know of no others

I got a new variety of "Da[ckua]" the other day so far as the cone went and it is the most perfect cone I have found in the Tropics. A native brought it out of the bush with him but without leaves or other parts – I send it to you – By & bye you shall have leaves & flowers

Faithfully yours
John B. Thurston

2 Theodor Weber, representative of Cesar Godeffroy & Sons in Samoa? See Davidson (1967), p. 46.

3 *C. nodosa* not in *IK*.

82.10.16 *From Sidney Johnson*[1]

RB MSS M12, Library, Royal Botanic Gardens Melbourne.[2]

Meredith[3]
S. School[4] No 1420
Octr 16th 1882

Baron F. Von Mueller –

Sir –

If, at your leisure you can send me the name of the accompanying flowers I shall be greatly obliged – As I get the names, I give object lessons on them in the school, as the bulk of the specimens are easily obtainable by the school-children. They take great pleasure in it.

I was unable to get the volume of your F. Australiensis[5] in Geelong. – At least only one Volume of the Descriptive portion – & unfortunately that did not contain any of the plants the names of which you have so kindly furnished me with. I shall try Ballarat next time.

With the hope that I am not proving a nuisance to you

I have the honour to be
Sir
Your obedient servant
Sidney Johnson

1 MS annotation by M: 'Answ 16/10/82'. Letter not found.
2 MS found with a specimen of *Chamaescilla corymbosa* (MEL 106922).
3 Vic.
4 i.e. State School = a primary school established by the government.
5 Bentham (1863-78)?

To James Stirling

82.11.02

Private hands.

Private

2/11/82

You will have found it strange, dear Mr Stirling, that I should not have earlier acknowledged your kindness of sending me, excellently colored and clearly indicated geologically, the geographic map of the Alps. The fact is, that the[1] parcel, when it arrived, got mixed up with some others, and only turned now up again in my small temporary office.

It does you infinite credit, to have worked out with Mr Howitts help, the geology of your district so well. In one respect I cannot help expressing regret, when looking on this map; it is the systematic manner in which all my early & toilsome work for the geography of the Australian Alps has become suppressed, even to the extent of the real Mt Hotham and to Mt Latrobe,[2] though they were named in a special despatch from Omeo as far back as 1854,[3] after my having ascended the two mountains, fixed approximately their position and measured their height near enough, though with the most scanty of instruments. That despatch was at the time at once placed before the then "Council" by Sir Charl Hotham, long before I came back to Melbourne printed, irrespective of my fuller account of the Alps in my annual Report of 1855,[4] written and printed before I went to Arnhem's Land. You will kindly understand, that I am to some extent aware, how these arbitrary and unjust changes in the nomenclature of several of our highest mountains occurred or were brought about long before your time; still it must be source of lasting regret, that in this way also the names of Sir Th. Mitchell & Sir Andrew Clarke, two of the three first

1 I should … that the *has been underlined.*
2 Vic.
3 M to W. Haines, Colonial Secretary, 16 December 1854, published as B55.01.01.
4 B55.11.01.

82.11.02 Surveyor Generals of our Territory became obliterated! Is there no means of remedying this yet?[5] Priority should also be respected in Geography[6]

But now let me say, as more to the point at present, that we must not attempt more in first instance, than to note the names of those plants, which are almost absolutely confined to marked geologic formations, except the gregarious kinds of trees, which constitute *real forests* and therefore great physiographic features. In this sense I have also written some time ago to Mr Howitt.[7]

By December the first volume with all "Vasculares" of Australia will be through the press[8] unless I should fall ill, and I certainly overworked myself for some months past. A copy will be sent you, and that will facilitate your insights also in the Flora there!

Regardfully your
Ferd. von Mueller.

There is no hurry for notes concerning the tracing of *particular* plants to strict geologic areas, as my next lecture will be delivered only after autumn. I would advise you, to continue your notes on elevations of various species, and I will look up my own data, mostly yet unpublished, on the same subject, though one of my fieldbooks was lost in one of the journeys, which happened to be one of particular sufferings and distress. Kindly keep your eye on minute & other *mosses*, where-ever in *fruit*, also on lichens. Among the *sedges* may also yet be novelty

Concise climatic statistics very acceptable indeed.

5 In one respect … this yet? *has been underlined.*

6 See also M to J. Stirling, 25 December 1882 [Collected Correspondence]. M had submitted a manuscript on this topic to the Royal Geographical Society in November 1876 [Collected Correspondence], but it was rejected for publication, Francis Galton, the referee, reporting that 'Baron Mueller's hope that a Geographical Congress could frame rules for Geographical nomenclature "to which the whole world would bow" seems to me too Utopian.'

7 Letter not found.

8 B82.13.16.

To George Bentham 82.11.08

RBG Kew, Kew correspondence, Australia, Mueller, 1882-90, ff. 38-9.

8/11/82

It is always with gladness, dear Mr Bentham, that I receive letters from you. If I have not often adressed you within the last few years, you must seek the cause in my reluctance of encroaching on your valuable time, which with your unimpaired energy you devote still to such important and difficult subjects in phytographic science. I renew my wish, that you may be spared for scientific purposes to an age like that of Humboldt, and longer still.[1] I look forward with deep interest to the part of your & Hooker's celebrated "genera plantarum", containing the Monocotyledoneae and the Supplements;[2] and still hope, that you and he will be induced to see the work really completed by the aid of great workers on "Evasculares", such as Berkeley, Agardh and other specialists before they pass away.

Unforunately I cannot avail myself of your & Sir Joseph's new observations on Monocotyledoneae in time for my Census of Australian plants, as that publication is under contract at a private printer and must be through the press, so far as Vasculares are concerned, by the end of this year.[3] This issue ought greatly to facilitate reference to Australian plants, and I hope to bring out the "Evasculares" (hitherto about 4000 species) in 1883,[4] with supplement to the Vasculares, because I have not been able to work up *all* the additional collections in my Museum, irrespective of such as may still arrive from new localities. My efforts, to keep up correspondence in many parts of Australia, are very great, and I leave the writing table seldom before midnight, there being also so much routine-work in my Department, without the former facilities for it.

1 Humboldt was almost 90 when he died.
2 Published in April 1883 in Part 2 of vol. 3 of Bentham & Hooker (1862-83).
3 B82.13.16.
4 This work was never published. The MS is at MEL.

82.11.08 In the preface to the Census you will find allusions pleasing to yourself, I trust.[5]

As regards an eight volume of the flora,[6] I merely wished to give the bare description of nearly 1000 vascular plants, added to my collections successively. The sequence would of course be precisely that of the 7 volumes, as otherwise confusion would arise; but I shall let it stand over til 1884,[7] as I expect yet some specific addenda from collectors in 1883. Your Flora Australiensis must ever be the full foundation for systematic native Botany in this part of the Globe; but there is no neccessity for alteration or geographic addition, though for specific addenda.

The Census is arranged in accordance with the method, in which my Museum collections are placed, and was originally not intended for publication, but merely for use in my herbarium-working. But the issue of it in print facilitates my transactions with Australian Correspondents.

I feel it quite a comfort to sort out collections in my Museum in the manner indicated in the Census, as thus I get out of the ambiguity of many Amyliferae or Curvembryoneae &c. Indeed the Census does not offer a new system, such as recently Eichler's,[8] J. Muellers,[9] Caruel's,[10] but merely brings the chaotic mass of Apetalae or Monochalmydeae into order; therefore my arrangement is that of Juss. & D.C.[11] with only one modification.

5 B82.13.16 is dedicated, 'as an appreciative tribute from a young colony in antipodal remoteness', to George Bentham, Joseph Hooker and Alphonse de Candolle 'who as heirs of great names worthily sustain world-wide ancestral fame, and who as leaders in phytography during more than a generation's time will be pre-eminent in biomorphic science through all ages'. Bentham did not like M's *Census*; see G. Bentham to M, 25 April 1883.

6 i.e. Bentham (1863-78). The '8th volume' presumably refers to a Supplement that M was expected to prepare; see G. Bentham to M, 27 June 1877. No such supplement was published.

7 1883 *deleted before* 1884.

8 Eichler (1875-8).

9 Relevant reference not identified.

10 Caruel (1881).

11 Jussieu (1789); A. P. De Candolle (1816-21).

Let me hope, that your health will continue firm, and that your interest in phytography will continue for many years yet a leading one, especially as your eyesight has remained unimpaired. 82.11.08

Very regardfully your
Ferd. von Mueller

The fire in the Gov. Printing Office here has prevented the continuation of the Eucalyptography[12] & the Fragm.[13] but both will go on again soon, & into the Fragm. I will collect, what became lately scattered.

To Lajos Haynald[1] 82.11.22

Magyar Természettüdományi Múseum, Budapest.

Melbourne,
22/11/82

Ganz mit Rührung, Eminenz, wurde ich erfüllt, wie ich Ihr gütiges Schreiben an S. Gnaden den Erzbischof von Mexico las und die Antwort dieses hochstehenden Prälaten. Gewiss wird diese herablassende Güte Ihrerseits und der örtliche Einfluss des Erzbischofs uns in den Stand setzen, auch der Pflanzen-Technologie Mexico's gerecht zu werden in der wissenschaftlichen Litteratur. Mit heutiger Post sende ich die Sydney Ausgabe der "select plants" an Seine Gnaden in Mexico, wodurch der Erz-Bischof in den Stand gesetzt wird, noch genauer unsere Wünsche kennen zu lernen. In-

12 See B79.13.11 and B80.13.14 for the first parts of *Eucalyptographia*; later parts appeared as B82.13.17, B83.13.07 and B84.13.19.

13 *Fragmenta phytographiae australiae*. One further part was issued in December 1882 (B82.12.03) but no more thereafter.

1 MS annotation: 'Muller fero 12 resp 15/1 1883' [Muller received 12 replied 15/1 1883]. Letter not found.

82.11.22 zwischen ist die deutsche Ausgabe von Dr Goeze bei Hr Th. Fischer in Cassel erschienen, u. ich habe den Verleger beauftragt Ihnen, Eminenz, ein Exemplar zur gütigen Annahme zu senden.

Auch hier überstürzt sich Alles im Aufbau der jungen Colonien, und da die Welt-Austellung von Melbourne meine Thätigkeit als einer der Commissäre und in 3 professionellen Abteilungen (die für Chemie eingeschlossen) als Preisrichter durch mehrere Monate erheischte, ist in meinem Departement gar vieles in Rückstand gekommen, zumal da ich noch in 1882 einen geogr. u chronolog. Census der Vasculares Australiens (etwa 8000 Arten) zu beendigen habe, denen in 1883, deo volente, die 4000 Evasculares folgen sollen. Dies Werk wird eine bequeme Übersicht liefern über die Gesammt Vegetation Australiens, und ich werde die Ehre haben, Ihnen ein Exemplar für Ihre Bibliothek zu senden, auch 1 für Ihre Akademie.

Unter einigen neuen eben angelangten Pflanzen, werde ich eine Art wählen, um Ihren hohen Namen bleibend mit der Vegetation des fernen grossen Südlandes in Verbindung zu bringen. Es war längst meine Absicht, Ihnen diesen wissenschaftlichen Tribut zu liefern, aber seit langer Zeit fand sich keine entschieden neue Pflanze schön genug, um Ihnen gewürdigt zu werden.

Mit S. Gnaden dem Erzbischof Dr Goold dieser Colonie habe ich die Ehre seit 30 Jahren persönlich befreundet zu sein. Im Jahre 1853 reisten wir durch die damals ungebahnten Wildnisse von Gippsland zusammen; in einem Ritt von 6 Morgens bis 10 Abends fast ohne Unterbrechung und ohne an irgend eine Niederlassung zu kommen, gelangten wir an einen Fluss, den wir durchschreiten mussten, geschwollen von Winter Regen in tiefer Dunkelheit, um das weit jenseits gelegene nächste Haus zu erreichen. Der Erzbischof selbst musste mit grösster Lebensgefahr fast bis zu den Schultern hinauf durch das strömende Wasser gehen!

Später nahm der Dr Goold auf meinen Rath für antimalarische Zwecke eine bedeutende Menge Eucalyptus-Samen von mir nach Rom, wie das erste grosse Conseil stattfand. Mehrere Jahre nachher, zur Zeit des 2t päbstlichen Rathes zeigten die Cardinäle unserem Erzbischof die bereits heranwachsenden Eucalypten zu Tre Fontana u anderwärts bei Rome an den Fiebersümpfen; u. dies war der Ursprung der ausgedehnten Eucalypten Pflanzungen, zu welchen ich auch die antiseptischen Tannen und Föhren Californiens u Ober Indiens aber noch nicht Mexico geliefert oder vielmehr zu liefern beginne, welche einen so grossen hygienischen Einfluss dort ausüben, u wegen welcher Massregeln ich noch immer mit dem Grafen Torelli in Verbindung bin, der im Italienischen Senat diese wichtige auch anderen Erdgegenden namentlich auch Spanien u. Cuba zukommende Massregel vertritt, welche auch in Dalmatien gefördert werden könnte, wie ich vor Jahren hier dem Feldmarschal Jochmus auseinander setzte. 82.11.22

Verehrungsvoll der Ihre
Ferd. von Mueller.

Ich unterstütze jetzt geldlich einen cathol. mir befreundeten Geistlichen, Rev. B. Scortechini, in einer botanischen Reise zu den S. östl. Gebirgen von Neu Guinea.

Bei gelegentlicher Zusammenkunft bitte ich Sie meinen ehrerbietigen Gruss den leutseligen Prinzen von Coburg-Gotha darbringen zu wollen.

Ich hielt bei der Schiller-Feier auch eine ex tempore Ansprache, unter anderem bemerkend, dass die Poesie von Schiller wie die von Homer noch nach Jahrtausende fortleben werde.

Bei der Jahresfeier des Geburtstags des Kaisers von Österreich hatte ich die Ehre zu präsidiren.

Melbourne
22/11/82

I was quite filled with emotion, Eminence, as I read your kind letter to His Grace the Archbishop of Mexico and the answer of this high-ranking prelate.[2] Certainly this condescending kindness on your part and the local influence of the Archbishop will put us in a position to do justice in the scientific literature to the plant technology of Mexico too. By today's post I am sending the Sydney

2 Letters not found. M sought Haynald's help in soliciting information about Mexican plants in M to L. Haynald, 24 January 1882 [Collected Correspondence].

82.11.22 edition of the *Select plants*[3] to His Grace in Mexico, by which the Archbishop will be better placed to understand our wishes. Meanwhile the German edition by Dr Goeze[4] has been published by Th. Fischer in Cassel and I have instructed the publisher to send you, Eminence, a copy for kindly acceptance.

Here, too, everything rushes ahead in the development of the young colony and since the International Exhibition in Melbourne[5] demanded my activity for several months as one of the commissioners and in 3 professional sections (including that for Chemistry) as judge, very much in my department has fallen behind, especially since in 1882 I still have to complete a geographical and chronological Census of Australian Vasculares (about 8000 species),[6] because in 1883, God willing, the 4000 Evasculares shall follow.[7] This work will give a convenient overview of the whole vegetation of Australia, and I will have the honour to send you a copy for your library, as well as 1 for your Academy.

Among some new plants just arrived, I will choose a species in order to bring your distinguished name permanently into connection with the vegetation of the distant great south land. It has long been my intention to give you this scientific tribute, but for a long time I found no decidedly new plant attractive enough to be worthy of you.[8]

For 30 years I have had the honour to be a personal friend of His Grace Archbishop Goold of this Colony. In the year 1853, we travelled together through the then untrod wilderness of Gippsland; in a ride from 6 in the morning to 10 at night almost without break and without coming to any settlement, we arrived at a river which we had to walk through, swollen with winter rain in deep darkness to reach the next house situated far on the other side. The Archbishop himself with the greatest danger to life had to go through the flowing water almost up to the shoulders![9]

3 B81.13.10.

4 B83.13.06.

5 International Exhibition, Melbourne, 1880-1.

6 B82.13.16.

7 Never published.

8 In B83.03.02, M named *Hibiscus Haynaldi* 'in recognition of his Eminence's researches into biblic phytology and of his strenuous endeavors to advance vegetable technology'.

9 There are no private letters in the Goold correspondence held at the Melbourne Diocesan Archives. Goold noted in his diary, 21 April 1853, that at an overnight stop on a trip to Gippsland, he met M, 'a German Doctor who introduced himself as Colonial Botanist'. The locality was somewhere south-east of Dandenong on the road to Western Port, probably the La Trobe Inn, Pakenham, Vic. See Darragh (2003).

Later, on my advice, Dr Goold took a considerable number of *Eucalyptus* seeds from me to Rome for antimalarial purposes, when the first great Council took place.[10] Several years later at the time of a 2nd pontifical council the Cardinals showed our Archbishop the already tall-growing eucalypts at Tre Fontana and elsewhere near Rome on the fever swamps; and this was the origin of the extensive plantings of eucalypts, to which I also supplied the antiseptic firs and pines of California and upper India, but not yet Mexico or rather am beginning to supply, which exert such a great hygenic influence there, and on account of which measures I am still in contact with Count Torelli,[11] who in the Italian Senate looks after these important measures, also reaching other regions of the earth especially also Spain and Cuba, which also could be promoted in Dalmatia, as I pointed out to Field marshal Jochmus here some years ago.[12] 82.11.22

Respectfully your
Ferd. von Mueller

Now I am financially supporting a catholic clergyman friendly with me, Rev. B. Scortechini, in a botanical expedition to the southeast mountains of New Guinea.

In the case of an accidental meeting I ask you to give my respectful greeting to the affable Prince of Coburg-Gotha.[13]

At the Schiller celebration I also delivered an *ex tempore* address remarking amongst other things that the poetry of Schiller, as that of Homer, will still survive after thousands of years.[14]

I had the honour to preside at the celebration of the birthday of the Emperor of Austria.[15]

10 i.e. 1869. See B80.13.14, entry for *Eucalyptus globulus*, and Barclay-Lloyd (1994).
11 See M to L. Torelli, 28 December 1876.
12 Jochmus visited Melbourne in 1870. See also A. von Jochmus to M, 8 March 1872 [Collected Correspondence].
13 Ernst II, Duke of Saxe-Coburg-Gotha. For M's earlier contact with the Duke, see M to A. Petermann, 26 August 1864 [*Selected correspondence*, vol. 2, pp. 257-62].
14 Attached to the letter is a clipping from the *Argus*, 11 November 1882, p. 9, reporting the Schiller comemoration at the Turn Verein in La Trobe Street, Melbourne. On this occasion M read Schiller's 'Rudolph von Habsburg'.
15 The birthday of Emperor Franz Joseph I (reigned 1848-1916) was on 18 August and a report of the celebratory dinner over which M presided appeared in the *Argus*, 19 August 1882, p. 11.

82.12.25 *To Archibald Liversidge*[1]

Royal Society of New South Wales, Sydney.

Melbourne,
Christmas 1882.

To Professor Archib. Liversidge, FRS., F.C.S., Honorary Secretary of the Royal Society of N.S.Wales.

Dear Professor

It is to me a source of infinite delight, that the Council of the Royal Society of New South Wales in so generous a spirit has bestowed on me the Clarke Memorial Medal for 1883, a distinction, which I prize beyond expressions in words adequate of my profound gratitude. Reflecting that the eldest scientific Society of Australia has selected me among the very first, on whom this token of scientific encouragement has been bestowed,[2] I may well be proud to be thus early enrolled on a list, which in the course of generations is sure to contain a long series of illustrious names, with which those of the first recipients of this honor will be brought historically into contact within the realm of science. For myself I treasure the Clarke-Medal all the more, as for more than a quarter of a century I had the privilege of scientific intercourse with its renowned founder, who while he left an enduring fame by his own

1 M's letter was read at the Royal Society of NSW on 2 May 1883, when the president, Christopher Rolleston, stated: 'At the Council meeting held 13th December, 1882, it was unanimously resolved to award the Clarke Medal for the year 1883 to Baron Ferdinand von Mueller, K.C.M.G., F.R.S., &c., Government Botanist, Melbourne.' (*Journal and proceedings of the Royal Society of New South Wales* (1883), vol. XVII, p. 212). By implication this was for his contributions to Australian natural history since the medal was established in 1878 'for meritorious contributions to the Geology, Mineralogy, or Natural History of Australia, to men of science, whether resident in Australia or elsewhere' (*Journal and proceedings of the Royal Society of New South Wales* [1882], vol. XVI, p. xlvii). M's letter was subsequently published in the Society's proceedings; see B84.13.15.

2 Previous recipients were Richard Owen, 1878; George Bentham, 1879; Thomas Huxley, 1880; Frederick McCoy, 1881; and James Dwight Dana, 1882.

great geologic researches in this part of the globe, was ever eager to promote the work of younger investigators entering successively the field of knowledge. The features of the Reverend and venerable sage on the Medal call vividly to my mind the several moments, when I met him in life, and render this medal a special souvenir. May I trust, that the opportunity will arise, to show in some tangible manner my gratefulness to the learned Society, which honored me so highly in preference to others, who had higher claims on this treasurable and lasting distinction. 82.12.25

Let me remain, honored Sir,
regardfully your
Ferd. von Mueller.

To Frederick Broome 83.01.02

Acc. 527 (1878-83), 522/123, State Records Office of Western Australia, Perth.

Melbourne,
2. Jan. 1883.

To his Excellency
Fred. Napier Broome Esqr, C.M.G.,
Governor of West Australia.

Sir.

In the interest of science your Excellency will excuse, when I as a stranger approach you, to seek a favor. I have been informed, that the Survey-party, which is to proceed to the northern territory of W.A., will be accompanied by a geologic observer and collector. To augment the scientific data, thus to be gained on W.A. resources, I would recommend for your Excellencys consideration, that also a botanic collector should be attached to the party, who would be able to collect the various grasses, saltbushes and herbage-plants, on which the pastoral capabilities of the Northern Districts there

83.01.02 depend. Such a collector would also obtain specimens of all other sorts of plants as material for extending the researches, which I personally instituted in the N.W. during the years 1855 and 1856 under the auspices of His Grace the Duke of Newcastle, then H. Maj.[1] Minister for the Colonies.

Should my proposal meet with your Excellency's approbation, I would draw attention to the ability of Mr Joseph Polak, a Gardener of superior training, now unemployed in WA. and therefore available for the duties indicated and that at a very moderate expense.

I have the honor to be your Excellency's obedient

Ferd. von Mueller, M.D.[2]

83.02.21 *From Charles Fawcett*[1]

RB MSS M7, Library, Royal Botanic Gardens Melbourne.

Kynnumboon. Tweed River,[2]
21 Feby /83

My dear Baron,

As I think I mentioned in a former letter[3] I am here doing duty, during his absense on leave, for a friend who is Police Magistrate and Land Agent for the Police district of the Tweed which in-

1 Her Majesty's.
2 MS annotation by Broome: 'C.S. Reply that this letter was handed to me by Mrs Forrest on the 13th inst. State botanical arrangements, if any, of survey expedition, & when it started. *F.N.B.* 14.6.83.' MS annotation in another hand: 'Baron Mueller written 15.6.83.' Letter not found.

1 MS annotation by M: 'Answ 7/3/83 FvM'. Letter not found.
2 All places named are in NSW. Marginal note by Fawcett: 'My address till I write otherwise'.
3 Letter not found.

cludes the Brunswick. I forward, in tin canister by same post, branch and fruit of small tree about 10 ft. When procuring timber specimens for the Government I noticed the tree near Ballina and the blacks informed me it bore a red fruit which they ate, but I could not get flowers or fruit then – I however sent you a branchlet such as you will find in the canister, though it may not arrive in a good state, as I have to take advantage of the post so as not to keep the fruit too long – The blacks' name is Bĭltarў as I booked it at the time – I have also the flower and leaf of what is *to me* a new nut. Having no spirit here I placed the two flowers brought to me in a bottle of Kerosene – one flower (the raceme) (fully developed) cut in 3 parts and the other in two. I have been told this tree did not make its appearance until the land was cleared (as some weeds have & the "Cape Gooseberry") that there is a thin red fruit over the stone which is not quite so hard as that of Macadamia ternifolia and that the kernel is good to eat but better roasted – and I was told by the person who brought it to me (one of the oldest residents) that the Blacks did not know it when it appeared first (not long ago) I have not yet been able to spare time to go to see one of the trees but I shall before I send the specimen off. On account of the Kerosene I cannot send by post, but can by the postman to Lismore, whence it can go by steamer to Sydney, &c – 83.02.21

With best wishes I am
yours truly,
Charles. H. Fawcett.

Baron F. von Mueller
&c &c &c.

83.02.26 *To Malcolm Fraser*

Acc. 527 (1878-1883), 1519/22, State Records Office of Western Australia, Perth.

26/2/83.

In first instance, dear Mr Fraser, let me congratulate you to the Premiership of W. Australia, which you with your energy and experience in local requirements will hold with great advantage to the colony.

Secondly let me thank you, for sending me a complete copy of the Report issued by you as Chairman of the Forrest-Commission.[1] As now all sides of this great local question are elucidated, the document ought to be valuable there for present and future forest-arrangements. Accept my best acknowledgment for the graceful manner, in which you alluded to myself in this respect.

As regards your question, whether Pines would be recommendable for forest-culture on the coast & islands there, I can emphatically encourage you in fostering such an object, particularly as you have no native trees fit for masts, spars and many other purposes, for which Conifers are indespensible. Should you not have the last edition of my "select plants for industrial culture and naturalization",[2] I would advise you, to ask your colleague in Sydney, the hon. the Colonial Secretary, for one or more copies, which the hon. Gentleman would send you for your patriotic purposes with the greatest readiness. In that volume (printed in 1881) I have marked all the best pines with asterisks. I only made a sending of pine-seeds lately again to the Italian Government for the sanitary plantations in the malarian districts of that Kingdom. So soon as you have completed some arrangements for Forest-nurseries, I will gladly procure fresh seeds of Himalaian, North American and European pines for your purposes also there. Sendings, sufficient for experiments, will be sent free. My last supplies went latterly to the Himalaian Kingdom.[3]

1 Fraser (1882).
2 B81.13.10.
3 Nepal?

A *few* more copies of your complete report would be acceptable for distribution abroad in the interest of your colony. 83.02.26

It would be well, to reserve some of your islands for Forest-culture; I recommended the same as regards Kangaroo-Island for South-Australia, and Fraser's Island for Queensland.

If I can in any way promote any of the rural interests in your wide territory, it will be done with much pleasure.

Regardfully your
Ferd. von Mueller.

To Messrs Watson & Scull 83.02.28

Unit 22, VPRS 1163/P1 inward correspondence, VA 1123 Premier, Public Record Office, Victoria.[1]

Copy

Melbourne, 28th February 1883

Messrs Watson & Scull
London

Gentlemen

By the P. & O. steamship Siam you will receive the large case with the Todea-fern, concerning which I wrote to you by last mail,[2] as intended for the Horticultural Exhibition of St Petersburg. As this international floral fete commences early in May, it is necessary, that the Todea should speedily be despatched to the final destination; but as in April still severe frosts may occur in St. Petersburg

1 MS is a 'transcript from the Departmental letter-book' prepared by Georg Luehmann; see M to T. Wilson, 3 April 1884 [Collected Correspondence]. It was registered as B84/3072 in the Chief Secretary's Department and then referred to the Premier.

2 Letter not found.

83.02.28 and even in England, care should be taken, to prevent exposure to low degrees of temperature. It is also undesirable to open the case in London, as it could not be properly closed again and as thereby the growth of young fronds would be disturbed. An assistant of mine has seen the huge case on board and safely lodged in the hold of the ship, so that under ordinary circumstances no damage through seawater, rats or frost can occur. The chief officer has taken a particular interest in this sending, and I would advise you, to see that gentleman personally on board, so that proper arrangements for safe reshipment are made. I deemed it best that the freight be paid in London[3] and have therefore sent a draft for £20 out of which the freight and the commission can be paid; and it seems best to defray also on my account the freight to St. Petersburg so that this exhibit may arrive there quite free, after the heavy expenses of dragging it here out of the ranges. Please telegraph at once after the receipt of this letter to His Excellency Dr. Von Regel, and see personally to the safe transit on board of the St Petersburg steamer, explaining to the officer in charge, that frost and seawater would be detrimental to this precious sending, and that a case weighing over 4000 lbs needs special arrangements for handling and for location in the hold of his ship.

Please be also very careful that you get the right case, as on the Siam among the Exhibition goods for Amsterdam is another case with a Todea, quite as large but not so heavy. It so happened that a second Fern-giant was discovered and so I sent it through the Victorian commissioners to Amsterdam. The case to Petersburg is distinctly addressed to Dr. Regel.

I require at once a special account, properly receipted, to be made out by you for freight and commission, so that it can be passed to the Treasury here.

Respectfully yours
(Signed) Ferd. von Mueller[4]

3 MS annotation by M: 'was eventually paid in Melbourne'.

4 See also M to G. Berry, 19 March 1883 [Collected Correspondence], in which M sought approval for charging the freight to his departmental account.

From George Bentham[1]

83.04.25

25 Wilton Place, London, S. W.,
April 25, 1883.

My Dear Sir, –

I have to thank you for your *Systematic Census of Australian Plants*,[2] received yesterday. The work is beautifully printed, and shows a great deal of laborious philological research into the dates of plant names (rather than of genera), which will be duly appreciated by those who occupy themselves with that subject, now much taken up; – but all that is not botany.

With regard to that science, it grieves me to think that you should have devoted so much of your valuable time to a work which, botanically speaking, is not only absolutely useless, but worse than useless. The interfering with established sequences of orders, without discussing in each instance the reasons for and against the doing so, is only producing confusion in the minds and collections of systematic botanists; and the wholesale amalgamation of genera, without any indication of the characters to be assigned to the new compound genus, or of its relations to allied genera retained as distinct, has no other effect than the unnecessary addition of many hundred names to the already over-loaded synonymy. I should much doubt whether you would find any one but a few of your immediate followers agree to placing Passifloreae and Cucurbitaceae between Rubiaceae and Compositae, and as to amalgamation of genera, there can be no great objection, beyond the multiplication of synonyms, in treating well-defined

1 Letter not found. For the text used here see Jackson (1906) pp. 253-4. Jackson does not give an archival source for the letter; since Bentham did not normally keep copies of his outgoing correspondence, Jackson's being able to quote it suggests it was never sent. There is no indication in M's surviving correspondence that he received it.

2 B82.13.16.

83.04.25 sub-tribes or subordinate groups as genera only, but in uniting *Olearia* and *Celmisia* with *Aster*, to the exclusion of *Erigeron*, would require for its justification a closer study of the numerous American, or Northern, nearly allied genera, than you have had the means of following up. If therefore, you wish to maintain the high position in which your name stands, let me entreat of you to give up the vain endeavour to attach the initials 'F. v. M.' to so many specific names, good or bad, as possible, and to devote your energies, your great abilities, and the splendid materials at your disposal, to the completion of such classical works as your *Eucalyptus* and similar monographs of the great Australian genera, to the supplemental volume or volumes of the *Flora Australiensis*,[3] or, above all, to a methodical digest of the copious and valuable data you have collected on the geographical distribution and relations of Australian plants.

With every wish that you may be enabled, like myself, to devote nearly sixty years of your life exclusively to botany, believe me, ever yours sincerely,

GEORGE BENTHAM.

Baron Ferdinand v. Mueller.

3 No supplemental volume was published.

To Thomas Wilson 83.05.07

Z83/4641, unit 1407, VPRS 3991/P inward registered correspondence, VA 475 Chief Secretary's Department, Public Record Office, Victoria.

Melbourne,
7. May 1883.

T. R. Wilson Esqr
Undersecretary.

Sir

In reply to your letter N. 1689[1] I beg to inform the honorable the Chief Secretary, that I shall in accordance with his wish endeavour to bring together as many industrial articles, as may be within my reach departmentally or otherwise for the Calcutta Exhibition.[2] In an establishment like that, administrated by me, vegetable resources would claim my attention. Thus woodspecimens in a presentable form, Tan-material, Gums, Resin, fibres and other products and educts of plants would be got together, so far *as they really*

1 James Thomson, Secretary to the Victorian Commission for the Calcutta International Exhibition, 1883-4, wrote to the Chief Secretary, G. Berry, on 28 April 1883: 'I have the honour by direction of the president, Joseph Bosisto, Esqr M.P. to state that a special effort is being made to ensure a thorough representation of the indigenous products of Victoria at the forthcoming international Exhibition at Calcutta, and the Commission will be glad to receive the assistance of the Government Botanist in the matter. It is proposed to form collections of seeds, dried plants, &c and at the close of the Exhibition to present them to various public institutions in India, with the view of obtaining analogous collections in exchange. The high scientific attainments and world-wide reputation of Baron von Muëller would combine to greatly enhance the value of any such collection which might be forwarded from the colony, and the Commission trusts therefore that you will be so good as to invite his co-operation in the direction indicated.' The Under Secretary, T. Wilson, immediately referred this letter to M with a covering letter. M returned the file on 11 May referring to the letter above (Y83/4315, unit 1407, VPRS 3991/P, PROV).

2 International Exhibition, Calcutta, 1883-4.

83.05.07 *represent articles or material for trade and export.* Thus I would not advise, that indiscriminately seed-collections are made up for the Exhibition, as only a very small percentage of such seeds would represent mercantile goods. The same remark applies to fibres, of which only such would be chosen by me, as could be got for practical purposes in large quantity cheaply, and as would readily yield to manufactural processes at rope-work, in paper-mills, on looms &c. Large numbers of samples might be prepared, if no restrictions are adopted, but such indiscriminate exhibits would be misleading to a great extent. I find it my duty, to point this out, as I should not aim at quantity, but *quality* on occasions like this, and as it would be unwise to form collections, when only *selections* are requisite. These are the principles, which guided me since I was a Commissioner and Exhibitor in 1854-1855 at the first Paris Exhibition.[3]

I would propose to exhibit also a large Todea-fern at Calcutta, as no species of that genus occurs in India; but it will not likely be possible to obtain such exceptionally gigantic specimens, as were sent to the Amsterdam international and the Petersburg Horticultural Exhibition.[4] A collection of dried plants could also be sent, or rather a *selection* of such, as would represent our best indigenous fodder-herbs, most nutritious grasses and leading technologic plants.

As I have no longer any laboratory or apparatus, not the labor of a carpenter and other former auxiliaries available, it may be worthy of the consideration of the hon. the Chief Secretary, whether my vote, which is the smallest of all departmental establishments, could have added to it a modest item for the Calcutta Exhibition specially, the means at my disposal during the present finance year having proved quite insufficient, to carry effectually

3 Exposition Universelle, Paris, 1855.

4 Internationale Koloniale en Uitvoerhandel Tentoonstelling, Amsterdam, 1883; International Horticultural Exhibition, St Petersburg, 1883. See M to Messrs Watson & Scull, 28 February 1883.

on the ordinary service along with the heavy extracalls on the departmental fund for the Amsterdam Exhibition.[5] 83.05.07

I have the honor to be, Sir, your obedient
Ferd. von Mueller.[6]

To Alfred Deakin[1] 83.05.11

Z83/4779, unit 1416, VPRS 3991/P inward registered correspondence, VA 475 Chief Secretary's Department, Public Record Office, Victoria.

Copy

Melbourne, 11th May 1883

Sir

In compliance with orders from the Honorable the Chief Secretary I have the honor of submitting to you herewith an estimate of expenditure for public works in the department of the Govt. Botanist for the finance year 1883/1884. I may be allowed to remark that since the last 10 years no special vote for my establishment has been obtained from the Public Works Department, although

5 On 9 May Wilson returned the file to M 'with the request that he will be good enough to state what he estimates the outlay will be of sending a collection to the Calcutta Exhibition'. M replied on 11 May: 'I would respectfully suggest, that £140 be extra provided on the new estimates for special preparation of articles illustrative of Victorian vegetable resources for the Calcutta exhibition through the Gov. Botanists establishment. If my original complete Department was still available for me, I would not venture to ask for any subsidy at all.'

6 Wilson sent the file to the Chairman of the Commission along with a report from W. Guilfoyle, Curator of the Botanic Gardens, on 17 May and the Secretary, J. Thomson, returned it on 23 May with no indication of what action, if any, was proposed.

1 MS written by G. Luehmann and signed by M. Deakin was Commissioner of Public Works at the time.

Figure 8. The old Melbourne Herbarium, showing the extension erected to house the Sonderian collection, photographed about 1890. The two men on the steps are probably Charles French Jnr and Gerhard Renner, M's assistants. Courtesy of the Library, Royal Botanic Gardens Melbourne.

I was favored on several occasions by aid out of the general vote for "fittings and furniture". The large increase of the botanic collections and material connected therewith since many years, and now also the acquisition of the great Sonderian collection by the Victorian Government renders some more accommodation at the Botanic Museum absolutely necessary. I have however not ventured to ask a large sum for the extension of the Museum, but in merely soliciting £450 I trust that the Government will not curtail that sum. Even if the Sonderian herbarium were finally to be placed elsewhere, it will be necessary to put such a very large and highly valuable collection into such form, as to render it publicly accessible to advantage, a process involving several years labor by myself at the present Museum of Botany. In proposing that an iron annex be built, I am anxious after the sad experiences lately of an other office building of an other department,[2] to render the collections as safe against fire as I can; moreover the Sonderian collections will arrive towards the end of this year, by which time no stone-building from new annual votes could be erected. An iron shed could at any time be taken down again and the material used for other purposes. 83.05.11

I have the honor to be Sir
Your obedient servant
Ferd. von Mueller,
Gov. Botanist.

The Honorable the Minister of Public Works

Estimate of Expenditure for public works in the establishment of the Govt. Botanist during 1883-1884.

Iron or other Annex to the Botanic Museum, with shelves
£450

Ferd. von Mueller.
11/5/83.

2 This may be referring to a fire at the Government Printing Office in 1881.

83.05.15 ## *To James Service*

P83/334, unit 4, VPRS 1163/P inward correspondence, VA 1123 Premier, Public Record Office, Victoria.[1]

Arnold Street.[2]
15/5/83.

The honorable Jam. Service, M.L.A.,
Premier and Treasurer of Victoria.[3]

In reply to your note of this day,[4] honored and dear Sir, I beg to say, that since the "fragmenta" were commenced to be issued, 25 years ago,[5] *a standing order of the hon. the Chief Secretary exists*, under which *100* copies are granted to my official establishment for literary interchanges and encouragement of collectors as each successive volume appeared. Unfortunately the edition was limited by myself to 500 copies, because I thought, that this strictly

1 M's letter is part of a large bundle of documents filed initially as P311, then attached to P334.

2 M's residence in the Melbourne suburb of South Yarra.

3 On 2 May 1883 Service asked the Government Printer, J. Ferres: 'On what principles are the works of the Baron von Mueller distributed? How many are printed & how are they disposed of.' Ferres replied on 8 May: 'Mueller's "Fragmenta", 500 [copies printed], [Distributed] To Departments, Visitors from other Countries, Horticultural Societies – on authorised requisitions, and by sale. "Botanic Teachings" 3000 To members of Parliament, Departments, Mechanics Institutes, Public Libraries, Horticultural Societies, Visitors from other Colonies – authorised requisitions, and by sale. "Native Plants" 3000 Ditto. "Eucalyptographia" 1000 Ditto'. In a letter of 11 May Ferres added: 'I find that none of the Newspapers are supplied with copies of the "Fragmenta" from this office, therefore, if they obtain it, Baron von Mueller must supply it. The Baron obtains 100 copies in parts, stitched, and 100 copies, bound in cloth. All his other works are supplied to the three papers from this office.' On 11 May Service asked: 'On what authority is the Baron supplied with these 200 copies – Please let me see it this afternoon'. Ferres replied on 11 May: 'I regret that the authority has been destroyed, but the copies as stated have been supplied for many years, if not from the commencement in 1859'. See also M to E. Thomas, 17 May 1883 [Collected Correspondence].

4 Letter not found.

5 B58.03.01.

professional work was or would be merely required for depart- 83.05.15
mental purposes in Victoria and abroad. But since the last quarter of a century so many new scientific unions sprung up, that the demand increased, and on *requisition, approved by the Minister of my establishment*, from time to time more copies were got as occasion arose. Moreover the "fragmenta" were by motions in Parliament, quite irrespective of any action of mine, distributed *several times*, so far as the volumes were available, to the honorable members. The free libraries and Mechanics Institutes sent also to Government in a good many instances for the work, through one or the other of the Ministerial Departments. Hence the supply of the earlier volumes became long ago exhausted; the *eleventh* volume appeared in 1881, and I hope, that the 12th will be concluded in 1884 (deo volente[6]). The "*systematic Census* of Australian plants" was by order of the hon. J. M. Grant published in 1882[7] at a private establishment without any extra-call on the finances of the country, out of the ordinary modest votes of my establishment; but I may passingly observe, that I find myself £60 short, as also heavy calls arose in 1882-1883 on my official fund through the international Amsterdam and horticultural Petersburg Exhibitions.[8] I have *not* ventured, to ask the hon. Graham Berry, though he has always been very kind to me, to place this £60 on the additional or supplemental estimates, as the expenditure beyond the vote, was not previously authorized. Indeed I may be allowed to add, that never a year passed in the 31, during which I had the honor of being a professional head of a Victorian establishment, without the necessity arising of augmenting the votes by my slender private means, which indeed are now – late in life – *all* sunk in my researches! of course I might have limited my literary work more than I did, fully 20 volumes having appeared departmentally, irrespective of several others, which either appeared on the aid of other Governments by Victorian ministerial sanction and

6 God willing. Only the first part of vol. 12 was ever issued (B82.12.03).

7 B82.13.16.

8 Internationale Koloniale en Uitvoerhandel Tentoonstelling, Amsterdam, 1883; International Horticultural Exhibition, St Petersburg, 1883.

Figure 9. Mueller, photographed in 1894, standing in front of his house in Arnold Street, South Yarra. Members of his staff, left to right, are George Parncutt, the brothers George French and Charles French Jr, and J. S. Lyng (the last two behind the fence). Courtesy of the Library, Royal Botanic Gardens Melbourne.

without monetary interest whatever to myself (Calcutta, Sydney,[9]), 83.05.15
or which were printed on my *private cost*. A good many copies of volumes of my works were purchased on my private expense for *not official* occasional distribution at the Gov. Printing office, or second hand from Melbourne and other Booksellers. This applies also to the seven volumes of the "Flora Australiensis" published by Bentham and myself,[10] to Miss Charsley's Atlas of Victorian flowers[11] &c, none of these purchases having ever been charged to departmental funds, my establishment, widely ramified as it is over the whole world, being the smallest endowed of all in Victoria. To return however to the principle of distribution, concerning which you desire to be informed, this was in each case dependent on the *nature of the particular volume*. Thus Mr Berry has most liberally placed at my disposal *250* copies of the *1000* copies of the "systematic Census", that work being also of use to any one for mere reference, without study. In response to a request of the Chief Secretary I submitted a proposition, how a portion of the 750 copies should be disposed of through the Gov. Printing office, which is the *only channel* through which copies of my works are *sold* or otherwise sent out, except the number of copies above indicated as reserved at once for reciprocal obligations and other strictly official distributions of my establishment. I omitted to mention, that of the *loose numbers* which constitute each volume of the fragmenta gradually, 25 copies are extraprinted, when ever each number appears, and given to me by Mr Ferres for use in my office and for sending to the principal amateurs at the time contributing unpaid to our collections. Each volume of the fragmenta requiring and having required from 2 to 4 years for its elaboration, with one or two exceptions.

Let me remain honored and dear Sir, regardfully your

Ferd. von Mueller.

9 San Francisco *deleted*. B80.13.07; B81.13.10.

10 Bentham (1863-78).

11 Charsley (1867).

83.05.23 *To William Brodribb*[1]

M 539, Royal Australian Historical Society, Sydney.[2]

Melbourne,
23 May 1883.

To the honorable W. A. Brodribb, MLA
&c &c &c

It needs not my assurance, dear Mr Brodribb, that I feel much honored with the offer of becoming President of the Australian Geographic Society, to be founded by you and your friends. Though probably only a short space of time remains left to me after my 35 years of uninterrupted toil in these far southern parts of her Majesty's dominions, I shall yet promote geographic interests as readily on the evening of my life, as during its earlier periods. In forming however a federal Society for Geography in the Australian Colonies, some difficulties present themselves to us; and I would like at the outset to offer a few observations for your further consideration. By starting the Society in the oldest Australian Colony, no jealousy could possibly arise in the other colonial territories here, it being understood, that the permanent Central Seat of the Society would remain in Sydney; but in such a case the Presidency ought to devolve on a Gentleman in New South Wales, unless it was to be held under any circumstances only for one year, as is the rule in our two medical Societies here. Perhaps it would also be well to consider, whether seniority in geographic research should claim precedence; in that case the hon. A. C. Gregory, M.L.C., C.M.G., of Brisbane would be entitled to the first years Presidency; and not only would he fill the office with dig-

1 MS annotation: 'Replied to [...] Letter Book 1 Folio 47'. See also E. Marin la Meslée to M, 6 June 1883 [Collected Correspondence].

2 For an abstract of this letter see ML MSS. 853/2 letter register, no. 2, Royal Geographical Society of Australasia (NSW Branch) papers, Mitchell Library, State Library of New South Wales, Sydney.

nity, but he would also have leisure to watch the interest of the new association. – Perhaps it might be advisable to elect a Vice President in each of the Australian Colonies; such a position I would be happy to fill as representant of Victoria; in this manner all the Colonies would be represented on the Council, and local cooperation would be secured in all parts of Australia. Whether however it would be possible for me, to attend any of the meetings in Sydney, remains very doubtful. My health has been fluctuating for some years, so that with much regret I was obliged to beg of the Field Naturalists Club and of the Turn-Verein to confer the honor of President on another Gentleman then[3] myself. Moreover I find late in life my professional work surrounded by far more difficulties than in former years; nor have I in any way prospered during my long and labourious career; therefore the completion of my literary works, for which I have made every sacrifice in life, has become more retarded than under former facilities; hence my time is now less free for extra duties, than it was in bye-gone years. I think therefore, that the Presidency of the new Australian geographic Society should devolve on a Gentleman, who by easier circumstances in life could give to so important a position the needful attention. By this arrangement I should not be prevented to write occasionally a short adress, or furnish some other communications, while it would remain to me a particular source of pleasure to advocate the cause of geography in an association like yours, as I have done independently for fully one third of a century. 83.05.23

Let me remain, honored and dear Sir, with regardful remembrance your

Ferd. von Mueller.

3 than?

83.05.25 ## *To Ebenezer Thomas*[1]

P83/334, unit 4, VPRS 1163/P inward correspondence, VA 1123 Premier, Public Record Office, Victoria.[2]

Melbourne,
25/5/83

To Er. Thomas Esqr
Private Secretary of the honorable James Service,
Premier and Treasurer of Victoria.

Sir

In reply to your note of the 22. inst.,[3] which reached me yesterday, I have the honor to state for the information of the honorable the Premier, that one copy of the "systematic Census of Australian plants"[4] was delivered addressed to the Editor of each of the following papers "Argus," "Age" "Daily Telegraph", "Herald" and "World" through a junior of my office on the 22 March, which was Thursday before Easter. These copies were brought by the Junior to each of the Offices personally, who asserts, that he has the clearest recollection on the subject, as he took the copies on the afternoon before the Easter Holidays. The copies not having been sent by post, they were merely entered in the alphabetic record of recipients of this work without date, but as the Junior of the Office is a very reliable person, I have not the slightest doubt that his statement is quite correct. The book of postal sendings bears this out indirectly, because many entries were made about the same time of copies forwarded by post, but as numerous copies were available by approval of the hon. the Chief Secretary for distribution, they could not all be sent out simultaneously. In each of the copies of the work, sent to the above mentioned newspapers I wrote

1 See also M to E. Thomas, 19 May 1883 [Collected Correspondence].
2 Filed initially as P311, then attached to P334.
3 Letter not found.
4 B82.13.16.

an inscription to the Editor, adding, to the best of my recollection, that the sending was "on behalf of the Victorian Government". 83.05.25

I have the honor to be,
Sir, your obedient
Ferd. von Mueller.

The name of the Junior, who delivered the copies at the newspapers is Mr Léon Henry; he adds, that in each instance he delivered the copy at the counter of the paper and obtained a promise from the person in attendence, that the book would be delivered to the Editor without delay.

F.v.M.[5]

5 On 26 May, Service noted: 'I send this to Mr Haddon as it seems circumstantial enough to cause wonder whether there is not some hitch in the Argus office – will be glad to learn the result of enquiry'. On 25 May Thomas wrote to the Government Printer, J. Ferres, for a statement of the supply of M's works to newspapers since 1 January 1881. Ferres replied on 30 May that part 8 of *Eucalyptographia* was supplied to the *Argus*, *Age*, and *Daily telegraph* on 4 April 1882 by messenger.

On 8 June Haddon commented on the letters of M and Ferres: 'I must apologize for not returning the enclosed official papers re Baron von Mueller's books earlier. Sheer want of time to write is my only excuse. Only a Premier can understand how incessant & engrossing are the demands on the time of an Editor of a big paper. As regards the Baron, I will deal first with Mr Ferres's memo. He is quite right in stating that a number of the Eucalyptographia was received at the office about the time he mentions. I had forgotten, when I wrote to you first, that this was one of the Baron's publications. But you will observe what Mr Ferres's memo. proves. It shows that from 1st Jany 1881 to May 30th 1883, nearly two years & a half, the Govt printer, at all events, supplied only 1 Part of a single work of the Baron's to the three Melbourne dailies! As regards the Baron's mere circumstantial account of the delivery of the "Census of Plants", at the *Argus* Office, I can only say that after the most minute & searching enquiry, not a trace of the book can be found. Of course, as we receive hundreds upon hundreds of books & parcels in the course of the year, it would be useless to expect the clerks in the front office to remember the receipt of this particular parcel but, as I told you before, special enquiry was made for it all over the Office by the Agricultural Editor of the *Australasian*, who had seen it noticed in some other journal. I also find gentlemen in our office, through whose hands books pass, remembering all about the publications of the Observatory, of Profr M'Coy, & of the Scientific departments of other colonies, but knowing nothing of works from the Govt botanist of Victoria, with the exception of the Eucalyptographia, which came from the Govt Printer. Beyond this I cannot go, but it certainly is remarkable that there

83.06.28 ## *To Graham Berry*

Y83/6227, unit 1416, VPRS 3991/P inward registered correspondence, VA 475 Chief Secretary's Department, Public Record Office, Victoria.

Melbourne
28 June 1883.

The honorable Graham Berry, M.L.A.,
Chief Secretary.

Sir.

I have the honor of soliciting your kind consideration of the circumstance, that the votes for my establishment during the financial year now closing have proved quite inadequate to the demands of the service, especially as for the Exhibitions of Amsterdam[1] and Petersburg[2] had to be provided and as the extensive Census of Australian plants[3] had to be printed.

Thus I have had no means to pay the juniors in the botanic Museum and in the office during the last quater of the finance-year from available votes, and it was my intention to defray the expenditure from my slender private means, rather than impair the progress of the service. I feel, that it is reproachable, if a Gov

should be such a consensus of opinion in the office that the Baron systematically ignores us.'

On 16 June Thomas suggested to Service that: '(1.) Call unexpectedly at his [i.e. M's] office & ask to see the record he refers to in his letter of 25 May (2.) Then, or afterwards, see & question Mr Leon Henry the alleged deliverer (3) Request the Baron, as the donations are made on behalf of the Colony to submit in every case a List for the Premier's approval of proposed recipients'.

On 3 July Thomas wrote to the Chief Secretary, G. Berry, intimating the desire of the Premier that 'no works printed at the expense of the Colony be bestowed on its behalf without authority in writing having first been obtained from the Premier'. The Under Secretary, T. Wilson, acknowledged the letter on 16 July on behalf of Berry, and informed Thomas that instructions had been issued accordingly (P83/334, unit 4, VPRS 1163/P1, PROV).

1 Internationale Koloniale en Uitvoerhandel Tentoonstelling, Amsterdam, 1883.
2 International Horticultural Exhibition, St Petersburg, 1883.
3 B82.13.16.

Officer exceeds the liabilities of his establishments, and I am therefore quite prepared to pay the wages of the three juniors as mentioned in the byefollowing memorandum myself, as indeed I have spent over and over in former years means of my own in promoting the efficiency of the branch of the public service entrusted to my care. I have however thought to appeal to your well known feeling of justice in this particular instance, though I may expose my administration to censure; but the last year has taxed our resources most severely, and I could really not help to exceed the expenditure, holding thereby myself as on many former occasions personally responsible. Should you be pleased to allow the sum of £60.9/- to go as "arrears for wages" on the new estimates, I shall of course not construe this as a precedent, and I may remark, that this is the first time during the 30 years of my administration, that I have made an application of this kind. 83.06.28

I have the honor to be,
Sir,
your obedient servant
Ferd. von Mueller.

Government Botanist.[4]
Arrears for the service 1882/3

Wages to the under-mentioned junior assistants in office and museum

L. Henry	from 2nd April to 30th June 1883	13 weeks at £2. 2/		£ 27. 6/ –
James Minchin		"	1.10/	£ 19. 10.
John Matthews		"	1. 1.	£ 13. 13.
				£ 60. 9/

Ferd. von Mueller,
Gov. Botanist.
28/6/83.[5]

4 The memorandum is written by Georg Luehmann and signed and dated by M.

5 On 29 June 1883 Berry minuted: 'Add to vote "arrears" but Dr Mueller must keep the expenditure within the votes for the future'. On the front of the file Berry has noted: '£70 added'.

83.08.06 *From Edwin Daintrey*

RB MSS M1, Library, Royal Botanic Gardens Melbourne.

Sydney Aug. 6th. 1883. –

My dear Baron,

I send you a plant sent to me from Warrah Liverpool Plains[1] with a request that I would name it. If not too much trouble would you kindly let me know whether it is not *Indigofera* brevidens or coronillaefolia – the description of the latter of which in the Flora[2] is not very definite. I am not able to send you the fruit as I would have wished. – In your Botanical Teachings[3] I observed that there is some doubt about the origin of the word "She Oak" the popular name for Casuarina. Sitting under one of these trees (that with pendulous foliage) many years ago among the Blue Mountains and listening to wail of the breeze through the tree it occurred to me that the name was a corruption of "Chiook" – an Onamatopoeia coined by the Australian Aboriginals from the sound above-mentioned, – or in other words that the name was coined on what Max Müller calls contemptuously the new word theory. I offer my derivation for your kindly consideration – We are sadly in want in N. S. Wales of a Book of Botanical Teachings with N. S. Wales Illustrations. I have been more than once asked for such a Book by friends in the Bush who wish to learn something of their Vegetable World and cannot get on because the[4] want this means of recognizing a few Species to start with and thence to extend their studies. – Apologising for trespassing on your valuable time and hoping you are in the enjoyment of good health I remain

My dear Baron
Yours very truly
Edwin Daintrey

Baron F. v. Müller.

1 NSW.
2 cf. Bentham (1864), p. 201.
3 B77.13.07, p. 34.
4 they?

To Asa Gray

83.08.25

Gray Herbarium Archives, Harvard University, Cambridge, Massachusetts.

25/8/83

The "Census",[1] dear Dr Gray, is to serve multifarious purposes; here as an index to the Australian portion of my Museum, as a framework of a new flora &c. &c. Allow me also to remark, what I have done already in a former letter, that since nearly 20 years I have no longer kept the Monochlamydeae (or as now unnecessarily called Incompletae) together, that I implored Bentham & Hooker, when they commenced the "genera"[2] to do away with that unnatural portion of the Jussieu system,[3] changed by D.C.[4] from Apetaleae to Monochlamydeae, to which pleading the answer was, that D.C' system must be adhered to in its entirety, because it was every where in use! According to *that*, also Habit does not debar Cucurbitaceae to be brought near Compositae, both gamapetalous. The grand climber, Senecio scandens, covers many a wall or bower here. Acanthosicyos horrida is an almost leafless but spiny bush without tendrils. The flowers of some Melothrias are smaller than those of many Composites (Cynara Scolymus etc.) The contact given by me to Orders cannot be applied to all other countries, as not *all* orders are represented in Australia; but I *do* think, that I have for the first time given to Elatineae, Plumbagineae, Plantagineae and some others their exact systematic position. The Curvembryonatae are all inseparable.

Pardon me, when I remark, that from the very commencement of my using the nat. system, more than 40 years ago, I felt that the wording Polypetaleae and Monopetaleae are *misnomers*, as in the great majority of cases their flowers are oligopetalous. The terms

1 B82.13.16.
2 Bentham & Hooker (1862-83). See M to G. Bentham, 24 January 1862 [*Selected correspondence*, vol. 2, p. 127].
3 Jussieu (1789).
4 A. P. de Candolle (1819), p. 247.

83.08.25 Choripetaleae and Synpetaleae are publicly with me in 1866,[5] long before Eichler,[6] who however was not aware of this fact.

Jackson in his "guide" writes also *Linné*.[7] DC writes Linne.[8] I have nearly all L.'s[9] works, but not the three "Resa",[10] so that I do not know, how L spelled his name in his native tongue. The accent is to show merely that the end e is not a silent one.

I see however in my library, that Schrebers edition of the Materia medica in Latin (1773) says Caroli a *Linné* mat. med.![11] As he was a disciple of L., this ought to be some authority. In Germany I never saw L.'s name otherwise written; indeed not only Brockhaus writes it so in the Encyclopaedie,[12] but also Chambers in the *English* Encyclopaedia.[13] It seemed to me best to adhere to one mode of spelling the names of Botanists, thus I kept to l'Ecluse, Bock[14] &c

It is with deep sorrow, that I hear of Mr. Bentham's failing health; after some repose he is likely to become invigorated again, unless some form of organic disease set in. We shall miss that great man all very much, after he passed away; my wish to meet him once in life will not likely be realized.

The use of the seeds of Abrus in Pannus and Trachoma will interest you much, as it depends on a bacterian fermentive principle, and supersedes so much more rationally and controlably the use of other inflammatory secretions, such as the antiquated blennorrhoeal. Some of the Australian Olearias are almost herbaceous,

5 See B66.13.05.
6 Eichler (1875-8).
7 Jackson (1881), p. 6.
8 A. P. de Candolle (1819), *passim*.
9 Carl von Linnaeus.
10 Linnaeus (1745); also (1747) and (1751).
11 Linnaeus (1773).
12 Brockhaus (1866), p. 480.
13 *Chambers's encyclopaedia* (1868), vol. 6, p. 142. M also wrote to J. Agardh seeking clarification (letter not found) and then published his reply in *Proceedings of the Linnean Society of New South Wales* (1883), vol. 8, pp. 532-3. Agardh explained that when Linnaeus was ennobled, he adopted the spelling Linné.
14 Charles de L'Ecluse and Hieronymus Bock were herbal writers. M owned L'Ecluse (1601) and (1605), see B65.10.01, but there is no indication that he owned any of Bock's works.

for instance O. ciliata; at best the woody stem could as little be generic as in Senecio & numerous other Composite genera Australian and Extraaustralian, so far as I venture to judge. Let me hope that you are now quite well again and with ease and happiness continue your glorious researches. Regardfully your 83.08.25

Ferd von Mueller

In sorting out plants into the great divisions, it is now quite a comfort to me, not to be troubled with the Monochlamydeae. Two European University-Professors sent me lately a Polycarpaea each as a Gomphrena; this is very pardonable, but shows, that a system cannot be in all respects natural, which keeps Caryophylleae near the commencement of Dicotyledonea and Amarant.[15] near their end. I admit, that a difficulty exists about the term Calyceae, especially so long as we have not yet settled the value of Lodiculae etc in Grasses, unless we adopt implicitly the views of Hackel. But to call the floral *lobes* of Orchids, Amaryllideae &c sepals and petals, is utterly inconsonant to calyx lobes and petals in Rosaceae, Myrtaceae &c

Linné's system should have never been abandoned. The Monochlamydeae may still drag on til the end of the century, but will not likely be maintained far into the next secular epoch! I need not mention to so leading a naturalist as yourself, to what extremity B. & H.[16] are driven, by being obliged to place actually the Loranthaceae into the Monochlamydeae.

If we maintain the name Polypetalae on priority, we must keep up also Monopetaleae.

15 Amaranthaceae.
16 Bentham & Hooker (1862-83).

83.08.28 *From William Dobson*[1]

RB MSS M65, Library, Royal Botanic Gardens Melbourne.

Hobart –
28 August 83

My dear Baron,

I was discussing with His Excellency, Sir George Strahan, a few evenings ago the question of introducing new industries. Amongst others he suggested Cork growing. I told him that we had the cork oak growing here, and that I had a young one growing in my garden. He was not satisfied that it was the true cork producing tree and asked me write to Sir Joseph Hooker for information. I told him that we need not go to the Northern Hemisphere for Botanical information for that you would probably know as much, or more about the matter than Sir Joseph himself, and that you would also know whether our climate is well adapted to its cultivation

I send you a twig of my plant so that you may be able to identify it

I promised His Excellency that I would venture to write to you on the subject, my only apology for troubling you is that I have always found you so ready to afford me information when I have been in difficulty as to the nomenclature of any fern (e.g. "Aspidium Hispidum", until then new to Tasmania) or other plant, that I look upon you as one of the scientists willing to aid the uninitiated when they seek information

I am, dear Baron,
Faithfully yours
W. L. Dobson.

Sir F. Von Mueller K.C.M.G.
Melbourne.

1 MS annotation by M: 'Answ 3/9/83'. Letter not found.

To William Thiselton-Dyer[1]

83.09.07

RBG Kew, Kew correspondence, Australia, Mueller, 1882-90, f. 69.

7/9/83

If my recollection carries me rightly, dear Mr Dyer, you wanted for Kew seeds of Boronia megastigma, Clianthus Dampierii[2] and Grevillea robusta. These I now send; if however I am in error about these supposed wants of yours, then the seeds now transmitted wil not come amiss for your interchanges.

I was at the grand Return Ball given to the Mayor of Melbourne last night, and as I shared much in the dancing, and came home only at 4, – I feel tired to day.[3] So I will conclude this brief epistle.

Regardfully
your
Ferd. von Mueller.[4]

1 MS annotation by Thiselton-Dyer: 'An[swere]d Oct. 26/83.' Letter not found.

2 *Clianthus dampieri*?

3 The *Argus*, 10 September 1883 (Monday), has a report of the Ball 'last Thursday night'; some 1,100 people attended and 'dancing was kept up with spirit until an early hour'.

4 MS annotation by John Baker: 'Baron von Mueller Two Selaginellas received M[ar]ch 28 1884 are both flabellata'. MS annotation by Thiselton-Dyer: 'Informed 5/4/84.'

83.10.20 *To Ralph Tate*

Barr Smith Library, University of Adelaide.

20/10/83

This day, dear Professor Tate, I received the proof-slips of your *very able* article on Kangaroo-Island,[1] and as I was not pressed with official work this (Saturday) afternoon, I at once revised the proof. The article is most varied in its information, even geographical (a new Cape &c[2]), so that it will be read even with interest at the geographic Society, and if you will in due time send me a separate copy in print, I will, if you like, propose you F.R.G.S.[3] – If the article was not already in type, and therefore types became locked up, I would advise to keep it in type for a few weeks, til you had revisited the island, as likely *lots of annuals* would readily come under your discriminating view there.

I hope you will not be displeased, that I took some liberty with the commata, but I was tempted to strike out a number and put in others &c. You will kindly understand, that this is not meant as a special correction up on your particular manuscript, but merely an upshot of a warfare, which I have been weighing for years against English interpunctation &c, but in which attacks I have as a rule been beaten by the English typographers, print-readers &c.

Now, in no language in the world except the English, a comma is put alongside such an ordinary conjuction word as "and", when it simply combines two substantives, in which "and" and the comma are equivalent though the comma does come with the "and", when the latter unites two sentences, each having a verb. To put commata before and behind such words as "however" disturbs the flow of thoughts in reading, and some English Renov-

1 Tate (1883a).

2 On p. 122, Tate refers to 'an unnamed cape, three miles to the west of the mouth' of the Hog Bay River.

3 Tate was never elected a Fellow of the Royal Geographical Society, and there is no evidence that M ever proposed him.

ators in language, feeling all this anomaly, went as far as to abolish all commata, thus falling from one extreme into an other, and depriving us of commata &c altogether. Perhaps you are not inclined to enter into linguistic philology; but to a foreigner this peculiarity of the English language comes very strange. I have succeeded in abolishing such words as Eucalypti for Eucalypts, photographs for photograms, membranaceous for membranous &c &c. 83.10.20

The hyphen is much through me coming into use to unite two substantives, instead of putting one into the genitive or adjective, exempli gratia Murray-River or Murray's River, West-Australia or Western Australia

Peron, speaking of Cassowaries, must have seen Emus in Kang. Isl. It is quite refreshing, to read such a good sense from such a man so long ago, especially as he was working under Baudin so disadvantageously. The Echidna from Kangaroo-Island (or rather Trachyglossum) seems quite a distinct species; at all events it is different from the two S.E. kinds.

Would you not allude to the desirability of K. I. for forest-reserves, where timber could be shipped? The importance of the island for cereal crops in favorable places, and the readiness of *naturalizing* Luzerne, clovers & other *leguminous* fodder-plants as *feeders* on *lime* is apparent, and the humid air would make them grow probably everywhere!

I do not understand all your remarks on Xanthorrhoea; can there be some confusion? Cakile must have come with Ballast. It has been *gradually* spreading on its own accord along our coasts, on places where it formerly not existed. Once here the seeds would be more and more washed about every year.

Could Cassytha have been drifted across by sea? The saltwater *possibly* would not injure the berries! Of course its germination on soil would be much easier than that of Loranthus on trees! Perhaps birds took it across undigested.

How strange that RBr.[4] did not collect more on K. I.; but he probably did not gather a second time, what he might have had from the mainland before.

4 Robert Brown.

83.10.20 Euc. cosmophylla on the upper Mt Lofty ranges[5] is *not* a plant of humid ground.

I have some suspicion, that Papaver aculeatum came to us from S. Africa, ever since I was the first to identify the African & Australian plants 25 years ago as one. Mt McDonnell 984' high![6] did you ascend it? Euc. cosmophylla I learn has been found 35' high.

Have you my translation of Wittsteins "organic constituents of plants" with addenda by me?[7] I could offer you a copy. If you gave an order to your bookseller, he would be able to get for you a copy of the new and *enlarged* American edition of the "select plants for industrial culture and naturalisation"[.][8] I shall probably have but few copies; it is inexpensive, and you would find it useful for reference. It will be out in Detroit, Michigan, at the end of 1883 in Mr George Davis publishing establishment. This is the *seventh* edition. I had no financial interest in any of them.

Should Solanum nigrum, Lepturus incurvatus and Gnaph luteoalbum[9] not be considered as Alliens?

I hope your health is better; my cough is still very violent, and seems not inclined to abate! With one, who descended from phthisic parents, this is a matter of seriousness. However as I am *alone* in the world, if I pass away, I shall not be missed much, unless it be by a few sterling scientific friends

Regardfully your
Ferd. von Mueller.

Centipeda orbicularis ought to be in Kang. Island; it is more prostrate than C Cunninghami; both grow often together.[10]

5 SA.
6 Kangaroo Island, SA.
7 B78.13.06.
8 *editorial addition.* B84.13.22.
9 *Gnaphalium luteoalbum*?
10 MS annotation: 'Herbert Robinson c/o Mrs Moon Kermode St N. Adelaide'.

To Graham Berry 83.10.24

Z83/10272, unit 1417, VPRS 3991/P inward registered correspondence, VA 475 Chief Secretary's Department, Public Record Office, Victoria.

Melbourne,
24. Oct. 1883.

To the honorable Graham Berry, M.LA.
Chief Secretary.

Sir

Observing in the "Age" of this day a notice, favorable of Bromus sterilis as a useful fodder-grass of fattening properties,[1] I deem it my duty to give my opinion to you, that the growth of this particular grass should not be encouraged. It is a common British grass and not liked there or elsewhere by the farming communities; in North-America it is only immigrated, the native range of this Brome-grass being through Europe and Western Asia. This grass is objected to, because it spreads readily into cultivation, interfering particularly with the growth of clovers and luzerne and overpowering also many other culture-plants, especially on friable and less fertile ground, by its

1 The file includes a clipping from the *Age* of the notice referred to: 'The Minister of Lands has received the following memorandum from Mr. Guilfoyle, in reference to a new native grass, to which Mr. W. Madden made reference in the Assembly last week. Mr. Guilfoyle states: – "As regards the grass which is known in the Bridgewater district, as 'Jack Hollis,' as soon as I received your memo., requesting me to furnish some information concerning the plant, I telegraphed at once to Mr. Macdonald, the curator of the public gardens, Colac, for a flowering or seeding specimen, in order that I might be able to determine the botanical name, &c. In answer to my telegram I have just received specimens of two species of grass, which I recognise as Ceratechloa uniloides (some time known as bromus uniloides), the common 'prairie grass' and bromus sterilis, as the ceratechloa, or prairie grass, is known to most farmers and graziers. I presume, in fact I feel almost sure, that bromus sterilis is the real 'Jack Hollis.' The grass is indigenous to North America, and is a tussocky annual, growing in loose soils to 18 inches or 2 feet in height. It has long been known in Queensland and New South Wales as a useful fodder and fattening grass. It is no stranger in this colony. It has been common for many years in the gardens and the domain and many parts of the suburbs. I remember the grass many years ago on the slopes of the Abercrombie and Lachlan Rivers, and more recently on some farms on the banks of the Clarence, Richmond and Tweed Rivers in New South Wales."'

83.10.24 rapid growth in spring; but being an annual grass, it fails when pasture-feed in a clime like ours is most needed. Irrespective of this, stock would browse on Bromus sterilis only before it comes into flower, as the long awned spikelets are too prickly for feed. Indeed sheep-owners also in Victoria complain, that the awns of this Bromus enter the fleece, and are even apt to penetrate the skin, causing losses in the flocks. The nutritive property of this grass even in its young state is far inferior to that of many other grasses; hence I have not admitted any recommendatory note on it into my volume on "select plants for industrial culture".

I have the honor to be,
Sir,
your obedient servant
Ferd. von Mueller.[2]

83.11.05 *From Sarah Brooks*[1]

RB MSS M1, Library, Royal Botanic Gardens Melbourne.[2]

Miss Brooks presents her compliments to Baron Sir Ferdinand von Mueller, and begs to state that in consequence of a paragraph in the "West Australian" newspaper,[3] she has dried, and now forwards, some plants; which she hopes may prove useful[4]

Waratah.
November 5th/883

Israelite Bay W. A.

2 Included in the file is a letter from R. McFarland to M, 22 October 1883 [Collected Correspondence], seeking the name of a specimen sent to M and identified by him as *Bromus sterilis*, warning that though 'good feed while young', 'the seed when ripe is fatal to sheep by penetrating through the skin into the flesh'.

1 MS annotation by M: 'Answ 29/11/83 FvM'. Letter not found.
2 MS found with a specimen of *Opercularia spermacocea* collected by 'Miss Brooke, Israelite Bay, 1883' (MEL 2267191).
3 See M to the Editor of the *West Australian*, July 1883 [Collected Correspondence].
4 See Archer & Maroske (1996).

From William Allitt[1] 83.11.26

RB MSS M1, Library, Royal Botanic Gardens Melbourne.

Tyrendarra[2] Nov 26th – 1883

Baron von Mueller,
Government Botanist Melbourne,

My Dear Sir,

It is many years, since I have had the pleasure of communication with you. In those years I hope that you have been prosperous in your undertakings, and health.

A few years back I started a small nursery at my place, but at present it does not pay, but I trust that it will do so in time, having to trust to Melbourne for receiving Plants and Bulbs from time to time for my nursery, and also to send away. I find now that it is not safe to send anything that I do not know out, for feer that they are rong named, untill I see them flower, for having duplicats of them I find that I receive at times a number of names with Bulbs, and when they flower they are all or a number of them the same.

From your long experience with the growers, I feel sure that you can give me the names of a few that I can depend on, also the name of some Sydney firm that I can get single Camelias from to graft, and an American firm, for which I shall feel much obliged.

There is also a Mr Bull a nurseryman in England that I should like to correspond with If I could get his address, he was under me for the first year he started to learn the nursery work in the hard wood department, at Mr Hendersons nursery St Johns wood

1 MS annotation by M: 'Answ 10/12/83 FvM.' Letter not found.
2 Vic.

83.11.26 London – I have just now in flower (childanthus fragrans[3]) with these names. (Amaryllis lutea) (Pancratium maritimum) (Phaedranassa chloracea[4]) (Sternbergia clussiana[5]) also a number of Amaryllis spec the same varieties or spc with various names, and as my Botanical knowledge is only slight, and my Library only small, I have a job to rectify the nurserymens mistakes if mistakes they are. – The enclosed flower I received many years ago from Sydney under the name of (Ixia pyrimidallis gracillis) I have often examined it but can never make it out to be an Ixia. Law Sumner[6] tells me that it is now called (Sporaxis Pulcherrima[7]) neither can I make it out to be a (Sporaxis) and shall feel much obliged if you will tell me its name, it is almost a evergreen Bulb, I must now beg to apologize to you for the troubl that I am giving you, and thanks for all past favours.

I am dear Sir yours most respectfuly
W Allitt

P.S.
I have this season risen a number of Chrysanthemums from seed, if any turne out good, are they named by you, in England in my time Dr Lindly regerstered all good seedlings plants.[8]

3 *Chlidanthus fragrans*?
4 *Phaedranassa chloracra*?
5 *Sternbergia clisoana*?
6 i.e. the Melbourne nursery suppliers Law Somner & Co.
7 Sparaxis pulcherrima?
8 John Lindley, presumably in his role as Secretary of the Horticultural Society.

To George Doughty Jr[1]

83.12.00

RB MSS M106, Library, Royal Botanic Gardens Melbourne.

A Christmas Greeting.
I wish you ev'ry happiness
Old Christmas can bestow!
A bright New Year, thy path to cheer,
Exempt from grief and woe!

AUSTRALIAN FLOWERS

To George F. St. H. Doughty
from his uncle
Ferd. von Mueller
Christmas
1883.

From William Fisher[1]

83.12.22

A84/1453, unit 8, VPRS 3992/P inward registered correspondence, VA 475 Chief Secretary's Department, Public Record Office, Victoria.

Dehra Dun,[2]
22nd December, 1883

My dear Sir,

Dr Schlich has written to me about an exchange of the 'Indian Forester' for your Eucalyptography & other papers from Aust-

1 MS is a card showing a card with the printed greetings lying on a bunch of Australian wildflowers. M's greeting is on the reverse.

1 See also M to G. Berry, 8 February 1884 [Collected Correspondence], with which this letter is filed. In due course the Premier, J. Service, approved the exchange proposed.
2 India.

83.12.22 ralia. Of course it will be very greatly to the advantage of the Indian Forester & I need not say that I shall be very much obliged to you & very glad to get an occasional paper on Australian Forestry. I have therefore directed the Publishers to send you the Indian Forester in future regularly, & will also send you as many back nos as I have in hand.

Owing to the large increase in the circulation of the I. F. consequent on its becoming a monthly, instead of a quarterly periodical, I am afraid that I have not a complete set of Vol IX available, but I will send you what I can.[3]

In India we have so much executive work to do, that litterary composition is only possible in spare moments, so that you will find the Forester merely a collection of practical suggestions from men who have hastily written down their impressions during the few hours they can spare from their work. I was very sorry to hear that Forestry in America has sustained a great check in the collapse of the American Journal of Forestry for want of support, & in the resignation of Dr Hough, the head of the Dept.

I hope that it will take up firmer ground in Australia

Yours truly
W. R. Fisher
Editor I. Forester

To The Baron von Müller
&c &c
Victoria

3 Vol. IX was the volume of the current year, 1883. The set of the journal at the Royal Botanic Gardens Melbourne lacks nos 6-8 of this volume.

To George Brown 83.12.30

A1686/23, Mitchell Library, State Library of New South Wales, Sydney.

30/12/83

The last mail, rev. and dear Sir, brought me the enclosed letter[1] from the great Vienne Linguist, Professor Müller,[2] to whom I had sent your most valuable publications[3] on the language of the Duke of York Islanders.[4] Prof Müller bestows on you and your collaborator the recognition so richly due to you. I enclose his letter

With my best wishes to you for a happy new year

your regardful
Ferd. von Mueller.

To George Bentham 83.12.31

RBG Kew, Kew correspondence, Australia, Mueller, 1882-90, ff. 74-5.

Drysdale, near Port Phillip-Head,[1]
Newyears eve, 1883.

Your kind but mournful letter,[2] dear Mr Bentham, reached me here on the coast, where I am since a few week to try, whether the pure sea air and the rural tranquillity will free me of a protracted bronchial inflammation; as both my parents died of Phthsis, I must look on my present severe illness with serious apprehension. – This I mention, that you see, how many months now I have been suffering, and thus had neither strength nor courage to write beyond

1 Letter not found.
2 Presumably Friedrich Müller (1834-98), professor of linguistics at Vienna.
3 G. Brown (1877)?
4 New Guinea.

1 Vic.
2 G. Bentham to M, November 1883 [Collected Correspondence].

83.12.31 what urgent official duties demanded from me. To give however a sign of life, I sen[t] you a paperknife of Siphonodon australis, thinking that the trifle might interest you, it being from a tree named by yourself.[3] From the remnants of the Calcutta-Exhibition[4] I also sent a paper-knife to Prof. Owen, the article being from one of the Owenias, and in writing back,[5] he kindly said (though I also sent no letter to that illustrious man), that he used it daily for cutting up new books, and that each time it came to him like a handshaking from me! –

I am deeply concerned to hear of your failing strength! – We all have been for these many years so accustomed to see you make uninterrupted and gigantic strides for the elaboration of the systema of plants of the world that I at the far distance can least of all realize that you have discontinued to advance the knowledge of the vegetable empire; and altho' you seem to have such gloomy forebodings, I still hope, that you after your overexertions of late years will rally, and will give the world some further litterary treasures. Prof Chevreul in Paris is now, I think, 96 years of age; and if he does perhaps no longer deliver a course of lecture at the University, he certainly did so still 2 or 3 years ago. Humboldt worked also on til 90. So I hope you will be cheered again![6] What you say of your long and illustrious career reminds me, that your name first became very familiar to me, when I cultivated in my widowmothers garden a little plot, on which I raised some of the beautiful Polemoniaceous annuals named and described by you,[7] plants which will ever hold their own in the horticulture of ordinary people and among them for all times commemorate your name!

Your attention was so absorbed in the "genera plantarum"[8] for the last few years, that I did not even wish to disturb you by sending the trifling prints, on additions to the Australian Flora. In 1883 the additional species from all parts of Australia obtained as the

3 Bentham (1863), p. 403.
4 International Exhibition, Calcutta, 1883-4.
5 Letter not found.
6 Bentham died on 10 September 1884, shortly before his 84th birthday.
7 Bentham (1833), (1845).
8 Bentham & Hooker (1862-83).

aggregate from numerous correspondents amounts to only 40 species. So the flora of Australia may be regarded as specifically almost completed. altho' many observations on the plants established, and so exhaustively detailed by you, will have to be made for centuries to come. Even this day I had to correspond[9] on your Acacia implexa as the most important tan-yielding tree in one part of New England. 83.12.31

This is the last letter I write in the byegone year! May the new one to you be one of serenity and joyous contemplation of what you have done for all times to benefit the world, should providence even not allow you to resume actively your great labors.

Ferd. von Mueller

From Joseph Hooker[1] 84.01.17

P84/759, unit 22, VPRS 1163/P1 inward correspondence, VA 1123 Premier, Public Record Office, Victoria.[2]

Jany 17/1/84

My dear Baron

I am much concerned to hear of your illness & do indeed hope that your fears may not be realized, & that a sojourn in the mountain air of Victoria will prove a complete restorative.[3]

9 Letter not found.

1 MS annotation by M: 'Answ 11/3/84. F.v.M'. See M to J. Hooker, 11 March 1884 [Collected Correspondence]. See also M to T. Wilson, 3 April 1884 [Collected Correspondence], in which M drew the Victorian Government's attention to Hooker's comment about the exploration of New Guinea and the New Hebrides.

2 This letter was registered in the Chief Secretary's Department as B84/3072 and referred to the Premier.

3 See M to J. Hooker, 3 December 1883 [Collected Correspondence], written from Mt Macedon, Vic., where M had gone seeking relief from a protracted bout of bronchial catarrh.

84.01.17 Thanks many for your vindication of the rights of nomenclature of the Gen Plant.[4] There is much difference of opinion upon the matter here, & I naturally hold aloof from saying what should be done in the new Nomenclator[5] which Mr D. Jackson is at work upon at Kew on Mr Darwin's munificent fund. I propose that Mr Dyer, Ball, Oliver & [...] should form a committee to decide,[6] & I will assent to whatever they may rule. I am not myself very enamoured of rules, which must be arbitrary & which applied to the letter rather than to the spirit of the law is the mark of the small fry of botanists who make more fuss about a wrong specific synonym than over a plant they put in a wrong [nat] Order!

Mr Gray in his last contribution to this subject[7] winds up by saying that those who readjust genera are bound to name the Species. This is all very well for those who do an Order or two, but when it is a matter of a Genera Plantarum it is impossible for two lives to overtake it. Especially if the unlucky authors have like myself a family of 7 to work for, & [a] laborious official duties to perform.

I am very averse to making the new Nomenclator an exponent of Kew, or Candollean or other "laws" so called.

Mr Bentham is exceedingly weak & quite incapable of any work – he now cannot walk across the room, & spends his whole time in a chair before the fire – he is unable to read more than a few minutes at a time or write more than a line or two – his is not the lusty old age, but a loss of physical power that never can be recovered; he is exceedingly attenuated too. He takes no interest in Science & sees hardly any one but myself, I see him weekly & yesterday I

4 Bentham & Hooker (1862-83). See B84.13.03 and M to J. Hooker, 3 December 1883 [Collected Correspondence].

5 *Index Kewensis* (= Jackson [1893-5]).

6 In his account of the compiling of *Index Kewensis* Jackson (1924), p. 226, wrote that 'a committee, consisting at first of Sir Joseph Hooker, Professor Daniel Oliver, and John Ball the alpinist, used to meet occasionally to discuss knotty points and take stock of progress'.

7 Gray (1883).

gave him your message,[8] for which he sends his thanks. I told him of your discovery of *Centunculus*. 84.01.17

The Herbarium at Kew advances rapidly of late [in] plants of Madagascar abounding in novelty, & from the Malay Peninsula where Perak seems to be a very rich and novel field. From Central Africa too we get small consignments, but they are not so interesting as I would have expected. We are sending an expedition to Kilimanjaro[9] which may do good service

We are also about to get up a list of all that is known of Chinese Botany[10] which will be useful as ground work In fact there are many countries which want cataloging prior to elaborating.

I hope that you will work New Guinea from your continent & also the New Hebrides & indeed the Pacific generally – Australia should send out a Scientific exploring Expedition to some of the Groups. I wonder that it has not fitted out an Antarctic one![11]

I am very busy with the Flora of British India,[12] a new Edition of the British Flora,[13] & a new guide to the Garden & the Arboretum.[14] We are also getting out a new Museum Guide[15] which Dyer undertakes & the Report for 1884 very delayed by press of other work & the Curators[16] 3 months illness followed by 6 months absence on leave. We will have a very heavy year of it.

Ever sincerely yr
Jos D Hooker

8 In his letter of 3 December 1883, M had asked Hooker to pass on his best wishes to Bentham.
9 Mt Kilimanjaro, Tanzania.
10 Probably Forbes & Hemsley (1886-1905).
11 See Home et al. (1992).
12 J. Hooker (1875-97), probably vol. 4, part 12.
13 J. Hooker (1884).
14 Royal Botanic Gardens, Kew (1885).
15 Royal Botanic Gardens, Kew (1886).
16 John Smith (1821-88).

84.02.07 *To Graham Berry*[1]

A84/1390, unit 7, VPRS 3992/P inward registered correspondence, VA 475 Chief Secretary's Department, Public Record Office, Victoria.

Melbourne,
7/2/84.

The honorable Graham Berry, M.L.A.,
Chief Secretary.

Sir,

I have the honor to report with deep regret, that Mr C. Groener, one of the Assistants of the Gov. Botanists establishment, died *yesterday*.[2] The particulars of his death are not yet known to me, as I only learnt this sad event through a telegram from Port Adelaide, he being on his way to King George's Sound,[3] where he was to collect plants for the Herbarium.

I have since fully 25 years been accustomed to send employées of the establishment occasionally to various parts of Australia for collecting purposes, and thought that a stay of Mr. Groener for some time in field-engagements at King Georges sound and its vinity,[4] where the clime is so salubrious, would tend to the improvement of his impaired health, while by these means desirable additions would be made to the departmental collections. I would

1 The file contains a separate note, in an unknown hand:

On salaries vote

Clerk – Luehmann –	£250
Assist. Groener –	£200
Paid out of Incidentals	
Henry –	9/6 per diem
Minchin –	6/6 – " –
French –	3/6 – " –

2 yesterday *is marked with double lines in the margin and annotated* 6/2/84.

3 WA.

4 vicinity?

beg at once to plead, that you, Sir, will give your kind consideration to the bereaved widow, who is left with very little of worldly means, so that by your generous action she may receive such compensation or gratuity, as is generally allowed to widows of Gov. Officers. She accompanied him on his last journey; he was ten years strictly in the Gov. Service, his salary being latterly £200 a year. Indeed he was nearly ten years more in the establishment as my private valet and at my private cost, though a large share of his time even then had to be given to the Gov. service, in which he thus spent the best part of his life.[5] 84.02.07

I have the honor to be,
Sir, your obedient servant
Ferd. von Mueller.

5 On 13 February the Under Secretary, T. Wilson, minuted: 'To the Government Botanist to note the Chief Secretary's approval, and to furnish particulars with respect to Mr Groener's length of service'. On 19 February Wilson fowarded M's reply, dated 14 February, to the Commissioners of Audit for verification. Their reply, 22 February, stated that Groener was employed from 1 July to 7 August 1873 and 1 July 1874 until his death on 6 February. They added: 'There appears no trace of him from 8th August 1873 to 30th June 1874'.

Wilson referred the file to M four days later 'with the request that he will be good enough to state in what capacity the late Mr Groener was employed & out of what vote or fund he was paid during the period from 8th August 1873 to the 30th June 1874, inclusive'.

M replied the next day, 27 February: 'Mr Groener was from the 8. Aug. 1873 til the 30 June 1874 entirely employed as second Assistant in the Gov. Bot. Department, but as only £300 were allowed for the whole finance year irrespective of my own Salary, and as it was impossible to carry on the departmental work with one Assistant only, I paid Mr Groener out of my private means; but in justice these eleven months ought to count service.' No response to M's action or suggestion was minuted.

84.02.14 *To Thomas Wilson*

A84/2889, unit 16, VPRS 3992/P inward registered correspondence, VA 475 Chief Secretary's Department, Public Record Office, Victoria.[1]

Melbourne,
14 Febr 1884.

T. R. Wilson Esqr.
Undersecretary.

Sir

In reply to your questions of yesterday I have the honor to state

1, The duties of the late Mr Groener[2] were those of a Superior Assistant as well at the office as in the Herbarium, besides attendance to town-work and occasional engagements for collecting plants.

– The duties of Mr Léon Henry are those of Junior Office-Assistant, aiding in clerical work, labelling bot. specimens at arrival, also minor mechanical duties.

– The duties of Mr James Minchin are those of Junior Herbarium-Assistant, sorting dried specimens of plants, treating Herbarium with preservatives and minor mechanical duties.

– The duties of Mr Charles French[3] are those of second junior Herbarium-Assistant, aiding thus generally in Herbarium-work and keeping the Building clean.

2, The present applicant[4] for the position of the late Mr Groener is the father of the young person above named.[5] He is twenty years in the Gov. service, 10 of which he spent under me in the bot. Garden, where he is still engaged, at present with an income of

1 The letter is numbered B84/1631 within the file.

2 See M to G. Berry, 7 February 1884.

3 i.e. Charles French Jr.

4 Charles French Sr.

5 See C. French to M, 8 February 1884.

about £150, having besides the free use of a 6 roomed cottage there, which latter he would have to give up, should he be appointed in Mr Groener's place; he will not gain monetarely by this transfer, but his scientific taste makes him desirous, to join my branch of the public service; besides I would beg to point out, that I should require a trustworthy and efficient person like Mr French (senior), who would need to have special knowledge and experience, such as he possesses. He is 41 years old. The three juniors in my establishment have not sufficient knowledge and are rather too young to be promoted. The three juniors are: 84.02.14

Mr Léon Henry, engaged since 1 July 1881; his present wages 7/6d daily.

Mr J. Minchin, engaged since 28 febr 1879; his present wages 6/6d daily.

Mr Ch. French (junior) engaged since 12 July 1883; his present wages 3/6 daily.[6]

I have the honor to be,
Sir, your obedient servant
Ferd. von Mueller.

6 The Chief Secretary, G. Berry, initialled his approval and on 19 February 1884, H. Moors for the Under Secretary minuted: 'To the Government Botanist to note the Chief Secretary's approval to Mr Charles French (senior) undertaking the duties performed by the late Mr Groener, and to report the date from which he entered upon them. I shall be glad to be informed of the actual date of Mr Groener's death.'

M replied on 26 February: 'Returned with best thanks for the hon. the Chief Secretary's kind action in this matter. Mr French will enter on his new duties at the first of March.'

On 28 February the Under Secretary, T. Wilson, reminded M that he had not notified the Chief Secretary's Office of the 'actual date of Mr Groener's death. The fact, and date of his death should be duly verified.' M replied on 1 March that Groener had died on 6 February 1884, and that a certificate could be obtained from Port Adelaide if necessary.

MS annotation: 'Chas French appointed, 5/3/84, Assistant in office of Govt Botanist, @ £200, from 1/3/84'.

84.03.14 *From Joseph Hooker*[1]

RB MSS M3, Library, Royal Botanic Gardens Melbourne.

Royal Gardens Kew[2]

March 14 /84

My dear Baron

I was vastly amused at finding my features engraved in the Melbourne paper & I thank you very much for your kindly notice of me, all too flattering though it be.[3]

There are one or two little inadvertencies in it – I was born at Halesworth in Suffolk, three years before my father went to Scotland & I am only [an] Foreign Associate of the French Academy.

I took your kind message to Bentham, who I found to be rather better – he can just walk across the room with a stick, & complains most of shortness of breath. He & I are sorry to hear that you are so poorly & earnestly hope that the mountain air will do you good.[4]

I was indeed surprised to see the Catalogue of the Melbourne Garden[5] published without an allusion to the former of the collection & the former head of the Establishment & its founder in short. – I have written thanking the donor for the copy & adding that "the absence of any allusion to your predecessor, & the former of

1 MS annotation by M: 'Answ 3/5/84. F.v.M.' See M to J. Hooker, 4 May 1884 [Collected Correspondence].

2 Embossed letterhead.

3 A full-page article about Hooker, including a portrait, was published in the Melbourne *Leader*, 3 June 1882, suppl., p. 1. As M explained in M to J. Hooker, 23 January 1884 [Collected Correspondence], he had supplied, at the editor's request, the notes on which the article was based, but had then mislaid the copy of the article he had intended sending to Hooker.

4 M had been suffering for some time from 'bronchial catarrh'. He had previously spent time at Mt Macedon, Vic., and at the seaside, without finding a cure. Now, he had told Hooker, he was going to try a visit to the Eucalyptus forests of Gippsland.

5 Guilfoyle (1883). M had alerted Kew to the publication of the catalogue; see M to W. Thiselton-Dyer, [October] 1883 [Collected Correspondence].

the collection, both surprized & pained me" – nothing more. I cannot understand any one behaving so 84.03.14

I am told that there are parties in Melbourne who sell dead trunks of *Dicksonia Antarctica* for the purpose of growing Ferns &c upon them. Can you tell me whether this is so or not. We have a series of pillars in our Fern house formed of dead stems of Tree ferns, covered with living Ferns, & we should be glad to get some dozen or two of trunks five feet long & upwards at a modest price.

Your great *Todea* is in splendid condition – it is impossible to count the fronds upon it – it stands in a saucer of cement with loose stones about the base, & one side covered with stones &c, the other exposed, & it goes on getting finer & finer every year. Of course it is in a shady spot. The big *Dicksonia* too is in splendid order.

Doryanthes excelsa is flowering –

Every most sincerely your
Jos D. Hooker.

To Ralph Tate 84.03.15

Barr Smith Library, University of Adelaide.

Drouin,[1] 15/3/84.

The two excursions,[2] of which you sent me an account,[3] dear Prof. Tate, must have been to the highest degree interesting. The one reminds me vividly of a tour of mine to Villunga,[4] during which

1 Vic.

2 In November 1882 Tate went to Mt Gambier, and in January 1883 he went to Kangaroo Island, both SA; see Tate (1883), p. 35 and (1883a), p. 134.

3 Letter not found.

4 Willunga, SA?

84.03.15 I followed the Onkaparinga valley[5] down through then widely uninhabited country, getting benighted without food and fire.

I will examine the Hydrocotyle[6] more closely on my return. The stay here has done me very little good, as the locality is too cold for me, and the weather was rainy and boisterous on many days of my stay. If I do not get worse during the winter here (much colder than in Adelaide) I will in early spring make a tour to the S.A. borders from Lake Hindmarsh, there being now coach-lines from where the Railroads cease. I hope that the German Residents will have patriotism enough left, to protest against the alteration of the name of their village Grünthal.[7] It is poor gratitude to those, who as early pioneer formed the settlement, and came on their own expense, to obliterate the records of their part of colonisation there on the geographic map. The German Members of Parlament should not allow this. The Giant Eucalyptus ought at Grünthal to be fenced and protected. Is the stem solid?

Let me offer my best felicitation to your 44th Birthday, and express a hope, that you will *double* that number of years in due time. Prof Chevreul of Paris lectures yet at the age of 98!

Regardfully your
Ferd. von Mueller

5 M may be referring to his trip in the Onkaparinga Valley in November-December 1848. He was also in the region in August 1850.

6 See also M to R. Tate, 2 March 1884 [Collected Correspondence].

7 The name Grünthal was retained until 1917 when, as part of the general wave of anti-German feeling aroused during the First World War, many German place-names in SA were given non-German names; Grünthal became Verdun.

To Robert Watson

84.03.30

No. 1192, unit 63, VPRS 425 Engineer-in-Chief inward correspondence, VA 2876 Victorian Railways, Public Record Office, Victoria.

30/3/84.

To the Chief Engineer of the Victorian Railways.

Sir

Would you oblige me by kindly informing me, what Eucalyptus-wood chiefly is used in your Department for Railway-Buffers? I should like to obtain this information for a new Victorian edition of my work on select plants for industrial culture.[1] You could add to the obligation, under which I shall be placed by you, if you would communicate to me a few brief notes from your local experience on the respective value of various native woods in Railway-construction; such information would be fully acknowledged in the work as obtained from you.

Respectfully
your
Ferd. von Mueller.

1 In B85.13.06, p. 145, M noted that *Eucalyptus rostrata* was used for buffers, among many other uses. However, he did not give a source for the information.

84.04.08 *To Graham Berry*

B84/3203, unit 17, VPRS 3992/P inward registered correspondence, VA 475 Chief Secretary's Department, Public Record Office, Victoria.

Melbourne,
8. April 1884.

The honorable Graham Berry, M.L.A.,
Chief Secretary.

Sir,

I have the honor to solicit, that you will kindly allow me, to do my Office-work for a few weeks at the residence of Dr Alex. Buettner, who built a large house at the higher part of Victoria Parade[1] in a very healthy position, and who as my principal medical adviser is anxious, that I should for some time be under his frequent and immediate care, my distressing cough having not abated. Dr Buettner offers me a large room for my sole use; the daily corresondence would be forwarded on to me, and I could do the ordinary office work (like in Gippsland and also at the sea-side) at the Doctors place, where my Assistant, when necessary, could personally call on me for instructions.[2]

I have the honor to be,
Sir, your obedient servant
Ferd. von Mueller

1 Melbourne.

2 Berry approved M's request on 16 April and the decision was immediately forwarded to M who returned the file 'with best thanks'.

To [Bernhard Perthes][1] 84.05.03

Briefsammlung, Archiv, Justus Perthes Verlag, Gotha, Germany.

3/5/84.

Mit tiefem Schmerz, habe ich, hochgeehrter Herr, die Nachricht von dem Tode des Herrn Professor Behm vernommen. Mit dem Hinscheiden dieses vortrefflichen Gelehrten hat die geographische Wissenschaft einen grossen Verlust erlitten, der wohl für Ihre herrliche Anstalt ein unersetzlicher sein wird, über wie schöne Arbeitskräfte Sie auch immer gebieten mögen.

Mir kam die Trauer Botschaft ganz unerwartet, denn obwohl ich kürzlich durch den Hr. Assistenten Behm's erfuhr, dass er erkrankt sei, so gab ich mich in keiner Weise dem Gedanken hin, dass wir den prächtigen Mann so schnell verlieren sollten. Erst vor wenigen Wochen liess er mir sagen, dass er sich sehr über die Reise des Herrn Winnecke ins centrale Australien gefreut habe, da doch auf diese Weise auch einmal wieder ein Deutscher ins Entdeckungsfeld Australiens gerückt sei! Mir selbst war Herr Dr Behm, wie es früher Herr Professor Petermann gewesen, ein besonders gütiger Freund, u. ich werde sein Andenken stets mit Rührung und Liebe bewahren!

Ehrerbietig der Ihre
Ferd. von Mueller.

Wollen Sie gütig der Familie des Herrn Dr Behm meine tiefe Condolenz ausdrücken.

Wäre es nicht wünschenswerth in den "geogr. Mitth." auf einer eignen Karte einfach in farbigen Linien darzustellen, was *deutsche Geographen für die Erforschung* Central Africas gethan? Ich erinnere nur an Barth, Vogel, Beurmann, Heuglin, Rolfs, Nachtigal u.

1 No addressee is indicated on the letter, but the contents suggest it was almost certainly directed to Perthes.

84.05.03 Schweinfurth, anderer Reisender unserer Nation nicht zu gedenken. Vielleicht würde das grosse Deutschland doch auch noch politische Vortheile aus diesen Errungenschaften ziehen, sei es durch Protectorate, sei es durch Colonisationen. –

3/5/84

With deep pain, highly esteemed Sir, have I heard of the news of the death of Professor Behm. With the demise of this excellent savant, geographical science has suffered a great loss, that probably will be an irreplaceable one for your splendid institution, whatever good employees you may have command over.

The sad news came to me quite unexpectedly, for although I recently learned through the assistant of Behm that he was ill, I did not in any way indulge in the thought that we were to lose the splendid man so quickly. Only a few weeks ago, he was able to say to me[2] that he was very pleased about the journey of Mr Winnecke into Central Australia,[3] for indeed in this way a German had taken to the Australian field of discovery once again. Dr Behm, like Professor Petermann previously, was a particularly kind friend to me, and I will always cherish his memory with emotion and love!

Respectfully your
Ferd. von Mueller

Would you kindly express my deep condolence to the family of Dr Behm.

Would it not be desirable in the *Geographische Mittheilungen* to show on its own map simply in coloured lines, what *German geographers* have done *for the exploration* of Central Africa? I only call to mind Barth, Vogel, Beurmann, Heuglin, Rolfs, Nachtigal and Schweinfurth, not to think of other travellers of our nation. Perhaps the great Germany would also still draw political advantages from these achievements, be it through protectorates, be it through colonisation. –

2 Letter not found.

3 See B84.08.02.

Last Will and Testament

84.06.17

No. 62/802, unit 256, VPRS 7591/P2, Public Record Office, Victoria.

This is the last will and testament of me, Ferdinand Jacob Heinrich von Mueller, K.C.M.G., M. & Ph.D., a Baron in the Kingdom of Wuerttemberg, at present living in Arnold-Street, South-Yarra. I appoint as executors of this my last will James Rudall Esq., FRCS., of Melbourne, Alexander Buettner Esqr, M.D., F.R.C.SE., of East-Melbourne, Hermann Buettner Esq., Gentleman, East-Melbourne, William Haig Esq., M.D., South-Melbourne and the Reverend William Potter, F.R.GS., of South Melbourne. – I request the Executors of my will to return through his Excellency the Governor my decoration of K.C.M.G. to the Right Honorable the Secretary of State for the Colonies, – to return my other decorations through the respective Consuls to their Excellencies the Ministers of foreign affairs of the various Governments, from whom I received these distinctions, except my decorations of St. Jago and of Isabella the Catholic,[1] which with the Patent of my Baronial or Freiherrns Rank are to be kept as an unsaleable and sacred heirloom in the families of my two sisters. I wish to give to the eldest son[2] of James Rudall Esqr the miniature diamond-star of the Legion of Honor,[3] and to the eldest daughter of Dr Buettner the chain of miniature-decorations, which I possess; – and I furthermore request all the executors of my will, to select from my moveable property such articles for souvenirs of their children, as they may deem fit.[4] – But from this selection are to be excludet[5] the five caselets of Jaspis-

1 M's medal of the Order of Isabella the Catholic is now at the Royal Botanic Gardens Melbourne.

2 J. F. Rudall.

3 This is presumably the star, said to have been M's, that is now held by the Dixson Library, State Library of NSW.

4 On the basis of this clause, the regalia (red sash, sash badge and breast star) of the Grand Cross of the Order of Christ of the Kingdom of Portugal that had been bestowed on M were gifted by the executors of his estate on 26 February 1900 to William Potter's daughter Beatrice. The regalia, together with the deed of gift, are now at the Powerhouse Museum, Sydney (catalogue registration number N15547).

5 excluded?

84.06.17 articles, which were presented to me by His Majesty the late Emperor of Russia, as these Jaspis-Articles are to be given to the youngest daughter of the honorable Dr David Wilkie of East Melbourne. The clock, presented by the employées of the botanic Garden to me, is to be placed as a gift into the botanic Museum.[6] I further direct, that all my furniture and other household-articles, carriage and any other domestic property be sold, as well as my ground and cottage in Arnold-Street, South-Yarra, and likewise any town-allotments, which at the time of my death I may possess in Perth, West-Australia, concerning which the honorable John Forrest, C.M.G., Commissioner of Crownlands there, will be able to afford information; and any other property, which I may hereafter acquire is to be disposed of in the same manner. My private and unofficial correspondence (kept ever apart from official letters) together with my Diplomas and manuscript journals I bequeath to the Baroness Maria Negri, daughter of Baron Christophoro Negri of Turin, the whole to be forwarded free of expenses.[7] I further wish the executors of my will to issue at my cost not less than 500 copies of reprints in book form with lithographic plan of the departmental reports, made by me from 1853 til 1874 as Government Botanist and also Director of the Botanic Garden of Melbourne to Parliament here, such copies to be distributed to various botanic Gardens and scientific institutions over the globe; – and I request the executors of my will furthermore, to cause for similar distribution an issue of not less than 500 printed copies to be prepared also on my expense of a collection or rather selection of favorable articles on the Melbourne botanic Garden, such as during my Directorship appeared in the metropolitan journals here, the transcription being ready at my residence.[8] I desire the botanic portion of my private library, con-

6 This clock is now at the Royal Botanic Gardens Melbourne. It bears the inscription: 'Presented to Dr F. Mueller F.R.S. as a token of respect By the employees of the Botanic Gardens January 17 1866'.

7 Maria Negri predeceased M. After many years' delay, in November 1910 the surviving executors of M's estate obtained a court ruling that his unofficial correspondence and associated materials should be placed at the disposal of his surviving immediate family.

8 Neither of these collections was ever published. The collection of newspaper articles appears not to have survived.

taining the books, purchased by myself since 1866 to be offered to the Victorian Government at a fair valuation; – and I wish also, that the Government be informed of the justness of making some monetary allowance to my heirs for the botanic collections, which I accumulated by private means from 1840 til to the present time, and which became united with those portions of the Museum-plants, which were acquired at Governments expense, – the fixing of any such allowance to be left to the judgement and good will of the Government, and the fulfillment of this solicitation under no circumstances to be excuted[9] by any legal means whatever. The same injunction I hereby give concerning any recovery of the extensive outlays, which in the eagerness of advancing the interest, utility and fame of my Department I spent out of my private means for official purposes since 1853 in defraying all personal travelling expenses myself, in incurring much private expenditure for foreign interchanges, agencies, freight, office-light, fuel, frequent purchases of seeds, having also provided since 1857 an office-keeper continually on private expense. Whatever proceeds thus may arise, I direct to be deposited in Government Saving-Banks with a view, that the proceeds of the Capital be given annually to my sister, Mrs Dr Wehl near Millicent, South-Australia, and after her death to her heirs and those of my late sister, Mrs George Doughty of Mount Gambier, but without any right whatever of any of such heirs to sell the benefit of such annual interest. From the investments above indicated is however to be set apart the sum of one hundred £, the interest of which is to be devoted to protect my burial place, and to keep it continually neat and planted with flowers. 84.06.17

I have subscribed my name to this my last will on the seventeenth day of June in the year eighteen hundred eighty four

Ferd. von Mueller

Signed by the above named Ferdinand Jacob Heinrich von Mueller in the presense of us, who on his request and in the presense of each other have here unto subscribed our names as witnesses

J. G. Luehmann
Charles French

9 executed?

84.07.14 *From Robert Johnston*[1]

RB MSS M1, Library, Royal Botanic Gardens Melbourne.

2 Davey st Hobart.
14 July 1884

My dear Baron

I send you enclosed two species of Ferns for your examination both of which appear to me to be new to Tasmania. One, the smaller, closely agrees with the description given of *Hymenophyllum marginatum*. Hook et Grev. Port Jackson, N.S.W. The other is so densely matted with scales that I am not sure of its alliance. They both exist in abundance in the Western region of Tasmania where the rocks are of Silurian age. Locality Honeysuckle Hill. Queeres' River,[2] in the vicinity of the Huxley, Tyndall and Owen range of mountains. The Collector, Mr Moore,[3] is one of our intelligent Surveyors who is engaged at present in opening up this little known district, and possibly other novelties may soon be placed in your hands. Should one of the ferns be new to science perhaps you will kindly encourage Mr Moore by associating his name with the discovery. You will recollect that our last addition to the ferns of Tasmania was obtained from this same region.

With regard to Baron von Ettingshausen's paper[4] on the Tertiary Flora of Tasmania I am pleased to find that the Author has very kindly forwarded a copy to me gratuitously and unsolicited This together with your paper to the Royal Soc of Tas,[5] will be of great advantage to me.

You will observe from a paper (advance copy enclosed) that I have discovered and described a new species of *Lepidostrobus* from

1 MS annotation by M: 'Rec & answ 23/7/84. FvM. 30/7/84, FvM'. Letters not found.

2 Queen River, Tas.?

3 Thomas Bather Moore.

4 Probably Ettingshausen (1883), in which he named and described *Alnus muelleri*, and which includes a brief section explicitly devoted to the tertiary flora of Tas.

5 B85.13.27.

our coal measures.[6] It is the only complete *Strobilus or cone* of this fossil genus, so far as I know discovered in the coal measures of Australia – certainly the only one from Tasmania. I have taken the liberty of naming [it][7] after yourself 84.07.14

When the drawing is finished in lithograph I will forward a copy.

Bye the bye, who is Mr Rasmussen who is now in Launceston?[8] He has recently come from Queensland and has got himself into bad odour in Launceston by lecturing once or twice on very elementary subjects the greater part of one of them having been plagiarised from various well known authors

Recently he has betrayed much ignorance upon botanical matters, abuses the systematic binominal nomenclature of botanists & otherwise, making a parade of the possession of elementary knowledge in botany.

You will observe I have taken notice of some of his observations in one of the scraps send to you herewith.[9] I cannot conceive how anyone who knows the necessity and the value of the existing bi-nominal system of nomenclature could possibly write in the way which Mr Rasmussen does. It is evident to me that he must possess very little knowledge of scientific method. Has he in any material way contributed to the knowledge of Australian Botany as far as you know?

With great regard
believe me
Yours sincerely
Robt M Johnston

Baron Ferd. von Mueller, K.C.M.G: M.D: F.R.S. etc etc,

Government Botanist
Melbourne

6 See Johnston (1884), in which Johnston described and named *Lepidostrobus Muelleri.*

7 *editorial addition.*

8 Possibly the herbalist Hans Peter Rasmussen (1861-1937); see Gibbney & Smith (1987), vol. 2, p. 201.

9 Items not found.

84.08.16 ## *To William McCrea*

Private hands.

16/8/84

It seems to me best, dear Dr M'Crea, to send you in first instance a copy of my volume on "select plants",[1] which little book I beg you to accept as a private gift for your own library from me. You will see at page 370 the genera mentioned of such plants, as will consolidate loose coast-sands.[2] To facilitate your selecting the best plants, available here, I draw your attention to Elymus arenarius, which Dr Williams raised at Queenscliff from seeds supplied by me. From the Lighthouse ground doubtless now roots and seeds could locally be supplied, without resorting anew to the introduction of seeds. Then I would advise, to get the so-called Buffalo-Grass, Stenotaphrum Americanum, which I introduced in 1858, and which is now in vast quantity available from numerous lawns, garden-edgings &c almost anywhere. This grass should be planted from rooted shoots. The Cynodon Dactylon could also readily be obtained from nurserymen at a cheap rate, either in seeds or in cut rooted shoots; but the sowing should be effected before hot weather sets in. If seeds of Agrostis alba var. Stolonifera can be obtained from the seedsmen, a good help would be gained in your operation, so far as grasses can be chosen for it, though any other quick growing and vigorous grass might also be tried; – and so is it with a variety of herbs mentioned by me under the generic headings. Cuttings of Tamarix, Poplars, Willows, Buddleyas, all cheaply and easily to be got, should be stuck in before the season advances too far. On ground, where the soil is not altogether loose and quite sandy the seeds of the Seacoast-Pines, reared on the European coasts, namely Pinus Haleppensis[3] and Pinus Pinaster should be

1 B81.13.10.

2 B81.13.10, p. 370 lists 68 genera M recommended as 'sand-coast plants'.

3 *Pinus halepensis*?

sown,[4] the seeds being now inexpensively got from the thousands of trees now abounding around Melbourne much through my early action. other Pines, mentioned in my work would be available, but their seeds would be far more expensive – Your success will however much depend in this first trial on the operations being entrusted to an experienced and zealous gardener. Indeed it would be recommendable, that such an artisan should be purposely engaged, to carry out not only the operations indicated, but also to watch over the plantations, collect the seeds of the "Sandstay" Bush, Leptospermum laevigatum and others of the best of our native Sandcoast-plants, and to transfer them also in the cool season as seedlings to the bare ground. Furze-seeds and lots of others could be then also at spare time collected in masses, so the seeds of the Golden Wattle, Acacia Pycnantha, which will succeed on sandy coast. The Melaleuca parviflora occurs still in vast numbers near Queenscliff; and it will live where the ground becomes too loose and shifting for the Wattle. The growth of what was sown this spring or planted then would show next autumn, which particular plants should receive preference on a large scale next season, irrespective of additional trials then.[5] 84.08.16

Regardfully your
Ferd. von Mueller

4 other Pines, mentioned in my work would be available, but their seeds would be far more expensive *is a marginal note with an asterisk indicating that the passage should be inserted after* should be sown, *but the text is placed here for ease of reading.*

5 Strangely, M does not mention Marram Grass (*Ammophila arenaria*), even though at his instigation this had been used successfully in the Warrnambool-Port Fairy region to stabilize drifting sand dunes in that part of Vic.; see Heathcote & Maroske (1996).

84.08.27 *To Thomas Wilson*

B84/7546, unit 38, VPRS 3992/P inward registered correspondence, VA 475 Chief Secretary's Department, Public Record Office, Victoria.

Melbourne,
27 Aug. 1884

T. R. Wilson Esqr
Undersecretary.

Sir,

I have the honor to acknowledge the receipt of your letter of yesterday, conveying the Chief Secretary's approval of Mr Gerhard Renner's appointment of junior Assistant in the Gov. Botanists Department vice Léon Henry resigned.[1]

Pray express to the hon. Graham Berry my best thanks for having sanctioned this proposition, and kindly inform the hon. Gentleman, that the kindness, thus shown by the Chief Secretary will be brought on my request by Mr Renner with Mr Berry's name at once under the notice of her Imperial Highness, the Crown princess of Germany,[2] who caused on her private expense this orphan to be educated during a period of six years. I have no direct communication from H. Imp. Highness; but letters carried by Mr. Renner from his uncle and granduncle in Germany, both divines, bear out the patronage thus shown in a most generous and condescending manner by the Princess Royal of England to young Renner. If I was wrong, in not mentioning this at once in my submitting his name, then I can only say that I really did not know how to act because I feared, it might appear that I had merely recommended his appointment on account of high patronage, not in consequence of merit.

I have the honor to be
Sir, your obed.
Ferd. von Mueller.

Mr Renner will enter on his duties to morrow, 28. Aug. 1884.

1 See also L. Henry to M, 25 July 1884; M to T. Wilson, 25 and 28 July and 15 August 1884; and M to G. Berry, 20 August 1884 [all Collected Correspondence], regarding Henry's resignation and replacement.

2 Princess Victoria Adelaide, eldest daughter of Queen Victoria and wife of Crown Prince Friedrich of Prussia.

To Thomas Wilson 84.09.06

E86/7716, unit 139, VPRS 3992/P inward registered correspondence, VA 475 Chief Secretary's Department, Public Record Office, Victoria.[1]

Melbourne,
6/9/84.

T. R. Wilson Esqr.,
Undersecretary.

Sir

In continuation of my last letter,[2] concerning the distribution of all Governments publication through the public Library, I have the honor to draw the attention of the Government to the desirability of such arrangements being made by the trustees, as will secure not only a *speedy* forwarding of the Victorian publications, but also an early return from the recipients. I beg respectfully to point out, that in our litterary engagements for *progressive* science, it is of the utmost importance, that our own publications should reach their destinations by earliest post, and that we here should also become as quickly as possible aware of the discoveries of those, with whom we are directly connected in our special branches of research, as otherwise lamentable retardation in scientific progress would arise and often also *clashing* of observations. I may further be allowed to state, that the Smithsonian Institute, with which my establishment effected interchanges since the last quarter of a century,[3] is made by no means a vehicle of *all* Government publications of the United States of America, and that but few of the none-Governments Institutions there avail themselves of the facilities, offered by the Smithsonian Institute, to promote or effect interchanges. Moreover often long delays have arisen in the arrival of such and other sendings at their destination, either by detentions in London or perhaps elsewhere, or by waiting til

1 Letter registered as A84/8087.
2 See M to G. Berry, 15 August 1884 [Collected Correspondence].
3 See J. Henry to M, 25 October 1869 [*Selected correspondence*, vol. 2, p. 522].

84.09.06 sufficient publications accumulated for forwarding them as ordinary freight-goods.

I would respectfully still further observe, that the Gov. *frank-stamp* of this colony carries our publications free to any part of the United Kingdom and United States, a favor of which doubtless the Public Library here will avail itself to the fullest extent; and if this system of Governments postal reciprocity could be extended to other countries, an immense advantage would be gained to the scientific and litterary intercourse of the different Governments Establishments in various parts of the world.[4]

I have the honor to be,
Sir, your obedient servant
Ferd. von Mueller.

84.09.08 *To Joseph Hooker*[1]

RBG Kew, Kew correspondence, Australia, Mueller, 1882-90, ff. 115-16.

Private

8/9/84.

In first instance, dear Sir Joseph, let me thank you for the copy of the new edition of your "Students Flora",[2] which renders a splendid work still more perfect; it reminds me of Koch's Taschenbuch,[3] with which and Kittels two pocket-volumes[4] I commenced to make myself acquainted with the north-german and south Danish flora, so much like that of the lowlands of Britain, more than 40 years

4 The file was forwarded to the Public Library on 17 October 1884 and returned on 22 November.

1 MS annotation by Hooker: 'An[swere]d Oct 27/84.' Letter not found.
2 Hooker (1884).
3 Koch (1844).
4 Kittel (1843).

ago. This sending was the first sign of thoughts regarding me for 3 Governments mails, 3 having passed without my hearing of you or Mr Dyer. I do not think, that so long a break in the Kew-correspondence or Kew communications did occur since I took office under Mr Latrobe, except when in the earlier years I was out on long exploring-expeditions. However I myself did neither write much lately, though I never missed any mail, to send something, and manifested also in other ways my continued loyalty to Kew. That I was not in the spirit of writing much, you will understand, when I say, that for more than 3 months this winter I was not able to leave the sickroom on account of my severe cough, though I continued to keep til midnight's hours on the writing table. 84.09.08

Under the grace of divine providence I feel however so much better, that I could leave the mansion of my generous hosts, Dr and Mrs Büttner and as the Apricots commence to flower now and as mild vernal air is setting in again, I can hope for fuller recovery, and trust that I may be spared for a few years more work, though at best I cannot hope for many more! – Let me hope, that Mr Bentham regained sufficient strength, during summer, to enable him with joyful contemplacy the onward-march of his favorite science, with which through the *unrivalled extensiveness* of his *phytographic* researches he must remain identified for all ages![5]

It was my intention, to proceed to the dry desert-country beyond the Wimmera[6] for some field-work, as one stretch of country there was formerly not accessible to me, especially as all my medical friends agree, that the warm dry spring air (away from the cold Antartic winds here) would do me good; but the *rabbits* are now swarming there in the sandy country by the hundreds of thousands; and I learn, that in some places all herbaceous vegetation is fed down by them, so that it may require an alteration of my plan, it having been my wish, to see the numerous spring-annuals on the S. Austral boundary-line for the completion of records on the Victorian flora; but they seem now doomed to extinction by

5 M hoped in vain; Bentham died on 10 September 1884.

6 Wimmera River, Vic.

84.09.08 the excessive multiplication of Rabbits in a winterless clime like that in the hotter parts of this colonial territory.

I was still too ill, when the new Governor, *your late chief*,[7] arrived, to attend the Levee; however I waited on his Excellency a few days ago, and was much inspired with *new confidence into my future* by the gracious manner, in which I was received; indeed Sir Henry Loch reminds me much of Sir Henry Barkly; and I feel sure, he will act to me in the same manner as the latter would, if he was here. Her Ladyship I have not yet seen. This interview with the new Governor has alleviated my mind of much embarrassement, as he is likely to take some interest in my fate and my work. The *anxiety*, which I had for some time, was this: what was to be the fate of my collections in the event of my death! In *confidence* I may add, that a hostile Minister would have the power to *force* me out of the Gov. Service on a *small* pension, when I get to the age of 60 years; but as I was appointed before the Civil service laws came here in force, the former rights are secured, and I have every reason to believe, that the Ministry will leave me undisturbed in possession of the herbarium and library so long as my eyesight and my other physical and mental strenght will last. It would render me still more miserable, if *any* kind of invasion took place on me, my natural temper not being of a kind, to yield to curtailing of administrative power and *competition* in work, though the plea might be set up, as was the case in reference to my loss of the garden, as Mr Casey put it: to "relieve me of drudgery".[8] Kindly remember, how wisely the British Ministry and the House of Commons acts. Sir Richard Owen ceded from the active Directorship of the Nat. Hist. Museum at 80. Your never to be forgotten Parent was called to the Directorship of Kew when he was nearly my own present age; and we all rejoice, that yourself remain undisturbed in your directorial functions and emoluments at Kew.[9] As regards monetary affairs of my own, I become abjectly poor since I had the Garden resources withdrawn from me. I feel sure, you *above all*

7 Henry Loch was a British Commissioner for Woods and Forests and Revenue, 1882-4.

8 See Cohn & Maroske (1996).

9 Hooker retired as Director at Kew in November 1885.

will not countenance any interference should such arise with me, while I am still able to carry on my administration, even should extra monetary provision be made for additional staff, which is quite unlikely. My present Ministerial Chief, the hon. Graham Berry, has the *best of feelings towards me!* 84.09.08

Regardfully your
Ferd. von Mueller.

Mr Berry has allowed me, to employ Mr Renner as an amanuensis in my Museum, where a vacancy happened to occur.[10] I have not only the *moral responsibilty* to initiate him into a life of earning, but also into an occupation of *learning* something for the benefit of his future; crude work elsewhere would mean loss of time in young life to no purpose.

The seeds of Manihot Glaziovi[i] kindly sent by you, have produced plants of extraordinary rapidity.

To Georgiana von Hochstetter 84.10.03

Private hands.

Melbourne
3/10/84

Die trauervolle Pflicht liegt mir ob, hochgeehrte Frau Hofräthin, Ihnen und den Ihrigen meine tiefe Condolenz auszusprechen, über das Dahinscheiden Ihres tiefbetrauerten Gemahls, des unvergleichlichen Gelehrten und liebevollen Freundes, dessen Güte und namenlosen Werth auch ich zu erkennen das Glück hatte in directem wissenschaftlichen Verkehr.

Wie innig wünschten auch wir hier in diesem entferntesten Theile der Erde, dass es ihm vergönnt sein werde, einen langen serenen Abend eines vielbewegten u ruhmreichen Lebens in sei-

10 See M to T. Wilson, 27 August 1884.

84.10.03 ner Grössten Schöpfung, dem herrlichen Wiener Staats Museum zu geniessen, und inmitten seiner blühenden Familie die reichen Triumphe seines vielseitigen Wirkens zu feiern!

Für Sie, edle Frau, muss es ein grosser Trost sein in Ihrer tiefen Betrübniss, dass der Name des Herrn von Hochstetter in der Wissenschaft fortleben wird, so lange die jetztige Welt bestehen mag, und dass in beiden Erdhemisphären geographische Monumente für den ausgezeichneten Gelehrten erstanden sind, welche so dauernd sein werden wie die gegenwärtige Schöpfung!

Mit tiefer Ehrerbietung
der Ihre
Ferd. von Mueller.

Melbourne, 3 October 1884.

I have the mournful duty, esteemed Lady, to offer you and your family my deepest sympathy at the passing away of your deeply mourned husband,[1] the incomparable savant and loving friend, whose kindness and inexpressible worth I, too, had the good fortune to know in the course of direct scientific communication.[2]

How sincerely we, too, here in this most distant part the world wished that it might be granted to him to enjoy a long and serene evening to his eventful and glorious life in his greatest creation, the magnificent Vienna State Museum,[3] and to celebrate in the midst of his flourishing family the great triumphs of his multi-faceted activities!

For you, dear lady, it must be a great consolation in your deep sadness, that the name of Mr von Hochstetter will live on in science for as long as the present world may endure, and that in both hemispheres of the earth geographical monuments have been erected to the excellent savant,[4] which will be as enduring as the present creation!

With deepest respect
your
Ferd. von Mueller.

1 Ferdinand von Hochstetter died on 18 July 1884.

2 M corresponded with Ferdinand von Hochstetter from the time the latter stayed in M's home during his sojourn in Melbourne in 1859; see Darragh (2001).

3 Hochstetter was appointed Intendant of the new Kaiserlich-königliches naturhistorisches Hofmuseum in Vienna in 1876, and worked thereafter to bring the museum into being. It did not open until 1889.

4 Hochstetter Hill, WA; Lake Hochstetter, NZ; Hochstetters Forland, Greenland.

To Franz Stephani[1]

85.01.23

Conservatoire et Jardin botanique, Geneva.

23/1/85

Nehmen Sie meinen besten Dank hin, hochgeehrter Herr Stephani, für die Benennung der austral. Jungermanniaceae. Sie werden inzwischen einige andere erhalten haben; auch werde ich mich freuen, Ihnen künftig Anderes aus derselben Pflanzen Ordnung senden zu können.

Würden Sie mir die Güte erweisen, bei unserem Freunde Herrn Professor Lüerssen vorzusprechen, um sich zu erkundigen, ob die von mir gesandten Farnstämme (mehreren Arten) in tauglichen Zustande und *frei* bei ihm ankamen; auch möchte ich gern wissen, ob ihn einige Filices in getrockneten Exemplaren erreichten. Fast fürchte ich, dass dieser rastlose Gelehrte sich überarbeitete u. krank wurde; denn seit langer Zeit hat er gar nicht mehr geschrieben. Vielleicht ist er auch missvergnügt, dass ich die Benennung der Pflanzen der Frau Dietrich noch nicht vollendete. Mein wiederholtes Kränkeln, unvorhergesehene neue Amtsarbeiten, die fast alljährigen Extra-Arbeiten für Weltausstellungen, haben mir nicht Musse gelassen, die Bearbeitung oberwähnter Pflanzen bisher ganz durchzuführen; dazu ist die Sache eine unerquickliche, denn oft fehlen die Früchte der Arten, u. Gartenpflanzen fremder Länder sind indiscriminirt den einheimischen zugemischt.

Heute sende ich ein nettes Paket jüngst angekommener Papuanischer Farne durch die Post an Hr Prof Lüerssen ab, hoffend dass solche ihm nützen werden u er selbige bald bearbeiten werde.

Sie ehrend u Ihnen Alles Gute wünschend
Ferd. von Mueller.

Dankbrief für sein letztes schönes Werk ging nach Leipzig.

Die 10te Decade meiner Eucalyptography ging auch an Hr Dr L. ab.

1 MS annotations by Stephani: '8/4' and '11/3 an Luerssen' [11/3 to Luerssen].

23.1.85.

Accept my best thanks, highly respected Mr Stephani, for the naming of the Australian Jungermanniaceae. You will in the meantime have received some others;[2] also I will be pleased to be able to send you others from the same Order of plants in the future.

Would you show me the kindness of calling on our friend Professor Luerssen to ask whether the tree ferns (several species) sent by me reached him in good condition and *free of cost;* also I would like to know whether some ferns in dried specimens reached him. I am almost afraid that this tireless savant has overworked himself and become ill, as he has not written to me at all for a long time. Perhaps he is also displeased that I have not yet completed the naming of the plants of Mrs Dietrich.[3] My repeated indifferent health, unexpected new official work, the almost yearly extra work for the international exhibitions have not left me leisure to quite carry out the working up of the above mentioned plants up till now; in addition the matter is an unproductive one, since the fruits of the species are often lacking, and garden plants of foreign countries are indiscriminately mixed in with the indigenous.

Today I will send a nice packet of recently arrived Papuan ferns to Professor Luerssen by post, hoping that they will be of use to him and he will soon work them up.

Respectfully and wishing you all good things
Ferd. von Mueller.

A thank-you letter for his last fine work went to Leipzig.[4]

The 10th Decade of my *Eucalyptographia*[5] also went off to Dr L.

2 See M to F. Stephani, 5 October 1884 [Collected Correspondence].

3 M described some of Dietrich's specimens including, for example, *Acacia dietrichiana* that he named in her honour. However, despite his promise to Luerssen, through whom he had acquired a large collection of Dietrich's material, he never published a systematic account of her collections. See also M to R. Tate, 16 December 1882, and M to W. Thiselton-Dyer, 22 June 1886 [both Collected Correspondence].

4 Letter not found.

5 B84.13.19.

To Louise Wehl

RB MSS M100A, Library, Royal Botanic Gardens Melbourne.

26/2/1885.

Though honoring the confidence, my dear niece, with which you approach me, and altho' I appreciate the kindly feelings, which you show to improve my household, – yet it would be a great disappointment to you, to come to this great city. I have lived here now more than 30 years, and have found that in most cases young Ladies could much easier establish a domesticity for themselves in the country than in the city. You would not believe it, unless you saw it yourself here, what a number of young Ladies have grown old here and remained alone in life, though highly educated, charming in their manners and endowed with personal beauty! How is this? – because in a great city the calls on every one are so exorbitant for support of endless things, from charities to pleasures, from churches to arts, that seldom engagements are made, unless the independence of family-life is secured by *property* already existing. This want of prosperity deters most people to add to their responsibility in great centres of population. Very different it is in country-districts; and the less such are populated, the better it is for the young people. So, I act in due regard to your likely welfare, if I most strongly advise you, not to give up the humble independence, which you so honorably established for yourself. – Even if I was endowed with worldly riches, I would give the same advise. But let me recommend to you, to petition Mr Todd for a position near your dear mother;[1] you will then not feel so lonely, can see her on sundays, and she can aid you in accomplishing the best for your worldly welfare. It was wrong of the Lady, who counselled you, to disturb the peace of your mind! As regards my own habitat-

1 M's sister, Clara Wehl. Louise Wehl may have already been working as a postmistress, the position she eventually came to occupy.

85.02.26 ion, it is only an office with four small rooms, all of which are overfilled with books, collections and office-concerns. So I keep only a groom in my simple household. I never invite any one, and only one meal daily is prepared, as there is no space for domestic concerns. I wanted for some time to add one room, so that at all events one private room might be in the poor place; but I have not even the means for that, as only lately I paid off the mortgage on the building. My income is so heavily taxed by my scientific intercourse with all parts of the world, that I have nothing whatever either to build or furnish any place. Had I received the slightest help from any one, when I "struggled for existence" in the bot. Garden, I could have had the happiness of some of my near relatives visiting me. In the garden was at least after 17 years a good two-storied House completed by me. You must also remember, my dear niece, that at my time of life I have very seldom any time for any pleasure; so I could not go out or even come to see you except at long intervals, if you stayed with friends in Melbourne; and if I had any spare-means, I feel sure you would like me rather to give them to your poor mother or for the education of your younger sisters and brothers.[2] Think me not unkind, but allow me to speak to you from a lifelong experience with the same candor, which you evinced towards me.

With my sincerest wishes for your welfare

Ferd. von Mueller.

I am sure, I have [bowed] to any young Ladies for many years.

2 Clara Wehl had twelve children who survived to adulthood.

To Heinrich Fayenz 85.04.01

P85/1279, unit 40, VPRS 1163/P1 inward correspondence, VA 1123 Premier, Public Record Office, Victoria.[1]

Melbourne
1/4/85.

Capt. Fayenz &c,
Commander of the Imperial Austrian Corvette "Saida".

Sir

I have the honor to inform you, that I have been instructed by the hon. James Service, the Premier Minister of Victoria, to transmit to you on behalf of the Victorian Government the following contributions towards the collections of the Imperial scientific Institutions of Vienne, from which kindred Gov. establishments of the Colony Victoria are in literary relation and material interchanges:

Two growing Todea-ferns, weighing about 1000lb each.
12 living ferntree-stems of Dicksonia Billardierii.[2]
one case of Victorian woodspecimens, worked up into technical articles.
one case of dried Australian plants.

In making this presentation by order of the Government of this Colony, I beg to assure you, that it will be always gratifying to the scientific establishments of this Colony, to remain in communication with the great Institutions of the Austrian empire for the advancement of learning and applied knowledge. – For the maintenance of this scientific relation the visit of his Imperial Austrian Majesty's Ship "Saida", under your able Command, afforded also particular facility; and I feel, that similar auspicious visits of Ships

1 First registered as D85/3471 in the Chief Secretary's Department, then referred to the Premier.
2 *Dicksonia billardierei*?

85.04.01 of the Imperial Austrian Navy in future years will here always be hailed with delight, as particularly calculated to connect also more and more the scientific and industrial interests of her Britannic Majestys young Australian colonies with those of a great continental empire, in which arts and sciences florished through very many centuries.

I have the honor, Sir,
to be your obedient servant
Ferd. von Mueller.[3]

85.04.07 *From Joseph Hooker*
RB MSS M3, Library, Royal Botanic Gardens Melbourne.

Royal Gardens Kew[1]

April 7 /85

My dear Baron

I have yours of 15/2/85[2] enclosing Scortechini's interesting letter.[3] As I told you, I had written to him at once on receiving your letter requesting me to do so.[4] My letter crossed one of his to me, & that also I have answered. I have implored him to describe the Palms, Pandaneae Musaceae, big Aroids & all such plants on the spot, for all the materials he will find here will only confuse him; & he will be bitterly disappointed to find that an Order, to the working up of which here, he will look with keen expectation, will prove utterly refractory.

3 See also M to G. Berry, 9 April 1885, and H. Fayenz to M, April 1885 [both Collected Correspondence].

1 Embossed letterhead.
2 Collected Correspondence.
3 Letter not found.
4 See M to J. Hooker, 23 October 1884 [Collected Correspondence].

What I am to do when I come to the Indian Palms (if ever I do) is a mystery. I can only take [so][5] much from Scortechini. I have evidence of 5 or 6 species of Phoenix in continental India & *not one is well known* not even P. *sylvestris*, if, as I believe, the Ceylon plant is not the Indian! & no one has proved the wild *sylvestris* of the Sth Deccan to be the right thing. But what I dread most in regard to Scortechini, is his not coming here at all! or in such bad health that, like Thomson, Falconer, Anderson, Brandis, Wight, Wallich, & a host of others – he will be unfitted by reason of previous fevers &c & exposures, *to work when he gets here*! This is the normal condition of Indian botanists. Clarke[6] was the exception that made the rule. – Even Beddome does nothing! 85.04.07

So to tell the truth I am more anxious to see S. well out of the Malay jungles[7] than continuing researches & collecting that may result, like Griffith's & Maingays in huge piles of unreported posthumous materials shovelled into Kew – full of novelty & interest, but wanting a master hand to sort with judgement.

Well one must hope for the best, – I have seen Sir Hugh Low about him who [promises] all the needful funds & I have through Dyer addressed him an offical letter suggesting the sort of work that Scortechini should bring out. – The Colony has plenty of money for the moment, & the sooner that S. is home & at work the better.

– Now as to Tenison Woods whose non election to the Royal you deplore. I should premise that for the last 4 or 5 years I have been far too busy to take my part in the R. S. affairs, & I have not been on the Council till this year, when I went on much against my will, – I was pressed to do so because Huxley (President) is ill & away on 6 months' leave, & it was thought that a past President would under the circumstances be a useful man. So I really have taken no part whatever in the Societys affairs till this year, when I will see see what I can do for him. I observe that his "recommendations" are very meagre and that except Davidson & Woodward no English Scientific man has put his name to his paper of candidat-

5 *editorial addition* – word omitted?

6 C. B. Clarke.

7 Scortechini died in Calcutta in November 1886.

85.04.07 ure from knowledge of himself or works – It is clear therefore that he wants a good backing up from Geologists – to pull him in amongst the 15 out of no less than 67 candidates.![8]

I see that his paper has not been put up this year but do not know whether this is an omission, or because Sr H Barkly may have forgotten it. It is however no matter for surprise that 2 years should have elapsed before he got in – 3 years is a very usual time to wait except for case of commanding attainments or labors. Of the 61 candidates for this year's election

1	has been up for	13	years!
1	– – –	7	–
2	– – –	6	
2	– – –	5	
3	– – –	4	
5	– – –	3	
22	– – –	2	

As to your suggestion of canvassing through Scortechini – pray be careful, it may do harm, as I said before any attempt at making Father Woods' claims known should be modulated by some one of *high position* in his own science.

The selection of candidates in Council is most carefully & thoroughly conducted, & the opinions of real judges from personal knowledge of the candidates work alone have any weight – no one for instance would be influenced by what I would say, except in the case of a Botanist. – Nothing can however be done this year, for his name is not up & it is too late to put it up. There should be an addition to his paper of recommendation pointing out his more important work. As a member of Council I am precluded from using any influence out of Council, I cannot even sign a Certificate.

I am very very busy, & all my spare moments go to Flora of India.[9] I have done Polygonum with 60 species, a thankless job. I find I must make a species of that little Tasmanian & Victorian so called *Strigosum* with spinate flowers; – the same occurs in the

8 Tenison Woods was never elected a Fellow of the Royal Society of London.

9 J. Hooker (1875-97).

mountains of Ceylon & the Khasia & always quite distinct from *Strigosum* (horridum of India). I propose to call it *P. praetermissum*. 85.04.07

I am now at Piper, soaking & dissecting – a most dreadful task wh[ich] will take weeks. Miquels[10] work is poor & C. D C's[11] *atrocious*.

By the way I have often intended to tell you that your long letters with specimens come torn & crumpled, frightfully, with the specimen often smashed across & reduced to powder

Ever sincerely your
J D Hooker

From Gerhard Renner 85.04.15

C85/4136, unit 72, VPRS 3992/P inward registered correspondence, VA 475 Chief Secretary's Department, Public Record Office, Victoria.

Botanical Museum of Melbourne
April the 15th 1885.

Baron Sir F. Von Mueller
K.C.M.G., MD. etca etca[1]
Government Botanist

Sir!

I have the honour to report for your information that on the night of the 13th and 14th I was awoke by a noise and heard somebody knocking gently at one of the windows of my room. When it was repeated I got up and when looking through the half opened blind I saw a rather suspicious looking person, apparently listening if any person was inside. I struck a light and when looking again

10 Miquel (1843).
11 Casimir de Candolle; see Candolle (1869).

1 etcetera.

85.04.15 through the window I saw the man had left. Hearing at the same time a knocking at the front door of the mainbuilding I went into the same and to the door, just in good time to see a man, of different stature than the above mentioned one, leaving the platform and disappearing. The key of the door had been thrown out of the door apparently to allow the use of a skeleton key and upon examining the door I found that one had tried to take out the glass. When I went outside I saw two shadows dissappear. Entering the Museum again by the front-door I heard knocking and shaking again on the backdoor and voices softly speaking. I thought it now the right time to make an end and to call assistance from the police station. In order to have at least a witness, should anything happen to me; and that the place might not be deserted but watched during my absence I went, protected by the darkness, to the Observatory to ask Mr. Burley (Senior messenger and caretaker of the Observatory) for assistance. He also advised me to report it to the Police at once. On the way to the Police Station and when approaching the back-door of the Museum again I saw a man busy to examine its locks. I called to stand and at the same time fired into the air to show I was armed. The man turned and suddenly disappeared, the heavy rain and profound darkness made it impossible to follow him

I am Sir
your most obedient servant
Gerhard Renner.
Custodian of Bot. Museum[2]

2 See also M to G. Berry, 16 April 1885 [Collected Correspondence], that M sent with Renner's letter, and in which he recommended installing a telephone line from the Herbarium to the nearby Melbourne Observatory. This was not done and in 1908 A. J. Ewart, the Government Botanist, was still asking for a telephone to be installed in the Herbarium (Cohn [2005]).

To Ferdinand von Krauss[1]

85.04.21

Staatliches Museum für Naturkunde, Stuttgart.

21/4/85

Diesmal möchte ich Ihnen nur mittheilen, edler Freund, dass ich Ihnen in nächster Woche durch das Agenturhaus von Mr Tate *frei*, ein Kästchen senden werde, welche einige in Alcohol conservirte Eidechsen u eine kleine Schildkröte enthält, welche alle nahe Cape York (Nordost-Australien) gesammelt wurden. Die kleine Schildkröte mag vielleicht neu sein, u in dem Falle wäre es gut, solche *bald* zu beschreiben, da jetzt mehrere Zoologen in Queensland u New South Wales alles Neue das in deren Hände fällt, gleich bearbeiten.

Allerlei Anderes grösseres liegt bereit für Sie, aber nicht genug um eine Kiste zu füllen.

Sie ehrend u Ihnen Alles Gute wünschend
Ferd. von Mueller

21/4/85

This time I would just like to inform you, noble friend, that I will send you a small box in the next week *free* through the agency of Mr Tate, which contains some lizards and a small tortoise preserved in alcohol, which all were collected near Cape York (north-east Australia).[2] The small tortoise may perhaps be new and in that case it would be good to describe it *soon*, since now several zoologists in Queensland and New South Wales immediately work up all new things that fall into their hands.

Various other larger things lie ready for you but not enough to fill a box.

Respectfully and wishing you all the best
Ferd. von Mueller.

1 MS annotation by Krauss: 'empf. 2 Juni Rep. 20. Aug' [received 2 June Replied 20 Aug]. Letter not found.
MS annotation by Krauss: 'Die 2 beigeschlossenen Schmetterlingene waren ganz zerdruckt, abgerieben' [The 2 enclosed butterflies were completely squashed, shredded].

2 See also M to F. von Krauss, 28 April 1885 [Collected Correspondence].

85.05.15 *To Otto Tepper*[1]

RB MSS M198, Library, Royal Botanic Gardens Melbourne.

15/5/85

Schon vor einigen Tagen, geehrter Herr Tepper, schrieb mir unser Freund Mr Molineux, wegen der Benennung der X. Tatei. Ich setzte ihm auseinander, dass bei der Benennung neuer Arten man zunächst das Recht des *ersten Finders* berücksichtigen muss, gleichviel ob Selbiger die Art als neu erkannte oder nicht.

Mir ist nicht bewusst, dass vor Prof. Tate's durch mich angeregte erste Reise nach der Känguruh-Insel überhaupt die Vertretung des Genus X. in der Käng. Insel wissenschaftlich bekannt war.

Wie mich Mr Somerville vor einiger Zeit besuchte, wurden die Charactere (die äusseren) der *ihm* bekannten X. Arten weitläufig erörtet, u. ich mache es sonach, die X. Tatei zu definiren in einem Artikel, den ich vor mehreren Wochen nach Wien sandte für die Zeitschrift des österreich Apotheker Vereins, dessen Ehren Mitglied ich seit vielen Jahren bin. In diesem Artikel sprach ich über alle Arten des G. Xanth. Ich *erinnere* nicht, dass ich Materialien von Ihnen hatte, um die K. I. Art zu definiren, wenigstens nicht genug, zumal da auch Capseln u reife Samen nöthig sind. Ob wirklich X. Tatei von X. australis, die westlich bis zum Glenelg Flusse doch schon vordringt, wirklich *specifisch* verschieden sei, ist noch weiter zu ermitteln. Wenn ich bei solchen Gelegenheiten in die einzelnen Umstände des Verfolgs der Entdeckung einer neuen oder seltnen Pflanze nicht immer eingehe, so müssen Sie bedenken, dass ich in meinen ganz engen und unzureichlichen Räumlichkeiten das litterarische zerstreute Material ohne grossen Zeitaufwand nicht zu finden vermag, u. in *meinem* Alter mit meiner Zeit sehr geizen muss. Bedenken Sie, dass ich als ein junger Doctor schon vor Ihrem Dasein in Südaustralien ankam, 1847!

1 MS annotation by Tepper: 'Reply 7.6.85 with "Lawn grass", Native Plants & [Taft] Animals'. Letter not found.

Figure 10. Otto Tepper, autographed for Mueller in 1879 and sent from Ardrossan, SA, contained in Mueller's photograph album. Courtesy of the Library, Royal Botanic Gardens Melbourne.

85.05.15 So habe ich es denn auch neulich übersehen, dass Prof. Tate zuerst die Styphelia costata von der K. Insel brachte.

Auch ich habe vom Secretair der L.S. nichts von Ihrer schönen Abhandlung gehört, Ihre neuen offenbar wichtigen Schriften werde ich in einer späten Nachtstunde zu lesen suchen. Ich habe alle für Lesen Abhandlungen u Journale nur spät im Schlafzimmer Zeit, nehme Ihre Schriften aber immer zuerst vor. Sie haben wohl keine Idee, dass ich fünf medicinische, zwei chemische, 4 horticulturistische, zwei geologische, etwa 1/2 Dutz geographische u ebensoviel botanische Journale regelmässig lese!

Freundlichst der Ihre
Ferd von Mueller.

15/5/85

Already some days ago, esteemed Mr Tepper, our friend Molineux wrote to me concerning the naming of the *Xanthorrhoea tatei*. I explained to him that in the naming of new species one first must consider the right of the *first discoverer*, no matter whether he recognised the species as new or not.

I am not aware that the representation of the genus *Xanthorrhoea* on Kangaroo Island[2] was scientifically known before Professor Tate's first journey to Kangaroo Island, prompted by me. When Mr Somerville visited me some time ago, the characters (external) of the species known to *him* were discussed at great length, and I did so afterwards to define *X. tatei* in an article that I sent several weeks ago to Vienna for the *Zeitschrift des österreich. Apotheker Vereins*,[3] whose honorary member I have been for many years. In this article I spoke about all species of the genus *Xanthorrhoea*. I do not *remember* that I had material from you to define the Kangaroo Island species, at least not sufficient, especially since capsules and ripe seeds are necessary. Whether *X. tatei* is really *specifically* different from *X. australis*, that certainly already penetrates westward up to the Glenelg River,[4] is still to be further ascertained. If on such occasions I do not always go into the detailed circumstances of the course of

2 SA.

3 M described *X. tateana*, collected by Somerville & Wilks on Kangaroo Island, in B85.07.01.

4 Vic.

discovery of a new or rare plant, you must consider that I am not able to find the scattered literary material in my quite cramped and insufficient premises without great waste of time and at *my* age must be very sparing with my time. Consider that I arrived in South Australia as long ago as 1847 as a young doctor before your existence![5] 85.05.15

Thus I have also recently overlooked that Professor Tate first brought *Styphelia costata* from Kangaroo Island.

Also I have heard nothing from the secretary of the Linnean Society of your fine paper.[6] I will try to read your new obviously important writings in a later night hour. I only have time for reading papers and journals late in my bedroom, but I will always attend to your writings first. You probably have no idea that I read five medical, two chemical, four horticultural, two geological, about half a dozen geographical and just as many botanical journals regularly!

Most amicably your
Ferd. von Mueller.

To Adolpho Möller 85.05.18

Arquivo da Correspondência do Dr Júlio Henriques, Biblioteca do Departamento de Botânica, Universidade de Coimbra, Coimbra, Portugal.

18/5/85[1]

Vor einigen Wochen, geehrter Herr Möller, sandte ich Ihnen durch die Agentur des Mr Tate ein Packet getrockneter Pflanzen, welches wohl vor diesem Briefe angekommen sein wird. Die Fracht ganz bis Coimbra war bezahlt.

Ich werde Ihnen demnächst Holzproben senden, auch mancherlei zool. Exemplare und vielleicht *Schädel*. Solche sind jetzt

5 Tepper arrived in South Australia in 1847 aged six.

6 Neither the Linnean Society of London nor that of New South Wales published any papers by Tepper.

1 18 *over* 12.

85.05.18 nach dem Aussterben der Eingebornen in naheren Gegenden schwer zu erhalten, denn die alten Grabstätten sind ausgebeutet.

Affen giebt es in Australien nicht. –

Mehr trockne Pflanzen können nach u nach folgen, wenn überzählige Exemplare hier vertheilt werden. Ist es nicht der Mühe für Sie werth, sich mit Jemand in Rio de Janeiro in Verbindung zu setzen, ob die brasil. Regierung nicht das Erscheinen einer portugiesische Ausgabe meiner "Select plants" nach Ihrer Übersetzung übernehmen würde vielleicht in der Gouvernements Imprimerie von Rio, zum Vertheilen an öffentliche Anstalten in dem ganzen grossen Lande, auch zum Verkauf. Jedenfalls möchte ich aber *keine* solche Anfrage in Lissabon.

Mit den besten Wünschen für Ihr Wohlergehen
Ferd. von Mueller

18/5/85

Some weeks ago, respected Mr Möller, I sent you a parcel of dried plants through the agency of Mr Tate,[2] that will probably have arrived before this letter. The freight was paid to Coimbra.

I will soon send you timber samples, also various zoological specimens and perhaps *skulls*. Such are now hard to get in closer regions after the extinction of the Aborigines, because the old burial grounds are depleted.

There are no monkeys in Australia. –

More dried plants can gradually follow, when surplus specimens here are distributed. Is it not worth the trouble for you to communicate with someone in Rio de Janeiro, whether the Brazilian Government would not undertake the publication of a Portuguese edition of my *Select plants* from your translation,[3] perhaps in the Government Printery of Rio, for distribution to public institutes in the whole large country, also for sale. At any rate, however, I would *not* like such an inquiry in Lisbon.

With best wishes for your welfare
Ferd. von Mueller

2 See M to A. Möller, 11 February 1885 [Collected Correspondence].

3 A Portuguese translation, attributed however to Julio Henriques rather than to Möller, was eventually published in 1905 (2nd edn 1931, but dated 1929 on the title page); see B05.13.01, B29.13.01.

To William Thiselton-Dyer 85.05.25

RBG Kew, Kew correspondence, Australia, Mueller, 1882-90, ff. 148-9.

25/5/85

By this post I am sending you, dear Mr Dyer, fragments of leaves and antheriferous and ovuliferous scales of a Macrozamia, to which I have given your honored name.[1] Concerning this Zamia a correspondence has been gone on for several years, til at last I got sufficient material from Esperance Bay,[2] to prove its distinctness from M. Fraseri. This shows also, that in measuring my share in the Flora Australiensis among many other facts the enormous efforts must be taken into consideration, made by me through nearly 40 years to bring such unexampled rich material (and elaborated too at once) together. I do not exaggerate, when I say, that I wrote about 100 000 letters during the last 1/3 century.

I have suggested to the Exhibition Commissioners at Swan River,[3] to take timely measures for procuring a fresh stem of M. Dyeri for their Court, the stem to be reared into foliage at Kew on an understanding, that the growing plant remains at your grand establishment after the exhibition as well as the fruitspike.

I have to thank you for sending me the most interesting new edition of Kew Guide,[4] also for returning to me a copy of the print of my rural adress.[5]

Sir Joseph wrote to me about the none-election of the Rev. Jul. Ten. Woods into the RS after his second candidature.[6] Rev. Scortechini can give, if he should not stay 2 more years in India, personal

1 M published the species as *Encephalartos dyeri* (see B85.06.02, p. 12) but the specimens available to him were poor and he was evidently uncertain as to its status, referring to it as 'this new Zamia or Encephalartos'. It was reassigned to *Macrozamia* by Gardner (1930-1).

2 WA.

3 i.e. the Commissioners responsible for mounting the WA exhibit at the forthcoming Colonial and Indian Exhibition, London, 1886.

4 Royal Botanic Gardens, Kew (1885).

5 Presumably B80.13.09.

6 See J. Hooker to M, 7 April 1885.

85.05.25 information about Woods If Sir Henry Barkly thinks proper, to put Mr Woods name up again for the third time,[7] I shall willingly allow me[8] name as that of a supporter also for the third time, not because that I think it nice to continue a candidature for three years in succession, but because I *may not live* an other year to attach my signature, my pulmonary sufferings not being subdued.

I suggested,[9] when sending my mite to Dr Masters for the placing of the likeness of the great Bentham in Kew, that the Linnean Society out of the fund bequeathed to it by the lamented great phytographer[10] should create a Bentham-medal for phytography, as that would give an additional impetus to systematic phytology and bring annually the name of one of its illustrious presidents prominently before the Society. In these times, when phytologic anatomy and physiology and morphology drive taxology so much into the background, such a medal as that suggested by me would help to keep phytography in its proper dignity, as so few even of enlightened people do reflect, that horticultural, geographic, medicinal technologic and rural Botany rest mainly or entirely on phytography not anatomy and physiology or evolutional research, however valuable they assuredly are. But We have to look to the great "*bread-affording*" branches mainly, and all physiology &c would have no basis, if we have not a progressive system of descriptive phytology. Someone in an educational position as a teacher spoke recently here of Botany as regarded by many as a science of hard names in the usual stereotypic manner of expression; but if we are not to have the greek and latin names, it will be best to shut out from our high schools all classi[cs][11]

Regardfully your
Ferd. von Mueller.

7 Tenison Woods was a candidate for election to the Royal Society of London, always unsuccessfully, in 1883-4 and 1886-8, but was not a candidate in 1885.

8 my?

9 Letter not found, but see M to J. Hooker, 6 and 28 November 1884 [both Collected correspondence], where the suggestions for a medal are made.

10 Bentham willed £1000 to the Linnean Society, expressing his desire that it should be used to open a Library Fund. M was perhaps unaware that Bentham had expressed such a wish.

11 *editorial addition.* – Text obscured by binding.

I wrote also to Mr B. D. Jackson[12] about a Bentham-Medal, and trust, that the proposition is not rejected because it arose in Australia. 85.05.25

Woods is now geologizing in China[13]

To Joseph Hooker[1] 85.05.30

The Linnean Society Guard Book, 1885-93, section 5, London.

30/5/85.

The byefollowing letter, dear Sir Joseph, explains the sending of the byefollowing skeletonized leaves. The Lady, who prepared them is the wife of one of our medical practitioners here, Dr Lewellin, and mother of an accomplished family, one of her sons being the principal resident Medical Officer of the Melbourne general Hospital. She has read of Sir John Lubbocks researches on leaves, and thinks the byefollowing specimens will interest that distinguished Baronet.[2]

You may remember that in the Eucalyptography with E. ptychocarpa I gave sections of the leaves of four species, so that the layers

12 Letter not found.

13 Tenison Woods travelled extensively in south-east Asia in 1883-4, and then from May 1885 spent several months in China; see O'Neill (1929).

1 MS annotation: 'spec. exhib. 3 Nov 1885.' The leaf skeletons were exhibited at a meeting of the Linnean Society on 3 November 1885, and the salient points of this letter were summarised in *Proceedings of the Linnean Society, 1883–6*, p. 117.

2 Mrs Lewellin thought the leaves might be used by Lubbock 'for his lectures "On the forms of leaves"'; see G. Lewellin to M, 29 May 1885 [Collected Correspondence]. See Seward in Duff (1924) for a discussion (pp. 179–82) and citations (pp. 194–5) of Lubbock's work on leaves. Lubbock (1886), pp. 97–147 contains the republished lectures. Lubbock read a paper 'On forms of leaves' at the meeting of the Linnean Society on 16 April 1885 (*Proceedings of the Linnean Society, 1883-6*, p. 76).

85.05.30 of cells might be noticed, the number of layers being comparatively definite, and stand probably in relation to the separation of the leaves in layers by the skeletonizing process.[3] It may also be worthy of Sir John Lubbock's notice, that Mrs Lewellin found while macerating many kinds of leaves for artistic work of former exhibitions, that Eucalyptus leaves in decay produce no bad odor, which results almost from all other foliage when soaked for skeleton-leaves. This shows the futility of M Riviere's recent attempts (in the Bulletin de la Societe nationale d'acclimatation de France)[4] to demonstrate that Bamboos are as good to subdue Malaria as Eucalypts! Certainly at a gutter along my poor office-place I have placed Arundo Donax to soak up stagnant foul water, there being no space to plant Eucalypts, but while the Bamboos and Reads[5] simply lay dry a shallow place of humidity their leaves have not the antiseptic properties of myrtaceous foliage dependent on the volatile oil, as also shown by Sir Joseph Lister in surgical treatments. Besides Eucalypts soak up moisture as quickly as Willows, Poplars, Bamboos, &c, &c. Perhaps all these observations may interest the worthy President of the Linnean Society while engaged in his present study of leaves.

Regardfully your
Ferd. von Mueller

3 B80.13.14, decade 5.
4 Rivière (1885).
5 Reeds?

To Ferdinand von Krauss[1]

85.06.08

Staatliches Museum für Naturkunde, Stuttgart.

8/6/85.

Heute haben Sie mich aufs Innigste erfreuet, theuerer Herr, obwohl es mich quält zu hören, dass Sie an das Ende Ihrer segensreichen Laufbahn schon denken. Chevreul hat 97 Jahr alt noch kürzlich seine Vorlesungen an der Universität Paris forgesetzt. Humboldt war ja auch zum 90stn Jahre geistes frisch u körperlich stark; Bentham u Hooker arbeiteten aufs Anstrengendste zum 84 Jahr, so können ja auch Sie Muth fassen, nur müssten Sie sich *Ruhe* gönnen, u. behaglich auf das Geschaffene Ihrer schönen Anstalt hinabschauen u. Ihre Kräfte schonen. Wahrlich *Sie* haben genug gethan, u. können nun doch auch sich selbst u den Ihren leben!

Dass der arme Dr Wilkie plötzlich in Paris starb anfangs April, ist Ihnen vielleicht bekannt geworden, da eine verheirathete Schwester der Frau Dr Wilkie vor einigen Jahren in Stuttgart lebte mit ihren Kindern. Er opferte sich offenbar seiner Familie auf, da die 3 prächtigen Töchter in Paris der Studien der Musik, des Gesangs u. der Malerei sich zum Vergnügen widmeten, was den hoch betagten seit nahe 50 Jahren an das austr Clima gewohnten Mann in die kalte Nacht-Luft Nordfrankreichs zu oft gebracht hat, so dass er der Lungen Entzündung unterlag! Mit ihm ist einer meinen edelsten Freunde, doch in dessen Haus ich seit mehr denn 30 Jahren aus u ein ging, dahin! Die Wittwe u. Töchter werden nun wohl bei Verwandte in Schottland für Erste bleiben.

Herrn Messner, als besonderen Landsmann und einen der angenehmsten Menschen den man sich nur denken kann werden Sie ebenso lieb gewonnen haben als ich ihn hochschätze. Ich freue mich schon recht auf seine Rückkehr! –

1 MS annotations by Krauss: (1) 'Empf. 27. Juli Rep. 20. Aug.' [Received 27 July Replied 20 Aug.]. Letter not found. (2) 'Als Geschenk durch die Linn. Soc. in London v. v. Müller Index perfectus ed C. Linnei species plantarum Collatore Ferd. de Müller. Melbourne 1880.' [As presentation through the Linnean Society in London from von Mueller ...]. See B80.13.04. (3) 'Eucalyptographia 1, 2.5.7 erhalten [received]'.

85.06.08 Dass Sie mich durch die neue *Sarcophila* oder Pennatula mit dem weltberühmten Anatomen u Physiologen Hr. Prof Kölliker in Berührung brachten, erregt meine hohe Freude, u bin ich Ihnen u dem verehrten Manne aufs dankbarste verpflichtet meinen Namen auch in der Gruppe der Meerfeder-Thiere vereinigt zu haben. Wenn ich diesem erhabenen Gelehrten in irgend einer Weise eine Freude bereiten könnte, so soll es gern geschehen!

Wenn ich bedenke, wie von der allgemeinen Welt Darwin, den auch ich als *Beobachter* verehre, förmlich *vergöttert* wird, so fühle ich die Ungerechtigkeit gegen solchen Herren, in der Natur Wissenschaft als Kölliker u Owen, deren solides Forschen bleibenden Werth hat u unumstossbare Resultate giebt, während doch nur ein kleiner Theil des Darwin's Theorien haltbar bleiben. Der 80 jährige Owen ist noch so frisch, so gütig, so geistigstark noch; mir hat er in *inniger Freundschaft*, obwohl wir uns im Leben nie trefen, stets nahe gestanden.

Ein Skelett nach *einem* Knochen herzustellen, ist die *Errungenschaft* eines Cuvier, eines Owen, eines Kölliker, das überstrahlt blosse Theorien.

Auch Hr Dr Fischer bitte ich für seine neue Dedication, diesmal unter den Ophidea, meinen besten Dank zu sagen, u freundlichsten Gruss zu entbieten. In diesem Jahr giebt es besonders schwere Arbeit bei mir. Die Obliegenheiten für die London Colonial Ausstellung mit ganz für mich unzulänglichen Mitteln, die Hülfe für Organisation der Neu Guinean Expedition; die Präsidentur unserer geogr. Gesellschaft hier, die Herausgabe einer neuen erweiterten Ausgabe des Werkes "select plants", die Vollendung der mit 80 Quarto Tafeln illustrirte Monographie der Myoporinen, die Ausarbeitung eines Schulbuchs für Benennung der mehr denn 1800 Vascular-Pflanzen Victorias, strengen mich neben der enormen Correspondenz aufs Übermennenste an; so dass Sie mir guter u edler Freund, nicht zürnen dürfen, wenn ich *Ihnen* gegenüber, dem ich aufs Tiefste verpflichtet bin, so lässig erscheine. Im nächsten Jahr, deo volente, werde ich wohl etwas freier sein, u da der liebe Hr Messner mir noch genauer sagen kann, was Sie gebrauchen, so soll Ihnen dann recht viel zukommen

Was ich eben an wanzenartigen Hemipteren auftreiben kann, soll Ihnen gleich per Post zukommen, auch bald die fehlenden Hefte der Eucalyptographie u manches Andere. 85.06.08

Nun halten Sie sich ja lange noch frisch u munter, u bleiben Sie fürder gewogen Ihrem

Ferd. von Mueller

Die Coleopteren &c *kaufte* ich von Mr French.

Die Holz-Ex. hoffe ich auch bald absenden zu können

Ein Wuerttemberger, Hr Bäuerlin, ist von mir für die Neu Guinea Expedition als Pflanzen Sammler gewählt

Von Sarcoptilus bisher nur 1 Ex. gefunden[2]

8/6/85

Today you have pleased me greatly, dearest Sir, although it torments me to hear that you already think about the end of your highly beneficial career.[3] Chevreul has just recently continued his lectures at the University of Paris at 97 years of age. Humboldt was certainly even in his 90th year intellectually fresh and strong in body; Bentham and Hooker worked most strenuously at 84 years, so you can certainly also pluck up courage, but you would have to allow yourself *rest* and look down contentedly on the achievements of your fine institution and conserve your strength. Truly *you* have done enough, and are certainly able now also to live for yourself and your family!

That poor Dr Wilkie suddenly died in Paris at the beginning of April is perhaps known to you, because a married sister of Mrs Wilkie lived some years ago in Stuttgart with her children. He obviously sacrificed himself for his family, since the 3 splendid daughters in Paris devoted themselves to the pleasures of music study, of singing and of painting, which has brought the highly aged man,[4] who was used to the Australian climate for nearly 50 years, into the cold night air of northern France too often, so that he succumbed to inflammation of the lungs! With him has died one of my most noble friends, indeed in whose house I went to and fro for more than 30 years! I suppose the widow and daughters will now remain at first with relatives in Scotland.

2 Von Sarcoptilus bisher nur 1 Ex. gefunden *is written on the flap of an envelope.*

3 Letter not found.

4 Wilkie was 69 when he died.

85.06.08 You will have become as fond of Mr Messner[5] as a special compatriot and one of the most pleasant men whom one can think of as I highly respected him. I am already looking forward very much to his return!

That you have brought me in contact with the world famous anatomist and physiologist Professor Kölliker through the new *Sarcophila* or *Pennatula*, arouses my great pleasure and I am most gratefully obliged to you and the esteemed man for having united my name with the group of sea pen animals as well.[6] If I could give pleasure to this eminent savant in any kind of way it shall gladly be done!

When I consider how Darwin, whom even I admire as an *observer*, is positively *idolized* by the world in general, I feel the injustice towards such men in the natural sciences as Kölliker and Owen, whose solid research has lasting value and gives incontrovertible results, whilst only a small part of Darwin's theory remains tenable. The 80-year-old Owen is still so bright, so kind, so intellectually vigorous; to me he has always stood in *intimate friendship*, although we have never met.

To reconstruct a skeleton from *one* bone, is the *achievement* of a Cuvier, of an Owen, of a Kölliker, that outshines mere theories.

Also please give Dr Fischer my best thanks for his new dedication this time among the snakes[7] and present most friendly compliments. In this year there is particularly heavy work at my place. The obligations for the London Colonial Exhibition[8] with quite insufficient funds for me, the assistance for organisation of the New Guinea expedition, the presidency of our Geographical Society here, the publication of a new enlarged edition of the work *Select plants*,[9] the completion of the monograph of the Myoporinae illustrated with 80 quarto plates,[10] the preparation of a school text for naming of more than 1800 vascular plants of Victoria,[11] together with the enormous correspondence fatigue me most overpoweringly, so that you, good and noble friend, will not be angry with me, if I seem so remiss towards *you*, to whom I am obligated most deeply.

5 Frederick Messner arrived in Victoria in April 1863 aboard *Great Britain* and made a successful career in business. In 1884 he decided to re-visit his home town, Stuttgart, and M provided him with an introduction to Krauss; see M to F. von Krauss, 20 August 1884 [Collected Correspondence].

6 Kölliker named *Renilla muelleri* in M's honour; see Kölliker (1872), p. 144. No subsequent naming of *Sarcophila* or *Pennatula* by him in M's honour has been located.

7 J. G. Fischer named *Hoplocephalus muelleri* in Fischer (1885). See Darragh (1996).

8 Colonial and Indian Exhibition, London, 1886.

9 B85.13.26.

10 B86.13.21.

11 B88.13.03.

In the next year, deo volente,[12] I will probably be somewhat more free and, since dear Mr Messner can tell me even more exactly what you need, shall then send very much to you. 85.06.08

What I can just find of bug-like Hemiptera shall be sent to you immediately through the post, also soon the lacking parts of the *Eucalyptographia* and some others.

Now keep hale and hearty for a long time and remain favourably disposed towards your

Ferd von Mueller

I *purchased* the Coleoptera from Mr French.[13]

I hope to be able to send the wood specimens soon.

A Wuerttemberger, Mr Bäuerlin,[14] is selected by me as plant collector for the New Guinea expedition.

Up to now only 1 specimen of *Sarcoptilus* found.

To Otto Tepper[1] 85.06.10

RB MSS M198, Library, Royal Botanic Gardens Melbourne.

10/6/85

Die 2 Pflanzen, welche Sie geehrter Hr Tepper mir so freundlich senden, sind Helipterum anthemoides u. Viola betonicifolia.

Ich lese Ihre Notizen immer mit grossen Interesse im F. & G. Ihre Sprache ist sehr schön, u. es treten auch immer Original-Ideen hervor, oft von grossem practischen Werth. Ich werde demnächst über cosmopolit. Naturalisation von Pfl. eine Abhandlung liefern. Eine neue erweiterte Ausgabe der "Select plants" geht nächster

12 God willing.

13 Some years earlier, M offered French's collection to the British Museum, but there is no evidence of its having been purchased; see M to R. Owen, 30 October 1878.

14 W. Bäuerlen.

1 MS annotation by Tepper: 'Reply 9.7.85 with Orchids & Ludwig's Digitalis'. Letter not found.

85.06.10 Monat zur Presse, um für die London Colonial Austellung fertig zu sein.

Für das Supplement des Census von 1885 ist auch schon manches notirt, aber es *gebricht* recht sehr an Sammlungen aus *vielen* Gegenden *Central* Australiens, von woher wir noch sehr viel zu lernen haben über die geogr. Verbreitung der Arten.

Auch bei Eucla sollte tüchtig gesammelt werden, da sich dort die S. u. die Westl. Vegetation begegnen. Sie können wohl aus Ihrem Museum u Ihrer Häuslichkeit nicht fortkommen; sonst würden in einem Monat am besten im October *Grosses* für Pflanzen u Insekten-Kunde da zu erreichen sein, u Gelegenheit dahin ist doch wohl nun zu Schiff oft genug. Würden die Trustees *Sie* nicht bei Eucla einige Wochen diesen Frühling sammeln lassen, um neue Tausch-Materialien zu erwerben. Wird die Museum Directur dort ganz eingehen?

Freundschaftlich Ihr
Ferd von Mueller.

10/6/85

The two plants that you, respected Mr Tepper, so kindly sent me are *Helipterum anthemoides* and *Viola betonicifolia*.

I always read your notes in the *Field and Garden* with great interest. Your language is very good, and also original ideas always emerge often of great practical worth. I will soon produce a paper on cosmopolitan naturalisation of plants.[2] A new enlarged edition of the *Select plants* will go to press next month to be ready for the London Colonial exhibition.[3]

Much is also already noted for the supplement of the Census of 1885,[4] but we very much *want* collections from *many* regions of *Central* Australia, from which we still have very much to learn about the geographical distribution of the species.

Also near Eucla should be efficiently collected, because there the southern and western vegetation meet. I suppose you cannot get away from your museum and your family life; otherwise it would be a *great thing* to reach there for

2 No such publication has been identified.

3 B85.13.26; Colonial and Indian Exhibition, London, 1886.

4 B85.13.19.

a month at best in October, for plant and insect studies, and there is probably opportunity to go now often enough by ship. Would not the Trustees let *you* collect for some weeks this spring near Eucla to acquire new exchange material? Will the museum directorship there be completely abandoned? 85.06.10

Amicably your
Ferd. von Mueller.

To Alfred Moloney 85.06.24

F86/6828, unit 136, VPRS 3992/P inward registered correspondence, VA 475 Chief Secretary's Department, Public Record Office, Victoria.[1]

Melbourne
24 June 1885.

To his Excellency Capt. Alfr. Moloney, C.M.G.,
Governor of Gambia &c.

Sir

I have the honor to acknowledge your Excellency's communication, dated 16. April,[2] in accordance with which I have procured on behalf of the Victorian Government for transmission through my ministerial Chief, the hon Graham Berry, Chief Secretary of Victoria, a collection of fresh seeds of such Eucalypts, as deserve trial-culture in Gambia and could be procured just now, the sending being a gift of the Government of Victoria. As regards the saline nature of the ground-water in many parts of your colonial territory I would beg to remark, that all Eucalypts are [...]y[3] to brackish soil; but doubtless many localities in Gambia would still be available for the culture of these important trees.

1 Letter registered as B84/1631.

2 See A. Moloney to M, 16 April 1885, and M to G. Berry, 24 June 1885 [both Collected Correspondence].

3 *illegible* – MS smudged.

85.06.24 To provide timber and fuel for places, where the humidity is saline, it will be best to choose species of Melaleuca, a few of which will grow in brackish water, for instance M. Leucadendron, M. ericifolia, M. linearifolia.[4] Of these I will procure seeds purposely, and send them for transmission to the hon. the Chief Secretary here. The Melalecuas like the Eucalypts have great anti-malarian power, hence would also [be][5] for sanitary purposes a valuable acquisition to your colony and other parts of Western Africa. Some other kinds of Eucalyptus-seed I hope to send later, when fresh procured. May I ask your Excellency, whether without much trouble a small quantity of fresh and well matured seeds of the Coapim-grass (Panicum spectabile), which is so famous for its rapid growth and nutritive property, could be obtained from your colony for test culture in Victoria

I have the honor to be
your Excellencys obedient
Ferd. von Mueller.

`85.06.26 *From Ferdinand Wehl*[1]

RB MSS M197, Library, Royal Botanic Gardens Melbourne.

Ehrenbreitstein, Millicent[2]
26th June 1885

My dear Uncle

As your birthday is approaching I write to wish you "Many happy returns of the the day" and that good health may be yours. I am now at home again having finished my survey work at Mt Gam-

4 *M. linariifolia?*

5 *editorial addition* – word omitted.

1 MS annotation by M: 'Answ 12/7/85.' Letter not found.

2 SA.

bier & am busy reading up to enable me to pass my examination for a License under the Real Property Act without which a Surveyor can only obtain limited employment here 85.06.26

I notice a newspaper report is current that a white man has lately been seen with[3] the blacks at the MacArthur River in the Northern Territory if this is true it is quite possible that he is a survivor of the ill fated Leichardt's party, I am happy to note that enquiries are to be made & if there is any truth in the report I presume a search party will be at once sent out. I should very much like to accompany such a party; I enclose a clipping giving details.[4]

We are having a very wet & cold winter here this year with severe frosts which cut off any herbage of a tender nature; the Buffalo Grass which you sent us some time ago is doing very well & seems to be very hardy. We are beginning to suffer now from the rabbit plague, the bunnies have been increasing very fast and I believe we shall have great difficulty in getting rid of them here in consequence of the numerous Wombat holes in which they shelter; Bisulphide of Carbon will have to be employed & that will entail considerable expense.

Bertha is now in Sydney Geo Harris having obtained an appointment in the Survey Department there,[5] our family is gradually getting scattered far and wide. I wish we were nearer Melbourne so that be[6] could see more of you, when we get the Railway through I suppose we shall be able to go & see you occasionally as the journey then will not take so much time.

Hoping to hear from you soon
I am, dear Uncle
Your affectionate nephew
F. E. Wehl.

3 I notice … seen with *is marked in the margin with a line.*

4 Clipping not found.

5 M's niece Bertha Wehl married George Harris in 1874.

6 we?

85.06.27 *To Graham Berry*

F86/3516, unit 121, VPRS 3992/P inward registered correspondence, VA 475 Chief Secretary's Department, Public Record Office, Victoria.[1]

Melbourne,
27 June 1885.

To the honorable Graham Berry, M.L.A.
Chief Secretary

Sir.

I have the honor to report, that on the 30th June 1885 I shall reach my 60th year; but as I am able to continue my duties of Governments Botanist quite well and have directed the whole studies of my life for that duty, I beg to *solicit*, that I may be allowed to continue in the service of the Government in the position of Governments Botanist.[2]

I have the honor to be,
Sir, your obedient servant
Ferd. von Mueller.

1 The letter is numbered C85/6411 within the file.

2 According to the rules of the Public Service Act, M, who was about to turn 60, was required to retire; however, under the rules applying when he had joined he was entitled to yearly extension at the discretion of the Governor in Council. On 6 July 1885 Berry recommended that M 'continue to perform his duties until it shall be further ordered'. The Public Service Board replied on 22 July, saying that M's continued employment 'must be determined not upon individual claims but solely upon public grounds'. The public grounds, sent to the Board on 28 August, included M's 'altogether unique and exceptional knowledge of Botany', which, along with his many previous and ongoing publications, enabled him 'to point out from time to time such plants, trees, &c as can be usefully introduced and grown here for economical or other purposes: ... He is also continuing in the interests of the Colony his extensive correspondence with the Scientific Botanists all over the World, ... [and] he is still quite fit to continue in the discharge of his duties.' On 15 September the Board recommended that his services be retained until 30 June 1886 (C85/7423, unit 170, VPRS 3992/P, PROV).

To William Thiselton-Dyer

85.06.29

RBG Kew, Kew correspondence, Australia, Mueller, 1882-90, ff. 150-1.

29/6/85

I am greatly beholden to you, dear Mr Dyer, by sending me your excellent photogram, which I shall value all the more, as it is the first likeness I ever have seen of you, and as we not likely will ever meet personally in life. This is an other mark of your generous sentiments towards me, of which you have given me so many proofs. I shall give this beautiful photographic picture of a sterling scientific friend a prominent position in the Album of illustrious correspondents.[1]

My photgraphic card will ere long be offered you in return for kind acceptance; but I have none at hand and have not been for years in an "atelier"; indeed I had never yet a *large* card-picture done for all my life!

I am somewhat out of the run of ordinary mortals; never bought a watch in my whole life; used in travelling the compas[2] instead, even chronometer-watches becoming unreliable for longitude in land-exploration. Never brought a mirror, only one a few spare[3]-inches large being in my poor place. My Hottentot-life so long in the "bush" has made me indifferent about the ordinary luxuries in the world, even if my money had not gone in travels, books, charities &c Still, as there was no Lady in the poor place at any time it did not matter much – The whole last 2 weeks we had here cloudy skies and rains; hence the photgram[4] of ♀ and ♂ of Macrozamia

1 Photograph not found. Thiselton-Dyer's portrait is not amongst the images included in M's photograph album held at MEL.

2 compass?

3 square?

4 photogram?

85.06.29 Dyeri[5] did not get ready by this mail. Next post I shall also be able to send N. vii of "Papuan plants"[6]

Regardfully ever your
Ferd. von Mueller

Lots of extra-work for London Colonial Exhibit.,[7] but I cannot do it to advantage, and only very partially.

Will try to get more seeds of the large Byblis.[8]

85.07.11 *To Johan Lange*

Lange papers, Botanisk Centralbibliotek, Copenhagen.

Private

11/7/85.

This morning, dear Professor Lange, I packed up for you the 5th & 10th vol. of the "fragmenta" out of my private library, as the 5th is no longer available from Government direct. In reference to your desiring, to obtain the complete Eucalyptography, allow me to suggest to you 1, to ask the *honorable James Service*, Premier for a *complete copy*, as it would appear strange that you received some only; the fact is, I had never enough exemplaria to provide all my Correspondents. Your present decades you could then utilize in interchanges elsewhere Be particular to adress the Premier *exactly* as I underlined it, and if you could send to the honorable James Service some *few* publications by post at the same time for the

5 See *Encephalartos dyeri* in B85.06.02, p. 12 and B85.13.11, p. 225.

6 Probably B85.06.03. Part vii (B86.02.03) was not published until the following year.

7 Colonial and Indian Exhibition, London, 1886.

8 will try ... Byblis *is written on a separate sheet and may belong to another letter.*

Melbourne public Library, it would be good. I return your letter, as it will be necessary, to write to him *direct*, *not* through me. 85.07.11

Please do *not* mention in your letter about the Todea, as it will be a *private* Gift of mine; if I was to send it through the Department, it would involve the necessity of your making a *return-sending* and necessitate also various official correspondence. All the Governments Departments here are much restricted in their actions.

Could you kindly extract out of Prof. Örsted's memoire[1] any *technologic* notes on Oaks and Pines of Mexico for the new edition of my select plants? Surely he made some allusion to the particular species which yield the best timber and most powerful tanbark. Perhaps you could oblige me yourself from your rich experience by notes on the value of such Scandinavian and Spanish trees, fodderherbs, pasture-grasses &c, as are not yet mentioned in the "select plants".

A new edition appeared last year in Detroit, Michigan, at Mr G. Davis's Establishment for N. America.[2] It is much enlarged; and perhaps the Librarian of your Agricultural Academy will send for it. It is not expensive; I have no copies to give away, and have no monetary interest in the sale and the issue of the work. The next Edition is to appear soon in Victoria,[3] and I add to it, – whenever I can, but adress myself only to *breadwinning people* in *extra-tropical* countries. About the respective value of arctic grasses and pasture herbs I know very little; yet notes on them would be of importance for our Australian Alps and other alpine countries, altho' we have not the long arctic summers.

Can I do anything for your Academy? Do you like me to send more Australian plants to your Herbarium in interchange? I am afraid I am in your debt yet for so many beautiful spanish specimens, highly important in connection with your & Willkomm's splendid "Flora Hispanica".[4]

1 Liebmann (1869).
2 B84.13.22.
3 B85.13.26.
4 Willkomm & Lange (1861-80).

85.07.11 The successive "Exhibitions" in various parts of the world gave much extrawork for years; and in this pushing young country so much else is to be done also in my department, that little time is left to effect herbarium-interchanges. When you see Prof. Japetus Steenstrup, Pray give this celebrated Savant my homage.

Regardfully your
Ferd. von Mueller

I met Prof Steenstrup in Kiel *1846*!

What do you think of my simplification of D.C.[5] syst., as given in the "systematic Census of Australian plants[6]

I have neither Liebmanns nor Örsteds papers on Mexican plants[7]

85.07.13 *To Adolf Engler*

Staatsbibliothek zu Berlin, Preussischer Kulturbesitz, Berlin.

Privat 13/7/85

Die letzte Post, hochgeehrter Herr Professor, brachte mir Ihren freundlichen Brief v. 31 Mai u. auch das begleitende Circular-Schreiben. Es bedarf wohl nicht meiner Versicherung, dass ich Ihrem schönen Unternehmen gern nützlich sein werde, soweit ichs vermag. Pflanzen-Familien, die Australien ganz oder fast ganz angehören, will ich gern bearbeiten, z.B. die Myoporinen, Tremandreen, Pittosporeen, Candolleaceen, Stackhousiaceen. Die neue Ausgabe meiner "select plants", welche für die Colonial-Ausstellung in London erscheinen wird, könnte ja auch Ihrem Werke zu gut

5 De Candolle.

6 B82.13.16.

7 Liebmann published several works on Mexican plants. Örsted also edited Liebmann's posthumously published book on Mexican oaks; see Liebman (1869).

kommen; u ich würde es auch erlauben, unter Angabe der Quelle, dass Abbildungen aus meinen hiesigen Werken in Xylography copirt würden, wie es bei Schnitzleier schon begann. An *einem* Felsen scheitern manche derartiger Unternehmungen, daran dass der Plan *zu grossartig* und zu *weitgreifend* gefasst wird. Mir scheint, ein Werk wie Lindley's vegetable Kingdom wäre das beste Vorbild mit Ausdehnung der Illustrationen etwa wie bei Maout u Decaisne. Ein Werk wie Baillon's ist schon veraltet ehe es zu Ende kommt (êntre nous); u *wann* kommt es zu Ende? Ist es besser, ein kleineres Werk zu liefern, und es durch neue Auflagen neu zu erhalten? *Zu* grosse Werke sind auch viel zu kostbar für das allgemeine Publicum u nicht tragbar auf Reisen. So wird es mit Prof Kirchhoff's geogr Werck "Unser Wissen von der Erde" sein. –

Mir scheint, dass für genera doch wohl Jeder B. & H. oder auch noch den immer ausgezeichneten Endlicher zu Rathe ziehen würde.

Eine besondere Schwierigkeit entsteht, *diese*: welche Sequenz soll beobachtet werden in der Anordnung der Familien? Ich habe nach vielen Nachdenken und endlosen Vergleichen zu keinem anderen Resultat kommen können, als dass das Syst von D.C. oder das umgekehrte Original-System von Jussieu das *beste* bleibt, nur müssen die Monochlamydeae oder Apetalae vertheilt werden anderweitig ausser Coniferen. Ich kann mich auch noch gar nicht mit dem Gedanken befreunden, dass die Coniferen den Filices nahe stehen. Man beobachte doch nur die Keimung. Hat sich die Beobachtung über die Entwicklung eines prothallus mit Archegonien im Embryoschlauch wirklich bestätigt? Liegt nicht vielleicht eine sehr verzeihliche Missdeutung vor, wie bei der Idee, dass die Lichenen aus einer Zusammensetzung von Algene u. Pilze bestanden? *Sie* sind gewiss in der Lage gewesen, *selbst* diese Beobachtungen zu prüfen; mir hat es an Zeit u besonderer Übung zu *solchen* Untersuchungen, ja auch den besten Instrumenten gefehlt. Geben Sie mir doch gütig Ihre Ansicht auch hierüber, denn ich bin immer offen für Belehrung, u. auch gern bereit meine Anschauungen zu ändern nach zu länglichen Beobachtungen.

Sehen Sie sich, bester Freund, doch auch *noch einmal ruhig meinen Census an*, u. sagen Sie mir gütig, was der darin angenommenen Anordung entgegensteht; bis zum gewissen Grade ist sie loyal für

D.C. Wäre nicht etwas Ähnliches im Embryosak anderer Dicotyledenen zu entdecken?

Sie haben mich ausserordentlich durch die Original-Belege der Aroideen verpflichtet; *die* haben Werth für Alle Zeiten. Demnächst soll auch wieder ein Packet au[s]erlesener australischer Pflanzen an Ihr Museum abgehen.

Die Ermittlung der Gattungen nach einem kurzen Schlüssel nützt nicht viel, da man dann zum Stillstand kommt, u. nichts über die Art weiss, wie man dies beim Gebrauch von Baillon's Monographien empfindet. Die Berechnung der Arten ist auch eine schwierige u. gewagte, seit die meisten Naturforscher überhaupt gar keine Stabilität von Arten anerkennen. In der That wäre es gut, wenn man die Fixität der Arten nicht mehr anerkennt auch den Ausdruck von Species ganz aufzugeben u. nur von Formen zu reden, oder Bastarden oder Varietäten. Welche Divergenz in der Calculation der Arten der Meliaceen selbst in wenigen Jahren, seit H. & B. u dann A de Candolle die Arten-Zahl berechneten! H. & B. schrieb ich von Anfang an, dass Species ... *notae* gesagt werden müsse; denn in vielen Gattungen wird der Zuwachs ja noch enorm sein, z.B. aus Central Africa den Sunda-Inseln, Neu Guinea, selbst Brasilien u andern central-amerikanische Ländern.

Ein *Vorläufer* eines grösseren Werks wäre – so scheint es mir – am Besten *jetzt*. Am Ende Jahrhunderts oder in Beginn des nächsten, könnte dann eher etwas *Erschöpfendes* geliefert werden

Ganz der Ihre
Freundschaftlich
Ferd. von Mueller.

Würden Sie nicht gelegentlich ein Schema der Verwandtschaft der Pflanzen-Familien nach *Ihren* eignen reichen Erfahrungen geben? Ich habe auch Baillon um die Aufstellung eines Systems nach seinen eignen Ideen gebeten. Absolut lässt sich freilich die Stellung mancher Familien nicht fixiren in einreihiger Linie, wie solche für Bücher nöthig ist, da ja die Verwandtschaft radial bleibt. Das ist aber das Schöne der Anlage eines bot. Systems in einem Garten, wie ich es B. de Juss schon 1857 hier nachahmte, dass man den vielseitigen Affinitäten gerecht werden kann. In meinem "Census" werde ich gelegentlich noch einige kleine Änderungen machen in Bezug auf die verwandtschaftliche Stellung einiger Familien; aber ich denke doch, dass im Allgemeinen ich nicht viel zu ändern habe, u. an der Verschmelzung der perigynae u epigynae halte ich fest. Schade, dass B. & H. meinen Rathe in 1862 nicht folgten, die Monochlamydeen den andern Familien zuzustellen nach respectiven Erfordernissen. Man sieht die Nothwendigkeit dafür recht bei den Curvembryonaten (od. Amyliferen).

Würden Sie gütig einige Körner frisch von Psamma in Ihrem nächsten Brief beifügen, Elymus arenarius *habe* ich hier naturalisirt.

Auf geldliche Zahlung für etwaige Beitrage wurde ich keine Ansprüche machen.

Soll die Synonyme der Gattungen auch vollständig berücksichtigt werden

13/7/85

The last post, highly esteemed Professor, brought me your friendly letter of 31 May and also the accompanying circular letter.[1] It probably does not require my assurance that I will gladly be of use to your excellent enterprise, as far as I am able. I will gladly work up plant families, which belong entirely or almost entirely to Australia, for example the Myoporinae, Tremandreae, Pittosporeae, Candolleaceae, Stackhousiaceae. The new edition of my *Select Plants*,[2] that will appear for the Colonial Exhibition in London,[3] could certainly also be beneficial to your work and I would also permit, with acknowledgement of the source, illustrations from my local works to be copied in xylography,[4] as it has already begun at the engraver's. Several undertakings of such a kind foundered on *one* rock, in that the plan was conceived too *grandiosely* and too *far reachingly*. It seems to me a work such as Lindley's *Vegetable Kingdom*[5] would be the best model with expansion of the illustrations somewhat like Maout and Decaisne.[6] A work such as Baillon's[7] is already obsolete before it comes to an end (entre nous) and *when* will it be finished? Is it better to produce a small

1 Letters not found.
2 B85.13.26.
3 Colonial and Indian Exhibition, London, 1886.
4 I.e. wood engraving.
5 Lindley (1847).
6 Le Maout & Decaisne (1868).
7 Baillon (1867-95).

work and to keep it updated by new editions? *Too* large works are also much too costly for the general public and not easy to carry on journeys. So it will be with Prof. Kirchhoff's geographical work *Unser Wissen von der Erde*.[8]

It seems to me that for genera each one would probably consult H. & B.[9] or even the always outstanding Endlicher.[10]

One particular difficulty arises, *this*: what sequence is to be observed in the arrangement of the families? After much thought and endless comparison I have been able to come to no other result than that the system of D.C.[11] or the inverted original system of *Jussieu*[12] remains the *best*, only the Monochlamydeae or Apetalae must be divided in some other way without the Conifers. I can also still not reconcile myself at all with the idea that the Conifers stand near the Filices. After all only the germination was observed. Is the observation about the development of a prothallus with archegonium in the embryo tube really confirmed? Is there not perhaps a very excusable misinterpretation, as with the idea that the lichens consist of a combination of algae and fungi? *You* certainly have been in the position to check these observations *yourself*. I lack the time and the special experience for *such* investigations, indeed even the best instruments. Do kindly give me your opinion about this, as I am always open to instruction, and also gladly prepared to change my views according to prolonged observations.

Do *calmly look over my Census*[13] *again*, best of friends, and kindly tell me what is opposed to the arrangement adopted in it. Up to a point it is loyal to D.C. Would not something similar be discovered in the embryo sac of other dicotyledons?

You have put an extraordinary obligation on me by the original documents of the Aroideae;[14] *which* have merit for all time. Shortly a packet of selected Australian plants is to again go off to your museum too.

The determination of genera according to a short key is not much use, because one then comes to a standstill and knows nothing about the species as one feels in using Baillon's monograph.[15] The determination of species is also a difficult and bold one, since the majority of naturalists recognise absolutely no stability at all of species. In fact it would be good if the fixity of species

8 Kirchhoff (1884-93).
9 Bentham & Hooker (1862-83).
10 Endlicher (1836-40).
11 Possibly De Candolle (1816-21).
12 Jussieu (1789).
13 B82.13.16.
14 Possibly Engler (1884).
15 Not identified.

was no longer recognised even giving up the expression of species completely and only speaking of forms, or bastards or varieties. What a divergence in the calculation of species of Meliaceae even in a few years, since Hooker & Bentham and then A. de Candolle determined the species number! I wrote to Hooker & Bentham from the beginning that species ... *notae*[16] must be said because in many genera the growth will still be really enormous, e.g. from Central Africa, the Sunda Islands, New Guinea, even Brazil and other Central American countries.

A *precursor* of a larger work would – it seems to me – be best *now*. At the end of the century or at the beginning of the next, a somewhat *exhaustive one* could then be produced.

Entirely yours
Ferd. von Mueller

Would you not give some time a scheme of relationship of plant families according to *your* own rich experiences? I have also asked Baillon[17] to put forward a system according to his own ideas. Absolutely of course the place of many families does not allow fixing in a straight line, as is necessary for books, because the relationship certainly remains radial. But that is the beauty of the layout of a botanical system in a garden, as I imitated B. de Jussieu here as early as 1857, that one can do justice to the manifold affinities. In my *Census* I will at times make in addition some small changes in respect to the relational position of some familes, but I do think that in general I have not much to change and I hold fast to the merging of the perigynae and epigynae. A pity that Bentham & Hooker did not follow my advice in 1862 to distribute the Monochlamydeae to the other families according to the respective requirements.[18] One sees the necessity for this properly in the Curvembryonateae (or Amyliferae).

Would you kindly add some grains fresh from Psamma in your next letter. I *have* naturalised *Elymus arenarius* here.

I would make no claims for monetary payment for any possible contributions.

Shall the synonyms of genera be completely considered as well.

16 i.e. known.

17 Letter not found.

18 See M to G. Bentham, 24 January 1862 (*Selected correspondence*, vol. 2, pp. 124-7).

85.09.10 *From N. Samwell*[1]

RB MSS M1, Library, Royal Botanic Gardens Melbourne.[2]

Georgetown[3]
Sept 10th 1885

Baron Von Müller

Dear Sir

I hereby forward you under seprate cover, a herb, also a piece of the same made up into a compound. The Compound – (which I believe is made from this herb) is used by the Chineese at Gilberton[4] for curing snake bites, for which purpose I have seen it used with good results the compound was pounded, and mixed with brandy, and then given internally, wound scarified & washed with water, then the same mixture applied – I have seen a horse bitten by a black Snake cured in this manner. I have also seen the wounds on two Chinamen who were bitten by a tiger Snake who said they were cured in this manner.

If the herb is so – it might become of some good – The herb might be a very common one for all I know, for "I am not much of a herbist." However if you should discover anything you could kindly let me know the name of the herb etc.

Hopeing that you receive it safetly and can make something out of it, also that it is not to scarious. I remain –

Dear Sir
Yours Very Truly
N. Samwell

Address –
N. Samwell
Georgetown, Etheridge Gold Field
North Queensland

1 MS annotation by M: 'Answ 16/10/85 FvM'. Letter not found.
2 Letter found with a specimen of *Blumea mollis* (MEL 545065).
3 MS annotation by M: 'Queensland'.
4 Qld.

To Eduard von Regel 85.09.21

MS 163, folio 21808, d. 76, op. 1, fond 335, Archives, Academy of Sciences, St Petersburg.

21/9/85.

In einem registrirten Brief an Sie, edler Freund, ist mit dieser Post ein Dankschreiben an Sr Ex. den Minister der Domainen abgesandt, obwohl es mir schwer ward, die Innigkeit meiner Gefühle der Erhebung in Worten auszudrücken. Die Diamanten sind von reinstem Glanz u. das Ganze unaussprechlich prachtvoll. Diese Ural-Diamanten sind wunderbar schön, u die Arbeit des Schleifens und die Ringfügung auch diejenige Künstler ersten Ranges.

Ich werde in meinem Testament durch eine besondere Clausel feststellen, dass stets dies Kleinod in höchsten Ehren gehalten wird. Ich selbst werde wohl nur noch kurze Zeit leben, um mich an diesem Kaiserlichen herrlichen Geschenke zu erfreuen. Es ist so ausserordentlich werthvoll, dass ich nicht weiss, wie ich mich dem Throne Russlands dankbar bezeigen soll. Würden Ihnen Eucalyten-Samen solcher Arten, als etwas Frost ertragen für die milderen Regionen Turkestans und für das Amur Gebiet zum Vertheilen willkommen sein? In Manchuria, wo Sie ein Japanisches Clima haben, könnten die besten Eucalypten für Bauholz auch der Kaiserl. Marine von hohem Nutzen werden. Vielleicht könnten Sie einen kleinen Experimental-Garten am Amur angelegt bekommen.

Verehrungsvoll der Ihre
Ferd. von Mueller.

Im Februar u nun wieder im Sept habe ich durch Tate's Parcel-Agency einige Museum-Pflanzen nach St Petersburg gesandt. Sollten solche an Sie adressirt werden? Die Akademie hat ja auch ein Herbar.

85.09.21 21/9/85.

In a registered letter to you, noble friend, by this post, has been sent a letter of thanks to His Excellency the Minister of the Imperial lands, though I found it difficult to express in words the intensity of my feelings of exaltation. The diamonds are of purest brilliance and the whole indescribably magnificent. These diamonds from the Ural are amazingly beautiful and the cutting and setting of the ring are the work of a first class artist.[1]

I shall make provision in my will by a special clause, that this jewel will always be held in highest esteem.[2] I myself will probably live for only a short while longer to enjoy this wonderful Imperial gift. It is so exceedingly valuable, that I do not know, how to express my gratitude to the throne of Russia. Would *Eucalyptus* seed of such species as would be able to withstand frost be welcome for distribution to the milder regions of Turkestan and for the Amur region? In Manchuria, where the climate is the same as in Japan, the eucalypts with the best timber for building could also be of great use to the Imperial Navy. Perhaps you could arrange for a small experimental garden to be planted in Amur.

Devotedly Yours
Ferd. von Mueller.

In February and again now in September I have sent some herbarium plants to St Petersburg through Tate's Parcel Agency. Should such be addressed to you? The Academy has a Herbarium also.

1 M formally acknowledged the gift through the Governor's office; see M to F. Traill, 15 September 1885 [Collected Correspondence].

2 There is no codicil to this effect to the will that M had already made (dated 17 June 1884).

To Edward Strickland 85.10.10

ML MSS.2134/1, Royal Geographical Society of Australasia (NSW Branch) papers, Mitchell Library, State Library of New South Wales, Sydney.[1]

10/10/85.

Last evening, dear Sir Edward, we held here a Council-meeting,[2] and this reminds me, that I still owe you an answer to your last kind letter.[3] We here will *endeavour* to get the vote renewed, and with some *care may* succeed in this for *one* year more.[4] Three days ago I received a first letter direct from Mr Forbes,[5] and this I laid yesterday before the council here.

We have reason, to set great hope on him! He opened the correspondence himself, which is easily understood, as he is much versed in the knowledge of plants himself, and as Sir P. Scratchley gave his own collector over to Mr Forbes's Expedition.[6] I had merely sent my little work on "Papuan plants"[7] to that distinguished Gentleman, and so he sent me a letter of acknowledgement, out of which the Council here ordered such passages to be published in the daily

1 For a summary of this letter see ML MSS.853/2 letter register, no. 802, The Royal Geographical Society of Australasia (New South Wales Branch), Mitchell Library: 're publications'. For another summary see ML MSS.774/1 General and Administrative Council, minute book, 10 November 1885, Royal Geographical Society of Australasia (NSW Branch), Mitchell Library: 'Baron von Mueller Re Forbes & another vote £1000'.

2 i.e. of the Victorian Branch of the Royal Geographical Society of Australasia.

3 Letter not found.

4 The Victorian Branch of the Royal Geographical Society of Australasia had been receiving a subsidy of £1,000 per annum from the Victorian Government.

5 Letter not found.

6 Forbes had been appointed to lead an expedition to New Guinea, jointly sponsored by the British Association for the Advancement of Science and the Royal Geographical Society. He arrived in Cooktown, Qld, in August 1885 and from there travelled to New Guinea with Scratchley, who had been appointed Commissioner for the new British Protectorate there.

7 Probably the first five sections, plus appendices, of M's work, issued between 1875 and 1877. Later issues commencing in 1885 were often treated as a second volume of the work.

85.10.10 journals here, as would interest the general public.[8] He goes evidently about in the best and safest way, advancing on the Owen Stanley's Range by the Sugaire line.[9] He writes, that the first set of his plants is to go to the British Museum, the second to the geographic Society of Australasia. As your Council so considerately ordered the bot collections of our Explorers to be handed for examination and subdivision over to me, it would be important that any sendings from the Expeditions should be forwarded to me *at once* after arrival in Sydney, so that we may with this part of the Expedition's work as early as possible be before the public. Mr Forbes writes, that at the British Museum the officers are so overwhelmed with work, that he looks to me for the elaboration of his plants. I will therefore communicate with Mr Carruthers and Mr Britten, so that a clear understanding is arrived at, whether they wish me to cede to them any particular families of plants for elaboration.[10] Mr Carruthers is a specialist in ferns and therefore I shall leave them entirely to him. Mr Forbes further remarks, that the Kew Establishment has *not* claimed any portions of his collections, which is in accordance with a former expression of Sir Joseph Hooker, that the elucidation of the Papuan flora, (as so much connected with that of North Australia) ought to fall to my share. Mr Forbes writes in a manner, as a true son of science would; and as our correspondence will be mainly phytological, I hope you will not there object to my remaining in direct letter communication with him.

Let me hope, that in this mild spring weather the excellent Mr Maiden will become quite well again. His intention, to come to Victoria, seems not to have been followed up; our cold antarctic

8 A brief report of Forbes' departure from the coast for the interior of New Guinea was published in the Melbourne *Argus*, 14 October 1885, p. 8.

9 The Sogeri District, where Forbes made his base camp at Saminumu.

10 See M to W. Carruthers, 21 October 1885 [Collected Correspondence], and M to W. Carruthers, 20 June 1886. In his introductory remarks in B90.05.01, M refers to 'a valuable enumeration' by H. N. Ridley [i.e. Ridley (1886)] of the monocotyledons collected by Forbes. M too described new species collected by Forbes in various publications, and in B90.05.01 he promised to devote the following, 10th, part of this series to plants collected in New Guinea by Forbes and Bäuerlen. However, though proofs of some pages exist, the promised work was never published.

breezes may also be against him in the *South*. He intended to bring Mr Bäuerlen's plants, sent from Thursday-Island;[11] so it would be best if they were forwarded by rail now.[12] 85.10.10

Regardfully your
Ferd von Mueller

What is your idea there, dear Sir Edward, about the issue of all our publications?[13] Would it not expedite matters, if each branch looked after its own issues, but that in Melbourne, Brisbane and Adelaide a fixed number of extracopies, all printed in conformity, were prepared for transmission to Sydney, where they could be bound up into the yearly or halfyearly volumes of the Society.

If at any time I should be guilty of any short comings to the Central Executive Council, or if an apparent want of loyalty should occur, you must be assured, my honored friend, that such will not be real and intentional shortcomings; but in my large and ramified Department, I get often *overpowered* by work, and may thus be utterly unable to be so attentive as I wish. The day has only 24 hours for all of us; I wish often it had 48! and in that *little time of life likely left me*, I like to carry out a little *progressive* work in science yet, irrespective of routine-work. So I hope, that I may become the Society's expounder also of some portion of the Papuan Flora.

Let me hope, dear Sir Edward, that you are happy and well, and remember me kindly also to our excellent Colleagues of the Sydney-Council.

11 North Qld.

12 J. H. Maiden was ill for much of the second half of 1885 and the early part of 1886, apparently with typhoid fever. M's fears notwithstanding, Maiden visited Melbourne in October 1885, carrying Bäuerlen's Thursday Island plants with him, and a second time in November; see M to E. Strickland, 29 October 1885 [Collected Correspondence] and Gilbert (2001), pp. 112-14.

13 i.e. of the Royal Geographical Society of Australasia.

85.10.15 *To David Lindsay*[1]

ML MSS.200/3, folder 1, Mitchell Library, State Library of New South Wales, Sydney.

15/10/85

Herewith, dear Mr Lindsay, the seeds promised for your expedition.[2] The wattle-seeds should be soaked for a night at each camp, where as a mark of your presence you may sow some. What else you sow may turn out edibles for the natives or other explorers hereafter, as doubtless some of the plants from these seeds would become naturalized. Could you reserve one or two of your new geographic localities for naming after some eminent persons in Europe?

I hope, you will take ample paper with you, (best a ream), so that you may have the means of drying a large number of different sorts of plants in flower or in fruit; the larger the material, the more extensive my Report afterwards; moreover I would willingly purchase the specimens.

Should you find traces of Leichhardt, please telegraph then to me direct, so that I may at once transmit the information to Prussia.

Wishing you a safe return and glorious success

your
Ferd. von Mueller.[3]

1 See also D. Lindsay to M, 26 September 1885 [Collected Correspondence].

2 In late 1885, Lindsay led an expedition that explored the path of the Finke River in central Australia.

3 See also D. Lindsay to M, 21 October 1885 and H. Dittrich to M, December 1885 [both Collected Correspondence].

To John Buchanan

85.11.03

A644, Mitchell Library, State Library of New South Wales, Sydney.

3/11/85

Since some time do I owe you, dear Mr Buchanan, an answer to your kind letter;[1] the correspondence has grown quite over my head; indeed I often wished the day had 48 instead of 24 hours, or that life was to each of us twice as long as it is. This year the extrawork for the London Exhibition,[2] New Guinea Exploration[3] and industrial culture[4] has been enormous.

Let me hope, after your many years toil you can enjoy on a fair competency and in serenity the rest of your life, which I hope will be to you a long and joyful one. Your help as a naturalist and artist will be very much missed in the Wellington Department.

Would you allow me, to ask you *privately* the question, whether the Auckland and Campbell islands are permanently inhabited by any one; also whether on Macquarie and Emerald Island[5] any landing can be effected, or if they have any safe harbours?

Why I ask the question, and why I ask it in *confidence* I will explain. As President of the Geographic Society here I have in *3 or 4 weeks* to give my annual adress, in which I wish to refer to antarctic exploration also. If this question was much talked of, in all probability articles on the subject would appear in the local press, *before* I had even myself a chance of my saying here. It seems that my remarks in last years adress led to the appointment of Hooker, M'Klintock and Nares as an antarctic explorat. Committee.[6]

Regardfully your
Ferd von Mueller.

for any informat on the antarctic islands I shall be very grateful

1 Letter not found.
2 Colonial and Indian Exhibition, London, 1886.
3 The expedition led by Forbes; see M to E. Strickland, 10 October 1885.
4 A new edition of M's *Select extra-tropical plants readily eligible for industrial culture* (B85.13.26) was published in 1885.
5 Southern Ocean, south of NZ.
6 M's 1884 Presidential address to the Victorian Branch of the Royal Geographical

85.11.28 *To William Thiselton-Dyer*

RBG Kew, Kew correspondence, Australia, Mueller, 1882-90, f. 162.

28/11/85

The hardy Agave, enduring the clime of S. England, dear Mr Dyer, cannot be a Fourcroya, as I found the species of the latter genus always suffering from night-frosts even here, which is not the case with the Agaves.

Your brother in law[1] is most successful at the smelting works, for extracting gold by means of fusing with lead.

His life is at present a rough one, but in our clime quite a healthy one, and the reminiscenses will be pleasing and even jocular to him in future life. I send him news papers.

This evening I have to make a presentation on behalf of the Melbourne School of Music to Mr Vogrich, a composer and grand pianist, before an audience of probably thousands of people in our great townhall.[2] He was married last year to a young Victorian Lady, probably one of the most exquisite singers in the world. I am sorry Brian is not here, to hear her, as he is so musical. I took him to one concert during his short stay in Melb.

Regardfully your
Ferd. von Mueller.

Next week my things will be packed for the Exhibition.[3]

Society of Australasia (B85.13.25) is generally seen as marking the start of a campaign for an Australian expedition to Antarctica. In his 1885 address (B87.05.03), delivered on 18 January 1886, he urged the colonization of Auckland and Campbell Islands 'with Macquarie Island as an outpost', as a preliminary to an assault on Antarctica itself. Meanwhile, the British Association for the Advancement of Science at its 1885 meeting, in response to a paper by Admiral Sir Erasmus Ommanney, set up a high-powered committee 'for the purpose of drawing attention to the desirability of further research in the Antarctic Regions'. See Home et al. (1992).

1 Joseph Hooker's son Brian.
2 M made the presentation and gave 'an elegantly designed address' (*Argus*, 30 November 1885, p. 6) at the farewell concert, held in the Melbourne Town Hall, by Max Vogrich and his wife, Miss Alice Rees, prior to their departure for Europe.
3 Colonial and Indian Exhbition, London, 1886.

With the loss of the garden, laboratory, cart, carpenter &c &c my wings are clipped; so I cannot fly far either in this direction, and I trust you will bear that in view as well as other jurors in fairly reporting on my things. I shall write to Dr Masters particularly also.[4] 85.11.28

To Mary Kennedy 85.12.12

Am 27/7, Mitchell Library, State Library of New South Wales, Sydney.

12/12/85

It is a long time, dear Madam, since any Lady went to so much trouble, to send me a collection of plants, such as I received lately from you. The value of the specimens is much enhanced by the native names you so carefully ascertained. All such plants are preserved in the great Gov. Collection, which I founded, now about half a million specimens; and with some ordinary care these collections can be preserved for centuries, so that your notes of the aboriginal names can be consulted with these plants, long after the Darling-tribes[1] have passed away.

Moreover the plants will continue in my Museum for ages to come the testimony of the original vegetation of each district, after changes through culture and settlement will have obliterated many of the indigenous plants largely and even in some cases entirely.

May I beg of you, dear Mrs Kennedy, to continue your searches after plants, especially as you need not trouble to send large specimens (however valuable they are), such as in this instance Mr Gayer so obligingly brought. Small specimens (in flower or fruit) closely packed into little packages would come quite well by post as parcels, if the edges are kept up, and merely a cross-string or tape is tied to hold the parcels. Successive posts would bring them. Perhaps you have friends far N. West, any where towards Central Australia to

4 Letter not found.

1 Darling River, NSW.

85.12.12 ask to collect also, whose sendings would be from entirely new ground. Even your collections contained some rarities, the new locality to be noted in my works under your honored name. All specimens are labelled, the name of the finder attached, and kept for perpetual reference in my Department.

With regardful remembrance your
Ferd. von Mueller

85.12.14 *To Alexander Macdonald*

A38 Royal Geographical Society of Australasia (Vic. Branch) papers, Mitchell Library, State Library of New South Wales, Sydney.

14/12/85.

This morning, dear Mr McDonald, I received the consent of Col. Disney,[1] for our Council[2] to join bodily the procession in poor General Scratchleys funeral.[3] We will need 2 two-horse Broughams, as all the corporate bodies are sure to have all fine carriages, drawn by two horses. This I explained also yesterday to the Rev. W. Potter, who will bring this morning the adress to lady Scratchley for your approval; it seems to me best, that all members of Council should sign it. Perhaps you will kindly telegraph to Capt Pascoe,[4] to come to your office for early signature.

With regardful remembrance
your
Ferd. von Mueller.[5]

1 See T. Disney to M, 12 December 1885 [Collected Correspondence].
2 i.e. of the Victorian Branch of the Royal Geographical Society of Australasia.
3 Scratchley contracted malaria while in New Guinea as special commissioner for the new British protectorate. He died at sea between Cooktown and Townsville, Qld, on 2 December 1885. His body lay in state in his Melbourne home before his funeral on 16 December (ADB).
4 C. Pasco.
5 MS annotation: 'Baron Von Mueller Dr Bride F. Scarr J. McD. Larnach A. C. Macdonald W. Potter'.

Will you kindly draw for me, when the pictures are raffled. 85.12.14

Just received your letter of Saturday.[6] Will be happy to make an hour or two free for spending at your friendly home with the French geographer[7] on Tuesday evening after 8.

From Karl Kirchhoff 85.12.21

A39 Royal Geographical Society of Australasia (Vic. Branch) papers, Mitchell Library, Sydney.

Halle, den 21. Dez. 85

Hochgeehrter Herr Doctor!

Gleichzeitig mit diesen Zeilen lasse ich einen Sonderabdruck meiner Einleitung zum 1. Band der "Länderkunde der fünf Erdteile" an Ihre Adresse abgehen.

Sie mögen aus meiner Widmung entnehmen, dass das gross angelegte Werk, dem Sie ein so hervorragendes Interesse entgegengebracht haben, nun wirklich ins Leben getreten ist. Und ich wage nur von neuem zu bitten, dass Sie Ihre *höchst* willkommene Unterstüzung der Bearbeitung von Australien auch fernerhin fortsetzen wollen. Dankbar habe ich noch letzhin eine neue Gabe von Ihnen zu verzeichnen, nämlich das inhaltreiche Werk über Australiens Fauna von Macleay.

Ist denn nun die Kontroverse über den höchsten Berg des australischen Festlandes entschieden? Darf man demnach den Mount Kosciuszko als solchen str[ei]chen?[1]

Noch in einer anderen Beziehung möchte ich um Ihren gütigen Rat bitten. Unser hiesiger Verein für Erdkund[e,] der die Ehre geniesst Sie unter sein[e] Mitglieder zu zählen, hat bisher vergeblich versucht ein Schriften-Austauschverhältnis mit der australi-

6 Letter not found.

7 Not identified.

1 *editorial addition* – word obscured by binding. All square brackets in the following text have this meaning.

85.12.21 schen ge[ogr.] Gesellschaft anzubahnen. Wir wissen allerdings kaum, ob der Hauptsitz dieser Gesellschaft in Melbourne oder in Sidney sich befindet, und wer die Präsidentenwürde zur Zeit inne hat. Wenn Sie mich darüber unterrichten wollten, so würden Sie aufs neue zu Dank verbinden

Ihren
hochachtungsvoll ergebenen
Alfr. Kirchhoff.

Halle, 21 December 1885

Highly esteemed Doctor,

At the same time as these lines I am also sending an offprint of my introduction to the first volume of the *Länderkunde der fünf Erdteile*[2] to your address.

You may see from my dedication that this grand scale work, for which you have shown such splendid interest, has finally been brought to birth. And I venture again to beg you, that you will continue your *most* welcome support in the treatment of Australia. Further I gratefully acknowledge another recent gift from you, the significant work on Australia's Fauna by Macleay.[3]

Has the controversy over which is Australia's highest mountain been decided by now?[4] Accordingly, should one strike out Mount Kosciuszko as such?

I would also like to ask your kind advice in another matter. Our local geographical Society, which has the honour to count you among its members,[5] has tried, so far unsuccessfully, to start a literary exchange with the geographical society of Australia. Admittedly, we hardly know, whether the headquarters of this society are in Melbourne or in Sydney, or who is the current President.[6] If you could advise me on this, you would once again place me under an obligation of gratitude.

Your respectfully devoted
Alfr. Kirchhoff

2 Kirchhoff (1886-93). This offprint is at MEL inscribed by Kirchhoff: 'Herrn Baron Dr. Ferd. von Mueller dankbar und hochachtungsvoll zugesandt vom Verf.' [Sent to Baron Dr Ferd. von Mueller with gratitude and respect by the author.]

3 Macleay (1885).

4 The controversy was whether Mt Kosciusko or Mt Townsend was the higher.

5 M was elected a corresponding member of the Verein für Erdkunde at Halle on 14 April 1880.

6 The Royal Geographical Society of Australasia had a federal structure, with largely autonomous branches in the various colonies, the largest of which were in Sydney and Melbourne. The president of the NSW branch was by convention president of the Society as a whole; Sir Edward Strickland occupied this position at this time.

From Julian Thomas

86.01.18

A39 Royal Geographical Society of Australasia (Vic. Branch) papers, Mitchell Library, State Library of New South Wales, Sydney.

MEMORANDUM.

Melbourne Address
YORICK CLUB
From "The Vagabond,"
Argus

Jany 18 1886
To Baron Von Mueller K.C.M.G
&c &c

My dear Baron,

I am very sorry that owing to an accident which has confined me to the house for three weeks I shall be unable to be present at the meeting of the Geographical Society this evening.

As I had the misfortune to write Captain Everill's obituary[1] I ought to join with the other members of our Society in congratulating him on his return

I am my dear Baron
Right truly yours
Julian Thomas

1 Everill commanded *Bonito* in an expedition up the Fly River in New Guinea in 1885. In November of that year, alarming reports were received that the entire party had been killed by the natives, but these proved to be unfounded and the expedition arrived safely back in Brisbane on 29 November. The initial reports, however, prompted a long article by Thomas, under his pseudonym 'The Vagabond', in the *Australasian*, 14 November 1885, supp., p. 1.

86.02.17 *To William Thiselton-Dyer*

RBG Kew, Kew correspondence, Australia, Mueller, 1882-90, ff. 167-8.

[17 February 1886][1]

Learning from your letter of the 20 Dec,[2] which reached me only this day, dear Mr Dyer, that you are no[3] duely installed into the Kew-Directorate, I offer you my earliest felicitation, and add the expression of my hope, that you will in unfailing health and unimpaired by adversity enjoy the highly honorable position for very many years, and thus add laurels to the achievements of the distinguished Hooker-family, to which you also belong. –

It needs not my assurance, that any little support, I can give you from here, will be gladly afforded.

I had also a very kind letter from Sir Joseph,[4] to which I will also briefly reply this evening.[5] Pray mention to him, that I did not advise his son[6] to take up pastoral pursuits at present, but think, it might be kept in view for his future. This would by no means necessitate his giving up his special profession; but all professions are unsafe or unlasting staffs to walk by, as some of our best medical men here have found also out. The best of mines get finally worked out; our splendid copper mines are at a standstill on many places, on account of the price of the metal having sunk so much just now. A subordinate position on any mine is a precarious one, though the young Gentleman gained already the honorable post of Acting Manager.

1 *editorial addition* – see M to J. Hooker, 17 February 1886 [Collected Correspondence].
2 Letter not found.
3 now?
4 Letter not found.
5 M to J Hooker, 17 February 1886 [Collected Correspondence].
6 Brian Hooker.

Be not surprised, that I sent some New Guinea ferns to Prof. Luerssen; but he most generously presented my Department with the completest (collection) series of Mrs Dietrichs[7] Queensland-plants,[8] for which I undertook to supply in return ferns from New Guinea and Polynesia in return,[9] as material for his forthcoming work.[10] 86.02.17

Next month my Ministerial Chief, the honorable Graham Berry, proceeds to England as Agent-General of this colony. He has been very friendly to me, and on my solicitation has left a minute in the Chief Secretary's Office, favorable to a new grant of £1000 to our geogr. Society here on the next estimates. This should have some weight with his successor. Please, take Mr Berry to the next meeting of the Royal Geographic-Society and Roy. Col. Institute, and introduce him to some of the leading members, and let them kindly *know* that he was favorable to the continuation of state-means for geograph. explorations.

The enclosed note[11] will put you likely into possession of any vegetable products for Kew from the collections in the Victorian Court.[12] The Albums would be of no use to Kew. Perhaps some of the younger members of the Royal family might like the Albums. It would be well, if you presented the note *early*, so that you may not be forestalled and promises given in other directions by the commissioners.

7 Amalie Dietrich.

8 be not surprised ... Queensland-plants *is marked with a line in the margin.*

9 See M to F. Stephani, 23 January 1885.

10 The letter is accompanied by a memorandum from D. Oliver, 13 April 1886, f. 169: 'Sir F V Mueller's letter 17/2/86. It is rather unfortunate about these New Guinea & Polynesian Filices going to Luerssen without fragments, at any rate, coming [here] Could Mueller lend us his own set with Luerssen's [names] if he [has] retained any? Also: Luerssen described in the [...] Dietrichian Queensland Colln to which Mueller refers Hemitelia Godeffroyi & Asplenium Dietrichianum both unknown to Mr Baker & not identified. Perhaps he wd lend us these at [some] time.'

11 Note not found.

12 i.e. at the Colonial and Indian Exhibition, London, 1886. The enclosed ... Court *is marked with a line in the margin.*

86.02.17 In Mr Bailey's supplement of the QL. Flora[13] 2/3 or 3/4 of the Phanerogams are from *me*, though it is not stated so. I have many additional Cryptrogams from Q.L.[14] besides those enumerated now by him, but cannot manage to get the 12th vol of the fragmenta completed.

Regardfully your
Ferd. von Mueller.

I directed some time ago Mr Morris's attention to the desirablity of examining the Antillan Species of Boxwood. Indeed the suggestion was made already to his predecessor by me.

Best thanks for Saharumpore[15] "ua" seeds.

I am sure you will clearly distinguish in the Australian Courts between the articles, which are of *real* rural, commercial or technologic value, and those which are mere curiousities or playthings.

Of course the percentage of fibres fit for looms, papermills and rope factories is very small.

Mr Clem Markham should see my halved Koilospheres[16] in the Exhibition (private property)

13 Bailey (1886).

14 Qld.

15 Saharanpur? (Uttar Pradesh, India.)

16 The following listing appears in Division B ['Education and instruction, apparatus and processes of the liberal arts'], Class 16 ['Maps and geographical ... apparatus ...'] p. 50, Colonial and Indian Exhibition, London 1886. Catalogue of Exhibits in the Victorian Court: 'Mueller, Baron Ferdinand von, ... A Geographical Koilosphere dimidiated.' There is no entry for 'Koilosphere' in the on-line edition of the OED (viewed 24 February 2004).

To Jean Müller 86.03.05

Conservatoire et Jartin botanique, Geneva.

5/3/86

Besten Dank, edler Freund, für die Güte, soviel Ihrer Zeit meinen letzten Flechten zugewandt zu haben! Möge die göttliche Vorsehung Ihnen Kraft und Gesundheit durch viele Jahre noch erhalten, so dass Sie Ihre grossen Arbeiten beenden können zum Nutzen aller Zeiten! Ich kam nicht dazu, am 12tn Band der Fragmente weiter zu arbeiten, da neben meinen gewöhnlichen Amtspflichten mich die Arbeiten für die London Ausstellung sehr in Anspruch nahmen, zumal da ich auch als Präsident der Vict. Abtheilung der geograph. Gesellschaft von Australien seit vielen Monaten Manches zu thun hatte, besonders für die Förderung der Erforschung Neu Guineas. Von dort sollten doch auch nach u nach viel Flechten kommen, besonders aus den höheren Bergregionen. Haben Sie dort auch eine geogr. Ges.? Dann könnte ich einen Abdruck der Präsidenten Rede senden (etwa 20 Seiten Druck) welche ich neulich hielt.

Hr Prof Lojka, der mich um Flechten gebeten hatte, erhielt eine Kleinigkeit von mir. Da Sie aber so munter fort arbeiten, soll Ihnen in Zukunft Alles wieder zugehen, es sei denn das Allergewöhnlichste aus der Nachbarschaft ausgenommen.

Sie ehrend und Ihnen Alles Gute wünschend
Ferd von Mueller

Endlich sind auch die 76 Quartplatten der austr Myoporinen gedruckt. Ich habe 4 X mehr Arten als AD.C. im prodr vor 40 Jahren verzeichnen konnte.

5/3/86.

Best thanks, noble friend, for the kindness, in having devoted so much of your time to my last lichens! May Divine Providence preserve your strength and health for many years yet, so that you can finish your great work to the good of all times!

Figure 11. Jean Müller of Aargau. Published in *Bulletin de l'Herbier Boissier* (1896), vol. 4. Reprinted courtesy of the Library, Royal Botanic Gardens Melbourne.

I happened not to carry on working on the 12th volume of the *Fragmenta*,[1] 86.03.05
because together with my usual official duties the work for the London Exhibition[2] occupied me, especially since I also had much to do as president of the Victorian branch of the Geographical Society of Australia for many months, particularly for the promotion of exploration of New Guinea. From there many lichens certainly should also come by and by, particularly from the higher mountain regions. Have you also a geographical society there? Then I could send an offprint of the presidential address (about 20 printed pages) that I recently gave.[3]

Professor Lojka, who has asked me for lichens, received a trifle from me. But because you continue working so vigourously, all shall go to you again in the future, with the exception of the most common of all from the neighbourhood.

Respectfully and wishing you all good things
Ferd. von Mueller.

At last the 76 quarto plates of the Australian Myoporinae are also printed.[4] I have fourfold more species than A.D.C[5] could record 40 years ago in the *Prodromus*.[6]

To Paolo Dattari[1] 86.04.14

MSS Tordi 547, 48, Biblioteca Nazionale Centrale, Florence.

14/4/86.

In reply to your note,[2] dear Mr Dattari, let me say, that when botanic works are written, we must draw the distinction between popular and strictly professional ones, and must adopt for the former the

1 See M to J. Müller, 26 August 1885 [Collected Correspondence].
2 Colonial and Indian Exhibition, London, 1886.
3 M's '1885' presidential address, delivered on 18 January 1886, was published as B87.05.03.
4 B86.13.21.
5 Alphonse de Candolle.
6 A. P. de Candolle (1823-73).

1 MS envelope front: 'P. Dattari Esqr [Kent] Cottage Station Street North Carlton'. Envelope dated 14 April 1886.
2 Letter not found.

86.04.14 most easily understood terms, while for the latter the precise scientific terms are used. In this sense we can in popular works speak of "fruits" or "fructifications" of ferns, whereas the term sori would be used in exact scientific writing for fruit-masses of ferns. However different the development of fruits of ferns, Characeae or any other acotyledonous order of plants may be from that of cotyledonar plants, they are still fruits in the wide popular sense of the words. In Hookers[3] and other standard works the term fruit has not been altogether excluded from ferns; the sporangia are often called capsules, altho' that designation rigorously defined could only be used for certain kinds of fruits of Mono and Dicotyledoneae. We do not wish to frighten beginners with using too many strictly botanic terms; thus I substituted in popular writings of mine "stalklet" for pedicel, fruitlet (a word coined by me) for carpel &c.[4] Let me advise you to wait for a few days with your publishing on Vict ferns, as you could then see how I have treated them in the explanations of the woodcuts, now all printed for the "Key".[5] Probably some early copies of the 200 woodcuts of the "Key" will be distributed at the annual meeting of the Club.[6] The word "seed" should never be used for cryptogamous plants, altho σπορα and σπερμα are both used in greek for what in latin is semen in english 'seed'

Regardfully your
Ferd von Mueller

What bears spores must be fertile as from them arise the young plants again

3 W. Hooker (1846-64).
4 See also B89.13.05.
5 B86.13.01.
6 Field Naturalists' Club of Victoria.

To Francis Barnard

86.05.22

FNCV 035-011 Archives, Field Naturalists Club of Victoria, Melbourne.

Melbourne,
22 May 1886.

It is my most pleasing duty, dear Mr Barnard, to express my deep appreciation of the generosity of the great Field Naturalists Club of Victoria, for having raised me to one of its Patronships.[1] Among the many marks of distinction, with which I have been honored in life, I value this one as among the highest, because it is a tribute from that country, in which I spent most of my years and with which my labours are most directly identified. To me it is in connection with this new dignity also a highly pleasurable thought, that work, which I commenced in Australia nearly 40 years ago, will in various directions be carried on by young workers, whom I met personally at your meetings, and who can be guided and can be encouraged by what was accomplished at my time; while they in their turn far on in the next century can inspire a younger generation, thus linking one century's scientific work to that of an other!

I feel particularly beholden to the accomplished and genial Professor Lucas, for having been my sponsor on this auspicuous occasion of alloting to me so high and permanent position at the Club; and I feel grateful also specially to these Gentlemen, who supported his proposition.

It is my hope, that in the brief space of time, which divine providence may yet graciously allow me for my earthly career, I shall be able to make many more hours free for active cooperation with the Club, than during the last few years, so very labourious to me.

Wishing your now widely ramified Institution, which is exercising such an elevating influence on social tone, on healthful recreat-

1 M had previously refused the presidency of the Club; see M to D. Best, 30 April 1884, and M to F. Barnard, 10 April 1885 [both Collected Correspondence].

86.05.22 ion, on practical education-developments and on progressive nature-sciences among us, also in all future a brilliant success, I remain

regardfully your
Ferd. von Mueller.[2]

86.05.23 *To Alexander Magarey*[1]

Collection of the Royal Geographical Society of Australasia (SA Branch) Inc., State Library of South Australia, Adelaide.

23/5/86

Dear Sir,

Our alert Hon Secr.[2] here handed to me a letter of yours,[3] in which the approval of your new Code for our different geographic branches is sought. Kindly send me also a copy of the proposed rules, and while the other circulates among my Colleagues of the Council, I will carefully make any annotations, which may seem to me desirable.

As regards the proposition of uniting with your branch the *historic Society* there, I must confess, that I should strongly disadvise such complications, – nor do I think that it would be just to your Royal Society there, who would more properly in its wide scope embrace other branches of science. History of geography is in reality only geography itself, just as history of medicine is only a part of medical Science.

2 See also M to F. Barnard, 17 June 1886 [Collected Correspondence].

1 MS envelope front: 'To the Hon. Secretary of the S.A. branch of the Austral Geograph Society Adelaide'.
2 Alexander Macdonald.
3 Letter not found.

If we encumber our Society with incongruous collateral obligations, we shall make our work still more onerous, get our attention diverted from the legitimate objects before us, and involve ourselves in additional responsibilities, while indeed the great science of geography will claim in a new part of the world, like ours, *all* the attention which any one of us can possibly bestow on it.[4] 86.05.23

Regardfully your
Ferd. von Mueller

To William Carruthers 86.06.20

RBG Kew, Archives, Miscellaneous correspondence, Mueller.

20/6/86

Since I last wrote to you, dear Prof. Carruthers, a set of Mr Forbes's plants has reached me, by far not so large as the one, which this distinguished traveller despatched to you; but it contains many new plants, some of great interest. In sorting them, I found two vaccineaceous plants, which I at once examined, as I was eager to find out, whether they threw any additional light on my genus Dimorphanthera, described from the Rev. James Chalmers collections in february last.[1] I had taken an interest in Vacciniaceae before, as I discovered Wittsteinea[2] myself in the Australian Alps 26 years ago and then defined it. – It seemed to me best, that I should note the results of my examination of Mr Forbes two additional

4 See A. Margarey to M, 29 May 1886 [Collected Correspondence], in which Margery responds that M has misunderstood his committee's intentions.

1 *Dimorphanthera amblyordinis*, described in B86.02.02 which, however, was not published.

2 *Wittsteinia*? B61.02.02, p. 136.

86.06.20 plants, his collection containing Dimorphanthera Moorhousiana[3] also. Accordingly I now beg to send you the descriptions of these two plants, one a Dimorphanthera (or sect. of Agapetes), to which Mr Forbes name might be given.[4] The other forms a highly remarkable new genus, connected with Oxycoccus in some respects, but with Pyrola in some other regards. I propose for it the name Catanthera;[5] and as Mr Ridley is interested in Ericacea and its allies, I would suggest, that he should single out these Vaccineae for early description, as he could add to the notes his own observations. The genus and the new species, in justice to him as well as to me, might be published under our *joint* authority.[6] Your Colleague, Mr Britten, would doubtless be glad to make these highly interesting plants known at once in his Journal.[7] You all will see the necessity of this being *early* done, as Dr Hollrung[8] may find the same plants on the Finisterre Range, and may send them early to the Continent for publication. It is rather perplexing, that on Papuan plants is actually worked *now at four* places; Beccari elaborates some at Florence; I here, and Eichler and others in Berlin and other places of Germany, while you and your coadjutors have them before you in London; – and yet only you and I have come to some understanding as regards our respective share of work.[9] – Now, it is *not* jealousy, that prompts me to point to this difficulty; but it will be a sad *loss of time* to any one of us, should we find (as is unavoidable), that identical plants had engaged simultaneously attention at different places.

There is so much material for all of us, that clashing of our researches should be avoided. I am the *Senior* and can at best only

3 B86.13.10, p. 163.
4 *Agapetes forbesii*, B86.10.01, p. 290.
5 M erected Catanthera (*C. lysipetala*) in B86.10.01, p. 289.
6 M is given as the only authority in *IK*.
7 i.e. *Journal of botany*.
8 Udo Hollrung, exploring in Kaiser Wilhelms Land (German New Guinea), 1886-7. He visited Melbourne in January 1888 on his way home. MEL holds a set of his specimens. See Schumann & Hollrung (1889).
9 See M to E. Strickland, 10 October 1885, and M to W. Carruthers, 21 October 1885 [Collected Correspondence].

work very few years longer; so I more than others have to grudge the sacrifice of time in doing what others, unknown to me, may also do elsewhere. In the set, which Mr Forbes sent me, (to be divided between Sydney, Brisbane and Melbourne), are very few Monocotyledoneae and ferns; on these, by our agreement, I shall not work. Lest Mr Forbes plants should not have yet reached you, when this letter comes to hand, I enclose fragments of the three Vaccineae, so that you and Mr Ridley can compare this original material with my descriptive notes. 86.06.20

The enclosed [short][10] list of names I made while hurriedly sorting the plants.[11] I will sent you the continuations, as my elaborations here proceed.

Always regardfully your
Ferd. von Mueller

I sent you a printed slip of Agapetes (Dimorphanthera) Moorhousii[12] some months ago

10 *editorial addition* – paper damaged, apparently by insects.

11 List not found.

12 *Agapetes moorhousiana*? See B86.02.02.

86.07.06 *To Henrietta Wehl*

No. 440, pp. 165-7, MS 1946 Charles Daley papers, National Library of Australia, Canberra.

6/7/86.

I should have thanked you, dear Ettie, before today for the elegant present, sent by you in remembrance of my birthday, and which token of your attachment, I all the more value as it is the product of your own cleverness, both in design and in working. I shall put it with my best treasure, and often thus be reminded of you.

It was also very pleasing for me to receive such kind letters from your dear mother and yourself.[1] The day was more than usually festive to me as several friends called, though the Marquis and the Marchioness Foverina could after 6 years this time only send me a gratulation from Malaya. Dr and Mrs Buettner called also in their splendid carriage and brought rich presents, as did some members of Dr Bardell's family, with whom I spent part of the evening and part with Dr Buettner's family. Dr Haig sent me also a splendid present, so Mrs Roberts and her daughter,[2] while my assistants also did not allow the day to pass without some kindly mark of remembrance, so Dr Llewellyn's family. I do not know how to thank all these kind people for their generous sentiments towards me.

You express a hope that I may meet you at your own domicile again; but really my chest sufferings forbid me to make long tours; indeed I have had a distressing cough since cool weather came on in autumn, nor was I free of pulmonary troubles even during the summer, the southern winds being even then so cold.

You live such a long way away and a journey to you even in summer would subject me to the influence of much cold air. Moreover the engagements in my Department have been most pressing

1 Letters not found.

2 Presumably Miss Edith Roberts, to whom M presented an inscribed vanity box, now held in private hands, on 15 July 1881.

for years past, particularly through the extra work for the Exhibitions, as there has been one every year somewhere since 1880,[3] for which my Department had to furnish lots of things and this under so very great difficulties, since I have no longer the facilities and larger means which for such purposes were available to me then in the Botanical Gardens. 86.07.06

I feel very proud when the Marquis of Lorne lately referred to me in such a laudatory manner in his Presidential address at the Royal Geographical Society of England.[4] I am often very sorry that my worldly means are so narrow and that my scientific relations all over the world involve so many pecuniary sacrifices, that I can do very little for the comfort of my own household, and thus also so little for anyone of yours there. I keep not even a single horse and only one man and a servant, yet it is impossible in such a position as mine to save anything, especially as the numerous charities of so great a city as Melbourne put us all here under such endless contributions. Let me hope that you are all happy and well. Health is next to a good conscience, the greatest treasure in the world unless family love.

With regardful remembrance to all of you,
Ferd. von Mueller.

3 e.g., International Exhibitions in Sydney, 1879-80, Melbourne, 1880-1, Amsterdam, 1883, Calcutta, 1883-4, London, 1884, St Petersburg, 1884 and the Colonial and Indian Exhibition, London, 1886.

4 See Lorne (1886), p. 443: 'With Baron von Müller as President of the Melbourne branch of the Australasian Geographical Society, we may be sure that the scientific aspects of the investigation of this magnificent new field [New Guinea] will not be overlooked, and if a moving interest be not already felt in the subject, it may well be excited by the admirable paper with which the Baron lately opened the session of that association.'

86.08.21 ## *To Alfred Deakin*[1]

187/6693, unit 191, VPRS 3992/P inward registered correspondence, VA 475 Chief Secretary's Department, Public Record Office, Victoria.

Melbourne, 21st August 1886

Sir

Observing in the Argus of this day a letter in reference to the second part of the "Key to the system of Victorian plants",[2] I deem it but right to offer some explanatory remarks on the supposed shortcomings of this work, as it is a Government's publication.[3]

In first instance I would observe, that the providing of woodcuts of an elaborate kind involves under any circumstances very considerable expenditure, even in European countries; Dr. Baillon

1 The letter is numbered F86/8274 within this file. MS written by G. Luehmann and annotated and signed by M. Some of these annotations are indicated, but changes to the sense of a sentence have not been noted. The file also contains a typewritten copy of this letter. M's letter was published in the *Argus* (Melbourne) on 28 August 1886 (B86.08.04).

2 On 21 August 1886, the *Argus* published (p. 5) a letter signed 'Student', dated 15 August, in which the author observed, after seeing Part 2 of the *Key*, that 'the Government may as well save further expense for all the use the book will be in our state schools'. 'Student' criticized the illustrations for having been used before and for not being of common plants. 'I do not deny that the illustrations are of value to the advanced botanist, but that is not what was wanted. It was desired, I believe, to have a text book in the simplest possible form that the Baron would condescend to write it, and illustrated by representations which any of the elder children in our state schools would recognise at a glance ... It is a sad disappointment to some who had looked forward to having a thoroughly practical key, but after reading the Baron's *Botanic Teachings* [B77.13.07] and *School Botany* [?B79.13.08] neither of which works has, I believe, been a success as a generally useful educational work, it was, perhaps, too much to hope for better things for the key. ... It is a remarkable fact that, out of the 152 illustrations, no less than 67 represent plants which the Baron was the first to describe, so that the well-known initials "F.v.M." become a weariness to the eye. The Baron seems to be so much in love with those plants, of which he is, so to speak, the godfather that he had selected them, rather than plants which we may all find easily, as illustrations for a text-book for botanical beginners.'

3 See also M to T. Wilson, 21 August 1886 [Collected Correspondence], in which M seeks permission to send his reply to the *Argus* for publication.

thus[4] in a letter to myself by last mail[5] refers to the costliness of the woodcuts of his "Dictionnaire de botanique", now under progress in Paris.[6] In adopting xylographic illustrations therefore, it is always an object, to re-use the woodblocks, so as to render the illustrations comparatively cheap in the end; consequently 57 blocks, which were prepared formerly for the little elementary school-book "Introduction to Botanic Teaching",[7] were employed again for the present "Key"; but as the number of plates for the latter amounts to 210, (two in many instances being devoted to one plant,) the majority had to be prepared additionally for this Key. In a large number of cases the drawings were taken from the quarto-works "Plants of Victoria",[8] because the choice then made for thus illustrating *orders* and *genera* rather than any particular species, has been considered by fair critics a good one, and saved the cost of preparing[9] new original drawings ample analytic dissections always being given.[10] Thus an alpine Buttercup (Ranunculus) illustrates the genus and order, to which it belongs, as well as any lowland-species; and I would further add, as particular objection was taken to this plate, that the several kinds of "Buttercup", occurring in the lowlands, are mostly very different in appearance from each other, one being a submerged floating aquatic; besides all our common Ranunculi have been previously illustrated in other works elsewhere.[11] It was out of the question, to figure any large share of common plants, as the number of Victorian species is altogether 1852, leaving even[12] Mosses, Lichens, Fungi and Algae out of account. Moreover of a good many orders, which are illustrated, only one or two species occur in Victoria, hence there was not much or no choice. Among the illustrations will be found many species of plants, to be found in or almost in the vicinity of Melbourne: 86.08.21

4 for instance *deleted by M and replaced with* thus.
5 Letter not found.
6 Baillon (1876-92).
7 B77.13.07.
8 Presumably B62.03.03 and B63.13.06.
9 the cost of preparing *added by M.*
10 ample ... given. *added by M.*
11 elsewhere *added by M.*
12 even *added by M.*

86.08.21 thus among such the Sundew, Pelargonium, Sneezeweed, Misletoe,[13] Milkwort, Sheoak, various coast Saltbushes, Pigface, Raspberry, two Gumtrees, Honeysuckle, Cypress-pine Statice, Waterlily, Gungang and some other aquatics[14] a few Rushes and numerous Ferns, while very many others represent *genera* occurring in the neighbourhood of the metropolis. If my name through early discoveries under endless dangers and difficulties became largely connected with the Victorian Flora, then the same can be said of R. Brown, and for Europe, and even here of Linnaeus without need of evoking disparing[15] remarks, particularly from any one in Victoria[16] In bringing out illustrations for any kinds of botanic works, it is always desirable, to strive at originality, and not to choose plants for delineation, which have been illustrated before[17] elsewhere. I freely admit that, some oversights have occurred in the part of the Key already issued; thus the quotations of the respective numbers of the last woodcuts were omitted, and a few plants were accidentally left out altogether in the list; this will be remedied in the part of the Key now under preparation, when also the additional species will be inserted, which may yet be discovered this season as new for Victoria, particularly in remoter regions. The larger work "Native Plants of Victoria", of which a portion appeared some time ago early to be continued,[18] and to which "Student" alludes, is in its scope and elaboration very similar to Sir Joseph Hooker's excellent[19] "Students Flora of the British Islands",[20] except that ours is illustrated, and that far less of strictly botanical terms are used, and that synonymy is omitted;[21] – whereas the little unpretensive work "Botanic Teaching" follows somewhat the lines of the admirable elementary publication, issued in Britain by the Revd. Prof-

13 Mistletoe?
14 and some other aquatics *added by M.*
15 disparaging?
16 If my … Victoria *added by M.*
17 in Britain or *deleted by M after* before.
18 B79.13.08. early to be continued *added by M.* No further part of this work was published.
19 excellent *added by M.*
20 J. Hooker (1878).
21 and that synonomy is omitted *added by M.*

essor Henslow for the very beginners.[22] Furthermore it seems an 86.08.21
ample allowance, when it is shown,[23] that about half the species, illustrated in the Key, occur in the *southern* region, to which Melbourne belongs, especially as a book of this kind is expected to serve *all parts* of the colony, each division having allotted to it a fair share while what is common at Melbourne is not always frequent elsewhere.[24] The whole however is tentative, and was not urged for the state-schools, but more particularly for the Field Naturalists Club, and somewhat[25] against my own opinion as to the plan, – the dichotomous method, – adopted by the late meritorious Revd Mr. Spicer for the flora of Tasmania,[26] mainly from works of Hooker, Bentham and my own. Nor is Spicer's book specially written or intended for school-children. Moreover the dichotomous method is not readily applicable to the natural system nor to large areas, therefore hardly now anywhere[27] in use in any European country, and at best it is only a *key* to larger works, for which also, – so far as Victoria is concerned, – ample provision has been made long ago by the Flora Australiensis[28] and other extensive publications.

I have the honor to be
Sir
Your most obedient servant
Ferd. von Mueller,
Govt. Botanist.[29]

The Honorable Alfred Deakin, MLA,
Chief Secretary

22 Henslow (1880).
23 when it is shown *added by M.*
24 while ... elsewhere. *added by M.*
25 somewhat *added by M.*
26 Spicer (1879).
27 now anywhere *added by M.*
28 Bentham (1863-78).
29 'Student' replied to M's letter on 7 September: 'The baron's defence of the illustrations in "Part 2" is based on two grounds, namely, expense, and the somewhat strange ground that it was not desirable to illustrate the book with examples which have been already figured in other books. When I look at almost any text book of

86.08.21 botany, I find the plants figured are almost invariably those which are the most easy to be obtained.' 'Student' drew attention to Lindley's *School botany* [Lindley (1862)], Balfour's *Manual* [Balfour (1875)] and Brown's *Manual* [Robert Brown (1874)] as examples. 'Instead of selecting for the woodcuts – (Why will the baron insist upon calling them "Xylographs"?) – plants which every child could find on its walk to school in the country, the illustrations, as a rule, represent the rarest specimens of our flora. ... Part 2 in its present form will be almost useless in the field to a beginner, as he will so rarely find figured the plant which he is striving to identify. As a supplemental volume, to be kept on the shelf for occasional reference, Part 2 will no doubt be of service, but beyond this I am too ignorant to perceive its merits. I see clearly that it will be necessary for some practical man to publish a useful series of illustrations after Part 1 makes its appearance. ... I believe that photography might be used in the illustrations, and that the cost might be thereby lessened. ... "Part 2." principally paste and scissors, bears too evident marks of hurried work. The baron speaks of some omissions. I could point out four myself.' (*Argus*, 10 September 1886).

On 11 September P. Dattari, as a member of the Field Naturalists' Club of Victoria, wrote a reply to 'Student'. Referring to 'Student's' first letter, Dattari suggested that the keys in Spicer's *Handbook of the plants of Tasmania* were misleading and contradictory. 'Moreover, the Tasmanian key is not of any use to beginners, but only for advanced botanists, who, knowing the peculiar character of the orders and genera, may perhaps by that artificial method discover the name of the plant. In reference to the woodcuts, I think the Baron quite justified in illustrating some of the plants not to be found near Melbourne, as the key is intended for the use of the whole colony of Victoria, and not for a portion thereof' (*Argus*, 13 September 1886).

'Student' answered Dattari's letter on 14 September by rising to the defence of Spicer's book. He took a copy into the *Argus* office and spoke to a sub-editor: 'On entering his room I saw, to my delight, a pot of fuchsia in flower on his mantelpiece, and I asked him if he would like to be made a botanist in five minutes. He very courteously fell in with my views, and, making him use Spicer's book himself, he traced his fuchsia to the order, Onagreæ, to which it belonged.' 'Student' held to the view that M's choice of woodcuts was poor, asserting that Dattari has misunderstood the criticisms made. 'Mr. Dattari has just as much right to express his opinion as I have to express mine, and I must leave it to the Chief Secretary and the intelligent public to decide which opinion is entitled to the greater weight. ... I am not without a gleam of hope that the baron himself may be convinced, by the plain arguments which you have allowed me to make public, to cancel Part 2 and re-cast it upon less scientific but more commonsense grounds.' (*Argus*, 21 September 1886).

On 21 September Deakin requested that the Minister of Education be asked 'if he would wish to utilise a simple handbook of Botany such as is here proposed by "Student"', and that the cost of preparing the plates required be ascertained. Deakin was informed that the cost of the woodcuts would be from £1 to £2 each. It was also suggested that M be asked for a copy of Spicer's handbook for inspection; see M to T. Wilson, 29 September 1886 [Collected Correspondence].

From Frederick Hagenauer[1] 86.10.04

Natural History Museum, London, General Library, Owen correspondence, vol. XIV, ff. 206–7.

Ramahyuck, Aboriginal Mission Station,
Gippsland,[2] Oct. 4, 86

Baron F. von Mueller, K.C.M.G F.R.G.S. M.D. PhD.
etc etc. etc.
Melbourne

My Dear Baron,

You will be pleased to learn, that after many years of zealous endeavours, to supply you for the venerable Professor, Sir R. Owen, with a new born little Platypus, which I have been successful at last to find through one of Aboriginals, with the mother Platypus in the nest a few days ago. As I had a long conversation with you on the subject a few days ago, I need not state the matter over again in these lines, but it gives me still greater pleasure to inform you, that this morning I went myself with my black boys to take the nest with a view of sending it to you for further investigation but to our greatest pleasure we have been successful in finding another nest complete, with the mother and two new born Platypi in it. The burrow was on a high bank of the River Avon, the hole from the level of the water narrowed into a passage large enough for one animal and rose gradually to about eight feet from the water at a distance of nine feet, then turning to the right to a distance of 14 feet and to within one foot from the ground surface, so that even the highest flood could not reach it. Here in a really very clean and comfortable, round burrow was the nest and in it the old Platypus with its young ones. The old mother was sitting on them or at least was covering both the little ones with her fur. The nest is constructed of grass and the centre with the leaves of the

1 MS annotation: '[*three illegible words*] 15/10/86'.

2 Vic.

86.10.04 Eucalyptus. We have taken great care to take nest and young ones with the mother, and in order to put you in full possession of all the important facts, I have packet up the ne[x]t[3] with the little ones and shall send the whole by the first train tomorrow morning to Melbourne to the Director of the Zoological Gardens with the request to at once communicate with you and give you the *nest* the *mother* and *one young one* for sending it to Professor Owen, to make further investigations on the subject. – As we now have three young ones with two mothers, I hope that the problem in question may be solved through it. You are aware, that ever since that time (I think at least 25 years ago) when Professor Owen through your endeavours, settled the question of the classification of these peculiar creatures as "Vivimammalia" I have taken the deepest interest on the subject and when two years ago Mr. Caldwell made the other discovery that they lay eggs and classified them under Ovifarous[4] class[5] I was still more interested, though not at all satisfied. Not being sufficiently at home in Natural Science I felt greatly puzzled and of course, rejoiced when a few days ago we discovered the nest with the mother and her young one. This puzzle, however, has been very much increased today, when we found the nest with the mother and two very young Platypi and not a sign of any egg shell or skin in the nest, neither any indication that the mother suckles her young ones, who evidently cannot live without food being supplied by the mother.

I trust that I have not been tiresome to you for giving you such a long account of my discovery, and as now, you are able to send Mother, Young one and nest to the great Naturalist in England,[6]

3 nest?

4 Oviparous?

5 Caldwell had confirmed in 1884 that Monotremes were egg-laying mammals. His conclusions were announced in a telegram, 'Monotremes oviparous, ovum meroblastic', to Archibald Liversidge in Sydney, who transmitted the news to the British Association for the Advancement of Science meeting in Montreal. See Caldwell (1888), p. 464, and also Moyal (2001), pp. 149-57.

6 There is no entry in the Natural History Museum Zoological Accession Register corresponding to these specimens, but Owen (1887) described a young platypus 'received ... from the Baron', with details from 'the Rev. Pastor Hagenauer ... to whose influence with the natives science is indebted for the acquisition' (p. 391).

and as likewise two specimens can remain in Melbourne either with you, or Professor McCoy and also Mr. Le Souef of the Royal Park,[7] I hope, that all will be made clear that has hitherto been a great problem.[8] 86.10.04

With very Kind regard
I remain
My dear Baron
Yours faithfully
F A. Hagenauer
Missionary.

To Thomas Whitelegge[1] 86.11.03

National Herbarium of New South Wales, Royal Botanic Gardens, Sydney.

3/11/86

I would be much beholden to you, dear Mr Whitelegge, if you could spare me any of your large specimens of Claudea Bennettiana. Perhaps what you collected shows Keramidia and other organs of multiplication, which I would be glad to describe, and of which you may have made yourself already microscopic sections and drawings. Copies of the latter would be very welcome.

Long ago I pointed out to Agardh, Schmitz and other European Algologist that the Claudea Bennettiana was not really congeneric with C. elegans, nor could be placed with Vanvorstia in

7 Le Soeuf was Director of the Melbourne Zoological Gardens at Royal Park.

8 See also R. Owen to M, 27 September 1887 [Collected Correspondence].

1 MS found with a type specimen of *Claudea bennettiana* Harvey (1859) = *Vanvoorstia bennettiana* (Harv.) Papenfuss (1956). The species has not been collected since 1886 and is now presumed extinct; see *New South Wales Fisheries* (2002).

86.11.03 one genus. I mentioned to them, that I had called it as a distinct genus *Sonderia*, in honor of Dr W. O. Sonder, one of my earliest friends in life, a great author on S. African plants[2] with whom I was for a third of a century in frequent phytologic correspondence, and to whom as a specialist most of my algologic collections were confided.[3] Four days before his rather sudden death he still worked on new Algs I had sent him, and the last letter, he wrote in life, was to me.[4]

It would be the aptest botanic genus, that could be dedicated to such a man, this one. The Sondera of Lehmann[5] among Droseraceae held not out. Perhaps you might mention this *incidentally* at the L. S. of N. S. W.[6] – So you can also see, that I am interested in this particular Alga with reference to its fruit. Dr Ramsay kindly sent small specimens,[7] but not in fruit.

Regardfully your
Ferd von Mueller

2 Harvey & Sonder (1859-65).

3 Agardh eventually published the new combination in *Sonderia* in Agardh (1890), p. 117. See also T. Whitelegge to M, 12 November 1886 [Collected Correspondence], reporting that he was as yet unable to supply specimens in fruit. No such letters written 'long ago' have been found. However, in a letter to Agardh, 15 November 1886 [Collected Correspondence], M reiterated the request: 'As stated I would like to name this alga with you, if the genus really is new, *Sonderia*'. In his reply, 25 December 1886 [Collected Correspondence], Agardh agreed that the newly found item was identical with the species described by Harvey as *Claudea bennettiana*, but that it agreed more with *Vanvoorstia* than with *Claudea*. However, in the absence of fruit it could not be decided whether it represented a new genus as M thought might be the case.

4 Letter not found.

5 Lehmann (1844), p. 44.

6 Linnean Society of NSW.

7 These E. P. Ramsay collections are held at MEL.

To William Thiselton-Dyer[1]

87.01.10

RBG Kew, Kew correspondence, Australia, Mueller, 1882-90, ff. 213-16.

10/1/87.

The mail just brought, dear Mr Dyer, the new fascicle of the icones plantarum, containing ferns, for which I beg to express my best thanks. I see in it Diplora, which came from the collections of the late Will. Sharp Macleay, which rich set of chiefly Polynesian ferns was gained for Kew on a solicitation suggested in first instance and supported by me.

It is delightful, to see hundreds after hundreds of specific forms thus illustrated. How the means of recognition of the species is facilitated for the next generation! and the younger workers now. Old people like myself cannot have much benefit from these great exertions in iconography any more.

Speaking of ferns, let me say, that I feel much beholden to the excellent Mr Baker for his generosity of offering, to let me name the new Trichomanes, gathered by my emissary, Mr Sayer. But it is evidently right that Mr Baker should name it,[2] as he worked out the specific details, altho' I certainly did not recognize it as known Australian.[3] I would however suggest, that it be named after Mr Sayer,[4] as he incurred not only endless toil in a season of unusually heavy tropical rains, but nearly lost his life through the hostility of the black savages, who are quite unsubdued and very numerous in the seclusions of the Bellenden-Ker Ranges.[5] Moreover he just

1 MS is stamped 'Royal Gardens Kew 21 Feb 1887', and annotated by Thiselton-Dyer: 'An[swere]d. 23.2.87'. Answer not found.

2 Named under joint authority as *Gymnogramme sayeri* (see B87.14.01, p. 163); M later listed it in B89.13.12, under *Grammatis sayeri* (p. 234), and *Trichomanes sayeri* (p. 230).

3 MS annotation by [Thiselton-Dyer]: 'Mr Baker', and by Baker: 'Answer at end'.

4 be named after Mr Sayer *is underlined presumably by Dyer.*

5 Qld.

87.01.10 writes,[6] that through the swollen state of the Russell-River,[7] his boat capsized, and he and his companions, though both good swimmers were nearly drowned, and lost everything except their revolver and one blanket, having to go almost naked to the nearest settlement. I shall bring some incidents of their ascent of Mt Bellenden-Ker before the Royal geograph Soc here this month,[8] when I shall have as President to aid also in a new enterprise for Mr Giles into Central Australia. The antarctic question will then also come up again.[9] Speaking of ferns, let me ask, *are the Todeas* out of the *South Austral* Court *of any size*? During 1847-1852, when I explored in S. Austr on my private expense, I met only in two vallies (near Mt Lofty) any Todeas at all, the largest of which would weigh only a few ctw.[10] Of course an other place for Todeas *may* have been discovered there since. I was the first to export Todeas.

How sad poor Scortechini's death, just when he wanted to go to Kew with his immense treasures.[11] It was myself, who placed him first with Sir Joseph Hooker in correspondence, though it was the Rev Jul. Ten. Woods whose scientific zeal and particular influence with Governor Weld secured for his lamented colleague the chance of exploring in the Malay-Peninsula, every facility for the same purpose being offered to me, when I returned with Mr Gregory from the discovery of the Kimberley-district in 1857[12] by the then Judge of the Peninsula, G. Windsor Earl.[13] My accepting the Directorship of the Melbourne bot Garden frustrated this scheme.

I really trust, that the Rev. Jul. Ten. Woods is not again in vain seeking the fellowship of the R.S., for which poor Scortechini in London would have interested himself personally out of gratitude

6 Letter not found.
7 Qld.
8 See M to A. Macdonald, 4 January 1887 [Collected Correspondence].
9 See Home et al. (1992).
10 cwt?
11 B. Scortechini died in Calcutta, India, on 4 November 1886.
12 WA, during the North Australian Exploring Expedition, 1855-6.
13 Letter not found.

to his friend Woods. The latter, after an other *three years field-work* in India, particularly for geology, is just returned with "impaired eyesight and benumbed hands."[14] It has been quite an enigma to me, that *such a man*, so varied in his accomplishments, so active in science since 30 years, so labourious in *field-work* under great danger (not mere commodious and safe house-work for science in settled places) could not even a couple of years ago be elected F.R.S.[15] I have written to Sir Henry Barkly[16] to seek for this worthy the powerful aid of Sir Joseph Hooker, were it only out of respect to the "Manes" of Scortechini! 87.01.10

Regardfully your
Ferd. von Mueller

Unfortunately we are again at a standstill in picking out the particular specimens, yet wanted for the Kew-Museum. The manual Assistant here[17] is ill; and anyhow it is a tedious undertaking to select odd pieces out of about 200 000 sheets of Australian plants; but the matter shall have early attention, so early as ever possible

It will be best in future, to send always *at once* a specimen of any newly discovered plant; but then, there is often *only one* specimen, as you may readily imagine. *This* year I actually *worked* from the 31 Dec. into the first January. Could not spare time to go to friends.

What does Mr Baker make of the Schizaea, figured by M[r]. Guillemin at plate 20?[18]

I am grieved to learn also of the death of Dr Wil[helm] Hillebrand, a friend of my youth,[19] who only lately sent me on my

14 See Press (1994), ch. 14.

15 J. Tenison Woods' candidature for the Royal Society was no more successful in 1887, or in 1888, than on earlier occasions, after which he was not nominated again.

16 Letter not found.

17 J. Minchin.

18 *This sentence is marked in the margin with a line.* For *Schizaea dichtoma* see Guillemin (1827), pl. 20, and also pp. 13-14.

19 Hillebrand met M in Adelaide in 1849.

87.01.10 request splendid specimens of the Hawaian Myoporum.[20] Perhaps I am the next, whose necrologe[21] is to be written![22]

Woods Essays are numerous, many on fossils.[23] His 2 large volumes on Austral explor. appeared 1865.[24]

Sir Rich Owen would be sure to support Scortechini's[25] candidature, if asked.

87.01.25 *To Henry Ranford*

Acc. 541, AN 3, 271/1887, Surveyor General's Department, State Records Office of Western Australia, Perth.[1]

25/1/87

The plants, which you so kindly collected, dear Mr Ranford, were brought to me yesterday by Mr Nyulasy, and I beg to send you herewith the list of names,[2] as you may wish perhaps to append it to your report. It is delightful, gradually to see the vast extent of Australia also phytologically explored, and it is a great credit to your honorable chief, that he fostered this line of extra-researches also in his Department, and that his officers are interested in this work also

West Australia is the richest part of Australia in numerical amounts of different species of plants, and large additions to the

20 Hillebrand died on 13 July 1886.

21 necrologue?

22 MS annotation by Baker: 'I am sending a note on the new Trichomanes to Journal of Botany and have called it T. sayeri, F.M. & Baker. *Schizae dichotoma*. We have a specimen precisely like Guillemins tab 20 from Moreton bay *Walter Hill* I call it dichotoma I find I have note of about 700 new fern-names published since last edition of Synopsis Filicum J G. B.'

23 A bibliography of Woods' scientific writings is in O'Neill (1929), pp. 399-406.

24 Woods (1865).

25 Woods'?

1 MS annotation: '5 Feb'.

2 List not found.

records will have to be made yet, while surveys and settlements proceed. I wish very much, that your colony and South-Australia would combine, to send Mr Giles out, to explore the country N. W. of Lake Amadeus,[3] where large tracts are likely to yield a revenue to the two colonies; and it seems a pity, that such country should remain unknown and so long be withheld from utilisation. If each of the two colonies, who are directly interested in this, would grant £500, the exploration could be carried out *this* season, as Sir Thomas Elder with his usual generosity would likely give the loan of Camels. Could you kindly bring this proposition under the favorable notice of the W.A. Government?[4] It would save S. A. & W. A. much expense, sooner or later to be incurred anyhow.[5] When you write to the Hon. John Forrest, pray, give him my best salutation.

As you so thoughtfully ask for suggestions, how collecting of plants during the Survey Expeditions might be made most valuable for science, I would beg to observe, that the aim should be, to get together *as many different* species as ever possible; the minutest plants count for science-purposes as much as the largest, the insignificantly looking as much as the showy. Then *fruiting* specimens are as valuable as flowering ones, and often more so. Waterplants should also be included; indeed from regions, not yet carefully traversed by Botanists, *any* kind of plant is valuable for record of locality. To expedite drying plants, put into sheets of papers, they (plants and paper) should be spread in thin sets on the air, and the sets occasionally *halved*, which brings the wet inside outside, and dry outside inside quickly. On Sundays the specimens (so far as dry) can be closely packed, to get much of the paper empty again and each parcel be kept by itself, probably not too large to be sent by successive mails.

Regardfully your
Ferd. von Mueller

3 NT.
4 Could you … Government? *has been underlined.*
5 The WA Government considered M's proposal but decided not to take it up; see M. Fraser to M, 7 March 1887, and J. Brooking to M, 10 March 1887 [both Collected Correspondence]. See also M to J. Brooking, 20 March 1887.

87.02.11 ## *To Edward Ramsay*

ML MSS.562, Letters to E. P. Ramsay 1862-91, Mitchell Library, State Library of New South Wales, Sydney.

Private

11/2/87

The fungus, kindly sent by you, dear Dr Ramsay, is a spec. of Battarea. I will give you the specific name bye and bye.

Do not think me illiberal, as regards Mr Bevans Expedition;[1] but privately I am quite *poor*, and what little I had for collecting purposes officially, this year, has been expended already. The ascent of Mt Bellenden Ker[2] and the bot exploration of the adjoining regions cost £100 during the last nine month (not £500 as stated in the papers); then I had some little expenditure, to complete the investigations on the flora of East Gippsland for the "Key"[3] – I could not possibly go to the ministry to ask for special additions to my funds, as every item must *before* be *voted* by Parliament.

For the £30, concerning which I telegraphed yesterday, I am *personally* responsible. I am much *discouraged* also in these arrangements; by past experiences! usually the return has been very small. Even at this moment there is a dispute monetaryily about some collections (not extensive) made by Lt Dittrich during Lindsay's Expedition.[4] On one occasion I gave out of my private means £50 to Goldie and in return I got a *handful* of plants.

Regardfully always your
Ferd von Mueller

1 Bevan left Sydney in February 1887 on his fourth expedition to New Guinea. See Bevan (1890).

2 Qld.

3 B86.13.01, B88.13.03.

4 See M to D. Lindsay, 14 and 21 December 1886, and D. Lindsay to M, 17 and 29 December 1886, 21 January 1887 and 11 February 1887 [all Collected Correspondence]. See also Lindsay (1889).

My own experiences moreover as regards obtaining collections of dried plants from N.G. by private expeditions are most discouraging. A collector on board of a vessels has so little space and so little facility for drying plants, – that the returns for his salary, outfit &c likely would be quite out of all proportion in value. 87.02.11

Of course, *this* expedition may prove an *exception* to the rule.[5]

Would it not be possible, to let Mr Moore's Department and Mr Maiden's Department[6] be responsible *each* also; also for £30; then you would have £90, and the specimens could be subdivided in *Sydney* into 3 sets, I getting one.

It will likely lead to dispute, if I were made responsible, to take all bot specimens at their value to *any extent*. Who is to fix the value? Collectors have – as I found with Edelfelt[7] in N.G., and with others – an *exaggerated* idea of the value of dried plants, as they are almost sure to bring mainly ferns (almost without novelties), neglect small not showy plants, pass aquatic weeds, don't trouble to get flowers from trees much seldom bring a fruiting specimen of any kind, generally think it too much trouble to wrap up a leaf and inflorescence of a palm or Bamboo, and regard it too much labor, to dry such or other bulky specimens properly.

With regardful remembrance,
dear friend, your
Ferd. von Mueller

I telegraphed that I will take specimens to the value of £30

5 M described a small collection of plants from Bevan's expedition to New Guinea in B87.11.01.

6 i.e. respectively the Botanic Garden and the Technological Museum, Sydney.

7 i.e. E. G. Edelfeldt.

87.03.02 ## *To Charles Moore*

4/7577 letters received, pp. 125-8, Colonial Secretary's Department, Archives Authority of New South Wales, Sydney.

Melbourne
2/3/87.

It needs not my assurance, dear Mr Moore, that I will be happy, to aid in the elaboration of the Flora of New South Wales; but in order that no clashings or contradictions occur in naming, characteristics and systematic disposition, it would be necessary, that my *Census*[1] should be the basis of operation. We will then be clear about everything, and the newest knowledge will be brought to bear on the work. It is of course *not* at all necessary, to follow the arrangements and nomenclature adopted for the Flora Australiensis,[2] because hardly any one, who will use the special work on New South Wales, will use the seven volumes of all Australia. The idea of keeping the Monochlamydeae by themselves, becomes more and more discarded; – even by last mail the leading Botanist of Western North-America[3] expresses himself on his own accord bound, to adopt my alteration of the DC.[4] system. Furthermore if all the naming of orders genera and species had to be strictly adopted from the Flora Australiensis, the rule of priority, which finally must prevail, would be carried out very imperfectly; and if the limitation of genera & species had also to be exactly in accordance with the Flora, all the research of the last 25 years (not only by me but also of European Botanists) would be lost sight of, and the work be so much *behind the times*!

1 B82.13.16 and supplements.
2 Bentham (1863-78).
3 Sereno Watson? Letter not found.
4 De Candolle.

If Mr Betche under your direction will prepare or[5] copy of the manuscript of order after order in the sequence of the Census, the manuscript could easily be sent to me for revision, when I could add also localities and *perfect* the whole in some other ways, before portion after portion goes to press. This can only be an advantage to yourself.[6] 87.03.02

I would however strongly advise, that the printing not be commenced much before the *beginning of next year*, so that efforts may be made, in which Mr Maiden wishes to share by help of his Department, to get the plants of the *remotest N. W. of N. S. Wales* next Spring.[7] I am satisfied, that many genera and a very large number of species from thence are yet to be added to the Flora of N. S. W.; but the search should commence in August and be carried on til October inclusive; if this is not done timely, the new Flora of N. S. W. will at once be very incomplete.

Always with regardful remembrance your
Ferd. von Mueller.

I should be glad if this supplement to your census[8] could be soon printed and some slips be sent to me. Look for instance on my plates of myoporinae[9] and see, whether there is any difference between Pholidia and Eremophila.

5 a?
6 In the preface to C. Moore (1893), Moore declared that 'the systematic arrangement adopted here is that of Baron von Mueller'; however, he made no mention of M's having reviewed the manuscript as M here suggests.
7 See J. Maiden to M, 4 May 1887 [Collected Correspondence].
8 C. Moore (1884).
9 B86.13.21.

87.03.20 ## *To John Brooking*

Acc. 541, AN 3, 271/1887, Surveyor General's Department, State Records Office of Western Australia, Perth.

Melbourne,
20/3/87.

Herewith, honored Sir, I beg to send for your kind acceptance a few lithographic plates, illustrative of W.A. Acacias from an Atlas of these kind of plants,[1] for which about 60 species are already lithographed, and in which the numerous species of your wide territory will also be largely contained. May I ask kindly to find out from flowering and fruiting specimens, whether it is the Acacia heteroclita, which yields about Geographe Bay and Cape Leewin[2] the peculiar Gum, well known as different from the ordinary Acacia Gum of W.A. I should like to know this for the new edition of my select plants.[3]

Altho Cape Leewin was discovered and named, as you will be aware, already in 1622, the vegetation of the vicinity is even now only imperfectly known.

For my work on the Australian Acacias my material is also yet very incomplete, particularly as regards the far eastern species, of many of which I have flowering but not fruiting specimens, and until the latter are gradually acquired, they cannot be treated in the Atlas. Indeed collections of *all sorts* of plants even the minutest weeds, rushes, aquatics &c are particularly wanted from your *far* eastern districts, not so much for actual novelty, but for tracing the distribution of the species. I am thankful for all the aid, already afforded by the survey-department, – and if the eastern settlers would all give a little local help, great strides could be made for completing the elucidation of the W.A. Flora.

1 B87.13.04, B88.13.01.

2 Cape Leeuwin? Both WA.

3 M's note on *Acacia heteroclita* in the next (1888) edition of *Select extra-tropical plants...* (B88.13.02, p. 7) was unchanged from earlier editions.

I am greatly beholden to you for allowing Mr King to furnish me with brief notes on the W.A. surveys of 1886 for my next presidential address;[4] and I feel also much indebted to you for recommending for his Excellency's favorable consideration the joint Exploration by W.A. & S.A. of the country N.W. of Lake Amadeus.[5] Should this not finally be carried out, it would still be a great advantage to W.A., if Mr Giles's services were secured for one of the *far inland* districts as Crownlands Commissioner, as by his connections and influence in the eastern colonies he could speed settlement much 87.03.20

Regardfully your
Ferd von Mueller

Is it not a pity, that a Gold-Medallist of the R.G.S. and a leading Australian Explorer should not be in his element again!, one who has done so much for W.A. particularly.

To William Thiselton-Dyer 87.03.30

RBG Kew, Kew correspondence, Australia, Mueller, 1882-90, ff. 56-9.

30/3/83[1]

Your letter of the 9 Febr.,[2] dear Mr Dyer, arrived this day; and as I wished anyhow to write to you, I will answer it at once. Before doing so, let me remark, that the Xanthorrhoeas will likely be sent

4 H. S. King. See B87.05.03, p. 31. In B87.04.03, M described the plants collected by King during his survey.
5 NT. See M to H. Ranford, 25 January 1887.

1 MS is stamped: 'Royal Gardens Kew 16 May 1887'. The date on the MS is clearly '83', but the Kew date stamp and internal references suggest an error on M's part. See notes below.
2 Letter not found.

87.03.30 by one of the next steamers.[3] The continuation of sendings of the sp. of dried plants, added to the Fl. Austr. (now over 1000 sp.) will have reached you, as one parcel went by last mail.[4] I have just a Carpenter to make additional repositories in the iron annex, which interferes with selecting more specimens; but your establishment will bye & bye be fully provided, so far as the material admits, which often is scant and fragmentary.

In two days Mr Al. Davidson, who accompanied Mr Sayer in the ascent of Mt Bellenden Ker,[5] proceeds to England again; he has promised, to look after the small case with well-rooted plants of Correa Lawrenciana, the red-flowering Variety. Mr Findlay potted the cuttings on the Hume-River,[6] and I have had them in my little culture-ground for several months. So they are well established, and ought to reach you well. Mr Davidson can give you an account of the exploit.

Our time is precious, and should not be vasted[7] in fruitless discussions, especially as *I* have so little left of my life; but as Kew is the principal establishment for Phytography in the world, I feel, that I ought not to leave your remarks on nomenclature unanswered. I take a far wider view of the question than probably most phytographers. To me the mere use of a name during even a century gives it no sanctity, because there will be *centuries after centuries*, when names will have to be used. So we must have *firm laws* for nomenclature. A. de Candolle himself says in his latest writings, that it would be *best*, to *have no exceptions to priority*. In this many leading phytographers concur. What ever views any one of us holds on this question is however a *mere individual view*,

3 '4 Xanthorrhoea australis' were received at Kew on 16 July 1887 (RBG Kew, inwards book 1884-7, p. 423). See also M to W. Thiselton-Dyer, 3 May 1887 [Collected Correspondence].

4 'Plants selected for Kew. Post. 25/2/87' (RBG Kew, plant lists 28, Australia/New Zealand/Polynesia, 1829-95, f. 155). It is annotated by M: 'Continuation of typical specimens to be followed up soon. F.v.M 24/2/87'.

5 Qld. Sayer and Davidson undertook this expedition in 1886-7. See M to A. Macdonald, 4 January 1887 [Collected Correspondence].

6 i.e. Murray River, NSW.

7 wasted?

and as phytographers may never fully unite, on where changes are 87.03.30
to commence in naming, and where to end, we *should mutually respect each others views*, though we do not agree.

Let us consider quite a recent instance. Mons. L. Pierre shows, that Stixis of Loureiro is Roxburgh's Roydsia.[8] Now – if this change cannot be adopted, than[9] it would be *anomalous* if not all other Loureiro's names, restored even in the genera of B & H,[10] be abolished again! So it is with lots of Aublets and others, which even in B & H genera have superseded well known generic names, fully a century in use.

In reference to Candollea, it must be remembered, that *Candollea among Dilleniaceae,* like Polanisia among Caparideae, *has fallen to the ground*; *that* greatly affects Candollea as a stylidaceous genus. If I make the change, others need not adopt it, but there is no cause for litterary or epistolar incivility on the subject

The remarks on my poor "Census",[11] concerning Stylidium a small genus in culture as yet[12] alias Candollea, in the Gardeners Chronicle, (from the initials attached) were not Prof. Oliver's.[13]

As regards Monochlamydeae, there is no doubt on my mind, that they should be inserted, each order, where its *real affinity* lies; otherwise we can have *no natural system.* I need not point out, that *all of them* (except Gymnosperms – not really Monochl.) are merely *reduced forms* of other families, just as Alchemilla stands against Rosa or Dodonaea against Aesculus, not to speak of hundreds of other examples. If we are to have no changes of any kind, then there can be no progress. –

8 Pierre (1887).

9 then?

10 Bentham & Hooker (1862-83).

11 B82.13.16.

12 a small genus … yet *is written in the margin of the MS, its position in the text indicated by asterisks.*

13 The article was signed 'W.B.H.'; i.e. Hemsley (1883). Of M's *Census*, Hemsley wrote that he approached it 'with mixed feelings, for although it embodies a vast amount of work, and will be exceedingly useful for reference, there are some features in it that we are sorry to see' – especially M's insisting on 'absolute priority in names', using M's treatment of *Stylidium* as an example. See also M to W. Thiselton-Dyer, 10 September 1881, especially n 37.

87.03.30 The *calm study* of nomen*clature* is not derogatory in biomorphic science, – but the endless multiplication of species is.

It is most cheering that Sir Joseph's Health is firm; I hope he will be able to give a new edition of the genera towards the end of the century. it is marvellous, that his eyesight is so good; but myopic eyes last usually longest.

As regards the giving up by your brother in law his *Gov.* position for one under a Committee, however excellent our friend Dr MacGillivray and his colleagues are, is a matter of grave consideration, especially as the income is rather less, though that may be increased.[14]

And now, dear Mr Dyer, let this letter bring our controversy about priority and nat. systems *for ever* to a close, and let me renew my assurance, that I shall always take a deep interest in the great national establishment of Kew.

Regardfully your
Ferd. von Mueller.

Your Bulletins will keep your industrial and technologic observations nicely together.[15]

I will try to get you the seeds of the Bluefloweringe Andersonias, also Acacia &c. When the Andersonia seeds come, they should be sown in soil, in which Calluna grows, but less sandy and not too dry. Sphagnum-soil does *not* exist in W.A.

Prof. M'Coy has returned, but I have had not yet leisure to see him.[16] I was not even able to attend this evening our branch of the Brit. Medical Association, when Phthisis had to be discussed, but sent my opinion in writing.

14 See M to J. Hooker, 10 March 1887 [Collected Correspondence], for further details about Brian Hooker's possible move to the Bendigo School of Mines, Vic.

15 *Bulletin of miscellaneous information* (1887), [RBG] Kew, vol. 1, p. 1: 'It is proposed to issue from time to time, as an occasional publication, notes too detailed for the Annual Report on economic products and plants'.

16 McCoy was in England on leave for several months from October 1886, following the death of his wife.

To Henry Moors 87.04.28

I87/8133, unit 197, VPRS 3992/P inward registered correspondence, VA 475 Chief Secretary's Department, Public Record Office, Victoria.[1]

Melbourne
28 April, 1887.

H. Moors Esqr,
Acting Under Secretary.

Sir.

In compliance with the request of your memorandum of yesterday,[2] I have the honor to report, that the preparing of each of the plates of Acacias,[3] first by

pencil-drawing costs with extras	£	2.10
The lithographing and naming	£	2.10
The lithographic paper for 1000 copies	£	1. 1
	£	6. 1
Therefore 1000 copies of each decade cost in the Gov. Botanist's Establishment	£	60. 10
The printing and binding of 1000 copies of each decade accord. to Gov. Printers estimate	£	34.6.8
Total	£	94.6/8d[4]

Thus each copy of a decade does cost not fully 1/11d.

But as the expenditure in my Department is defrayed out of the *ordinary item* of publishing of plants, and as the portion of the work, done in the Gov. Printing Department is done by the *ordinary staff* and not by extra-hands, no special expenditure beyond ordinary departmental means is incurred by this publication. In

1 The letter is numbered I87/4007 within this file.

2 See M to C. Pearson, 29 March 1887 [Collected Correspondence] and the notes thereto. M sought Pearson's approval for a list of institutions and individuals that he enclosed to receive copies of his work on Acacias. This attracted the attention of the Premier, D. Gillies, who requested information about the cost of preparing the work.

3 Eventually published as B87.13.04 and B88.13.01.

4 *sic.* The total should be £94.16.8.

87.04.28 distributing the decades, the Governm. Property in the Departments becomes pari passu[5] enriched through the litterary interchanges, while for educational and industrial purposes these decades will remain of importance *for all times*. Moreover there should be a *considerable sale*; – and here I might instance, that the last edition of my volume on "select plants"[6] was sold in seven months from the Gov. Printing Office, covering not merely the whole printing expenditure, but rendering a portion of copies available for Gov. distribution free at the time. I may be allowed to add, that thus widely and permanently knowledge on utilitarian plants for the benefit of the rural and industrial population became dispersed *without any loss* to the Government monetarely, and I would much *recommend, that a new edition be brought out soon*, numerous enquiries for this work on "select plants for industrial culture" occurring, and not a single copy being by purchase or otherwise obtainable.[7]

In conclusion I may observe, that I am and have always been guided by the strictest principles of economy also in the litterary branch of my establishment also; thus the decades of the Acacias are brought out *uncolored*, as a chromo-lithographic issue of such a work would have *doubled* or *tripled* the expenditure.[8]

I have the honor to be, Sir,
your obedient servant
Ferd. von Mueller,
Gov. Botanist.

5 simultaneously and equally.

6 B85.13.26.

7 A new edition was published the next year (B88.13.02).

8 The file was forwarded to the Premier's Office on 3 May (P87/1264). It was returned on 17 August with the approval 'of the List submitted' (i.e. the proposed distribution list for M's works on Acacias that M had submitted with his letter to Pearson) and an apology for the 'unavoidable' delay. The next day it was sent on to M who returned it on 19 August 'with best thanks'.

On 8 September 1887 E. Thomas, Secretary to the Premier, asked that 'the file P87/1264', concerning the 'cost of preparation of each plate of the Acacias for Lithographic Illustrations of Australian Acacias' again be sent to the Premier's Office. On 12 September Thomas returned the file, noting that 'the approved list of distribution has been withdrawn for reference in the Office' (J87/8133, unit 281, VPRS 3992/P, PROV).

From James Dickinson[1]

87.05.11

RB MSS M1, Library, Royal Botanic Gardens, Melbourne.[2]

Hepburn[3] May 11th 1887

My dear Baron;

An excuse to delay answering your letter[4] will be found when I inform you that I was determined to make a trip or two to give you information relative to the scarce plant that you enclosed and to elicit as much as possible relative to it.

Know then I have had several tours in search of any more of that species in vain. I have been at the place where I first got it. It was at the mouth of a desirted gold mine on the summit of one of the highest hills in the neighbourhood the mine is perpendicular, eight feet by four: many mines were around, and it with great difficulty that I obtained it. It was growing prostrate straight down the pit. I had to tear all the plant out. I forget whether it had two or three flowers.[5] I went around half-a-dozen other pits, but failed to discover any more. I send it back according to your directions but I retain a few of the leaves and shall take them to identify with any any other that I may find. By some mistake you have omitted to insert the generic name although you have given the characteristic letter S. in several other species of the genus.

The busy time that has ocupied me has caused me to defer sending the seed I have for you. For I have been ocupied with the one plant, but another I have not found. I am paying particular attention to plants and aquatics and swamp plants, but I must inform you that there are no pools of water, or swamps; yet I I[6] hope to succed

1 MS annotation by M: 'Answ 12/5/87'. Letter not found.

2 MS found with a specimen of *Ozothamnus cuneifolius* (MEL 1588820) collected by Dickinson 'near Hepburn'; the original label refers to *Helichrysum backhousii.*

3 Vic.

4 Letter not found.

5 An attached note by P. Short says that this does not describe the specimen found with the letter.

6 I *repeated.*

87.05.11 when I take longer stroles. you will please to allow me a week or ten days further before you expect their transmission to you

You write relative to the payment I shall require. Poor and economical as I am, my zeal in the business of "looking through nature up to natures God"[7] compels me to moderate my desires; I propose that one pound per month, *when I am actively employed* will be ample payment. The packet of seed I will leave the remuneration to yourself. I am now lonely. You, my very dear Baron, is the only sympathy that I have left. I have proved the strength of that friendship The Poet Thomas Young declares

"Can gold gain friendship? Impudence of hope!
"As well mere man an angel might beget.
"All like friendship, few the price will pay
"This makes friendship such a miracle below
Love and love alone is the price of love[8]

Deep this truth impress my mind
Through all his works abroad,
The heart benevolent and kind
The most resembles God! Burns.[9]

Affectionately, yet respectfully
yours truly
James Dickinson

7 Alexander Pope, *Essay on man*, epistle iv, line 331.

8 The quotation is from Edward Young's 'The complaint: or night thoughts' (1743-5), the section entitled 'Night II. On time, death, and friendship'. The lines are quoted out of order, being respectively lines 551, 552, 556, 557 and 553, and in the original read as follows:

Can gold gain friendship? Impudence of hope!
As well mere man an angel might beget.
All like the purchase, few the price will pay
And this makes friends such miracles below
Love and love alone is the loan of love.

9 The final four lines of Robert Burns' poem 'A winter night' (1786). The first line in the original is: 'But deep this truth impress'd my mind'.

To Henry Ridley 87.05.17

RBG Kew, Archives, Miscellaneous correspondence, Mueller.

Private 17/5/87.

This day, dear Mr Ridley, I have sent to you the only specimen exstant of Microstylis Bernaysii, a plant named in honor of a brother of Prof Bernays of St. Thomas-Hospital.[1] I adressed it to Mr Carruthers, so soon as it turned up in some mislaid parcel of plants here. As it is an only specimen, I should like it returned, after you have done with it; some flowers and leaves of it might be retained for the Brit-Museum Collections. The only plant in culture at Brisbane is lost.

Let me thank you for your information on the synonymy of Dendrob. longicolle.[2] Your allusion to other Dendrobiums refers, I presume, to D. Macfarlanei, renamed lately by G. Rchb,[3] who evidently overlooked my description of it in the "Papuan plants,"[4] and then gave the same name to the previously described D. Johnsoniae.

Pray, ask Mr Carruthers, to excuse, as you must kindly do, my being still so much behind with Mr Forbes's plants.[5] The elaboration of the "Key" to the system of Vict. pl.[6] proved so time-taking beyond all calculation, that I shall only be able to finish it in June or July; – and as for the early issue of it a demand was made in the Parliament here,[7] I am bound to finish that first. I mentioned already, that if you

1 i.e. L. A. Bernays. See B78.03.01, p. 21.
2 Letter not found.
3 Reichenbach (1882).
4 B76.04.01, p. 29.
5 M and Carruthers had come to an agreement as to how the work of describing Forbes's plants should be divided between them; see W. Carruthers to M, 1 March 1886 [Collected Correspondence]. See also M to H. Ridley, 12 January 1887 [Collected Correspondence].
6 B86.13.01, B88.13.03.
7 In July 1886, F. Dobson asked in the Legislative Council, as a question on notice, when the *Key* would be published; see M to A. Deakin, 28 July 1886 [Collected Correspondence].

87.05.17 liked to elaborate Mr F's Euphorbiaceae and Urticeae, to speed the work, I should be very glad.[8]

Some unpleasantness has occurred here quite lately between Mr F. & our geogr. branch here. For *your* and *Mr Carruthers* special information, but as a confidential remark, I like to say, that the Council of the Vict. geogr. Society, after the Rev Mr Chalmers inability to take the field,[9] decided, lately, to send out a special expedition from Victoria, so that all the reports, diaries, maps and collections might become in first instance available to the colony, which provides the fund, – particularly so, as we have *no prospect of a repetition of the vote*. on the above condition only our £1000 would this season have become available to Mr Chalmers, if he could have gone out for us. Mr F would perhaps have been willing to enter on a similar arrangement with us, – but as he estimates the cost of *one seasons expedition at £2000* (therefore only for portion of a year!), and as the Special Comissioner wants only – so far as hitherto arranged [–] to contribute out of the £15000 (voted for N.G. by the three eastern colonies of Austr, Q, N.S.W. & V.) either £500 or perhaps £1000, for forming a track through a gap of the ranges from Pt. Moresby to Acland bay[10] *across the peninsula*, we could not entertain here any proposals of Mr F, as we have only £1000 and we want *highland* –, not lowland – exploration. Mr Douglas is, so Mr F. says, not likely back from the Louisiades[11] til july, therefore too late this season, to withdraw – if he thinks proper – the condition on which his support depends.

I have for about a dozen of years in writings & prints of mine expressed a hope, that the elaboration of the highland plants of NG. should fall to my share; and as I *feel*, that my worldly work is drawing to a close, I am sure no one will grudge me my *swan-song* on the alpine and subalpine plants of N.G., more particularly so, as in Europe ample materials exist from other parts of the globe,

8 See M to H. Ridley, 21 September 1886 [Collected Correspondence].

9 Chalmers was in England from August 1886 to July 1887; he left Sydney for Port Moresby, New Guinea, on 13 September 1887 (*Flora malesiana*, vol. 1, p. 103).

10 Dyke Ackland Bay?

11 An archipelago in the Coral Sea.

to occupy every competent phytographer, and all the younger ones can work for decennia after me! I hope, to do a vast amount of work on Mr Forbes's and Mr Bäuerlens plants during the last half of this year, and some of the msc. can go to Mr Britten.[12] 87.05.17

When the "Key" is out, I must devote some time for selecting more plants for the Brit Museum.

Confidentially I may add, that I have asked a particular friend, who is this year "Rector magnificus" of one of the Continental Universities, to confer the honorary degree of Ph.D. on Mr Carruthers. I shall rejoyce, if my solicitation is successful.[13]

Regardfully your
Ferd. von Mueller.

To Charles Riley 87.05.21

Box 65, entry 2, RG7, Records of the Bureau of Entomology, National Archives, Washington DC.[1]

Melbourne,
21 May 1887.

C. V. Riley Esqr,
Entomologist to the Department of Agriculture,
Washington

In reply to your letter of the 15th December,[2] dear Mr Riley, I beg to inform you, that the Icerya Purchasi, (or a closely alied species) altho occurring on Acacia mollissima and some congeners in the

12 Some descriptions from Papuan collections were published after this date in the *Journal of botany* of which Britten was editor, for example, B91.06.02.
13 Letter not found. No evidence of such award has been found.

1 For published extracts of this letter see B87.13.26, where it is incorrectly dated 21 March 1887.
2 Letter not found.

87.05.21 colony Victoria, has not attacked here, (so far as I can learn or had occasion to observe) destructively the Orange-Orchards. I will however make further enquiries as well in this colony as in New South Wales, South Australia, New Zealand, and let you know the result.

Possibly the Icerya develops more readily in a moister clime than that of Victoria, and thus becomes more mischievous in California than here.

The introduction of this destructive insect into your states by means of Acacias seems to me very unlikely, because the various species of Acacias are so easily raised from seeds, that no one will think to introduce them by living plants. Moreover it could not have been the Acacia latifolia, which was the host of the Icerya because that species is a native solely of the North-Coast of Australia and as yet nowhere existing in horticulture. Acacia armata certainly is grown for hedges, but always raised from seeds, chiefly obtained from South-Australia. It seems therefore more likely, that when Acacias are grown anywhere, they would afford – particularly in humid climes, – a favorable opportunity for the Icerya to spread. A similar circumstance occurred in Ceylon and an other in some parts of Brazil, where an indigenous insect-plague became aggravated, when Eucalypts, on which that insect preferentially seized, became reared always from seed. Whether the Icerya was originally an inhabitant of Victoria or merely immigrated, I will endeavour to ascertain; but such a subject of inquiry is surrounded with difficulty now after half a century's existence of the colony, particularly as the Icerya drew no attention here by any extensively injurious effects on any cultural plants, though it may have caused on some places minor or transient injury. Allow me, honored Sir, to congratulate you to your recovery from a long and severe illness, and let me express a hope, that your grand talent and unrivalled experience will long advance science yet and rural industries.

Regardfully your
Ferd. von Mueller

To Ferdinand von Krauss[1]

87.06.11

Staatliches Museum für Naturkunde, Stuttgart.

11/6/87.

Endlich, edler Freund, ist es mir gelungen, des Balges eines Macropus rufus habhaft zu werden, u soll solcher mit dem Ende dieses Monats von hier abgesendern P. & O. Steamer frei an Mess. Watson & Scull zur Weiterbeförderung an Sie gesandt werden, u allerlei Anderes werde ich hinzufügen. Sie sehen denn doch, dass ich Ihnen nicht ganz abtrünnig geworden bin. –

Es ist aber nicht so leicht, wie früher, zoologische Gegenstände zu erwerben, namentlich Hochwild in Australien. Man gönnt den armen Thieren das wenige Gras, Kraut u Gebüsch nicht, was sie zur Nahrung bedürfen, denn die Weideländereien sind mit Heerden nun fast überall erschöpfend besetzt, und so werden die Känguruhs u auch Emus erbarmungslos vernichtet. In nicht gar langer Zeit wird manche australische Art grosserer Thiere durch Menschenhand aus der Schöpfung gestossen sein! So werden die armen Geschöpfe auch hier das Schicksal des Didus ineptus, der Dinornis-Arten, und Ihnen näher der Alca impennus theilen, von der in so rührender Weise Blasius berichtet, dass das letzte Paar in Island erschlagen sei, in dem die braven unglücklichen Geschöpfe Ihr Nest vertheidigten! Es ist zu verwundern, dass die Zoologen Dänemarks nicht zeitig schützend eintreten, damit die Brütung fortbestehe. Wir hier schützen jetzt den seinem Untergang nahen Leiervogel.

Immer der Ihre
Ferd von Mueller.

1 MS annotations: ' Empf. 25. Juli 87 Rep. 2. Nov. 87.' [Received 25 July 1887, reply 2 November 1887.] (reply not found); 'Kiste 15. Sept. 87 angekommen. Bericht über seine Sendung u. Dank' [Box arrived 15 Sept. 1887. Report about the consignment and thanks]; 'Woher sind die Eier?' [Where are the eggs from?]; 'Anzeige dass [*two illegible words*] jetzt 840. frägl.' [Indicate that [...] now 840. doubtful].

11/6/87.

At last, noble friend, I have been successful in getting hold of a skin of *Macropus rufus*,[2] and it will be sent with the P. & O. steamer despatched from here at the end of this month free to Messrs Watson & Scull for forwarding to you and I will add all kinds of other things.

You will then certainly see that I have not been quite disloyal to you. –

However, it is not so easy, as previously, to acquire zoological objects, especially big game. They begrudge the poor animals the little grass, herbs and bushes that they need for food, because the pasture lands are now almost everywhere exhaustively occupied with herds, and so the kangaroos and also emus are exterminated mercilessly. In no time at all many Australian species of large animals will be expelled from the creation by the hand of man! So the poor creatures even here will share the fate of *Didus ineptus*,[3] the *Dinornis*[4] species and nearer to you the *Alca impennus*,[5] of which Blasius reported in such a touching way, that the last pair in Iceland was killed, in which the brave unfortunate creatures defended their nest![6] It is amazing that the zoologists of Denmark did not early intervene protectively, so that the hatching continued. Here we now protect the lyre bird that is nearing its extinction.

Always your
Ferd von Mueller

2 Red kangaroo.
3 Dodo.
4 Moas.
5 Great Auk.
6 Blasius (1883). The last known pair of Great Auks was killed on 3 June 1844 on the island of Eldey, off the coast of Iceland, by a landing party sent by a bird collector to obtain specimens of what was already recognised to have become a very rare species.

To William Thiselton-Dyer 87.09.16

RBG Kew, Miscellaneous reports 6.5, New Guinea, Fiji and Pacific Islands 1850-1928, ff. 23-4; RBG Kew, Kew correspondence, Australia, Mueller, 1882-90, ff. 228-9.[1]

16/9/87

In answer to your very considerate last letter,[2] dear Mr Dyer, let me say, that I had *not* the *slightest knowledge* of Mr Forbes's arrangements, concerning the mode [of][3] distribution of his plants in Europe, *nor* did I exercise *any influence* either directly or indirectly on the distribution. The Rev W. G. Lawes of Port Moresby Mission Station acted kindly as Agent for the R.G.S. of Austr, and through his influence, the second best set (therefore not a complete series) came to Melbourne, which set must be *subdivided* between the bot. institutions of the three colonies, from which the votes were obtained, out of which [o]n [r]ig[ht][4] proportion £500 were given to Mr Forbes, this second set of bot specimens being the only return, as we did not even get a sketch-map or any extracts from diaries. Now, my generous friend, you must consider the *wide difference* for collecting fund from *private institutions* and obtaining fund from the *public exchequer through Parliament* In the one case, there is no responsibility, in the other the greatest possible *responsibility*, and here under the vigilance of a very exacting press! Englishmen in Australia, and I consider myself one of them since my 40 years naturalisation, are not unreasonable; all we wanted here, feeling our responsibility to the local Parliament, and the equally watchful local journals, that we should *fairly share* in the proceeds of F's Expedition. That the arrangements with him could not renewed, was because he wanted £*2000 for one season* for the lowlands or uplands; the High Commissioner had promised a subsidy

1 The letter as transcribed here comprises two parts, filed in different series at Kew. The internal evidence that suggests that the two parts belong to the one letter is given below, at the end of the first part of the letter.

2 Letter not found.

3 *editorial addition* – MS damaged.

4 *editorial addition* – MS damaged.

87.09.16 for the lowlands track across the peninsula, while *we* here wanted highlands exploration, and we *had simply not £2000.*[5] I will do *all I can* also as regards New Guinea supplies for the great Kew Department, but give me a little more time, as the Field Naturalists here press me for the "Key to the Syst of Vict. Plants"[6] for use still this spring, and altho' the printing *has* commenced, the work cannot appear for some weeks yet. When Cuthbertson's and Sayer's plants arrive, I will methodically go through the whole Papuan Collections, and make up as good a set for you as I can.

Our new expedition will give us the plants of the third and 4th zone, but not yet that of the fifth, which we must try to grasp next year, and the elucidation of all this shall be my swan song![7]

Kindly tell Sir Joseph, that we push antarctic exploration, because we wish to increase the *revenue*, trade and manufactures of these colonies, and see the now greatly unsuccessful whaling in the north through steamers carried to the promising fields of the S. It was on *my* proposition that an antarct. Committee was here formed, and as mover of the resolution, I could have claimed the Presidency, but on my own free impulse I moved a celebrated Naval Surveyor into the chair, our senior naval Officer,[8] an accomplished Astronomer and a religious man, whose father as flag-Lieut. of Nelson at Trafalgar hoisted the celebrated and ever memorable signal, and whose son is one of the most brilliant of navigating Lieutenants in the Royal Navy. The idea however of using in nightless time and calm weather Balloons for reconnoitering at

5 The volume includes a number of press cuttings from Australian papers which report the formation of the expedition, and a number of letters from officers of the Royal Geographical Society and the British Association for the Advancement of Science concerning their support for and later dissatisfaction with Forbes. There are also letters from Forbes to Thiselton-Dyer.

6 B88.13.03.

7 The letter filed in Miscellaneous reports 6.5 ends here without valediction. The remainder of the letter as transcribed is filed in Kew correspondence, Australia, Mueller, 1882-90. The two sections are united on the basis that the date '16/9/87' has been added by Thiselton-Dyer at the top of f. 228 which begins without salutation but contains a valediction, and, importantly, M's return to discussion of Forbes at the end of the letter. However, the paper of the two parts is different, the first is on unruled and the second on lined paper.

8 Crawford Pasco.

Photo. by Elliott & Fry.

SIR WILLIAM TURNER THISELTON-DYER
K.C.M.G., C.I.E., F.R.S.

TO WHOM VOLUME CXLII.
IS DEDICATED

KEW, DECEMBER 1, 1916.

Figure 12. William Thiselton-Dyer. Published in *Curtis's botanical magazine dedications* (1931) p. 354. Courtesy of the Library, Royal Botanic Gardens Melbourne.

87.09.16 various longitudes for triangulation beyond the ice-barrier arose with me. Only yesterday I urged on the Premier of N.S. Wales with my Colleagues a subsidy, which Sir Henry Parkes kindly promised. So we hope to see Sir Allen Young here out yet this season.

Of course, you will have read what I said in my two annual Presidential Discourses.[9] The suggestion of Mr Goeschen being asked for some help was also emanating from me, his firm, Frühling & Goeschen then, transacted for me transits to the Continent in the 40-[10] and -50 years of this century. We want thus to get profitable engagements also for the Scotish whalers. Returning once more to the Forbes-affair, I must say in justice to my Colleagues, that no incivility was shown him, though some sternness in our administrating public fund. He has no reason to be so severe upon these colonies, particularly while he is an Officer of the High Commissioner, and thus solely sustained by the three colonies, which hitherto found the fund for the administration of British New Guinea.

With all good wishes for all at Kew your
Ferd von Mueller.

Should you meet the most generous of all the generous, Sir H. Barkly, will you kindly explain about the antarctics and about Forbes. I find, that I shall have no time for a long letter to him, til the Key[11] is out, for the Cherry trees &c are now here in full flower already.

I am to be at private luncheon at the Viceregal palace[12] tomorow, when antarctic[s] and New Guinea will also be the topic.

I have written within the last few days written[13] as a byework over hundred various letters with my own hand to leading colonist, so that Giles may finish off geographically Central Austr. within our first century.[14]

9 B85.13.25, B87.05.03.

10 18 *deleted before* the 40-.

11 B88.13.03.

12 i.e. Government House, Melbourne.

13 written *repeated.*

14 For an example of a response to his appeal, see F. Sargood to M, 19 September 1877 [Collected Correspondence].

To Henry Forbes 87.10.26

A36 Royal Geographical Society of Australasia (Victorian Branch) papers, Mitchell Library, State Library of New South Wales, Sydney.

Copy[1]

Melbourne.
26. Oct. 1887.

In my whole career, dear Mr Forbes, I had never an occasion, to write an unfriendly letter to any man of science, and I do not wish to deviate from that principle now at the late autumn of my life. Instead therefore of entering into the tone of your last letter to me,[2] I adress you in terms of civility, and assure you, that I have quite friendly feeling towards you, as evinced in my pleading for you with H. Excell. the Special Commissioner[3] in allowing you, to proceed with explorations in New Guinea, so that you would still enjoy support from the three Australian colonies, who vote the funds of the Protectorate.

In my usual candor I must say, that I felt discouraged to write to you, after you applied in an article through a daily journal a Shakespearian highly undeserved quotation to the Council of the R.G.S.A, who simply as Custodians of public funds, entrusted to their administration, endeavoured to secure a fair return for the disbursements. If any shortcomings towards yourself occurred in our transactions with you, then be so just to remember, that we have not a single man of leisure on our Council, that in the stern duties of daily life we might overlook details in connection with our honorary work for the R.G.S.A., and that a single letter to us, asking plainly, what in justice to yourself required to be done, would have been better, especially as we gave you half a thousand

1 The copy is written in an unknown hand with some corrections by M.
2 See H. Forbes to M, 5 September 1887 [Collected Correspondence], in which Forbes demands a retraction of 'the extremely inaccurate statements' made by A. C. Macdonald at a Council meeting of the Victorian branch of the Royal Geographical Society of Australasia and subsequently published in the *Argus*.
3 J. Douglas.

87.10.26 £, when[4] attacking us without any previous notice in the public press. Your letter, now received, in which you demand from the Council here some apology on points in dispute, I have handed to the Hon. Secret.,[5] who will place it before the Council at its next meeting, and I feel convinced, that my honored colleagues will do you justice, if it can be shown, that inadvertently we omitted any thing due to you. That we, after being unable, to provide the £2000, asked by you for highland-exploration, made other arrangements for an expedition within our means, arose from our desire not to loose the season, while at least one other expedition was proceeding to New Guinea; – and that the Rev. J. Chalmers, to whom alone I had pledged myself in the choice for the leadership, was not waited for, arose from an expression of the then last letter from the Rev. Gentleman, in which he held out no hope of his taking the field this season.[6] In fairness to us here you will allow me to remind you, that every readiness was shown here by us, to support any Expedition of yours under the auspices of the High Commissioner and therefore on the expense of the Australian Colonies additionally so far, as to purchase a series of any specimens from such expedition at the highest price, that according to value could be placed on them. That you should have recently stated, no specimens of plants should ever be sent to me any more by you, I cannot think to be correct, because it would be unfair to me as the one, who for a dozen years urged highland exploration in New Guinea, – because it would be ungrateful to me as the one, who mainly advocated the former grant to you, – and because it would cause here public discontentedness. I frankly concede, that you have cause to be disappointed in not having yet received more names of your former and only sending, not quite as many species as you publicly stated; but I had no idea that the dichotomous method, demanded for the "Key to the system of Victorian plants",[7]

4 than?

5 A. Macdonald.

6 The Council contracted instead with W. Cuthbertson; see M to W. Cuthbertson, 28 May 1887 [Collected Correspondence]. See also M to W. Thiselton-Dyer, 16 September 1887.

7 B88.13.03.

would take up so very much more time, than I estimated, when I had the pleasure of meeting you here. It is therefore not want of faith, but simply miscalculation of time required for an important public duty, which has thrown back my working on the collections of Dicotyledoneae from you. I hope however to finish the "Key" at last next month; and working on Papuan plants will then be my main-engagement in all hours which can be rendered free from urgent official duties.[8] Meanwhile I have sent, – indeed many months ago, – your Sapindaceae to Prof. Radlkofer of Munich, who is engaged on a monography of that order,[9] while Mr Britten has undertaken the elaboration of the Urticeae and Euphorbiaceae.[10] Of some of your plants I have sent you long ago the print, and shown a generosity of spirit not only by dedicating specimens to you, but even by adding your honored name in some instance to the authority.[11] As a Naturalist you will be aware, that the critical elaboration of a new species is usually a good day's work; and you must also know, that Dr Beccari has not even yet finished to describe his Mt Arfak-plants after a dozen of years. A full list of plants is not fitted for a concise report, such as we hoped from you, and to such a document I could add here at any time a brief appendix, so that your forwarding or completing your report need not be delayed on the ground of yet unfinished detail examination of your plants. 87.10.26

With my best wishes for you.
signed Ferd von Mueller

8 It appears that M never published a comprehensive account of Forbes's plant collections; some, however, were recorded in B90.05.01 after being described by M elsewhere (see n. 11).

9 Radlkofer (1890).

10 No such publication by Britten has been identified.

11 M described various plants collected by Forbes in B86.08.03 (including *Sterculia oncinocarpa* credited jointly to M and Forbes), B86.09.02, B86.09.05, B86.10.01, B86.10.02 and B87.01.02.

87.10.26a *To Rudolph Virchow*

NL Virchow, Nr. 1499, Akademiearchiv, Berlin-Brandenburgische Akademie der Wissenschaften, Berlin.

Melbourne
am 26 Oct. 1887.

Mit der deutschen Post, edler Herr, habe ich Ihnen eben ein neues Werk meines Freundes, Mr E. Curr, des hiesigen Gouvernements Heerden-Inspectors, über die australische Race zugesandt für freundliche Entgegennahme und als eine Gegengabe für die wichtigen u. zahlreichen anthropologischen Hefte, welche unter Ihrer regen Fürsorge hervorgehen. Von der grausigen Tiefe, welche die australischen Autochthonen in der menschlichen Cultur einnehmen, giebt Mr Curr glaubwürdig lebhafte u. erschütternde Darstellungen. Ich möchte inter alia hinzufügen, dass unlängst ein noch nicht dreijähriger Knabe eines Halbcivilisirten im östlichen tropischen Australien seine jüngere Schwester absichtlich tödtete, damit ihm wieder allein die ganze Aufmerksamkeit seiner Eltern zukomme; ob dieser Frevelthat lobte der Vater den schrecklichen Jungen, da er schon jetzt beweise, wie brav ein Krieger er später sein würde; – und als vor einiger Zeit ebenfalls im tropischen Australien ein Sammler von Schädeln einem Schwarzen ein ganz kleines Geschenk bot für einen craniologischen Beitrag, lockte der Bösewicht einen prächtigen civilisirten Knaben ins Dickicht, schnitt ihm den Kopf ab und brachte solchen zu dem Sammler mit einer Mien[e], als ob dies ganz gehörig wäre! Mit diesem Jahrhundert wird der Cannibalismus doch wohl selbst in Neu Guinea aufhören, und in der nächsten Generation es doch wohl durch Mission u. Civilisation keinen wirklichen Wilden mehr geben.

Mit tiefer Ehrerbietung und Bewunderung Ihres grossen u. vielseitigen Wirkens

Ferd. von Mueller.

Melbourne 26 Oct. 1887. 87.10.26a

With the German mail, noble Sir, I have just sent you a new work of my friend, Mr E. Curr, the local government stock inspector, on the Australian Race[1] for kindly acceptance and as a gift in return for the important and numerous anthropological fascicles that emerge under your active care. Of the horrible depth, which the Australian autochthones occupy in human culture, Mr Curr gives authentic, lively and deeply moving depictions. I would like to add, *inter alia*, that recently a boy not yet three years old belonging to a half-civilised man in eastern tropical Australia deliberately killed his younger sister, so that the whole attention of his parents belonged to him again; the father praised the terrible child for this outrage, because he already now proved how worthy a fighter he later would be; – and as some time ago also in tropical Australia, a collector of skulls offered a Black a quite small present for a craniological contribution, the scoundrel decoyed a splendid civilised boy into the jungle, cut off his head and brought it to the collector with a look as if this was quite proper! Cannibalism will certainly cease this century probably even in New Guinea and in the next generation there will be probably be no more true savages through missions and civilisation.

With deep respect and admiration of your great and many faceted work

Ferd. von Mueller

To Henry Parkes 87.11.01

A925 Parkes correspondence, vol. 55, pp. 253-4, Mitchell Library, State Library of New South Wales, Sydney.

Melbourne,
1/11/87.

The honorable Sir Henry Parkes, K.C.M.G.,
Premier of New South Wales.

Allow me, dear Sir Henry, to introduce to you a relative of myself, F. E. Harris Esq. of Silverton,[1] who particularly asks me to recom-

1 Curr (1886-7). See also E. Curr to M, 25 October 1887 [Collected Correspondence].

1 Possibly a relative of George Adolphus Harris, who married Ottilie Bertha Wehl in 1874, the oldest child of M's sister Clara. Or did M mean to refer to his nephew Ferdinand Edward Wehl, who had been living in Silverton, NSW?

87.11.01 mend him to your kind consideration, as he is eager to carry through some business with the Department for Mines in Sydney, which objects he feels sure your own great influence would facilitate.

Let me remain,
dear Sir Henry,
regardfully your
Ferd. von Mueller.

87.11.14 *From William Macleay*

A39 Royal Geographical Society of Australasia (Vic. Branch) papers, Mitchell Library, State Library of New South Wales, Sydney.

LINNEAN SOCIETY
N.S.W.

Sydney 14th November 1887

My dear Baron

I think I will be able to undertake the identification & description of the New Guinea Insects you wrote about[1] but before positively undertaking the task I should like to see them. Could you manage to send me the entire collection of all orders of the Insects. If I find that I cannot satisfactorily undertake the description of them I shall return them without delay & if I do describe them I shall return them with as little delay as possible. Do you wish the descriptions published here in the Linnean Proceedings or sent to you in Melbourne for publication elsewhere?[2] I shall of course be glad to accept for my own Museum specimens of any of the Insects of

1 Letter not found.

2 No such publication appeared in the *Proceedings of the Linnean Society of NSW.* In Macleay (1887), p. 204, Macleay wrote that in view of the fact that others were working on New Guinea insects, he himself was 'deterred' from further work on them, lest he describe the same species.

which there are duplicates, but I make no claim for any, & will leave that to be arranged by you after the return of the Insects. 87.11.14

I am my dear Baron
yours very truly
William Macleay

From William Tietkens[1] 87.11.24

A39 Royal Geographical Society of Australasia (Vic. Branch) papers, Mitchell Library, State Library of New South Wales, Sydney.

Adelaide
Novr 24th 87.

My dear Baron

I thank you very much for your kindly expressions of congratulation[2] upon my appointment to the charge of the Lake Amadeus Expedition and with regard to Mr Giles I would wish to point out that I have never at any time put myself forward in this matter I was asked a year ago as to who would be the best man to take the party & I unhesitatingly said Giles and Giles was applied to and for a while it was an understood thing that he was to take it but as there seemed no immediate prospect of Funds being collected he entered into an engagement to go to the Kimberly & I feel much hurt that he did not write to me before he started, for I know nothing whatever of his whereabouts this is certain, Giles knew perfectly well that I wished to go and *very generously withdrew* his name from the undertaking but I hope sincerely hope that he has not expressed feelings of disappointment to you for I assure you my dear Baron that I have not once put myself forward, the Geograph Sec wrote to Giles repeatedly without receiving any reply and then they appealed to me, So that I am somewhat puzzled when you tell me of his telegram from Port Darwin expressing

1 MS annotation by M: 'Ackn end of Nov see letter book'. Letter not found.
2 Letter not found.

87.11.24 disappointment for it [...][3] agree with his generously with drawing his name a few months previous:

I would wish for an opportunity to write to him but his Sister in a letter to a friend of hers said that he had gone to Hong Kong & from there he was going to Kimberly it seems to be a fairly good season in the Interior & we may hope that not more than the usual amount of difficulties may present themselves during the trip

With every expression of kindliest regard
I remain dear Baron
yours Sincerely
W H Tietkens

87.12.00 *To [Philip Sclater]*[1]

RBG Kew, Kew correspondence, Australia, Mueller, 1882–90, ff. 236-40.

Private

As you are on the Council of the R.S., I would venture to ask, whether you see your way clear to interest yourself on my behalf at the next distribution of the medals.[2] This asking of mine may seem an unusual one, but I beg you to consider, 1, that I have no

3 *Word illegible.*

1 Probably written in December 1887 or early January 1888, apparently to P. Sclater (see P. Sclater to W. Thiselton-Dyer, 14 February 1888, that accompanies M's letter in the file, where Sclater asks Thiselton-Dyer to look at the 'enclosed extraordinary letter fr. Baron v. Müller'). A version of M's letter was published in Powell (1978).

2 Sclater was not then on the Council of the Royal Society and indicated to Thiselton-Dyer (see n. 1) that he would write to that effect to M. He also informed Thisleton-Dyer that '[Mueller] seems to me to have done some good work'. Maxwell Masters reported in a letter to J. Hooker, 8 March 1888 (RBG Kew, Archives, miscellaneous reports, Melbourne, Mueller, vol. 1, p. 128), that M 'is burning with desire' for the Royal Medal, 'and writes to me frequently about it'. Maxwell judged that M 'has certainly deserved well not only on botanical but on geographical grounds'. The awarding to M of one of the Royal Society's two Royal Medals for that year was announced at the Society's Anniversary Meeting on 30 November 1888; see M to G. Stokes, 1 January 1889.

opportunity to meet any member of the Council in person; 2, that my works coming from such a distance and extending in volumes over more than 30 years would likely to a large extent escape notice of most members of the Council; 3, that in the very late autumn of my life I may not be destined by divine providence to *live through an other year*, while it would infinitely be joyful to me, having before I pass away my name enrolled on the honor list of the R.S. 4, that I am *senior* to men in science of extensive working in this part of the world, and as such should set the highest value on being the *first* in Australia, on whom the honor is conferred; – 5, that Kew surely would *allow* such a distinction to be extended to these portions of H. M.'s dominions, particularly in the year of the first centenary jubilee here; 6, that I was the first in Australia, who published original and ample observations for science in three languages; 7, that I am the first Australian science-author, of whom works (as regards one particular volume) have been translated into two languages and are soon to pass into two more; 8, that I am the first, who published lithograms of plants in the southern hemisphere by the hundreds (about 600 plates now chiefly quarto) always with analyses from the bud to the embryo; 9, that I commenced my *various* researches in Denmark and N. Germany as far back as 1840; that I attended the German Soc. of physicians & naturalists as an *active* member as far back as 1846, meeting then Steenstrup, Oersted, Forchhammer, Schleiden, Kunze, D'Alton, Rammelsberg, Volger, Roeper, and other illustrious men, now nearly all numbering with the dead; – 10, that I uninterruptedly worked for *geography* and for phytology in all its branches since 1847 here, having taken out my Doctor-degree at the University of Kiel before I left Europe, – 11, that I represent *besides* chemistry, having worked in Pfaffs and Himly's laboratory for two years, some alkaloids, glycosids and organic acids of Austral. plants having been discovered by me. also geology as regards particularly veget. palaeontology, two decades of pl[i]ocene[3] fruits for the geologic survey here. also further geography, medicine especially therapeutics, being thus President of the Vict branch of the R Geogr. Soc. of Austr., and examiner 87.12.00

3 pleiocene?

87.12.00 in therapeutics and phytology at two Austral. Universities; – 12, that I have contributed to Zoology in all its branches – as you will be aware – by extensive collections through many years; 13, that the total of my volumes extends to about *40* irrespective of my share in Benthams flora Australiensis, 2/3 of its material (nearly all elaborated by myself) coming from me; – 14, that I established over 100 new genera (well acknowledged) of plants, described more than 2000 new species, and transferred over 1000 to better generic positions, adding over 1000 well founded spec to the phytography of Australia since the seven volumes were published by the great Bentham; 15, that the Royal Soc. of Victoria founded *by three persons* I being the only survivor in my room prior to my proceeding in 1855 with the two Gregorys to what is now known as the rich pastoral Kimberley-District (almost as large as Britain) in an expedition, which kept us away *17* months at a stretch from any settlement or any communications with the civilized world; – that under my Presidency of the Roy. Soc. of Vict. its Hall was built, and her Majestys patronage granted; 16, that my *observation-lines* in Australia on *foot* or *horse* extend to 30,000 (nearly) English miles, including many of the first triangulations of the Australian Alps, when some hypsometric notes were also by me taken, even the remotes[4] 30 miles of the Yarra River, emptying at Melbourne, being discovered from the Alps by me. – in many of the explorations (from 1847 til 1853 on my private expense) the danger of crossing unbridged rivers, breaking through waterless country and encountering the fury of the aboriginal natives being extreme; not to speak of hunger and thirst and excessive and parching heat under the canopee of heaven – on one occasion no food for five days! – one of the six votes of special promoters of geography at the Congress of Venice being accorded to me; that my exertions and influence called forth numerous expeditions those of M'Intyre, Giles, Mc[Kinlay &c][5] lately that also of Cuthbertson in New Guinea when the [th]ird mountain-zone was reached for the first time, and causing also the likely *revival* of *antarctic exploration* on my urging 3 years

4 remotest?
5 Partly obscured by binding.

ago, and on my suggesting an appeal also in the interest of steam whalers at home to the Right Hon. Mr Goeschen, to whom I am known since the fourtier and fiftier years. – 17, that my efforts in a cosmopolitan spirit have given first and mainly the Eucalypts and best tan-Acacias to all warm temperate zones, have brought extensively the Conifers and endless numbers of other utilitarian plants first and largely to Australia in the interest of forest-culture and other rural industries; 18, that from the empire of plants I have always endeavoured to increase technologic industries not only here but also elsewhere, having been already one of the 12 local Commissioners for the first Paris Exhibition of 1855 here, the Right Hon. Mr Childers, General Sir Andrew Clarke and General Pasley being among my Colleagues; I being again one of the Commissioners for the Melbourne Exhibition of 1888-1889 now. – 19, that the generous spirit of men of Science throughout the world had raised me to the position of hon. or corresp. member of more than 150 science-institutions, my facility of the use of several languages giving me the means of ready communication. 20, that at the celebration of the 25 annual jubilee of the marriage of their Maj. the King and Queen of Wurttemburg I was as a *representative of science* raised to an equivalent of peerage of that Kingdom, and at the 50 years reign of Her Brit Majestys jubilee I became a Grand Cross of one of the S. Eur. Kingdoms, over which a Sovereign out of the House of Coburg-Gotha reigns, a Commander-Order from one of the sons in law of her maj. Queen Victoria, being graciously conferred on me also at that glorious festival. – remarks offered here with feelings of gratitude not vanity, – but also in explaining my *status* in approaching through you (and the kind Dr Schuster perhaps) the Council of the RS. 21 that the Governm herbarium formed by me here, and founded on the *free gift* of my own private collections comprises now about *half a million* of sheets of named plants for reference in all futurity; – that Dromedaries, Dear[6] of various species, Angoras, Llamas, Alpacas, Salmon-trout, carp, and song-birds foreign ducks &c with many other kinds of other animals for the enrichment of the Austral. Fauna were first introduced here

6 Deer?

87.12.00 while I held (til my Garden space became too small) the Directorship of the zool. Garden also and for about a dozen years the Vice presidency of our Acclim. Soc. 22, that for 17 years I founded and directed the bot-Garden of Melbourne, built the first large Conservatories in Australia, grew first the Vict. Regia here and had at the time the largest collection of species of ornamental & utilitarian plants under cultivation in any part of Australia, and maintaining a most extensive interchange, so that no Glass-house of any pretensions *anywhere* is likely without plants from me, including also the ferntrees, the export of which entirely originating with me. 23, that as Vice President of the greatest Liedertafel I support the noble music art for the benefit, which it exercises on the tone of Society, development of mental faculties and of one of the best enjoyments of family life. 24, that as Patron of the Field-Naturalists Club I aid in collecting the scientific-forces here for united action and mutual learning; – 25, that as Patron of the Turn-Verein I promote the gymnastic exercises of the youths in our metropolis, and the *rational* enjoyments connected therewith; – but above all; 25,[7] as Patron of the young christian Gentlemens Assoc. of one of the Great Presbyterian Churches and at all other occasions I seek to advance the blessing of our holy religion for ennobling us during our earthly career and for leading us to a hope of a sanctified eternity!

Should you desire to consult anyone on this wish of mine, then I advise you, to communicate with Sir H. Barkly, Sir Rich. Owen and Sir John Lubbock particularly, and I feel also sure, that besides my special Colleagues in Britain also Prof Gilbert, Dr Gladstone, Mr Ball, the Righ Hon Mr Childers, Dr Günther, Prof. Maskelyne, Prof Judd, Prof Etheridge and others would aid any movement in my favor.

As I cannot present myself personally, such a step as consulting them, would be quite justifiable in my case, so exceptional from antipodal distance. With some of those, mentioned by me, you are probably befriended personally.

[Ferd. von Mueller][8]

7 26?
8 *editorial addition.*

To Rudolf von Fischer-Benzon 87.12.16

Cb 66, Nachlass Rudolf Fischer-Benzon, Schleswig-Holsteinische Landesbibliothek, Kiel.

16/12/87

Ihre Zuschrift, geehrter Herr Doctor, ruft in meiner Erinnerung wieder jene Zeit besonders wach, in welcher das Leben mit seinen Hoffnungen mir meist noch vorlag, und in der ich mit jugendlicher Begeisterung die Wunderwerke der Pflanzenwelt, soweit solche in freier Natur und in Garten-Culturen sich mir darboten, zu erforschen begann. Jene Jahre gehören zu den herrlichsten meines Daseins! Gern entspreche ich Ihrem Wunsch, für den schönen Zweck, welchem Sie und Herr Dr Prahl sich widmen, u. zu dem meine ersten Arbeiten im Felde der Wissenschaften auch vielleicht etwas beitragen, einige Notizen über meinen Lebenslauf zu liefern; u wenn ich schreibe vielleicht zu umfangsreich in dieser Hinsicht, so können Sie ja das, welches Ihnen als Zweck entsprechend erscheinen mag, auswählen.

Verwaist trat ich als Eleve in das pharmaceutische Geschäft des Herrn A. B. Becker in Husum Ostern 1840 ein, im Alter von noch nicht ganz 15 Jahren, nachdem ich meinen ersten Unterricht am Gymnasium in Rostock, meiner Vaterstadt erlangt, und nach dem Tode meines Vaters, der dort erster Zoll-Controlleur war und als Freiwilliger 1813 mit bei Seestedt focht, mit meiner Mutter und drei Schwestern zu den Grosseltern in Tönning zog, wo ich die höhere Schule von 1836-1840 besuchte.

In dem Becker'schen Hause hatte zum Anfange des Jahrhundert auch Prof. Forchhammer, der scandinavische Geologe, seine erste Bildung für die Naturwissenschaften erhalten, u. ich hatte das Glück, diesem edlen Manne in der Naturforscher-Versammlung zu Kiel in 1846 zu begegnen, bei welcher für mich höchst anregenden Gelegenheit ich auch Oersted, Dalton, Schleiden, Roeper, Steenstrup, Kunze, Waitz, Rammelsberg, Hoppe u viele andere hervorleuchtende Männer der Wissenschaft persönlich kennen lernte.

Wie 1842 Hr. A B. Becker starb, ging die Apotheke als Erbschaft an dessen Neffen, Hr E. G. Becker, einen besonders tüchti-

87.12.16 gen Chemiker über; und hier will ich es wieder spät im Leben anerkennen, dass beide sich des verwaisten Knabens mit fast väterlicher Zuneigung annahmen. Ja, mit der Witwe des vor einigen Jahren in Berlin dahingeschiedenen Hr E. G. Becker bin ich noch jetzt in brieflichem Verkehr.

In 1845 bezog ich die Universität zu Kiel, wo unter den noch Lebenden Prof. Himly u Dr. Mein, meine Lehrer waren, denen ich meinen ehrerbietigen Gruss übersenden möchte! durch Sie theils um das pharmac Staats Examen zu bestehen, theils u. besonders um mich den Natur-Wissenschaften zu widmen, da meine Ausflüge in die Pflanzen-Fluren schon damals einen innigen Wunsch erregt hatten, in andern Welttheilen selbständig zu forschen, zu welchem Begehr ich besonders noch durch das Lesen von Humboldt's Reisen in die Äquinoctial-Gegenden der neuen Welt noch weiter angeregt wurde, u für welchen Zweck ich auch mich damals schon etwas der Medicin zuwandte.

Nachdem ich 1847 promovirte als Doctor der Phil. wanderte ich aus, trotz zweier für den Jüngling schon glänzenden Anerbietungen in Schleswig-Holstein, denn bei meiner ererbten Disposition zur Pulmonar-Phthisis fühlte ich, dass ich den nächsten Winter in Nord- oder Mittel-Europa nicht überleben würde; u da auch meine älteste Schwester an jenem Leiden erkrankt war, u diesem später als Opfer verfiel, wanderte ich nach Süd-Australien aus, wohin der Zug der deutschen Emigranten sich damals besonders wandte.

Von Ende 1847 bis Mitte 1852 bereiste ich das Territorium von Südaustralien bis zur Grenze von Victoria, u bis nahe zum Lake Torrens auf eigne Kosten, zu einer Zeit wo es kaum eine Brücke oder Landstrasse gab, u. jede Tour zu Fuss oder auf dem Reit-Pferd durchgeführt werden musste, die Rast-Stätten fast immer unter freiem Himmel u wie noch jeder Reisende in der Gefahr stand, den Feindseligkeiten der Eingebornen zu erliegen. Diese Untersuchungs Reisen erstreckte sich damals schon lineal über 4000 engl. Meilen. Im Jahr 1852 nahm ich einen Ruf vom Gouverneur Latrobe an, die in Melbourne entstandene Stelle eines Regierungs Botanikers zu übernehmen, während für die des Regierungs Geologen Mr Selwyn, (jetzt Director des Museums in Canada) berufen wurde. Ich bereiste phytologisch die meisten Theile des Landes (Victoria) von 1852 – Mitte 1855, einschliesslich dreijähriger Sommer-Reisen in den damals *pfadlosen* austral. Alpen, bestimmte gleichzeitig die Position vieler der Alpenhöhen u unternahm auch hypsometrische Beobachtungen. Auf Urlaub von Mitte 1855 – Mitte 1857 begleitete ich als Naturforscher Mr A. C. Gregory in der Expedition, die vom Herzog von Newcastle als damaligen Staats Minister der Colonien für die Erforschung des tropischen Australiens ausgesandt wurde, war einer der vier Entdecker der grossen Weide-Region, jetzt Kimberley-District genannt, u war auch einer derer, die durch die ganze Zeit im Felde waren, 17 Monate lang von allem Verkehr mit der civilisirten Welt absolut ausgeschlossen, und wiederholt in Collision mit den Eingebornen, durch Monate mit meinen Gefährten von örtlich getrocknetem Pferde-Fleisch unserer armen Packtiere lebend. 87.12.16

Der enorme Aufschwung der austral Colonien während der glänzendsten Gold-Zeit machte die schnelle Entwicklung der ruralen Interessen auch zur Notwendigkeit. So übernahm ich 1857 unter dem Gouverneur Sir H Barkly das Directorat des kürzlich begonnenen botanischen Gartens von Melbourne, u knüpfte Horticultur-Verbindungen mit vielen Plätzen in der civilisirten Welt an. Ich baute die ersten Gewächshäuser dann gleich in Australien, zog die Victoria regia u schuf die damals grösste Sammlung lebender Garten Pflanzen in diesem Theil der Welt, u begründete 1857 das bot Museum, es aber auf getrocknete Pflanzen beschränkend, jetzt ziemlich 1/2 Million Bögen umfassend. Da für die Erfordernisse der schon damals grossen Stadt ein technol. Museum auch für Pflanzen-Producte neben der öffentl. Bibliothek errichtet wurde, da der bot Garten in ziemlicher Entfernung von Melbourne liegt. In 1873 zog ich mich von der Garten-Directur zurück, da ich der Anforderung, die Mittel einer solchen Anstalt in einer jungen Colonie hauptsächlich für Rasenculturen, Blumenbeete und anderer unproductiven Schau nicht opfern wollte. Durch einige Jahre war ich auch Director des zoolog. Gartens, bis solcher nach der entgegen gesetzten Seite der Stadt verlegt wurde, weil es bei mir an Raum gebrach für Weide der Nutztiere, wie Dromedare, Angoraziegen, Llamas, Alpacas, verschiedene Hirsche u Rehe &c.

87.12.16 Der Einführung der Sperlinge u Kaninchen aber widersetzte ich mich; behielt aber im Garten das Aviarium, und im Garten-See, wo ich die Inselchen u die Geiser-Fontaine (80' hoch) schuf, manche der Wasser-Vögel

[...][1]

Zwar werden Sie wohl nur in ganz kurzen Zügen ein Lebensbild von mir entwerfen wollen; da ich aber nicht weiss, nach welcher Richtung hin Sie meine Lebens Ereignisse schildern möchten, habe ich den 11ten Band der fragm. gesandt, aus dessen Titel-Blatt Sie meinen Status bis zum Jahre 1881 ersehen können. Seitdem ist das Werk nicht fosgesetzt,[2] da andere Arbeiten dringender wurden während der mir karg zugemessenen täglichen Studien-Stunden in einer jungen Colonie, wo *angewandte* Wissenschaft Alles gilt! Vielleicht nützt es Ihnen aber doch zu wissen, dass zu meinen Würden in letzteren Jahren die eines Grosskreuz des Christus-Orden von Portugal hinzugekommen ist Bei Gelegenheit des 50 jähr. Jubiläums der Königin von England.[3] u die eines Commenthurs des Philip's Ordens von Hessen auch bei derselben grossen Feier, u die eines Ritters des Oldenburg Verdienstordens. Dann bin ich Präsident der geograph Gesellschaft von Vict seit den 3 Jahren der Existenz dieses Vereins, der einen Zweig der Royal geogr. Soc of Austr bildet; auch habe ich die Ehre Patron des Vict. Field-Naturalists Club zu sein u. Vice Präs. der grossen Melbourne Liedertafel; Giles McIntyre u Andere waren geographisch zuerst meine Emissäre. Seit 1881 bin ich auch noch als Ehrenmitglied in manche wissenschaftliche Vereine gewählt; in London (Ehrenm. Roy. Hort Soc), Brescia, Gent (Ehrenm. der agric u hortic Gesellschaft), Boston, Edinburgh (geolog Ges.) Danzig, London (geol. Gesellsch. u auch Society of arts), Brüssel, Bordeaux, Brisbane, Launceston, Dresden (Isis), Gotenburg, New York (chemische Gesellschaft) Melbourne (ordentl Mitglied der Brit medical Association, da meine Studien durch viele Jahre hier auch eine medicinische Wendung nahmen.

1 pp. 13-16 missing from MS.

2 fortgesetzt?

3 Bei ... England. *is a marginal insertion by M.*

In einem Clima der Orangen-Zone halten selbstverständlich die 87.12.16
Pflanzen der Mittel-Europäischen Kalthäuser im Freien aus, so dass es für viele der Palmen, Bambusaceen u anderer scenischer Pflanzen hier keines Schutzes bedarf, u so ist es entsprechend mit den Nutzthieren.

Um aber nicht prolyx zu sein, muss ich mich kürzer fassen, zumal da ich als einer der Commissäre auch mit Extraarbeiten neben den immer grossen Departements Arbeiten gerade jetzt sehr bedrängt bin; aber dies führt zu der Bemerkung, dass ich mit dem späteren Kanzler von England Childers und den jetzigen Generälen Sir Andr. Clarke u Sir Charles Pasley einer der 12 hiesigen Commissäre schon für die Pariser Welt Ausstellung von 1855 war. (Zur Zeit der ersten, durch den *grossen Prinzen* hervorgerufenen war ich weit im Innere abwesend).

Ich habe zu Fuss oder zu Pferde etwa 28000 engl. Meilen *Beobachtungs Linien*, die auch für die Geographie und andere Zweige des Wissens bedeutungsvoll waren, zurückgelegt, habe 40 Bände unabhängiger Original Werke geschrieben, abgesehen von meinem Theil an der Bearbeitung in Benthams Flora Austr., für welche ich in 70 grossen Kisten etwa 70000 getrocknete Pflanzen-Exemplar *fast ganz durchgearbeitet* lieh, etwa 2/3 des ganzen Materials bildend!

Ich war der Erste, welcher ein grosses Werk botanisch schrieb in diesem Welttheil, war der erste Australier, von dessen Werken eins in die deutsche Sprache durch Hr Dr Goeze und in die französ. durch M. le Dr Naudin de l'institut überging; – der erste Australier, von dem ein Band wissenschaftlichen u ruralen Inhalts in 7 Ausgaben in engl Sprache erschien (Select plants for industrial culture and naturalisation), selbst eine Edition für die Vereins Staaten v. N. Am. u eine für britisch Indien; war der Erste der extensiv Lithogrammata naturhistorisch in Australien lieferte, etwa 600 Platten von Pflanzen, fast stets mit Analysen *von der Knospe* bis zum *Embryo*. Ich war einer der *drei* Ersten in Australien, welchen von der Königin von England so gnädig der Orden von St. Michael u St. George verliehen wurde, wie dieser aus dem Mittelalter herrührende Orden, der auf die Länder des Mittelmeers beschränkt war, auch auf die britischen Colonien ausgedehnt wurde.

87.12.16 Ich bitte Sie aber, geehrter Herr, irgend etwas diese Notizen in *dritter nicht erster* Person zu geben, schon deshalb, weil mein Gefühl stets *gegen* Biographien *Lebender* gewesen ist, u ich *hier* wiederholt rund das Schreiben von biograph. Notizen auf mich selbst bezüglich ablehnte, *Ihnen* aber doch bereit entgegen kommen wollte. –

Und nun zu Schleswig Holstein! Es ist nicht bloss der Süd westlichste Theil, den ich kenne; ich ging auch nach Grafenstein, durchwanderte auch die schöne Landschaft Angeln, der das mächtige England als Anglia seinen Namen verdankt, kam auf meinen Zügen bis Heiligenhaven,[4] sammelte bei Altona, Wandsbeck, besuchte Föhr u Sylt, u fand auf letzterer Insel manche merkwürdige Pflanze, darunter die Gentiana Pneumonanthoides in einer Strandform nur etwa 1 Zoll hoch, welche ich 1846 an Fürnrohr, Sonder u Andere Correspondenten sandte, mit letzterem in innigem Freundschaftsverkehr stehend bis zum Ende seiner Tage, so dass sein *letzter* Brief, wie auch der letzte von Lindley an mich gerichtet war.

Meine Correspondenz nach den Departements Eintragungen hier erforderten während des letzten Drittel des Jahrhunderts etwa 100,000 Briefe aus meiner eignen Hand!

Was *Sie* besonders wohl zu wissen wünschen ist der Umstand, dass auch Dr Ecklon's Pflanzen vom Herzogthum Schleswig nach dem Tode des auch mir befreundeten Dr Steetz von meiner Anstalt erworben wurden, so dass ich Ihnen über zweifelhafte Notizen seinerseits authentisch Auskunft geben kann. Ich selbst fand Epipogium Gmelini in einem Buchen-Walde zwischen Kiel u. Heiligenhafen wieder auf, – sammelte Trifolium spadiceum reichlich auf feuchten Wiesen bei Ihrer Stadt, auch Hieracium virescens in einem Gehölz dort, u *gab* davon ein Ex. an Sonder, der es nicht selbst fand.

Mein plötzliches durch schwere Krankheit veranlasstes Auswandern, verhinderte mich diejenigen seltnen Pflanzen Schleswig-Holsteins zu veröffentlichen, welche nicht in das Gebiet des Breviariums gehören, u nachdem ich in die Forschungen eines *winterlosen*

4 Heiligenhafen?

Continents, fast von der Grösse Europas, gezogen wurde, blieb mir nicht Zeit, auf das in der Jugendzeit-Gesammelte zurück zu kommen. Ja, meine Reise-Journale seit 1847 liegen noch unveröffentlicht. 87.12.16

Auch blieb einiges selbst von Husum als unerledigt zurück, da ja auch andere Wissenschaften mich in Anspruch nahmen; – so glaube ich *zuerst* als eine selbstständige Art ein Potamogeton erkannt zu haben, welches rigider ist als der normale Zustand von P. pectinatus u in brachischen Gräben ganz nahe bei den Häusern am Haven[5] Husum vorkommt; es ist besonders dadurch auch merkwürdig, dass es scheu ist, Früchte zu bilden, nach welchen ich vergebens spät im Herbst suchte, selbst bis die Gräben vereisten. Wenn ich mich nicht täusche ist seitdem dieselbe Art aus Schweden beschrieben. Rücksicht zu Prof Nolte, mit dem ich seit 1841 correspondirte, hielt mich ab diese Art unabhängig von dem *damalig* besten Potamogeton-Kenner der Welt vor die Öffentlichkeit zu bringen.

Und schliesslich, wir armen mangelhaften Menschen, sollen auch immer der göttlichen Macht u Huld gedenken, u. so füge ich als am Wichtigsten vor Allem hinzu, dass ich Patron des Vereins christlicher junger Männer in einer der grössten schottisch. Kirchen Melbournes bin, u dort gelegentlich freie Vorträge gebe, u nie eine Gelgenheit versäume zum Anhalt an die Kirchen zu ermahnen, den erhebenden Gebeten u Predigten zu lauschen u in die Hymnen mit einzustimmen! Wie viele Thränen würden nicht fliessen, wieviele Qualen der Reue vermieden, wenn *Jeder* sich ans Kirchenleben eng anschlösse.

Sie ehrend der Ihre
Ferd. von Mueller

Das Breviarium war bereits 1845 vollendet, aber durch meine Auswanderung verzögerte sich bis 1853 die Veröffentlichung.

Mein Einfluss auf die Geographie Australiens (u jetzt auch auf Neu Guinea) ist so bedeutungsvoll erachtet, dass im Geogr. Con-

5 Hafen.

87.12.16 gress zu Venedig eine der fünf Dank-Voten für Promotoren der Geographie mir zufiel.

Etwa ein Dutzend Werke in mehreren Wissensch. sind mir dedicirt.

Für die technol. Industrie u für die Pflanzen-Chemie habe ich hier in einem eigenen Laborator auch manches Neue geschafft.

Manches aus der Fauna Australiens ist zuerst durch mich bekannt geworden.

Eine grosse Höhe in Spitzbergen, ein grosser Cataract in Brasilien, ein Berg in Neu Guinea, ein Hauptfluss in Australien, ein Gletscher in Neu Seeland führen meinen Namen.

Die Eucalyptus Culturen in allen anderen Theilen der Erde wurden meist von mir hervorgerufen u die franz Regierung [...][6] mich darüber schon vor fast 30 Jahren.

In meinen Key to the syst of Vict pl habe ich die Thymel. zu den Rosaceen u vor Planchon die Vinif. zu den Araliaceen gestellt.

Baronisirt wurde ich durch die grosse Huld Sr Maj. des Königs von Württemberg bei der Feier S. M. silbern Hochzeit mit der Schwester der verst Czarin.

Sie u Hr. Dr Prahl werden meine Vereinfachung des Juss-D.C. System aus dem Census kennen lernen

Es freuet mich ausserordentlich, dass Sie u Hr Dr Prahl die Flora Schl Holst. *erschöpfend* bearbeiten wollen.

16/12/87

Your letter,[7] dear Doctor, awakens memories especially of those days, when life with its hopes lay still largely before me, when I began to explore with youthful enthusiasm the wonders of the plant world as far as they presented themselves to me in open nature and in garden culture. Those years belong to the most wonderful times of my life! I am happy to respond to your wish to furnish you with a few notes about my life for the splendid purpose to which you and Dr Prahl are

6 *illegible word.*

7 Letter not found.

devoting yourselves,[8] and to which my first labours in the field of science may perhaps contribute a little as well. If I write perhaps too voluminously about it, you are free to select whatever seems suitable for your purpose. 87.12.16

Orphaned, I entered the pharmaceutical business of Mr A. B. Becker in Husum as a pupil at Easter 1840, at not quite 15 years of age. I had received my first education at the classical grammar school in Rostock, my home town, and after the death of my father, who was First Customs Controller there and in 1813 had fought as a volunteer near Seestedt,[9] I moved with my mother and three sisters to my grandparents' in Tönning, where I attended the secondary school from 1836-1840.

At the beginning of the century Professor Forchhammer, the Scandinavian geologist, had also received his first education in natural history in Becker's house. I had the good fortune of meeting this noble man during the meeting of natural scientists held in Kiel in 1846. On this for me most stimulating occasion I also made the personal acquaintance of Oerstedt, Dalton, Schleiden, Roeper, Steenstrup, Kunze, Waitz, Rammelsberg, Hoppe, and many other prominent men of science.

When Mr A. B. Becker died in 1842 the pharmacy went as an inheritance to his nephew Mr E. G. Becker, a particularly able chemist; and here I want to acknowledge late in life that both cared for the orphaned boy with almost fatherly affection. Yes, I still correspond today with the widow of Mr E. G. Becker, who died a few years ago in Berlin.

In 1845 I went up to the University of Kiel – where from among my teachers Prof. Himly and Dr Mein are still living, and I would like to pass on to them my respectful regards through you – partly to pass the pharmaceutical state examination, partly and particularly to devote myself to the natural sciences, as my excursions into the open fields of plants had already then aroused the fervent wish to explore independently in other parts of the world. This desire was further stimulated by my reading of Humboldt's travels[10] in the equinoctial areas of the New World and for this purpose I applied myself a little to medicine already even then.

After I had graduated with a Doctorate in Philosophy in 1847 I emigrated, even though I had received two splendid offers for a young man in Schleswig-

8 Prahl (1888-90). The second part of this work includes a history of floristic research in Schleswig-Holstein, with biographical and bibliographical notes on those who have worked on the subject. M's entry is on p. 39.

9 i.e. at the battle of Sehestedt, in Schleswig-Holstein, during the Napoleonic wars.

10 It is unlikely that M would have had access to a full 24-volume set of Humboldt's account of his travels in South America; see Humboldt (1805-34). Most likely he read one of the many abbreviated editions published in German.

87.12.16 Holstein. With my inherited disposition to pulmonary phthisis I felt that I would not survive the next winter in northern or central Europe. As my eldest sister[11] had also become ill with this disease and later fell victim to it, I emigrated to South Australia, where the stream of German emigrants then mainly flowed.

From the end of 1847 to mid-1852 I travelled at my own expense in the territory of South Australia, as far as the border with Victoria and to near Lake Torrens, at a time when there was hardly a bridge or a road, and each journey had to be made on foot or on horseback, camping places nearly always under the open sky, and when every traveller was in danger of falling victim to the hostile natives. These exploration journeys already covered a distance of more than 4000 linear miles. In 1852 I accepted the appointment from Governor Latrobe to fill the position of Government Botanist, then created in Melbourne, while Mr Selwyn (now Director of the Museum in Canada[12]) was appointed Government Geologist. I explored phytologically most parts of the country (Victoria) from 1852 to mid-1855, including summer journeys for three years to the then *pathless* Australian Alps, at the same time determined the position of many alpine summits, while also conducting hypsometric observations. On leave from mid-1855 to mid-1857 I accompanied as naturalist Mr A. C. Gregory's expedition,[13] sent out by the Duke of Newcastle as the then State Minister for the Colonies to explore tropical Australia, was one of the four discoverers of the vast pasture regions now called the Kimberley District, and was also one of those who remained in the field through the whole time, completely cut off from all contact with the civilised world for 17 months, repeatedly in collision with the natives, living with my companions for months on the locally dried meat of our poor packhorses.

The enormous boom of the Australian colonies during the peak of the gold-rush made the rapid development of rural interests a necessity. Thus I accepted in 1857 under Governor Sir H. Barkly the directorship of the recently-created Botanic Gardens of Melbourne, and established horticultural contacts with many places in the civilised world. I soon built the first glasshouses in Australia, raised the *Victoria regia*, and created the then largest collection of living garden plants in this part of the world. In 1857 I founded the botanical Museum, restricting it to dried plants, which now comprise about half a million sheets, the reason [for the restriction] being that for the needs of the already large city a technological museum also for vegetable products was built adjoining the public library, as the Botanic Garden is a fair distance from Melbourne.

11 Bertha Müller.
12 Selwyn was Director of the Geological Survey of Canada.
13 North Australian Exploring Expedition, 1855-6.

In 1873 I resigned from the Directorship of the Garden, as I did not want to sacrifice the means of such an institute in a young colony largely for the cultivation of lawns and flowerbeds and other such unproductive show.[14] For some years I was also Director of the Zoological Garden, until this was moved to the opposite side of the town, because I was lacking the space for pastures for the useful animals such as dromedaries, angora goats, llamas, alpacas, various stags and deer, &c. But I was strongly opposed to the introduction of sparrows and rabbits. I did, however, retain an aviary in the Garden, and in the Gardens Lake, where I created the islets and the geyser fountain 180 feet high, many of the water birds 87.12.16

[...][15]

You will probably want only a very short biographical sketch from me, drawn in very broad strokes; but as I do not know from which point of view you might want to recount the events of my life, I have sent the 11th volume of my *Fragmenta*. From its title page you can see my status up to the year 1881. Since then the work has not been continued,[16] as other work became more urgent during the meagre study hours given to me in a young colony, where *applied* science is everything! But perhaps it will help you to know, that the Grand Cross of the Christus Order of Portugal has been added in recent years on the occasion of the golden anniversary of the Queen of England[17] and that of a Commander of the Philipp's Order of Hesse[18] also during the same great celebration, and that of a Knight of the Oldenburg Order of Merit.[19] Furthermore, I have been President of the Geographical Society of Victoria for the 3 years since its inception which forms a branch of the Royal Geographical Society of Australasia. I also have the honour to be patron of the Field Naturalists Club of Victoria and Vice-President of the large German Liedertafel. Giles, McIntyre and others were first my geographical emissaries. Since 1881 I have been elected to many scientific societies as an honorary member: in London (hon. member of the Royal Horticultural Society), in Brescia, Ghent (hon. member of the Agricultural and Horticultural Society), Boston, Edinburgh

14 M is stretching the truth here concerning the circumstances of his leaving the Botanic Garden; see C. Hodgkinson to M, 31 May 1873, and subsequent correspondence [*Selected correspondence*, vol. 2, and Collected Correspondence]. See also Cohn & Maroske (1996).

15 Four pages of text missing in the MS.

16 M did publish a first fascicle of vol. 12, pp. 1-26, in December 1882 (B 82.12.03), but the volume was not completed and remained without a title page.

17 Queen Victoria succeeded to the British throne on 20 June 1837 and was crowned on 28 June. M was appointed to the Portuguese Order of Christ on 28 April 1887.

18 M was appointed to the Order of Philipp the Magnanimous, Hesse, on 21 June 1887.

19 M was appointed to the Oldenburg House and Service Order on 17 January 1887.

87.12.16 (Geological Society) Danzig, London (Geological Society and also Society of Arts), Brussels, Bordeaux, Brisbane, Launceston, Dresden (Isis), Gothenburg, New York (Chemical Society), Melbourne (ordinary member of the British Medical Association, as my studies here over many years also took a medical direction).[20]

In a climate comparable to the orange-zone the plants of the central European cold-houses grow, of course, in the open, so that many of the palms, Bambusaceae and other scenic plants do not require any protection, and it is the same with the useful animals.

But in order not to be too prolix, I must cut it shorter, especially seeing that, as one of the Commissioners,[21] I am pressed with extra-work just now, apart from the always large workload of the Department. That leads me to remark that I was one of the 12 local commisioners for the Paris International Exhibition of 1855, together with the later Chancellor of England Childers and the present Generals Sir Andrew Clarke and Sir Charles Pasley. (At the time of the first Exhibition, constituted by the *great Prince,*[22] I was absent far away in the interior.)

I have covered on foot or on horseback about 28,000 English miles, making observations, which were also of value to geography and other branches of science. I have written 40 volumes of independent original works, apart from my share in the work of Bentham's *Flora australiensis*,[23] for which I lent about 70,000 dried plant specimens in 70 large cases, *almost completely worked out*, which comprised about 2/3 of the whole material!

I was the first to write a large botanical work in this continent, – was the first Australian who had one of his works published in the German language through Dr Goeze and in French by M. Dr Naudin of the Institut;[24] – the first Australian who had a work with scientific and rural contents published in 7 editions in the English language (*Select extra-tropical plants, readily eligible for industrial culture and naturalisation),* with an edition even for the United States of America and for British India.[25] I was the first to produce extensive natural history lithographs in Australia, about 600 plates of plants, nearly always with analytical drawings *from the bud* to the *embryo*. I was one of the first *three* in Australia, who was so graciously appointed to the Order of St Michael and St George from the Queen of England,[26] when this medieval order, previously restricted to Mediterranean countries, was also extended to the British colonies.

20 See Appendix B for a list of M's memberships and appointments.
21 For the Centennial International Exhibition, Melbourne, 1888.
22 The 1851 Exhibition in London, promoted by Prince Albert.
23 Bentham (1863-78).
24 i.e. the Institut de France. See B83.13.06 and B87.14.06.
25 B84.13.22 and B80.13.07.
26 See Royal Warrant, 24 April 1869 [*Selected correspondence*, vol. 2, p. 497].

But I ask you, dear Sir, to publish anything from these notes in the *third, not in the first* person, if only because my feelings have always been *opposed* to biographies of the *living*, and because I have refused repeatedly the writing of biographical sketches of me out of hand *here;* but I did want to oblige *you.* 87.12.16

And now to Schleswig-Holstein! I know not only the south-western part; I also went to Grafenstein, wandered through the lovely countryside of Angeln, from which mighty England derives its name, went as far as Heiligenhafen during my excursions, collected near Altona, Wandsbeck, visited Föhr and Sylt, and found many a strange plant on the latter island, among them *Gentiana pneumonanthoides* in a coastal form only about 1 inch high, which I sent to Fürnrohr, Sonder and other correspondents in 1846. With the latter I remained in contact as a close friend to the end of his days, so that his *last* letter was written to me, as was Lindley's.

My correspondence according to the Department's registers here required about 100,000 letters written in my own hand during the last third of the century!

What *you* are probably particularly interested to hear is the fact that Dr Ecklon's plants from the Duchy of Schleswig were purchased by this institute after the death of my friend Dr Steetz, so that I am in a position to give you authentic information on any doubtful notes by him. I myself rediscovered *Epipogium gmelinii* in a beech forest between Kiel and Heiligenhafen, collected *Trifolium spadiceum* plentifully on damp meadows near your town, also *Hieracium virescens* in a wood there, and *gave* a specimen of it to Sonder, who did not find it himself.

My sudden emigration due to a severe illness prevented me from publishing those rare plants, which did not fall within the area covered by the 'Breviary',[27] and once I was drawn into the exploration of a *winterless* continent almost the size of Europe I lacked the time to return to what I had collected during my youth. Yes, my travel journals since 1847 are still lying here, unpublished.[28]

Even a few things from Husum remained unfinished, as other branches of science occupied me as well. Thus I believe I was the *first* to have recognised a separate species of *Potamogeton*, which is more rigid than the normal form of *P. pectinatus* and occurs in brackish ditches quite close to houses near the harbour of Husum; it is remarkable particularly in that it is reluctant to form fruit, for which I searched in vain late into the autumn, even till the ditches were iced over. If I am not mistaken this species has since been described from

27 B53.08.01, B53.08.02.

28 M's journals disappeared, along with most of his files of correspondence, early in the twentieth century.

87.12.16 Sweden. Consideration for Nolte, with whom I corresponded from 1841, prevented me from bringing this species before the public independently of the *then* best *Potamogeton* specialist in the world.[29]

And finally, we poor imperfect human beings should always remember God's power and kindness, and so I add as most important before anything else, that I am patron of the Christian Association of young men in one of the largest Scottish churches in Melbourne,[30] where I give occasional free lectures and never miss an opportunity to urge involvement with the Church, to listen to the up-lifting prayers and sermons and to join in the hymn singing. How many tears would remain unshed, how many tortures of remorse could be avoided, if *everybody* took an intimate part in the life of the Church.

Respectfully your
Ferd. von Mueller.

The 'Breviary' was already completed in 1845, but due to my emigration publication was delayed until 1853.

My influence on the geography of Australia (and now also on New Guinea) was deemed so significant, that during the Geographical Congress in Venice I received one of the five votes of recognition for the promoters of geography.[31]

About a dozen works in several fields of science have been dedicated to me.

I have produced much that is new for technological industry and vegetable chemistry here in my own laboratory.

Much of the fauna of Australia first became known through me.

A high peak in Spitzbergen, a large waterfall in Brazil, a mountain in New Guinea, a major river in Australia and a glacier in New Zealand bear my name.

The *Eucalyptus* culture in all other parts of the globe was almost entirely instituted by me, and the French Government [honoured] me for it almost 30 years ago.[32]

29 M's judgement of Nolte is greatly coloured by his obvious respect for the man, but was not shared by later critics of Nolte's importance or achievements (see R. von Fischer-Benzon, 'Geschichte der floristischen Erforschung des Gebietes', in Prahl (1888-90), part II, pp. 40-4). Elsewhere, M, too, was more critical; see M to J. Hooker, 28 July 1866 [*Selected correspondence*, vol. 2, pp. 368-70].

30 i.e. at the West Melbourne Presbyterian Church; see M to R. Tate, 20 August 1890 [Collected Correspondence].

31 See *Report upon the third international geographical congress and exhibition at Venice, Italy, 1881 ....* (Washington, 1885). p. 22.

32 M was appointed to the French Légion d'Honneur in 1863.

In my *Key to the system of Victorian plants*[33] I have placed the Thymeleae before the Rosaceae and before Planchon the Viniferae to the Araliaceae. 87.12.16

I was made a Baron through the great kindness of His Majesty the King of Württemberg during the celebrations of His Majesty's silver wedding anniversary with the sister of the late Czarina.[34]

You and Dr Prahl will get to know my simplification of the Jussieu-De Candolle system through my Census.[35]

I am extraordinarily pleased that you and Dr Prahl will work exhaustively on the Flora of Schleswig-Holstein.

To Asa Gray 88.01.26

Gray Herbarium Archives, Harvard University, Cambridge, Massachusetts.

26/1/88.

When the centennial anniversary of the independence of the U.S. was celebrated, my honored and venerable friend, you wrote me a letter, which I treasure among my best epistolar possessions.[1] Now it is my turn, to write you on the most memorable day of Australia,[2] when the second century of its settlement is commencing! You wrote, that you would devote such a day of honor to correspondence with science-friends; and this sentiment I share so completely, that I also shall devote this day to communications with the learned, who honored me with their friendship. I telegraphed my felication to Sydney, whence I was invited;[3] so I have done thus far my homage also on this grand historic occasion! –

33 B86.13.01.

34 See Karl I, King of Württemberg, to M, 6 July 1871 [*Selected correspondence*, vol. 2, pp. 580-2].

35 B82.13.16, with several later supplements.

1 Letter not found.

2 The centenary of the establishment of the first British settlement on 26 January 1788.

3 Telegram not found. The principal celebrations of the centenary took place in Sydney, where the first settlement was established.

88.01.26 Hardly any special question is before me, concerning which I should adress you, though thousands of objects interest us mutually, but could not be discussed by letter-writing.

I am grateful, that divine will has spared me, to live into the second century of Australian civilisation, and to witness what the energy and enterprise of the British nation has accomplished so gloriously within such a space of time also in her Majesty's Australian Dominions!

To myself only a very brief period of worldly existence can be left; but it is with some pride, that I look back now to the results of more than 40 years uninterrupted toil in endeavouring to advance the interests of these great southern lands, *humbly though*, in applied geography, rural development and abstract science.

To the fullest extent do I concur with your views of the indesirability of superseding the first specific name in a correctly chosen genus; it was one of the reasons for the chronology of Austral plants in the "Census",[4] to subdue that practice, to which Bentham even adhered yet in the earlier volumes of the Austral. Flora[5] – he then even yet writing to me, that if he made not such changes, to which I was opposed, *others would* – to drag forward a name for a species wrongly placed before as regards its genus. On this and many other points, concerning the systematic key of Vict. plants,[6] I am just writing an essay;[7] and I crave of you, to withhold any review of that work, until this essay will be before you.[8]

Ever regardfully yours
Ferd. von Mueller.

4 B82.13.16 and subsequent supplements.
5 Bentham (1863-78).
6 B88.13.03.
7 B89.13.05.
8 Gray died four days after this letter was written. M's *Key* (B88.13.03) was reviewed by G. L. Goodale, *American journal of science*, no 221, May 1889, pp. 416-7.

To Adolf Engler 88.02.22

Staatsbibliothek zu Berlin, Preussischer Kulturbesitz, Berlin.

22/2/88.

Bereits durch eine der letzten Posten, edler Freund, theilte ich Ihnen die Schwierigkeiten mit, *bald* mich bei der Mitarbeit an den natürlichen "Pflanzen-Familien" zu betheiligen; die Hindernisse, welche entstanden sind die Extra-Arbeiten für die Welt-Ausstellung hier, die ausserordentlich Zeit nehmende Bearbeitung der Pflanzen Victorias nach der dichotomischen Methode zum *ersten mal ohne Aufopferung* der *naturlichen Verwandtschaft*! und das lange Zurückstehen dadurch der Materialen, welche ich so mühsam erwarb für die Flora von Neu-Guinea.

Das Departement hier erfordert ohnehin alljährlich das Schreiben von zwischen 3000 u 4000 Briefen von meiner eignen Hand, der vielseitigen anderen Amts Pflichten in einer jungen practischen Colonie gar nicht zu gedenken. Ich müsste Sie u den Herrn Verleger also lange warten lassen, bis ich Ihnen gerecht werden könnte, u. bitte Sie daher, dem Letzteren die jetzige Sachlage mizutheilen.[1]

Mit meiner gewöhnlichen Offenheit möchte ich mir dann auch noch erlauben, Ihnen noch zwei andere Gründe mitzutheilen, welche mich bestimmen möchten, von meiner Identificirung mit dem Werke abzustehen. Die eine dieser Ursachen ist eine wissenschaftliche, die darin besteht, dass auf eine ganz *ungewöhnliche* Weise die Illustrationen aus anderen Werken entlehnt sind, ein Verfahren das den natürlichen Pflanzen-Familien die Originalität grossentheils entzieht, selbst wenn die Erlaubniss für diese Entlehnung vom Verleger erlangt sei, wie ich annehme. Nach meiner Meinung wäre es viel besser, den Preis etwas zu erhöhen und das Erscheinen lieber etwas zu verlangsamen, als Angesichts der deutschen Kunst und Original-Kraft, das Werk so ausgedehnt auf die artistischen Resultate einer früheren Zeit zu stützen.

1 mitzutheilen?

88.02.22 Den zweiten Anlass zu meinem jetzigen Bedenken, an diesem schönen deutschen Zusammen-Wirken in *einer* der Wissenschaften, welche ich vertrete, Theil zu nehmen, ist freilich nur ein fast persönlicher, obgleich es mir als dem Schöpfer und Administrator einer bedeutenden Anstalt auch der Regierung von Victoria gegenüber obliegt, die Interessen meines Departements zu wahren. Ich fühle nämlich, dass meine eignen litterarischen Arbeiten (von etwa 40 Bänden) in den "Pfl. Fam." ganz unbeachtet geblieben sind, was unter den besonderen Schwierigkeiten und unnennbaren Aufopferungen meinerseits (während der 41 Jahre meines Lebens durch die erste Generation der Colonie Victoria) mich sehr entmuthigen muss, selbst eingreifend das Opus, von welchem ich rede, zu fördern, obgleich es mir stets Freude machen wird, den wirklichen Autoren dafür Materialien von hier aus dafür zu liefern. Weder meine "select plants" (trotzt einer deutschen Ausgabe durch Hr D Goeze), noch der "Census" sind in Gebrauch gezogen für die Pfl. Fam.; weder der "Fragm. phytogr. Austr" noch meiner paläontolog. Arbeiten (von welcher letzteren *Sie* eine so gütige Beurtheilung gaben), sind in den Pfl. Fam gedacht; und ich sehe fast voraus, dass auch meine Eucalpytographia (der *umfangsreichsten* Monographie, welche überhaupt je geschrieben), wohl gar keiner Notiz gewürdigt werden wird, weil sich das Werk auf ein genus nur beschränkt, u meiner Iconography der Acacien wird es vielleicht nicht besser ergehen, obwohl ich in kaum über *ein* Jahr 130 Original Platten mit ebenso viel Arten (mit Analysen von der *Knospe bis zum Embryo*[)][2] geliefert habe, *keine* derselben vorher abgebildet. Vielleicht würde bei der Nennung der Litteratur der Atlas der Myoporinen Beachtung finden; aber wenn ich bedenke, dass in der Kritik des Werkes in Deutschland Niemand analisirt hat, dass es auf *40* jährige Original Beobachtungen beruht, so mache ich mir auch keine besondere Vorstellung, dass es zur gebührenden Ehre kommen wird bei dem Autor, der nun diese schöne Ordnung für Sie u Hr Prof Prantl dort übernehmen mag. Dass *das* Buch u. so die meisten anderen meiner Werke den deutschen Phytologen nicht zugänglich seien, ist unmöglich; denn nicht

2 *editorial addition.*

nur gebe ich jährlich etwa £50 allein aus meinen *Privat*-Mitteln zum Verschenken meiner Bände her, sondern auch wird es mir möglich, etwa 100 Ex. durch die Liberalität der hiesigen Regierung, im Austausch über die Erde zu vertheilen. 88.02.22

Verehrungsvoll der Ihre
Ferd. von Mueller

Der "Census" giebt leicht u. schnell Nachweis über die Ansichten, welche ich in den elf Bänden der Fragm. entwickelte.

22/2/88

Already through one of the last posts, noble friend, I informed you of my difficulties of *soon* participating in the collaboration on the *Natürlichen Pflanzen-Familien*.[3] The obstacles which arose are the extra work for the International Exhibition here,[4] the extraordinarily time-consuming preparation of the *Plants of Victoria* according to the dichotomous method for *the first time without sacrificing the natural relationships!*[5] and for this reason the long setting-aside of the material which I so laboriously acquired for the flora of New Guinea.

The department here in any case necessitates annually the writing of between 3000 and 4000 letters from my own hand, not to mention the manifold other official duties in a young practical Colony. Thus I would have to let you and the publisher wait a long time until I could do justice to you and so I ask you to inform the latter of the present state of affairs.

With my usual frankness then, I would also like to take the liberty of informing you of still two other reasons, which would induce me to forgo my identification with the work. One of these causes is a scientific one, which lies in the fact that in a quite *unusual* way the illustrations are borrowed from other works, a procedure which takes away the originality of the *Natürlichen Pflanzen-Familien* to a great extent, even if permission for the borrowing be obtained from the publisher, which I assume. In my opinion, it would be much better to raise the price somewhat and, in view of German art and original power, preferably delay the appearance somewhat than basing the work so extensively on the artistic results of an earlier time.

3 Engler & Prantl (1887-1915). See M to A. Engler, 25 December 1887 [Collected Correspondence].

4 Centennial Exhbition, Melbourne, 1888-9.

5 B88.13.03.

88.02.22 The second cause of my present misgiving in taking part in this fine German co-operation in *one* of the sciences, which I represent, is admittedly only an almost personal one, although it is also my duty to the Government of Victoria as the creator and administrator of a significant institution to protect the interests of my department. You see I feel that my own literary works (of about 40 volumes) have remained quite unnoticed in the *Pflanzen-Familien*, which under the particular difficulties and unspeakable sacrifices on my part (during the 41 years of my life through the first generation of the Colony of Victoria) must dishearten me very much, even taking action to promote the work of which I speak, although I will always be pleased to provide material from here for it to the actual authors. Neither my *Select plants* (in spite of a German edition by Dr Goeze),[6] nor the *Census*[7] is made use of for the *Pflanzen-Familien*; neither the *Fragmenta phytographiae Australiae*, nor my palaeontological works (of which latter *you* gave such a good assessment), is considered in the *Pflanzen-Familien*; and I almost expect that even my *Eucalyptographia*[8] (the *most extensive* monograph actually ever written) will probably not be acknowledged by any mention at all, because the work is only restricted to a genus, and my *Iconography of the Acacia*[9] will perhaps not fare better, although I have produced in scarcely over *one* year 130 original plates with just as many species (with analyses from the *buds to the embryo*), *none* of them previously illustrated. Perhaps the *Atlas of the Myoporinae*[10] would find attention in the literature entry; but when I consider that in the review of the work in Germany no-one has recognised that it is founded on *40* years of original observations, so I get no particular idea that it will come into due favour with the author, who may now take over this beautiful order for you and Professor Prantl there. That *the* book and also most of my other works are not accessible to the German phytologists is impossible, since not only do I annually part with about £50 alone from my *private* means to give away my volumes, but also it is possible for me through the liberality of the government here to distribute about 100 copies in exchange around the world.

Respectfully yours
Ferd. von Mueller

The *Census* will easily and quickly give information about the views that I developed in the eleven volumes of the *Fragmenta*.

6 B83.13.06.
7 B82.13.16 and supplements.
8 B79.13.11, B80.13.14, B82.13.17, B83.13.07, B84.13.19.
9 B87.13.04, B88.13.01.
10 B86.13.21.

To William Lawes

88.03.19

A36 Royal Geographical Society of Australasia (Vic. Branch) papers, Mitchell Library, State Library of New South Wales, Sydney.

copy

Melbourne 19/3/88.

To the Rev. W. G. Lawes, F.R.G.S.,
Representative of the R.G.S.A. for New Guinea &c.

It was the intention of the geographic Council of Victoria, reverend Sir, to pay you some homage during your present stay in our metropolis; but the shortness of your visit to this colony together with the multifarious engagements and duties, devolving on you in your exalted position, have left us not sufficient time, to call the Vict. branch of the R.G.S.A. purposely together, to do you special honor, as we intended. My colleagues of the Council here have desired me therefore, to write on its behalf a valedictory letter to you, expressive of a hope, that Gods divine providence may spare you for very many years with all your energies, talents and high-mindedness for continuing your great labors in that cause, to which you have devoted the best part of your life so successfully and with so much selfsacrifice. I am certain, that I express further the sentiments of my honored geographic Colleagues here, all of whom I cannot see before your departure, when I say, that we trust you will live to see fully realized all your intentions and hopes, that in the great papuan Island Christianity with all its blessings, – first and largely spread by yourself, – will become universal, and that it may thus be destined for you and your renowned and noble coll-aborator, the Rev. J. Chalmers, to celebrate triumphs at all events through British New Guinea similiar to those, which the Rev. S. King[1] achieved in so touching a manner for Samoa already through a still earlier call into the mission-field. We as members of the geographic council here cherish also the hope, that your efforts and

1 Rev. Joseph King? *copyist's misreading*?

88.03.19 those of your honored Colleagues and your numerous disciples for securing substantial benefits to the Papuan Autochthones through civilisation, sped by you, will be crowned likewise with brilliant success, so that in a manner, alike to that adopted recently by the independant tribes of the Moaries[2] in New Zealand, the territory of each of the settlements in New Guinea may be permantely secured to the native inhabitants and may by surveys and legal enactments under British Sovereignty be permanently alloted to each family in just proportions for peaceful homesteads of modern comforts and for prosperous and largely enriched rural estates as heirlooms to their descendents. Thus, we trust, the spiritual and worldly welfare of the natives in New Guinea will be alike advanced collateral to the requirements of an unencroaching colonisation, so that a large and peaceful dominion may be added to the Great British empire, and this in its turn contributing also to the general blessing and ever hoped-for universal happiness and religious unity of the world!

With deep reverence
your Ferd von Mueller,[3]
President of the Vict. Branch of the R.G.S.A.

2 Maori.

3 Copyist's annotation before M's name: 'signed'.

To Paul Maistre[1]

88.04.18

Institute for the Study of French-Australian Relations, Archives, Department of French and Italian Studies, University of Melbourne.[2]

Melbourne,
18 Apr. 1888.

Mons. P. Maistre,
Acting Consul General of France &c.

Honored Sir,

In continuance of former correspondence I beg to inform you, that as yet I have been not very successful in obtaining the seeds of the Grasses and Saltbushes from the interior, desired by his Excellency the Minister of foreign affairs of your great country, – owing partly to the advanced season and partly to the devastations caused by the Rabbits widely over the interior of Australia.

For these two causes it became difficult, to send purposely a collector far inland, so that I had to rely on the aid of friends in the interior, to obtain the desired seeds. Every prospect however exists now, to get supplies from different localities, and it seems best, that such sorts, as do come in, be at once fresh despatched, as the arrival of the various kinds will extend over considerable time.

Accordingly I now have the pleasure of sending seeds, just received, of the valuable Chloris truncata, which grass will prosper in the driest regions, and which grass is particularly mentioned at page 203 of the "Manuel de l'Acclimateur" of Prof. Naudin & myself.[3]

1 MS annotation: 'Eé. No. L12 le 19/4/88. Accusé reception le 23me Transmis avec rapport le 25 Avril' [Registered no. L12, 19/4/88. Acknowledged on the 23rd. Sent with report on 25 April].

2 The text is taken from a photocopy, the original letter having been returned to the French consular archives, Nantes. For a printed version of this letter see Home & Maroske (1997), pp. 18-19.

3 B87.14.06.

88.04.18 Whenever more sorts of Grass-seeds or of other pasture-plants, adapted for North-Africa, shall have arrived, I will always forward them at once to you.[4]

I have the honor to be, Sir, your obedient
Ferd. von Mueller

88.05.24 *To Samuel Davenport*

Collection of the Royal Geographical Society of Australasia (SA Branch) Inc., State Library of South Australia, Adelaide.

24/5/88

I am delighted, dear Sir Samuel, that the exploration of Central Australia beyond Lake Amadeus is proceeding this season under the auspices of the S.A. Branch of the R.G.S.A.,[1] and I feel sure, that important results will be attained not only for geography, but for rural and mining interests also. Only a dozen years of the century are left, and good use will have to be made of each season to get the exploration of our continent completed within that time, as surely we all desire.

While congratulating your branch on the sending out of this Exploring party, into a region, which remains as one of the most promising to be explored, I would like to plead specially also yet for the cause of phytologic researches; and I trust that some special arrangements are made, to secure botanic specimens if even

4 See M to L. Dejardin, 1 February 1892 [Collected Correspondence], where M sends Saltbush seeds for North Africa.

1 See W. Tietkens to M, 24 November 1887. Tietkens led the expedition sponsored by the SA branch of the Royal Geographical Society of Australasia into the region west of Lake Amadeus. The expedition was, however, delayed for want of funds and did not take place until 1889.

fragmentary only. Perhaps not many actually novel forms will be discovered, but from the whole unsearched regions beyond Lake Amadeus nothing will come amiss, because the geographic distribution of the numerous species of W.A. plants will have to be traced further, and your Expedition will probably be in a good position to determine the limits of many tropical plants southward and vice versa, while likely also the extent of the western so very peculiar vegetation can be ascertained Eastward. If the party would only daily trus[ts][2] sprigs of *any* kinds of plants, whether flowering or fruiting, into envelopes indiscriminately, valuable material for botanic records would be obtained and a good insight from such likewise be gained into the capabilities of the soils. I would have written earlier on this subject, but I learnt only last week at our geographic Meeting, that this new Expedition had been formed. Perhaps it is not yet too late, to telegraph to the party, should not special arrangements have been made, to render so splendid an opportunity without additional cost or toil profitable for the branch of science which I professionally represent. 88.05.24

With regardful salutation to Lady Davenport and yourself

Ferd von Mueller

2 thrust?

88.08.01 ## *To Thomas Wilson*

K88/7448, unit 247, VPRS 3992/P inward registered correspondence, VA 475 Chief Secretary's Department, Public Record Office, Victoria.

Melbourne,
1 Aug. 1888.

T. R. Wilson Esqr,
Undersecretary.

Sir.

I have the honor to report, that last evening I met with an accident by falling from a tramcart; but altho' this will prevent me from leaving my dwelling for some days, it will not hinder me in performing my office-duties.[1]

I have the honor to be,
Sir, your obed. servant
Ferd. von Mueller,
Gov. Botanist.

1 Wilson submitted M's letter to the Chief Secretary, A. Deakin, on 3 August and Deakin signed it on 6 August. That day Wilson minuted a draft reply: ' I much regret to hear of the occurrence and trust he will soon be able to resume active duties'. A letter was sent to M on 6 August but it has not been found.

To Walter Gill 88.09.18

RB MSS M33, Library, Royal Botanic Gardens, Melbourne.

18/9/88.

It needs not my assurance, dear Mr Gill, that I will be happy to name any plants for you, even mosses, lichens and fungs; and with this view I would advise you, to send specimens, numbered consecutively, and to keep back a set correspondingly numbered. What you may send need not be large, so long as it is instructive, by bearing flowers or *fruits*. Small parcels come best by successive mails, and can better be attended to at once, than largely accumulated lots. If you include in your collection also the minutest spring-weeds, little annual rushes, saltbushes, floating and submerged plants, novelties and many rarities would be likely secured.

With best compliments
your Ferd. von Mueller.

Of course I can only through your excellent Chief[1] communicate with you, and shall always send the name-lists &c through him.[2] The seeds of Acacia iteaphylla[3] will be quite a boon, as it would flower so early in European Conservatories Envelopes, opening at the narrow end, are very convenient, to trust[4] plants into, even indiscriminately, so that a dozen or more sorts particularly of small kinds can be trusted together into an envelope, which need not be opened again, but can be placed in sun-light or near a fire-place for drying. A mere pocket-handkerchief, knotted to gether at its four corners will take half a hundred different plants while you are[5] any *friends* of yours are travelling, and these specimens could, when any camp is reached, be placed into envelopes, surplus-branches being broken off before

1 John Ednie Brown, Conservator of Forests in SA.
2 MS is folded over but has not been sealed, and on the outside M has written: 'Walter Gill Esqr &c Chief Forester &c *Wirrabarra*'.
3 Not in APNI or IK.
4 thrust?
5 and?

88.11.02 *From Jacob Agardh*[1]

RB MSS M108a, Library, Royal Botanic Gardens Melbourne.

Lund, Schweden 2 Nov. 1888.

Hochgeehrter Herr Baron!

Sie werden Sich erinneren, dass mit Ihre Genehmigung, und ich glaube sogar auf Ihre Anregung, ich schon früher (ich glaube im Jahre 1883 und 1884) Exemplare der von Ihnen mir zugesandten Algen als *Algae Muellerianae* vertheilt habe, und zwar an diejenigen Algologen, denen Solche Samlungen von Interesse seyn mögten. Die seitdem mir zugekommene Algen Remissen – wenn ich von den schönen Samlungen von Bracebridge Wilson ganz absehe – obgleich quantitativ weniger bedeutend, enthalten doch einige früher mir kaum vorgekommene Formen, mit deren Auslegung zu eine neue Distribution von Algae Muellerianae ich letztens beschäftigt war. Für Ihre eigene Samlung habe ich heute ein kleineres Päckchen an die Post abgegeben. Es enthält besonders Algen von besondere Localitäten (Neu Zeeland, Norfolk Isl etc), sonst einige Seltenheiten so wie "Kritische Arten" die vielleicht nicht immer gut bestimmt vorkommen mögten.

Andere Samlungen werden nach Kew, nach Trinity-College in Dublin, wo die Samlungen von Harvey sind, nach Bornet in Paris, nach Grunow in Wien, nach Hauck in Trieste, Askenasy in Heidelberg, Kjellman in Upsala, der die arctischen Algen beschrieben hat, Reichs Museum in Stockholm, welches die bedeutende Samlung von J. E. Areschoug jetzt besitzt – und so weiter – abgegeben. Es war eine bedeutende Arbeit alle diese Exemplare zu vertheilen und etiquettiren.

Sonst bin ich in der letzten Zeit mit eine genauere Bestimmung der Sargassen beschäftigt gewesen; und ich hoffe eine specielle Arbeit über die australischen Sargassen Ihnen einmal Senden zu können.

1 MS annotation by M: 'Beantw 29/12/88' [Replied 29 December 1888]. Letter not found; but in a brief letter to J. Agardh, 31 December 1888, M acknowledged the receipt a few days earlier of a parcel with algae.

Ich habe aber dabei sehr bedauert, was auch schon Harvey sagte "that few of my obliging correspondents in the Colonies are careful to seek out and preserve the brown or Fucoide Algae sufficiently"; und was besonders die Sargassen betrifft, dass gewöhnlich nur Bruchstucke davon gesammelt werden, so dass errathen werden muss welche Obertheile zu seinen Untertheilen gehören. 88.11.02

Ich fürchte dass Bracebr. Wilson entweder die Algen, oder mich persönlich vergessen habe. Wenigstens habe ich von seiner Samlungen im letzten Jahre nichts gesehen, was ich viel bedaure, da in diesen Samlungen neue oder seltene Arten öfters steckten.

Mit besondere Hochachtung
zeichnet ergebenst
J G Agardh

Lund, Sweden 2 November 1888.

Highly esteemed Baron!

You will remember that with your approval and I believe even at your suggestion, I had already (I think in 1883 and 1884) distributed specimens of algae sent to me by you as *Algae Muellerianae*, and in fact to those algologists, to whom such collections may be of interest. The remittances of algae sent to me since then – if I completely disregard the fine collections of Bracebridge Wilson – although quantitatively less significant, do contain some forms that previously were scarcely met with by me, with the interpretation of which for a new distribution of Algae Muellerianae I was recently occupied. Today I have delivered to the post office a small parcel for your own collection. It contains particularly algae from special localities (New Zealand, Norfolk Island etc.), otherwise some rarities as well as "critical species" that perhaps may not always occur properly determined.

Other collections are delivered to Kew, to Trinity College in Dublin, where the collections of Harvey are, to Bornet in Paris, to Grunow in Vienna, to Hauck in Trieste, Askenasy in Heidelberg, Kjellman in Uppsala, who has described the Arctic algae, Royal Museum in Stockholm, which now has the important collection of J. E. Areschoug – and so on. It was a considerable task to distribute and label all these specimens.[2]

2 The labels of the 'Algae Muellerianae' were initially handwritten, but soon printed, with Agardh filling in details by hand though he often omitted to give the collector and the date.

88.11.02 Otherwise recently I have been occupied with a more exact determination of the Sargassae; and I hope to be able to send you a special work on the Australian Sargassae[3] some day. I have, however, regretted very much what Harvey has also already said, "that few of my obliging correspondents in the Colonies are careful to seek out and preserve the brown or Fucoide Algae sufficiently"; and what concerns the Sargassae particularly, that usually only fragments are collected of them, so that it must be guessed what upper part belongs to its lower part.

I fear that Bracebridge Wilson has forgotten either the algae or me personally. At least I have seen nothing of his collections of late years, which I regret very much, because new or rare species are often in these collections.

With particular respect signs devotedly
J G Agardh

89.01.01 *To George Stokes*

Royal Society, London, Miscellaneous correspondence vol. 15, 1889-92, letter no. 1.

Melbourne,
Newyear, 1889.

To Professor G. G. Stokes,
M.A., D.C.L., L.L.D., &c &c
President of the Royal Society of London.

It devolves on me, honored Sir, the most gratifying task, to express to yourself as President and to the distinguished Councillors of the Royal Society my feelings of profound gratitude for the grand distinction, of the bestowal of which you all held me worthy in so generous a spirit.[1] Indeed inexpressibly do I feel touched

3 Agardh (1889).

1 The announcement that M was awarded a Royal Medal of the Royal Society 'for his long services in Australian exploration, and for his investigations of the flora of the Australian continent' was made in the Anniversary Address of the President, 30 November 1888, *Proceedings of the Royal Society of London*, vol. 45, pp. 47-72. M's citation is on pp. 55-6. For M's lobbying for the award, see M to [P. Sclater], December 1887. For his first reaction to news of the award, see M to W. Thiselton-Dyer, 7 November 1888 [Collected Correspondence].

by this condescendence of the greatest forum of science, – all actions of which are surrounded by a halo of plurisecular renown, – to connect my humble name as a link in the chain of glory, unitedly formed from among the leaders of progressive thought through these brilliant awards. This grand mark of consideration, shown me by you, Sir, who occupies the most exalted place in British Science, and by your celebrated Colleagues, is in its value to me still further enhanced by the circumstance, that this is the first time, at which an Australian name became enrolled among those of the illustrious Elite, successively singled out for this ever glorious distinction in the world of knowledge; thus also at the verge of the first century for colonisation of the fifth continent, Australian researches become still more encouragingly recognized at the eldest centre of learning in the great British Empire. 89.01.01

I am however conscious, that this unsurpassable gift is not won by me through what I may have endeavoured to accomplish by individual exertions, but that it rather should be considered as a recognition of bravery, shown through a whole century by science-votaries in these great Australian dominions of her Majesty, – I merely happening to be one of their seniors. Undoubtedly this graceful act of the Royal Society – let us say – towards all Australia will have here a long lasting and immensely cheering effect on science-work in our next secular epoch! In the late autumn of my life, after multitudes of science-honors, – I fear but scantily deserved – have been almost showered on me, I shall always feel anew elevated by this grandest prize of all, and it will inspire me to new endeavours and stimulate me into increased exertions for contributing to the cognizanze of God's wondrous world for the benefit of mankind through such time, as the grace of divine providence may yet allot to my worldly career.

Most reverently your
Ferd. von Mueller

89.01.27 *To Isaac Balfour*

Royal Botanic Gardens Edinburgh, I. B. Balfour correspondence, M (Mi-Mu).

27/1/89

If you allow me, dear Prof Balfour, I will send you from time to time fresh seeds, – when obtainable –, of any Australian Everlastings, with a view to you kindly causing them to be raised, and in spring some seedlings to be planted on the grave of the honorable Dr Wilkie, who died in Paris 3 years ago during the winter cold from acute pulmonary inflammation while he was with his family on a visit to Europe after 40 years active practice here. – He was buried in Edinburgh,[1] where his father about the middle of the Century was a renowned presbyterian preacher.[2] I enjoyed for more than 30 years the intimate friendship of this worthy man here, and it would be a sad pleasure to me, if during the summer-warmth a few Australian flowers could ornament his grave.

Regardfully yours always
Ferd. von Mueller.

1 Greyfriars cemetery.
2 Rev. Daniel Wilkie (1782-1838).

To Malcolm Fraser 89.01.28

Colonial Secretary's Office, acc. 527, no. 453/89, State Records Office of Western Australia, Perth.

28/1/89

Allow me to enquire, dear Sir Malcolm, whether the Government of WA. would like to avail itself for a year or two of the services of Dr von Lendenfeld for the investigation of the lower marine animals of your coast.[1] This energetic and accomplished Naturalist has just completed his grand works on the Medusae and the sponges of other parts of the Australian shores,[2] yours remaining extensively uninvestigated. So important are his researches, that the Royal Society of London voted £300 towards the cost of printing his illustrations. His employment there, would enrich much your Museum, and place the Colony more in communication with men of sciences abroad. Dr v. Lendenfeld is just free of engagements; but may any moment get a call to an University. For the completion of his last works he has been staying with an uncle, who is a nobleman in Austria. A few hundred £ would suffice for a years salary; instruments he has himself. So the whole cost would not likely exceed £500 for a years work on your shores.

Regardfully your
Ferd. von Mueller.[3]

1 Earlier M had recommended von Lendenfeld for a position in Sydney; see M to A. Stephens, 25 November 1885 [Collected Correspondence].

2 Lendenfeld (1887), (1888), (1889).

3 On 13 February 1889 Fraser sent a minute to the Governor, F. Broome: 'no doubt it would be very interesting to have further knowledge of the "lower marine animals". We have some good specimens of the "Teredo navalis" in the sternpost of the "Meda" – I believe – but whilst putting Baron Von Mueller's letter before Your Excellency I am unable to advocate any action. Geology and Mineralogy are two branches of Science useful immediately to us, but a grand work on Medusae and Sponges will not profit us at this date.' Broome replied on 16 February: 'Reply courteously that we are not in a position at present to avail ourselves of this offer. F.N.B.' A reply was sent to M on 18 February; letter not found.

89.02.05 ## *To Alfred Deakin*

N89/5627, unit 291, VPRS 3992/P inward registered correspondence, VA 475 Chief Secretary's Department, Public Record Office, Victoria.[1]

Melbourne,
5 febr. 1889

The honorable Alfr. Deakin, M.L.A.,
Chief Secretary

Sir.

I have the honor to submit an application from Mr Renner[2] for a week's leave of absense on condition, that this time of absense be *deducted* from his three weeks usual leave. In recommending this application to your favorable consideration, I would beg to advise, that this present leave be granted on the *above* condition. The Museum of plants is situated on a rise away from buildings in a healthy place, and there is nothing in Mr Renner's departmental occupation to make him liable to illness. As the working hours are comparatively short and often holidays intervene, it would render the average-working time still shorter, if repeated absense on account of indisposition is granted without reducing the annual furlough.[3]

I have the honor to be,
Sir, your obedient servant
Ferd. von Mueller.

1 The letter is numbered N89/1613 within the file.

2 See G. Renner to M, 5 February 1889 [Collected Correspondence].

3 The Under Secretary, T. Wilson, approved M's request on 7 February and the file was forwarded to M who returned it the next day 'with best thanks'. See also G. Renner to M, 20 March 1889, in which Renner requested further sick leave, and M to T. Wilson, 20 March 1889 and the notes thereto [both Collected Correspondence].

To Ralph Tate

89.03.17

Barr Smith Library, University of Adelaide.

17/3/89

Let me assure you, dear Professor Tate, that I appreciate most highly the *honor* of your connecting my name now also with conchology,[1] and that I also most gratefully recognize the *sentiments*, which induced this dedication. Quite with admiration I have glanced over your list of the plants from the two rich regions, recently visited by you, and I am further charmed by the *splendid* manner, in which you at once turn all these observations to account for the geologic history of Australia! Indeed you are the *only one*, who possesses within himself the united knowledge of three branches of science sufficiently, to generalize effectually on the past history of the Australian Continent in *all* respects, you also only having with these comprehensive views gone personally over wide tracts of Australia with grand facilities now existing.

The Melb. meeting of the Austr. Assoc.[2] has been definitely fixed for *Jan 1890*. I hope, this will give you the opportunity of seeing the Austral. Alps also This year the railway will be finished to Bright,[3] from whence (at a drive of a few hours) elevations of 6000 feet can be reached. I have recommended to the executive Committee, that an effort be made, to induce the Government, to fix at that height some tents, so that the members of the Assoc. and their Ladies can have shelter and refreshments there amidst some of the grandest alpine scenery of the world! This tour would be made in the second week of the meeting The party can then return late in the afternoon to Bright, the whole tour taking up three days from Melbourne Of course under the tent-shelter a camp can be kept up for a few days, so that professional scientists, like yourself, can make excursions to the not very distant elevations of

1 Probably *Semicassis muelleri*, Tate (1889).
2 Australasian Association for the Advancement of Science.
3 Vic.

89.03.17 7000', which first were scaled in 1854 (by myself). According to the rules of the Assoc. only Delegates of the former meetings can be members of the executive council. Thus I have not a seat on it, and have had very little influence on the choice[4] of the Presidents Vice-Presidents and Secretaries of the sections, which seems however to be a happy one, altho' I should have liked *you* to have been in a very prominent position. But you are sure, to be President, when the Assoc. meets in Adelaide,[5] and then something can be done for you in London also.[6]

Ever regardfully your
Ferd. von Mueller

I am trying to bring a new edition of the Census out in 1889.[7] Ceratophylleae will then be placed near Ranunculaceae; – Thymeleae next to Rosaceae, Viniferae near Araliaceae &c How many distinct creations and migrations of living vegetation of Australia can be discovered according to *your* views?

4 Thus ... choice *is bracketed in the margin and marked* Private! *by M.*

5 Tate was indeed President when the Association first met in Adelaide, in September 1893.

6 Tate had been an unsuccessful candidate for election to the Royal Society of London in 1884-6, but was never proposed subsequently.

7 B89.13.12.

From William Armstrong 89.05.09

RB MSS M1, Library, Royal Botanic Gardens Melbourne.[1]

Waranga[2]
May – 9th /89–

Baron von Mueller – Esq –

Sir in reference to the Noxious weed known As the Chinee weed i will give you The required knowledge as far As is known in the District in Which it Grows The oldest residents And relyable ones At Rushworth[3] Says –19. years Ago there were But one Plant At the Chinees Camp Rushworth it has spread Rapidly since i Have had Nearly the same Time Experience In this District & i know it To be spreading rapidly Up to The Present time = Sir = you would Oblige if you would Give Me the knowledge of the Properties it contains As it will Burn Green or Dry And i was thinking it Contained An oil if you would Furnish me with the Name of the Plant you Would oblige And if it Would sute[4] you To Publish Through the Age Paper you Are At liberty

yours Sir Truely
Wm Armstrong

Address
William E. Armstrong
Farmer
Waranga
Rushworth – P – O –

1 MS found with a specimen of *Cassinia arcuata* (MEL 221137), coll. W. E. Armstrong, Lower Goulbourne [i.e. Goulburn] River, 1889.

2 Vic.

3 Vic.

4 suit?

89.05.12 *From Ferdinand von Krauss*

Staatliches Museum für Naturkunde, Stuttgart.

Stuttgart 12. Mai 1889.[1]

Verehrtester Herr Baron u. Freund!

Heute erlaube ich mir wegen des "Statuts für das von Dr Ferd. v. Müller in Melbourne gestiftete naturhistorische Reisestipendium v. 18. Mai 1869" mich an Sie zu wenden.

Wie ich Ihnen schon einigemal schrieb, hat sich seither kein ernstlicher Bewerber um Ihr so reiches u. wohlwollendes Stipendium beworben, obwohl es pflichtlich jedes Jahr am schwarzen Brett in Tübingen den Studirenden öffentlich bekannt gemacht worden ist u. jetzt rund 800 Mark jährlich beträgt. Dieser Mangel ist wohl hauptsächlich dem Umstand zuzuschreiben, dass Studirende der Medicin, auf welche sich das Stipendium beschränkt, nach den neueren Vorschriften über ihren Studiengang ihr Studium nicht gern unterbrechen u. sobald als möglich ihren eignen Herd gründen wollen. Auch existirt eine "erste medicin. Staatsprüfung" in früheren Sinne nicht mehr.

Es ist desshalb in Erwägung gezogen worden, ob Ihnen nicht der Vorschlag gemacht werden soll, den Kreis der Berechtigten u. das Reiseziel zu erweitern.

Dr. Eimer, der gegenw. Prof. d. Zoologie u. Vorstand des zool. Instituts in Tübingen, hat nun in dieser Richtung vor*geschlagen, das Stipendium "für naturwissenschaftliche Reisen überhaupt"* zu vergeben, wobei er namentlich die zoologische Station in Neapel in Auge hat, um seine Studenten mit Unterstützung dahin schikken zu können. Dieser Vorschlag geht nun aber offenbar zu weit, denn nach Ihrer sehr dankenswerthen Bestimmung wollen Sie

1 MS annotation: 'Abschrift des Briefes von Krauss an Freiherr Ferd. v. Müller' [Copy of the letter of Krauss to Baron Ferd. von Mueller]. The file includes this letter and M's reply dated 22 June 1889, a copy of a memo to the Royal Administration, 17 October 1889, by Krauss and a copy of suggested changes to the regulations of the Mueller Foundation, dated 17 October 1889 [all Collected Correspondence].

durch Ihr Stipendium nicht allein dem Studierenden die Kosten für wissensch. Reisen ausserhalb Europas erleichtern, sondern was wir sehr hoch anschlagen, zugleich auch dem K. Nat. Kab. in St. durch solche Reisen naturhistorische Gegenstände aus fernen Ländern zu seiner Verwahrung zuwenden. Diese würde aber, wenn das Stipendium wie vorauszusetzen hauptsächlich für Neapel Verwendung fände, nicht erreicht, denn das K. N. Kab. hat seither von solchen, welche den würt. Sitz der Station von Neapel benutzen, noch nie einen nennenswerthen Gegenstand erhalten, auch können Mittelmeer-Objekte durch die Station selbst in vortrefflicher Zubereitung käuflich erworben werden. Auch bezüglich der Qualifikation der Bewerber müssen doch bestimmtere Voraussetzungen vorgeschrieben werden. 89.05.12

Die Direktion der wissensch. Sammlungen (Präsid. Dr. v. Silcher) hat nun in Uebereinstimmung mit mir Vorschläge zu einer Modification der Bestimmungen des Statuts Ihrer Stiftung entworfen, welche ich Ihnen im Auftrage des K. Kult-ministeriums im Anschlusse mitzutheilen mich beehre mit der Bitte, dieselben an der Hand des Statuts v. 1869 zu prüfen u. mir Ihren Entschluss bald gef. mittheilen zu wollen.

Nach diesen Vorschlägen würde der bisherige Zweck Ihrer Stiftung in der Hauptsache beibehalten u. nur der Kreis der zuzulassenen Bewerber in so weit erweitert, dass auf eine regere Betheiligung an derselben zu hoffen wäre, nutz würde, wenn sich trotzdem kein Bewerber für aussereuropäische Reisen finden sollte, ausnahmsweise u. mit Beschränkung auf 1/4 des Jahresertrags das Stipendium auch für Reisen innerhalb Europas vergeben werden können, übrigens unter ausdrücklicher Berücksichtigung der Interessen des Natur. Kabinets.

Meinen Brief v. 9. Mai mit der gewünschten Zusammenstellung der bis jetzt bekannten lebenden Arten der Thiere werden Sie nun in Händen haben.

Mit hochachtungsvollen Grüssen
Ihr dankbar ergebener
F Kr.

89.05.12 Stuttgart 12 May 1889.

Most honoured Baron and Friend!

Today I take the liberty of consulting you concerning the "Statutes of the natural history travel grant endowed by Dr. Ferdinand von Mueller in Melbourne of 18 May 1869".[2]

As I have written to you already several times, no serious applicant has applied since then for your so substantial and benevolent grant, although it has dutifully been made known publicly to the students each year on the notice board in Tübingen and now amounts to around 800 Marks annually. This shortcoming is probably mainly to be ascribed to the circumstance that students of medicine, to whom the grant is restricted, do not willingly interrupt their study following the new regulations concerning their course of studies and want to found their own home as soon as possible. Also a "first medical state examination" no longer exists in the previous sense.

We therefore have considered whether we should suggest to you that you enlarge the circle of those entitled and the purpose of travel.

Dr Eimer, the present professor of zoology and chairman of the Zoological Institute in Tübingen, has now suggested that the award be granted "for *natural history travel generally*", by which he especially has the Zoological Station in Naples in mind, to be able to send his students there with support. This suggestion, however, obviously goes too far, since according to your very commendable regulation you intended to alleviate the costs for scientific travel outside Europe not only for the student, but what we value very highly, at the same time also to obtain natural history objects from far lands for the Royal Natural History Cabinet in Stuttgart through such journeys for its keeping. However, this would not be achieved, if the grant as presupposed had use mainly for Naples, because the Royal Natural History Cabinet has never received an object worth naming from those who used the Württemburg seat at the Naples Station, also Mediterranean objects can be obtained by purchase through the Station itself in excellent preparation. Also respecting the qualification of the applicants more definite requirements must certainly be prescribed.

The Director of the Scientific Collections (President Dr von Silcher) has now in agreement with me drafted suggestions for a modification of the statutes of your Foundation,[3] which I have the honour to communicate to you by instruction of the Royal Cultural Ministry in connection with the request to examine them with the Statutes of 1869 at hand and to kindly inform me of your decision soon.

2 See *Selected correspondence*, vol. 2, pp. 763-7.

3 Marginal annotation: 'Siehe Beil' [See enclosed].

According to these suggestions the former purpose of your Foundation would be kept in the main and only the circle of the applicants to be admitted enlarged only so far that it would be hoped for a more lively participation in it, if nevertheless no applicant should be found for travel outside Europe, exceptionally and with restriction to 1/4 of the annual proceeds the grant could also be awarded for travel within Europe, however, under the express consideration of the interests of the Natural History Cabinet. 89.05.12

You will now have in hand my letter of 9 May with the requested compilation of the living species of animals known to date.[4]

With respectful greetings
Your grateful devoted
Ferdinand Krauss

To Eduard von Regel 89.05.29

MS 177, folio 21808, d. 76, op. 1, fond 335, Archives, Academy of Sciences, St Petersburg.

29/5/89

Erlauben Sie mir, edler Freund, Ihnen eine Bitte vorzutragen, die Sie direct oder durch Vermittlung Ihres ausgezeichneten Sohns, des Stabs-Arztes, leicht erfüllen können. Herr Dr Feoktistov in St. Petersburg hat zahlreiche Beobachtungen in letzter Zeit über das Ophidian Gift angestellt, und hat damit gewissermassen den Untersuchungen des Dr Aug. Mueller in Yakandandah weiteren Halt gegeben, da dieser Victorianische Arzt die Wirkung des Schlangen Giftes auf Paralysis der Motor-Nerven zurückführt, nicht auf eine Gewebe-Änderung sondern nur auf dynamische Action, obwohl wir freilich die etwa herbeigeführten Molekular-Alterationen noch nicht kennen. Einfach nach therapeutischen Grundsätzen wendet nun Hr Dr Aug. Mueller das Strychnin als

4 Letter not found; but see M to F. von Krauss, 24 March 1889 [Collected Correspondence], in which M requested an estimate, that he could use in his forthcoming presidential address to the Australasian Association for the Advancement of Science (B90.13.01), of the number of living species of the different classes of animals.

89.05.29 Antidot (durch hypodermisch Injection) an u zwar in den wenigen unlängst beobachteten Fällen mit dem glänzendsten Erfolge. Ich habe diese seine Methode berührt in meiner Eröffnungs Rede der therapeutischen Section des Congresses austral Ärzte Jan. 1889, u auch eine Mittheilung darüber schon vorher an den mir befreundeten Geheimrath Virchow gemacht, welche in seiner Zeitschrift zur Veröffentlichung kam. Da nun Hr Dr Feoktistov in dieser Angelegenheit ein reges Interesse haben muss, möchte ich Sie oder Ihren geehrten Herrn Sohn bitten, die Methode Dr A. Muellers zu besprechen. Ich verfolge diese Sache einfach vom wissenschaftlichen Standpunkt, da mein Namens Genosse mir nicht verwandt ist, möchte aber doch von Ihrem gelehrten Mitbürger gern eine Notiz in dieser Hinsicht haben für die Präsidentur Rede in der austral Association for the Advancement of Science, welche im Jan. 1890 Ihre Versammlung in Melbourne halten wird, gerade wie die Versammlung deutscher Naturforscher u Ärzte alljährlich zusammen kommt, u mich 1846 mit vielen hervorragenden Gelehrten in Contact brachte.

Was denkt der Herr Decan der medic. Facultät in Petersburg von meiner Idee, dass ein Rundschreiben an alle Universitäten ergehen solle, uns für den nächsten medicinischen Congress, der 1891 in Sydney gehalten werden soll, je einen Abgeordneten zu senden.

Sie ehrend der Ihre
Ferd. von Mueller

Ich weiss recht wohl, dass Sir Joseph Fayrer in Calcutta in seinen Strychnin Experimenten mit dem Gift der schrecklichen Naja negative Resultate in Bezug auf Strychnin erziehlte, aber er hatte nicht Gelegenheit die Wirkung an Menschen zu prüfen, u die Experimente mit Hunden sind auch gar nicht zahlreich gewesen

29/5/89

Permit me, noble friend, to place before you a request, which you directly or through your excellent son, the medical officer, can easily grant. Dr Feoktistov of St Petersburg has made numerous observations on snake poison in recent times[1] and in so doing has, in a way, given further credence to the investigations of Dr August Mueller in Yackandandah.[2] This Victorian doctor attributes the effects of snake poison to paralysis of the motor nerves, not to a change in the tissue, but solely to dynamic action, though admittedly we are not yet aware of any possible molecular changes which may occur. Based on purely therapeutic principles Dr Aug. Mueller uses strychnine as antidote (by hypodermic injection), and did so in the few most recently observed cases with the most brilliant success. I touched on this method in my opening speech[3] before the therapeutic section of the Congress of Australian doctors in January 1889, and also sent a notice to my friend Privy Councillor Virchow prior to that, which was published[4] in his journal. As Dr Feoktistov evidently must have a keen interest in this matter, I would like to ask you or your learned son to discuss this method of Dr A. Mueller. I pursue this matter solely from a scientific point of view, as my name-sake is not related to me, but I would like to have a notice from your learned compatriot for my next presidential address before the Australasian Association for the Advancement of Science,[5] which will hold its meeting in Melbourne in January 1890, just as the congress of German naturalists and doctors meets annually and in 1846 brought me into contact with many eminent scientists. 89.05.29

What does the chairman of the medical faculty in St Petersburg think of my idea, that a circular should be addressed to all universities, asking them to send us one delegate each to our next medical congress,[6] to be held in Sydney in 1891?

With my respects
your
Ferd. von Mueller.

I know very well, that Sir Joseph Fayrer in Calcutta obtained negative results with his strychnine experiments with the venom of the terrible Naja,[7] but he did not have the opportunity to test its effects on humans, and his experiments on dogs were not at all numerous.

1 Feoktistov (1889).
2 Vic. See A. Mueller (1888); also A. Mueller (1893) and *Australian medical journal*, (1893) new series, vol. 15, pp. 472-4.
3 Not in the printed version, B89.13.16.
4 MS annotation by Regel: 'nicht gefunden' [not found]. See A. Mueller to M, 14 February 1888 [Collected Correspondence], published in A. Mueller (1888a).
5 B90.13.01. M made no such reference in his presidential address.
6 MS annotation by Regel: 'medisch. Academie senden' [send to medical Academy].
7 The Indian cobra, a highly venomous snake.

89.07.12 *To Friedrich Franz III, Grand-Duke of Mecklenburg-Schwerin*[1]

Signatur: 4630, Grossherzogliches Kabinett III, Mecklenburgisches Landeshauptarchiv, Schwerin.

Melbourne,
am 12 July 1889.

Sr. Königlichen Hoheit
dem Grossherzoge Friedrich Franz III
von Mecklenburg-Schwerin &c &c &c

Es ist meine ehrenvolle Pflicht, Ew. Königlichen Hoheit gnädige Zuschrift vom 24 Mai anzuerkennen, durch welche mir tiefgerührt eben die Kunde wird, dass Ew. Königliche Hoheit in herablassendster Güte sich bewogen fühlten, die grossherzoglich-mecklenburgische Goldmedaille, – "den Künsten und Wissenschaften" gewidmet –, mir zu verleihen. In tiefster Dankbarkeit und sinnungsvollster Würdigung betrachte ich dies Ereignis als einen Glanzpunkt meines Lebens, – um so mehr als dadurch auch Ew. Königlichen Hoheit allergnädigstes Interesse an einen in weitester Ferne weilenden Sohn Ihres schönen Reiches bekundet wird; und es ist wahrhaft für mich erhebend, meine schwachen Bestrebungen, auch hier Glanz auf deutsche Wissenschaft zu werfen, von dem edlen Herrscher meines engeren Vaterlandes, meiner geliebten Heimat, in einer so hervorragenden Weise belohnt und ermuntert zu sehen! Besonders empfindungsvoll gedenke ich dabei auch des hohen Ahnen Ew. Königlichen Hoheit, welcher dies prächtige und bedeutungsvolle Ehrenzeichen stiftete, des Grossherzogs Friedrich Franz I, der meinen nach dem Gefecht bei Seestedt (1813) zurückkehrenden theueren aber früh dahingeschiedenen Vater, damals einem jungen Manne aus dem Kaufmanns Stande, der Übertragung der ersten Kontrolleur-Stelle in Rostock werth hielt, und

1 MS annotation: 'Präs.: Jgdhs. [Präsentiert: Jagdhaus] Gelbensande, 26. August 1889' [Presented: Gelbensande Hunting Lodge, 26 August 1889].

ihm so aufs allergnädigste die Gelegenheit gab, sein häusliches Glück zu begründen, und später mir dort die erste Erziehung geben zu lassen und damit den Anstoss zu meiner künftigen Laufbahn. Dann ist es noch für mich eine besondere Erhöhung, dass selbst Ew. Königl. Hoheit in freudiger Huld davon Kenntnis nahmen, dass jene Gesellschaft, in welcher Newton einst präsidirte, mir eine so seltene Auszeichnung zu erkor, – so dass nun auch durch einen mecklenburgischen Namen eine Spange gebildet wird, in der Anteil derer der wenigen Sterblichen, welche während der letzten zwei Jahrhunderte durch die Gold-Medaille der Royal Society in unmittelbaren Contact gebracht sind innerhalb der Geschichte der Wissenschaften! 89.07.12

Möge es mir gestattet sein an meine Worte unaussprechlichen Dankgefühls die innige Hoffnung zu knüpfen, dass die göttliche Vorsehung Ew. Königliche Hoheit und deren erhabenem Hause alles Erdenglück zu Theil werden lasse weit ins nächste Jahrhundert. So verharre ich als Ew. Königlichen Hoheit ehrerbietigster und unterthänigst Ergebenster

Ferd. von Mueller

Melbourne,
12 July 1889.

His Royal Highness
Grand-Duke Friedrich Franz III
of Mecklenburg-Schwerin, &c &c &c.

It is my honourable duty to acknowledge your Royal Highness's gracious communication of 24 May,[2] by which I received the deeply touching news that Your Royal Highness in condescending graciousness felt inclined to bestow on me the Grand-Duke of Mecklenburg's gold medal, dedicated to the 'Arts and Sciences'. In deepest gratitude and genuine appreciation I consider this event to be one of the highlights of my life, all the more so as it testifies to Your Royal Highness's most gracious interest in a son of your beautiful realm who lives in a most distant place. It is truly edifying for me to see my feeble efforts to reflect

2 Letter not found.

89.07.12 glory on German science here rewarded and encouraged by the noble monarch of my closer fatherland, my beloved homeland, in such an eminent way. On this occasion I remember also with particular sentiments the exalted forebear of Your Royal Highness, who founded this magnificent and significant medal, Grand-Duke Friedrich Franz I, who considered my dear but early deceased father, then a young man of the merchant class returning from the battle of Seestedt (1813),[3] worthy to be given the position of First Controller in Rostock,[4] thus enabling him to found his domestic happiness and later there to provide me with my first education and with that the starting point for my future career. Furthermore it is especially elevating for me that even Your Royal Highness took note with ready favour, when that Society, over which Newton once presided, bestowed on me such a rare distinction, so that now through a Mecklenburg name a tie is also formed with those select few mortals, who during the last two hundred years were brought into direct contact within the history of the sciences through the gold medal of the Royal Society.[5]

May I be permitted to add to my words of inexpressible gratitude the sincere hope, that Divine Providence may grant Your Royal Highness and Your exalted House all earthly happiness well into the next century. I remain Your Royal Highness' most respectful and most obediently devoted

Ferd. von Mueller.

3 Sehestedt, Schleswig-Holstein, scene of a battle during the Napoleonic wars.

4 M's father was appointed a *Strandvoigt* or customs officer in Rostock in 1818, responsible for controlling the flow of goods through several of the city's gates.

5 In 1888 the Royal Society of London awarded M one of the two Royal Medals it bestowed each year. See M to G. Stokes, 1 February 1889.

From James Thomson 89.07.26

A36 Royal Geographical Society of Australasia (Vic. Branch) papers, Mitchell Library, State Library of New South Wales, Sydney.

Royal Geographical Society of Australasia.[1]

ADDRESS QUEENSLAND BRANCH.
HON. SECRETARY & TREASURER,
ROYAL GEOGRAPHICAL SOCIETY OF AUSTRALASIA.
BRISBANE, QUEENSLAND, AUSTRALIA. *Brisbane*. 26. 7. 1889.

Honoured Sir,

On behalf of the above-named Society I beg to acknowledge receipt of your esteemed letter dated the 3rd inst.[2] addressed to our ex-President, & to inform you that it was duly placed before our Council & carefully considered at its meeting held today.

As your letter, in its conclusion, solicits an expression of opinion upon the substance of its own contents, which, are in effect, the deliberations of your distinguished Council, I am directed in the first instance to place before you a brief summary of previous action in connexion with the proposed establishment of a Fellowship, then to convey to you the opinion of our Council in accordance with your request. –

About twelve months ago this branch recognising in our Australasian Society the absence of a stimulating influence for the reward of meritorious labours & the encouragement of latent talent which would in no way limit individual operations, &, observing that although our learned colleagues representing our sister branches had, in conference assembled, adopted no active means to make such provision further than mere reference thereto, also observing

1 MS annotation by M: 'Received evening 1/8/89 F.v.M. An[swere]d 5/8/89'. See A. Macdonald to J. Thomson, 5 August 1889 (ML A36, f. 199, Mitchell Library).

2 Thomson got the date wrong; see M to I. N. Waugh, 2 July 1889 [Collected Correspondence].

89.07.26 no practical & united efforts by individual or collective branches in that direction, unanimously adopted a resolution at its annual meeting then held, providing for this want. This resolution, which also invited united action & solicited an expression of opinion from our sister branches, was subsequently forwarded to them for their respective consideration. Our Sydney branch acknowledged the receipt & promised attention by its Council. Our Adelaide confreres never acknowledged that nor any subsequent communication, a circumstance which indicates inactivity. Our learned colleagues of the Melbourne branch with their characteristic courtesy duly acknowledged receipt by letter dated August 10. 1888.[3] This letter, which, singularly enough, not merely acknowledged but actually promised co-operation, also stated that the importance of the matter had been previously brought before the geographical conference held in Adelaide by the Victorian branch. Although subsequently written about we heard nothing more of the question till about a month ago we quite unexpectedly received a letter from Melbourne dated May 24th,[4] accompanied by a copy of Diploma of Membership, which, in a slightly altered form we were asked to order from your printer for presentation to all of our members. The compliance with your request, which practically admitted of no choice,[5] for the order & adoption by us of these Diplomas of membership, which, be it marked, had previously been adopted by you, & actually issued to your members entirely unknown to us, was duly authorised by our Council. – In consideration of this subsequent arrangement for the issue of Diplomas to all classes of members, our Council, which assembled especially for the purpose, considered it necessary to reconstruct its previous resolution, & again submit it to our members for conformation.[6] This reconstructed resolution having been previously prepared & transmitted to our sister branches, was brought forward at our last annual meeting held on the 15th inst. & I have

3 Letter not found.
4 Letter not found.
5 MS annotation by M: 'reply to this'.
6 confirmation?

great pleasure in informing you, unanimously adopted. We are also informed by letter that our Sydney branch has heartily co-operated by unanimously adopting our resolution at its last annual meeting. From the foregoing you will observe that ample time was afforded to all our sister branches for the full consideration of the whole subject. 89.07.26

Having thus summarised past actions, I am directed to convey to you our warm felicitation, & in reply to the suggestions contained in your letter our Council expresses the following opinions. (1) That as no affiliation or obligation exists between the Royal Geographical Society of London & our own we are in consequence, in the future as in the past, dependent upon our own individual or collective resources for development of energy or application of forces, &, that as progressive science & art is entirely dependent upon individual originality & force of character & not by the adoption of current opinion, so will the career of our Society be entirely influenced by the condition of its administrative policy.

(2) That although no distinction may be drawn between members & Fellows in name, the Royal Geographical Society of London possesses in addition to its distinction of honorary & corresponding members various classes of medals, grants, & testimonials which it awards annually as distinctive attributes for distinguished services. –

(3) That the want of these stimulants in our Society is already felt by many who flag in interest, & by others who transfer their labours elsewhere. –

(4) That our distinction of corresponding member is only available to non-residents, & that of Hon. member usually to those who have completed a life of usefulness, both of which preclude activity in the affairs of the Society.

(5) That our adopted resolution anticipates no additional grade of membership. –

Taking into consideration the sum of the foregoing conditions, the Council begs to submit that the time is fully ripe for the generalisation of united & concentrated action by the several branches of our Society in making the necessary provision for the establishment of a limited Fellowship, as a distinctive attribute

89.07.26 analogous to the medals, grants, & testimonials awarded annually by our sister societies elsewhere, &, to be as unlike the distinction of Hon. & corresponding membership, as to leave the possessor free to act as an active agent in all matters connected with the Society, to provide a stimulus to the Society, & to foster unity & social intercourse. –

In conclusion I am also directed to state that in so far as this branch is concerned & that of our N. S. Wales sister action is complete, consequently we are unable to further consider negotations[7] on the subject. Our Council will therefore be glad to receive an intimation that you have taken speedy & hearty co-operative action by the unanimous adoption of the resolution referred to, copy of which is in your possession. –

I am,
Honoured Sir
Yours faithfully
J. P. Thomson
Hon. Sec. & Treasurer.[8] –

Baron Sir F. von Mueller K.C.M.G., F.R.S. &c.
President
Vic. Branch R.G.S.A.
Melbourne

7 negotiations?

8 In his reply (see note 1), Macdonald reported that the Council of the Victorian branch was unanimously of the view that 'the question of establishing even a limited number of Fellowships, issue of an annual medal etc. is one that should be unanimously agreed to by all the branches', and suggested that the proposals be discussed by the Geography Section of the forthcoming congress of the Australasian Association for the Advancement of Science, at which all the branches would be represented. In response to Thomson's complaint about being confronted with a *fait accompli* in regard to the issuing of membership diplomas, Macdonald said he had 'merely suggested the advisability of your Council adopting a form of *Certificate*, which is quite different from a Diploma of Fellowship'.

To Arthur Lucas

89.08.10

FNCV 014-034 Archives, Field Naturalists Club of Victoria, Melbourne.

10/8/89

Having learnt, dear Prof. Lucas, that some members are eager, to get the name changed of the Field-Naturalists Club, I think, that I shall not step out of my proper position, when I advise, that such a change *not* be made. I offer such objections, as occur to this (according to my views) in writing, as my severe cough does not allow of my going out, particularly at late evening-hours in this cool season. The very word "Club" implies an union without rigerous ceremonies, a freer coming together, than in abstract science-societies, as evinced also by the membership of our Field-Naturalists Club being happily open to Ladies. By the change of the name, as far as I can see, nothing would be gained for our particular work, while much to us in our free scientific intercourse and in our unrestrained field-operations might be lost. I further have heard, that some members of the Club are anxious to establish *grades* in our union, according to greater or lesser accomplishments and experiences. This proposition came up formerly in more than one science-society of Australia, but I gave my advise against such a measure fully thirty years ago. The question in such case arises, who is to adjudicate? The member, to be raised to a higher grade, can as regards his merits only be judged by the few, who are engaged in the same speciality, – whereas the whole of the Club would be voting on such a proposition. But this is not the gravest ascect[1] of the question. The danger is of envey and discord being generated, and a harmonious union, working for a common purpose, being split up into factions. We have already the power to reward very distinguished workers in the Club with an honorary membership, which need not be restricted to persons beyond the colony. This seems to me quite sufficient for an organisation of ours. If still more is demanded and a separation-system is to be introduced, then the Royal Society offers very con-

1 aspect?

89.08.10 siderately for such special aspirations the best scope among us, without the Candidates for higher grades thereby being lost to us for united work, the sectional system of the R.S.V.[2] being exactly what would meet the case.

Regardfully your
Ferd. von Mueller.

89.09.01 *From Mordecai Cooke*[1]

RB MSS M31, Library, Royal Botanic Gardens Melbourne.

146 Junction Road
London N

My dear Sir/

According to the terms of agreement between the Departments of State my official connection with the Herbarium at Kew ceases at the expiration of the present year, and I am not aware of any intention to extend it[2]

Under these circumstances I shall be prepared to enter upon any other eligible arrangement which will produce some little remuneration, and occupy my time. It has been proposed to me more than once to enter upon the preparation of a volume on the fungi of Australia, but I have felt no desire to enter upon such a project at my own expence. If the Australian colonies will combine for this purpose it can be carried out

2 The Royal Society of Victoria at this period had several special-interest Sections – including one for microscopy, formed when the Microscopical Society of Victoria merged with the Royal Society; see Kernot (1888). The Field Naturalists Club of Victoria later also introduced Sections.

1 MS annotation by M: 'Answ 17/10/89'. Letter not found.

2 Cooke was employed at Kew under an arrangement with the India Office. His contract, which was to expire in December 1889, was extended for a further three years; see English (1987).

The production of 500 copies of the book in 1 vol. of upwards of 500 pages and 48 plates, illustrating the Genera and subgenera would cost about £420. If this sum were to be advanced for the purpose, and all the copies sold at 25/- or 30/ each I think the whole amount would return to the Exchequer, but I have not the money to spare to invest £400 in such a venture at a time when I shall be wholly without offical employment and only a small pension to depend on 89.09.01

Under these circumstances I have submitted my scheme, with estimates, and should be glad if you would take steps to ascertain the feeling of the governments of the several colonies and as early as possible communicate to me the result[3]

I am my dear Sir
yours most obliged
M Cooke
1st Sept. 1889[4]

From Jemima Irvine 89.09.30

RB MSS M1, Library, Royal Botanic Gardens Melbourne.

Corona Station[1]
Septr 30th /89.

Baron von Mueller.

Dear Sir

By this mail I am sending you a box of plants. My last from here I fancy as I start on my homeward journey this day week. I would

3 M followed up Cooke's proposal, which led to the publication of Cooke (1892); see M to W. Thiselton-Dyer, 1 July 1891.

4 MS is contained with a bundle of letters to M about the publication of Cooke's work. The letters are inside a folded sheet annotated by M: 'Drawings of Fungs by G. Massee 1889'. No drawings found. See also F. Bailey to M, 17 October 1889 and G. Massee to M, 7 May 1888, 22 April 1889 and 12 August 1889 [all Collected Correspondence].

1 In the Barrier Range, north of present-day Broken Hill, NSW.

89.09.30 be glad to know if you receive the box – so please address a line to me at the Silverton Post Office. I can call there for it. here it would arrive some time after I had left. On my arrival in Melbourne I should much like to see you – so would call at the Botanic Museum if I knew your hours. please name them – I am anxious to know if you care for any of the things I have sent. *One* value they have, many of them have not been seen for years, and Mr. Kennedy who knows this country so well, told me it might be ten years before the like showed themselves again. When I come in with arms full of things my son always says, "if you had been here this time last year you would not have found a green leaf" so I have been most fortunate in the time of my visit. I fear from what Mr. French[2] said that my box of lovely everlastings reached you in very bad condition – I received one box myself crushed flat. I am sorry, for I packed them most carefully – and wished you to see all their beauty. I have collected a lot for seed this afternoon but fear they are not quite ripe – so will get some the day before I leave. I almost doubt their growing away from their own rocky hill tops – great branches grow out of a little split in a rock, where you would hardly think they could find root hold – there are some seeds of all the different kinds – all bleach white as they get ripe. I am sending a bit of three or four different kinds of acacia – may be something new to you. I *fear* not. Have you decided about the *holly plant*? is it any thing new? – With the £10 you gave me for my Western Australian flowers I bought Block 10 shares at 32/- each, they are now £5-5- so I was lucky – Should I find any thing more worth sending you shall have it.

I remain dear Sir
Yours truly
Jemima Frances Irvine

2 Presumably Charles French Sr.

To Thomas Wilson[1] 89.12.07

P89/5045, unit 215, VPRS 1163/P1 inward correspondence files, VA 672 Premier's Office, Public Record Office, Victoria.

Melbourne, 7. December 1889

Sir

I have the honor to report, that the printing of the "Second Census of Australian Plants",[2] for which the Honorable the Chief Secretary kindly granted permission, has now been completed at Messrs. McCarron Bird & Co's establishment, and I beg to forward an impression herewith. I further beg to submit list of those persons and societies, whom I consider, through literary interchanges or other services rendered to the Govt. Botanist's Department, entitled to receive this new publication in the usual official course through the Public Library. But I would consider it a particular kindness, if the Government would allow me a moderate number of copies of this edition (the impression being 1000) for direct use in the Govt. Botanist's office, such number to be fixed as the Government may deem proper,[3] especially as the work was printed by a private firm at the expense of my branch of the Service, at a

1 MS written by G. Luehmann and edited and signed by M.

2 B89.13.12.

3 T. Wilson, the Under Secretary, asked M how many copies he wanted, to which M replied, 23 December 1889: 'Could 25 or 50 copies be granted of this inexpensive work at a time, when after such supply being exhausted a return of, how disposed, would be sent in.' This was referred to the Premier's Office where an amount representing the value of the gifts was asked for. When this message reached M via the Chief Secretary's Department he replied, 30 January 1890: 'Fifty pounds Sterling would cover the annual expenditure, which from next year might be defrayed from the ordinary vote.' The next day M added: 'Best thanks for information afforded; will the copies be sent out according to the proposed list, by the public Library.' The information referred to seems to have been given orally, since there is no indication on the file of what it was. The file was returned to the Premier's Office and on 20 March 1890 the Secretary, E. Thomas, ordered 50 copies to be sent to M 'for which an account is to be rendered for payment'.

89.12.07 very moderate cost through being without illustrations. For opportunities arise frequently when the work is quickly required in the exigencies and to the advantage of the Service, such copies to be distributed always on behalf of the Government, and a record kept in my office about such distribution, a return of which might quarterly be submitted to the minister of the Department, similar to an arrangement existing in the Govt. Statist's office as regards Mr Hayter's Year Books.[4] I may be allowed respectfully to remark, that in cases of urgency I have for several years past purchased out of my private means the various publications of my own at the Govt. Printing Office, to expedite the departmental service on many occasions, the expenditure having amounted to about £50. in each of those years, with no means of reimbursement. The Government might perhaps also, as generously done during the Medical Congress with regard to the "Select Plants",[5] grant the favor of placing a certain number of copies of the Second Census of Australian Plants at the disposition of the biological section of the Australian Association for the Advancement of Science, which is to assemble early in January in our metropolis, wheras such a meeting is not likely to be held here again for many years. The remaining copies might as usual be sent to the Govt. Printing Office for sale.

I have the honor to be
Sir
Your obedient servant
Ferd. von Mueller,
Govt. Botanist

T. R. Wilson Esq &c
Under Secretary

4 i.e. *Victorian yearbook*, published annually by the Government Statist, H. Hayter.

5 Intercolonial Medical Congress, Melbourne, January 1889. A copy of M's *Select extra-tropical plants* (B88.13.02) was presented to each delegate to the Congress by the Victorian Government; see M to T. Wilson, 22 December 1888 [Collected Correspondence].

Proposed distribution of Second Census of Australian Plants. 89.12.07

Academy of Science		New York	US.
"		San Francisco	"
"		Chicago	"
"	Professor Meehan, Vice President	Philadelphia	"
"		Munich,	Germany
"		Goettingen	"
"		Buda Pest	Austro Hungary
"	Leopold. Carol.	Halle a/S.	Germany
"		St Petersburg	Russia
"		Lisbon	Portugal
"		Madrid	Spain
"	dei Lincei	Rome	Italy
"	Athenaea	Brescia	"
"		Bruxelles	Belgium
"		Liége	"
"		Amsterdam	Netherlands
"		Leyden	"
"		Utrecht	"
"		Stockholm	Sweden
"		Upsala	"
Agricultural Department		Washington US.	
Agri-Horticultural Society		Calcutta	
Acclimatisation Society		Paris	
Cosson, Dr.		Rue de la Boëtie, Paris	
Apotheker Verein (Chemical Society)		Vienna, Austria	
Baillon, Professor H.		Rue Cuvier Paris	
Beccari, Dr. Ed.		Florence	Italy
Bailey, F. M.		Govt. Botanist Brisbane	
Botanical Society		Edinburgh	
"		London	
"		Hamburg,	Germany
"		Copenhagen,	Denmark
"		Goerlitz,	Germany
Caruel, Professor		Florence	Italy
Colonial Institute		London	
Davis, Geo. Editor Therapeutic Gazette,		Detroit, Mich.	US.
Director Botanic Garden		Calcutta	
Engler, Professor Director Bot. Garden		Berlin	

89.12.07	Forrest, Hon. John, CMG	Perth,	West Australia
	Geologic Survey Dept.	Washington	US.
	Goeze, Dr. E.	Greifswald,	Germany
	Henriques, Professor	Coimbra,	Portugal
	Heurck, Dr. H. van	Antwerp	Belgium
	Holtze, M., Govt. Gardens Port Darwin		
	Horticultural Society, Royal	London	
	Editor of the “Queenslander”	Brisbane	
	Linnean Society,	Burlington House London	
	”	Sydney	
	Forest Department	Adelaide	
	Govt. Botanist’s Office	Melbourne	
	Luerssen, Professor	Königsberg,	Germany
	Masters, Dr. M. T., Editor Gardeners Chronicle,		London
	Moore, Chas. Direct. Bot. Garden	Sydney	
	Hooker, Sir Joseph. KCSI	Royal Gardens, Kew, London	
	Kirk, Professor Thos.	Wellington,	N.Z.
	Museum of Natural History	Mexico	
	Naudin, Professor	Antibes (Var)	France
	New Zealand Institute	Wellington	
	Regel, His Excellency Dr. E.	Botanic Garden St. Petersburg	
	Royal Society	London	
	” of Canada	Ottawa	
	”	Sydney	
	”	Melbourne	
	”	Adelaide	
	”	Hobart	
	”	Brisbane	
	Royal Irish Academy	Dublin	
	School of Mines	Ballarat	
	”	Sandhurst[6]	
	Smithsonian Institute	Washington	US.
	Society of Natural History	Boston	”
	” ”	Stuttgart,	Germany
	” ”	Danzig	”
	” ”	Bützow	”
	Physiographical Society	Lund,	Sweden

6 Now Bendigo, Vic.

Société des sciences physiques & naturelles		Bordeaux	89.12.07
Société Impériale des Naturalistes		Moscow	
Société des Amis de l'Histoire Naturelle		Moscow	
Society of Natural History	Rockhampton QL.[7]		
Society of Natural History	Bremen,	Germany	
Field Naturalists Club	Melbourne		
Solms-Laubach, Professor Count,	Strassburg,	Germany	
Treub, Dt. Director Bot. Garden,	Buitenzorg,	Java	
University of	Cordoba,	Argentina	
"	Cairo,	Egypt	
"	Melbourne		
"	Sydney		
"	Adelaide		
Wittmack, Professor,	Invaliden Strasse, Berlin		
Virchow, Professor,	University Berlin		
Woolls, Revd Dr.	Burwood, Sydney		
Planchon, Dr.	Montpellier, France		
Fowler, Professor	Kingston,	Canada	
Dawson, Sir W.	Ottawa	"	
Kerner, Professor	Bot. Garden Vienna, Austria		
Heldreich, Professor	Athens, Greece		
Fitzgerald, R. D. FLS	Hunter's Hill, Sydney		
Editor of the Chemist & Druggist,	Cannon Street, London		
Torrey Botanical Club	New York		
Society of Natural History,	Braunschweig, Germany[8]		

7 Qld.

8 The list was approved by the Premier and the books were forwarded by the Public Library by 18 May 1890.

Figure 13. 'The Baron Discourses'. One of three sketches accompanying an article about the visit of the Australasian Association for the Advancement of Science to the Australian Alps, 15-18 January 1890, *Illustrated Australian news and musical times* (Melbourne), 1 February 1890, pp. 16-17. Reprinted courtesy of the State Library of Victoria, Melbourne.

To William Thiselton-Dyer 90.01.28

RBG Kew, Kew correspondence, Australia, Mueller, 1882-90, ff. 299–300.

28./1./90.[1]

It is with much regret, dear Mr Dyer, that I learn of the failing of the first attempt, to introduce the two splendid Verticordias.[2] They moreover arrived there in winter, and must have suffered from frost, being subtropic plants. They were to have been shipped several months earlier, but delays arose in finding out the precise localities, as through depasturing and burning off of the scrubs such plants have become very scarce in their native haunts. What I now shall do, is to recommence at once operations in W.A, so that the plants become better established, earlier sent in the season and are more carefully attended to during the voyage. But I would beg you particularly, *to leave this introduction in my hands*. If you communicated on the subject with any party, connected with establishments here, my ideas and plans, in the usual style, would be taken up, *not the slightest mention be made* of the originator of the concern, and it be sought also in this instance to take credit of other peoples labour quite unscrupulously and unblushingly.

I will communicate with you from time to time on the progress of the second trial.

Regardfully your
Ferd von Mueller

My alpine tour with a party of the Austral. Association[3] was the first since *six years*, since which time I have not been away *a single day* from my little cottage through sheer *want of time*. Indeed I

1 MS is stamped: 'Royal Gardens Kew 10. Mar. 90'.

2 Report of failure not found. See M to W. Thiselton-Dyer, 28 August 1889, 31 January 1890, and for examples of repeated attempts to introduce the species to British horticulture, M to W. Thiselton-Dyer, 29 August 1892 and 25 June 1895 [all Collected Correspondence].

3 By train to the NE Victorian ranges, see M to W. Thiselton-Dyer, 18 April and 4 May 1889 [both Collected Correspondence].

90.01.28 have all that time hardly been out of door, unless occasionally at church and in the Liedertafel. The fresh bracing mountain-air has done me much good. To give you an idea of the magnitude of the work here, my letter-book shows about 3000 letters annually from my own hand.

90.03.24 *From Hans Gundersen*

A36 Royal Geographical Society of Australasia (Vic. Branch) papers, Mitchell Library, State Library of New South Wales, Sydney.

CONSULATE
FOR
SWEDEN AND NORWAY,
MELBOURNE.

Melbourne, the 24 March 1890[1]

Dear Sir,

Baron Nordenskiold's address,[2] delivered before the Royal Swedish Academy of Science on the 8th of January last, gives the history of the plan to a Swedish-Victorian Antarctic Expedition, referring first to a letter of 1887 from the Agent General for Victoria in London, with a telegram from the "Antarctic Exploration Committee of the Royal Society of Victoria" to the following effect: "Antarctic. Communicate direct with Baron Nordenskiöld. Would he feel disposed to cooperate local Society, each furnishing one vessel, Baron Nordenskiold commanding? etc[3] and also with a program for such expedition, made out by the said Committee, – thereafter mentioning your speech in the Royal Geographical Society

1 MS annotation by M: '(Postmark 27th March)'.

2 Nordenskiöld (1890).

3 to cooperate ... commanding? etc *is marked by double lines in the margin.*

of Australasia the 2nd September last,[4] and my subsequent letter to the Swedish Foreign Department. – He thereafter shortly exposes, what benefit and scientific results may be hoped from such an expedition, in the solution of different geodetic, hydrographic, meteorologic, magnetic and other geophysic questions, also alluding to practical results for hunting, whaling and fishing. – Finally he states, that having deliberated the matter with Baron Oscar Dickson, this nobleman has promised his pecuniary assistance to covering half the costs, on condition that the other half, not above £5000, be forthcoming from Australasia. And herewith Baron Nordenskiold concluded, recommending the Antarctic Expedition, that he calculates will sail from Sweden in 1891, to the Royal Swedish Academy of Science, which will, he hopes, give to this Expedition the same amount of valuable help and assistance as to the previous Arctic Expeditions. (This assistance has in former cases been very important, consisting in loan of scientific instruments, charts, books, necessaries for gathering and conserving plants and animals, exhibitions to young men of science etc. etc.). 90.03.24

Supposing that it will yet take some considerable time, before the Royal Geographical Society of Australasia can give a definite answer to the Swedish proposal, I venture to ask, if it might be possible for you, already at this stage of the affair, to give me, preliminarily, an idea of the prospects for realisation of the plan or your privat opinion as to the probability of obtaining, from any source, the desired Australian subsidy? Although it is nowhere expressly said, I am strongly under the impression that the Swedish now calculate upon this subsidy as a matter of course and are already commencing their preparations for the Antarctic Expedition.

I have the honour to remain,
Sir,
yours most respectfully,
H. Gundersen

To Baron Ferd. von Mueller,
K.C.M.G. M & Ph. D. F.R.S. etc. etc. etc.

4 B89.10.01.

90.04.03 *To Francis Barnard*

FNCV 014-019, Archives, Field Naturalists Club of Victoria, Melbourne.

3/4/90

It is with regret, dear Mr Barnard, that I find it impossible to make this evening free for any of the *three* meetings, to which I am invited including that of the Med. Society, on account of pressure of work for the outgoing European mail. I feel honored with being made a member of the Committee of the Field-Naturalists Club for preservation (in apt localities) of the indigenous vegetation and marsupials as well as various birds. I have however held from the commencement of this movement, that we could not possibly induce the Government, to cede so large an area for that purpose as the whole of Wilson's promontory;[1] the distance from the metropolis would also be too great for the multitude of the people, to derive an adequate advantage from such reservations. In my opinion our *first attention* should now be given, that not all the most picturesque vallies get defaced and alienated from the crown. Thus an application might be made to the hon the Minister of Lands *at once* for withdrawing from selection the best of the *Waratah-Vallies* in Eastern Gippsland, also all places in which large cataracts or cascades exist.

Prof Spencer and his companions of the E. Gippsland-tour,[2] made a year ago, would be able to describe these vallies and cascades as regards *precise localities*, so that the district-surveyors might become instructed, to keep these glorious spots intact, and perhaps some arrangements might be made thus far also, to prevent shooting in these reserved localities. Places at Mt Baw Baw, the Buffalo-Ranges and towards Cape Otway might also be protected.

Regardfully your
Ferd von Mueller

1 Vic.

2 From 28 December 1888 to 20 January 1889 Spencer and four other members of the Field Naturalists' Club of Victoria made an expedition to the Croajingalong area, Vic.; see Spencer & French (1889).

From Thomas Elder[1] 90.07.02

ML MSS.2134/1, Royal Geographical Society of Australasia (NSW Branch), Mitchell Library, State Library of New South Wales, Sydney.

Copy[2]

Knock Castle Largs, N.B.[3]
2nd July 1890.

My dear Baron von Mueller,

I was delighted to receive and have had much pleasure in carefully perusing your inaugural address to the Australian Association for the advancement of Science[4] in Melbourne. The paragraph in your address, referring to geography, has revived the interest, which I have always taken in Australian exploration, and as you say, that "Talent, Enthusiasm and Experience" are available at present, I cannot but agree with you, *that it would* be *almost* a *"reproach"*, *to permit the opportunity to pass for completing, what you properly describe as the main-work of Australian Land exploration.* You say, that this work in the past has *devolved on nine travellers only*, and *that space seems left now for only one more great explorer to rank with the nine.* This being the case, *I would like to furnish the tenth exploring Expedition;*[5] and if you will take the matter energetically, as you have done successfully on former occasions, I will hold myself responsible for the funds, so that no unnecessary delay will take place. Everything of course will depend on the Leader of the

1 The file includes a covering letter from A. Macdonald, Secretary of the Victorian Branch of the Royal Geographical Society of Australasia, to J. Maiden, his opposite number in the NSW Branch, 20 August 1890, reporting that M had brought Elder's letter before the Council of the Vic. Branch, and seeking the co-operation of the NSW group in arranging the proposed expedition. For printed versions of Elder's letter see both Melbourne *Age* and *Argus*, 23 October 1890, and the *Transactions of the Royal Geographical Society of Australasia (Victorian Branch)*, vol. 8, pt 2 (1891), 24-5.

2 MS is marked '*Copy*' in a different hand.

3 North Britain (i.e. Scotland).

4 B90.13.01.

5 The passages indicated have been underlined in another hand.

90.07.02 Expedition and his party; but being on the spot, and connected as you are with the various Geographical Societies, not only in the Australian Colonies but elsewhere, you will have no difficulty I should think, in helping me, to succeed in this scheme. What I would like you to do, my dear Baron is this – viz: to intimate to the Melbourne Exploration and Geographical Societies and to other kindred societies in the *Australian* Colonies, that I am willing to bear the whole charges of this proposed final expedition, provided a scheme is formulated and submitted for my approval. I have already referred to the importance of finding a thoroughly competent leader, as upon that much of the success of the expedition will depend; He ought to be a man not merely of pluck, courage, energy, influence over his men, and possessed of all the required physical qualities, but of such scientific attainments, as will enable him, to report advantageously on the topographical, geographical, botanical, geological and other features of the tracts of land, which he may travel over; but I need not expatiate at length on this branch of the subject, as no one knows better than yourself, the special qualifications, required by the party conducting such an expedition, if it is to be thoroughly successful. What I would like you to do is this: viz to intimate to the Melbourne Exploration and Geographical Societies, and to other kindred Societies in the *Australian* Colonies "*that I am willing to bear the entire cost.*" I hope that in the midst of your important and engrossing pursuits you will be able to help me in the way pointed out – Let a scheme be carefully prepared in concert with the best experts for the final important work of Australian Land-exploration, and transmitted to me, when, if approved, I shall immediately take steps, to have the scheme realized at my own charges. Have the goodness to write me at my present address and

Believe me
with sincere regard & every good wish
Yours sincerely
(signed) Thos. Elder

From Daniel McAlpine 90.07.18

P90/7868, unit 359, VPRS 3992/P inward registered correspondence, VA 475 Chief Secretary's Department, Public Record Office, Victoria.

5 Wallace St
Toorak,[1] 18th July /90.

My dear Baron,

I had no opportunity, when seeing you last, of speaking upon a very important matter in connection with Vegetable Pathology, so now write you on the subject.

The Hon. Mr Deakin, Acting Minister of Agriculture, sent for me last Thursday & had a long interview regarding my department. He recognised, first of all, that instead of being a temporary appointment, as it is under the Bonus system, it ought to be a permanent one. You understand that I am appointed Consulting Vegetable Pathologist under that heading of the Bonus Regulations where it is stated that "The Governor in Council may from time to time engage persons temporarily & during his pleasure", & at present I am only supposed to devote my spare time to the subject, but Mr Deakin considers it, quite as entitled to an independent position, & even more so, than the office of Government Entomologist.

The next question that came up for consideration was, as to what department the Vegetable Pathologist ought to be connected with. I considered that it properly belonged to the Government Botanists Dept & it is on that matter that I now specially write you. Mr Deakin here spoke of your invaluable services to the Colony & hoped that you would long be spared to serve the Government further. I may explain that the Hon. Mr Deakin is personally very favourable to me, & can hardly believe that it is six years since I delivered my letter of Introduction to him, & that the Colony had only begun to utilise my services, at least in a Government capacity. Also I would like to point out that I have already

1 Melbourne.

90.07.18 applied for a Laboratory & Museum, as well as a Library in connection with Vegetable Pathology, & their desirability & utility has been conceded. The buildings for these purposes to be located where the Old Training College (Mr Topp's) now stands.[2]

Bearing this in mind, it was suggested by Mr Deakin (in a very friendly & private way) that if you were to apply for an *Assistant* to attend to Vegetable Pathology, it would simplify matters considerably & he (Mr Deakin) could then proceed with the permanent appointment. You understand that he could not originate a matter like that without coming from yourself. I also mentioned the subject of Seed Control to him as being possibly assigned to the same officer.

Now, if you are of opinion that the subject of Vegetable Pathology ought properly to belong to your Department – just as Dr Cooke acts under the Director of Kew Gardens – & if you think, as I do, that a Museum of Vegetable Pathology & Economic Botany could go well together & not be scattered, say one building at the Botanic Gardens & so on, then I shall be happy to see you at any time which suits your convenience, to confer upon the matter. Many points could then be discussed which cannot be fully done justice to in a letter.

I feel that your support has obtained for me the position which I now hold as Consulting Vegetable Pathologist, & I am sure you will do what lies in your power to make the office of Vegetable Pathologist a permanent one, as its importance demands.[3]

I remain
my Dear Baron
Yours faithfully
D. McAlpine

2 Victoria's first Teachers' Training College, directed by C. A. Topp, was located in Spring Street, Melbourne. The College moved to new purpose-built accommodation in Carlton, adjacent to the University, at the end of 1889.

3 See M to A. Deakin, 19 July 1890.

To Alfred Deakin 90.07.19

P90/7868, unit 359, VPRS 3992/P inward registered correspondence, VA 475 Chief Secretary's Department, Public Record Office, Victoria.

Melbourne,
19 July 1890.

The honorable Alfred Deakin, M.L.A.
Chief Secretary.

Sir.

I have the honor to submit a letter from Mr D. McAlpine,[1] just received, in reference to his intended appointment of permanency as Vegetable pathologist, and feel it my duty to give respectfully my opinion, that such a position could most advantageously be connected with the Departm. of the *hon. the Minister of Agriculture* for the following reasons.

1, As it is understood, that Mr M'Alpine's services should be in *practical application to the crops* of the colony, the position must necessarily be closely connected with the agricultural and pastural interests here, just as that of the Gov. Entomologist; and the building, in which Mr M'Alpine is to operate, when not engaged in the field, will be close to the Office of Agriculture. Here, with me, is not even working space enough for the special work of the Gov. Botanist, and even if additional buildings were provided, it would cause continued complications, if Mr M'Alpine's work was constantly passing through *two Departments*.

2, The engagements of Mr M'Alpine would require frequent *cultural tests*, for which the Gov. Botanist's establishment has no ground whatever, while the Agricultural Department has all the facilities for the purpose through its *experimental* farms.

1 See D. McAlpine to M, 18 July 1890.

90.07.19 3, The aid, to be afforded to Mr M'Alpine from the establishment here, would be similar to that in chemistry through Mess. Blackett and Newbery, to that in geology through the Mines-Department, to that in meteorology by the astronomic observatory.

4, In the *United states* the position of Veg. Patologist has also been always under the hon. the Commissioner for Agriculture.

5, From the Gov. Botanists Department long since the botanic Garden and Seeds Magazine were withdrawn; my laboratory and its appliances went to the Department of Agriculture, the vegetable products and educts were transferred to the technologic Museum at the public Library so that my establishment is reduced to correspondence, to litterary work and the herbarium; it would be of the utmost difficulty *now* to extend or restore it to its original scope for purposes just discussed.

6, The Gov. Veg. Pathologist would, by the vast interests involved, require a particular library and special appliances, which are not to the required extent contained in my branch of the service.

7, Dr Cooke's position at Kew-Gardens is merely engagement of a small share of his time for general knowledge of fungi in connection with dried specimens, not for pathologic researches on crops which is outside of the functions of the great Kew Establishment. Indeed Dr Cooke is in a private position almost entirely as a litterary Gentleman, and depends mainly on the issue of his works for public sale. No Assistant is specially here required in the herbarium for the maintenance of the fungus-collection.

It needs not my assurance, that I will gladly aid at any time the Gov. Veg. Pathologist with advise and information at the Agricultural Department, where so much work will devolve on him in travelling, observing, sketching and recording, that it would be advisable to make his *appointment a full one*, so that he may devote his whole time for the important special duties, expected from him there; – but if I am allowed, to ask a favor in this instance, it would be this, *not to enlarge my establishment and burden additional obligations on it*, so that at this very late time of my life I may quietly continue and finish that particular work, which I have laid out for myself during the remainder of my earthly career, while strength still may last me.[2] 90.07.19

I have the honor to be,
Sir, your obedient servant
Ferd. von Mueller,
Gov. Botanist.

May I add, that if Mr McAlpine was even pro forma only placed into the Gov. Botanists Department, the real and probably harrassing *responsibility* for his work would necessarily fall on myself, whereas evidently the control over his work ought to be exercised by the Department for Agriculture.

To Jacob Agardh 90.08.01

Handskr. Avdl., Universitetsbibliotek, Lund, Sweden.

1/8/90

Eben, verehrter Herr Professor, kommt wieder ein Packet Algen für Sie bei mir an von unserem braven Freunde J. Br. Wilson. Das Packet ist zu gross, um mit der Buchpost gesandt zu werden; so soll es mit der nächsten Packet-Post, am 5 Aug, an Sie abgehen; es ist hier *nicht* geöffnet.

2 The Under Secretary, T. Wilson, submitted both M's and the enclosed letter from McAlpine to the Chief Secretary on 23 July 1890. Deakin minuted on 25 July: 'Mr McAlpine's letter attributes his own views & opinions, to which I only assented in a qualified & conditional manner, to myself – Inform him that the report of Sir F. Von Mueller is unfavourable to his proposition'. Wilson wrote to McAlpine on 29 July: 'In reply to the letter addressed by you to Sir Ferd Von Muller K.C.M.G. &c. in which you state that in your opinion your office of Consulting Vegetable Pathologist should be attached to the department of the Government Botanist, I am directed to inform you that the report received from that gentleman on the subject is unfavorable to your proposition.' A copy of this letter was forwarded to M on 11 August and he replied the next day: 'Very grateful for the information afforded and the decision arrived at by the honorable the Minister.'

90.08.01 Erlauben Sie mir die Vorfrage, ob Ihre Universität auch gelegentlich verdiente Gelehrte honoris causa zu Ph.D. promovirt? In diesem Falle möchte es Ihrer Erwägung zu empfehlen sein, ob Sie unseren Freund in Geelong auf diese Weise ehren und *überraschen* können. Er weiss nicht, dass ich diese Vorfrage mache; wenn also nichts geschähe, so wäre ja niemand compromittirt.

Sie ehrend und Ihnen alles Gute wünschend Ihr
Ferd von Mueller.

Mr Wilson ist seit 1857 Rector der Hochschule in Geelong

1/8/90

Respected Professor, another parcel of algae has just arrived at my place for you from our worthy friend J. Br. Wilson. The parcel is too large to be sent by book post, so it is to go off to you with the first parcel post on 5 August. It has *not* been opened here.

Permit me the preliminary enquiry, whether your university also occasionally confers PhDs on deserving savants *honoris causa*. In this case might I recommend to your consideration whether you could honour and *surprise* our friend in Geelong in this way. He does not know that I am making this enquiry, thus if nothing happens, no one would be compromised.[1]

Respectfully and wishing you every good thing
Ferd von Mueller

Mr Wilson has been the principal of the high school in Geelong[2] since 1857.

1 M repeated this request in M to J. Agardh, 4 July 1893. The degree was not awarded.
2 Geelong Church of England Grammar School, Vic.

From John MacKay 90.08.27

RB MSS M1, Library, Royal Botanic Gardens Melbourne.[1]

Clover Creek via Bourke
N.S. Wales
August 27th 1890

Baron Von Mueller Ph & M.D.

Dear Sir

Yours of 8th inst, in answer to my enquiries about the Burr Daisy,[2] duly reached me. – and I thank you much for the information so kindly supplied by you about it. – By this mail I am sending to you,[3] under separate cover, two more samples (No 2) of the "Eremophila Bowmanii", in Company with some other plant specimens, one of these samples has the fruit of it ripe, the other one is in bloom, hereabouts its habitat is on white clay flats and it grows to the height of from 2 1/2 feet to 3 feet, in the vulgar nomenclature it is called the blue bush, at one time it was very plentiful but is now getting rather scarce, as Stock are partial to it.

The Saltbush plants hereabouts are not at present in seed but when they are so I will get all the varieties that I possibly can of them collected and sent to you and as you wish I will also ask my neighbours round here to assist me in doing this for you – All the saltbush plants growing in this quarter are eagerly fed on by Stock and they are not at all so plentiful as they once were, indeed in many places where they formerly luxuriated their absence has now to be supplied by giving Stock Rock Salt. – It is a great pity that more care was not taken to prevent their being eradicated, to the extent that they have been. Another grand fodder plant "*Kochia,*" in common nomenclature "*Cotton bush*" which formerly was very plentiful is also now getting very Scarce and I believe, that, unless

1 MS found with a specimen of *Eremophila bowmanii.*
2 Letters not found.
3 Letter not found.

90.08.27 the Governments of the different Colonies take means to preserve specimens of all the most useful native fodder plants, by cultivating them in growth, – some of the best of them will soon become extinct, – which would be a great national loss. –

Yours faithfully
John MacKay

90.08.28 *From Friedrich Schmitz*[1]

RB MSS M108b, Library, Royal Botanic Gardens Melbourne.

Greifswald 28.8.90.

Geehrtester Herr Baron!

Die Sendung Meeresalgen-Material, die in diesen Tagen mir zukam, erinnert mich daran, Ihnen endlich einmal meinen verbindlichsten Dank auszusprechen für die verschiedenen Sendungen, die Sie mir seit meinem letzten Briefe haben zugehen lassen. In den beiden Algen-Sendungen, von Port Phillip und von der Armstrong-Bai, habe ich verschiedene Exemplare gefunden, die mir für meine Untersuchungen von Wichtigkeit sind, und die mir daher sehr willkommen waren. Vor allem aber habe ich Ihnen meinen besten Dank zu sagen für die Zusendung Ihres Syst. Census of Austral. Plants. Sind meine eigenen Studien zur Zeit auch nicht gerade speziell den Phanerogamen zugewandt, so hat doch das Buch mein Interesse in hohem Grade wach gerufen, so dass ich Ihnen aufs Beste danken muss für die freundliche Zusendung. Hoffentlich werden Sie recht bald dem ersten Theile die Kryptogamen nachfolgen lassen, von denen die Algen mich, wie Sie ja wissen, ganz besonders interessiren.

1 MS annotation by M: 'Beantw 5/10/90' [Replied 5 October 1890]. Letter not found.

Dabei muss ich Ihnen dann ein Versehen bekennen, das mir bei der Aufstellung meiner Florideen-Übersicht begegnet ist. Ich hatte leider übersehen, dass eine Gattung Müllerella bereits unter den Flechten existirt. Meine Gattung Müllerella muss ich daher anders benennen. Da bin ich nun aber einigermassen in Verlegenheit wegen der Wahl des Namens. Der Name Müllera ist ja bereits vergeben (von Linné). Es bleibt mir daher nur noch die Namensform Mülleromia zur Wahl übrig, eine Namensform, von der ich nicht weiss, ob dieselbe Ihnen willkommen sein würde. In jedem Falle aber möchte ich hierdurch einmal bei Ihnen anfragen, wieweit Ihnen diese Form Mülleromia genehm ist, eventuell aber Sie freundlichst bitten, mir eine andere Namensform vorzuschlagen, welche die Gattung als Ihnen gewidmet kenntlich macht. 90.08.28

Meine ausführlichere Florideen-Arbeit ist in diesem Jahre bisher nur sehr langsam vorwärts gerückt. Seit dem leichten Influenza-Anfall, den ich Ende des Jahres 1889 durchmachte, kränkele ich fast ununterbrochen, bald an diesem, bald an jenem Übel. Auch jetzt bin ich noch immer Patient und darf noch nicht wieder an eine volle Arbeitsthätigkeit denken.

Hoffentlich darf ich annehmen, dass Sie bisher von dieser allverbreiteten Influenza verschont geblieben sind. – Mit grösstem Interesse habe ich Kenntnis erhalten von Ihrem jüngst hier eingetroffenen Bilde, dass mir Herr Garteninspektor Dr Goeze gestern vorzeigte.

Mit nochmaligem verbindlichstem Danke und mit besten Grüssen

Ihr ergebenster
Fr. Schmitz

Greifswald[2] 28.8.90

Most esteemed Baron!

The consignment of marine algal material that arrived recently reminds me again to express my most obliging thanks to you for the various consignments that you have forwarded to me since my last letter. In both the algal consign-

2 Mecklenburg.

90.08.28 ments from Port Phillip and from Armstrong Bay[3] I have found various specimens that are of importance to me for my researches and which were therefore very welcome. However, above all I have to give you my best thanks for sending your *Systematic Census of Australian plants*.[4] Even if my own studies at present are not just especially devoted to the Phanerogams, the book has awakened my interest to a high degree, so that I must thank you very much for so kindly sending it. It is to be hoped you will be able to send the first part of the Cryptogams very soon,[5] of which the algae, as you know, interest me quite considerably.

At the same time I must confess a mistake to you, that has happened to me in the drawing up of my Floridaceae overview. Unfortunately I had overlooked that a genus *Muellerella* already exists among the lichens. I must therefore rename my genus *Muellerella*. Even so, I am now to some extent embarrassed on account of the choice of the name. The name *Muellera* is, you see, already taken (by Linné). Therefore the form of the name *Muelleromia* probably only remains as a choice for me, a form of name that I do not know would be welcome to you. However, in that case I would first like to ask you to what extent this form *Muelleromia* is acceptable to you, or ask you most kindly to suggest another form of the name to me, which would make the genus recognisably dedicated to you.[6]

My detailed Floridaceae work in this year has hitherto moved only very slowly forwards. Since the slight influenza attack that I experienced at the end of 1889, I have been unwell almost constantly, sometimes from this, sometimes from that illness. Even now I am still a patient and cannot yet think of a full working activity.

I hope I may assume that you have hitherto been spared from this universally distributed influenza. – I was informed with the greatest interest about your portrait that recently arrived here, that Garden Inspector Dr Goeze showed me yesterday.

Again with obliging thanks and with best greetings
Your most devoted
Fr. Schmitz

3 Not identified.

4 B89.13.12.

5 M's census of the Cryptogams was never published.

6 Schmitz renamed the genus *Muellerena* in Schmitz & Hauptfleisch (1897), p. 496.

From Wilhelm Bäuerlen[1] 90.08.30

RB MSS M21, Library, Royal Botanic Gardens Melbourne.[2]

Monga
30/8/90.

Geehrtester Herr Baron.

Vorgestern langte ich wieder hier an u. fand Ihren Brief mit einliegender Remisse, für welche ich herzlichst danke, u. schliesse nun die Quittung ein, so dass dieselbe mit der ersten Post abgeht, denn wir haben hier die Post blos zweimal in der Woche.

Die Eucalypte, die Sie in mitgehendem Packet erhalten, scheint sich mit keiner deren in Ihrer "Eucalyptographia" illustrirten übereinstimmen zu lassen, wesshalb ich Ihnen sogleich einiges Material mit Notizen zusende, u. wenn erforderlich oder erwünscht kann ich Ihnen mehr u. grössere Ex. später senden. Diese Art fand ich vorgestern auf einer Höhe von etwa 4,000 f. als ein kleiner dünner Strauch von ungefähr 5'-10' f Höhe u. 2 Zoll im Durchmesser. Gestern nahm ich mir vor den Strauch abwärts zu verfolgen, u. so fand ich dass er weiter unten malleeartig wächst, d. h. mehrere Stämme kommen von einem Wurzelstollen, u. noch tiefer hinab wird er baumartig von ungefähr 40 f Höhe u. 6"-10" Zoll im Durchmesser, immer mit einem starken dicken Wurzelstollen, so dass die Art von 4,000 f bis ungefähr zu 2,500 f hinabsteigt. Blüthe, Frucht, Knospen u. Blätter bleiben sich auf den verschiedenen Elevationen immer gleich. Die Art wächst auf sehr steilem Boden, zwischen wilden Felsblöcken u. losem Steingeröll. Die sehr steile Bergseite auf welcher sie vorkommt schaut nördlich, u. auf dem südlichen Abhang des Berges konnte ich bis jetzt keine Spur derselben entdecken. – Zwar bin ich immer etwas schüchtern Ihnen Eucalypten Ex. zu senden, doch denke ich dass die mitgehende

1 MS annotations by M: 'Beantw 7/9/90 F.v.M' [Answered 7/9/1892 F.v.M.] and 'N. 321. (1884) vielleicht var von E. viminalis' [No 321 (1884), perhaps a variety of *Eucalyptus viminalis*].

2 MS found with specimen of *Eucalyptus baeuerlenii* F. Muell. (MEL 733006), 'coll. W. Bäuerlen, Sugarloaf Mountain at the Clyde River, NSW, 1890'.

90.08.30 interressant ist, u. Ihnen nicht die Zeit unnöthigerweise raubt zumal da der Standort selbst schon merkwürdig ist, indem derselbe früher schon zwei neue Arten lieferte, namlich Eriostemon Coxii u. Hakea Macraeana. Auf der obersten Elevation kommen die drei Arten miteinander vor. Eriost. hört aber schon nach ungefähr 200 f Niederung auf, während die Hakea die Eucalypte ganz hinunter begleitet. Auch ist diese Eucalypte die einzige die jetzt in dieser region in Blüthe steht. Die Rinde ist glatt u. bräunlich von Farbe, sie streift sich in schmalen Bändern ab. Ich füge einige Notizen bei, welche an gepressten Ex. vielleicht nicht auffallend sind u. entgehen möchten. Der Kelch ist innen oft, jedoch nicht immer ganz blutroth oder rosenfarbig. Auch der Griffel ist oft ganz roth u. manchmal die eine Hälfte roth u. die andere grün, manchmal aber auch ganz grün. An der Aussenseite der Blüthe läuft oft, da wo die Stamen an dem Kelch sitzen ein schöner rother Ring herum, welcher dann den Blüthen ein liebliches Ansehen gibt. Es mag wohl blos Zuffall sein, aber an zwei Blumen bemerkte ich, dass sich der Griffel der Länge nach zu spalten anfing; die Theilung aber schien in der Mitte, d. h. der halben Länge des Griffels zu beginnen u. nicht ganz nach unten oder oben auszulaufen. Die Blätter sind ohne Glanz u. von gleicher Farbe an beiden Seiten. Die Blattkanten sind meistens roth, u. die äussere (convexe) Kante ist gewöhnlich röther, denn die innere (concave) Kante. Blätter des jungen Zustandes liegen bei. Die Krone des Baumes ist sehr licht, sogar unter den Eucalypten.

Mit besten Wünschen bleibe ich Ihnen stets ergeben

Wilhelm Bäuerlen
C./o. Mr George MacRae
Clyde Road
Brentwood.

Monga[3] 90.08.30
30/8/90.

Most esteemed Baron,

The day before yesterday I arrived back here and found your letter with the enclosed remittance, for which I thank you most sincerely. I herewith enclose the receipt, so that it leaves with the first mail, because we have the mail here only twice a week.

The eucalypt, which you will receive in the accompanying parcel, does not seem to agree with any of those illustrated in your *Eucalyptographia*. For this reason I am sending you some material at once, together with some notes. If required or desirable I can send you more and larger specimens later on. I found this species the day before yesterday at an altitude of approximately 4,000 feet as a small sparse shrub, about 5-10 feet high and 2 inches in diameter. Yesterday I decided to follow the shrub downhill. I discovered that further down it grows mallee-like, that is, several trunks arise from the one rootstock, and still further down it becomes tree-like, to about 40 feet high and 6-10 inches in diameter, always with a strong thick rootstock. Thus the species descends from 4,000 feet to about 2,500 feet. Flowers, fruit, buds and leaves remain the same at the different elevations. The species grows on very steep ground, between rough rock masses and loose boulders. The very steep side of the mountain on which it occurs faces north, and so far I have not been able to find any trace of it on the southern slope of the mountain.

I am always a bit shy about sending you *Eucalyptus* specimens, but I think that this present one will be of interest and will not waste your time unnecessarily. Even the locality itself is remarkable in that it already yielded two new species in *Eriostemon coxii*[4] and *Hakea macraeana*.[5] The three species occur together at the highest elevation. But *Eriostemon* disappears altogether 200 feet further down, while the *Hakea* accompanies the eucalypt all the way down. This *Eucalyptus* is also the only one at present in flower in the area. The bark is smooth, brownish in colour, and is shed in narrow strips. I add a few notes, of things which may not be very conspicuous in the pressed specimens and could be missed. The interior of the calyx is frequently but not always blood-red or rosy red. The style is also often completely red, sometimes half red and half green, but sometimes also completely green. At the level where the stamens are attached to the calyx a lovely red ring often runs around the outside of the flower, which gives the flowers a very pretty appearance. It may be just coinc-

3 NSW.
4 B84.12.02, p. 62.
5 B86.11.02, p. 430.

90.08.30 idence, but I noticed that on two flowers the style had begun to split lengthwise. The split seemed to start in the centre, i.e. halfway up the style, and did not continue all the way to the top or the bottom. The leaves are without gloss, with both sides the same colour. Leaf edges are usually red, and the outer (convex) edge is generally redder than the inner (concave) edge. Juvenile leaves are enclosed. The crown of the tree is very sparse, even among the eucalypts.

With best wishes I always remain

your devoted
Wilhelm Bäuerlen
c/o Mr George MacRae
Clyde Road
Brentwood[6]

90.11.11 *To Edward Ramsay*

AMS 355, Holograph letters from noted scientists and individuals, item no. 46, Australian Museum, Sydney.

11/11/90.

In reply to your letter of the 8 Nov.,[1] dear Dr Ramsay, I beg to point out, that the Expedition, which the munificence of Sir Thomas Elder is calling forth,[2] must necessarily be a *light party,* as in all probability vast stretches of waterless country will have to be traversed, and as great privation and toil are likely to occur. To maintain under such circumstances the needful discipline and unity of purpose, all members of the party must be under the absolute control of the Leader, and all must share in the general work of the field. Any deviation from this well recognized principle would introduce a disturbing element into the enterprise, which may prove disastrous to the whole.[3] The opportunities for collecting can in

6 NSW.

1 E. Ramsay to M, 8 November 1890 [Collected Correspondence].
2 See T. Elder to M, 2 July 1890.
3 After this M wrote 'enterprise' and then 'expedition' and finally crossed out both.

an expedition, the mainobjects of which are geographical, also only be very limited, so that no delays occur, and that the lines of the exploration may cover within the season as much of the blank spaces on the map as possible. Let it be considered also, as fully pointed out to Mr Gipps of the N.S.W. geographic Council, that the mere Salary of any one in the party is only a small fraction of the expenditure, which will arise to the noble originator and sustainer of the enterprise for each individual member of the party, to provide for locomotion, protection and sustenance. You will deem it also but just that the first series of specimens of any kind, resulting from the expedition, should fall to the share of institutions in that colony, which through one of its leading Residents provides the whole means for the contemplated exploit, in this instance S.A., with which from 1847 to 1852 I was as a settler also identified. The monetary question is however not raised, but equality of footing as regards all members of the party insisted on. The geographic Council of Victoria has already shown itself very favorable to the appointment of Mr Helms as a zoologic Collector, provided that he enters as an ordinary officer of the Expedition, subject to all the conditions imposed on the other members of the party. But you will understand, that no definite engagements can be made in any direction, until Sir Thomas Elder has approved of or altered the propositions, which have been submitted to him in a letter, sent two weeks ago.[4] 90.11.11

With regardful remembrance your
Ferd. von Mueller

Mr Vogan's well sustained aspirations for a position in the party will also be fully considered at the right time

All Documents, emanating from the Expedition and so all collections are necessarily the property of Sir Th. Elder, who alone will dispose of them

4 Letter not found.

90.11.14 *To Léon Dejardin*

Institute for the Study of French-Australian Relations, Archives, Department of French and Italian Studies, University of Melbourne.[1]

Melbourne,
14 Nov. 1890

To the Chevalier De Jardin,
Consul General for France &c &c

Allow me, dear Consul, to approach you on a subject, which is very painful to me, and which – from what I *suddenly* learn this day – requires some explanation of my own. I notice just from a weekly journal, that at the geographic meeting on Friday last[2] or subsequently some offensive remarks occurred, of which I never heard before. If they were made at the meeting, I *did not hear* them, otherwise I would at once have risen, and pronounced such expressions as highly improper. After the Rev. Mr. Macdonald, who unfortunately against my special request drifted into politics, had read his remarks, I at once rose; and emphaticallly gave it as my opinion, "that I was convinced, the French Government would with sanctity see the convention maintained, entered into with Britain, and that I felt also sure, the Government of the French Republic would remove any difficulties and any causes, which might lead to injustice or hinder progress of settlement, if clearly pointed out."

1 The text is taken from a photocopy, the original having been returned to the French consular archives, Nantes. Published in Home & Maroske (1997), pp. 28-9.

2 Two papers were given at the meeting of the Royal Geographical Society of Australasia (Vic. Branch) on 7 November 1890. The first was by J. Lindt on his ascent of the Tanna volcano in the New Hebrides and the second was by the Rev. D. Macdonald on the condition of affairs in that group. The *Argus* (Melbourne) reported that Macdonald 'advocated that France should either join with Britain in framing and enforcing equally upon all Europeans all necessary New Hebrides laws as to land, labour, and trade, or yield any claim to the islands in consideration for the concession of territory in some other part of the world. In the meantime, he urged that Australia should insist upon the spirit of the existing convention or joint protectorate being wholly observed, so that in every respect British subjects in the New Hebrides should be placed on an equal footing with the French, and that France must either grant Australia her rights or recede from the convention.' See *Argus* 8 November 1890, p. 9.

As usual the extempore remarks of Speakers are not reported by the press, as they are too long for record, and thus you and your compatriots can not be aware of the position, which in justice to France I took at the meeting. 90.11.14

It was only *this morning*, that I learn of the subjects, brought forward by Mr. Macdonald, having become matter of *official correspondence*,[3] and I take the earliest oppportunity of expressing to you as the dignified representative of France my sorrow, that anything, arising from the last geographic gathering should have hurt your and your compatriotes feelings in any way.

It was only expected, that the Missionary Macdonald, as he happened to be in Melbourne, should supplement the *itinerary* observation [...][4] Lindt.

I spoke myself to the Rev. Gentleman, insisting that no political opinions should be brought before any geographic meeting; and had I been able, to see his manuscript, I would have struck out some of the passages, as I have done on a former occasion, and so far as hurtful they certainly shall not appear in the proceedings of our geographic branch society here.

Be so kind, to convey these sentiments to the Government of your great country, and allow me to reiterate my assurance, that I shall always endeavour, as I have done during the last 36 years, to promote in my small professional and departmental way, also rural and scientific, the interests of the great nation, in which I have so many science-friends, and from which I experienced so many acts of generosity!

Regardfully your
Ferd. von Mueller

3 M's anxiety may have been prompted by a report in that morning's *Argus* (p. 9) of a discussion the previous day at the General Assembly of the Presbyterian Church of Victoria concerning the situation in the New Hebrides. Here, reference was made to agitation in the Victorian Legislative Council by James Service 'to induce the French, German and American Governments to come into the same arrangements with regard to the prohibition of these articles' (i.e. guns and intoxicating liquors) as the British Government already imposed on its citizens, and to responses received from those governments.

4 *illegible.*

91.01.21 *To Joseph Hooker*

RBG Kew, Kew correspondence, Australia, Mueller, 1882-90, ff. 331-5.

21/1/91.

On my return voyage to Melbourne from N. Z.,[1] dear Sir Joseph, I have the last opportunity quietly to write, and so like to say a little about my movements in those islands, where at *almost every step* the vegetation reminds of *you!* The universal regret has been, that you could not come personally at this scientific festival to the scenes of your earliest triumphs. – This year I could less easily take out a month of my time for absense from my Department, than could likely be done for twice or thrice that time in any other year. But the duties for getting the new Central Austr expedition[2] ready, which is to map for the first time portions of the two blanks, one as great, the other greater than Britain, call me greedily back, so that the party may be organized in time, for being at the outskirts of the settlements at the beginning of the cool season, these two tracts of country being in all likelihood extensively waterless. Sir Thomas Elder, from whom I had a London telegram last month,[3] will also be back in a few weeks, so that I must not delay the fitting out of the new Expedition. It is an outcome of the Melbourne-meeting of the Austr. Assoc., just as most probably the N. Z. meeting will bring about the getting of the remaining means for Nordenskiold's contemplated antarctic enterprise, which should tend largely to increase subsequently the revenue of all the Australian colonies, but particularly that of N. Z.[4] I spoke emphatically on the objects of these antarctic plans in the geographic Section, and must own, that my appeal met with an enthusiastic response, especially as Sir James Hector and Prof Kernot delivered also vigorous

1 M had attended the meeting of the Australasian Association for the Advancement of Science held in Christchurch, January 1891.

2 Elder Scientific Exploring Expedition, 1891-2.

3 Telegram not found.

4 See Home et al. (1992). Nordenskiöld's proposed expedition did not proceed.

speeches on the subject,[5] and as Mr Griffith, the President of the geographic Section, made antarctic concerns the main-subject of his inaugural adress.[6] As the voyage from Port Phillip to N Z takes one week going and one week coming, and as the Assoc. meetings required one week more, I had only one week for tours, and devoted this to a journey into the vicinity of Mt Cook. This left me only one day for the glaciers,[7] to be in time for the opening meeting at Christchurch. During that day I crossed the huge boulders of the Hooker-River, which are overlaying the glacier, and shifted gradually forward by the huge ice-masses, which press on them from the flank of the alps above. Vegetation however does not exist among these rock-fragments, as they are slowly moving and afford therefore no permanent footing to plants. The Hooker-River, where my companions and myself crossed it under the guidance of Mr Huddlestone is a mile wide and 2500 feet above sea-level; and the aspect of the alps, so close by, is magnificent beyond description. but on account of the vast extent of icy mountains in long chains the cold descending currents render at the Hooker-River the vegetation quite alpine, among other plants the glorious Ranunc. Lyalli being conspicuous. The plains and lower ranges towards the alps coming from the eastern side are absolutely treeless, and the two great lakes towards Mt Cook near the Hooker River[8] are also devoid of any fringe vegetation, so that I will push your Eucalyptus Gunnii and E. urnigera into this region for embellishing the landscape and for affording fuel and timber. Many Pines should succeed also there, but it is too cold for Pinus insignis and Cupressus macrocarpa, the two quickest growing of Conifers. 91.01.21

Your Handbook[9] has delt 25 years ago almost exhaustively already with the N. Z. flora, so that I saw no novelties, unless it is a

5 Neither M's comments, nor those of Hector and Kernot, were published in the report of the meeting. No doubt all three spoke in the context of the presentation of the report of the committee (of which M was a member) appointed at the previous Congress 'to consider the question of Antarctic Exploration'.

6 Griffiths (1891).

7 The Tasman, Hooker and Mueller glaciers near Mt Cook.

8 Lakes Tekapo and Pukaki.

9 J. Hooker (1864-7).

91.01.21 Viola near V. filicaulis, but more stoloniferous and with a yellow tinge towards the base of the lowest and the lateral petals. It was to me however of high interest to notice the varied consociations of plants in N. Z. and to see those in a living state, which had I had not in cultivation, while at the bot. Garden.

The presense of Prof Goodall,[10] the President of the American Assoc. for Advancement of Science at the N. Z. meeting shed a particular lustre on it, he bringing a greeting across two oceans![11] A more genial and generous man, so learned, and yet so modest, can not be imagined.

Sir James Hector intended, to take me with others through the North Island, of N. Z., but my duties call me back to my Department by the end of January.

Ever regardfully your
Ferd. von Mueller

It may interest you for the suppl. of the gen. plantarum[12] that Baillon points out to me the affinity of Macgregoria to Floerkea and Limnanthes[13]

I have no Pachystoma (Apaturia) from Australia (Arnhem's Land)

Sir James Hector kindly offered to take me through the northern Island but I could not delay my return

10 G. Goodale.
11 For Goodale's remarks see Goodale (1891).
12 No supplement to Bentham & Hooker (1862-83) was published.
13 Baillon's letter not found.

To Jane Gray 91.01.24

Gray Herbarium Archives, Harvard University, Cambridge, Massachusetts.

24/1/91.

On my return-voyage from New Zealand, dear Madam, where I had by the rules of the Austral. Association for Advancement of Science to instal Sir James Hector as my successor in the Presidency, I avail myself of spare-hours for correspondence, and as the genial Professor Goodale, to whom I mentioned two remarkable passages from letters of your lamented Consort, desired me to communicate them to you, I feel it a particular but sad privilege to do so, – altho' I write only from memory.

The one passage constituted the whole of a letter, written to me at the time of the dreadful fratricidal civil war,[1] when the arms of the Southern States were in the ascendence. Thus he wrote "My dear Dr. Mueller. I am distracted, but do not forget you, and now send you a parcel of plants from Texas" Yours Asa Gray. The other letter, to which I refer,[2] was also of laconic but significant briefness: "My dear Baron. This is the Centennial day, and in the streets it is noisy joyous, but I get a quiet day for my work." This shows with what close application your renowened husband devoted himself to science-considerations.

Prof. Goodale's presence at the N.Z. meeting of the Australian Association[3] shed a great lustre on it. In my short opening speech, referring to the American president, I paid him homage for having crossed two oceans to bring us a greeting from antipodal distance of your great Science-Union, and added, that the highest praise, we could bestow on him, was to recognize in him the Successor of Asa Gray![4]

Ever, dear Mrs. Gray, yours with profound regard

Ferd. von Mueller.

1 The U.S. Civil War, 1861-5. Letter not found.

2 Letter not found.

3 Third meeting of the Australasian Association for the Advancement of Science, Christchurch, NZ, January 1891.

4 These comments were not included in the published version of the speech; see B91.13.08, pp. xxvii-xxviii.

91.01.24 It was so kindly thoughtful of Prof Asa Gray to invite me specially to the American Assoc. meeting, when he was President,[5] but in the administrative Departments in Victoria never vacation-time arises, and as I was then extra-taxed also with unusually heavy work, I could not make the needful time free, and thus one of my greatest worldly wishes, to meet Asa Gray, remained unfulfilled. When I think of him, I like to express myself in slightly altered words of Caroline Pichler, written at the time of the death of Koerner,[6] a companion under arms in 1813 of my father, "Also stand er hoch vor Wissen's Söhnen, Weckte mächtig mit des Wortes Tönen, Die Begeisterung die ihn durchglüht"![7]

91.03.12 *To Thomas Wilson*

R91/2419, unit 397, VPRS 3992/P inward registered correspondence, VA 475 Chief Secretary's Department. Public Record Office, Victoria.

Melbourne

T. R. Wilson Esqr 12/3/91.
Under Secretary.

Sir

In reply to your letter of yesterday, N. 1252,[1] I have the honor to inform you, that I will readily wait with getting the additional room at the Botanic Museum til *early in the next finance-year*, altho' it is much needed. Not only is the space in the Main building of the museum much over burdened, but also the necessity has arisen for utilizing a *detached new room for employing the Bisulphid*

5 American Association for the Advancement of Science. Gray was president in 1872. Invitation not found.

6 The poet Karl Theodor Körner (1791-1813) was killed while fighting against the French in the Mecklenburg-based Lützow Corps.

7 'Thus he stood high before the sons of knowledge, mightily with resounding words awakening the enthusiasm which burns in him.'

1 See T. Wilson to M, 11 March 1891 [Collected Correspondence].

of Carbon as a preservation of the botanic collections, because for the greater safety of the botanic Museum it has been arranged, that the Junior, lately appointed, *sleeps in the building*, where the inhaling of the noxious preservative at closed door during nights would become *prejudicious to his health*. I will therefore defer the use of the Bisulphide of Carbon altogether til the new room can be erected by *revoting* the needful amount in the next finance-year, that being the *only* vote for public works in the Gov. Botanists establishment. The error, as regards the heading of the vote this year, called *Repairs*, arose in the public works Department.[2] 91.03.12

I have the honor to be,
Sir,
your obedient servant,
Ferd. von Mueller,
Gov. Botanist

To Edward Ramsay 91.05.13

ML MSS.562, Letters to E. P. Ramsay 1862-91, Mitchell Library, State Library of New South Wales, Sydney.

13/5/91

I think, dear Dr Ramsay, that I sent you a book on nebular theory, of which the author J. Spottiswoode Wilson, a former travelling companion of mine, appointed by Sir Rod. Murchison sent me some Copies.[1] Uncautiously I sent them out, before I had read them, and now find to my horror, that the book contains some irreligious passages in the worst of wordings. Should you therefore have received this highly objectionable book, please destroy it. You will be aware that I am a strict adherent to christian faith.

Regardfully your
Ferd von Mueller

2 MS annotation by G. Langridge, Chief Secretary: 'noted GDL 18/3/91'.

1 J. S. Wilson travelled as geologist on the North Australian Exploring Expedition, 1855-6. The book in question was presumably Wilson (1890).

91.06.22 ## *To Thomas Wilson*[1]

R92/2768, unit 454, VPRS 3992/P inward registered correspondence, VA 475 Chief Secretary's Office, Public Record Office, Victoria.[2]

Melbourne
22 June 1891,

The Undersecretary.

Sir.

As I shall have attained my 66th year at the end of this month,[3] I have the honor to solicit, that the honorable the chief Secretary will allow me, to continue my services as Gov. Botanist also in the coming finance-year.[4] I feel fully capable, to carry on the duties of the Department as in former years, and would like to render the long and unique experience, which I possess available also in future, so long as my mental power and physical strength may last. As mentioned on former occasions, I have sunk all the best years of my life and all my worldly means in the obligations of this establishment, formed by myself, or in researches connected with my scientific position, and would much wish to do so also in future.[5]

I have the honor to be,
Sir, your obedient servant
Ferd. von Mueller,
Gov. botanist.

1 See also M to G. Berry, 27 June 1885.

2 Within the file this letter is registered as R91/5970.

3 MS annotation on the front of the file: 'Date of Birth – 30/6/25'.

4 See H. Moors to M, 22 July 1891, and M to A. McLean, 28 November 1891 [both Collected Correspondence].

5 The file contains several notes, one of them in shorthand. One note is the draft of a letter to the Chairman of the Public Service Board, dated 30 November 1891, and identified as 're Mueller & Hayter's [the Government Statist] [retirement *deleted*] retention': 'Ref[errin]g to a letter add[e]d by me by the C. S. dir[ection]s on the [*space left in MS*] to the Secy to Prem Dept recom[men]d[in]g that the services of Baron V. M. & Mr Hayter be further cont[inue]d, I am now directed to inform you that it has been decided [*four lines deleted*] to abolish the Govt Botanists branch of this Dept on the 31st March next'. A further note states: 'Mr Hayter's retention not yet dealt with by Cabinet. Awaits Premier's return to town 5 January 92.'

To Walter Gill

91.07.00

RB MSS M33, Library, Royal Botanic Gardens, Melbourne.[1]

213, Ixodia achilleoides, RBr.
The narrow-leaved form
214, Verbascum Creticum.
(Introduced.)
216, Helichrysum Blandowskianum, Steetz.
217, Darwinia micropetala, Bentham.
218, Helichrysum adenophorum, F.v.M.
219, Patersonia longiscapa, Sweet,
unless it is a dwarf plant
220, Calocephalus Brownii, F.v.M
215, Acacia[2]
This I cannot at once identify. Does the fruit really belong to the same species as the flowering specimen. Where were the seeds obtained? Perhaps this is a new species. It is however near A. melanoxylon
222, Styphelia concurva, F.v.M.
223, Apium prostratum, Labill
224, Scaevola aemula, R. Brown
225, Helichrysum retusum, FvM
226, Acacia retinodes, Schlecht.
227, Leptospermum lanigerum (var.)
228, Acacia Oxycedrus Sieber

The Goodenia, dear Mr Gill, is a *variety* of G. ovata, described by R.Br.[3] as G. acuminata.

Just as in the human species so in those of plants, some forms may rise to an almost complete constancy at particular places, but that constancy will break down in other places. Such forms can

1 Identifications 213-215 are on a single lined foolscap sheet, while the remainder of the letter – if it is in fact part of the same document – is on an unlined double foolscap sheet from which the top half of the front page, with identifications 222-228, has been detached.

2 See also M to W. Gill, 2 September 1891 [Collected Correspondence].

3 Robert Brown.

91.07.00 never be regarded as *true* species. An enormous number of varieties have been erroneously described as species. The occurrence of fertile hybrids sometimes impair the specific discrimination To recognize a *real* species in its permanent value, it must be studied over wide areas. The variability of species is greater in our clime than in Britain. A *variety* can of course have a distinct name as such, but the appellation can never be specific in the true sense. It will require yet studies through long times to settle the exact limits of *all* species.

Regardfully your
Ferd. von Mueller

My own *field*-studies have extended over 40 years in Europe, Australia and for a short time in S. America.

When I commenced to study plants I admitted far more species than later with longer experience and mor[4] matured judgement[5]

91.07.01 *To [William Thiselton-Dyer]*

RBG Kew, Kew correspondence, Australia, Mueller, 1891-96, ff. 7-8.

1/7/91[1]

Private

For some time, my honored friend, I intended to write to you specially in reference to the sendings of Fungs for elaboration in England. After the lamented death of Berkeley, they went to Dr Cooke, usually direct, with an understanding, that a series of the specimens, so far as available, should be left at Kew, and a series be returned to my establishment. This has worked satisfactorily to all of us hitherto; but I think it right now to inform you

4 more?

5 The position of the postscript on the sheet shows that it was written after the list of identifications 222-228 was detached.

1 MS is stamped 'Royal Gardens Kew 19. Aug. 91'. MS annotation by Thiselton-Dyer: 'An[swere]d 19/8/91'. Letter not found.

officially of what you likely have learnt already indirectly, that Dr Cooke is engaged to write a descriptive volume on the mycology of Australia for which his thus far now unrivalled experience and his facilities at Kew he is specially suited.[2] 91.07.01

The subject was discussed some few years ago in correspondence between him and me, when I proposed, that the estimated cost of £400 should be borne by N.S.W., V. QL[3] and South-Australia on a similar subsidial arrangement as that effected for the issue of the 7 volumes of Flora Australiensis.[4] I wrote at once to Mr Ch. Moore and Mr F. M. Bailey, but neither of them could procure the needful £100 each, whereas I made myself responsible for Victoria to that amount. I had no fund in the Department to which I could charge so large a sum as £400, and times were not propitious here to get from the Ministry a special grant, particularly also as a clamour had arisen likewise for a bryologic, lichenologic and phycology volume, the number of Australian Algs for instance, now known being *double* that recorded by the never to be forgotten Harvey.[5] This year the Agricultural Department here, on which Mr M'Alpine (much through my influence) received an appointment as veget. Pathologist, while Mr C. French became Entomologist, moved in the revival of the subject, to get through Dr Cooke the proposed systematic enumeration of the about 1500 of the Austr. Fungs done and printed in London; this succeeded by communicating not with the bot Gardens but with the agricultural Departments of the 4 colonies, and when I was consulted about it I gave it of course strenous support.[6] Having now duely brought this arrangement under your notice, I have resolved to send in future for this volume any new material accumulating in my Department, *always to you*, and would beg, that it be arranged to give

2 Cooke (1892). See M. Cooke to M, 1 September 1889.

3 Vic., Qld.

4 Bentham (1863-78).

5 Harvey (1858-63).

6 strenuous? For correspondence relating to the proposal see M. Cooke to M, 1 September 1889 and F. Bailey to M, 17 October 1889 [Collected Correspondence]. On its title page, Cooke (1892) was stated to be published 'for the Departments of Agriculture in Melbourne, Brisbane, Sydney, Adelaide, Hobarton'.

91.07.01 the name of each finder with the addition ("FvM or BvM"[])[7] in brackets when the specimens came from or through me,[8] and this should also be done with the Kew material from me since 1857 as I must watch the interest of my Department, and as it would be but the *barest justice* to myself, as in many cases the material is from paid collectors or obtained in interchange. More over in hardly any case any one, who sent to me fungs would through this long series of years have done so, had it not been for my special inspiration Omitting my name would make it appear, as if all these fungs had gone from the Collectors *direct* to Dr Cooke. I have treated that Gentleman with every consideration, not followed the example of Mr Tepper of sending the material to Saccardo, Winter and others, have allowed him to keep a set for his herbarium, have slightly remunerated him for naming material, have subscribed to all his publications, have paid for many illustr. plates of the Grevillea.[9]

Regardfully your
Ferd von Mueller

Kindly acknowledge this letter at receipt and let me have any remarks of yours thereon

7 *editorial addition.*

8 Cooke did not accede to the request in Cooke (1892) where, instead of giving collectors' details under each individual species, he made a general acknowledgment to collectors – 'Messrs. F. M. Bailey, Dr. Berggren, Mrs. Flora Martin, Baron F. von Mueller, F. Reader, Schomburgck, and others' (p. v) – with a more specific thanks to 'the perseverance of Mr. F. M. Bailey, Mrs. Martin and Baron F. von Mueller, in continuing to secure and forward specimens to England for identification' (p. vi). However, from vol. xvii of *Grevillea* on, Cooke indicated in his series on Australian fungi the species described from specimens sent by M by adding an asterisk before the species name, as well as giving details of the collector in parentheses at the end.

9 Cooke (1872-92).

To James Stirling

Private hands.

91.07.27

27/7/1891.

It needs not my assurance, dear Mr Stirling, that I will gladly support your candidature for a geologic officership in the forthcoming Swedish-Australia Expedition,[1] and I will place your letter, as a timely application, before the Antarctic Committee on Monday of next week, when its monthly meeting will be held. Let me however remark that you should not be influenced by any prospect of discovering living plants, unless perhaps sea-weeds, because terrestrial vegetation ceases southward at Macquarie Island. It is however probable that, in your geologic researches, you may discover some vegetable fossils. We are, as Antarctic Committee, now in communication with Barons Nordenskiold and Dickson as regards final arrangements; until the main plans are agreed on, no appointments can be made. But I will watch your interest at the right time. At this stage nothing definite can be done. You must kindly however remember that the other fund-contributing Colonies may also put forward scientific candidates for positions in the expedition.

Regardfully your,
Ferd. von Mueller.

Have you as the head of a family well considered the dangers to be braved in approaching the south polar regions?

1 The proposed expedition to Antarctica under Baron N. A. E. Nordenskiöld did not eventuate.

91.08.14 *To Jacob Agardh*

Handskr. Avdl., Universitetsbibliotek, Lund, Sweden.

14/8/91

Diesmal schreibe ich, edler Freund, um anzuzeigen, dass ich Ihnen eben per Post ein Packet Algen von Cape Flattery, in Nordost-Australien, gesandt habe. Es mag nichts Unbekannter[1] darunter sein, aber die Sammlung ist doch insoweit wichtig, als wir von diesem besonderen Standorte bisher noch keine Algen hatten. Es mag so ein Zuwachs asiatischer doch noch für die Liste der australischen Algen gewonnen werden.

Es ist eine geraume Zeit, seit welcher ich *keine Rücksendungen* für die Staats-Sammlung hier erhielt. Dies setzt mich als öffentlicher Beamter in *Verlegenheit*, denn jetzt sind viele von mir gesandte Arten veröffentlicht, von welchen keine Repräsentanten im Herbar hier sich befinden. Es fangen nach u. nach auch die jungen Australier an zu studiren die Meeres-Flora, und wenn viele Lükken in den entsprechenden Museum-Sammlungen fort bestehen, wäre ich Tadel öffentlich ausgesetzt. Wenn aber Unica untheilbar sind, erwarte ich nichts zurück, möchte dann aber hier eine *Abbildung* vorlegen können, z. B.[2] von Amansia mammilaris. Wenn von neuen oder seltenen Arten auch nur Bruchstücke mit der *Musterpost* zurückkämen, könnte ich hier doch etwas zeigen.

Sie ehrend stets der Ihre
Ferd. von Mueller

14/8/91

This time I am writing, noble friend, to report that I have just sent you per post a packet of algae from Cape Flattery[3] in northeast Australia. There may be nothing unknown among them but really the collection is important in so far as

1 unbekanntes?
2 zum Beispiel.
3 Qld.

previously we had no algae from this particular locality. Thus there may yet be an increase of Asiatic algae obtained for the list of Australian algae. 91.08.14

It is some time since I received any *return consignments* for the state collection here. This *embarrasses* me as a public official, because now many species sent by me are published of which no representatives are to be found in the herbarium here. Gradually young Australians are also beginning to study the flora of the sea, and if many gaps in the corresponding museum collections endure, I would be publicly reprimanded. If, however, unica are indivisible, I expect nothing back, but would then like to be able to exhibit an *illustration*, e.g. of *Amansia mammilaris*. If only broken pieces of new or rare species came back with the *sample post*, I could indeed show something here.[4]

Respectfully allways your
Ferd. von Mueller

From Richard Helms 91.10.04

PRG 40/21, Mortlock Library of South Australiana, State Library of South Australia, Adelaide.

Fraser Ranges 4ter Oct. 1891

Herrn Baron F. von Mueller

Ew Hochwohlgeboren

Die freundlichsten Grüssen von hier sendend erlaube ich mir Ihnen anzuzeigen dass in etwa 14 Tagen ein Wagen von der Station hier in der Nähe abgehen wird womit meine Sammlungen bis an die Küste befördet werden um weiter nach Adelaide verschifft zu werden. Da die botanischen Objecte doch jedenfalls sofort an Sie hochgeehrter Herr weiter gehen so erlaube ich mir einige Bermerkungen über die selben. Ich möchte besonders hervorheben dass

4 In a letter to Agardh, 18 November 1891 [Collected Correspondence], M apologised for writing too hastily. He had now found that a collection of algae had been received from Agardh, including a piece of *Amansia mammilaris*, while M had been away in NZ, but had not been drawn to his attention when he returned.

91.10.04 seit wir die Everard Ranges hinter uns hatten das ganze Land unter der fürchterlichsten Dürre litt und wahrscheinlich schon lange vorher so gelitten hatte und dass ausserdem grosse Strecken von der Eingebornen abgebrannt waren die sich noch nichts wieder erholt hatten. Diese Zustände verhinderte mich oft am sammeln oder vielmehr am finden von Pflanzen und ganz besonders bezieht sich diesen auf die kleineren und zahrtere Gewächsen. Ich möchte deshalb in dieser Beziehung um Ihre gütige Nachsicht bitten. An Mühe habe ich es mir nicht fehlen lassen und soweit ich konnte habe ich den Auftrag der Geo. Gesellsch.: Möglicherweise 4 bis 6 "representative collections" zu sammeln zu erfüllen versucht was unter den manigfalltigen Schwierigkeiten hat vielleicht die Zubereitung zu wünschen übriggelassen. Doch da Sie ja nur zu gut hiermit bekannt sind bin ich überzeugt dass auch ohne meine Entschuldigung diesewegen, solcher Mangel bei Ihnen berücksichtigt werden würden.

Die Localitäten sind oft als Camps so und so angegeben und hat mir der Surveyor versprochen in Kürze die genaueren Positionen derselben zu geben und werde ich dieselben Ihnen dann übermitteln. Bis Camp 18 incl. ist zu den Everard Ranges und fängt von dort eigentlich die Exploration an weshalb die officielle Nummer Camp 1. meine Camp 19. hätte sein sollen. Ich bin aber wieder von dort mit No 1 angefangen und mit den officiellen Nos fortgeschritten um nicht in Confusion zu gerathen. Camp No 1. bis 18 incl. kommt deshalb zweimal vor doch sind die Daten natürlich verschieden

Wir haben eine schwierige Reise von den Barrow Ranges bis hierher gehabt und haben die Kamele Wunder geleistet. Etwa Halbwegs zwischen Mt. Squires und der Victoria Spring fanden wir in zwei sich folgenden Camps einen üppigen Wuchs von Mesembrianthemum woran die armen Thiere sich gütlich thaten und hat uns dieses und das glücklicher Weise verherschend kühle Wetter gerettet. Ohne wirklich guten Trank sind die Thiere 34 Tage gewesen obwohl sie am Wege je eine Bucket voll schlammigen übelriechenden Wassers kriegten und wieder bei der Victoria Spring dieselbe Quantität dessen Qualität auch nicht viel besser war denn es war schlammig und roch nach vermollesten Pflanzenresten.

Wie ich höre ist das Land zwischen hier und dem Murchison auch fürchterlich dürre und werden wir wohl noch wieder eine schierige[1] Reise haben wenn es nicht bald regnet. Hier hat es nicht geregnet einigen kleinen Schauern in Mai abgerechnet, seit Febr 1890. Am Murchison werden wir jedenfalls nicht vor Mitte December anlangen was meine Hoffnung auf eine gute Insecten Ausbeute zu Schanden macht und mir sehr unlieb ist denn bisher habe ich in der verdorrten und abgebrannten Region durch welche unser Weg führte nur wenig habhaft werden können. Leider hat man meinen Gehalt theilweise von dem Erfolg meiner entomologischen Beute abhängig gemacht. Ich übernahm diese Bedingung ohne viel zaudern da ich mir mehr versprochen hatte als ich anderswo in Australien begegnet bin wo es mir nie scher[2] geworden tüchtig was zu sammeln bin aber in meinen Erwartungen über alle Maase getäuscht und komme nun noch dazu mehrere Monate später am Murchison an wo noch was gutes zu erwarten war, wenn die beste Saison für Insecten schon vorüber ist. 91.10.04

Obwohl ich fast keine Stunde zum lesen habe finden können so habe ich doch ein Handbuch der Aust. Botanie sehr vermist. Wenn ein solches existirt so würde es mir sehr willkommen sein.

Mich Ihren gütigen Wohlwollen weiter empfehlend
verbleibe ich in aufrichtiger
Hochachtung
Ew Hochwohlgeboren
ergebenster Diener
R. Helms

Fraser Ranges[3] 4th October 1891

Baron von Mueller

Your Honour

In sending the friendliest greetings from here, I permit myself to notify you that in about 14 days a wagon from the station nearby will leave in which my

1 schwierige?
2 schwer?
3 WA. Helms was the collector on the Elder Scientific Exploring Expedition, 1891-2.

91.10.04 collections will be forwarded to the coast to be shipped further to Adelaide. However, because the botanical objects will immediately go on to you, respected sir, I permit myself some remarks about them. I would especially like to stress that once we had the Everard Ranges behind us, the whole country suffered from the most terrible drought and probably had already suffered long beforehand and that in addition large stretches were burned by the aborigines, which had not yet recovered again. These circumstances often hindered me in collecting or much more in finding plants and particularly in respect to the smaller and more delicate plants. Therefore in this respect I would like to ask for your kindly forebearance. I have spared no trouble and as far as I could tried to fulfil the instruction of the Geographical Society: as far as possible to collect 4 to 6 "representative collections", which under the manifold difficulties perhaps the preparation has left something to be desired. However, because you are certainly only too well acquainted with this, I am convinced that even without my excuse on that account, such deficiency will be allowed for by you.

The localities are often stated as Camp so and so and the surveyor has promised me soon to give the exact positions of these and I will then transmit them to you. Up to Camp 18 inclusive is to the Everard Ranges and the exploration really starts from there, so the official number Camp 1 should have been my Camp 19. But I have started again from there with No. 1 and continued with the official numbers in order not to cause confusion. Camp No. 1 to 18 inclusive occur therefore twice, but of course the dates are different.

We have had a difficult journey from the Barrow Ranges to here and the camels have performed wonderfully. About half-way between Mt Squires and the Victoria Spring, we found a luxuriant growth of Mesembrianthemum in two successive camps, which the poor animals enjoyed. This and the fortunately prevailing cool weather saved us. The animals were without a really good drink for 34 days, although on the way they each got a bucket full of muddy, nasty smelling water and again at the Victoria Spring the same quantity whose quality was also not much better, because it was muddy and smelled from decayed plant remains.

As I hear the country between here and the Murchison is also terribly dry and we shall probably again have a difficult journey if it does not rain soon. With the exception of some small showers in May, it has not rained here since February 1890. At any rate we shall not arrive at the Murchison before the middle of December, which will frustrate my hope of a good insect collection, for which I am sorry because previously in the dry and burnt-out region through which our track led I was able to get hardly anything. Unfortunately my salary was made partly dependent on the success of my entomological catch. I accepted this condition without much hesitation, because I had expected more than I had encountered elsewhere in Australia, where it has never been difficult for me to collect a great deal, but I am disappointed in my expectations beyond measure and as well now will arrive several months later on the Murchison, where again good things were to be expected, when the best season for insects is already past. 91.10.04

Although I have scarcely been able to find any hour to read, yet I have missed a handbook of Australian botany very much. If such a thing exists, it would be very welcome to me.[4]

Further recommending myself to your kindly benevolence,

I remain in
sincere respect your honour's
most humble servant
R. Helms.

To Otto Tepper[1] 91.12.25

RB MSS M198, Library, Royal Botanic Gardens Melbourne.

Weihnacht
1891.

Nehmen Sie meinen besten Dank hin, lieber Herr Tepper, für die freundlichen Wünsche, welche Sie mir auch diesmal wieder zur Festzeit senden. Möge die neue annuelle Spanne Zeit, in die wir nun hineinleben werden, auch für Sie und die Ihren eine recht freudevolle sein. Im Lauf des Januars hoffe ich das Vergnügen zu haben, Sie in Adelaide persönlich begrüssen zu können.

Es war mir final nicht möglich, im Nov. oder Dec. Zeit für Reisen frei zu machen. Ihr Wunsch, dass ich in den Ruhestand gehen solle, befremdet mich etwas, obwohl er so gütig gemeint ist. Das Unglück hat es gewollt, dass selbst manche meiner besten Freunde dachten, dass die Wegnahme des Gartens von mir eine Wohlthat

4 M forwarded this letter to Samuel Davenport in Adelaide; see M to S. Davenport, 4 November 1891 [Collected Correspondence].

1 MS annotation by Tepper: 'Reply 1.2.92.' Letter not found.

91.12.25 wäre, u doch leide ich täglich noch dadurch. Wenn ich nun meine Wissenschaft der Pflanzen die ich noch im engeren Sinn auszuführen suche, aufgäbe, nachdem ich *Alles* was ich im Leben hatte, – meine Zeit, mein Vermögen, meine geistige Kraft in meine Sammlungen, Bibliothek, Reisen u Studien gesenkt, was bleibt mir dann? Ich habe keine oder fast gar keine Geldmittel, nicht einmal Familie auch nicht den geringsten häuslichen Luxus. Andere würden sich in die Anstalt drängen, meine Collectionen, Bücher Manuscripte u.sw. in Besitz nehmen, und wahrscheinlich, wie es im Garten ist u war, den Aufbau einer grossen Anstalt nach meinen *unsäglichen* vieljährigen Bemühungen u Aufopferungen *gegen mich* gebrauchen! u ausnutzen. Es wundert mich, dass Ihr Sohn, das welches nach u nach noch neu sieht, nicht mehr sammeln will. Da die Liste, welche Sie senden, doch *meistens* nur Pflanzen enthält, welche im tropische Austr eine weite Verbreitung haben, u dazu nur eine Fraction der Pflanzen seiner Umgebung behandelt, wäre es am Besten, wenn solche eine etwaige *allgemeine Skizze* seiner Gegend begleiten; ich möchte aber doch der Sicherheit wegen rathen, nur solche Arten aufzunehmen, über welche Sie *gewiss* sein können. Vollständig kann eine solche Liste ja doch nicht sein. Die Pflanze, welche Sie diesmal senden, ist doch zu unvollständig, um etwas damit zu thun.

Sie ehrend und Ihnen alles Gute wünschend
Ferd von Mueller

Hoffentlich erfreuen Sie sich einer festen Gesundheit, so dass Sie Ihre eignen schönen Forschungen fortsetzen können.

Besten Dank für Ihre schöne Weichnachts Karte

Hätte Hooker nicht seinen Schwägersohn in Kew anstellen können als Nachfolger wäre er sicher nicht aus der Anstalt getreten, dann ist er doch wenigstens wohlhabend.

Christmas 1891. 91.12.25

Accept my best thanks, dear Mr Tepper, for the friendly wishes that you have sent me again for the festive season.[2] May the new annual span of time in which we will now live also be a very joyful one for you and yours. In the course of January I hope to have the pleasure to be able to greet you personally in Adelaide.

It was not possible for me finally to make time free in November or December for travelling. Your wish that I should go into retirement displeases me somewhat, although it is so kindly meant. Misfortune has willed it that even some of my best friends thought that the removal of the gardens from me would be a benefit, yet I still suffer daily as a result. If I now gave up my science of plants that I still seek to carry out in the narrower sense, after sinking *everything* that I had in life, my time, my funds, my intellectual power into my collections, library, travels and studies, what remains for me then? I have no or almost no money at all, not even family, also not the slightest domestic luxury. Others would enter into the institution, take possession of my collections, books, manuscripts and so on, and probably, just as it is and was in the garden, use and exploit the construction of a great institution after my *unspeakably* many years' efforts and sacrifices *against me*. I am surprised that your son no longer wants to collect that which in the course of time still seems new. Because the list that you send does contain *mostly* only plants that have a wide distribution in tropical Australia, and moreover deals with only a fraction of the plants of his surroundings, it would be best if it accompanied a possible *general sketch* of his region; but still I would like to advise for safety's sake only to include such species about which you can be *certain*. Such a list really cannot be complete. The plant that you sent this time is really too incomplete to do anything with it.

Respectfully and wishing you every good thing
Ferd von Mueller

It is to be hoped you enjoy constant health so that you can continue your own fine researches.

Best thanks for your beautiful Christmas card.

If Hooker had not been able to appoint his son-in-law[3] in Kew as successor, he certainly would not have resigned from the institution, and he is indeed at least well off.

2 Letter not found.

3 W. Thiselton-Dyer.

92.01.12 *To Nicholas Holtze*

GRG 19/391, State Records of South Australia, Adelaide.

12/1/92.

This I write, dear Mr Holtze, on board of the Steamer, which brings me back from Hobart, where I had to instal the Governor[1] as President of the Australian Association for Advancement of Science. It is too late to felicitate you properly to the new year, but let me trust, that the new annual span of time, on which we have entered, may be replete with happiness to you. In answering your letter of the 17 Dec.[2] I have no books for reference before me; but I can at least in part offer information, such as you are seeking. Answer about Utricularia albiflora[3] has not yet come from London. In using the name Terminalia latifolia before for your collection, I made a memory mistake. I used that name in my exploratory diaries 1855-1856,[4] but on return found that it was preoccupied, so I changed it to T. platyphylla. As regards Mimulus Uvedaliae and M. debilis, the perplexity may by you perhaps be solved, by limiting M. debilis, as I did in first instance to yellow flowered plant. I saw it only once myself; so I can not bring your extensive local experience to bear on it. Let me however remark, that I never in the hundreds of cases, when I could watch here in the south M. gracilis found it vary with yellow flowers, and so it is here neither with M. repens. Both however grow very much larger in wet soil, than in drier localities.

That I not recognized your Zizyphus as belonging to that genus, you must adscribe to the fact, that I have often to name specimens at late evening-hours, when after the toils of the day my visual power becomes dim. Oryza I certainly regard as indigenous in trop. Australia. I never saw it, until I came to the Upper Vict

1 Sir Robert Hamilton, Governor of Tas.

2 Letter not found.

3 See also M to M. Holtze, 8 February 1891 [Collected Correspondence].

4 During the North Australian Exploring Expedition, 1855-6. M's field diaries have not survived.

River, and to Sturt's Creek;[5] it is nowhere in Australia a coast-plant, so far as I am aware. Consider also, what a multitude of other Grasses are Australian as well as Indian In 1855 & 1856 Oryza could not have reached the places where I saw it, through Malayan advents. If migratory water birds brought it, then we have to regard it as indigenous, for that would apply to many places in South-Asia as well. 92.01.12

Monochoria[6] I recommend to your special local study. M. cyanea is the only species which has all the stamens of equal, and is exclusively Australian whereas the narrow-leaved spec, discovered by you is identical with an Indian, which has one anther different to the rest. It is however quite possible, that M. cyanea occurs also in a narrow-leaved form. A glance of yours on the flowers of the living plant will show at once, whether you have before you M. cyanea or an other species

Phaseolus Max is the oldest name of the species, though long discarded. I have restored it in the Census[7] and Select plants.[8] I may not have named hastily this plant on subsequent occasions, because several allied species exist, and I do not like to commit myself by trusting to my over burdened memory, and have not always time to refer on such subjects at once to authorities. It is very pleasing that you will still further enlarge the material, so thoughtfully provided for the fuller study of your many Utricularias, and thus look forward also with particular expectation to the fruits of the narrow-leaved plant, allied to Clerodendron. Is it always dwarf. It is possible, that an error occurred in naming and describing Sida Holtzei It may be a Malachra. I did not think of that hitherto extra-australian genus, until its characteristics, well known to me, from earlier days flashed again across my mind When I am back at my humble little dwelling and working place, I will reexamine the Sida Holtzei[9] so also the Melodorum which I have

5 WA.

6 See also M to M. Holtze, 2 June 1892 [Collected Correspondence].

7 B82.13.16.

8 B76.13.05.

9 For M's efforts to gather more information about *Sida Holtzei* see also M to N. Holtze, 16 January and N. Holtze to M, 7 March 1892 [Collected Correspondence].

92.01.12 since if I rightly remember transferred to an other genus, still as new. I yet set great hopes on you to obtain from your youthful enthusiasm & experienced search many plants additional for Australia.

If not unforseen hindrances intervene, I shall see your worthy parents in Adelaide[10] during this month, as I intend to consult personally with Sir Thomas Elder about the measures yet required for the forth coming south-polar expedition of Baron Nordenskiold.[11]

With regardful remembrance
your Ferd von Mueller

92.01.16 *To Thomas Wilson*[1]

U92/692, unit 663, VPRS 3992/P inward registered correspondence, VA 475 Chief Secretary's Department, Public Record Office, Victoria.[2]

Melbourne,
16/1/92.

T. R. Wilson Esqr
Under Secretary.

Sir,

I have the honor to submit a memorandum, conveying my views in reference to the advantages of placing the Herbarium under the control of the trustees of the public Library and Museum,[3] on whose behalf I would most willingly exercise over these extensive and unique collections *an honorary custodian-ship*, so that the cause

10 Maurice Holtze became director of the Adelaide Botanic Garden in 1891.
11 See H. Gundersen to M, 24 March 1890.

1 See also T. Wilson to M, 6 January 1892 [Collected Correspondence].
2 The file also contains a typed copy of this letter.
3 See also M to A. McLean, 1 January 1892 [Collected Correspondence], responding to the announcement that it was proposed to abolish M's department.

of the science of plants may suffer as little as possible here by the abolishment of the Gov. Botanists Department, and my connection with these treasures not be severed during the little time, which divine providence may yet destine for my worldly career and progressive researches. May I avail myself of this opportunity, to draw the favorable attention of the hon. the Chief Secretary to *two points*, concerning the position, which I had the honor to hold, namely 92.01.16

1, that I commenced my collections in March *1840*, that all through life I sunk my whole resources, therefore also all my private means, in the Herbarium and the studies, travels and science-intercourse connected therewith, so that after 7 years special professional work in Europe and after five years exploring in Australia on my private expenses (1847-1852), when I was called to create the Gov Botanist's Office early 1853 by Governor Latrobe, I made my *then already extensive collections* the foundation of the grand Herbarium, which in Melbourne thus arose, never sparing my income and time to enlarge it by my private means and very long daily working hours without any stipulations in this free gift to my Department!

To appreciate fully the importance of this herbarium also for all future, I may be allowed to mention, that it ranks in vastness as *one of nine* in the world, the others being at Kew, Paris, Leyden, Berlin, Florence, Petersburg Boston and Geneve, but for Australian material the Melbourne Herbarium is larger, than all the others taken together

2, When in 1873 I ceded from the Directorship of the bot Garden, I fear much to the impairments of the rural and technic interests of the colony, I had an emphatic though only verbal assurance from the hon Mr Ramsay, that my position towards the Herbarium as administrative should never be disturbed so long as I might be able to work.

I have the honor to be, Sir,
your obedient servant
Ferd. von Mueller

I will gladly continue as an honorary duty the *very extensive professional correspondence* and afford all other special information as

92.01.16 before gratuitously whereby by[4] scientific honor in the world of learning will than also be saved.

Memorandum concerning the future of the Herbarium of the Gov. Botanists establishment.

If it is really the final intention of the Government, to abolish the Gov. Bot. establishment as a distinct institution after all its achievements and its intended future work, then it would be most advisable, to place the Herbarium under the *control of the Trustees of the public Library and Museums* for the following reasons because

1, the trustees have already all other Gov. Scientific collections under their supervision.

2, The trustees will be able to place this great and largely unique herbarium into the new museum, soon to be erected *near the public Library*, where it will be more *readily accessible* than any where else.

3, The herbarium can then be used easily in connection with the respective books, already contained in the public library at close proximity.

4, The herbarium will then also be located near to the *University*, the *working Mens College*, the College of Pharmacy, and also the Gov. Offices in the city, while Amateurs in any branch of the science of plants can most conveniently utilize it.

5, The great *technologic Museum* for the further study of the vegetable products would have the herbarium close by.

If the Herbarium was located in the bot. Garden, it could be only of extremely limited use after the Gov. Botanists Department ceased, as the Garden is simply a horticultural establishment, whereas the herbarium is to serve also the systematically descriptive, rural, medicinal, forestral, technologic and any other interests connected with plants.

To place the Herbarium into the University or the agricultural Department would render it available only for *restricted purposes*,

4 The typed version has 'my'.

and it could not be accessible to the general public, and would be too far away from the centre of the metropolis.[5] 92.01.16

Ferd von Mueller,
Gov. Botanist

15/1/92

To Allan McLean 92.01.17

T92/2768, unit 454, VPRS 3992/P inward registered correspondence, VA 475 Chief Secretary's Department, Public Record Office, Victoria.[1]

South-Yarra[2] 17/1/92

With the greatest of readiness I beg to certify, that Mr G. Luehmann has been an Officer of my Department for 23 years, from 1869 to 1873 as clerical Assistant in the botanic Garden and since then my first Assistant in the Office of Governm. Botanist. He is a Gentleman of superior education, zealous, inostentatious and persevering in the performance of his duties, and he earned all throughout his connection with my establishment the fullest confidence of myself and the subordinates. Moreover he acquired under me botanic knowledge to a considerable extent, and it is with much regret that I become deprived of his assistance now when the De-

5 The decision to abolish M's department prompted considerable public outcry and numerous letters and editorials in the press, copies of which are filed at 92/2768, unit 454, VPRS 3992/P, PROV. In the end, M's department was continued, though on a much-reduced basis. For further stages in the resolution of the issue, see M to T. Wilson, 23 and 27 January 1892 [both Collected Correspondence].

1 The letter is numbered U92/692 within this file. A note in a non-standard Pitman-style shorthand, headed 'Mr Killen', dated '23 May 92' and referring to Luehmann, is attached to the letter.

2 Melbourne.

92.01.17 partment of Gov. Botanist as a distinct institution is to be discontinued. I trust that after his long and faithful services during the best years of his life, the Government will place so able and reliable Officer in a position, commensurate to the fair claims he can advance from the past.[3]

Baron Ferd. von Mueller
Gov. Botanist.

92.02.05 *From George Murray*[1]

RB MSS M108b, Library, Royal Botanic Gardens Melbourne.

BRITISH MUSEUM (NATURAL HISTORY),
CROMWELL ROAD,
LONDON: S.W.[2]
5 February 1892.

Dear Baron von Mueller

You have been so very kind to us in sending valuable Algae that I venture to ask you to do me a great favour in procuring a few *spirit* Specimens. A most interesting research on *Scaberia Agardhii*, *Notheia anomala* & *Hormosira* (on which it grows) on *Myriodesma* and on *Xiphophora* = *Fucodium gladiatus*[3] Labill. and on *Sarcophycus potatorum* is now in progress partly in my own hands &

3 See also M to A. McLean, 18 January 1892 and G. Luehmann to M, 19 January 1892 [both Collected Correspondence]. Luehmann's position with M was not abolished.

1 MS annotation by M: 'Answ 6/6/92'. Letter not found.
2 MS also bears the embossed insignia of the British Museum (Natural History).
3 *Fucodium gladiatum*?

partly in those of a pupil. We have carried the matter so far as our dried material permits and now spirit Specimens are necessary to enable us to solve a problem of the very greatest morphological importance. In our Herbarium we have *Sarcophycus* from Port Fairy Victoria and *Notheia anomala* on *Hormosira* (both needed) also from Port Fairy – as well as from Georgetown Tasmania. *Xiphora* (Fucodium gladiatus also from several points on the coast of Victoria & Tasmania and *Myriodesma* from Geelong & Brighton Beach.[4] Perhaps Mr Bracebridge Wilson will add to his former kindness to me by helping you in this. We have *Scaberia Agardhii* collected by yourself in Streaky Bay[5] & from others King Georges Sound[6] Bass Strait, Adelaide (R. Brown) Tasmania etc. 92.02.05

I know how heartily you encourage research and I trust to your interest in this work and kindness of heart to enable to settle beyond doubt this most absorbing question of the development of the conceptacles of these Algae.

In about two months Messrs Dulau will issue British Museum Phycological Memoirs Part I.[7] which I am sure will interest you. May I ask you to make Part II interesting as well!

Yours very truly
George Murray.[8]

4 Both Vic.

5 SA. M never visited Streaky Bay; the specimen in question must have been collected by someone else and passed on to him.

6 WA.

7 See Dulau & Co. to M, 12 April 1892 [Collected Correspondence].

8 Murray (1892-5). See also G. Murray to M, 18 November 1892 [Collected Correspondence].

92.02.20 *To Otto Tepper*[1]

RB MSS M97, Library, Royal Botanic Gardens Melbourne.

20/2/92

Es ist am Besten, lieber Herr Tepper, Ihnen gleich über die eben angelangten Pflanzen einige Auskunft zu geben, obgleich bei den gestörten Departements Verhältnissen ich mich noch nicht für zusammenhängende Arbeiten sammeln kann. 3/4 der Verwaltungs Gelder sind in diesen schlechten Zeiten der Anstalt entzogen!

Die Epiphyt-Orchiden ist Cymbidium canaliculatum Die Acacia will ich demnächst genau untersuchen. Möglicher Weise wird sie für eine der Platten des Suppl. des Acacien hier das Material liefern.

153, Persoonia falcata, RBr

152, Euc. Terminalis, FvM kleine Form.

Sollte diese Art für Feuer Holz gefällt werden, so waren Früchte leicht erlangbar u so auch die bald ausfallenden Samen. Gerade für Samen in einiger Quantität hat diese Art bei mir besondere Verwerthung! Ich weiss freilich nicht, was ich Ihrem Sohn als Gegengabe senden soll, wenn Sie mir nicht dabei etwa Rath geben. Einstweilen habe ich Ihm mein lithographirtes Portrait mit autographischer Dedication gesandt. Das Solanum muss genau verglichen werden. Könnte der junge Sammler auch reife Beeren herbeischaffen? Da 700 *haltbare* Arten bekannt sind, u. diese auch vielleicht in S.W. Asien vorkommen mag, muss man in der Benennung recht vorsichtig sein.

Hier haben wir jetzt ein wenig Regen gehabt. Ob es wohl je gelingen wird der Atmosphäre extensiv Regen abzuzwingen durch Dynamit-Explosionen in bedeutenden Höhen von Ballons? Es könnte wohl nur bei trüber Luft geschehen, und dann müsste Staub mit dem Nitroglycerin gemischt sein, damit sich die Regentropfen daran bilden können.

Ihr

Ferd von Mueller.

1 MS annotation by Tepper: 'Recd 23.2.92 Sent names to Otto by P.C. Reply 29.2.92. With No 122 Roebuck Bay'. Letter not found.

Es ist unmöglich, Gingins verantwortlich zu machen für eine Hybanthus-Art, da er längst gestorben war, wie ich den übersehenen älteren Namen Jacquins wieder herstellte für Ventenats Ionidium. Wohl weiss ich den bedauerliche Gebrauch, auch in der Zoographie, einem Autor binäre Namen zuzuschreiben, die er nie gebraucht oder anerkannt hat. Selbst in Parenthese ( ) zu schreiben ist missleitend u unaccurat. Wenn Gingins Namens Gebung bleiben soll, so könnte es nur in diesem Falle so geschehen Hybanthus suffruticosus (Ionidium Gingins) der Kürze wegen ziehe ich englich[2] "from", deutsch, "nach" vor, dann braucht man die verworfenen Namen nicht mehr durchzuschleppen u wird doch dem ersten *specifische* Namensgeber gerecht 92.02.20

Wenn Amarantus auch Exolus einschliessen soll, dann muss auch Claytonia in Portulaca aufgehen! Ich habe im Census alles dies Jahr sorgfaltig erwogen. Selbst J. Hooker nimmt manche der Benenungen Benthams nicht an in Bezug auf Genera! Es wäre schade, wenn die Namen der Pfl. in Australien nicht einförmig blieben. Wir haben hier ein besseres System als anderswo. Neulich z. B. veröffentlicht Maiden Echinocarpus (*Sloanea*) australis, nachden[3] *alle* neueren Autoren, die Unhaltbarkeit von Echinocarpus eingesehen. Hätte er gedruckt Sloanea (Echinocarpus) so wäre doch etwas Sinn darin gewesen. Mir scheint es wäre am besten alles Zweifelhafte u Unvollständig ganz auszulassen, denn solche stellt doch wohl nur 1/3 od 1/4 der Veg. dar. Ich habe allein von der Nachbarschaft von Nickol Bay etwa 400 Arten veröffentlicht.

Amarantus pallidflorus ist ein achten[4] Amarantus.

Echium vulgare scheint bei Ihnen sich eingebürgert zu haben.

Algen giebt es an der Roebuck-Bay wohl nur wenige.

Ich muss ja auch vielen anderen Correspondenten gerecht werden, u der Tag hat doch nur 24 Stunden.

Können Sie noch ein Stück der fraglichen Pfl. senden? Ich erinnere nicht welche sie meinen.

2 englisch?

3 nachdem?

4 aechter = echter?

20/2/92

It is best, dear Mr Tepper, to give you some information immediately about the plants just arrived, although in the troubled departmental circumstances I still cannot collect myself for coherent work. 3/4 of the administrative funds are withdrawn in these hard times of the institution!

The epiphytic orchid is *Cymbidium canaliculatum*. I will investigate the *Acacia* carefully soon. Possibly it will supply the material for one of the plates of the supplement of acacias here.

153, *Persoonia falcata* RBr

152, *Euc. terminalis* FvM 1, small form.

Should this species be felled for firewood, fruits would be easily attainable and so also the soon dropping-out seeds. This species has particular value for me just for the seeds in some quantity! Admittedly I do not know what I am to send your son as a return present, if you do not give me some advice about it. Meanwhile I have sent him my lithographed portrait with autographic dedication. The *Solanum* must be carefully compared. Could the young collector also get ripe berries? Since 700 *tenable* species are known and these also perhaps may occur in southwest Asia, one must be very careful in the naming. Here we have now had a little rain. I wonder whether it would ever be possible to wrest extensive rain from the atmosphere with dynamite explosions at considerable height from balloons? I suppose it could probably only happen with cloudy air and then dust would have to be mixed with nitroglycerine so that the raindrops could form on it.

Your

Ferd. von Mueller

It is impossible to make Gingins[5] responsible for a *Hybanthus* species, since he died long ago, while I established the overlooked older name of Jacquin's again for Ventenat's *Ionidium*.[6] Well I know the unfortunate use, also in zoology, of ascribing binomial names to an author that he has never used or recognised. Even writing in parentheses ( ) is misleading and inaccurate. If Gingins's name-giving is to remain, it could only so happen in this case *Hybanthus suffruticosus* (*Ionidium* Gingins) for the sake of brevity I prefer English "from", German "nach", then one no longer needs to struggle through the rejected names and yet do justice to the first *specific* name giver.

5 Gingins de la Sarruz (1824), p. 311.

6 See B89.13.12, p. 10.

If *Amaranthus* is also to include *Euxolus*, then *Claytonia* must also merge into *Portulaca*! I have carefully pondered everything in the *Census*[7] this year. Even J. Hooker did not accept some of Bentham's naming in reference to genera! It would be a pity if the names of the plants in Australia did not remain uniform. We have a better system here than elsewhere. Recently for example, Maiden published *Echinocarpus (Sloanea) australis*, after *all* recent authors saw the indefensibility of *Echinocarpus*. Had he printed *Sloanea (Echinocarpus)* there would really have been some sense in it.[8] It seems to me best to completely leave out everything doubtful and incomplete, because such probably only represents 1/3 or 1/4 of the vegetation. I have alone published about 400 species from the neighbourhood of Nickol Bay.[9] 92.02.20

Amaranthus pallidiflorus is a true *Amaranthus*.

Echium vulgare seems to have become naturalised at your place.

There are probably only few algae at Roebuck Bay.[10]

I must indeed also do justice to many other correspondents and the day does have only 24 hours.

Could you send another piece of the doubtful plant? I do not remember which you mean.

7 B89.13.12.

8 See Maiden (1889), pp. 420-1. For *Sloanea australis* see B64.05.01, p. 91 and for *Echinocarpus australis* see Bentham (1863), p. 279.

9 WA.

10 WA. Tepper's son, also Otto, was based at Roebuck Bay.

92.03.15 *To Otto Nordstedt*

Handskr. avdl., Universitetsbibliotek, Lund, Sweden.

15/3/92.

Die letzte Post, geehrter Herr Professor Nordstedt, brachte mir Ihren freundlichen Brief über den Fortschritt der Characeen-Monographie; aber Sie werden aus meinem letzten Schreiben ersehen haben, dass ich die Subsidie nicht weiter zahlen kann, obwohl ich für das, welches *schon fertig* geworden ist, geldlich verantwortlich bin. Privat-Vermögen habe ich gar nicht, denn ich habe Alles in meine Wissenschaft geschickt.

Da die Revenue des Jahres 1891 so sehr den Erwartungen des Ministeriums hier zurückblieb, entschloss sich die Regierung bedeutende Einschränkungen im Staatsdienst zu machen, und so mussten von den 5 Gehülfen, welche ich in den verschiedenen Abtheilungen meines vielseitigen Departements beschäftigte, drei gekündigt werden. So ist es auch mit den anderen Verwaltungs Geldern in meiner Anstalt. Nur 1/4 ist geblieben, 3/4 ist fort! So kann künftig auch kein lithographisches Werk mehr von mir erscheinen!

Mit der Post in nächster Woche werde ich Ihnen wieder £10 senden.

In Bezug auf Ihre Bemerkung, dass Subspecies anzunehmen wären, theile ich Ihre Ansicht nicht. Es scheint mir der unglücklichste Ausdruck in der ganzen Naturgeschichte zu sein, und ich habe ihn nie gebraucht in meinen Werken. Was immer die Ansichten der Naturforscher sein mögen über die Beständigkeit oder Inconstanz der Arten, wir müssen Alle zugestehen; dass zwischen Species die Kluft viel grösser ist als zwischen Varietäten. Subspecies können nur Varietäten sein, und jener Name ist verwirrend. Manche sogenannte Subspecies mag aber berechtigt sein, zu einer besonderen Art erhoben zu werden.

Sie ehrend der Ihre
Ferd. von Mueller

The last mail, respected Professor Nordstedt, brought me your kind letter[1] about the progress of the Characeae monograph;[2] but you will have gathered from my last letter[3] that I can no longer pay the subsidy, although I am responsible financially for that which is *already finished*. I have no private means at all, because I have put all into my science.

Since the revenue of the year 1891 fell so much behind the expectations of the ministry here, the Government resolved to make significant reductions in the state service, and so three of the five assistants whom I employed in the different sections of my extensive department had to be dismissed. So it is also with the other administrative funds in my institute. Only 1/4 has remained, 3/4 has gone! So for the future no more lithographic work can be published by me!

I will send you £10 again with the mail next week.

In reference to your remark that subspecies would be adopted, I do not share your opinion. It seems to me to be the most unfortunate expression in the whole of natural history, and I have never used it in my works.[4]

Whatever the opinions of natural scientists may be about the permanence or inconstancy of species, we must all admit that the gap between species is much larger than between varieties. Subspecies can only be varieties, and the former name is confusing. Some so-called subspecies may, however, be entitled to be raised to a separate species.

Respectfully your
Ferd. von Mueller

1 Letter not found.
2 Nordstedt (1891); only one part was published. M had suggested that Nordstedt prepare such a monograph and later agreed to pay a subsidy of £10 for each Decade of the work; see M to O. Nordstedt, 21 January and 25 December 1890 [both Collected Correspondence].
3 See M to O. Nordstedt, 31 January 1892 [Collected Correspondence].
4 In a letter to Nordstedt, 8 December 1891 [Collected Correspondence], commenting on the proofs he had just received of Nordstedt (1891), M urged Nordstedt to drop one of the two plates he had included of *Chara leptopitys*, depicting the subspecies *subebracteata*, and substitute a plate of a different species. The request was not heeded.

92.04.27 *From David Lindsay*

ML MSS.200/3, folder 5, Mitchell Library, State Library of New South Wales, Sydney.

Hutt Street, Adelaide
27th April 1892

Baron Von Mueller K.C.M.G. Ph D MD. &c &c &c &c
Melbourne

Dear Sir,

I have intended writing you for sometime past, and have really had so much writing to do that I have put it off from time to time until I feel that now an apology is due to you for want of courtesy shown you.

I hope ere long to have an opportunity of chatting with you on the subject of the breakup of the Elder Expedition,[1] a result which I know has caused you a great deal of pain, and disappointment – to me it has been the severest blow I ever experienced – and I hope you clearly understand it is no fault of Sir Tho Elder – he has not changed his mind or regretted the expenditure, nothing of the sort. I do not care to write what I could tell you, but I hope what you have heard from Mr MacDonald[2] will have given you a fair idea of the trouble.

When you have the whole of the evidence before you, you will see the almost *impossible* part I had to fill, and what bad material my party was composed of. I am extremely glad to be able to say that Mr Helms[3] collections of botanic specimens could not have been eclipsed by any one for he was an indefatigable worker – his

1 The Elder Scientific Exploring Expedition, 1891-2, broke up in disarray near Geraldton, WA, in January 1892 when all but one of the scientific members of the party resigned. See Lindsay (1893) and Feeken, Feeken & Spate (1970), pp. 200-2.

2 A. C. Macdonald?

3 Richard Helms, surveyor and third in charge on the expedition. See also R. Helms to M, 4 October 1891.

whole heart was in the work but dear me what a fearful temper he has, the least thing would send him off into a frightful passion. Still that could be dealt with better than the underhand sneaking actions of Leech[4] and the Dr.[5] 92.04.27

Had we been permitted to continue our work, 18 months more would have completed the grandest exploration ever made in Australia.

You sir must be terribly disappointed as I know your dearest wish has been to discover some relics of Leichardt, and MacPhee's account[6] renders it probable enough that something might have been found between the Oakover[7] and Lake MacDonald.[8] I too would have given much to have settled the mysterious and interesting question as to what became of that brave but unfortunate explorer.

The winding up of the expedition was carried out with such indecent haste that there was no opportunity of altering the decision. We have travelled about 2250 miles in new country and mapped down about 80,000 square miles so that the results of the expedition are by no means insignificant.

Hoping that you are still blest with good health and sympathising with you in the worry caused by the impecuniosity of the Victorian Govt

I am faithfully yours
David Lindsay

4 F. W. Leech, second in command on the expedition.

5 Presumably F. J. Elliott, medical doctor and photographer on the expedition.

6 Alexander McPhee reported meeting Aborigines at Lagrange on the NW coast of Australia who told him of a party of white men who had perished one by one in the sandy wastes, some days' walk SE of Joanna Spring. See D. Lewis (2006).

7 Oakover River, WA.

8 WA.

92.05.16 *To Thomas Sisley*

MS7593, Box 586/1(a), La Trobe Australian Manuscripts Collection, State Library of Victoria, Melbourne.

16/5/92.

Feel much honored, dear Mr Sisley, with the kind invitation of the Vict Artists Society to their annual festival, and will so arrange my work, that the evening of the 26 May remains free for this great enjoyment.

Regardfully
your
Ferd. von Mueller

92.08.05 *From Frederick Bailey*

Letter press copy book 2, p. 151, Queensland Herbarium, Brisbane.

Colonial Botanists Office[1]
August 5th [189]2

Dear Baron

The dates of my Botany Bulletins[2] are the time when the manuscript is given to the printer, while in his hands I keep adding to the publication information as such come to hand. The issue of the work is often delayed, but I often obtain [slips] which are sent out, – small matter like [...] [...][3] are kept in type so that at any time, to [...] very last, matter can be inserted. Thus matter from the June number of the Victorian Naturalists was sliped[4] in. I have no given

1 Brisbane, Qld.
2 Bailey (1892).
3 *illegible* – MS damaged. All [...] in the following text have this meaning.
4 slipped?

time for issuing the publications, but have matter printed when sufficient is ready.[5] Will you allow me as an old friend to advise you to be less dogmatic? I consider that what I have done regarding Tribulus occidentalis and Millettia Maideniana is quite justified. I found little in your herbarium to assist me with regard to plants. I might have done better if you had afforded me a little of your time during my weeks stay in Melbourne, but I was not even met by you at the herbarium and a very large proportion of specimens which I wished to consult were not there. With regard to the Millettia, I find nothing in Endlicher's description of Pterocarpus australis[6] or in your reference to it in Journ. of Bot. 22 – page 290[7] to lead me to the conclusion that it was the plant now named by me after Mr Maiden. You might see from my description that it & M. Megasperma F.v.M. are quite distinct from each other.[8] With regard to your Census[9] my work is so totally ignored in this publication that now it is of little moment whether plants of my naming appear in it or not. There is no mans work so perfect that fault cannot be found with it. I do my best to assist Queenslanders with a knowledge of the plants of their colony pointing out all the economic features of both the indigenous and naturalised species, and you throw discredit upon my work. This should [...] it will not aid the cause of [bota]ny. I have no specimens of Solanum orbiculatum *Dun*. I went by the description in the [...] [...] [...] S. oliga- 92.08.05

5 M published the new species *Endiandra exostemonea* and *Polyosma reducta* in B92.07.05. Bailey promptly included these Qld species in his 5th *Botany bulletin* (1892), on pp. 13 and 24. Since the preface of this work was dated May 1892, even though it was not published until July (see F. Bailey to M, 9 July 1892 [Collected Correspondence]), readers were likely to gain the impression that Bailey had priority.

6 Endlicher (1833a), p. 94.

7 B84.10.02.

8 In Bailey (1892), pp. 7-8, Bailey prefers *Tribulus occidentalis* to *T. hystrix*. Both are Robert Brown's names (see Sturt [1849], appendix, p. 69) so it is unclear what M's objection might have been. On pp. 12-13, Bailey published *Milletia maidenia* as a new species, believing the specimen he had was different from *M. megasperma* published in Bentham (1864), p. 211.

9 B89.13.12.

92.08.05 canthum *F.v.M.* in Bulletin.[10] Of this latter I will send you specimens which you will doubtless find is wrongly determined. I received but a small flowering specimen of the Acacia melanoxylon *R.B.* from Gladfield, and as I thought it most probably a form of that species gave the description from [...] of A. melanoxylon[11] in bulletins to assist persons to identify, if it was again collected when in fruit –

Yours very truly
F. M. Bailey.

Baron Ferd von Mueller, K.C.M.G &c
Government Botanist for Victoria

92.08.06 *To [Adolf Engler]*

Staatsbibliothek zu Berlin, Preussischer Kulturbesitz, Berlin.

6/8/92

Die letzte Wochen-Post brachte mir von Ihnen, hochgeehrter Herr Professor, ein Druckblatt mit Fragen über Prioritäts Rechte, u. sende ich nun kurze Antworten, ohne mich dadurch irgendwie einstweilen binden zu wollen, denn um eine möglichst einheitliche u weitreichende Entscheidung herbeizuführen, bedarf es nicht nur eines neuen internationalen Congresses der Phytographen jetziger Zeit, sondern auch einer Einholung schriftlich gegebener Ansichten derjenigen, die nicht persönlich erscheinen können

10 Bailey did not consider (in Bailey [1892], p. 21) that these two species of *Solanum* were distinct, notwithstanding M's having carefully separated them while describing the latter of them in B55.13.03, p. 19. Bailey accepted M's name but provided what he described as 'the first full description which has been published of this interesting species'.

11 Bailey (in Bailey [1892], p. 13) republished, word for word, the description of *A. melanoxylon* from Bentham (1864), p. 388.

Eigentlich ist es nicht nothwendig, dass ich mich in diesen Beziehungen aufs Neue erkläre, denn in meinen Schriften, namentlich auch im 1st. u 2t. Census habe ich die Ansichten entwickelt, welche mich beim Pflanzen-Benennen geleitet haben, u. ich möchte auch in Zukunft *mein eignes Gerechtigkeits Gefühl* in dieser Hinsicht ausüben. Ich bin auch überzeugt, dass Alle, welche an einem Congresse nicht Theil nehmen können, sich frei halten werden, in der Ausübung von deren Ansichten, selbst wenn Notizen für den Congress eingereicht wären. 92.08.06

Um aber doch gleichmässige Namens Wahlen in der Pflanzen-Kunde mit anzubahnen, gehe ich doch kurz auf Ihre u Ihrer Freunde Fragen ein, möchte aber gleich vorher bemerken, dass die 4 Thesen ja mehrere der hochwichtigen Fragen gar nicht berühren, z. B. *dass nur derjenige Binal-Name*, *welcher zuerst* im *richtigen Genus* angewandt wurde Vorrecht haben kann, was sich ja stets aus der *Jahreszahl* ergiebt, sonst: "you get the *shadow* instead of the substance" wie das englische Sprichwort sagt.

Infolge der Behandlung von Entada Pursaetha im 2t. Band der Fl. Austral. sprach ich mich über das ebengenannte Prinzip entschieden aus, und überzeugte Bentham, dass es das einzige richtige sei, so dass er stets *nachher diese* Regel anwandte. Es liegen noch mehrere andere wichtige Fragen vor, welche Ihre 4 Propositionen nicht erledigen, aber in einem Briefe wäre es zu weitläuftig, darauf einzugehen, u soll Alles in der Vorrede zum 3t Census 1893 oder 1894 besprochen werden von meinem Standpunkte.

1, Selbstverständlich kann ein *Genus* erst Geltung erlangen, nachdem es durch systematische specifische binäre Anwendung legitimirt ist. Solche binäre Namen können aber sehr wohl *vor-Linneische* sein, Vide "Papuan Plants" vol. I denn L. ist selbst durch Bauhin u Andere zu den Doppel-Namen geleitet. Der Anfang legitimer *Gattungen* ist aber durchaus *nicht mit L.* Das war ja eine ausserordentliche Ungerechtigkeit für den hochverdienten Tournefort u manche Andere!

Es giebt allerdings nur wenige binäre Arten-Namen vor 1753, aber sehr viele richtige und wohl motivirte Gattungs Namen aus *früherer Zeit* als L. sp plant.

92.08.06 2, Da das *Arten-Recht* für *Abbildungen* nicht bestritten wird, so wäre es arbitrar solches für Genera zu verbieten. Eine *gute* Abbildung erfordert viel mehr Mühe als eine kurze Genus- oder Species-Definition, namentlich solche, wie im vorigen Jahrhundert und im ersteren Theil des Jetzigen üblich waren. Wenn aber Aublet, Loureiro u Andere gleichzeitig 2 *Arten eines Genus* unter zwei verschiedenen Gattungs Namen abbildeten, dann werden nach meiner Ansicht recht wohl beide Genera *invalid*, u Schreber oder dem Nächstkommenden gebührt der Vorzug in der Annahme von deren Genera Wenn einem Phytographen ein in Prinzip *richtiger* binärer Pflanzen-Name in einer Sammlung oder sonstwie *bekannt* geworden ist, hat er *nicht das Recht* solchen zu verwerfen, einfach um *selbst* einen neuen Namen zu geben.

3, Diesen Satz habe ich in den transact. of the Royal Soc. of N.S.W. vor etwa 3 Jahren besprochen und dieselbe Ansicht geäussert, hinzufügend dass durch eine kleine Änderung e.g. ia anstatt ea oder vice versa gleichlautende zoographische von phytographischen Genus-Namen ohne eigentliche Störung geschieden werden könnten.

4, Ich erlaube mir hier *die* Genus Namen zu besprechen, in Bezug auf welche ich von Ihren Ansichten abweiche oder weitere Winke oder Erklärung geben möchte für Ihre dortigen Berathungen.

Ich bekenne aufrichtig, dass mir Adanson's Werke nie zugänglich waren; daher weiss ich nicht ob er sich auf L's species bezog. Wäre es so, dann gebührt ihm das Gattungs Recht, wenn es nicht etwa durch etymologische Gründe ungültig wird wie "Gans-Blum", was er aus dem Deutschen genommen haben muss.

Wenn Scopoli Gattungen aufstellte, z. B. Wilkia u solchen auch nicht eine benannte Species als Wilkia hinzufügte, so gebührt ihm doch die Priorität für die Gattung, denn er fügte Linnés Arten als Beispiele hinzu. Es wäre Pedanterie, darauf zu bestehen, dass er neben Linnés Art nicht auch *seine* Benennung drucken liess, was er bloss aus Bescheidenheit unterliess, da die Sache ja selbstverständlich war. Daher sollte auch *derjenige*, welcher ein vernachlässigtes gutes Genus zuerst wieder hervorzieht, der *Autor aller derjenigen Arten* sein, welche unverändert als *wohl bekannt* u wohl verstanden in das Namensveränderte Genus übergehen. Dem späteren Monographen gehören *nur* solche zweifelhafte Arten, als er zuerst richtig erklärt, oder als *nach* der generischen Namens Umstellung hinzu gekommen sind! Ich nenne es keine treue Wissenschaft, einfach hurtig eine *erste* Namens Liste drucken zu lassen, wenn die Umstellung längst vorher, wie z. B von Baillon geschah, denn diejenigen Arten Namen waren damit schon *gedeckt*, welche sich auf zweiffellos begründete Arten beziehen. *Baillon* u Andere hielten es nicht für nöthig, *gleich* eine Namensliste in den Druck zu bringen von allen bekannten Arten u es wären ja denn doch auch nur "*Nomina nuda*". Wer wird überhaupt in *späterer* Zeit eine *solche* Besitznahme anerkennen? 92.08.06

Die Behandlungen, welche RBr in der ersten Hälfte dieses Jahrhunderts von den Botanikern ohne Ausnahme erfuhr in dieser Hinsicht nach seinen Andeutungen im Prodromus war eine *würdige*.

Ionidium ist bereits durch Hybanthus absorbirt. Die Scrophularien Gattung Calceolaria stützt sich schon seit *1725* auf Feuillé sowohl in Beschreibung u *Abbildung*

Tissa u Buda fallen als *widersprechend gleichzeitig* hervorgegangen.

Malviopsis scheint mir unantastbar gut.

Wenn willkürliche Ausnahmen gemacht werden, wo sollen solche beginnen, wo enden. Es wäre unmöglich gerade in *dieser* Hinsicht eine Meinungs-Einigung zu erzielen. Das chronologische Urtheil ist unabänderlich.

Cola Caspar Bauhin brauchte schon diese Namen, was nicht unbeachtet bleiben kann u zeigt, dass wir nicht exclusiv mit 1753 beginnen können.

Aphora hier muss die Weise, wie Necker dies und andere seiner Genera vor bringt entscheiden, so ist es mit Heisters Meibomia, aber es wäre *bedauerlich* zu einem solchen Genus zurückkehren zu müssen, wenn es wirklich von *Arten-Namen* begleitet war. Ich werde Desmodium beibehalten.

Placus. Man muss bedenken dass jetzt eine Menge Genera von Loureiro spätere verstossen haben, sogar die weit bekannte Myriogyne, Sponia, Dithyrocarpus &c. Entweder wir müssen *alle* fehlerlosen Genera Loureiros *annehmen* oder alle verwerfen. Glücklicherweise wird sein genus Callista werthlos schon dadurch, weil

92.08.06 er für das, was wir Dendrobium nennen, auch noch ein *zweites* Genus hat

Linnés Lobelia (auf L Dortmanna wohl gestützt, erschien schon *1737*.

Statice; dies Genus gehört L. Wie kann Willdenow dafür Autorität werden, weil er einfach Armeria abtrennte? –

Herpestis fällt vor dem unanfechtbaren Bramia Lam.

Isopogon u seine anderen neuen Proteaceen Genera brachte RBr *1809* schon vor der LS. zur Öffentlichkeit in der persönlichen Gegenwart von Salisbury. Ich habe daher im Census auch 1809 quotirt. So sinken die meisten von Knight & Salisbury Genera, wie aber Linkia übersehen u Persoonia bleiben kann ist nicht leicht fassbar. Viel andere Genera, eben so lange in Gebrauch als Persoonia sind von A. de Cand, Benth., Hooker und anderen *Leitern* der Pflanzen-Systematik umgestossen, ohne dass Jemand appellirt hat. Banksia L.f. sollte *bleiben*, denn die Forsters hatten deren Genus Banksia auf Murrays Antrieb *eingezogen*. Ich halte dies für genügenden Grund in einem solchen Fall das bisherige Genus Banksia unangetastet zu lassen, verkenne aber damit die freundlichen Gefühle nicht, welche Dr O Kuntze veranlassen, meinen Namen damit verbinden zu wollen, worauf ich ganz stolz sein würde, wenn es ginge.

Für Exocarpos kann Xylophyllos nicht substituirt werden, denn der letztere Ausdruck ist ganz *unpassend*.

Gyrostachys ist schon desswegen nicht annehmbar, weil die Anordnung der Blüthen keine kreisige oder circinnate ist, sondern spiral – was die *beste* Interpretation des griechischen Wortes ist.[1]

Bulbophyllum stammt schon aus dem Jahre 1822, da dies Genus des Thouars *angenommen ist*, kann man ihn nicht dessen verlustig erklären, weil er keine Arten Namen publicirte denn das Genus wurde zeitig angenommen. Alles dies zeigt, dass die Behandlung der Genera *keine schroffe* sein kann, chronologisch u.s.w.

R.Br legte schon am 17 *Jan 1809* seine grosse Abhandlung über Proteaceae der L. Gesellschaft vor.

1 – was … ist *is an interpolation by M. The sense suggests he has misplaced it and that it schould precede* sondern spiral.

Pinalia. Auf eine *Monats* Berechnung können wir uns nicht einlassen, wenn ein lange bekanntes Genus umgestossen werden soll, welches in *demselben* Jahr erschien. Daher erkenne ich auch Swartz genus Stylidium nicht an, u der schwedische Forscher hätte *selbst* gleich aus Ehrfurcht für DC sein Stylidium einziehen sollen. Ich bezweifle das Wohlthun, Loureiros Stylidium mit Alangium zu vereinen. Freilich beruht die Auffassung der Genera auf individuelle Ansicht, denn die *Natur* schaft nur Species, nicht Genera oder Ordungen, die einfach uns den Überblick erleichtern und unseren Gedächtniss helfbar sein sollen. Die Verwandschaft von Loureiro's Stylidium zu Alangium erkannte ich allerdings selbst schon an, als ich 1860 das Genus Pseudalangium aufstellte. Dass Labill. Candollea der Dilleniaceen *ganz unhaltbar* ist, habe ich wiederholt nachgewiesen nach mehr *vollständigen* Material als irgend einem Anderen zu Gebote stand 92.08.06

Ich kann in Libertia nichts Anderes als eine Section (*schwach* dazu) von Sisyrinchium.

Astelia wird schwerlich zu retten sein. Diese und änhliche Schwierigkeiten habe ich auch im Census angedeutet.

Draco der Name ist geradezu *sinnlos*.

Chlamysporum wird doch wohl "den Kampf ums Dasein" überleben! Was ist aber mit Thysanella (1806) zu thun. Wird sich fürs Vorhandensein zu enthschuldigen haben! –

Pandanus ist durch Rumph schon hinlänglich geschützt.

Manisuris, Rottboellia ziehe ich vor geschieden zu lassen

Chamaeraphis wurde von RBr nicht im Sinn von Setaria aufgesetzt

Nageia muss unzweifelhaft wieder hergestellt werden, so kann Podocarpus für Phyllocladus bleiben.

Sie ehrend der Ihre
Ferd von Mueller

Du Petit Thouars Genera *leider* fallen theils, weil er keine Arten-Namen gab.

In der Bot. Zeitung u dem Centralblatt sind auch Aufsätze von mir über Prioritäts Rechte.

Last week's mail brought me a printed sheet[2] from you, highly esteemed Professor, with questions about right of priority and now I send you short answers, without thereby wishing to commit myself in any way for the present, because to produce the most uniform and far-reaching decision as possible, it requires not only a new international congress of phytographers of the present time but also the procurement of written opinions of those who are not able to appear personally.

Actually it is not necessary that I explain myself anew in this respect, since in my writings, particularly in the 1st and 2nd *Census*,[3] I have developed the views that have guided me in plant nomenclature, and I would also like in the future to follow *my own sense of justice* in this regard. I am also convinced that all who cannot participate in a congress will feel free to follow their own views even if notes were submitted for the congress.

But in order to initiate stable choices of names in plant studies, I will certainly for a moment accept your and your friends' [questions], but would like to remark immediately beforehand that the 4 theses certainly do not touch at all on several of the highly important questions, e.g. *that only that binomial name which at first* was used in the *correct genus* can have priority, which certainly always follows from the *date*, otherwise "you get the *shadow* instead of the substance", as the English proverb states.

As a consequence of the discussion of *Entada Pursaetha* in the 2nd volume of the *Fl. Austral.*[4] I expressed the opinion firmly about the above-mentioned principle and convinced Bentham that it is the only correct one so that *afterwards* he always applied *this* rule. There are several other important questions that your 4 propositions do not deal with, but in a letter it would be too lengthy to go into them and all shall be discussed from my standpoint in the preface to the 3rd *Census* in 1893 or 1894.[5]

1, Of course a *genus* can attain validity only after it is legitimized through systematic specific binomial use. Such binomial names can, however, be very

2 M's speaking of a 'printed sheet' suggests that he is referring to Ascherson et al. (1892). The '4 propositions' that M goes on to discuss also appear in Ascherson (1892), together with a list, on which M here offers comments of conserved names decided upon by the 1892 International Botanical Congress.

3 B82.13.16, B89.13.12.

4 Bentham (1864), p. 298, described *Entada scandens*, basing the name on Linnaeus's description of *Mimosa scandens*. He also referred to de Candolle's later description of the species as *Entada Pursaetha*. M in his *Second census* used de Candolle's name.

5 This promised work was never published.

probably *pre-Linnaean*, see *Papuan Plants* vol. 1,[6] since L[7] himself was led to the double name by Bauhin and others. The beginning of legitimate *genera* is, however, by *no means with L*. That was indeed an extraordinary injustice for the highly deserving Tournefort and several others! 92.08.06

Admittedly there are only few binomial species names before 1753, but very many correct and well-justified generic names from *earlier times* than L. *sp. plant.*[8]

2, Because the *species law* for *figures* is not disputed it would be arbitrary to forbid such for genera. A *good* illustration requires more trouble than a short generic or species definition, particularly those that were customary in the previous century and in the first part of this. But if Aublet, Loureiro and others at the same time figured *2 species of a genus* under two different generic names, then in my opinion both genera very well become *invalid*, and the priority in the acceptance of such genera belongs to Schreber or the one next coming after. If a binomial plant name *correct* in principle has become *known* to a phytographer in a collection or some other way, he does *not have* the right to reject it, simply to give a new name *himself*.

3, I have discussed this sentence in the *Transact. of the Royal Soc. of N.S.W.*[9] about 3 years ago and expressed the same opinion, adding that by a small change, e.g. ia instead of ea or *vice versa*, similar sounding zoographical or phytographical generic names can be separated without real disturbance.

4, I take the liberty here to discuss *those* generic names respecting which I diverge from your views or would like to give further suggestions or explanation for your discussions there.

I frankly acknowledge that Adanson's works were never available to me, therefore I do not know whether he referred to L's species. Were it so, then the generic law entitles him, if it does not become invalid on etymological grounds such as "Gans-Blum", which he must have taken from the German.

When Scopoli erected genera, e. g. *Wilkia*,[10] and also did not attach a named species to such as *Wilkia*, then the priority for the genus does belong to him, because he added Linne's species as examples. It would be pedantry to insist that he had not had printed *his* nomenclature beside Linne's species too,

6 B76.06.22, pp. 37-40, where M discusses Linnaeus's use of binomials of earlier botanists.

7 Linnaeus.

8 Linnaeus (1753).

9 B89.13.05.

10 *Wilckia* was published by Scopoli (1777), p. 317. M in his *Census* published the orthographic variant *Wilkia* (B82.13.16, p. 5).

92.08.06 which he merely omitted from modesty, because the matter was certainly obvious. Therefore should also *those* who pull out an ignored good genus again be the *author of all those species* which transfer unchanged as *well known* and well understood into the genus with the name change. *Only* such doubtful species as he at first correctly defines belong to the later monographer, or as are added *after* the change of generic name! I regard it no true science simply to have printed quickly a *first* list of names, when the change occurred long ago, as for example from Baillon, because those species names were already *covered* by it which refer to doubtfully based species. *Baillon* and others did not regard it as necessary to *immediately* bring a list of names into print of all known species and they would certainly then also be only *Nomina nuda*.[11] Whoever in *later* times will acknowledge *such* an act of taking possession?

The treatment which R Br experienced in the first half of this century from botanists without exception in this regard according to his hints in the *Prodromus*[12] was a *deserving one*.

Ionidium is already absorbed by *Hybanthus*.[13] The scrophularian genus *Calceolaria* is based on Feuillé as long ago as *1725* both in description and *figure*.[14]

Tissa and *Buda* fall as arisen *simultaneously* and *contradictorily*.[15]

Malveopsis[16] seems pretty unassailable to me.

If arbitrary exceptions are made, where shall such begin, where end. It would be impossible just in *this* regard to obtain a unanimous opinion. The chronological judgement is unalterable.

Cola Caspar Bauhin already used this name, which cannot remain unnoticed and shows that we cannot begin exclusively in 1753.

Aphora Here the manner must decide just as Necker proposes this and others of his genera, so also with Heister's *Meibomia*, but it would be *regrettable* to have to return to such a genus, if it was really accompanied by *species names*. I will retain *Desmodium*.[17]

11 'Naked names', unattached to a formal description.

12 R. Brown (1810).

13 *Hybanthus* was published by Jacquin (1760), p. 2, and *Ionidium* by Ventenat (1803-4), t. 27.

14 See Linnaeus (1771), pp. 143, 171 for *Calceolaria* which he based on Feuillé (1714-25), vol. 3, p. 12, plate 7.

15 See Adanson (1763), vol. 2, p. 507.

16 Presl (1844), p. 449.

17 For *Aphora* see Necker (1790), vol. 3, p. 50; for *Meibomia, see* Heister ex Adanson in Adanson (1763), vol. 2, p. 509; and for *Desmodium* see Desvaux (1813-4), vol. 1, p. 122, t. 5.

Placus. One must consider that a number of genera of Loureiro have later 92.08.06
been rejected, even the widely known *Myriogyne*, *Sponia*, *Dithyrocarpus* &c. Either we must *accept all* faultless genera of Loureiro or reject all.[18] Fortunately his genus *Callista* becomes valueless already for this reason because he also has a *second* genus for that which we call *Dendrobium*.[19] Linne's *Lobelia*[20] (probably based on *L. Dortmanna*) appeared as early as *1737*.

Statice; this genus belongs to L.[21] How can Willdenow be the authority for it, simply because he separated off *Armeria*?[22]

Herpestis falls before the indisputable *Bramia* Lam.[23]

R Br brought *Isopogon* and his other new Proteaceae genera before the public at the L. S.[24] as early as *1809*[25] in the presence of Salisbury. Therefore in the *Census* I have also quoted 1809. Thus the majority of Knight & Salisbury's genera sink into synonomy, but how *Linkia* can be overlooked and *Persoonia* remain is not easily understandable.[26] Many other genera just as long in use as *Persoonia* are set aside by A. de Cand., Benth., Hooker and other *leaders* of plant systematics, without anyone appealing. *Banksia* L. f.[27] should *remain*, because the Forsters had *withdrawn* their genus *Banksia* on Murray's inducement. I regard this as sufficient reason in such a case to leave the former

18 Loureiro's names *Centipeda*, *Trema* and *Floscopa*, introduced in Loureiro (1790), are now preferred to *Myriogyne*, *Sponia* and *Dithyrocarpus* respectively that were introduced later by other authors, but not his name *Placus* (Loureiro [1790], p. 496) which = *Blumea*.

19 *Callista* Loureiro (1790), p. 510 = *Dendrobium* Swartz (1799), p. 92. Loureiro's second genus referable to *Dendrobium* was *Ceraia* (Loureiro [1790], p. 518).

20 Linnaeus (1737), p. 267.

21 Linnaeus (1753), vol. 1, p. 274.

22 Willdenow (1809), p. 333.

23 *Herpestis* C. Gaertner (1805-7), p. 186; *Bramia* Lamarck (1783-1808), vol. 1, p. 459.

24 Linnean Society, London.

25 Robert Brown read a paper on Proteaceae at the Linnean Society on 17 January 1809. Brown's paper was published in 1810 (R. Brown [1810a]), but prior to this, a work on the same subject was published by Joseph Knight (1809) who acknowledged drawing heavily on the manuscripts of Richard Salisbury; indeed, the publication has often been regarded as Salisbury's. Salisbury was present at the reading of Brown's paper, and the work published in Knight's name included, without acknowledgement, several of Brown's descriptions. See Britten (1886).

26 *Linkia* was named by Cavanilles (1798), p. 61, published on 14 March, while *Persoonia*, though published two months later by Smith (1798), p. 215, on 24 May, was preferred as the conserved name in Ascherson's list.

27 Linnaeus fils. *Banksia* (Proteaceae) was described by Carl von Linné (1781), pp. 15, 126. *Banksia* (Thymelaeaceae) described by J. R. & J. G. A. Forster (1776), p. 7, is now *Pimelia*.

92.08.06 genus *Banksia* untouched, but thereby do not fail to see the friendly feelings which prompted Dr O. Kuntze to wish to couple my name with it, for which I would be quite proud, if it happened.[28]

For *Exocarpos, Xylophyllos* cannot be substituted, because the latter expression is quite *unsuitable*.[29]

Gyrostachys[30] is already for that reason not acceptable, because the order of the flowers is not circular or circinate, which is the *best* interpretation of the Greek word, but spiral.

Bulbophyllum originated as early as 1822;[31] because this genus of Thouars *is accepted*, one cannot declare him to have forfeited it because he published no species names, since the genus was accepted early on. All this shows that the discussion of genera cannot be abrupt, chronological and so on.

As early as 17 *January 1809* RBr submitted his great paper on Proteaceae to the L. Society.

Pinalia.[32] We cannot enter into a *month's* calculation, if a long-acknowledged genus is to be rejected, which appears in *the same* year. Therefore I also do not recognise Swartz's genus *Stylidium*, and the Swedish researcher should, *himself*, have withdrawn his *Stylidium* out of respect for DC. I doubt the value of uniting Loureiro's *Stylidium* with *Alangium*. Of course the interpretation of genera depends on individual opinion, because *Nature* creates only species, not genera or orders, which simply are to facilitate the general view for us and are to be helpful for our memory. Admittedly I acknowledged the relationship of Loureiro's *Stylidium* to *Alangium* myself, when I erected the genus *Pseudalangium* in 1860.[33] That Labill. *Candollea* of the Dilleniaceae is *quite untenable* I have repeatedly established from more *complete* material than is available to any other person.[34]

28 In 1891 Kuntze published 74 new combinations of *Banksia* in the family Thymelaeaceae, rejecting the genus *Pimelia*, but none are named after M; see Kuntze (1891), p. 583.
29 *Xylophylla* Montrouzier (1860), p. 250 = *Exocarpus* Labillardière (1800), vol. 1, p. 155.
30 Persoon (1805-7), vol. 2, p. 511.
31 Du Petit-Thouars (1822), tab. esp. 3.
32 Described by Don (1825), p. 31.
33 M erected *Pseudalangium* (*P. polyosmoide*) in B60.08.01, p. 84.
34 See M to W. Thiselton-Dyer, 10 September 1881. Loureiro (1790), p. 220, described *Stylidium* in Cornaceae, while Swartz published *Stylidium* (Stylidiaceae) in Swartz (1807), p. 48 (predated by Willdenow [1805], p. 146, as *Stylidium* Sw. ex Will.). *Alangium* was published by Lamarck (1783-1808), vol. 1, p. 174. M in B82.13.16, pp. 85-6, rejected the Swartz/Willdenow *Stylidium* in Stylidiaceae in favour of *Candollea*, which had been published by Labillardière (1805), p. 453, and on this basis made 86 combinations under this name. However, Labillardière

I can [see] in *Libertia* nothing other than a Section (*weak* besides) of *Sisyrinchium*.[35] 92.08.06

Astelia will be difficult to save. I have indicated this and similar difficulties in the *Census* as well.[36]

Draco The name is almost *senseless*.[37]

I suppose *Chlamysporum* will survive the "struggle for existence"! But what is to be done with *Thysanella* (1806). It will have to justify its existence![38]

Pandanus is sufficiently protected already by Rumph.[39]

I prefer to leave *Manisuris* and *Rottboellia* separated.[40]

Chamaeraphis was erected by R. Br not in the sense of *Setaria*.[41]

Nageia must undoubtably be restored again, thus *Podocarpus* can remain for *Phyllocladus*.[42]

Regardfully your
Ferd von Mueller

Unfortunately Du Petit Thouars's genera fail partly because he gave no species names.

In the *Bot. Zeitung*[43] and the *Centralblatt*[44] there are also articles by me about right of priority.

himself ([1804-6], vol. 2, p. 33) had rejected his own *Candollea* in Stylidiaceae and re-used the name in Dilleniaceae; i.e. he seems to have accepted that *Stylidium* of Swartz/Willdenow was his *Candollea* and that their name had priority.

35 *Libertia* Sprengel (1825-8), vol. 1, pp. 127, 168; *Sisyrinchium* Linnaeus (1737), p. 273.

36 *Astelia* R. Brown (1810), p. 291. In his *Census* (B82.13.16, p. 117) M listed only one species. *Astelia* is now a conserved name.

37 *Draco* Crantz (1768), p. 13 = *Dracaena* Vandelli in Linnaeus (1767), p. 246.

38 *Chlamysporum* Salisbury (1805), plate 103, subsumed under *Thysanotus* R. Brown (1810), p. 282. M wrote the date incorrectly: *Thysanella* Salisbury (1866), p. 67, is now considered an illegitimate name and is also subsumed under *Thysanotus*.

39 *Pandanus* Rumphius ex Linné (1781), p. 64.

40 *Manisuris* Linnaeus (1771), p. 164; *Rottboellia* Linné (1779), p. 22.

41 *Chamaeraphis* R. Brown (1810), p. 193; *Setaria* Palisot de Beauvois (1812), pp. 51, 178.

42 *Nageia* J. Gaertner (1787-91), vol. 1, p. 191, was subsumed under *Podocarpus* Persoon (1805-7), vol. 2, p. 580. This meant that *Podocarpus* Labillardière (1804-6), vol. 2, p. 71, was invalid, being re-named *Phyllocladus* by Richard (1826), p. 129. If *Nageia was* accepted, *Podocarpus* Persoon was vacated, *Podocarpus* Labillardière was no longer invalid and *Phyllocladus* was no longer needed.

43 B79.08.01

44 B82.13.02.

92.08.19 *To William Thiselton-Dyer*

RBG Kew, Kew correspondence, Australia, Mueller, 1891-96, ff. 27-31.

19/8/92[1]

I am just preparing an inaugural adress for the Horticultural Students Society here, dear Dr Dyer, so that I can answer your kind letter only hurriedly, other extra-work devolving on me in the Department also just now. I thank you for the sympathetic words, offered by you in my present departmental depression, and for the support official also you always give me, when occasion arises. P*rivately* I may say, that your idea, to get support from the other colonies, could not be carried into effect, as I am already as a Victorian Officer, on whom the other colonies have no claim, often harrassed by parties beyond Victoria, and if I had any emoluments from their colonies, the demands and worry would still become greater and my time for original research still more be encroached on. Moreover each colony has now its separate Herbarium already of its own plants I have discontinued subscriptions to nearly all periodicals, even Just's &c, but gladly keep the botanic Magazine[2] on. Pray thank Sir Joseph for his genial interest in my fate.

Of course any novelties, which I successively elaborate, will also in future be sent representatively to Kew. I shall be very pleased to see the morphologically so remarkable Correa Baeuerleni figured by Prof Oliver.[3] In such a celebrated Serial as the icones, any one seeing a plant of his displayed, must be proud. I have cuttings of it now growing, and shall manage ere long to get it across to

1 MS is stamped 'Royal Gardens Kew 3. Oct. 92' and initialled by W. Hemsley.

2 *Just's botanische Jahresbericht*; *Curtis's botanical magazine*.

3 M described C. Bauerlenii in B85.13.24, p. 960. It was illustrated in *Hooker's icones plantarum*, xxiii, t 2245, 1892. In his accompanying text Daniel Oliver wrote: 'Sir F. von Mueller, to whom we are indebted for excellent flowering specimens ... suggests that this species is deserving of a figure ... as "of singular structural interest"'; see M to W. Thisleton-Dyer, 10 May 1892 [Collected Correspondence].

Kew. After 30 years exertions I have managed to get seedlings of Verticordia oculata and V. grandis. When they are well established some shall go also to Kew. V. oculata is to my taste the most beautiful plant in its own delicacy of all Australia. 92.08.19

Kindly ask the excellent Mr Hemsley, to look once more through my 13 Decades of Acacias and Albizzias.[4] As in each case the fruit is given, he will see, that on the *form* of fruit neither Pithecolobium nor Hansemannia can be maintained as genera, just as little as Tetracheilus.[5] With regard to the consistence of the fruits, we must remember that the pithy legume of Vachellia counts not for a good genus, and that in Gleditschia (not far off in affinity) we have species with pulpy and dry fruit. After my 46 years of the study of 300 species of Acacias, my judgement – I thought – should have some weight. Of course, genera, unlike species, are not natural, so that no absolute rule for their limitation can be laid down. In Trimens Journal of Botany 1872[6] I pointed out, that at all events the Pithecolobiums of the eastern hemisphere could not be kept apart from the Albizzias. The well known A. lophantha shows us, that the inflorescence does not help us so far for the genera, and we find quite similar variation of fruit-form in Acacia as in Albizzia incl. Pithecolobium.

I do not stand alone in uniting the two genera, as some recent authors followed me in this. In writing on the subject I do so, because the question came forward in the last Bulletin.[7] Sooner or later, whatever fate these genera may have, their position will become more settled, but I now write to show, that the union was effected only after long and careful study on my part.

Archidendron was established[8] when I knew only the species with numerous pistils, up to 15, as shown in the plate.[9] But when

4 B87.13.04, B88.13.01.
5 Tetracheilos?
6 B72.01.01.
7 came forward in the last Bulletin *is marked with a line in the margin.* See *Kew bulletin* (1892), p. 125 under *Hansemannia oblonga.*
8 B65.10.04, p. 59.
9 B88.13.01, decade 13, pl. 7.

92.08.19 I subsequently ascertained in other species, that the number came down to 2, I gave up the genus. Excuse the frankness of my remarks, but I am thrown – so to say – on my defense.

Always regardfully
your
Ferd von Mueller.

Prosopis contains also species with succulent and dry fruits.

I have *not* responded to the invitation from Berlin to subscribe to the 4 theses, and to the rights of a list of genera; contrarily I pointed out to Prof Engler,[10] that the 4 passus[11] do not cover some still more important laws; – further that many of the genera, of which he and his collaborators acknowledge priority, cannot claim such, e.g. Salisbury & Knight's genera of Proteaceae as RBr's belong to *1809*, not to 1810. This they could have seen in my Census.[12] As RBr. read his famous essay on the Proteaceae in *Jan* 1809, before the L.S., he must have finished the whole in *1808*.[13]

Private

Merely to show, how hopeless it would be, to work a Central Herbarium on the expenses of the several Australian Colonies, I instance the case of the Eucalyptography. Before I entered for Victoria alone on so *costly* a work, I applied to more than one of the neighbouring colonies for aid, but the result of a lengthened correspondence was *none*! They purchase some few copies perhaps *and use* the work for the sake of their own trees, without probably ever thinking of the sacrifices made for it by my Department. However I did *not apply to the other Governments direct*, as the preliminary enquiries were not encouraging from among the colonists most interested.

10 See M to A. Engler, 6 August 1892.

11 passus *is marked with a question mark in blue pencil, and annotated in lead pencil.* passus = steps.

12 B82.13.06, pp. 65-73; B89.13.12, pp. 111-24.

13 Brown's paper on Proteaceae was read at the Linnean Society on 17 January 1809. For further details, see M to A. Engler, 6 August 1892, n. 25.

Mr Fitzgerald, who death we just so much deplore,[14] had also no financial aid in his splendid "Australian Orchids"[15] from the other colonies. If I accepted any financial help, even if obtainable from the other colonies, I might – to speak figuratively – "be torn to pieces["][16] by inordinate claims raised. Therefore I accepted during the 40 years of my official service *never a single fee* nor a single present, though perhaps not a day elapsed, except on occasional travels, when I had not to write *professional* information *beyond* Victoria! The Work on the fungs of Australian Fungs[17] mainly arose under the hope that it would help to combat the extensive mycetous diseases in our *warm clime* and under other exceptional circumstances concerning which I need not write, as this subject is not one out of which much acknowledgement to me will arise. 92.08.19

Any how Dr Cooke had a substantial and disinterested support from me for many years.

Private.

The bot Garden has still after retrenchment £8000 or £9000 irrespective of heavy extra-expenses for Water-supply, the large Observatory Vote is also but slightly reduced, and Sir Fred. MacCoy[18] has still £400 a year for his decades of chromolithography.[19]

Can the doubts about the true Pittosporum ferrugineum Aiton be finally settled?[20]

14 Fitzgerald died on 12 August 1892.
15 Fitzgerald (1875-94).
16 *editorial addition* – obscured by binding.
17 Cooke (1892).
18 i.e. McCoy.
19 No new decades of McCoy (1878–90) were issued.
20 Can the doubts … settled? *is marked with a line in blue pencil.*

92.09.03 *To Josiah Cooke and Charles Jackson*

Letter book, vol. 10, p. 32, The American Academy of Arts and Sciences, Library of the Boston Athenaeum, Boston, Massachusetts.

Melbourne, 3/9/92

To Prof. Josiah P. Cooke &c,
President and Professor
Ch. Lor. Jackson, Corresp. Secretary of the American Academy of Arts and Sciences.

Honored Gentlemen.

It is with the deepest grateful emotion that I receive your communication of the 15 June,[1] apprizing me of the extreme honor, shown me by the American Academy of Arts and Sciences in electing me to the position, which became vacant by the lamented death of the illustrious C. J. Maximowicz, as a foreign honorary Member. This singular mark of generosity touched me all the more, as unlike my celebrated Predecessor, I had so insignificant claims on the consideration of the science of the western world, with which Prof. Maximowicz by his East Asiatic travels and researches became so intimately connected. I am therefore all the more beholden to your great Academy for raising me to the proudest postion attainable in American Science! Indeed I regard it especially from antipodal distance, as an honor, equalled only by one other bestowed on me during my 53 years phytologic studies, more particularly so, as it comes from the eldest Union in science of your great country. The bonds of progressive intellectual efforts, linked together all over the world will ever be fascinating, and will individually encourage me also always to share, if ever only so humbly in advancing knowledge, a privelege all the more proud under the aegis of the "American Academy of Arts and Sciences"

With profound reverence
your
Ferd. von Mueller

1 Letter not found.

Figure 14. Mueller at 68 years (1892/3). Courtesy of the Library, Royal Botanic Gardens Melbourne.

92.09.10 *From George Perrin*[1]

RB MSS M1, Library, Royal Botanic Gardens Melbourne.[2]

Forest Branch
Melbourne
Sept 10 1892

My Dear Baron

I must apologise for not replying to your kind letter in re naming the Eucalypt from Tasmania after myself. The fact is I have been away from Town and when I returned I found such a mass of correspondence that your letter was overlooked in the rush of it all.

You will allow me to thank you very much for the honor you have done me in giving to one of the leading genus of the Natural Order Myrtaceae a specific name after myself. I am truly glad that my deduction as shown forth at the Melbourne meeting of the Science Association has proved correct.[3]

I trust you will put the matter in re Moores Eucalypt[4] and Hodgkinsons Yellow string bark[5] right as to which should have the honor of your own name – Moore's tree if you are satisfied that it is distinct from E. urnigera of which I have been in doubt

1 Letterhead of the Department of Lands and Survey, Victoria, has been altered by hand to Department of Mines.

2 MS found with a specimen of *Eucalyptus perriniana* (MEL 1611957).

3 At the Melbourne meeting of the Australasian Association for the Advancement of Science in January 1890, Perrin had exhibited a specimen of a Tasmanian eucalypt that he suspected was a new species; see the *Report* of the meeting, p. 557. M evidently later named this *E. perriniana* in Perrin's herbarium but did not publish a formal description. It was subsequently reduced by H. Deane and J. Maiden to *E. gunnii*, var. *glauca*; see Deane & Maiden (1901), p. 135. See also Maiden (1903), p. 377.

4 T. B. Moore named *Eucalyptus muelleri* in T. B. Moore (1887).

5 Clement Hodgkinson had drawn attention to the potential usefulness of the timber of this tree, first described by A. W. Howitt under the name *Eucalyptus muelleriana*; see *Report of the Melbourne Harbor Trust Commissioners* (1890), pp. 18-20, and Howitt (1890), pp. 89-91. Perrin drew M's attention to the clash of names 'with a view of alteration in time to prevent trouble by and bye'; see G. Perrin to M, 25 August and 30 November 1892 [both Collected Correspondence].

myself[6] – as I saw E. urnigera on the S.E slope of Mt Wellington[7] 92.09.10
about 150 feet high whitish salmon col'd[8] bark with *red wood* not unlike "jarrah" in appearance a most excellent timber tree to my mind – the tree I saw had just been chopped down that is how I came to notice the color of the wood –

If you determine these trees before the next meeting of the Science Association at Adelaide, it would be a good opportunity to bring under notice these recent discoveries of Eucalypts[9]

Thanking you again for your kind thoughtfulness in giving my name to the new Tasmanian Eucalypt

I remain
Yours very faithfully
Geo. S. Perrin.

Baron von Mueller
K.C.M.G. &c. &c

6 M had expressed his doubts on this subject in B87.13.05.

7 Tas.

8 coloured?

9 M did not attend the Adelaide meeting of the Australasian Association for the Advancement of Science in September 1893 or submit a paper to it of the kind suggested.

92.12.10 ## *From William Potter*[1]

RB MSS M1, Library, Royal Botanic Gardens Melbourne.

Baron Sir. F. Von Mueller, K.C.M.G. F.R.S., M.D. &c &c.

Sunday, December 10th 1892.

My dear Baron –

As Sir George Dibbs was good enough to send me a Railway Pass for the Blue Mountains tour I left Sydney on Saturday and travelled as far as Lithgow then returned to Wentworth to visit the Falls. It is my intention to return to Sydney by the 6 a.m. Train tomorrow, in order that I may interview the Hon. Dr Norton and other members of the Council before the meeting takes place on Tuesday. The weather is delightful.

I deemed it wise to telegram you to write to the Hon. Mr King[2] and the President, because a timely word from you might stir them up to activity. Sir George[3] at once admitted the difficulties that beset the Antarctic Committee, and the danger of the vote being lost if it should have to come before Parliament again. He told me (but this is strictly *inter nos*) that if the information I had given him were officially brought before him, he should *cancel the vote*. Personally he did not regret Baron Nordenskjold's withdrawal, and thought the Committee should resolve to send out an Australian Expedition.[4] I mentioned to him the statement of Mr Crummer that it would be sharp practice and deceiving the Government to take Mr Reid's money[5] to make up the Geographical Society's sub-

1 Letterhead: 'Joseph Toll, Proprietor. Hotel Wentworth, Wentworth Falls [NSW]'.

2 Probably Hon. Philip Gidley King, MLC [see ADB].

3 Sir George Dibbs, Premier of NSW.

4 In February 1892, Barons Nordenskiöld and Dickson wrote to the Swedish Academy of Science (with a copy to the Australian Antarctic Exploration Committee), withdrawing from the proposed Swedish-Australian expedition to Antarctica on the grounds that the required Australian contribution to the funding was not on hand by the specified date of 1 January 1892. See Swan (1961), p. 75.

5 The NSW Government had agreed to contribute £1,334 if the NSW Branch of the Royal Geographical Society of Australasia raised £666 from other sources. The Victorian businessman and politician Robert Reid had subsequently agreed to give £1,000.

scription, and he laughed at it. He read me the conditions of the Government grant, namely that the £666 was to be collected from *private sources*. The treasurer (Mr Crummer) had only to show a voucher of the money having been collected, it mattered not to the Government who from or how, and the Grant would be paid over to Trustees named by the Geographical Society. 92.12.10

So you will see, my dear Baron, the difficulty of my position. If I should say to the Council here that the vote would be cancelled should an official communication be put before the Minister that Baron Nordenskjold had withdrawn, then some one antagonistic to the expedition or to the Victorian Society may at once get a question on the subject put in Parliament.

On the other hand, if I hold my tongue on the subject, the hostility to Victoria of some persons here, who profess that there is no such hurry for getting the money out of the Government, as I assert, then the very delay may prove fatal to the Societys interest, and the matter be brought before the Government through the press. For that reason I have kept away from all newspaper offices and Pressmen this visit.

At the Council meeting on Friday, I may as well inform you (in addition to what I wrote to Mr Macdonald) that Mr Crummer spoke bitterly against what he called Victorian interference, as if they in Sydney were not able to manage their own affairs. You may depend upon it that I did not let that misstatement go unrefuted. I showed the necessity laid upon the Antarctic Committee to state publicly, before long, what it intended to do with the moneys already collected seeing that the Swedish-Australasian Expedition had been abandoned, and also that the committee had all along carried on direct communication with the Governments of Australasia. As in Victoria, every one had to get away to catch his train, so the meeting was hurried through. After the meeting Mr Crummer and Mr Mann made arrangements for me to dine with them at the Australia on *Tuesday* at noon. We appear to be on the best of terms. I pressed him very earnestly to make an effort to get a good meeting on Tuesday, and share in the glory of getting an Antarctic expedition out into the field. He said he would do all that he could, but he must have a month in which to collect before he touched Mr Reid's

92.12.10 money. I pointed out that future co-operation would be endangered if the present available vote were lost through his views of a Victorian taint on money prevailing with the Society. Victorian geographers took broader grounds, and while they would not let the matter of Antarctic Exploration drop, because of a rebuf, the Committee would no doubt by and by make an independent effort.

When we meet at dinner on Tuesday (and I shall have seen several of the members of the Council privately by then) it is my intention to see whether or not Mr Crummer cannot be induced to move a resolution in the Council something to the following effect: – "That having heard the Statement of the Hon. Secretary of the Antarctic Committee, it is hereby resolved – (a) That all the funds raised in New South Wales for the Antarctic expedition be placed in the hands of Trustees in trust for Antarctic Exploration. (b) That the sum required to claim the Parliamentary vote be at once collected and the Premier be asked to pay over the vote to the Trustees appointed by this Council. (c) That the following gentlemen be appointed Trustees: Messrs. ------------- (d) That this Council believes it would be wise to fit the Antarctic Expedition out under the English-Australian flag, should sufficient funds be available."

Mr Crummer thought there was not any necessity to detain me for the Council meeting but I said the Antarctic Committee had incurred the expense of my visit, and it would be far more satisfactory to the Committee that I should wait and take back a direct answer to enable the Committee to have a basis for further and prompt action. I presume that I shall leave here (Sydney) on Wednesday morning). It is too expensive to stay long. Sir George desires to be kindly remembered to you. I believe he will write you a short note.[6] I will do my utmost to succeed,

Yours devotedly
Wm. Potter.

6 Note not found.

To Jacob Agardh[1] 93.00.00

RB MSS M200.38a, Library, Royal Botanic Gardens Melbourne.

CLARISSIMO PHYCOLOGO
J. G. AGARDH

INUENTE AETATIS SUAE ANNO OCTUAGESIMO
VIII DEC. MDCCCXIII MDCCCXCIII
GRATULANTES OFFERUNT AESTIMATORES

G B De Toni – W. G. Farlow – Prof. Chr. Gobi – Ferd. von Mueller – C. Cramer – [...] – F. R. Kjellman – O. Nordstedt – Dr Antonio Siccone – Reinke – Prof. Pringsheim – J. J. Rodriguez – A. Weber – L. Kolderup Rosenvigne – Fr. Schmitz – N Wille – Ed Perceval Wright – Paul Falkenberg – Eug. Warming – E. Bornet – Prof. Dr. Ferdinand Cohn – Aug Le Jolis – Herbert M. Richards – E. Strasburger – D. C. Eaton – Dr. P. Hauptfleisch – Dr M. Worowin – Luigi Dufour – J Bracebridge Wilson – T. Caruel – M. Foslie – Dr J. A. Henriques – J. Wiesner – A. Blytt – F. W. C. Areschoug – S. Berggren – T. Reinbold – G. Hiern – Prof. Dr. M. Möbius – L. Krug – Mrs F. A. Curtis – Prof. Geo. F. Atkinson – G. Lagerheim – Prof. E. Askenasy – F. Heydrich – Ch. Flahault – Joh Lange – Paul Magnus – Ernst Ljungström – N W Lagerstedt – Wittmack – H H Gran – L. M. Neuman – R. J. Harvey Gibson – B. Jonsson – Barthold Hansteen – J. M Norman – O. Borge – L. J. Wahlstedt – D. Bergendal – Axel Vinge – Aug. Hultberg – E. V. Cedervall – Veit Wittrock – Jakob Eriksson – George King – Dr. J. Brunchorst.

To *the most illustrious phycologist* J. G. *Agardh on attaining the age of eighty,* 8 December 1813-1893, *his admirers offer congratulations.*

1 MS is a decorative certificate, printed in Venice in 1893, on which the handwritten signatures have also been reproduced.

93.01.14 *To Veit Wittrock*

Centre for History of Science, Kungl. Vetenskapsakademien, Stockholm.

Melbourne 14/1/93.

Nachdem ich mit letztwöchentlicher Post Ihnen, hochgeehrter Herr Professor, meinen besten Dank dargebracht für die ausserordentliche Auszeichnung von der grossen altehrwürdigen Kgl. Schwedischen Akademie der Wissenschaften, – erreichte mich Ihr gütiger Brief vom 30 Nov., aus welchem ich ersehe, wie überaus hoch die Ehre ist, welche mir zu Theil geworden! Es rührt mich auch besonders, dass ich der Nachfolger von Regel bin! Durch 40 Jahre war ich mit ihm nicht nur in amtlicher sondern auch in befreundeter Correspondenz, u wenige Wochen vor seinem Tode schrieb er mir noch eigenhändig einen Brief den ich als den Abschied von ihm betrachten musste, da er offenbar sein Ende heranahen sah! Sein Andenken wird mir stets heilig bleiben. Für die Schenkung Ihres *prächtigen* Hort. Bergian. Bandes sprach ich Ihnen auch bereits schon meine beste Anerkennung aus.

Nachdem ich die Liste derjenigen meiner nun auch erhalten, welche sich bereits in der Bibliothek der Königl. Akademie befinden, ist es mir gleich möglich gewesen, etwas für Vervollständigung abzusenden, so den "Key to the System of Vict. plants", in welchem Werke ich eine Vereinfachung des Juss.-D.C. Systems entwickelte, u. in welchem es mir auch gelang eine Dichotomie ganz nach dem natürlichen System ohne irgend eine Unterbrechung durchzuführen vom Beginn der Ordnungen bis zum Ende der Arten.

Auch habe ich die neueste (und bedeutend erweiterte Ausgabe der "Select plants" gesandt. Mehreres Andere, wie die "Papuan plants" habe ich selbst nicht mehr noch ist es direct kaufbar, doch werde ich diese Bücher wohl gelegentlich antiquarisch herbei schaffen können.

Haben Sie Campbell's New Hebrides für welches kleine Werk ich einen phytographischen Anhang schrieb? Besitzen Sie dort auch vielleicht schon Wittsteins Chemie der Pflanzen-Theile, welche ich ins Englische übersetzte, und mit vielen Nachträgen vermehrte? Trockne Pflanzen-Exemplare will ich von Zeit zu Zeit in kleinen Päckchen durch die Post senden, was das Einfachste des Schickens ist. 93.01.14

Sie ehrend der Ihre
Ferd von Mueller.

Auch die fehlenden Platten der Fragmente gehen mit dieser Wochen-Post an Sie ab.

Die Todea wird Ihnen durch meine London Agenten, Mess Watson & Scull, Lower Thames-Street, kostenfrei zugehen.

Spätherbst (hier Apr. & Mai) wird die beste Zeit zum Senden einer Todea sein. Wenn ich keine mehr erlangen kann von bedeutender Grösse, so will ich 2 kleinen senden, welche dicht neben einander gestellt das Bild eines Riesen Exemplars geben werden

Melbourne 14/1/93.

After I offered my best thanks by last week's post,[1] highly esteemed Professor, for the extraordinary distinction from the great venerable Royal Swedish Academy of Sciences, your kind letter of 30 November[2] reached me from which I gather how exceedingly high is the honour that has fallen to my lot! It also moves me particularly that I am the successor of Regel! I was in correspondence with him for 40 years not only officially but also as a friend and a few weeks before his death he wrote me a letter[3] still in his own hand, that I had to regard as the farewell from him, because he obviously saw his end approaching! His memory will always remain sacred to me. I have also already expressed my best acknowledgement for the present of your splendid Horto Bergiano volume.[4]

After I also receive the list of those things of mine that are already in the library of the Royal Academy, it may be possible for me to send something to

1 Letter not found.
2 Letter not found.
3 Letter not found.
4 Wittrock (1891).

93.01.14 *complete it, such as the Key to the System of Victorian plants,*[5] *in which work I developed a simplification of the Jussieu-De Candolle system and in which I also succeeded in carrying out a dichotomy completely according to the natural system without any interruption from the beginning of the orders to the end of the species.*

Also I have sent the latest (and considerably enlarged edition of the Select plants.[6] *Several others, like the Papuan plants,*[7] *I myself have no longer nor is it directly purchasable, but I will probably be able to get these books second-hand as occasion offers.*

Have you Campbell's New Hebrides, for which small work I wrote a phytographic introduction?[8] *Do you perhaps already have there Wittstein's Chemie der Pflanzen-Theile, which I translated into English and enlarged with many addenda?*[9] *From time to time I intend to send dried plant specimens in small parcels through the post, which is the simplest way of sending them.*

Respectfully your
Ferd. von Mueller

Also the missing plates of the Fragmenta will go off to you with this week's post.

The Todea will reach you free of cost through my London agents Messrs Watson & Scull, Lower Thames Street.

Late Autumn (here April & May) will be the best time for sending the Todea. If I can no longer get one of considerable size, I will send 2 small ones, which placed close to one another will give the impression of a large specimen.

5 B86.13.01, B88.13.03.
6 B91.13.10.
7 B75.11.01 and subsequent fascicles.
8 B73.13.01.
9 B78.13.06.

To Paul Ascherson

93.02.24

Staatsbibliothek zu Berlin, Preussischer Kulturbesitz, Berlin.

Melbourne 24/2/93.

Eben, hochgeehrter Herr Professor Ascherson, langt Ihr Circular hier an, welches von der zu Genua mir erwiesenenen hohen Ehre die Kunde giebt, in das internationale Committee für systematische Pflanzen-Benennungen gewählt zu sein, und nehme ich mit grossem Dank für das mir erwiesene Vertrauen diese Wahl in die Commission nicht nur an, sondern werde mich auch besonders bestreben bei dieser wichtigen Angelegenheit meinen hohen Collegen wirklich nützlich zu sein. Diese Stellung giebt mir ja auch Gelegenheit, die Gründe noch weiter zu entwickeln und zu besprechen, welche mich nach so langer Erfahrung bei der Ausarbeitung meiner eignen phytographischen Werke leiteten.

In Bezug auf die besonderen Fragen, welche in Ihrem Genua-Circular gestellt sind, möchte ich vorschlagen:

1, dass ein Subcommittee für engere Berathungen erwählt werde.

2, dass dies Subcommittee bestehe aus M. A. De Candolle, Sir J. Hooker, Prof Baillon, und entweder Prof Saccardo (für Cryptogamie) oder Prof. Coulter (für Americanische Consultation) mit Ihnen selbst als Secretair. Mir scheint eine Subcommission von 5 genügend zu sein, denn mehr Mitglieder würden die practischen preliminaren Berathungen zu weitläufig machen.

3, dass dies Sub-Committee alle Mitglieder der Commission ersuche, eine volle Exposition [von][1] deren Ansichten über die zu berathenden Fragen einzusenden, dass nach solchen vollen schriftlichen Vorlagen ein Plan von der Sub-Committee ausgearbeitet werde, und dieser vorläufige Plan den 30 Mitgliedern für weitere Erwägung zukäme, wonach dann eine finale Ausarbeitung dem Sub-Committee möglich sein würde, um solche einem specialen botanischen internationalen Congress vorzulegen.

1 *editorial addition.*

93.02.24 4, dass die Wahl des Ortes, in welchem dieser Congress zu halten wäre, vom Sub-Committee bestimmt werde, so wie auch der passende Zeitpunct.

Um der Sub-Committee die Deliberationen und Ihnen die Correspondenz zu vereinfachen und zu erleichtern, möchte es wünschenswerth sein, den 30 Mitgliedern eine Reihe categorischer Fragen zu stellen formulirt und exemplificirt, z. B. Ob Tournefort, der eigentliche Begründer der Pflanzen-Genera, ganz zu übergehen sei, ob wenigstens als (T.) Anerkennung verdiene, oder ob wie ich mit Asa Gray, Baillon &c wünschte, als erste und älteste systemat. Gattungs-Autorität zu betrachten sei (e.g. Ranunculus, Anemone &c. Im letztere Falle bliebe die Annahme für die Species doch 1753, und nur solche Genera aus der Zeit vor L. würden angenommen werden, als durch die binare Benennung confirmirt wurden. So auch könnte eine Frage lauten, ob der erste Art-Name anzunehmen sei, obwohl solcher in den meisten Fällen auf Irrthümer in der Gattungs Auffassung beruht, und eine solche Annahme dem wahren Prioritäts-Recht widerspricht; e. c.[2] Nasturtium terrestre (1812) oder N. palustre (1821) obwohl Leysser 1761 und Withering erst 1776 schrieb.

Sie ehrend,
der Ihre
Ferd. von Mueller.

Ferner, ob es gerechtfertigt sei, dass bei der Transposition von Arten aus einem genus in ein anderes irgend einem Anderem, als dem der die erste Indication gab, die Autorität zufiele für Arten deren richtige Auffassung vollkommen bekannt waren, also selbverständlich in das neue Genus oder andere G. übergingen.

Andere Fragen möchten sein. Verwerfung von Gattungs Namen ohne Indication von Arten selbst in Zeiten nach Linné, wenn auch die Nothwendigkeit durchaus nicht vorliegt, dass der neue Doppel Name gedruckt sei, da solche Andeutungen manchmal wie Scopoli, R Br und andere solche gaben ganz genügen Es möchte auch zu erwägen sein, ob das Sub-Committee auch einige Zoographen

2 exempli causa.

zu Rath ziehen solle, um für die systematischen Benennungen in der ganzen lebenden Natur eine Gleichförmigkeit in den Principien anzubahnen. Vor Allem kommt es aber darauf an, dass alle Bestimmungen auf volles *Gerechtigkeits* Gefühl beruhen, wenn solche dauernden Bestand haben sollen. 93.02.24

Melbourne 24/2/93.

Highly respected Professor Ascherson, your circular just arrived here, which gives the news of the high honour shown to me at Genoa, to be elected to the International committee for systematic plant nomenclature, and I not only accept this election to the committee with great thanks for the trust shown to me, but also will strive especially in this important matter to be really useful to my distinguished collegues. This position also certainly gives me opportunity to develop even further and to discuss the reasons which guided me after such long experience in the elaboration of my own phytographic works.

In reference to the particular questions which are put in your general circular, I would like to suggest:

1, that a sub-committee be elected for closer consultation.

2, that this sub-committee consist of M. A. De Candolle, Sir J. Hooker, Professor Baillon and either Professor Saccardo (for cryptogams) or Professor Coulter (for American consultation) with yourself as secretary. It seems to me a sub-commitee of 5 is sufficient, because more members would make the practical preliminary consultation too lengthy.

3, that this sub-committee request all members of the committee to send in a full exposition of their views about the questions to be discussed, that after such complete written presentations a plan be worked out by the sub-committee and this provisional plan come to the 30 members for further consideration, after which it would be possible to submit a final elaboration by the sub-committee to a special international botanical congress.

4, that the choice of the place in which this congress would be held will be determined by the sub-committee as well as the suitable time. In order to simplify and make the deliberations of the sub-committee and the correspondence easier for you, it may be desirable to put a series of categorical questions to the 30 members, formulated and exemplified, for example, whether Tournefort, the actual founder of plant genera, is to be ignored completely, whether at least that (T.) deserves recognition, or whether, as I with Asa Gray, Baillon &c wish, he is to be regarded as the first and oldest systematic authority of genera (e.g. Ranunculus, Anemone &c). In the latter case the acceptance for species remains really 1753 and only such genera would be accepted from the time

Figure 15. Friedrich Krichauff. Courtesy of the Library, Royal Botanic Gardens Melbourne.

before Linnaeus as were confirmed by binomial nomenclature. So also could a question be whether the first species name be accepted, even if in most cases it is based on an error in the generic conception and such an acceptance contradicts the true right of priority, for example Nasturtium terrestre (1812) or N. palustre (1821) although Leysser 1761 and Withering 1776 wrote them first.[3] 93.02.24

Respectfully your
Ferd. von Mueller

Further, whether it is justified that in the transfer of species from one genus into another the authority for species whose correct conception was completely known passed to somebody other than he who gave the first indication, thus of course turned into the new genus or other genera.

Other questions may be: Rejection of generic names without indication of species even in times after Linné, even if the necessity is not by any means known that the new double name be printed, because such mentions sometimes like Scopoli, Robert Brown and others were given quite sufficiently. It may also be considered whether the sub-committee should also consult some zoologists, to initiate uniformity in the principles for systematic nomenclature in the whole of living nature. What matters above all is that all determinations are based on complete feelings of justice, if such are to have permanence.

To Friedrich Krichauff 93.04.15

PRG 715/2/5, Mortlock Library of South Australiana, State Library of South Australia, Adelaide.

15/4/93.

Dein Brief hat mich unendlich traurig gemacht, lieber Krichauff, denn er ruft wach die Erinnerung an längst vergangene hoffnungsvolle Tage! Deine nun betrauerte Schwester steht in Gedanken vor mir, wie sie mich lächeln Corydalis digitata, Tulipa silvestris und andere Seltenheiten auf der Schlossgarten-Wiese bei Husum sam-

3 *Nasturtium terrestre* was published by R. Brown in Aiton (1812), vol. 4, p. 110, while *N. palustre* was published in A.P. de Candolle (1816-21), vol. 2, p. 191. *IK* credits Crantz with an earlier publication of *N. palustre*, in 1769, but does not note any publication of *N. terrestre* prior to de Candolle's.

93.04.15 meln sah, und ihre kleinen Kinder waren die *schönsten* der Stadt! Jeder freute sich bei deren Anblick. – Jeder wird dieser edlen Dame ein herrliches Andenken bewahren!

Du selbst wirst uns gewiss als viel jüngeres Glied der Deinigen des Vaterhauses noch recht lange erhalten bleiben und so auch zum Besten Deiner Colonie. Du bist ja aus einer Familie von Longävität! – In Husum hast Du wohl gar keine Angehörige mehr.

Gewiss werde ich Dir und Deinem hochbetagden braven Schwiegervater und Deiner lieben Gemahlin und prächtigen Söhnen auch etwas von meiner Adelaider Zeit widmen. Du bist ja mein einziger Dutzbruder in Australien!

Stets freundschaftlich
der Deine
Ferd. von Mueller

15 April 1893.

Your letter[1] has made me infinitely sad, dear Krichauff, for it recalls memories of long-gone days filled with hope! I can see your now lamented sister[2] in my mind, standing before me, smiling, as she watched me collect Corydalis digitata, Tulipa silvestris, and other rarities in the meadows of the castle garden near Husum. Her young children were the most beautiful in town! Everybody was delighted by their appearance. – Everyone will retain wonderful memories of this noble lady!

You, as a much younger member of your family, will surely be preserved to us for a long time yet, and also for the benefit of your colony. After all, you come from a family with longevity! – you probably have no longer any relatives left in Husum.

I will certainly spend some time during my visit in Adelaide with you and your worthy old father-in-law,[3] and your wife, and your splendid sons. After all, you are my only 'Dutzbruder'[4] in Australia!

Yours in constant friendship
Ferd. von Mueller.

1 Letter not found.
2 Elizabeth von Kaup, née Krichauff.
3 D. L. T. Fischer.
4 In German only family members and close personal friends were addressed with the familiar 'Du'; the verb is 'duzen' = to address someone with 'Du', hence 'Duzbruder'. The more formal and general form of address is 'Sie'.

To Lord Kelvin

93.04.23

Royal Society, London, Miscellaneous correspondence vol. 16, 1889-92, letter no. 31.[1]

Melbourne 23/4/93

To the Right Honorable Lord Kelvin, D.C.L., L.L.D., P.R.S. &c &c[2]

May I approach your Lordship on a subject entirely in the interests of science, of which you as the present occupant of what once was Sir Isaac Newtons chair,[3] are the renowned leader in the British Empire. The favor, sought by me, is one, which I never asked for during the 30 years of high privilege of belonging to the R S.

Would your Lordship then take with your honored Councillors into Consideration for reward from the science-throne of Britain, the high claims of the renowned Physicist, the Geheim-Rath Prof Dr Knoblauch, the present President of the Acad. Caes. Leop. Carol. Nat. Curioso[r.][4] in Halle. It is purposeless to enlarge on the merits of this great man and on his elevated Position as representative of a Society almost as aged as the R.S.

Could you honor this great Investigator by some distinction either on the foreign list or as the Recipient of a medal, when the next opportunity arises? Prof Knoblauch is long beyond the zenith of life, so that perhaps divine providence has not destined him to continue long his earthly career, and thus to enjoy long the tokens of appreciations, yet to be bestowed on him.[5]

Reverently your
Ferd von Mueller

1 This letter is filed with one of M's calling cards and a three-page printed document, with handwritten additions by M, headed 'Liber Baro Fedinandus de Mueller' and dated 20 June 1892, listing, in Latin, all M's degrees, awards, honours and society memberships, presumably sent as an enclosure with this letter. See Appendix B below.

2 MS annotation: 'Officers'.

3 i.e. the presidency of the Royal Society of London.

4 Academia Caesarea Leopoldino-Carolina Naturae Curiosorum.

5 Knoblauch was neither elected a Foreign Member of the Royal Society nor awarded one of the Society's medals. See also M to W. Thiselton-Dyer, 25 October 1892 [Collected Correspondence].

93.04.23 It is too late now to offer your Lordship my felicitation directly in being raised by our gracious Sovereign to the Peerage.[6] But may I say, that when as Past-President of the Austral Assoc. for Advancem. of Science, I had to instal his Excellency Sir Rob. Hamilton, LLD., the late Governor of Tasmania, to the Presidency,[7] I alluded gratulatorily in my speech, to the event, then just telegraphed, that so worthy a tribute had been made by the Crown of Britain as to confer on your Lordship for the incalculable science-services rendered the dignity of hereditary Nobility!

Gladly do I share in the movements in memory of my late friend Sir Rich. Owen and in homage to Sir John Lawes and Prof Gilbert

93.05.20 *From William Thiselton-Dyer*[1]

RBG Kew, Kew Gardens, Colonial Floras, Kewensia, K9/655.59, ff. 105-6.

Kew, May 20, 1893.

Dear Sir Ferdinand,

I have your letters of Easter & 9. 4.[2] It is of course obvious that you are the person from whom a continuation of the Flora Australiensis would be naturally expected. I cannot imagine that if you

6 William Thomson was raised to the peerage as Baron Kelvin of Largs in the New Year's honours list of January 1892.

7 James Hector, the retiring President of the Australasian Association for the Advancement of Science, was ill and unable to attend the Association's meeting in Hobart in 1892. M therefore chaired the meeting of the General Council on 5 January, and installed Sir Robert Hamilton that evening as the new President. M's speech on this occasion was not published in the official report of the meeting.

1 This document is a typescript draft with minor manual corrections that have been incorporated in the transcription. MS annotation by Thiselton-Dyer: 'Sir J.D.H.'

2 M to W. Thisleton-Dyer, 2 and 9 April 1893; see also M to J. Hooker, 28 March 1893 [all Collected Correspondence]. See also Clements (1998).

definitely express your intention of undertaking the work Mr Bailey would wish in any way to conflict or compete with you. At the same time it seems to me that you cannot expect to keep the field indefinitely unoccupied and that it is a pity you did not obviate the present difficulty by announcing your intentions a good deal earlier. 93.05.20

Mr Bailey appears to me to have a real zeal for Botany and although no doubt one could wish his work to be better, one cannot altogether blame his ambition.

My own opinion is that a difficulty of this kind can be adjusted without unnecessary personal feeling by the judicious intervention of the Colonial Royal Societies. I am very glad to see that the Royal Society of Adelaide has taken up the matter.[3]

There are so few persons who occupy themselves with Systematic Botany that one would be sorry to stand in the way of anyone who is anxious to do work. I hope therefore that you will not be too hard on Mr Bailey in the matter.

I may remind you that the Australian Governments entered into an arrangement with one of my officers to prepare a Fungus Flora I believe with your approval but without in any way my cognisance.[4]

Although the work was confessedly based in great measure on the Kew Herbarium I was not in any way consulted about it nor have I to this moment been favoured with a copy of the published work. Although I think the proceeding was under the circumstances irregular I have not troubled myself about it. As long as useful botanical work gets accomplished I for my part am not disposed to stand on my dignity.

Yours sincerely
W.T.T.D.[5]

3 M reported this in his letter of 9 April 1893 [Collected Correspondence].

4 The officer was Mordecai Cubitt Cooke. See M to W. Thisleton-Dyer, 1 July 1891; also M to W. Thisleton-Dyer, 2 July 1891, 20 August 1891 and 30 August 1892 [all Collected Correspondence].

5 The initials are typed.

93.05.31 *To Frederick Bailey*[1]

RBG Kew, Kew Gardens, Colonial Floras, Kewensia, K9/655.59, ff. 108–9.

Copy[2]

Melbourne, 31st May 1893

Sir

in reply to your letter of the 25th inst.[3] I have the honor to inform you that Fagus Moorei was discovered some time ago by Mr R. Collins on high mountains towards Jamaroocau,[4] where it seems to occur only in one spot, and it is easily comprehended, that this discovery would interest the Natural History Society of Brisbane, to which I also belong. Had the finding of this tree in Q.L[5] been of any entomologic importance such would of course have been explained at the recent meeting of the Society.[6]

In reference to a circular in type writing,[7] sent several weeks ago by you without letter or further explanation to me just like to some others, announcing suddenly and unexpectedly your intention to supplement the Flora Australiensis "you are fully aware of my desire to keep published up to date an account of all the plants of Australia",[8] with the final view of my completing that work,

1 MS is stamped: ' Royal Gardens Kew 10. Jul. 93'.

2 The copy, marked by the copyist, 'F. M. Bailey F.L.S Colonial Botanist for Queensland', was enclosed with M to W. Thisleton-Dyer, 31 May 1893 [Collected Correspondence]. M's corrections have been incorporated into the transcription.

3 F. Bailey to M, 25 May 1893 [Collected Correspondence].

4 Not identified. Possibly the copyist's error?

5 Qld.

6 Bailey in his letter had noted that a Brisbane newspaper had reported M's sending a specimen to the Entomologist of the Queensland Museum, and requested that M give Bailey details of the locality.

7 M sent a copy of Bailey's circular with M to J. Hooker, 28 March 1893 [Collected Correspondence]; there is another copy at Kew in RBG Kew, Kew Gardens, Colonial Floras, Kewensia, K9/655.59, f. 99.

8 M is quoting F. Bailey to M, 25 May 1893, where Bailey uses the wording to refer to himself.

and you must also know well enough that I have been almost daily 93.05.31
making exertions for this purpose ever since the succesive volumes of Flora appeared and that Bentham himself publicly desired that additions should be brought up connectedly If any one should compile from my "Fragmenta" and other literary property of mine, outside what pertains to the colony in which he is officially engaged, it will be against my rights and against my permission, as such an act would inflict injuries on my status and department. My explorations have been variously and extensively carried on *in all* the Australian colonies since 1847, not like yours merely in one and only in much more recent times. Furthermore solely in my collections are the needful materials for a full elucidation the universal vegetation of Australia. I would however beg to advise that as early as possible a *full* flora of Queensland be written; and remains surprising that by far the largest portion of its territory has never yet been botanically traversed, notwithstanding the facilities long ere this offered by through railways, coaches and steamers existing widely through the colony for a number of years.

I have the honor to be
Sir
Your obedient servant
(signed) Ferd. von Mueller
Govt. Botanist for Victoria

93.06.03 *From Frederick Bailey*

Letter press copy book 2, p. 431, Queensland Herbarium, Brisbane.

June 3rd [189]3

Dear Baron Mueller

You seem to take it hard that I should prepare a volume to supplement Bentham's work the Flora Australiensis.[1] But the work is wanted and who is to do it? You have all along ignored the nomenclature & classification of Bentham and I have followed it in all my publication. Thus can you wonder at my being asked to supply the matter for a supplementary volume. You have studiously treated all plants named or determined by me with contempt, never could you possibly avoid it allowing such to a place in your works on the general flora. Therefore I cannot consider you capable of doing justice to the portion of Australia to which I have the honor to be the Government Botanist. In conclusion I beg to state that I never pushed myself forward to do the proposed work, infact when repeatedly asked to undertake it have steadily declined. But at last seeing how persistantly you worked to throw discredit upon the work of Bentham I consented to undertake its compilation. And when you received my circular[2] had you written me that you were preparing not a Muellerian volume but a supplementary volume following Benthams system & style I would have gladly left the work in your hands, and will do so even now if you assure me that such shall be done and that my work will not be ignored or tampered with –[3]

an early answer will oblige
yours very truly
F. M. Bailey

Baron Ferd von Mueller
Melbourne

1 See M to F. Bailey, 31 May 1893.
2 See M to F. Bailey, 31 May 1893, note 7.
3 See also F. Bailey to M, 14 June 1893 [Collected Correspondence].

To Pier Saccardo

93.06.20

Orto Botanico, Università degli Studi di Padova, Padua.

20/6/93

It was very gratifying to me, dear Prof. Saccardo, to received the two extraprints from the Hedwigia[1] so important in this instance as it shows the connection of Mylitta australis with a Polyporus.[2] This was the only case, in which this came under my notice in the 46 years of my uninterrupted stay in Australia. Prof Spencer shall receive one of the two extraprints, when he returns from England, where he is on a science-mission from here.[3]

By this post I send you two small parcels of Fungs. They appear all common species, which Dr Cooke must have had long ago, and they would only be of interest to your own collection for the sake of locality.

I had hoped to attend the medical Congress in Rome,[4] but at this depressed financial period I could not leave my post for the several months requisite. So I shall probably never meet you or any of my Italien Science-friends personally. I have however provided some notes on the therapeutic value of Duboisia and Homeria for the Rome Meeting[5]

With regardful remembrance your
Ferd. von Mueller

I was President of the Therapeutic Section of the inter colonial medical Congress of Melbourne.[6]

1 Saccardo (1893).
2 That the 'Native bread', *Mylitta australis*, is a Polyporus and not a Truffle was first published by Cooke & Massee in Cooke (1892a) p. 37, and independently soon afterwards by Saccardo, who based his description on specimens collected from Western Port, Vic., by Baldwin Spencer; see McAlpine (1904).
3 Spencer was in England during most of 1893 on sabbatical leave from the University of Melbourne.
4 International Medical Congress, Rome, 29 March-5 April 1894. The meeting had been announced for September 1893 but was postponed.
5 No paper by M was published in the proceedings of the Congress.
6 Intercolonial Medical Congress of Australasia, Melbourne, January 1889; see B89.13.16.

93.06.20 We like Delegates for our Australian Congresses from Europe

Can I be in any way of special use to your Societa Veneto-Trentina?

93.08.20 *To Annie Walker*

Am 27/1, Mitchell Library, State Library of New South Wales, Sydney.

Sunday, 20/8/93.

Some few days ago, dear Miss Walker, our kind friends at Brighton brought your superb drawings of fungs. The one, marked by you as a Lycoperdon is a Lysurus, and as it seems to me distinct from any of the few species, known of that genus, I have named it Lysurus Walkerae and will see it described.[1] Is it a very local species there? I suppose, you have no duplicates of these drawings, so that the[2] must be returned or be copied here. You can easily imagine, that I like to make the mycologic display in my two Herbarium-Halls (near the Observatory-Building) as large and instructive as possible. You deserve infinite credit for the talented zeal, with which you follow up these enquiries, which certainly in N.S.W. have the additional charm of novelty. The scarlet Polyporus is P. sanguineus (Meyer) You figured the typic form of Aseroe rubra, on which Labillardière founded the genus, the A. pentactina being only a variety.

May I advise, that you kindly number the figures of the drawings and correspondingly number the specimens also; then the danger of misreference will be averted. I will then do, all I can, to bring your researches to permanent scientific honor.

The yellowish Coral-Fungus is a species of Clavaria, allied to C. crispula, the other is also a Clavaria. There is in N.S.W. one region

1 No description was published.

2 they?

left, where among all sorts of plants yet new species might be found, 93.08.20
that are the sources of the Bellinger River and the country through which they send their upper waters. If *you* only could make some weeks stay there in a snug farm and get the youngsters out with baskets to bring you *all* sorts of plants, from branchlets of trees to the smallest weeds, what a glorious time you would have of it. You would see every day plants, never beheld by you before! Such a tour should not be very costly for any one, and during midsummer the temperature must be delightful. The scenery surpasses in some places even that of the Blue Mountains, particularly through the zone of the ever-green *true* Beeches (Fagus Moorei).

With regardful remembrance your
Ferd. von Mueller.

Kindly think in the season of the Thysanotus with short leaves (Fringe-Lily).

To Franz Stephani 93.08.21

Conservatoire et Jardin botanique, Geneva.

21/8/93.

Mit dieser Wochenpost, sehr geehrter Herr Stephani, sende ich Ihnen 2 Couverte gefüllt mit Jungermanniaceen (ich ziehe diesen Ausdruck dem von Hepaticae vor wegen der viel passenderen ausgedehnten Anwendbarkeit) Ich möchte Sie aber besonders darauf aufmerksam machen, dass aus dem westlichen Tasmanien kaum bisher Pflanzen dieser Ordnung gesammelt wurden. Mr J. B. Moore hat einen guten Blick für Unterscheidung der Arten, aber da die Exemplare im (südlichen) Winter gesammelt wurden, sind wenige in Frucht. Um ihn zu ermuthigen, zum Weitersammeln in seiner günstigen feuchten berg-waldigen Gegend, rathe ich, ihm doch eine Art, die sich als neu erwiesen haben wird, zu widmen. Dr Cooke hat manchen australischen Pilzsammler entfremdet dadurch

93.08.21 dass er sich stets weigerte den Sammlern u denen die zum Collectiren inspirirten irgend eine Species zu dediciren. Ein solcher Wunsch nach all den Anstrengungen, Ausgaben u Gefahren ist doch wohl eine verzeihliche Ambition. Das Material ist meistens reichlich diesmal. Beim Vertheilen lassen Sie es wohl kund werden, dass die Ex. aus meiner Anstalt kamen. Zurückbehalten habe ich *nichts*, sehe daher auch diesmal Rücksendungen entgegen.

Sie ehrend, der Ihre
Ferd. von Mueller.

Wenn Sie dem ausgezeichneten Vertreter der sächsischen Handels Interessen Hr M. F. Bahse begegnen, bitte ich ihm meinen besten Gruss zu entbieten. Wir hier bewahren sein Andenken aufs Freundlichste.

21/8/93.

With this week's post, very respected Mr Stephani, I am sending you 2 envelopes filled with Jungermanniaceae (I prefer this expression to that of Hepaticae on account of the much more suitably extensive applicability). However, I would like to make you aware particularly that hitherto plants of this order were scarcely collected from west Tasmania. Mr. J. B. Moore[1] has a good eye for the differentiation of species but because the specimens were collected in the (southern) winter there are few in fruit. In order to encourage him to further collecting in his favourably moist mountain forested region, I suggest dedicating a species to him that may have turned out to be new. Dr Cooke has alienated several Australian fungi collectors by always refusing to dedicate any species to the collectors and those inspired to collect. Such a request after all the efforts, expense and dangers is surely an excusable ambition. This time the material is mostly ample. On distribution I suppose you will make it known that the specimens came from my institution. I have kept back nothing, therefore I expect returns this time.

Respectfully your
Ferd. von Mueller

If you meet the excellent representative of the Saxon commercial interests Mr M. F. Bahse, please offer him my best greetings. We retain most friendly memories of him here.

1 While M appears definitely to have written 'J. B. Moore', he was presumably referring to Thomas Bather Moore who was active on the west coast of Tasmania at the time and after whom Stephani named several Hepaticae.

To Henry Deane 93.10.15

No. 548, folder 28, series 4, MS 610 Deane family papers, National Library of Australia, Canberra.

15/10/93.

In setting to work, dear Mr Deane, for adding notes to yours for the new part of Fitzgerald's Orchids[1] I find, that you likely not have my "Census of Australian plants"[.][2] I will send you the second edition,[3] because it will refer you to notes in the "fragm. phytogr. Austral", a copy of which is likely accessible to you. A second point for consideration is presenting itself, if you quote *verbatim* from Bailey's synopsis, which Mr Fitzgerald never has done, in such cases, where a description was given in the Flora or in the fragmenta, then these two works would become *superseded*, which would not be just to Bentham nor to myself. You have such excellent knowledge of the science of plants, that you could easily give a *description of your own* At all events the Flora and the Fragmenta ought to be quoted and *not solely* the synopsis, unless the species were first described in the synopsis. As this is the commencement of your Editor ship which – I trust – will continue through many years, I take the liberty of pointing out to you my views on this subject, and I have a sort of moral obligation also for watching Bentham's interest. It would undoubtedly be well, even if some slight delay occurred to do justice to Bentham. Of course, it would be necessary to adduce *discarded synonyms* in your work. Of Adelopetalum bracteatum I have never seen a specimen. Indeed our departed friend never communicated to me any dried Orchids of any kind

1 Following Fitzgerald's death in August 1892, Deane was commissioned to complete Part 5 of vol. 2 of his *Australian orchids*, 'using such notes as had been left by Mr Fitzgerald, and supplementing these from his own valuable experience and observations'. Deane evidently sought M's advice on what he had done, and his comments are reflected in some of the published notes. The new material was published in 1894.

2 *editorial addtion.*

3 B89.13.12.

93.10.15 Will you kindly try to see, when on the Tweed,[4] or ask your officers to do so, whether only Nymphaea gigantea (with serrated leaves) occurs there; or whether N. coerulea (N. stellata) can also be found (with entire leaves). Some other characteristics distinguish these two, which often grow intermixed, though Bentham saw not the distinctions notwithstanding my previous identification of the two.[5] I believe N. coerulea could be shown also to occur in NSW and so in lagoons Aldrovanda vesiculosa, of which I gave a figure in the transact LS. of N.SW.[6]

With regardful remembrance your
Ferd von Mueller

93.10.16 *From Eugene Hilgard*

C-B 972 letterpress copy books, vol. 19, Bancroft Library, University of California, Berkeley.[1]

October 16, 1893

Baron Ferd von Muller, Melbourne.

My dear Sir,

Your letters of july[2] and seeds accompanying them were duly rec'd and would have been answered long ago but that a case of severe illness in my family kept me from all my usual correspondence just about the time the last steamer was to leave; so I waited until another is soon to go.

4 i.e. Tweed River, NSW.
5 Bentham (1863), p. 61. M distinguished the two species of *Nymphaea* in B61.05.02, pp. 141-2.
6 *Transactions of the Linnean Society of New South Wales,* new series, vol. 2, 1887, p. 621, records a drawing of *Aldrovanda vesiculosa* being exhibited on M's behalf at a meeting of the Society on 28 September 1887. However, the drawing does not seem to have been published.

1 The letterpress copy books contain typed copies of the outgoing letters.
2 See M to E. Hilgard, 2 July 1893 [Collected Correspondence]; no other letters written in July 1893 have been found.

Accept thaks[3] for the seeds, some of which are specially interesting: we shall give the Med. orbicularis a particular trial also that melon, whose designation I at first took to be a hybrid superlative – "sweetissima". 93.10.16

As you have heard of my visit to Europe, let me say that it was to me an occasion of very great interest, as I had not been over long enough to more than glance at things for nearly 40 years. That is a long time in the fast-striding nineteenth fin de siecle, and accordingly I found things very much changed, and that to great advantage, as compared with the early fifties, when the revolutionary movement upset everything more or less and neither the old nor the new condition of affairs showed to advantage. The industrial progress of Germany has simply been stupendous; and despite all the grumbling that is heard about the army bill, the agrarians [&c.] it is not a country to be lightly abandoned. If I were foot-loose I would like very much to live there several years at least. If I am not mistaken you also have not seen the old country for a long time; if I am right, to[4] take a "jump" across at the earliest opportunity. – When I was going to Berlin I was told that the Berliners were snobs, from first to last, and the military men simply insufferable. I had no pleasant recollections of the Prussians about the time they reconstructed Baden in 1849-50 and was half way afraid it might be so. But I must say that three winter months I passed at Berlin was the most profitable time I had in my year of absence. As you may have noted, I was not left to rest in idleness there; and perhaps I had to work as hard as I do here, but it was con amore, with an appreciative public. I ought to say the same for Munich; and for London. I shall ever regret that an attack of grippe in Paris cut my stay there so short The three weeks I had I thoroughly enjoyed, the more as for our special benefit London had not a single fog during the time.

Of course I went to Chicago on the way back, but could stay only a few days, because of the illness of my wife which since our return – now four months – has kept me on the anxious seat. It

3 thanks?
4 do?

93.10.16 was a grand sight and nothing short of a month's stay could have done it justice.[5] Germany was pre-eminent in the industrial exhibits – renewing the impression I received when there – Since then I have had numerous visitors from the old world here – quite a number of those I saw on the other side, too; so I really feel as though I had established a fresh bond of brotherhood with them, and as if I had accomplished something by going. This feeling has since been accentuated by the bestowal upon me of the Liebig medal for Soil investigation, by the Munich Academy. It is a pleasant offset to the petty annoyances and the indifference of ignorance with which my best work is regarded by a majority of mankind in this free country.

I tell you all this because I think that, being so far away from the center of culture you, like myself, have had many a hard trial of spirit in your labors for the benefit of your adopted country and of science. It is very refreshing to come in contact once more with a thoroughly appreciative public after many years of this quasi isolation, and I trust you may have a chance before long to experience this personally. The thought occurred to me particularly as I read your brief but pregnant biography, kindly sent me; for it is not mentioned that you ever left Australia, and however highly appreciated are your labors in science everywhere, yet the personal renewal of contact involves a gratification far beyond any recognition "ad distans".

Your paper on thistles[6] has also greatly interested me, since the weed question is with us also a alarming one, [and h]as required legislation like that regarding insect pests. I perceive you have to keep out the "Canada thistle"; we have done so thus far by timely action, but are constantly on the lookout for an invasion.

But I must close this lengthy epistle. With best wishes for your continued health and success in your good work

Yours faithfully
[Eugene Hilgard][7]

5 Hilgard had evidently visited the World Columbian Exhibition in Chicago.

6 B93.13.01.

7 *editorial addition* – there is no signature on this copy.

To William Thiselton-Dyer 93.11.03

RBG Kew, Miscellaneous reports 7.2, Queensland Botanic Gardens etc., 1873-1919, f. 50, and Kew correspondence, Australia, Mueller, 1891-96, f. 53.[1]

3/11/93

Your letter of the 12 Sept[2] has just reached me, dear Dr Dyer, and I will answer it seriatim. By action and advise I have done for Mr Bailey what I could. I have sought the support of two Brisbane Physicians for him directly, Drs Bancroft and Lauterer, and I have pleaded his QL.[3] claims indirectly with other leading men of Science in Brisbane.[4] The civil service regulations here forbid my acting officially in his favor with his own Ministry or Parliament,[5] and as a proof of this I like to mention, that when the S.A. Government wished on its own accord to have my opinion of Mr Holtze's already a S.A. Garden-Director then (Pt. Darwin), capabilities and fitness for the Adelaide Director Ship, the Minister

1 The two sheets of this letter have been filed separately in different sections of the Kew archives. Both have been stamped with date of receipt: 'Royal Gardens Kew 11. Dec. 93'. The first has been annotated by Thiselton-Dyer: 'Sir F. von Mueller'. The second sheet commences: '2, As regards pre-linnéan plant-names'. It too has been stamped: 'Royal Gardens Kew 11. Dec. 93'.

2 Letter not found. Presumably Thiselton-Dyer's answer to M to W. Thisleton-Dyer, 4 August 1893 [Collected Correspondence], which announced the intention of the Queensland Government to abolish Bailey's post.

3 Qld.

4 See J. Thomson to M, 27 November 1893 [Collected Correspondence], and M to J. Hooker, 19 December 1893.

5 Thisleton-Dyer, however, wrote to Sir James Garrick, the Qld Agent General, urging Bailey's reinstatement. Garrick sent Thiselton-Dyer's letter of 16 September (draft at ff. 43-4) 'to the colony for the information of the Chief Secretary' (f. 45). Thisleton-Dyer also wrote on 15 December to Sir Robert Meade, Permanent Secretary at the Colonial Office, (draft at ff. 51-2). Bailey was reinstated at a reduced salary on 1 January 1894, and wrote on 7 February 1894 (ff. 53-4) to thank Thisleton-Dyer for his support: 'nothing which has ever been done by an Australian Government has caused such universal dissatisfaction, the press and the people from all parts of Queensland wrote and waited upon the members to try and prevent the abolishment of my office, but it was of no use until your letter was sent out by the Agent General'.

93.11.03 concerned wrote to my Minister here to be allowed to give my opinion.[6]

It would be very difficult to restore the Office of Gov. Botanist of Queensland after it had been suddenly and unexpectedly abolished by the local Parliament;[7] but if his friends make an united effort a pension might be secured for Mr Bailey just as for Mr Walter Hill before; or failing this – a compensation should be applied for. His son, I understand – is retained in the Herbarium, and he could enjoy probably various official advantages even after retirement while engaged in congenial work.[8] I mentioned already before, that Botany plain and simple does not present itself as needful to the Colonists in general;[9] hence I make the great and special efforts here, to apply it to rural and technologic interests as you do at Kew. But I am afraid to say almost anything in this sad affair, lest it be misconceived or misconstruc[t]ed.[10]

Mr Bailey's sedulity and cleverness we all recognize

2, As regards pre-linnéan plant-names I have fairly given my views many years ago in the "Papuan plants",[11] and if they are absolutely identical with those of Linné, then, I hold, it is a grave injustice to deprive the original authors of their rights. Linnaeus rested mainly on works through centuries of his predecessors and was not always mindful of them and their claims. It is futile to have any code of priority unless based like other laws on absolute justice. In reference to genera, I agree with many others, among them so undisputable authority as Asa Gray, that we should go back as far as Tournefort at all events, if they became by the spec. plantarum

6 Letter not found. M. Holtze was appointed Director of the Adelaide botanic garden in 1891 following Richard Schomburgk's death.

7 Bailey's retrenchment had been announced in October. Newspaper clippings on the debate about this in the Qld Legislative Assembly on 4 October 1893 are bound at ff. 48 and 49. Key sections have been marked in red ink.

8 In the newspaper report, the Secretary for Lands is quoted as saying: 'As to the museum of economic botany, it might not be known to hon. members that Mr Bailey's son – a young man of considerable attainments – was in the department; the museum was not going to destruction'.

9 For an early example, see M to W. Thiselton-Dyer, 1 January 1881 [Collected Correspondence].

10 misconstrued?

11 B76.06.01, pp. 37-40.

fidemised![12] Why for a familiar instance should Lindern be set aside as author of the genus Limosella? *Bentham* was *open* for reconsidering questions and I aided in he adopting the oldest species-name within the correct genus, after he had followed DC[13] for a long time in using the oldest species name from the wrong genus! This is still considered by the majority of Phytographers as heresy! 93.11.03

If you, my honored friend, had known all the circumstances of my relation to the Academy for bot. Geography of Le Mans,[14] you would have written in a different tone. I did not know its existence when I was elected Director *for a year*. I saw great names on the list, among them Duchartre's the *Nestor* of French Botanist[s][15] and the designed[16] Director for 1894. I all along urged that there should be a local Council I disadvised the creation of an Order, but then suggested a medal though regarding it far too early. I was against large numbers of medals and any payment e[ver] so small for them. I induced Dr Rougier the Cholera-Investigator and nephew of Pasteur to proceed personally to Le Mans, to argue with Prof Leveillé on points in which I did not concur. I had allotted to Prof Leveillé some discretionary power, which he thought he could use for classifying the gift of medals, for which I am *not responsible*, though my name is attached and his counter-signature wanting, nor do I share the responsibility of the list in N. 25 of Le Monde, to which however my signature does not appear.[17] I in-

12 A fusion of 'bona fide' and 'legitimized'? Linnaeus (1753).

13 De Candolle.

14 International Academy of Botanical Geography.

15 *editorial addition* – Text obscured by binding. All square brackets in the following text have this meaning.

16 designated?

17 A list of recipients of vermillion, silver and bronze medals awarded by the International Academy of Botanical Geography was published over M's name in no. 24 of *Les mondes des plantes* (1 September 1893), and no. 25 (1 October 1893) contained a further, much longer list of recipients of bronze medals without M's name attached. In the earlier of these lists, J. Hooker was among those to receive a vermillion medal. Issue no. 23 (1 August 1893) included a list of prices for different categories of award. A later issue (no. 27, 1 November 1893) announced, over M's name, that Duchartre had declined to become Director. His letter of refusal was published in full in no. 29 (1 December 1893), followed by an announcement that on account of Duchartre's withdrawal, M would continue as Director in 1894, until a successor could be elected.

93.11.03 sisted on recipients being asked before hand, whether they wished to accept the medal and compromised no one. I wrote you already 2 weeks ago, that I had resigned at the Academy of Le Mans before [effect of these.][18] Kindly show this letter to Sir Joseph.

Regardfully your
Ferd von Mueller

I did not like applications to various Governments for subsidy.
I was aware that no medal can be worn except military.

93.11.25 *To Henry Deane*[1]

No. 586, folder 30, series 4, MS 610 Deane family papers, National Library of Australia, Canberra.

25/11/93

Am thrown into deep mourning, dear Mr Deane, through the death of the only son of my late sister,[2] a young accomplished Gentleman in Adelaide. Through this sad event the calculation of my time was also deranged. I have several sendings from you for which I am thankful and which shall now soon have attention.

Regardfully
your
Ferd von Mueller.

18 Letter not found. M's resignation was not mentioned in *Le monde des plantes*. However, in no. 31 (1 January 1894) G. Rouy was listed as Director.

1 MS black edged; M's nephew, George Doughty, died on 19 November 1893.

2 Bertha Doughty.

To Joseph Hooker[1]

93.12.19

RBG Kew, Letters to Joseph Hooker, vol. 16, ff. 1-2.

19/12/93[2]

This is the first mail, leaving for Europe, dear Sir Joseph, since I received from Dulau & Co the grand Kew-Index. What a splendid opus, to have the whole synonymy of Phanerogams up to 1885 before us! It will be an incalculably valuable gain, when the whole work is out, and I have no doubt, that you and Mr Jackson will be able to supplement it by to the end of our century, when not much will be left to be done.[3]

I am glad, seeing you supersede also Citrosma by Siparuna – Aublet['s][4] names undoubtedly should stand, unless disqualified by creating 2 genera instead of one. There is far more labour, as you know from own lengthened practice to furnish a plate than a mere brief diagnosis. Thus I uphold Lamarck illustr.[5] also. Am delighted, that you did not abandon Tournefort, as so many of the Moderns or Novices would have it. Nomenclature must rest on an *unimbiased feeling of justice*. That has been my reply to many communications, which I had on primogeniture also in plant names. I have by this mail again a heap of prints on this subject, which lot I have not yet time to read; but my unalterable reply will be again, that even if we have ever so many congresses or conferences or

1 MS black edged; M's nephew, George Doughty, died on 19 November 1893.

2 M wrote this date in the margin of the following newscutting, which is pasted at the top of the folio: 'Queensland. Brisbane, Sunday. Mr. F. M. Bailey, one of the recently retrenched civil servants, has been reappointed colonial botanist, at a salary of £200 per annum.' Bailey was reinstated from 1 January 1894.

3 *Three exclamation marks have been added.* Volumes 1 and 2 of *Index Kewensis* were published in 1893, and the remaining two volumes of the original work in 1894 and 1895.

4 *editorial addition* – Obscured by binding. All square brackets in the following text have this meaning.

5 For an example of M's use of a Lamarck plate without description as evidence of nomenclatural priority see M to W. Thiselton-Dyer, 31 December 1881 [Collected Correspondence].

93.12.19 deliberations, they will be overthrown like all other unjust legislature, unless all arbitrariness is avoided. I often wonder, what they will think of us about the agitation now on priority of plants names, a hundred years hence. As I said, permanency and unanimity can only be secured by absolute justice.

Perhaps I may venture to point out occasional inaccuracy, unseparable from all human work and more especially so from a gigantic work like yours. Bass[ia] Erskineana is a Sapotaceous plant. I maintained the name Bassia at the time[6] yet for Sapoteae, calling Allionis genus Chenolea before.[7]

You will be gratified, that our exertions on Mr Bailey's behalf have been successful. I communicated with the hon. Aug. Gregory,[8] pointing out that "as Pres. Austr. Assoc. he had for the time charge of Australian Science, and calling on him in that capacity to exercise his great influence for the benefit of Mr B."

With best festive salutation your
Ferd von Mueller

Will you kindly let Dr Dyer know about the [...] &c.[9]

Of course no one is bound to accept decisions of Congresses which he did not attend

6 1885, when M published *Bassia Erskineana* (B85.04.01, p. 930).

7 M described *Bassia Erskineana* as a sapotaceous plant in B85.04.01, whereas in *IK* it was listed among the Chenopodiaceae in a second genus of *Bassia*, erected by Allioni in 1766, that the editors treated as superseded by *Chenolea* (erected by Thunberg in 1781). Though M thus maintained *Bassia* among the Sapotaceae, as he says, he added: 'The generic name Bassia might well be changed to Illippe, as given by Koenig, as long ago as 1771 (Linné mantissa altera 663 [sic]), inasmuch as Allioni five years earlier established already a genus Bassia among Salsolaceae'. Linnaeus (1771), p. 563, indicated that *Bassia longifollia* is Koenig's MS *Illipe malabarorum*. In B85.06.03 M himself transferred the species to *Illipe.* M had used *Chenolea* in B76.10.01, pp. 91–2. Allioni's priority in the use of the name *Bassia* is now recognised.

8 See J. Thomson to M, 27 November 1893 [Collected Correspondence].

9 *two illegible words.*

Figure 16. Sketch of Mueller in 1893 (aged 68), made on blotting paper by J. A. Panton. Reprinted courtesy of the Royal Historical Society of Victoria, Melbourne.

93.12.23 *To Alexander Macdonald*

A38, Royal Geographical Society of Australasia (Victorian Branch), Mitchell Library, State Library of New South Wales, Sydney.

23/12/93

It was my intention, dear Mr Macdonald, to pay you personally a visit for conveying my felicitation at these festive days. But my cough is so severe that I cannot venture out unless on a hot day. Indeed I lost almost my voice, and must *avoid speaking* as much as possible. So, excuse me when the little Christmas gift for your bright young Lady is not brought by my self. Perhaps you may have a spare-hour during the holidays to come over to my dwelling, as I like to consult with you on several geographic subjects. Kindly give me a description or sketch of your present abode.

With best wishes for the new year
to all of you
Ferd von Mueller.

To Thomas Wilson 93.12.25

V93/8741, unit 527, VPRS 3992/P inward registered correspondence, VA 475 Chief Secretary's Department, Public Record Office, Victoria.

25/12/93.

T. R. Wilson Esqr
Under Secretary.

Sir

I have the honor of transmitting herewith the memoranda, promised by me at my interview at your Office last week.[1]

I have the honor to be,

Sir, your obed. servant
Ferd. von Mueller,
Gov. Botanist.

Antarctic whaling and sealing.
The Victorian branch of the Royal geographic Society of Australia has for the last ten years advocated Whaling and Sealing by *Steam ships* under some bonus-system, similar to that by which in our colony the cultivation of new sorts of crops is encouraged. Hitherto we have had in the southern hemisphere only *sailing* whalers, which cannot move through the intricate masses of ice forming an irregular belt between our shipping sea and the waters at the permanent ice towards the south-pole, the fartest[2] southern waters being the most promising for the chase. Last summer however 3 Scotish and one Norwegian Steam-Whaler went to the south of America, and were without going far into the antarctic regions very successful in seal hunting; and this season a Norwegian Steam-Whaler is testing the antarctic waters under Australian meridians. If success-

1 MS annotation: '1 – Antartic [sic] whaling 2 – Cultivation of coast lands 3 – Yarra floods 4 – Alpine Rural settlement 5 – Phylloxera'. MS annotation beneath this: 'Type copy to C.S. [Chief Secretary] 29/12/93'.

2 farthest?

93.12.25 ful, we may expect quite a fleet of such steamers out in future years making their final start and first return from and to Australian harbours. Part of the cargos would be sought by Victorian factories, and part would fall to the share of our naval commerce.

Cultivation on hitherto unutilized coast-lands

Many of our shores are not producing any revenue, and might furnish superior pasture grasses and fodder-herbs for naturalisation, so also the New Zealand Flax Agaves, Yuccas and other selfprotecting plants for fibres, the Sand-Wattle for tan-bark and many kinds of Pines and other trees for resin and turpentine in future years might be disseminated at comparatively small expense and for permanent good. What can be done in this direction has been shown by the planting of many miles of drift-sand already from Warnambool[3] to Portland-Bay[4] with the Marram-Grass,[5] which even affords cattle-feed for some months in the year, and which ought to be established and naturalized on all our sand-coasts, to prevent the farm-lands behind to become sanded in. About the various plants, needed for the purpose indicated, my work on "Select plants for industrial culture"[6] affords information also. Perhaps occasionally this might be considered, when questions concerning the Unemployed, come under discussion.

Yarra-Floods[7]

As the discoverer of many miles of the Upper Yarra from its remotest eastern alpine sources on Mt Baw Baw in 1860 (at 5000 feet elevation) I may be permitted to point out, that the vast expenditure, which would be involved by daming[8] up the upper flood-waters or by canalisation in the lowlands, could be avoided, dams moreover in high localities being always a source of danger through bursting, if at a comparatively inconsiderable expense a slight *tapping*-system was adopted. Yarra-floods of severity occur mainly

3 i.e. Warrnambool, Vic.
4 Vic.
5 Heathcote & Maroske (1996).
6 B91.13.10.
7 The typed version of this section is annotated: '*See Davidson*'.
8 damming?

under four simultaneous conditions, melting of snow at the sources in Spring, heavy rains, southern gales and spring-tides. To allow time for the lowland-waters, to flow off, before the upper waters reach the lower Yarra, I would suggest that the level of the successive upper reaches be dimined[9] by boring above falls at the base of rocks, or by cutting through small ridges, the object being, to divert slowly and gradually much of the water into vallies adjacent for irrigation, so that still the great principle of "water-conservancy" be adhered to. As yet hardly any vested water rights have arisen on the upper Yarra, nor is the river navigable above Dights fall near Richmond, to require the water to be kept at the present level exactly. If the water is led off on many places for inexpensive new irrigation by simple gravitation, then much of the downpour from the upper regions of the Yarra-country will find space for storage not existing now, with the advantage also, that much less of the Yarra-waters would flow unutilized into the Bay and finally into the ocean. 93.12.25

Rural Settlements in the Victorian Alps to promote mining also.[10] When nearly 40 years ago I discovered, ascended and named Mount Hotham and traversed our then pathless snowy regions in many directions, I became already impressed with the importance of our alpine [regions][11] becoming fully developed. Permanent settlement is possible in our latitudes up to 4500 feet elevation, and during the summer-months pastural, cultural and mining operations are possible to the summits of our alps. We can have table-fruits and vegetables from one to two months later, if grown in our subalpine regions, and the railway, now already extended to Bright,[12] affords facilities of bringing such products to the markets of the lowlands, when otherwise out of season there. Hardy grain could also be raised at high elevations, so poultry be kept and depasturing by herds and flocks carried on in the high regions from Sept til March. This would give encouragement to mining prospectors for

9 diminished?

10 The typed version of this section is annotated: '*Unemployed*'.

11 *editorial addition.*

12 Vic.

93.12.25 more persevering in their searches, as resuply of provisions could be effected from farms cheaply and locally, instead of packing up on horses, as now, all that is necessary for sustenance from the lowlands. Doubtless vast riches of gold and other metals are yet to be discovered in our highlands and their vallies over a vast extent still uninhabited, the Morning Star hill[13] alone having in course of years yielded nearly a million pound Sterling worth in gold.

Highlanders from Scotland, Swiss and Scandinavian people among us would be particularly those, fit to be placed for rural purposes on our Alps, as they are accustomed to a cold clime. The produce of the farms would in many cases be more tasty than that of the lowlands. Surplus stock, when frosty weather sets in, could be brought to marked,[14] or as in the colder European countries is much done, salted or smoked and be sold in casks to town-customers for family supply as food more cheap and superior than that under ordinary circumstances available. Natural irrigation can be obtained for most of the alpine vallies also here.[15] Culture-plants are much less subject to diseases there than in the lowlands of ours.[16]

Phylloxera.

The Phylloxera vastatrix was shown to exist in Victoria 1877 at a time, when I was exploring in West-Australia, to connect my northern researches of that vast territory, instituted as an Officer of Gregory's expedition during 1855 and 1856,[17] with my southern explorations in 1867. An albe[18] Board then appointed under the auspices of the Government has stamped out the first Phylloxera invasion, and requires not my assistance personally, though I have visited

13 Near Woods Point, Vic.

14 market?

15 The typescript has 'herb' instead of 'here', and omits the full stop.

16 M made similar suggestions about settlement in the Alps in M to A. McLean, 19 December 1890 [Collected Correspondence]. At that time a report was sought from the mining surveyor James Stirling on the resources of the Alps, for which M prepared a botanical appendix. The report was circulated extensively within the relevant government departments but the proposal died when the Secretary for Agriculture declared that there was no vote available to support the developments proposed; see G29797, unit 22, VPRS 619, PROV.

17 North Australian Exploring Expedition, 1855-6.

18 able?

the Geelong-Phylloxera-area more than once by myself. Foreseeing however, that the dreaded Phylloxera might at any time be reaching our colony anew, I distributed in several years fresh seeds of several kinds of American Vines, as secured by me purposely, such as are phylloxera-proof, as regards their roots, so that the European Vine, which is never attacked above ground by the Phylloxera, could be grafted on American roots, – the process being described briefly in the latest editions of my "select plants",[19] – should such a measure at any future time be deemed necessary. I may add, that since some few years Victoria has a Gov. Entomologist, Mr Ch. French, whose able services could at any time be rendered available for information on the vital processes of the Phylloxera vastatrix, while I could be consulted if desired on any special points, which come in respect to the phylloxera-disease within my own professional reach. 93.12.25

From Sophie Hooker 94.01.05

MS 443/D, Archives and Manuscripts, Hocken Library, University of Otago, Dunedin, NZ.

Kilkivan.
Queensland.
Jan 5th/94.

Baron Sir Ferd. von Mueller, K.C.M.G.
&c. &c. &c.

My dear Baron,

I hope you will not be anoyed at my writing thus to you – I am Brian Hookers[1] wife, but I cannot hope that you remember me –

19 B91.13.10.

1 Brian was the son of Joseph Hooker.

94.01.05 My letter is to beg the aid of your influence in trying to procure an appointment for my husband.

He has been most unfortunate, & is still in New Zealand. He has been there for the last fourteen months, and cannot get back. He has not been idle, it is true; but his Salary was exceedingly small, and he has been compelled to take the last month or so out in shares in the Co[y], because it is short of money & cannot pay.

I am afraid it is quite impossible to get an appointment unless one has someone of influence to speak for one. May I ask you, dear Baron, to do that for him? – for the sake of your old friendship for his father.

I dread asking, but one must move if one must live; and I dare not ask anyone else. I have others, too, to think of beside myself.

I know how difficult it is to procure anything, but if you were only to speak on his behalf I am sure that it would benefit him.[2]

Sincerely asking your pardon for so troubling you.

Believe me
My dear Baron,
Yours faithfully
Sophie Hooker.[3]

2 M wrote seeking a position in Tasmania for Brian Hooker, but without success; see M to J. Hooker, 18 June 1894 [Collected Correspondence]. Brian eventually found a position on the goldfields in WA; see M to J. Hooker, 5 August 1894 and 2 March 1896.

3 MS annotations by M: 'Kilkivan.' 'Rec and answ 13/1/94.' Letter not found.

From Alphonse Milne-Edwards[1]

94.02.09

MS 2474, f. 555, Bibliothèque Centrale du Muséum National d'Histoire Naturelle, Paris.

Baron Von Müller

9 fevrier 1894

Cher Monsieur

J'ai recu votre beau livre sur les Eucalyptus et je le presenterai à l'Academie des Sciences qui [en] comprendra tout l'interet.

Les services que vous n'avez pas cessé de rendre à la Science en general et aux Naturalistes francais en particulier vous designent tout naturellement à leur attention et votre nom serait dignement placé à cote de ceux des Correspondants de notre Institut. Vous pouvez etre assuré de mon concours et je parlerai à mes collegues de la manière obligeante dont vous avez toujours aidé les études de ceux qui s'adressaient à vous, des materiaux de recherches que vous leur fournissiez avec la plus grande liberalité et afin de donner plus de poids à mon [...]onnement je vous serais bien [...][2] de m'envoyer une liste sommaire des principaux travaux que vous avez publiés – je serais aussi heureux de presenter d'autres [...] de vous à la bibliotheque de l'Institut. Nos correspondants sont repartis en plusieurs Sections et le nombre des membres y est limité – ceux de la Section de Botanique, etrangers à la France sont au nombre de 5

Sir Joseph Dalton Hooker à Kew
M N Pringsheim à Berlin
M Agardh à Lund
M Treub à Batavia
M M. [T.] Masters à Londres

Je ne doute pas que lorsque une vacance se produira vous n'ayez a l'academie des avocats très chalereux

1 Letter not found. The text is from a draft prepared by Milne-Edwards.
2 *illegible* – All [...] in the following text have this meaning.

94.02.09 Le Myrmecobius fasciatus et les oeufs d'oiseau que vous avez remis au Dr Rougier me sont arrivés en bon etat. Malheureusement les Corbeaux (Corvis australis) ne me sont parvenus [...] et je n'ai pas les remettre a M Geoff D'ailleurs ce dernier n'est plus Directeur du Jardin d'acclimatat. Le conseil d'administration de ce Jardin a voulu lui donner un caractère commercial qui ne [convient] pas au petit fils d'Etienne Geoff et au fils d'Isidor Geoff et il prefère cesser ses fonctions. Il continue a etre president de la Societe d'Acclim qui est tout a fait distincte et separe du jardin

Je presume que vous avez recu ma lettre vous remerciant des Dendrolagus et du [...] marsupial aveugle.

9 February 1894

Dear Sir

I have received your beautiful book on the Eucalyptus[3] and will present it to the Academy of Sciences which will take great interest in it.

The services that you have not ceased rendering to Science in general and to French Naturalists in particular brings you very naturally to their attention and your name would deservedly be placed alongside those of the Correspondents of our Institute. You may be assured of my support and I shall speak to my colleagues of the obliging way in which you have always helped the studies of those who have appealed to you, of the research materials that you have supplied to them with the greatest liberality and in order to give more weight to my [urging] I should be most [grateful] if you would send me a summary list of the principal works that you have published – I should also be happy to present other [works] of yours to the Institute's library. Our correspondents are divided into several Sections and the number of members in each is limited – those of the Botany Section who are foreigners number 5

Sir Joseph Dalton Hooker at Kew
Mr N. Pringsheim at Berlin
Mr Agardh at Lund
Mr Treub at Batavia
Mr M. [T.] Masters at London

3 i.e. *Eucalyptographia*. See M to A. Milne-Edwards, 1 January 1894 [Collected Correspondence].

I have no doubt that when a vacancy appears you will have very strong advocates at the Academy[4] 94.02.09

The Myrmecobius fasciatus[5] and the birds' eggs that you sent to Dr Rougier have reached me in good condition. Unfortunately the Ravens (Corvus australis) have not arrived […] and I have not passed them on to Mr Geoff.[6] Moreover the latter is no longer Director of the Acclimatization Garden. This Garden's administrative council wanted to give it a commercial character which did not [sit well] with the grandson of Etienne Geoff and the son of Isidor Geoff[7] and he preferred to give up his position. He continues as president of the Acclimatisation Society which is completely distinct and separate from the garden

I presume you have received my letter thanking you for the Dendrolagus[8] and for the [little] blind marsupial.[9]

4 See also C. Naudin to the Académie des Sciences, Paris, 24 June 1895, pp. 753-9 below. M was elected a Corresponding Member of the Académie on 1 July 1895.
5 Banded ant-eater or Numbat.
6 Albert Geoffroy Saint-Hilaire. M had indicated he was sending a pair of Australian ravens, *Corvus australis*, for the zoo in the Bois de Bologne, Paris.
7 Étienne and Isidor Geoffroy Saint-Hilaire.
8 Tree kangaroo.
9 In the letter to which Milne-Edwards is replying, M mentioned having sent, some months earlier, a marsupial mole from central Australia.

94.04.19 *To Walter Gill*

RB MSS M33, Library, Royal Botanic Gardens Melbourne.

Private

19/4/94.

You are probably right, dear Mr Gill, in regarding the Box-tree of one Central Austr. region to be Euc. microtheca. It is a valuable timber tree, and forests of it ought to be raised in any country, the climate of which would allow of its growing.

Am delighted with the glorious success, which you have with the Date-Palm. I alluded to it, when coming back from Central Australia (Sturt's creek) in a lecture, at which H.E.[1] Sir Henry Barkly was present,[2] as one of the most promising of plants on any oasis or moist place at our deserts. More than 20 years ago and later I sent Dates to the Mission-Station at the Finke River,[3] where this useful Palm exists now also in considerable quantity.

Am afraid, I have not acknowledged all your kind sendings, whether litterary or otherwise. I wrote 6000 letters or their equivalents last year with my own hand, so that I may easily miss writing any more, my work-hours daily are 16, Sundays 12, to allow attending evenings divine service.

Since the Rev Mr Kempe left the Mission-station on the Finke-River I very rarely hear from there. Now and then some seeds of any kind from the interior will be very welcome.

With friendship your
Ferd von Mueller[4]

1 His Excellency.

2 B58.05.02, p. 95. Barkly's presence at the meeting of the Philosophical Institute of Victoria on 30 September 1857 at which M read this paper is recorded in *Trans. Phil. Inst. Vic.*, vol. 2, p. xli.

3 i.e. Hermannsburg Mission, NT.

4 MS is accompanied by an envelope addressed to 'Walter Gill Esqr F.LS., F.R.H.S. Conservator of Forests, Adelaide'.

From Thomas Kirk 94.06.08

RB MSS M70, Library, Royal Botanic Gardens Melbourne.[1]

June 8. 1894

Victoria Terrace
Wellington

My dear Baron

I enclose herewith specimens of 5 plants which I should feel obliged if you would kindly compare for me & inform me of the results. The *Eragrostis* appears to be near E. *tenella* although it does not fully agree; *Euphorbia Drummondii* is I think correctly named but my specimens are not very good – too scrappy – A Cyperus and what appears to be a Polemoniaceous plant but only a scrap – These two I venture to ask you to return as it quite uncertain whether they will appear again. They were found in ballast supposed to have come from Port Philip[2] or Sydney & with them were found many other plants of interest as *Cenia turbinata*, *Xanthium Strumarium*, *Setaria imberbis Bowlesia tenera*. *Bromus arenarius*? –branched from the base – our N.Z. plant is always single stemmed – *Acicarpha tribuloides* and two or three others which I have not yet determined together with a host of common naturalized species. I do not think however that that many of these which we see here for the first time will prove permanent. *Xan[t]hium Strumarium* only developes a single head of male flowers at the apex of each terminal or lateral spike instead of several as in England.

The ballast must have come from old cultivated land, portions of it contained fragments of mortar, brick &c but the bulk of it was fairly good surface soil.

The Honourable Mr Seddon our premier wishes to have our new Flora illustrated with a coloured plate of each native and natur-

1 MS found with a specimen of *Pseudopanax ferox*.
2 Port Phillip Bay, Vic.

94.06.08 alized plant The cost would however be tremendous and our unemployed difficulty will absorb all the money that can be raised. – The Minister for Education is pressing the necessity of a new Flora on the Cabinet and it seems probable that something will be done this session.[3]

Please inform me if any varieties of *Solanum aviculare* exhibit spreading anthers with you – we have two strongly marked forms. 1. a rather coarse strong growing plant with compact anthers of the usual type, flowers of a dull lurid pale purple – 2. a slender & much smaller plant with deep violet coloured flowers & wide spreading anthers, an attractive plant. I have made repeated search for your *S. vescum*, without result.

Did I mention that amongst some scraps of Norfolk Island plants collected for me by Miss Gertrude Purchas during a short visit is a solitary specn[4] of *Corysanthes rivurlaris* a most interesting find – most of the other plants were of little interest – a large form of *Lobelia anceps – Samolus repens. Gnaphalium luteo-album* & the like

In your last[5] you enquire about Mr Buchanan's plants: his own collection consists of very small specn. mounted in folio volumes but not localized, but in fairly good condition. His memory is quite decayed and as no arrangements have been made for their disposal they will probably become the property of the man with whom he is lodging.[6] The Herbarium of the Colonial Museum which contains many of his type specimens is in a shocking state, every thing is being devoured by insects. The Museum Staff consists of the Director,[7] the Secretary who has not the slightest scienfific knowledge and a messenger – with a youth who is I believe paid by the Colonial Secretarys departt.[8] It is very sad.

3 Kirk (1899)?
4 specimen.
5 Letter not found.
6 Adams (2002), pp. 109-15, discusses the fate of Buchanan's collections.
7 James Hector.
8 department.

I enclose fresh seeds of *Metrosideros tomentosa* and *Pseudopanax ferox*. – the young leaves of the latter daily move [through] an arc of from 80° to 85°. 94.06.08

With all kind regards believe me to be

Yours truly
T. Kirk.

If quite convenient would you kindly favour me with specimens of *Erigeron conyzoides*, *E. linifolius* & *Ageratum conyzoides*

To John Shillinglaw 94.07.30

Box 246/2, Shillinglaw papers, La Trobe Australian manuscripts collection, State Library of Victoria, Melbourne.

30/7/94
Near Midnight

Was sorry to miss your visit, dear Mr Shillinglaw, which would have been a spright one! That a man of genius should carry for some miles an umbrella back I regard almost an unheard of affair. I was out on a *double* errand of *charity* even on the evening of the mail for the whole eastern hemisphere!

But to return to the umbrella! As I am often lost in thought, I am apt to leave such trivialities of the outer world behind; but the "moral" has been for years to me thus far to purchase the cheapest umbrella in the market only, as when one has gone astray, not all are so attentive as you.

Salve!
Ferd. von Mueller.

Among my posthumous memoires this may serve as a treatise on umbrellas.

94.08.05 ## *To Joseph Hooker*

RBG Kew, Letters to Joseph Hooker, vol. 16, ff. 14-16.

5/8/94.

The anxiety felt by you, dear Sir Joseph, concerning the well-being of your Son Brian and his family, will have been relieved before this by a letter of his and also one of mine,[1] apprizing you of his good fortune to get the Managership of a Gold Mine in Coolgardie, W.A. Indeed, to use his own words, he expects to become a Millionaire, and that is quite within the reach of possibility there, though such a happy luck falls to the lot of but few mortals anywhere. Coolgardie is the southern terminatio[n][2] of a vast tract of auriferous country, interrupted by wide spaces, under meridians at which, when I was with the two Mess Gregory in 1876[3] we discovered the northern probabl[e] termination at Sturts and Hooker's Creeks. But we had in a season of drought and travelling with horses only no time for methodic experimental "digging". We were 2000 miles on the one side and 3000 miles on the other side distant from the nearest settlements and the country either side unexplored. Had we made too much of the Gold-Indications then, numbers of "Prospectors", enticed by any uncautiousness of ours, would have fallen under the hands of the Cannibals! –

The generous offer of Lady Hooker to educate your eldest Australian grandson[4] with your youngest Son[5] should be eagerly seized under ordinary circumstances by Mrs Brian Hooker, but a mothers feeling may be such, that she may not perhaps like to be separated from one of her darlings, more particularly as all monetary distress seems warded off, though she could not live at the Goldfields, but could live very comfortably a long way towards

1 Letter not found.

2 *editorial addition* – Obscured by binding. All square brackets in the following text have this meaning.

3 1856?

4 Brian William Hooker, b. 1889, son of Joseph Hooker's son Brian; see Allan (1967).

5 Richard Symonds Hooker, b. 1885; see Allan (1967).

Coolgardie, somewhere east of York, a region known to me by my travels in 1877. I think I cannot do wrong in writing at once to Sir John Forrest or Lady Forrest, who are personally known to me, informing them of Lady Hookers Offer. It would be better, to do that than to write direct to your son or his wife, who would certainly not be in Perth. The Premier or his Lady could communicate with them and would doubtless see the little boy sent safely wth a family of passengers to your domicile, where he should arrive in the northern spring; this as a medical man you will see to be necessary. 94.08.05

I would like to write to you on hundreds of things, but am so overwhelmed with work at this *dire* period of financial distress, that I must reserve further writing to a later time.

Let me felicitate you on the rapid progress, you are making with the Index Kewensis.[6] You will be entitled to the gratitude of the whole bot. world for all times by this work most particularly. Every one of us must find it of to be of advantage to our working almost daily. No doubt a competing[7] edition will become due at the end of the century, when little of absolute novelty will remain to be discovered.

Always with friendship your
Ferd. von Mueller

Your resuscitation of the Banksian diaries is most remarkable and opportune.[8]

6 Vols 1 and 2 were published in 1893 and vol. 3 in 1894.
7 completing?
8 Banks (1896).

94.08.23 *From Rose Grainger*

RB MSS M22, Library, Royal Botanic Gardens Melbourne.

"Killala"
Caroline St.
S. Yarra.[1]

Dear Baron Von Mueller,

May I ask the favour of your patronage & presence at little Percy's concert on Monday September 10th.

I have not forgotten my promise of the photograph, & will send it very soon.[2]

Hoping you will grant my request.

I am, dear Baron von Mueller,
Sincerely Yours
Rose Grainger.

Aug 23rd 1894.

94.09.04 *To Charles Musson*[1]

ML MSS.2009/21, item 27, Linnean Society of NSW, Mitchell Library, State Library of New South Wales, Sydney.

5/9/94

As even my own bryologic collection, and library is [not] sufficiently large for *independent* working on Mosses here, dear Mr Musson, I cooperate with an European Specialist on these kinds

1 Melbourne.

2 Filed with the letter is a photograph of Percy Grainger inscribed: 'To Baron von Mueller with love from Percy. 1894.'

1 MS is accompanied by an envelope addressed to 'Ch Musson Esq F.LS Lecturer at the Hawkesbury Agricult. College, *Richmond*, New South Wales'.

of plants, and as I have done so since 1847 methodically and thus mainly built up the Bryology of Australia, I would expect *loyalty* to me from all bryologic workers in Australia, and I am sure from your former action[s] towards me I shall have this from you also now. If therefore you will send me a good full specimen of any of your mosses *numbered* correspondingly to your set, I will undertake to furnish in reasonable time the names. A direct communication from you to any Home-Bryologist would create an "Imperium in imperio"[2] 94.09.04

Mr Bracebridge Wilson for instance has passed *all* his algologic collections through my hands to an European Specialist to avoid what I just mentioned and this has been to this distinguished microscopist rather to advantage and never to disadvantage

As regards vascular plants, for the study of which I have all the original books and authentic collections, I am always willing to give *direct* and *early* information, if proper specimens in flower and fruit are furnished, what unfortunately often is not the case. As you hold an independent position at the local Agriculture-College, which to obtain I was happy[3] you are probably not obliged to follow the wishes of any one else, any superior in subjects of the kind now under discussion.

As regards German University degrees, they have been so altered gradually, especially since the last 20 years, that no degree can be obtained "in absentia" & in all cases a rigorous examination personally made is requisite.

Ready to render you any service so far as I can, I remain regardfully your

Ferd von Mueller[4]

2 Empire within an empire.

3 which to obtain I was happy *is a marginal note with its intended position marked with an asterisk.*

4 See also M to C. Musson, 17 September 1894 [Collected Correspondence], in which M acknowleged receipt of Musson's collection of mosses.

94.10.16 *From Georgina King*

RB MSS M13, Library, Library, Royal Botanic Gardens Melbourne.

Beresford Road
Homebush[1]
Oct. 16th /94

Dear Baron von Mueller,

I have just got some interesting specimens of "Sea-grasses", which I hasten to send to you, they are again from our "Harbour" I wonder if the spores contain the pollen that fertilises the flower (if they are flowers) in the accompaning specimen? I see how *requisite* your kind advice was about not jumping at conclusions, for I know so little about "Sea-grasses", never having studied them – and know nothing about those in Britain. If I am impatient to get the names of the marine fossils from Sir Frederick McCoy please forgive me, for I have a paper waiting for them that will be most useful to science just now!

Please accept my sincere sympathy in the loss of your friend the Dean of Melbourne[2] – I never forget that it was his son that introduced me to you. You have been such a valuable friend ever since – I always feel I can't do enough to show you how grateful I am for all your kindness to me.

I am, with kindest regards
Yours gratefully
Georgina King

1 Sydney.
2 H. B. Macartney.

From John Brooks 94.10.29

New Norcia Archives 04255, Benedictine Community, New Norcia, WA.[1]

Balbinia[2]
29th Oct 1894.

Baron Sir Ferdinand Von Mueller &c
Melbourne

Dear Sir

I did not for one minute contemplate the possibility of my humble letter to you attaining such an exalted position as thanks to your extremely kind offices it has, or I should have exercised much greater care in its compilation – when I think of its being read before that august assemblage of the learned in Brisbane after New Year bristling with Presidents & Vice Presidents whose names & fame are World wide I fairly shudder.[3] But when I read with my own eyes the gracious & very flattering encomiums it provoked from two gentlemen whose praise is so valuable as yourself & Major Boyd I am sure I felt quite 10 years younger & 1/2 an inch taller. it has been a source of great gratification to my dear mother & sister & a proud & happy man am I.

I have with shame now to inform you that I made a dreadful slip of the pen in my letter to you I think when writing of Pollocks Reef[4] the Southern extremity of the Recherche Archipelago I wrote "*Beware Reef* the dread of mariners &c" Beware Reef is just off Cape Conran near the mouth of the Snowy River on the Gippsland Coast[5] where I travelled just before I came on this Coast & I have

1 The text is a handwritten copy, in what appears to be Brooks' hand, of what was presumably sent to M.

2 Balbinia Station was established at Israelite Bay, WA, by the Brooks family c.1883.

3 J. Brooks to M, 25 July 1894 [Collected Correspondence]. The letter was read to the Geography Section at the sixth Congress of the Australasian Association for the Advancement of Science, Brisbane, January 1895.

4 WA.

5 Vic.

94.10.29 always had them both a little confused in my mind hence my mistake will you kindly correct it.[6]

Additional point has been just recently given to my remarks through the wreck off the ill fated SS. Rodonda.[7] her compasses not being sufficiently insulated were deflected by the proximity of machinery she was conveying for the Dundas gold mines so she was a good many miles out of her supposed course & struck on the Pollocks reef at 2 a.m. the majority of the passengers were a mean cowardly lot. though not without one or two brilliant exceptions as for instance when the Captain ordered all hands on deck the ruffians rushed the boats before they could be lowered from the davits the ropes broke & they were precipitated into the sea & 4 or 5 of them met the fate such cowards deserve two boats capable of carrying 30 passengers each were cut adrift while yet containing only 8 or 10 each the Captain's threats & entreaties to them to return as otherwise the people could not all be saved were of no avail so the mate sprang up on the bulwarks & after abusing them to the best of his ability without effect threw off his jacket & sprang overboard to swim after & capture them "Well Sir" said Mr Fitzgerald – a passenger – a splendid handsome man "You shall not go alone" & he sprang in too. & they succeeded in bringing the boats back & in landing all of the 196 souls on Southeast Island there was a lady among them with two little children She could not give them as much as a biscuit without some unmanly wretch snatching it out of their hands. it is not surprising that when the captain called for volunteers to man a boat to go to the mainland to cut the telegraph wire & so attract attention that none should be forthcoming whereupon Mr Fitzgerald guaranteed £100 to every man who who[8] venture he making one himself & after a deal of difficulty a crew was procured who were fortunately picked up by the "Grace Darling" & all were safely conveyed & landed by her at Point Malcolm where when they dropped anchor they gave 3 cheers to the amazement of Messrs Ponton & Sharps Establishment who

6 The correction has been made in the printed version of the letter.

7 The wreck of SS *Rodondo* attracted extensive coverage in the WA press; see for example *West Australian*, 9, 10, 11, 12, 13, 20 and 27 October, and 2 November 1894.

8 *word repeated.*

were there shearing – the cheers from the throats of 196 people producing a volume of sound never heard on this coast before. 94.10.29

Scarcely had this occurred when our small community was again convulsed by a affray with an Afghan Camel Caravan. A respectable young fellow named Knowles whose father belongs to the civil service in S.A. while camped at an outstation of Pontons named Wahganninya saw an Afghan performing his ablutions in the dam[9] Knowles went & expostulated with him telling which contains the only available supply there of water for travellers him to get a dish & wash in that so as not to defile the water for other people the Afghan replied nature provided the water & he would use it as he liked in defiance of "Englishmans" for whom he had no respect Knowles – a little handful of some 8 stone weight – then endeavoured to intill[10] a little into him with such effect that the Afghan yelled for his comrades to come to his assistance & 16 of them ran over & began pelting Knowles with sticks & stones which he dodged until a stone struck him on the cheekbone near his eye half stunned & maddened with pain he drew his revolver & shot the Afghan who caused the disturbances wounding him in the left shoulder, the other Afghans then rushed on Knowles & a scuffle ensued for possession of the revolver which accidentally went off & shot an Afghan dead the others overpowered Knowles & tied him up to a tree & beat him till he became insensible they then went & captured Pontons bullock driver & another young man & tied them up. when they returned to Knowles & beat him again. they then informed the other two that if the wounded Afghan died they would burn them all alive the head man of the Afghans named Anghwon Zada returned in the night & told Knowles if the wounded Afghan died he would kill Knowles. They kept them tied up till 3 p.m. next day when Knowles rode into Israelite Bay & reported to the police

9 'the water in which is all the only supply that travellers have.' *deleted by the copyist but seemingly in error. It would appear that it was the second occurrence of this clause that should have been deleted, leaving the sentence reading:* Knowles ... saw an Afghan performing his ablutions in the dam which contains the only available supply of water for travellers. K went & expostulated with him telling him to get a dish ...'.

10 instil?

94.10.29 – he is now under arrest & I sincerely trust no jury will be found to convict him for it is of such stuff that heroes are made

There is a man living near Eyres Patch[11] named Fred Schultz his father is or was a general in the Sweedish Army. he himself has been a lieutenant in the same, he lives with an aboriginal woman & made a lot of money kangarooing which he took to London & spent in rioting returning to Eyres patch he found the kangarooing line done & now makes a precarious living cutting sandalwood which is said to vary considerably from that to the Westward, he it was who first informed me that the Doombarry – (wild peach) & the Mulgarra (ab[12]) – (Anglici wild cherry) were of the true sandalwood genus. I did not credit him till I saw it also in one of your works. he might be of service to you as he knocks about in Country I may never see again & the points to which you may especially direct his attention are (1) the Sandy ground around [Hamps] Rockhole on top of the cliffs just above his house (2) the face of the cliffs especially the gulches & grassy patches on them (3) the recent formation or country between the beach & the cliffs (4) If he reaches so far inland the belt of Casurina[13] on the northern outskirts of the Nullabor plain.[14]

There remains for me now only to answer the points on which you desire information & I take them in the order you have given

(1) the Blackboy or Xanthorrhoea of this district is as far as I am aware identical with that of wide distribution in the westward specimens have been procured by me & forwarded to you by Miss Brooks[15] but I will obtain another & if possible forward it with this letter.

(2) the Cabbage or Xmas tree (Nuytsia floribunda) answers to your description & appears to me to correspond with the sample you […][16]

[John Paul Brooks][17]

11 A telegraph station on the south coast of WA.
12 aboriginal?
13 Casuarina?
14 Nullarbor Plain, WA.
15 Sarah T. Brooks.
16 An unknown amount of text missing.
17 *editorial addition.*

To Jacob Agardh

94.11.11

Handskr. Avdl., Universitetsbibliotek, Lund, Sweden.

11/11/94.

Jetzt sende ich Ihnen wieder, edler Freund, eine beträchtliche Anzahl Algen, welche in letzten Zeit von Mr O'Halloran an der Fowler's Bay und anderen Plätzen nahe der Great Bight gesammelt sind und mir eben zugehen. Es wäre gut, wenn dieser Herr *bald* Auskunft über seine früheren Sendungen erhalten könnte u gewis[1] werden Sie ihn ferner ermuthigen durch Dedication von neuen Arten. Sonder und Kützing waren in dieser Hinsicht immer sehr freundlich, Harvey viel weniger, wenigstens mir gegenüber. Mit dem Vater des Senders, dem Major O'Halloran, war ich in den *vierziger* Jahren des Jahrhunderts persönlich bekannt in Süd-Australien, wo er auch Mitglied des gesetzgebenden Rathes war. Der Sohn ist Land-Richter, der die entfernten Ortschaften and[2] der Küste von Süd-Australien periodisch als Judge zu besuchen hat.

Alle Ehre, dass er sich neben so schweren Obliegenheiten im Algensammeln mir so gefällig zeigt

Die Küste nach der Great Bight zu sollte für die Geographie der Algen wichtig sein, da dort sich gewiss manch W.A. u Ost Austr Algen begegnen. So fand ich es rücksichtlich der Phanerogamen

Gewiss werden Sie mich bald auch mich auch bald wieder mit weiteren Notizen erfreuen über die vielen Sendungen, welche Ihnen in 1894 zukamen.

Lassen Sie meine Hoffnung ausgesprochen sein, dass Sie sich wohl befinden. Zum Neuen Jahr auch Ihnen meine beste Felicitation.

Ihr Ergebener
Ferd von Mueller

1 gewiss?
2 an?

94.11.11

11/11/94.

Now I am sending you again, noble friend, a considerable number of algae that were recently collected by Mr O'Halloran[3] at Fowlers Bay[4] and other places near the Great Bight and just reached me. It would be well, if this gentleman could receive information *soon* about his earlier consignments and undoubtedly you will encourage him further by dedication of new species; Sonder and Kützing were in this respect always very kind, Harvey much less, at least towards me.[5] I was personally acquainted with the father of the sender, Major O'Halloran, in the *forties* of this century in South Australia, where he was also member of the Legislative Council. The son is a magistrate, who periodically has to visit the distant places and the coast of South Australia as judge. All credit that he shows himself so willing to oblige me in algae collecting besides such heavy obligations.

The coast towards the Great Bight should be important for the geography of algae, since undoubtedly many western Australian and east Australian algae meet there. I found it so with regard to the phanerogams.

Undoubtedly you will also please me again soon with further notes about the many consignments that came to you in 1894.

Let my hope be expressed that you are well. Also my best felicitation for the New Year

Your humble
Ferd von Mueller

3 T. J. S. O'Halloran.

4 SA.

5 For example, *Caulerpa muelleri* Sonder (1853), p. 661; *Codium muelleri* Kützing (1846-69), vol. 6, p. 34, pl. 95, fig. II.

To Benjamin Robinson

94.11.17

Library of the Gray Herbarium, Harvard University, Cambridge, Massachusetts.

17/11/94.

Am much touched, dear Dr Robinson, with Prof Goodale's and your kindness. It is most kind of your honored Chief, to send me such a large lot of Mexican plants, many of which will be specifically new to the Museum of dried plants here. I must endeavour to make a proper return sending, if even only by small successive lots.

I send you by this mail an impression of the second Census of Australian plants,[1] which work you have not yet received from me, altho it is in the Library of the Gray-Herbarium. May I recommend this to your kind consideration. It is the result of half a century's thought and practical applications. I felt 40 years ago that the Monochlamydeae could not be maintained and distributed the Orders pertaining to them since nearly 30 years. This I mention to you for a special reason, because a prominent bot. Professor in the eastern states in a recent letter to me[2] makes the remark "American Botanists are breaking away from the Candollean system" But by expressing his preference of a particular recent system he goes too far, because no other system can be so natural than that of Jussieu-DeCandolle, provided that the Apetalae of Juss are abolished

I see the danger, that many whose experiences are not large, may be fascinated by one or other of the newest arrangements, instead of showing loyalty to the sequences of D.C., so far as this is possible altho' his 4 Divisions of the Dicotyledoneae are not so good and clear as Jussieu's. In the third Census[3] I shall make only two alterations, bringing the Droseraceae near the Saxifrageae

1 B89.13.12.
2 Letter not found.
3 Never published.

94.11.17 where I place also Nepenthaceae and others and the Stackhouseaceae (minus Macgregoria) next to the Halorageae. Asa Gray, whom we all adore, would doubtless, had he lived, broke up also himself the Monochlamydeae for which "Incompletae" is quite a superfluous word.

Where American Phytographers can particularly help us all, would be to continue the publication of Asa Grays and Sprague's genera;[4] we should then get a better idea of Buckleya and a number of other genera, which are of instructive significance for their natural position.

Next year an enlarged edition of the "Select plants["][5] will appear (further enlarged), and I will send it then to you and Dr Goodale also. In the Census I have evolved the principles of Nomenclature also, such as lifelong studies tought[6] me to be just and correct; if we legislate phytologically and zoologically not with justice all our legislation like any other will be overthrown again.

With friendship
your
Ferd von Mueller

Best greeting to the genial and generous Prof Goodale.

4 Sprague & Gray (1848-9).

5 B95.13.02.

6 *M wrote* thought *then deleted the* h *but did not completely alter the word to* taught.

From Lancelot Lindley-Cowen 95.01.05

RB MSS M1, Library, Royal Botanic Gardens Melbourne.

Bureau of Agriculture of Western Australia.[1]

WELD CHAMBERS,
Perth, 5th January 1895

Sir,

Fibre Plants

I am sending you under separate cover samples of Fibre plant indigenous to this Colony.

Some twenty years ago it was discovered in our South Western Districts, & a local cordage manufacturer converted it into rope, & made some very passable headstalls for horses.

The Government, with a commendable desire to patronise local industry, gave the manufacturer a large order for halters. A survey party then starting out into the interior was furnished with these halters as part of their equipment. The horses were tethered in Perth the evening prior to the start of the expedition but in the morning it was discovered they had developed, during the night a taste for their halters, & had eaten all that was possible, apparently with the greatest relish.

Whether this was the cause which proved the death blow to the industry I cannot ascertain, but the manufacture of the fibre ceased soon after this untoward event.

The Bureau would be greatly obliged if you will give us the botanical name of the plant, & your opinion of its value commercially for rope making purposes.

I have the honor to be
Sir,
Your obedient servant.
L. Lindley-Cowen
Secretary[2]

1 Document bears the government insignia, and is a typed letter, signed by Lindley-Cowan.

2 MS annotation by M: 'Rec & answ 15/1/95'. Letter not found.

95.01.28 *To Francis Reddin*

B95/3533, unit 581, VPRS 3992/P inward registered correspondence, VA 475 Chief Secretary's Department, Public Record Office, Victoria.[1]

Copy[2]

South Yarra[3] 28/1/95

F. Reddin Esq
Secretary to the Public Service Board.

In compliance with the request of the Public Service Board,[4] dear Mr. Reddin, the two officers, Mr. Minchin and Mr. French, present themselves.

Without in any way endeavouring to influence the Commissioners in their independent judgment, I would beg to point out respectfully

1, that the officers are both trained through many years for the special herbarium work and that, as pointed out in former communications, there is ample occupation for them here;

2, that at the Herbarium at Kew (England) *eight* officers (all with considerable salaries) are regularly employed under the Director;

3, may I also venture to mention that on the present finance-year's Estimates Mr. Minchin and Mr. French are not mentioned as transferable, as they were in the year before;

4, that three years ago already enormous retrenchment was effected in my small Branch of the Public Service;

1 This letter is A95/1888 in the file.
2 Copy apparently made at Chief Secretary's Office.
3 Melbourne.
4 Letter not found; but see C. Topp to M, 7 March 1895 [Collected Correspondence].

5, that the Melbourne Herbarium is the only one of vast magnitude in the southern hemisphere, commenced by me in 1839 and comprising nearly a million sheets. 95.01.28

I will gladly wait on the Commissioners if desired.

Regardfully your
Ferd. von Mueller.[5]

To Jean Müller[1] 95.03.04

Conservatoire et Jardin botanique, Geneva.

4/3/95

Da ich nun in die 8te Decade meines Lebens trete, hochgeehrter Freund, muss ich eilen grössere litterarische Arbeiten abzuschliessen, und so soll denn 1896 der dritte Census of Australian Plants" erscheinen. Diesmal möchte ich denn auch die Evasculares mit aufnehmen, und so möchte ich Sie bitten, die Lichenes zu liefern. Ich ersuche Agardh um die Algen, Brotherus um die Moose.

Da Ihnen in Genf der 1ste u 2te Census zugänglich ist, werden Sie sehen dass in Conformität damit nur der Name, der Autor,

5 See also F. Reddin to M, 30 January 1895 and M to C. Topp, 8 March 1895 [both Collected Correspondence]. Both Minchin and French were transferred by the Public Service Board to the Department of Agriculture (see E. Duffy to M, 5 February 1895), on the ground that they had been declared to be in excess in the Government Botanist's office. Minchin returned to the Herbarium in April 1895, but French remained in the Department of Agriculture where he succeeded his father as Government Entomologist.

1 MS annotation by Müller: 'resp. Ja in einem Jahr nach Publ. meines Consp. od. 8-10 Mon. später, falls Wilson seine Sachen schickt' [replied yes a year after publication of my Conspectus or 8-10 months later if Wilson sends his things]. Letter not found. Presumably the work to which Müller refers was not his *Conspectus systematicus lichenum Novae Zelandiae* (1894), but a new work that he failed to complete before his death in 1896; 'Wilson' is probably the lichenologist F. R. M. Wilson (1832-1903).

95.03.04 sein Werk und etwa noch ein General Werk angegeben werden sollen, dann indication der Austral. Colonien und anderer Welttheile Die Arbeit würde also keine Synonyme noch Varietäten enthalten, u sollte daher weder eine schwierige noch langwierige sein. Kann Ihre unvergleichliche Meisterschaft dafür gewonnen werden? Ihr Theil könnte ja zuerst in Europa gedruckt werden und dann in meinen dritten Census übergehen.

Sie ehrend, der Ihre,
Ferd. von Mueller.

4/3/95

Because I am now entering into the 8th decade of my life, highly esteemed friend, I must hasten to conclude the larger literary works and so then in 1896 the third *Census of Australian plants* should be published.[2] This time I would like to include the Evasculares as well and so I would like to ask you to provide the lichens. I am requesting Agardh for the algae and Brotherus for the mosses.[3]

Because the 1st and 2nd Census[4] is available to you in Geneva, you will see that in conformity with it only the name, the author, his work and some other general work should be cited, then indication of the Australian colonies and other parts of the world. The work would hence contain neither synonyms nor varieties, and should hence be neither a difficult nor boring one. Can your incomparable mastery be won over to it? Your part could certainly be printed at first in Europe and then merged in my third Census.

Respectfully your
Ferd. von Mueller

2 This work was never published.

3 See M to J. Agardh, 11 February and 22 February 1895. Letter to Brotherus not found. See also M to F. Stephani, 15 February 1896 [Collected Correspondence], in which M asks Stephani to do the liverworts.

4 B82.13.16, B89.13.12.

Figure 17. Members of the Royal Geographical Society of Australasia, Victorian Branch, and their families on an excursion to 'The Hermitage', J. W. Lindt's summer retreat at Narbethong, Vic., 9-14 March 1895. Photographed by John Lindt. Reprinted courtesy of the Royal Historical Society of Victoria, Melbourne.

95.06.02 *To Veit Wittrock*

Centre for History of Science, Kungl. Vetenskapsakademien, Stockholm.

2/6/95.

Vor einer Reihe von Jahren, hochgeehrter Herr Professor wurde mir aus Stockholm der Wunsch Sr. Maj. bekannt, dass der König eine Steinaxt für seine Waffen-Sammlung wunsche.

Es wurde mir damals nicht möglich eine solche herbei zu schaffen, aber eben jetzt habe ich eine solche erlangt, und sollte S. Majestät noch wünschen diese seiner Sammlung zuzufügen, und geruhen sie anzunehmen, so will ich das schöne und seltene Exemplar sogleich nach Stockholm absenden. Wollen Sie also freundlichst Erkundigung am Königl. Hofe einziehen. Diese Axt wurde vor mehr als 40 Jahren von einem der ersten Missionäre aus der Cook-Insel gebracht. Das Steinstück ist gross und nur ein wenig beschädigt am einen Ende der Schärfung, was leicht abgeschliffen werden könnte. Der Stiel ist viereckig und sehr künstlich geschnitzt, und mag nur ein grosses Futteral sein für den Gebrauchsstiel. Dieses Utensil oder diese Waffe ist im Besitz eines früheren Chefs gewesen und solche Gegenstände werden schon lange nicht mehr angefertigt, da Europäische oder Nord-Amerikanische Waffen und Utensilien diejenigen der Stein-Zeit verdrängt haben. Ich würde die Axt durch meine London-Agenten Mess. Watson & Scull senden.

Ich habe Ihnen Vieles senden wollen, habe aber in diesen bedrängten Zeiten der Colonie keinen Sammler im Felde, und in meinem vorgerückten Alter, ich werde deo volente 1897 mein 50 jähriges Doctor Jubiläum feiern und bei der jetzigen Wucht *ruraler* Bureau-Geschäfte, ist mir nicht mehr die Zeit übrig im Felde persönlich zu wirken. Ich hoffe aber bei den sich hier jetzt bessernden Zuständen auch bald mehr für Sie thun zu können. Keimen bei Ihnen solche Samen, als ich zuweilen sandte, gut? Wie befindet sich die Riesen-Todea bei Ihnen, nachdem solche aus dem sonnigen Heim in die ausgedehnte Dunkelheit eines scandinavischen Winters gelangt ist?

Sie ehrend, und Ihnen alles Gute wünschend
Ferd. von Mueller.

2/6/95. 95.06.02

A number of years ago, highly esteemed Professor, the desire of His Majesty was made known to me from Stockholm that the king wanted a stone axe for his weapon collection.

At the time it was not possible for me to procure such a one, but I have just now obtained one and should His Majesty still want to add this to his collection and deign to accept it, I will immediately send off the beautiful and rare example. Would you therefore kindly make inquiries at the Royal Court? This axe was brought from the Cook Islands more than 40 years ago by one of the first missionaries. The stone piece is large and only a little damaged at one end of the sharpening, which could easily be ground down. The handle is rectangular and very artistically carved, and may only be a large case for the real handle. This utensil or this weapon has been in the possession of a former chief and such objects have not been made for a long time, because European or North American weapons and utensils have displaced those of the stone age. I would send the axe through my London agents, Messrs Watson & Scull.

I have intended to send you many things, but have no collector in the field in these hard times of the colony, and in my advanced age I will, deo volente,[1] celebrate my 50-year doctoral jubilee in 1897, and with the present load of *rural* office business there is no longer time remaining for me to work personally in the field. I hope, however, with the now improving conditions here also to be able to do more for you soon. Do such seeds as I sent occasionally germinate well at your place? How is the giant *Todea* at your place, after it arrived from the sunny home into the extended darkness of a Scandinavian winter?

Respectfully and wishing you every good thing
Ferd. von Mueller

1 God willing.

95.06.18 *To Jacob Agardh*

Handskr. Avdl., Universitetsbibliotek, Lund, Sweden.

18/6/95

Besten Dank, hochgeehrter Herr Professor, für die Übersendung der letzten Analecta als Extra-Exemplar. Ich hatte mein eignes bereits an Miss Hussey abgesandt, und behalte nun selbst die letzten Abdrücke. Es wäre unzart gewesen, hätte Miss Hussey Ihnen unaufgefordert und direct deren photogram gesandt. Sowohl sie als Miss King bat *ich*, mir deren Lichtbilder zu senden, da *ich* sie Ihnen gern anbieten möchte. Es ist auch noch eine Dame und zwar in West Australien, die anfängt Algen zu sammeln und noch eine andere in Süd-Australien, so dass Sie noch 2 mehr Photographien erwarten können. Ich denke mir, dass Ihnen so eine Freude bereitet wird, when[1] Sie junges Damentalent für Ihre Lieblings Wissenschaft erwachen sehen.

Ich denke auch, dass Ihr Lebens Abend durch Freund Wilsons und mein Wirken angenehmer gemacht wird. Sie werden in Gerechtigkeit zu uns anerkennen, dass wir Beide zusammengenommen Ihnen mehr Neues liefern als die ganze andere Welt zusammengenommen.

Sie ehrend der Ihre
Ferd von Mueller

Ich treffe eben hier die Gräfin Wachtmeister, die Mutter des jungen prächtigen Ihnen bekannten Edelmannes

18/6/95

Best thanks, highly esteemed Professor, for sending an extra copy of the latest *Analecta*.[2] I had already sent off my own copy to Miss Hussey, and now myself have the latest issues. It would have been indelicate, had Miss Hussey sent her photograph to you directly and without being asked. *I* asked both her and

1 wenn.

2 Agardh (1894-1903). The second part was published in 1894.

Miss King to send me their photographs, because *I* would like to offer them to you. There is also another lady, in fact in West Australia, who is beginning to collect algae[3] and another in South Australia,[4] so that you can expect another two more photographs. I imagine that it will give you such pleasure when you see young ladies' talents aroused for your favourite science.[5] 95.06.18

I also think that the evening of your life will be made more agreeable by friend Wilson's[6] and my work. You will in justice to us acknowledge that both of us together provide more new things to you than the whole of the rest of the world together.

Respectfully your
Ferd von Mueller

I have just met Countess Wachtmeister here, the mother of the splendid young nobleman[7] known to you.

From [Charles Topp] 95.06.28

A95/4969, unit 587, VPRS 3992/P inward registered correspondence, VA 475 Chief Secretary's Department, Public Record Office, Victoria.

28 June [18]95[1]

Sir,

I beg to inform you[2] that the Government has decided that from the 1st proximo your branch of this Department is to be amalgamated with this office and placed under one Division Adminis-

3 Clara Ryan? See M to C. Ryan, 29 August 1895.
4 Probably Eleanor (Nellie) Davey who collected on the lower Yorke Peninsula, SA; see Robertson (1986), p. 141.
5 None of these photographs are among Agardh's papers in Lund.
6 John Bracebridge Wilson.
7 Count Axel Wachtmeister.

1 *editorial addition* – MS is a letterpress copy.
2 A minute from the Chief Secretary's Office, dated 25 June 1895, that prompted this letter is attached to it: 'It will be best to inform Baron Von Mueller of the change to be made from 1st July in regard to payment of the salaries in his branch'.

95.06.28 tration and [benefit] in the Estimates for 1895/6 and that consequently the Salaries of yourself and your officers will be paid by this Office from the date above mentioned.

I have the honor to be,
Sir,
Your obedient servant
[Charles Topp][3]

Sir Ferd. Von Mueller K.C.M.G.

95.07.01 *From the Deutsche Verein von Victoria*[1]

RB MSS M200, Library, Royal Botanic Gardens Melbourne.

DER DEUTSCHE VEREIN
VON VICTORIA
HERRN
BARON SIR FERD. VON MUELLER K.C.M.G.
ZUM 70. GEBURTSTAGE
30. JUNI 1895

AN HERRN BARON SIR FERD. VON MUELLER K.C.M.G.
Sehr geehrter Herr Baron,
Hochgeschätzter Landsmann und Freund!

Ihre Landsleute in Melbourne können die Feier Ihres siebzigsten Geburtstages nicht vorübergehen lassen, ohne die Gefühle der Hochachtung und Verehrung, von denen sie für Sie beseelt sind, Ihnen kund zu geben.

3 *editorial addition* – the letter-copy is unsigned. Topp, at this time Under-Secretary to the Chief Secretary's Department, is the most likely signatory.

1 Sie frontispiece. MS is a leather bound illuminated address signed by the artist 'L. Lang del. Melb.' The heading is on the front cover; inside, the text is on the left and the signatures on the right.

Wenn auch Verdienste um die Wissenschaft nur von Berufsgenossen völlig gewürdigt werden können, so haben doch Ihre von reichem Erfolg begleiteten Forschungen auf neu erschlossenen Länder- und Wissens-Gebieten Ihren Namen weit hinaus über die eigentlichen Fachkreise getragen und ihn unter jene eingereiht, welche die Nation, als ihr zum Ruhm und zur Zierde gereichend, für sich in Anspruch nimmt. Fürsten, Universitäten und andere gelehrte Körperschaften haben die Bedeutung Ihres Schaffens erkannt und Sie mit Auszeichnungen überhäuft. Die australischen Colonieen aber sind Ihnen ganz besonderen Dank schuldig für Ihre rastlose und hingebende Thätigkeit in der Erforschung der heimischen Flora wie in der Einführung geeigneter Culturpflanzen aus anderen Erdtheilen, wodurch Sie einen wesentlichen und segensvollen Einfluss auf die Entwicklung dieser zukunftsreichen Gemeinwesen gewonnen haben. 95.07.01

Dass Sie neben unermüdlichen wisssenschaftlichen Arbeiten Zeit und Musze[2] für die Förderung der schönen Künste und zahlreicher Werke der Barmherzigkeit gefunden haben, darf nicht mit Stillschweigen übergangen werden. Für Ihre unerschütterliche Anhänglichkeit aber an unser gemeinsames Vaterland und die deutsche Sache sagen wir Ihnen unseren aufrichtigen und tiefempfundenen Dank.

Am heutigen Festtage wünschen wir von ganzem Herzen, dass die Vorsehung Ihnen noch eine lange Reihe durch Gesundheit, Freundschaft und wissenschaftliche Erfolge verschönte Jahre gewähren möge: der Menschheit zum Nutzen, dem deutschen Namen zur Ehre und Ihren australischen Landsleuten zu freudiger Genugthuung.

Der Deutsche Verein von Victoria,

J. G. Luehmann	D. Diercks
Präsident	Secretair

Melbourne
1ten Juli 1895

2 Muße.

95.07.01 WA Brahe
Imp German Consul
[Imperial German Consul]
Ch Pinschof
K. k. öster. ungar. Consul
[Imperial Royal Austro-Hungarian Consul]
Hermann Büttner
F. A. Hagenauer
H. W. Püttmann
M L. Enes
Miss Eva Meissner
E. H. Meissner
Alwin Caro
Ernst Hartung
Rudolf Himmer
N. Guckenheimer
Alfred Pfaff
Dr Büttner
W Mommsen
Emilie Himmer
Luise Wiedermann
Antonie Pfaff
Ida Hartung
Emilie Ampt
Elise Pinschoff
H. Elmhirst-Goode
S. Weinheim
Max Kronheimer.
G Damman
Sabina Peipers
Frau Arthur Mueller
Lulu Peipers.
Millie Benjamin
Annabella Puttmann
Katie Moody.
Doris H Lichlam
Martin Feldheim
Jo. Diercks
Arthur Mueller
Hermann Herlitz
Minnie Herlitz
August Wehl
C. H. Ernst Boese
R. Bernde
Charles Troedel
A W Enes
C Kussmaul
M. L. Kreitmayer
H. M. Kreitmayer
Mrs Charles Braché
J. E. Braché
Gustav Aron
Charles Braché
Elise Braché.
Jenny Law.
M. Damman
Lillie Damman
Annie Damman
Julia Troedel
Frau Emil Weber
Jules Simonsen
Kate Püttmann
Hermine Püttmann
Anna Mommsen
William Hirsch
C. Hesselmann
Edith M. Hagens.
Maude Luehmann
Edwin Louis Jacoby.
Hugo Kinderling
D. Diercks.
Paul Steinfeld.
Wm Wiesbaden

Präsident des M. D. Turn Vereins
[President of the Melbourner
Deutsche Turn Verein]
Frau Wiesbaden
D. Kussmaul.
A. T. Beker
Frau Anna Heyne
C B. Heyne.
S. J. de Beer
Carl Schmahl
Georgie Aron
Laura Wehl
Otto Zorn
Erich Wiegmann
A Zorn.
G Ampt.
T. G. Hagens
Max Joseph
Siegfried Lazarus
Siegfried Meyer
Hannah Jacoby.
Hilda Guckenheimer
Regina Lazarus.
Sigismund Jacoby.
H Roberth
J. Engländer
W. H. Junker
Mrs W. H. Junker
A. C. Jacoby.
Felix Petsch.
Martin Schmidt.
Hans. P. Rassmussen.
Friedrich Holtz
Lieutenant z. See W [...]aty
[Naval lieutenant]
Hugo Wertheim
W. H. Wischer J.P.

95.07.01

The German Club of Victoria to Baron Sir Ferd. von Mueller K.C.M.G. for his 70th birthday 30 June 1895

To Baron Sir Ferd. von Mueller K.C.M.G.
Very respected Baron, highly esteemed fellow-countryman and friend!

Your fellow-countrymen in Melbourne cannot let the celebration of your seventieth birthday slip by without expressing the feelings of deep respect and admiration with which they are filled for you.

Even if services to science can only be fully appreciated by colleagues, yet your researches on recently opened up fields of geography and science accompanied by bountiful success have carried your name far away beyond the specialist circles and ranked it among those whom the nation claims for itself as redounding to its glory and lustre. Princes, universities and other learned bodies have acknowledged the significance of your creative work and heaped you with honours. The Australian colonies, however, owe you quite particular thanks for your tireless and devoted activity in investigation of the native flora as in the introduction of suitable cultivated plants from other continents, by

95.07.01 which you have gained an essential and highly beneficial influence on the development of this community with its promising future.

That you have found time and leisure apart from tireless scientific work for the support of the fine arts and numerous works of compassion may not be passed over in silence. However, for your unshakeable devotion to our common fatherland and the German cause, we express to you our sincere and cordial thanks.

On this day of celebration we wholeheartedly wish that Providence may grant you yet many long years embellished by health, friendship and scientific success; for the benefit of mankind, for the honour of the German name and to the joyful satisfaction of your Australian fellow-countrymen.

The German Club of Victoria,

J. G. Luehmann — D. Diercks
President — Secretary

Melbourne
1 July 1895

[Signatures as above.]

95.08.01 *To Charles French*

RB MSS M99, Library, Royal Botanic Gardens Melbourne.

S. Yarra[1] 1/8/95.

It has been very pleasing to me, dear Mr French, that while you are entering on your new lines of duties under your excellent father, such sentiments, as you express towards me in your parting letter[2] are pervading you. I recognize with much gratification also your desire of preserving feelings of attachment for the Institution, in which we worked so long together, and I feel sure that your

1 Melbourne.

2 Letter not found. In January 1895 the Public Service Board transferred French and J. Minchin, another of M's assistants, to the Department of Agriculture. In April Minchin returned to the Herbarium, but French stayed at the Department of Agriculture where he later succeeded his father as Government Entomologist.

career here will be of lasting beneficial influence on your future service also under the Government. We must all strive, when in a public position, to seek honor in placing our duties towards the state always above private personal considerations. We will then be kindly remembered also in futurity. To you such conception of our positions is all the more a sacred one, as you have young desendents, to whom you doubtless like to leave a high public status as an inheritage. Let me thank you for the services so long rendered me often regardless of your own interests, often also tendered to me outside of the precincts of State duties, and let me hope that divine providence will grant you a long series of years replete with pleasure with pride of your successes and with happiness in your family life. 95.08.01

Ferd von Mueller

To Clara Ryan 95.08.29

Letters to Mrs Ryan of Eucla, acc. 2146A, Battye Library, Perth.

29/8/95.

It was most kind of you, dear Madam, to send me the nice lot of seaweeds (Algs), for which I express my best thanks. May I hope, that you will continue to collect seaweeds, as they are particularly valuable from *your* locality, where the eastern and the western species largely meet, so that your collections would be particularly valuable, as there from the geographic limits of various species may become fixed. Besides some novelties would you with the help of Lady friends would persevere to collect Algs. Perhaps you have friends on other points of the coast, whom you could also induce to collect seaweeds. I will gladly send you and friends of yours also seeds of various highly useful plants some will be despatched by this weeks mail already. Others can follow, as they may ripen gradually.

95.08.29 Indeed I should like also to get dried pressed specimens of the land plants of your locality also, particularly also *minute* weeds and other very small plants, which come out during spring. Among *them* may be novelties. But at all events the *new locality* for rare plants would always be mentioned under your honored name. I can utilize fruiting sprigs as well as flowering from your vicinity and other places near you, but the locality should always be marked in each lot. You can send one pound weight of specimens by post for some few pences. That is the simplest, quickest and cheapest mode of sending The postage can be repayed by me. Would aboriginal youths bring you specimens from distances in baskets fresh? You might send them a little[1]

Very regardfully
your Ferd von Mueller.

95.09.03 *To Veit Wittrock*

Centre for History of Science, Kungl. Vetenskapsakademien, Stockholm.

3/9/95

Durch diese Wochen-Post sende ich Ihnen, edler Freund, einige Diatomeae, welche ich von einem Specialisten hier erhalten habe Mr W. Bale, dessen locale Amateur-Forschungen ich unterstütze. So hat er als Darlehn von mir das grosse Werk des mir durch viele Jahre befreundeten Archdiaconus Schmidt, der jetzt 83 Jahr alt sein Amt niederlegt, aber geistig noch kräftig ist und so auch körperlich bis zu einem gewissen Grade.

Ich habe Mr Bale mit Herrn Prof Van Heurck in Correspondence gebracht, und könnte Ihnen auch manches aus den Diatomen Ihnen senden.

1 The final sentence is incomplete in the MS.

Seine Notizen über das jetzt Übersandte füge ich bei, aber das Gläschen N.12 zerbrach auf dem Wege zu mir, und der Inhalt wurde verschüttet. 95.09.03

Brauchen Sie und Ihre Wissenschafts Freunde dort eben jetzt etwas Besonderes von hier? Vielleicht könnte ich es herbei-schaffen.

Sie ehrend, der Ihre
Ferd von Mueller.

3/9/95

By this week's mail I am sending you, noble friend, some diatoms, which I have received from a specialist here, Mr Bale, whose local amateur research I support. So he has as a loan from me the large work of Archdeacon Schmidt,[1] who has been a friend of mine for many years, who now at 83 years of age has resigned his office, but is still sound in mind and also in body to a certain extent.

I have brought Mr Bale into correspondence with Professor Van Heurck and could also send you much from the diatoms.

His notes about those just sent I enclose, but the small glass number 12 broke on the way to me and the contents were spilled.

Do you and your scientific friends there need anything in particular just now from here? Perhaps I could furnish it.

Respectfully your
Ferd von Mueller.

1 Schmidt (1874-1959). At the time M wrote this letter, 200 plates with letterpress had been published in fifty parts. The Royal Botanic Gardens Melbourne holds a set of Plates 1-176 (1874-92) bound as a single, large volume; this was probably what M lent Bale.

95.09.26 *To Thomas Cheeseman*
Auckland Institute and Museum, Auckland.

26/9/95

In accordance with your wish, dear Mr Cheeseman, I send you some few typic specimens of Epilobium, as viewed by Bentham. But as it is now nearly 30 years, since we both worked on the 3th vol. of the Fl. Austral.,[1] a large lot of specimens also of Epilobium has accumulated, but my views remain unaltered, that for *continental* Australia only one genuine species can be admitted. No doubt, if the species are limited by such characteristics, as are adopted in the Fl Austral, they *are transmutable*.

Have *you* studied Epilobium in Europe? I *have* in nature from 1839-1846 in Denmark and Germany, besides species in bot Gardens there We have in Australia not the Section with divided stigma, nor the Section with unsymmetric flowers. In New Zealand the play of forms would not be so great as in Australia, because here Epilobium advances also into the *hot dry desert*. I can send you specimens of later Epilobiums (later than 1866 collected), but I have not named them, as I found no stability in the characteristics adopted by Bentham for this part of the world. As he never was personally in Australia, he could have no field experiences here, such as tracing Epilobium from the Austral Alps to the far inland-deserts.

I am much beholden to you, that you will send to the great Herbarium here a set of your typic Coprosmas. It is also pleasing that you will record your further observations on this genus.

From 1857-1873 I cultivated several N.Z. species of Coprosma in the bot Garden here, and my observations on these kinds of plants led me thus to advise you to take up the genus for study in nature at its principle home. In favor to myself you might briefly mention this, when you deal with the suppl. collections. You will on close scrutiny also find, that Nertera cannot be kept out of the genus.

With regardful remembrance your
Ferd von Mueller

1 *Epilobium* is treated in the third volume of *Flora australiensis* (i.e. Bentham [1866]), pp. 303-6.

From May Wise[1]

95.10.01

RB MSS M1, Library, Royal Botanic Gardens Melbourne.[2]

Sale. Vic.
Oct. 1st 1895.

Dear Sir Ferdinand,

I have to acknowledge receipt of both your letters[3] and to thank you very much indeed for your kindness in writing to me. I shall try to procure the ripe fruits of the Styphelias to send you, as I should like very much to be able to distinguish the two species. I am glad you were pleased with the Caladenia Cairnsiana and as we had collected a number yesterday I thought they might be useful for your correspondents and so forwarded them to you today.

We had a holiday from school and we spent it in a trip down the Port Albert Road collecting plants. My sister Lillie was fortunate enough to find a specimen of Pterostylis barbata and as it is the first of the species we have heard of being found here, she duly forwarded it to you, as it may be a new record for this district.

Caladenia Cairnsiana are exceedingly plentiful this season in all parts of Longford. Caladenia Patersoni are also plentiful, at least we call several varieties by that name, whether rightly or not. I enclosed two specimens with the C. Cairnsiana so that should it be a mistake you will I know with your extreme kindness correct it. We also found a very red variety about a fortnight ago.

We have only found one good specimen of Acianthus caudatus since Miss Bennett sent hers. We shall be delighted at some future time to publish a list of the species of plants indigenous to this district.

With very kindest regards
I am
Yours sincerely
May Wise.

1 MS annotation by M: Answ 6/10/95'. Letter not found.
2 MS found with a specimen of *Caladenia cardiochila* (MEL 712213).
3 Letters not found.

95.10.05 *To Jacob Agardh*

Handskr. Avdl., Universitetsbibliotek, Lund, Sweden.

5/10/95

Der beigefügte Brief von Miss Hussey für Sie, venerabler Freund, kam zu mir *ohne* eignes Couvert, eingeschlossen in ein Schreiben für mich, welches ich auch hinzufüge, u woraus Sie ersehen werden, welch eine prächtige junge Person Sie ist.

Sie hat inter alia auch eben eine ausserordentlich kleine Form von unserem europ. (hier eingewanderten) Plantago Coronopus, eine Art welche ich aus der Nähe der Nordsee u des baltischen Meeres von 1839-1846 lebend kenne, die aber im dortigen Clima einen *unterbrochenen* Entwicklungs Gang hat, also nicht diese Winterform oder Frühlingsform von *unglaublicher* Kleinheit und doch einiger Fruchtbarkeit nicht hervor zu bringen vermag wie in einem winterlosen Klima; für eine volle Auffassung des Species-Begriffs muss man innerhalb der Wende Kreise lange beobachtet haben. So können z. B.[1] in der winterlosen Zone Cardamine, Epilobium u manche andern Genera hier nicht nach denselben Prinzipien für wirklich[e] Arten-Definitionen behandelt werden, wie bei Ihnen, wie ich aus Feld Erfahrungen aus Dänemark u Deutschland weiss

Ich werde über das kleine Wunderding Plantago, welches Miss Hussey eben fand, u welches in anderen Gattungen der Landpflanzen manches Analoges hat, einen kurzen Aufsatz liefern hier demnächst.

Hoffentlich sind Ihnen die vielen Algen Sendungen von mir oder durch mich in Ihrer Winterbehausung alle richtig zugegangen. Würden Sie mir freundlich bald schreiben, was angelangt ist, so dass ich die manchen Amateure über die richtige Ankunft beruhigen kann. Das Herbeischaffen von so schönen u so umfangsvollem Material macht wohl viel mehr Mühe und Umsicht u Beharrlichkeit meinerseits, als Sie vielleicht denken. Dazu kommt

1 zum Beispiel.

nun auch noch, dass ich das Interesse der Sammler durch Zuschriften und Geschenke *wach* zu halten suche. Die Sammler haben ja auch selbst Ausgaben. 95.10.05

Ich sehe mit Erwartung auch einiger ermuthigenden *vorläufiger* Auskunft über den Inhalt der Sammlungen entgegen. Während der Winterstille werden Sie gewiss manches Merkwürdige darüber daraus entwickeln

Ich habe Miss Hussey nicht persönlich kennen gelernt. Ich glaube sie ist nur taub, durch Krankheit oder einen Sturz in ihrer Kindheit; sie ist also nicht stumm, wie ich früher hörte.

Mit besten Wünschen für Ihr Wohlergehen
Ferd von Mueller.

Sie setzen wohl gütig die Vertheilung der Algae Muellerianae fort Wenn Sie gelegentlich einige übrig haben, würden Sie auch ein Päckchen an meine alten Freunde Colmeiro in Madrid u Henriques in Coimbra senden?

Fräulein Hussey hat von mir meinen Key to the Syst. Of Vict plants (descriptiv) mit vielen ausfürlich. Xylographien.

5/10/95

The enclosed letter from Miss Hussey for you,[2] venerable friend, came to me *without* its own envelope, enclosed in a letter for me, which I also enclose,[3] and from which you will gather what a splendid young person she is.

She has *inter alia* also even [included][4] an extraordinarily small form of our European (introduced here) *Plantago coronopus*, a species which I know living from the neighbourhood of the North Sea and the Baltic Sea from 1839-1846, but which in the climate there has an *interrupted* development, therefore not[5] this winter or spring form of *unbelievable* smallness and yet of some fertility is not capable of bearing as in a winterless climate; for a full conception

2 Hussey's letter to Agardh, dated 4 October 1895, is preserved with Agardh's papers in Lund.

3 J. Hussey to M, 4 October 1895 [Collected Correspondence].

4 *editorial addition* – M omitted the verb.

5 *sic.*

95.10.05 of the species concept one has to have observed for a long time within the Tropic. So for example in the winterless zone *Cardamine*, *Epilobium* and many other genera here cannot be dealt with according to the same principle for true species definition like at your place, as I know from field experience from Denmark and Germany.

I will produce a short essay here soon on the small marvellous *Plantago* that Miss Hussey just found and which has many analogues in other genera of land plants.[6]

I hope that the many dispatches of algae from me or through me have reached you in your winter dwelling all correctly. Would you kindly write to me soon what has arrived so that I can reassure the many amateurs about their safe arrival. The gathering of such beautiful and such extensive material requires probably much more trouble and prudence and persistance on my part, than you perhaps think. On top of that now I try to keep the interest of the collectors *alive* through letters and presents. The collectors also indeed have expenses themselves.

I look forward with expectation to somewhat encouraging *preliminary* information about the contents of the collections. During the winter quietness you will undoubtedly develop from them many noteworthy things about them.

I am not acquainted with Miss Hussey personally. I believe she is deaf through illness or a blow in her childhood, she is thus not dumb, as I heard previously.

With best wishes for your well being
Ferd von Mueller

I suppose you are kindly continuing the distribution of the Algae Muellerianae. If you happen to have some remaining, would you also send a small packet to my old friends Colmeiro in Madrid and Henriques in Coimbra?

Miss Hussey has from me my Key to the system of Victorian Plants[7] (descriptive) with many detailed wood engravings.

6 No such essay appears to have been published.
7 B88.13.03.

Figure 18. Jessie Hussey, photographed by Stump & Co., Adelaide. Reprinted courtesy of the Library, Royal Botanic Gardens Melbourne.

95.10.25 *To Jessie Hussey*

State Library of South Australia, Adelaide.

509. Festuca bromoides L.
510. Heleocharis acuta RBr or a closely allied species Specimens with perfectly *ripe* fruit must decide this.

25/10/95.

You will have probably have learned, dear Miss Hussey by this time of the very sad event, which deprived us one of our most genial friends! Early this week poor J. Bracebridge Wilson passed away.[1] On Wednesday I was one of the pallbearers, having gone to Geelong to pay him the last wordly[2] homage. I knew, he was ailing, but was not aware til after his death, that he was seriously ill. This is an irreparable loss to us also. He was most loyal and generous to me. I induced him to extend his researches on the phytozoa also to the Algs, as he was very near to rich localities of oceanic plants. In 1859 I supplied him with many young Pines and other trees to plant around the stately buildings of the famous Geelong Grammar School from which many distinguished Legislators, Doctors, Divines and Jurists have emanated preparatorily. So one after [another][3] passes away of my eldest friends. On Monday last I delivered a short oration on Pasteur,[4] who as[5] one of those who supported my unanimous election into the Institut de France. – Roper, the last of the first celebrated Leichhardt Expedition died also this month.[6] So I must expect to be one of the next to be called away from my mortal career! –

Always regardfully your
Ferd. von Mueller.

1 J. B. Wilson died at Geelong on 22 October 1895.
2 worldly?
3 *editorial addition.*
4 B95.10.03.
5 was?
6 John Roper died on 15 September 1895.

To John Keltie

95.12.01

Royal Geographical Society, London, Archives, RGS correspondence 1881-1910 MOO-MUL, Mueller, F. von.

Private

Sunday, 1/12/95[1]

Your letter and the printed article from the "Times", dear Mr Keltie, reached me by last weekly mail,[2] and a similar communication came somewhat earlier from the honored President of the R G.S.[3]

To the latter, so I learn, a reply has already been forwarded by the Reverend W. Potter, FRGS., to the Times direct.[4] As the subject is mainly one between the Rev Gentleman, who acted honorarily as Agent for Capt Kristensen in geographic subjects here, and Mr Borchgrevink, it does not seem, that the antarctic Committee here is directly concerned in this more particularly personal dispute. I have therefore forwarded also *your* sending to Mr Potter, and if after his letter to the Times he thinks it necessary, he will

1 MS annotation: 'rcd [received] Feb 8 96'.

2 See J. Keltie to M, 17 October 1895 [Collected Correspondence]. Presumably the article was the letter from C. Borchgrevink that appeared in *The Times*, 9 October 1895, p. 11, claiming to correct Markham (1895). See also 'The International Geographical Congress', *The Times*, 2 August 1895, p. 8, which summarizes Borchgrevink's paper 'On the voyage of the Antarctic to Victoria Land' read to the Congress, and letters to the editor of *The Times* by Leonard Kristensen, who was in command of *Antarctic* on its voyage (6 September 1895, p. 4); Borchgrevink (7 September 1895, p. 7) in which he threatens legal action against Kristensen; and Clements Markham (11 September 1895, p. 6). As well as items specifically cited below, there is substantial related correspondence at the Royal Geographical Society, Archives, RGS correspondence block 1881–1910, Borchgrevink, C. E. and RGS correspondence block 1881– 1910, Bull, H. J.

3 Clements Markham. Letter not found.

4 Potter's letter was published in *The Times,* 6 January 1896, p. 14; a further letter, quoting a minute of the meeting of the Antarctic Committee of the Royal Geographical Society of Australasia, Victorian Branch, held on 2 December 1895, was published in *The Times*, 8 January 1896, p. 13.

95.12.01 doubtless communicate with you early.[5] I am reluctant to be involved in this public discussion, more especially so, as I am only an ordinary member of the antarctic Committee, altho' I was the mover of the resolution by which it was appointed, and could by virtue of this have been its Chairman But as I was already Pres. of the Vict. Branch of the RGSA., I desired Capt Pascoe, whose Grandfather was an junior flag-officer at Nelson's ship at Trafalgar, to be President of the antarctic Committee, he being the Senior here of the R.N.

If it becomes necessary to elicit statements from the antarctic Committee, then doubtless the venerable Capt Pasco, who is still quite bright in thought and physically robust will stand up in this case also for the antarctic Committee.[6]

As for myself I have shown the greatest attention due to Mr Borchgrevink here, but we must all feel, that Capt Kristensen, as the Commander of the "Antarctic" *could not be ignored.*

I gave Mr Borchgrevink introductions to Sydney, provided him with particularly valuable maps and helped him [o]n so far as I could consistently with the great objects of the science of geography.

We published in our transactions the lectures of Mr Bull and Mr Borchgrevink along with Capt Kristensens account.[7] What

5 MS copies of Potter's letters published in *The Times* on 6 and 8 January 1896 were enclosed with W. Potter to J. Keltie, 10 December 1895, with the following annotation on the copy of the letter that was published on 6 January: '*N.B.* The copies of the two letters to the Editor of *The Times* are herewith sent to the Secretary of the Royal Geographical Society so that in the event of the Mail by which the originals were forwarded being lost they may be taken to the Editor with this explanation.' (Royal Geographical Society, Archives, RGS correspondence block 1881–1910, International geographical congresses, Royal Geographical Society of Australasia.)

6 The committee met on 2 December and 'unanimously affirmed' the letter that Potter had sent; see M to J. Keltie, 3 December 1895 [Collected Correspondence].

7 M's statement was a little premature. The Baillieu Library, University of Melbourne, has a set of proofs, intended for the *Transactions* of the Victorian Branch of the Royal Geographical Society of Australasia but with a separate title-page, of Kristensen's account of the voyage of *Antarctic*, together with the texts of the lectures given by Borchgrevink and Bull at the public meeting in the Melbourne Town Hall on 19 March 1895 arranged by the Antarctic Committee to celebrate Kristensen's successful landing on the Antarctic continent. In the end, however, only Kristensen's journal was published (Kristensen [1896]).

astonishes me most is, that Mr Bull and Mr Borchgrevink are no longer befriended, as they were in such intimate alliance on the expedition-cause before. 95.12.01

Of The Account of the Expedition with map & illustrations was an early proof advance-impression sent on to you before we knew that Mr Borchgrevink who was then with Mr Bull lecturing in N.S. Wales, would attend the geographic Congress.[8]

I had a letter direct from Mr Borchgrevink telling me of his good prospects in London, to which I gladly replied.[9] It seems to me, that for him the best field for operation would be the *ice-plateau*, as he is accustomed to traverse the icefields of Norway on snow [shoes] and as his constitution is such as endures great cold, he moreover being a Gentleman of education and a keen observer. That the glaciers of Vict. Land *can* be ascended as I predicted a dozen years ago, we now know positively. It should not cost much to charter a Norwegian or Scotish Steam-Whaler for about a year, so as to provide not only for a summer-tour toward the magnetic pole but also for winter quarters through one season in the recently discovered harbour. The Admirality would during the present politic disquietude not likely be willing, to grant means for a large expedition, but two of the smaller and older ships might perhaps be told off at [comparatively] insignificant expense expense, particularly if – what telegraphically might be arranged, the British far southern possessions would contribute. I advise to interest Lord Loch also. I wrote to his Excellency while he was Governor of S. Africa on the facility of reaching Enderby Land as a basis of operation, with open sea more than once seen southward from that Locality. His Exc. brought the subject before his Ministers, who seemed not at all unfavorable to give some help, especially as the revenue was overflowing from Diamond and Gold-Discoveries; but the tide of adversity set soon subsequently in at several of the Australian Colonies, so that further antarctic exploration, in which Britain should not be forestalled, could not be advanced substantially from here. In any [i]nofficial conversation with the Right Hon. Mr Goschen use could be made

8 International Geographical Congress, London, July 1895.
9 Letters not found.

95.12.01 of my name, as very many years ago I stood in some slight business-relation with *one of* the firm, of which he was then one of the Principals.[10]

Perhaps our illustrious President Markham can see the first Lord of the Admiralty[11] for some few moments at some fête, to plead for an expedition of one or two seasons duration on a not very expensive scale, and so timely dispatched that Britain becomes not forestalled. Sir Henry Barkly would be sure to use as well his influence also as Lord Loch.

Very regardfully your
Ferd von Mueller.

95.12.03 *From Eugene Hilgard*

Bancroft Library, University of California, Berkeley.[1]

December 3, 1895.

Baron F. von Mueller, Melbourne.

My dear Sir,

Since I wrote to you[2] in response to your letter written shortly after your 70'th anniversary,[3] a long siege of illness, partly personal and not yet ended, partly involving life and death in the case of my wife, now slowly convalescent,[4] has kept me able only to attend,

10 The bankers and forwarding agents Frühling and Göschen. See also M to W. Thiselton-Dyer, 16 September 1887.

11 George Goschen.

1 The letterpress copy books in which this document is located contain typed copies of Hilgard's outgoing letters. Hilgard has signed this copy. The text is reproduced including the typing errors.

2 Letter not found.

3 Letter not found.

4 See also E. Hilgard to M, 16 October 1893.

as well as possible, to the demands of my daily official work. I am now again on my feet, though yet unable to take enough of the much-needed exercise, and with no immediate prospect of definite improvement. Mine is a bowel trouble of long standing, dating from the war, 30 years ago,[5] but as I grow older becoming less manageable – as such things will do! 95.12.03

I hope you receoved the papers noticing the celebration of your anniversary – one was in the "Rural Press" of S. F.,[6] and I sent your clipping to Dr. Behr who promised to make good use of them and to write to you. But having been unable to cross the Bay for months, I have not seen him since then.

It will be a satisfaction to your know[7] that at last there is a downright "boom" over here on the subject of your salt bushes, for the prsent notably A. Semibaccatum,[8] of which thousands of acres have been and are being planted in the alkali districts. As you have seen from our bulletin[9] which has reached you before this, we are distributing seed of the same amd also plants of your latest A. leptocarpum,[10] of whose success we do not yet feel quite so sure. The 200 lbs. seed we gathered from our plantings at the Tulare Station will not suffice for the demand, and some seedsmen are offering the seed at a dollar an ounce! Until now most people having alkali land have tried to hide the fact, so as not to injure possible sales; now all at once we hear of them everywhere, and all accounts agree as to the success and value of the Atriplex even upon the worst of our alkali soils. One enthusiastic eidtor says this single introduction is worth to the state all the College of Agriculture has cost! We have sent seed in all directions, nort in order to see how much cold it will stand, both here and in Russia; and at the request of Vilmorin we have sent 25 lbs to them for use

5 Presumably the American Civil War, 1860-5.

6 Presumably the *Pacific and rural press,* San Francisco.

7 to you to know?

8 *Atriplex semibaccata*?

9 Probably *University of California Agriculture and Experiment Station bulletin*, no. 105.

10 *Atriplex leptocarpa*?

95.12.03 in Algeria and Egypt. So you see that your recommendation "zu Ehren gekommen ist"[11] at last, although your first favourite – A. nummularia – is still scouted by our cattle, nearly as much as our native salt bushes. Is it not curious that with the enormous extent of alkali lands west of the Rocky Mountains, there should not be a single plant that is truly acceptable to stock like your salt bushes? – And by the way: Can you tell me what are the more promising plants of "karroo" vegetation of S. Africa? it seems to me that we mivht look there for additional useful plants for the alkali lands of this and other countries. SE Russia, Turkestan, Siberia and India also stand in need of them!

With the completion of our new plant house here we are entering upon wider possibilities in the way of acclimation and experiement, although for the moment our finances are disxouragingly low. Up to this time we have frequently lost our plants and plantings for want of proper space and protection; we are now trying hard to get up as full a collection as possible of the economic plants of the world. I will now also have some help from the dep't of Botany which under Prof. Greene's auspices devoted itself almost exclusively to the systematic botany of California, so that I had even to make my own determinations of any foreign plants. Prof. Setchell of Harvard, who succeeds Greene, is of a different mind and we will now be able to push the work more energetically.

I have not yet seen an announcement of the new edition of your "Select Plants",[12] but on the strength of your statement that it was going through the press at the time you wrote, I have put it on our new annual order for books. My copy is in such constant use that I often have to hunt for it in other rooms of the department, and we shall want several copies when we know it is out.

What with my illness and the increasing work put upon me, I do not progress at all in the wirting of my book on "Soils", in which I intend to lay special stress upon the interconnection of climatic and soil conditions with vegetation, and the interpretation of the

11 'has arrived at a place of honour'.

12 B95.13.02.

natural growth with respect to the adaptations of the soil. It may be 95.12.03
that if I reach my sevetieth anniversary I may have it completed![13]

Trusting that your own health continues good and will long so continue yet and leave you to serve both science and practice effectually as in the past, belive me

Ever sincerely yours
E. W. Hilgard

To Baldwin Spencer 96.01.17

Private Hands.

17/1/96

Herewith, dear Professor Spencer, I send you the description of the new scrophularineous genus, Elacholoma, which Prof Tate discovered, when you [w]ere out in Central Australia.[1] As Prof. Tate sent me all his rough notes on the Horn plants, I find that Elacholoma Hornii should be inserted between Buechnera and Limosella in the msc. with you.[2] It is the *best plant*, he discovered. Either to day or to morrow you will receive also a note on Helichrysum Ayersii, which Prof Tate, I see from the rough notes – indicates from Giles finding, on which I based the species The Professors specimen is somewhat aberrant as regards the pappus, but cannot be specifically separated.[3]

13 Hilgard (1906).

1 Horn Scientific Expedition to Central Australia. 1894. M's description was published in the report of the expedition, part 3 (1896), p. 190, i.e. Tate (1896), the name having previously been published in the *Victorian naturalist*, vol. 12 (May 1895), p. 14.

2 See Tate (1896), p. 172.

3 M's note on the specimens of *Helichrysum Ayersii* collected during the Horn Expedition was published in Tate (1896), p. 165.

96.01.17 On Sunday and in the early part of next week I will attend to two cyperaceous plants left yet unfinished by me for the Horn-plants; that will be in time, as they come in only at the close which will take several days yet.[4] Be so kind to send me the proofsheet, or proof slips, on which Elacholoma will appear, as I would be glad to revise the proof.

Very regardfully your
Ferd von Mueller

Could the R.S. here[5] publish the Elder-plants, if Prof Tate approves of it? They would not require very many sheets print.[6]

Through the urgent enormous rural work in my branch of the service, the phytographic writings get incessently interrupted.

The correspondence is something prodigeous.

96.02.04 *To Jacob Agardh*

Handskr. Avdl., Universitetsbibliotek, Lund, Sweden.

4/2/96.

Wieder habe ich Ihnen, verehrter Herr, eine nette Anzahl Algen gesandt frisch gesammelt von Miss Hussey, und zu welcher Sammlung Dr Verco (nicht Vercoe) auch auf einer Fahrt mit dem Schlepp-netz in der Encounter-Bay beigetragen hat Dr Verco practizirt in Adelaide, ist ein eben so ausgezeichneter Arzt für innere Krankheiten als für chirurgische Operationen, und wurde daher zum Praesidenten des ersten intercolonialen medicinischen Congresses ge-

4 A note by M distinguishing a variety of *Cyperus umbellatus* that he named *fasciculigerus* was included in the account of the Cyperaceae in Tate (1896), p. 181.

5 The Royal Society of Victoria. Spencer was Secretary of the Society.

6 The report by M and Tate on the plants of the Elder Expedition to NW Australia was published in the *Transactions* of the Royal Society of SA (B96.14.02).

whählt,[1] welcher – sei es beiläufig gesagt – *von mir zuerst und öffentlich* als wünschenswerth hier erklärt wurde, obwohl diese Idee zuerst in Adelaide zur Ausführung kam. Ich war im 2ten Congress, welcher in Melbourne gehalten wurde, und in welchem etwa *500* Aerzte persönlich Theil nahmen, President der therapeutischen Section; meine Eröffnungs Rede wurde zum Theil auch in der grossen London "Lancett" abgedruckt. Ich kenne Herrn Dr Verco persönlich. Er ist jetzt zum ersten mal auch in den Algen Pfad gezogen. 96.02.04

Seien Sie so freundlich auch bald wieder eine Series, so weit es geht, der "Algae Muellerianae" nach Melbourne zu senden, damit in dem von mir seit 1839 begonnenen und nun etwa *eine Million* Bögen umfassenden Herbarium, auch die *Algen-Localitäten* recht vollständig vertreten sind. Da ich allein bin im Leben, konnte ich meine enorme Privat Sammlung, welche mir im Lauf der Decennien mehrere Thausend £ kostete frei der von mir begründeten Anstalt *geschenkt*, doch trachte ich noch immer darnach das Ganze zu vergrössern, u Algen sind hier besonders gut in meinem Museum vertreten.

Dann liegt ja auf mir die *Responsibilität*, für Studien und Nachweise hier das Material so vollständig wie möglich zu haben.

Hoffend, dass Sie Ihre wunderbare Arbeitskraft bewahren, und sich im Triumph Ihrer Errungenschaften recht glücklich fühlen Ihr

Ferd von Mueller

Es würde mich freuen wenn Sie auch in Zukunft in den Analecti meiner gedenken wollen als des Sender. Ich habe ja das ganze Sammeln hervorgerufen, und es verursachte mir doch auch Mühe u. Kosten

Sie haben *mir* noch nie eine Alge gewidmet –

Im nächsten Jahr 1897 werde ich mein 50 jähriges erstes Doctor-Jubiläum feiern, sollte die göttliche Vorsehung mir die irdische Laufbahn so lange erhalten.

1 gewählt?

Figure 19. M in the 1890s wearing some of his medals. Lithograph printed by Charles Troedel. Mueller sent inscribed copies of this portrait to several people. Courtesy of the Library, Royal Botanic Gardens Melbourne.

4/2/96. 96.02.04

Again I have, respected Sir, sent you a handsome number of algae freshly collected by Miss Hussey, and to which collection Dr Verco (not Vercoe) also has contributed on a voyage with the dragnet in Encounter Bay. Dr Verco practises in Adelaide, is just as excellent a physician for internal diseases as for surgical operations and was therefore elected President of the first intercolonial medical congress, which – let it be said in passing – was announced *by me first and publicly* as desirable here, although this idea first was carried out in Adelaide.[2] I was president of the therapeutic section in the second congress that was held in Melbourne and in which about *500* physicians personally took part; my opening address was in part also printed in the great London *Lancet*.[3] I also know Dr Verco personally. He is now for the first time also drawn into the algae path.

Also be so kind to send again soon a series as far as possible of the "Algae Muellerianae" to Melbourne, so that in the Herbarium begun by me in 1839 and now embracing about *one million* sheets, the *algae localities* are also represented complete. Since I am alone in life, I was able to *present* my enormous private collection, which in the course of decades cost many thousand £, gratis to the institution founded by me, but I still endeavour to enlarge the whole, and algae are represented particularly well in my museum.

Then indeed the *responsibility* lies on me to have the material as complete as possible for study and information here.

I hope that you keep your wonderful capacity for work and feel really happy in the triumph of your achievements.

Your

Ferd von Mueller

It would please me if in future in the *Analecta*[4] you will also remember me as the sender. I have indeed given rise to the whole collecting and it did cause me trouble and expense

You have never dedicated an alga to *me*.

Next year, 1897, I will celebrate my 50-year doctoral jubilee,[5] should Divine Providence preserve my earthly career so long.

2 First Intercolonial Medical Congress of Australasia, Adelaide, 30 August-2 September 1887.

3 The second Intercolonial Medical Congress was held in Melbourne in January 1889. M's address was published as B89.13.16. No published extract from it has been found in *Lancet*.

4 Agardh (1892-1903).

5 M's PhD from Kiel University in 1847.

96.02.09 *To [William Thiselton-Dyer]*[1]

RBG Kew, Kew correspondence, Australia, Mueller, 1891-96, f. 81.

[9.2.96][2]

At a recent german festival here,[3] I in proposing the health of His Excellency the Governor stood up in dignified terms for the greatest of political greatnesses that of Albion[4] speaking in the English language and one of our leading British Colonist told me on his own accord subsequently that I sent "a thrill of delight" through him and other British byestanders by my wording and sentiment! I gave my oat[5] of allegiance to Her Majesty here in 1847,[6] when as a young Doctor I came to these dry winter-less regions to save my life, and when I brought some cash fortune, which enabled me mainly to travel for the first 6 years on my private expenses in Australia til in 1853 Governor Latrobe asked me to take Office.[7] I have never crossed the Equator again as I could not spare the time for recreation. As I have therefore spent *more than 2/3* of my life on British soil, and the other portion of my existence was divided between Germany and Denmark I ought to be considered a Britain![8]

[Ferd von Mueller][9]

1 Correspondent based on location of filed letter; although the letter might have been written to Joseph Hooker, whose post-retirement letters from M are filed in this volume. The letter may be incomplete.

2 *editorial addition* – The date '9.2.96' is marked on the letter, probably by W. Hemsley.

3 M proposed the toast to the Colony of Victoria, in English, at a banquet attended by some 700 people on 18 January 1896 to mark Germany's National Day. His speech was subsequently published in *The weekly times* on 25 January and in the *Record* (South Melbourne) on 29 February (see B96.02.01). A German translation was published in *Deutsch-Australisches Echo* on 23 January 1896, p. 7.

4 i.e. Britain.

5 oath?

6 M arrived in SA in December 1847 but did not take out British citizenship until August 1849.

7 As Government Botanist of Victoria; see W. Lonsdale to M, 26 January 1853 [*Selected correspondence*, vol. 1, p. 138].

8 Briton?

9 *editorial addition.*

From Joseph Hooker[1] 96.03.02

RB MSS M3, Library, Royal Botanic Gardens Melbourne.

March 2 /96
THE CAMP.
SUNNINGDALE[2].

My dear Baron

I have not yet thanked you for your [acceptable] letter of the New Year,[3] with its kind felicitations, which I cordially accept & respond to.

Only today I have received your Address[4] to the Geograph. Sect. of the Australn. Ass. for A. of Sc.[5] at Brisbane. I have read it with very great pleasure & [*hot*] interest. It is capital, & worthy of you; so full of sound matter & of sound sense, & all so well put, that one "runs as one accord"; & what a store of information it contains. The summary of a world's ways & means from an antipodeal point of view, & that a British standpoint, is verily refreshing, & makes one's blood course faster. Thank you very much for it *& its contents.*

You may guess how doubly welcome your letter was, when I tell you that it as yet the only intimation I hear of Brian's[6] discovery of mercury at Coolgardie.[7] I had heard from him only very shortly before (I suppose) the happy find. He told me that he had a good billet as manager of the "White Feather mine" with £600 a year.

I am still struggling with the Indian Grasses, & Stapf is preparing the same order for the Flora of Trop. Africa & of S. Africa; so we have much work in common. That benefits us both. The chaotic condition of the African Grasses in the Herbarium is inconceiv-

1 MS annotation by M: 'Answ 18/4/96'. Letter not found.
2 Hooker's home in his retirement.
3 See M to J. Hooker, 1 January 1896 [Collected Correspondence].
4 B95.13.01.
5 Australasian Association for the Advancement of Science.
6 Hooker's son. See also S. Hooker to M, 5 January 1894.
7 WA.

96.03.02 able, & I only wish that I could withhold publication till Stapf's work is over, for I can foresee that his work will throw great light on the Indian that cannot appear till his is completed.

I am also busy with Banks' narrative, which necessitates a great deal of work in detail.[8] Happily I am actively assisted in this by my son Reginald (now in the Agricult. Dept), as to which appt I thank you heartily for your kind congratulations.

I shall have excellent portraits of Banks from that in the R. Sy.[9] Rooms, & of Solander from that in the Lin Socy[10] Rooms. The valuable feature of the work is, the revealing[11] Banks' his right place as a "working naturalist," the pioneer of the illustrious band of Naturalist Voyagers of which Darwin is the culminant. It nowhere appears in the accounts of Banks' life & work that he was a bona fide naturalist, in which respect Hawkesworth[12] does him no justice. Banks was further the interpreter of the Expedition; the commissariat-officer so to speak; & the thief catcher to whose energy was due the recovery of the stolen Quadrants, but for which the Expedn would have been a failure. – In short but for Banks the results of Cook's voyage would have been confined to Geograph. discovery. – His subsequent position as the Maecenas of Science[13] has eclipsed hitherto all he did in his early days as a scientific worker. Had he but published his collections what a mark he would have made in the scientific world proper! As it was he gave every one liberty to make use of them, & except for the fragment of the Botany published by Brown,[14] there was nothing gained by those magnificent collections.

The conclusion of the "Index"[15] was indeed a relief, for which, at my age, I am more thankful than I can express: for I had always

8 Hooker was editing the journal kept by Joseph Banks during his voyage around the world with James Cook, 1768-71; see Banks (1896).

9 Royal Society.

10 Linnean Society.

11 giving *deleted and replaced by* revealing, *but* his *not then deleted.*

12 Hawkesworth (1773).

13 Banks was President of the Royal Society of London, 1778-1820.

14 In R. Brown (1810).

15 *Index kewensis* (i.e. Jackson [1893-5]).

the fear of Saturn's scythe[16] at my heels during the long period of its gestation. In one's 79th year the inevitable stares one in the face so long as unfinished work is in hand – & so it is with the Indian Grasses,[17] but as nearly half of this is printed I may live to see it concluded by midsummer. 96.03.02

I still go three days a week to Kew, & work here for the rest of the week. The medallion of my father goes by next mail.

Ever sincerely your
Jos D Hooker.

To Ellis Rowan 96.03.03

MS 2206 Ellis Rowan papers, National Library, Canberra.

3/3/96.

Your genial letter,[1] dear Madam, has reached me, the one of the 17 Jan, and I feel very much indebted for the generous sentiments, which you evince towards me, and for the spirited exertions, made by you in the antarctic cause. All the more grateful acknowledgement is due to you, as your glorious achievements in floral art are absorbing so much time and thought and you should thus not be withdrawn from your own splendid and important engagements. We can all be sure that you will win new laurels in the incomparably rich flower-fields of South-Africa.[2] I cannot express, how elevated I felt on your behalf, when you related the extraordinary

16 Roman god, represented with a scythe and an hour-glass and symbolizing death.
17 J. Hooker (1897).

1 Letter not found.
2 So far as is known, Rowan did not visit South Africa at this time, or ever for any extended period.

96.03.03 interest which our gracious Queen bestowed on your exquisite work,[3] and that her Majesty should have alluded in so condescending manner also to myself, is one of the greatest triumphs of my life! As you evince such vivid interest in the progress of antarctic discovery, being an Australian, let me say, that Britain by possessing more territory in the far south than any other nation, ought not to remain unrepresented in the new discovery voyages for which now preparations are made.

The American expedition contemplated seems postponed,[4] but Germany, Belgium and Norway seem to get ready with their expeditions for the summer 1896-1897 already.[5] Mr Borchgrevinks enterprise[6] will be, as perhaps first suggested by me as most adequate, one for traversing the icy table lands plateaux towards or beyond Mt Erebus and Mount Terror, so that it will not be naval exploration, nor do I feel sure, whether it can be considered strictly a British enterprise, – but it will be very important in its own way. President Markham[7] will not likely get together the very large fund, which a three years expedition by 2 steamers is estimated to cost, nor as you ascertained could in these politically disturbed times ships be obtained from the Royal Navy. My idea is, that a *British Naval Exped.* should still be got ready to operate in the farthest south next season (96-97) that its preparation and outfit be limited to 2 summers and one winter, so that only half the costs would be incurred, that Graham's Land, not so very far from S. Africa which has a good harbour and often open water, should be the basis of

3 Queen Victoria 'accepted three of her paintings' (ADB, vol. 11, p. 466).

4 No new American expedition to Antarctica took place for many years.

5 All these planned expeditions were delayed. The Belgian expedition under Adrien de Gerlache de Gomery was first in the field, in the summer of 1897-8, but then became beset by ice and spent twelve months drifting in pack-ice, the crew thereby becoming the first humans to winter south of the Antarctic Circle. A German expedition under Erich von Drygalski and a Swedish expedition under Nils Nordenskiöld both headed south in late 1901 and spent the winter of 1902 in the field.

6 A privately funded British expedition under Carsten Borchgrevink landed at Cape Adare in April 1898 and spent the following winter there. See Borchgrevink (1901).

7 Clements Markham, President of the Royal Geographical Society, London.

operation, that the South-African Government with which I communicated at the time of Lord Loch[8], should contribute to the expenditure, in which also probably N.S.W. and Victoria would share. All this could still be arranged preparatorily within the next few months, if S. Africa was communicated with by telegrams. Britain would thus not be forstalled! As for steamers the Scotish whaling and sealing vessels would serve well and have Commanders and Crews accustomed to ice-voyages. 96.03.03

I had a very kind letter also recently from President Markham of the Royal geogr. Soc.[9] who introduced to me last month 2 midshipmen from Lord Brassey's training ship.[10] Perhaps you could visit President Markham, to whom Lord Brassey has also referred me, or transmit this letter to that distinguished Geographer.

Britain will also derive new commercial and industrial advantages from these voyages of its own, and not be shut out from the scientific discoveries, having always since the time of Capt Cook stood leadingly or singly in antarctic researches.[11]

Ever regardfully your
Ferd. von Mueller

8 Letter not found.

9 Letter not found.

10 Lord Brassey sailed his private steam yacht *Sunbeam* from Britain to Melbourne to take up his appointment as Governor of Victoria in October 1895.

11 See also M to E. Rowan, 1 December 1895 [Collected Correspondence], in which M covers many of the same points.

96.03.11 *From Ignatz Urban*[1]

RB MSS M1, Library, Royal Botanic Gardens Melbourne.[2]

KÖNIGL. BOTANISCHES MUSEUM.

Berlin, W. den 11. März 1896.

Hochverehrter Herr Baron,

Beifolgend gestatte ich mir, Ihnen Abschrift dessen zu übersenden, was ich seiner Zeit über Boronia floribunda Sieb. gesagt habe und ein Fragment der Originalpflanze beizulegen. Hoffentlich finden Sie meine Meinung bestätigt.

Zu meinem grössten Bedauern kann ich Ihnen über die Veröffentlichung der Uebersetzung Ihres Werkes von Seiten meines Schwagers immer noch nichts mittheilen. Er hat mir darüber nie eine Zeile geschrieben, niemals auf meine Frage geantwortet. Da er seit einem Jahre wieder tüchtig arbeitet, so steht zu hoffen, dass er sich auch dieser übernommenen Verpflichtung erledigt.

Da unser Notizblatt auf Kosten des Gartens hergestellt wird, so können wir mit demselben tauschen. Lohnt es sich für uns, mit dem australischen Chemist and Druggist, in dem Sie so viele neue Arten beschrieben haben, zu tauschen? und welches ist seine Adresse? Viele Ihrer Werke erhielten wir, wenn ich mich recht erinnere, durch die Melbourner Nationalbibliothek. Müssen wir nicht auch das Notizblatt dahin schicken und unter welcher Adresse?

Mich Ihnen bestens empfehlend, zeichne ich

mit vorzüglichster Hochachtung
ergebenst
I. Urban.

1 The file also contains a copy of an extract from a publication by Urban annotated by him: 'Jahrb. des kgl. Botanischen Gartens zu Berlin II (1883) p. 391'.

2 MS found with a specimen of *Boronia floribunda* (MEL 250962).

Royal Botanical Museum 96.03.11

W. Berlin, 11 March 1896

Honoured Baron,

Allow me to send you a copy of what I said at the time about *Boronia floribunda* Sieb. and include a fragment of the original plant. It is to be hoped you will find my opinion confirmed.

To my greatest regret I am still not able to tell you anything about the publication of the translation of your work on the part of my brother-in-law.[3] He has never written me a line about it. Because he is working hard again for a year, it is to be hoped that he will also finish this assumed commitment.

Since our *Notizblatt*[4] is produced at the expense of the Garden, we can exchange with it. Is it worthwhile for us to exchange with the *Australian Chemist and Druggist* in which you have described so many new species? and what is its address? We receive many of your works, if I remember correctly, through the Melbourne Public Library. Do we also have to send the *Notizblatt* there and under what address?

Warmly recommending myself to you I sign with
most particular respect most humbly
I. Urban.

To Casimir de Candolle[1] 96.03.31

Conservatoire et Jardin botanique, Geneva.

Melbourne, 31/3/96.

This day, dear Prof de Candolle, I received the large volume on Bromeliaceae, which you so generously forwarded to me.[2] It is a superb volume of a splendid series! Mr Mez may prove a second Bentham and Boissier. He is in an exceptionally fortunate position to be rich, I believe, and therefore can devote to his favorite science all his time, his attention being in no way diverted by pub-

3 Not identified.

4 *Notizblatt des königlichen botanischen Gartens und Museum zu Berlin.*

1 MS annotation: 'rep. 3.8.96'. Letter not found.

2 Mez (1896).

96.03.31 lic duties, and as he is still young, he may – if his constitution is strong – do vast phytographic work far into the next century. This brings me to an idea, which long since I have formed, that the Universal Flora of the World ought descriptively remain under the sway of the Family of De Candolle! By the end of the century the vegetation of our planet will be almost completely known, as within the next four years the few remaining regions will become penetrated and thus also their main plant-wealth rendered known.

My views are, that at the commencement of the century a new "prodromus" (vegetationis totius mundi[3]) should be begun, much in the style of the earlier volumes of your illustrious grandfather. As numerous species become abolished as untenable the whole phanerogamic flora of our globe could be treated in 20 volumes briefly, and as most plants have now been carefully studied the condensation of their chief characteristics into short diagnoses should not be taking more time than 20 years for 20 volumes by 20 special authors, and as the volumes would not be illustrated, they would be not very expensive to print, and the sale would be most extensive all over the world, which could not be the case, when the first prodromus was published by Aug. Pyr. DC, Alph. D.C. and Casimir D.C Perhaps a son of yours can take part in the Editorship. What a splendid work this new "prodromus" (in true consonance to the name, would be. After 20 years or occasionally earlier could appear some supplemental volumes Tomus I ought to appear 1900.

Ever regardfully your
Ferd von Mueller

The third Census of Austral plants is ready for publication.[4]

I am meanwhile now pushing on descriptions of Papuan plants

Kindly give me your opinion on the proposition of a new prodromus.

The volumes must appear in the sequence of the orders of D.C. syst, except that Monochlamydeae be inserted among the Dichlamydeae, reserving only [Con]iferae and Cycadeae both certainly dicotyledonous

3 'of the vegetation of the whole world'.

4 This work was never published.

To Walter Rothschild[1] 96.04.12

Natural History Museum, London, Museum Archives, TM1/21/24 (Mueller).

Melbourne 12/4/96.

Hierbei habe ich die Ehre, edler Herr Baron Rothschild, Ihnen Auszüge eines Briefes vorzulegen, welchen ich in den letzten Tagen von meinem Sammler, Mr W Fitzgerald aus New Guinea erhalten habe, und hoffe ich dass es für Sie erfreulich ist ein Zusammen-Wirken eingeleitet zu sehen durch Ihren Agenten, um eine ornithologische Expedition von Ihnen mit meiner geographischen and[2] phytologischen in Cooperation zu bringen. Auf diese Weise wird ja die Parthie der Reisenden um so stärker, ohne weder der einen noch der anderen Hindernisse zu bereiten. Für Ihre Zwecke liegt noch der besondere Vortheil darin, dass Ihre Expedition das *Hochland* erreichen wird, ja vielleicht sogar die *Glacier*-Regionen, wo ja mit veränderter Vegetation auch ein anderes Vogelleben sich zeigen wird als im Unterlande.

Mr Fitzgerald is an Alpen-Reisen durch seine Touren in Tasmanien gewöhnt worden, und so entwarf if[3] für Ihn den Plan einer Reise in die [n]erforschten[4] höchsten Schneigen[5] Gebirge der grossen Papua-Insel, und habe ich dafür die Mittel durch Hülfe von Freunden zusammen gebracht. Die jetzige Reise ist die zweite in Neu Guinea, da er schon in der letzten Hälfte des vorigen Jahres hauptsächlich auf meine eigne Kosten den Gouverneur Sir Will. Macgregor auf einer Fahrt ins Innere zu Wasen begleitete. Er hat also jetzt auch schon Papuanische Erfahrungen, welche Ihrer Expedition mit zu gut kommt.

Ich verspreche mir viel von diesen vereinten Forschungen und wenn es den Reisenden gelingt bis zu Höhn von 16000 Fuss vor-

1 MS annotation by Rothschild's secretary, Ernst Hartert: 'Answ 22/5 E H'. See E. Hartert to M, 22 May 1896 [Collected Correspondence].

2 und?

3 ich?

4 unerforschten?

5 Schneeigen?

96.04.12 zudringen, dann wäre die *hoch-alpine* Flora und Fauna von NG. zuerst aufgedeckt.

Seine Excellencz Sir William Macgregor nach Proposition von mir selbst unternahm es, der Owen Stanley's Range bis zu seiner subalpinen culmination (13000') zu erklimmen und die vorkommenden Pflanzen zu sammln, über welche ich vor einigen Jahren eine grössere Abhandlung in den Transactions of the Royal Soc. of Vict, deren Hauptbegründer ich 1854 war zu veröffentlichen Sie ehrend und hoffend Ihr freundliche Antwort bald zu erhalten

Ihr
Ferd von Mueller

Melbourne 12/4/96

Hereby I have the honour, noble Baron Rothschild, to submit to you an extract of a letter I received in the last few days from New Guinea from my collector Mr W. Fitzgerald,[6] and I hope that it will be agreeable to you to see a cooperation arranged through your agent in order to bring an ornithological expedition of yours into cooperation with my geographical and phytological expedition.

In this way the party of travelers will become all the stronger, without either one causing obstacles to the other. There is another particular advantage for your purpose, that your expedition will reach the *highland*, indeed perhaps even the *glacier* regions, where with more diversified vegetation than in the lowland another birdlife will also appear.

Mr Fitzgerald has been used to alpine journeys through his travels in Tasmania and so I drafted for him the plan of a journey in the unexplored highest snowy mountains of the large island of Papua and I have raised the funds for it through the assistance of friends. The present journey is his second in New Guinea, where he already accompanied the Governor Sir Will. Macgregor on a journey into the interior to Wase in the last half of the previous year principally at my own expense. Therefore he now also has Papuan experience, which will be of advantage to your expedition too.

I have high hopes of these combined researches and if the travellers succeed in reaching the height of 16,000 feet, then the *high alpine* flora and fauna of New Guinea would be discovered for the first time.

6 See W. Fitzgerald to M, 13 March 1896 [Collected Correspondence].

His Excellency Sir William Macgregor himself on my proposal undertook to climb the Owen Stanley Range up to its subalpine culmination (13,000') and to collect the plants occurring there, about which I published a larger paper some years ago in the *Transactions of the Royal Society of Victoria*,[7] whose principal founder I was in 1854. 96.04.12

Respectfully and hoping to receive your friendly answer soon

Your

Ferd. von Mueller

To Ada Armytage

1-10/1234 Armytage papers, University of Melbourne Archives. 96.04.27

S. Yarra,[1] 27/4/96

When I receive, dear Miss Armytage, kind letters from Ladies, which is not very often, I feel always much cheered, and thus I was also particularly the case when I received[2] your thoughtful communication,[3] and the grass seeds as I on several occasions, when I called at Como,[4] you all were out or absent in the country. For some few months I shall be overpowered with professional work in my Department, but after that time I hope to call again on the few friends which I have. Perhaps next summer the Mitchell-Grass will be bearing grains, the last season having been one of drought through most parts of Australia But your mothers N. Queensland's Station[5] suffered less, as it is, I believe within the reach of the tropic summer-rains. I should like very much to know, whether the gigantic annual Teosinte Grass of Mexico and the creeping Wonder-

7 B89.13.11.

1 Melbourne.
2 particularly glad to receive *changed by M to* particularly the case when I received.
3 Letter not found.
4 The Armitage family home in South Yarra, Melbourne.
5 Afton Downs, Hughenden, Qld.

96.04.27 grass Coapim of W. Africa and the Burr Trefoils without burrs from Italy succeeded & become naturalized on the station.

Regardfully yrs
Ferd von Mueller

My best thanks for the fruiting specimens of the Mitchell-grass. (Astrebla elymoides another species occurs in N. QL[6] with denser spike)

96.06.14 *To John Shillinglaw*[1]

Box 246/2, Shillinglaw papers, La Trobe Australian manuscripts collection, State Library of Victoria, Melbourne.

Sunday, 14/6/96

Am much touched with your kind letter received yesterday,[2] dear Mr Shillinglaw, and I fully recognize the generous sentiments, evinced by yourself and by mutual friends of ours towards me. But I have repeatedly expressed my views to some of those "near and dear" to me, that public tributes, such as you so feelingly intend also for me, should only be bestowed after the recipient had passed away. Then I have further desired from the few with whom, as otherwise *alone* in the world, I am in confidence, that any token, which are also bodily to preserve the memory of myself, should *not be a bust*, but a half size painting. *This view I have ever held*, because of the lifeless appearance of marble- or bronze-busts or statues, on account of the want of visual expression, and I further suggested that any oil painting of which I possibly may be deemed worthy in a public place should be kept under Glass. My views clash with ancient aesthetics and time honored traditions and customs, and as it must be painful to anyone to foresee what was done should only

6 Qld.

1 MS annotation by Shillinglaw: 'Recd 15.VI.96'.

2 Letter not found.

come into use after his death, it will be best to leave all your friendly intentions for the present. If it is still then be preferred that a bust should be prepared, then the cast for it or the other necessary measure could be taken in my last illness and be sent to the Artist, whom also I highly admire. As regards a *place* for such a monument I have asked long ago confidential friends, who promised to what little requires to be done, for one, who has *no worldly property* whatever left, that any portrait shoud be put either in a place at the University, as I was there through many years an honorary examiner for the degree of M.D., or that the portrait should find a location in the "Turn-Verein" where English and German members have largely and so happily met during the last 30 years. On the public Library I have no claims; my regretable cessation from the bot Garden lessened my opportunities so much, to work for the technologic Museum, that any special remembrance of myself *there*, would be inconsonant with real claims and this view I have as a special wish expressed years ago and ought to prevail. It is very different with our never to be forgotten sterling Newbery, and it will also be quite different with Sir Fred McCoy and Colonel Ellery in what I trust will be a distant time to honor their special work by a bust. 96.06.14

Ever with regardful and grateful attachment
your
Ferd. von Mueller.

To Louise Matthiessen 96.07.02

RB MSS M100a, Library, Royal Botanic Gardens Melbourne.

2/7/96

Am quite delighted, dear Louise, with your kind letter and with the felicitation at the anniversary of my life[1] from all of you. I had also a kind letter from your dear Mother,[2] and she baked as in former

1 M's birthday on 30 June.
2 M's sister, Clara Wehl.

96.07.02 years for me the same little cakes which our own lamented Mother, whose name Louise you have from her, prepared for us children in the earlier part of the century. I send you a little book for your children, as the plant-pictures may inspire them to take notice of the flowers particularly in a humid year like the present. I could send you seeds for your Flower- and Kitchen-Garden if you have means for cultivating Herewith you receive also some envelopes, into which with the help of your little ones you might place plants obtainable near you. I want particularly *minute* plants, of all kinds, and of these Australia is richer than any other of the great divisions of the globe. Each envelope would cost only when filled to the weight of 2 ounces 1d[3] by ordinary post, 2d for 4 ounces. It is for tracing still further the geographic distribution of these miniature-plants, that I require them. Perhaps Ladies on Stations or Farms near you would also collect these *minute* plantlets. Next to them the most important are floating or submerged weeds.

I send you seeds of the most beautiful of all Australian Everlastings. Cattle, Horses and sheep like this West-Australian plant, and it gets easily naturalised. You find a note on it in the "Select plants" under Helipterum Manglesii. The newest Edition[4] is sent by this mail, and Mr Matthsien[5] will find [many] notes in it which he can turn to account on your estate. All land- or station-owner, should have this work, and you might show it to some of your Neighbours. The Government Printer here sells it as[6] cost price 5/ only and postage. I have no monetary profit from this or any other works of mine.

With best greeting to all of you
Ferd von Mueller

Have you my large lithographic portrait?

All senders of of[7] dried specimens get full credit in my works for any discoveries made.

Kindly give some of the Envelopes to people far inland

3 i.e. one penny.
4 B95.13.02.
5 i.e. Matthiessen.
6 at?
7 *word repeated.*

To Georgina King

96.07.14

ML MSS 2117/1, Mitchell Library, State Library of New South Wales, Sydney.

14/7/96

Your kind letter has reached me, dear Miss King, also the additional note on the hatched by the Rev Dr Gill,[1] to whom – please – convey my best thanks. I certainly will dedicate to him a new plant, but before the middle of August I shall have no time for phytographic labours, rural work absorbing all my attention til then and perhaps later. Here the season has been one of fogs and nebules almost continually and thus the danger of *rust* is greater in Wheat than in any with much clear sunlight, as I pointed out in 1865,[2] when I was President of the Rust-Committee (in a very bad year) of the Gov Board of Agriculture as constituted at that time.

Will Mr Moore return?[3] I fear not, as he not likely will live any longer in Sydney. Altho' we held very opposite views on the objects of *botanic* gardens, he always acted *honorably* towards me. English Gardening in an egyptian clime can only be carried on at fabulous expense, which sums in young colonies, so I always held, go largely for promoting rural interests for the benefit of the producing classes and thus for the breadwinning people and to *augment the* revenue. Now my counsels, for which I made so strenuous I might say heroic efforts here with slender means, seem to prevail also in your colony, and a reaction under the deep financial distress of these colonies, seems to have set in with you. Moreover I had never any water for extensive lawns and mosaic or carpet flower-beds, but I had the Victoria regia, the Geyser fountain, the aviary, the pine sloaps,[4] the first large glasshouse in Australia and a marvellously rich collection of varied plants, the industrial kinds

1 Letter and note not found.

2 B65.09.01.

3 Charles Moore visited Europe after his retirement as director of the Sydney botanic garden in May 1896; he later returned to Sydney.

4 slopes?

96.07.14 largely represented, and I supplied on an immense scale for 15 years as Director here in the then only commencing Victorian colonisation; the grounds of Churches, manses, Cemeteries, public reserves, railway enclosures with plants, so that when I ceded from my Directorship I was presented with an adress of thanks by the Clergy of the Church of England[5] If the 1/4 million sterling, which has been spent on the bot Garden here since I left had been under my administration, what vast utilitarian work would have been accomplished!

I see Sir Frederick[6] very rarely; we live so far from each other, our researches are in different directions, I never give up time for attending to mere formal festivals at the University, and have – for which I am thankful late in life, when my time becomes so precious, [no][7] further duties as an hon. University-Examiner which I was for therapeutics and materia medica for many years. Are the "Select Plants" often mentioned in your colony? I know they are in almost daily use there on[8] many public institutions, journals and by private colonists. I have worked on that book fully 30 years, and an occasional kindly word in an official Document would be cheering to me and aid in the support of my Department But I find so often in Australia that those who use my original works give themselves the air, as if they had known all already themselves. Let me hope, that you and those near and dear to you are all happy and well.

Ferd von Mueller

5 Address not found.
6 Frederick McCoy.
7 *editorial addition.*
8 in?

To Jacob Agardh[1]

96.07.18

Handskr. Avdl., Universitetsbibliotek, Lund, Sweden.

18/7/96

In Beantwortung Ihres letzten Briefes, hochgeehrter Herr Professor, spreche ich meine Bewunderung von Neuem aus, dass es Ihnen möglich ist noch as[2] Octogenar Algen-Dissectionen vorzunehmen und microscopisch zu verarbeiten Ich kann mir daher vorstellen, wie unerquicklich es für Sie sein muss, mit unzulänglichen Material zu thun zu haben. Die Sammler mit seltenen Ausnahmen in Australien wollen sich nicht die Mühe machen, vollständige Exemplare zu sichern wenn solche wirklich erlangbar sind. Welche Zeit habe ich im Lauf meiner Forschungen nutzlos verschwendet in den Versuchen blumenlose oder fruchtlose Phanerogamen-Arten zu benennen. Wenn die Collectoren das Bibel-Wort "An den Früchten sollt Ihr sie erkennen" auch in dieser Hinsicht in Gedanken behielten, so wären die systematischen Arbeiten leichter für uns Wenn ich bedauerliches Algen-Material dennoch an Sie abzuschicken wagte, so war die Idee, dass Sie *leicht* das Eine oder Andere dazwischen finden würden von phytogeographischem Werth oder was sonst ohne grosse Mühe benutzt werden könnte. Ich dachte, dass Sie etwa einen Studenten der Universität oder einen Amateur zu Hand hätten, welcher das schlechte Material doch präpariren möchte, wobei dann doch vielleicht not[3] etwas herausgearbeitet werden könnte. Das Material zurückzusenden wäre nicht rathsam, denn meine Zeit ist so besetzt, dass ich selbst nichts damit thun könnte, und nach J. B. Wilsons Tode beschäftigt sich Niemand in Australien untersuchend mit Algen beschäftigt. Material, welches Ihnen nichts nützt möchte noch immer Bornet, de Toni, Heydrich, Murray, Wright, Holmes, Falkenberg, Reinke, Cramer, Reinbold, Berthold, Kukuk oder sonst einen jüngeren Phycologen

1 MS annotation by Agardh 'Fragm.'

2 als?

3 noch?

96.07.18 Europas für deren Sammlungen willkommen sein, und ich überlasse es ganz Ihrer Wahl, wem Sie solches zuwenden wollen, wenn ich nur weiss wer es hat um Nachfragen der Sammler gerecht zu werden.

Stets Sie ehrend, der Ihre,
Ferd von Mueller

18/7/96

In answer to your last letter,[4] highly esteemed Professor, I express my admiration anew that it is still possible for you as an octogenarian to carry out dissections of algae and to work with the microscope. I therefore imagine how fruitless it must be for you to have to do with inadequate material. The collectors in Australia with rare exceptions do not want to take the trouble to secure complete specimens when such are readily acquirable. What time I have wasted uselessly in the course of my research in the attempts to name species of phanerogams without flowers or fruit. If the collectors would keep in mind the biblical words "And by their fruits shall you know them"[5] also in this respect, systematic work would be easier for us. Nevertheless when I dared to send off deplorable material to you, the idea was that you would *easily* find one or another among them of phytogeographical value or what otherwise could be used without great trouble. I thought that you would have had for instance a student of the university or an amateur at hand, who might prepare the bad material, whereby then perhaps something more could be worked out. To send the material back would not be advisable, since my time is so occupied that I could do nothing with it myself and after J.B. Wilson's death[6] no one in Australia is occupying himself investigating algae. Material which is of no use to you might still be welcome to Bornet, de Toni, Heydrich, Murray, Wright, Holmes, Falkenberg, Reinke, Cramer, Reinbold, Berthold, Kuckuck or such other younger phycologists of Europe for their collections, and I leave it completely to your choice to whom you wish to give it, if only I know who has it in order to do justice to the enquiries of the collectors.

Always respectfully your,
Ferd von Mueller

4 None of the letters Agardh wrote to M in 1896 have been found.
5 Matthew, ch. 6, verse 20.
6 Wilson died on 11 June 1895.

From Robert Etheridge, Jr

96.07.22

AMS 49/4, Curators' private letterbooks, pp. 244-5, Australian Museum Archives, Sydney.

22.7.96

My dear Baron,

Will you do me the favour to consider the following points, & give me the benefit of your opinion –

In your "Observations on New Vegetable-Fossils", Decade 2, 1883, p. 22,[1] you described fossil wood from the Haddon Gold-drift, assuming it to be that of your fruit genus *Spondylostrobus*. Now Schenk in Schimper's Palaeophytologie Vol. of Zittel's "Palaeontologie" p. [...],[2] describes wood from the Ballaarat[3]-Gold-drifts under the name of *Phyllocladus Mülleri*, which he says the "Botanical Collection of Leipzig owes to the kindness of Dr Ferd. von Müller", I presume meaning yourself. Of this he gives three figures, and on the authority of Kraus says the wood of *Phyllocladus* is distinguished from that of Cupressi-form Conifers by the large oval [pores], inclined to the left, that are found on the parenchyma of its medullary rays. Now, from the fact that the wood on which *Phyllocladus Mülleri* was founded, is said to have been sent to Leipzig by you, it has occurred to me are *P. Mülleri* & your *Spondylostrobus* wood one & the same organism? On the other hand, altho' the perforations of the medullary rays of your Fig. 3, of Pl. 22[4] (Spondylostrobus, op. cit.) are fairly large, still they do not appear to me to attain the size of those figured by Schenk in the corresponding part of his *Phyllocladus Mülleri*, nor are they oblique as in the latter, consequently the two forms may be different. It is on this point that I shall be glad to have your opinion.

I am,
Faithfully yrs
R. Etheridge

1 B83.13.04.
2 Space left in MS. The description appears in Schimper & Schenk (1890), p. 873.
3 Now Ballarat, Vic.
4 The reference should be to Plate 20.

96.08.16 *To Thomas Hart*

RB MSS M24, Library, Royal Botanic Gardens Melbourne.

S. Yarra,[1] 16/8/96.

In the quietude of the Sunday, dear Professor Hart, I bring up arrear correspondence, accumulated last week, when I was extraordinarily busy in the Department, and now first attend to you, offering my best felicitation to you at your auspicious promotion to the Ballarat Professorship.[2] You have with natural abilities also honorably striven, to win a prominent scientific position; and as you are yet far from the zenith of life, you can hope to work as one of our Leaders in knowledge far into the next century with the charm also of identifying yourself with the independent and progressive elucidation of forms in nature here at a time, when enveyably so much of organic and inorganic objects in the division of our planet can first yet be rendered known.

Let me advise you, to start at once a Field-Naturalists Club in your important and prosperous town. You can soon in youthful enthusiasm bring together some few bright spirits there, as all such organisations even the Royal Society of London commenced with the gatherings of merely some few members. You will then have locally at once a number of Amateur-Collectors, both Ladies and Gentlemen, and an immense impetus will be given to local research and elevating thoughts on Nature's riches there. Litterary Gentlemen, such as the hon. Mr Vale are sure to support such a movement. In the German Association of Ballarat you are likely to find also some, who will share in this new action. In two or three years, therefore before the end of the Century you can have already in a Florale of Ballarat the descriptive records of the several hundred of indigenous plants then [A] "Fauna of Vertebrata" could soon follow, – of Invertebrata later. Discuss the subject in the press, to test public feelings thereon. It will be a glorious initiative of a wide

1 Melbourne.

2 Hart had just been appointed science lecturer at the Ballarat School of Mines, Vic.

scientific career of yours there. When the warm weather sets in, I shall redeem a promise to the German Association there, to spend a day or two in Ballarat once more, but the weather is for me too cold yet to travel.

As regards your Pimelea, I think it is only a form of P. axiflora.[3] A similar variety I have from Mt Cole[4] which seems to be the N.W limit of this species. The bisexual flowers may be sterile The crucial test for the exact specific position of your plant will be the examination of the *perfectly ripe fruit*. Thus the allied P. drupacea, which I found only on one single place outside of Tasmania has the fruits outside black. The Bible word, ye shall know them by their fruits[5] is very applicable in plants science also!

I will be happy to aid you in the examining of any plants, which may remain to you doubtful after your own investigations. But now I like to invoke also your early aid for the promotion of special studies of mine. I am thus eager to get from as many places as possible in the *far interior* during *this spring* particularly *any* kinds of the *minutest plants*, in which Australia is richer than any other wide region of the globe. The[6] can be easily collected, put into flat- or bag-envelopes, such as carries this letter, and the postal arrangements are so excellent in all these colonies that 1 lb weight of plants by *sample*-post can be sent for 8d, and such package would convey many hundreds of specimens. The third "Census of Australian plants" is due next year,[7] so that I like to make that issue as full as possible up to the time, and as ordinary Collectors and Amateurs generally proved inattentive to wee plantlets, though a simple lense will reveal also in these ephemerous children of Flora very perfect floral structures, an ampler display of rarities for noting geographic distribution will yet be found among these smallest beings, than among any other forms of Australian vegetable life, although *of uncommon* other plants should neither be lost sight

3 MS annotation by M on one of two herbarium labels attached to the letter: 'Pimelia axiflora FvM Mt Buninyong 1896 Prof Th. Hart.'

4 Near Ballarat, Vic.

5 Matthew, ch. 6, verse 20.

6 They?

7 Never published.

96.08.16 of. As you are aware, every finder gets credit in my works for new localities of rare plants and for the discovery of novel forms. Will you kindly make soon free some few hours of your valuable time for pleading with distant thoughtful friends this cause of mine? As I am to celebrate my 50 years Dr Jubilee next year, my worldly career must soon come to a close, so that I may not live during other springs to repeat this solicitation

With regardful remembrance your
Ferd von Mueller

Can a small piece of meteoric iron be spared from the Mines-Departments Collection there?

I gave it as my opinion, when adressing the Field-Nat Club here at its last annual gathering that every town throughout Australia ought to have some such Association

The new locality of the Pimelia shall be recorded under your honored name.

Would you like to join the Linnean Soc. of London? I will gladly be your Sponsor.[8]

96.09.08 *To Édouard Bornet*

MS 539, No. 2099, Correspondance scientifique d'É. Bornet, Lettres adressées à É. Bornet, Laboratoire de Cryptogamie, Muséum National d'Histoire Naturelle, Paris.

8/9/96

With the greatest of gratitude, dear Dr Bornet, I have received the large parcel of Algs, which you so beautifully prepared from Dr Engelhardts and other collections & will send a set to the finders as

8 Would ... Sponsor *is written on the second of M's attached herbarium labels.* The Linnean Society has no record of receiving a nomination for Hart.

named by your high authority, and this will induce them to continue their researches But I am afraid that your *valuable time* is taken up too much for naming crude material, requiring resoaking and spreading out, and I fear also, that you find seldom species additional for your own great collection. But some interest is attached to localities as means to trace geographic ranges. Thus I know now Claudea elegans from 4 Australian colonies and six places. 96.09.08

I can send you still many more crudely dried specimens.

Prof Agardh wrote me recently that as an Octogenarian he cannot any longer devote time to naming specimens of unprepared crisp Algs.[1] So I can send to you many from *new localities* in future, by which means your collections might be enriched also.[2]

Very regardfully
your Ferd von Mueller

Entre nous[3]
Prof Maxime Cornu has not written to me for a long time. He is necessarily very busy with his students and the jardin des plantes. Still the idea has arisen with me that he may feel hurt at something unknown to me. I have sent him *numerous* small parcels of plants during 1895 & 1896 by post for the "Herbier", and in each instance some *very rare* species were among them. If he only sent occasionally a post-card, I would know whether he is pleased with such sendings. Numerous of the species must have been quite new to the "Herbier".[4]

1 Letter not found.
2 See M to J. Agardh, 18 July 1896.
3 Between us.
4 MS accompanied with an envelope addressed to 'Docteur E. Bornet Membre de l'Institute Quai de la Tournelle Paris'.

Letters about Mueller, 1876-1897

This section includes a selection of letters that make significant reference to Mueller. The letters are transcribed, and in some instances translated, in the same style as the letters in the main sequence.

Louis Smith to Graham Berry 77.07.11

L77/7970, unit 153 VPRS 3991 inward registered correspondence, VA 475 Chief Secretary's Department, Public Record Office, Victoria.

Melbourne
11th July 1877

Sir,

I have the honor to forward herewith the Report of the Board appointed "to inquire into the present position of Dr Mueller, in relation to his professional duties, with the view to advise, what alteration, if any, is necessary, to afford him reasonable facilities for the due discharge of his scientific labours."

I have the honor to be
Sir
Your most obedient Servant
Louis L. Smith
Chairman of the Board

The Honorable
Graham Berry M.P.
Chief Secretary and Treasurer
&c &c &c

77.07.11 We the undersigned, being a Board appointed,[1] "to inquire into the present position of Dr Mueller in relation to his professional duties, with the view to advise, what alteration – if any – is necessary, to afford him reasonable facilities for the due discharge of his scientific labours", have the honor to present the following as our report: –

We think it well to state in the outset, in order to guard against misapprehension that the complaint of Baron von Muëller, as laid before us by him, is not as to his personal remuneration, but it is, that while he holds an office of practical importance to the Colony, and of interest to the scientific world, he is deprived of the means of properly discharging its functions.

We find that the present position of Baron von Muëller, is that of Government Botanist; but that he is without a Garden; without a Laboratory for technical work; with an insufficient Museum for dried plants (an absolute requisite in Botanical science); and that he is without a sufficient staff and premises, for effectually carrying on his department.

A State Botanist in such a position as this, cannot perform the duties required of him: his proper place is in the midst of his living plants which he needs for daily study. An adequate and ample Museum for dried plants; a Laboratory and Apparatus; are also essential requirements in Botanical science.

We deem that an alteration of this state of things is most advisable and for the following reasons: –

1st Because the important work of observing and describing the Flora of Victoria is yet far from complete.

2nd Because it has nevertheless, under the assiduous labours of the Government botanist, arrived at such a stage as to have commanded the favourable notice of scientific men in all parts of the World.

3rd That to stay the work at such a juncture, would in our opinion be exceedingly unwise.

4th That this colony having the peculiar advantage of the services of one of the foremost botanists of the day, it would we think

1 The file includes letters to W. Odgers from L. Smith (22 June 1877), M. King (23 June 1877) and J. Bosisto (23 June 1877) agreeing to sit on the Board.

be a shortsighted policy, not to prosecute the work in question – which would stand to the credit of Victoria throughout all time. 77.07.11

5th That the labours of the Government Botanist relate not only to abstract, but also to practical science, demonstrating the utilitarian and commercial value of our indigenous vegetation as well as indicating what portions of the vegetation of other countries may be acclimatized here.

6th That although this might entail a further outlay to a small extent at the commencement, it would prove ultimately remunerative.

We therefore after due consideration, submit the following as our recommendations

1. That the duties of the Government Botanist be defined, and his title of Director restored; so that it should be clearly understood that the scientific functions of Government Botanist pertain to him. For instance we consider that letters of enquiry, as to the habits, growth, and culture of plants; and all communications bearing upon the science of Botany should be addressed to the Government Botanist, and that any not so addressed, should be transferred to him to be dealt with.

2. That the Government Botanist should have a Garden of acres in extent in close proximity to the City and the present Botanic Gardens, and that it be termed the "Scientific Botanical Garden", and that he be allowed three specimens each, of such plants as he requires from the present Botanic Gardens.

We refrain from recommending any particular spot, but would point out that there are certain lands apparently available, such as those adjacent to the Immigrants' Home and the site around the present Botanical Museum; or, the Albert Park. The first two are at present unsightly, and their cultivation would ornament the Government Domain, and improve at small cost one of the entrances to the City.

3. That the Government Botanist be allowed a sufficient staff.

4. That he should have a Laboratory with suitable appliances.

5. That the Botanical Museum should be enlarged or some suitable building be provided. To save outlay, we would suggest – after having inspected them – that the buildings of the present Immi-

77.07.11 grants' Home being in close proximity to the City and the Botanic Gardens, would be suitable for offices, Museum &c.

6. That part of one of the State Forests should be at the Government Botanist's disposal for the culture of trees and plants, – indigenous or acclimatized – for economic purposes.

All which we submit as our Report.

Louis L. Smith, Chairman

M. L. King
Joseph Bosisto. } Members

Melbourne
11th July 1877

90.01.22 *James Walker to Mary Walker*[1]

Archives of the Religious Society of Friends, University of Tasmania Library, Hobart.

22 Jany 1890

[...]

After-dinner speeches are usually a weariness to the flesh, but at this dinner tho' everyone spoke & necessarily, impromptu, nearly every speech was good & most laughable or witty. I wish I could tell you a few, but if I began, this letter would be double or treble weight. The President of the Association this year was Baron von Mueller, and he was quite the funniest feature of it. A little German, spare in figure & thin in features, with round blue eyes, & a

1 This is part of a longer letter describing the writer's experiences while attending the second Congress of the Australasian Association for the Advancement of Science in Melbourne in January 1890. Only the section relating to M is published here.

florid complexion unused to soap & water, his clothes ill fitting & set off by a dingy white comforter which would be much improved by a wash – he is a most grotesque figure & has a propensity for making little German English set speeches on every occasion. When Sah and I were presented he said "I am prahd Madam, to make your acquaintance; I am prahd you should have come from a so far country to dis association." After a similar greeting to another girl, at the reception where there were refreshments he said "Madam, I hope you have partaken of my *frugality*". At the big champagne lunch at Sunbury he replied to the toast of the Assocn & himself in one of the funniest speeches I ever heard. I wrote down a few gems on the spot. It was good for scientific men to come to these meetings "where each meets mit its oder". He enlarged on what the Assocn would come to "when our warty (worthy) host's children had grown up in anoder century" The Association would last, in fact "its permanency can never come to an end". The weather was bad & the pass wet; but this had its compensations for "it would have mitigated our pleasures if we had any great excessive heat". Referring to himself he said "I am prahd of the magnificent reception dat has been accorded to me, but dis reception is not due to my 'umble merits, but to dis Association's "*magnimosity*". That word is worthy of 'Alice in Wonderland'. 90.01.22

[...]

With my blessing –
Thine
JBW

94.05.23 *Kurt Lampert to Gustav von Silcher*[1]
Staatliches Museum für Naturkunde, Stuttgart.

Koenigl. Direction

beehre ich mich, von einigen wertvollen Bereicherungen der Sammlungen Kenntniss zu geben, welche das K. Naturalienkabinett in den letzten Monaten geschenkweise erhalten hat.

[...]

Baron Dr Ferdinand Müller in Melbourne, welcher unermüdlich für die Bereicherung der Sammlung thätig ist & von dem von Zeit zu Zeit mehr od. weniger bedeutende Sendungen einlaufen, hat kürzlich dem K. Nat. Kab. ein sehr wertvolles Geschenk gemacht durch Ueberschickung eines Exemplar des Beutelmull, Notoryctis typhlops Stirll. Das Tier wurde vor einigen Jahren entdeckt in Innern Australiens an der Telegraphenlinie zwischen Adelaide & Port Darwin; eine von Prof. Shirling[2] speciell zur Aufsuchung des Notoryctis ins Innere Australiens unternommene Reise führte nur zur Erbeutung von 6 vollständigen Exemplaren, sie gelangten durch Prof. Shirling zur Verteilung an das britische Museum, in London, das Roy. Coll. Surg in London, das Museum in Stockholm & das Museum in Cambridge. Nach Bekanntwerden der Entdeckung wendete ich mich an Baron Müller mit der Bitte die eventuelle Erwerbung eines Notoryctis für das Naturalienkabinett ins Auge fassen zu wollen & erhielt Anfang dieses Jahres von Baron Müller die Mitteilung, dass es ihm "zu seiner grossen Freude" nach jahrelangen Bemühungen gelungen sei, das Tier zu erwerben, für welches man "fabelhafte Preise" zahle; einige Wochen später traf das angekündigte Geschenk ein. Nachdem schon bisher, besonders durch die fortgesetzten Schenkungen des Baron Müller unsere Sammlung australischer Beuteltiere die reichhaltigste & vollständigste gewesen steht sie durch den Besitz des Notoryc-

1 This is part of a longer letter. Only the sections relevant to M are published here.
2 i.e. Stirling.

tis nun nicht nur den Sammlungen des Continents sondern selbst australischen Sammlungen voran. 94.05.23

[...]

Im Interesse der Sammlungen dürfte es erwünscht sein, wie den genannten 3 Schenkgebern & insbesondere dem Baron Dr. v. Müller in Melbourne von Seiten eines Hohen Kgl. Ministeriums der Dank für ihre wertvollen Schenkungen ausgeprochen würde.

Zugleich beehre ich mich, zu berichten, dass nunmehr die Sammlung gesichtet ist, welche Dr. Krauss von Tübingen von seiner im vorigen Jahr mit Gewährung des Freiherrl. v. Müller'schen Stipendiums in die Algerische Sahara unternommenen Reise zurückgebracht hat; dieses umfasst im Ganzen 431 Arten in c 1500 St., grossen Teils Insekten [...]

Verehrungsvoll
Prof. Dr. Kurt Lampert

Stuttgart den 23. Mai 1894

Royal Administration

I have the honour to give information of some valuable enrichments of the collections that the Royal Natural History Cabinet has received as presentations in recent months.

[...]

Baron Dr Ferdinand Mueller in Melbourne, who is untiringly active for the enrichment of the collection and from whom more or less significant consignments come to hand from time to time, has recently made a very valuable present to the Royal Natural History Cabinet by the sending of a specimen of the marsupial mole, *Notoryctis typhlops* Stirling. The animal was discovered a few years ago in the interior of Australia on the telegraph line between Adelaide and Port Darwin; one journey into the interior of Australia by Prof. Stirling undertaken specially on the search for the *Notoryctis* lead only to the capture of six complete specimens. They came by distribution by Prof Stirling to the British Museum in London, the Royal College of Surgeons in London, the museum in Stockholm and the museum in Cambridge. After the discovery becoming known, I turned to Baron Mueller with the request of having in view the eventual acquisition of a *Notoryctis* for the Natural History Cabinet and at the beginning of this year received information from Baron von Mueller that "to

94.05.23 his great joy" after years of effort he had succeeded in getting the animal, for which a 'fabulous price' was paid. Some weeks later the announced present arrived. Since our collection of Australian marsupials hitherto was already the most extensive and the most complete, especially through the continued presentations of Baron Mueller, it now stands at the fore not only of the continent but even of Australian collections through the possession of the *Notoryctis*.

[...]

In the interests of the collection it would be desirable that thanks for their valuable presentations be expressed on the part of the Royal Ministry to the above named three donors and in particular to Baron Dr von Mueller in Melbourne.

At the same time I have the honour to report that now the collection is sorted, which Dr Krauss from Tübingen has brought back from his journey in the Algerian Sahara undertaken last year with a grant of Baron von Mueller's scholarship. This comprises in total 431 species in about 1500 specimens, for the most part insects.

[...]

Regardfully
Prof. Dr. Kurt Lampert

Stuttgart 23 May 1894

Charles Naudin to the Académie des Sciences, Paris 95.06.24

Archives de l'Académie des Sciences, Paris.[1]

Notice sur les travaux de M. le B[ar]on Ferdinand Müller

Le B[ar]on Ferdinand von Müller comptera parmi les savants les plus laborieux de ce siècle. Arrivé jeune en Australie, il consacra toutes ses forces à l'étude des plantes de ce continent, où dejà quelques botanistes l'avaient précédé, mais auxquels la difficulté de pénétrer dans l'intérieur et le danger des explorations n'avaient guère permis de s'aventurer loin des côtes. Grâce à l'immigration croissante des européens, au recul des populations indigènes et à la fondation de colonies dont le développement a été rapide, grâce aussi à la bienveillance et aux encouragements des autorités locales, les recherches des explorateurs ont été graduellement facilitées et Fd Müller a su mettre à profit ces avantages pour pénétrer beaucoup plus loin que ses devanciers dans l'intérieur du pays. Sans arriver tout à fait à son extrémité septentrionale, il l'a scruté, avec un soin meticuleuse dans toute sa longueur de l'Est à l'Ouest, depuis le Queensland et la N[ouv]elle Galles du Sud jusqu'à la colonie de Perth, dans le sud-occidental du continent. On imagine sans peine que, dans une si vaste étendue de pays, et si peu explorée jusqu'alors, Fd Müller a dû faire de larges récoltes de plantes, la plûpart peu ou point connues, qu'il s'est empressé de partager avec les musées de l'Europe, fournissant par là aux monographes d'innappréciables ressources. Mais, tout en récoltant des plantes pour les herbiers, notre botaniste préparait déjà, par des observations faites sur place, les travaux de botanique systématique qu'il devait publier quelques années après.

Sa situation personnelle n'a du reste pas tardé à se consolider. Le Gouvernement anglais, en en faisant son botaniste attitré, le chargeait en même temps de la direction du Jardin botanique de

1 The file has a title page: 'Rapport sur les travaux de M. le baron Ferdinand Müller par M. Ch. Naudin (Seance du 24 juin 1895)' [Report on the works of Baron Ferdinand Müller by Mr Ch. Naudin (Meeting of 24 June 1895)].

95.06.24 Melbourne, la grande et florissante capitale de la Colonie de Victoria. C'est dans ce repos relatif qu'il a publié les travaux qui devaient lui faire prendre rang parmi les botanistes les plus en vue. Ils sont trop nombreux et trop divers pour que nous puissions les rappeler tous; il nous suffira de citer, entre les années 1860 et 1862, les deux volumes intitulés *Plants of the Colony of Victoria*, accompagnés d'un troisième volume de figures; en 1882 le *Systematic Census of Australian Plants*, recensement général de toutes les plantes d'Australie connues à cette date, présenté sous forme dichotomique, avec des compliments géographiques et chronologiques; de 1885 à 1888 une véritable flore de la Colonie de Victoria, en deux volumes, avec une carte botanique du pays, sous le titre de *Key of the system of Victorian Plants.* Peu avant cette dernière publication, l'auteur avait été chargé par le Gouvernement d'aller inspecter la nouvelle colonie de Perth et d'en faire connaître les ressources forestières et l'état du commerce des bois. Le résultat de cette exploration a fait la matière d'un grand volume in-folio, contenant, outre une minutieuse description des principaux arbres du pays, une série de planches très bien exécutées qui les représentent. La lecture en est instructive sous plus d'une rapport.

La flore australienne est si particulière, et si nettement différente de celle des autres régions de la terre par la prédominance de certains types d'organisation, que l'idée devait naturellement venir à Fd Müller d'entreprendre la monographie des groupes de plantes les plus caractéristiques du continent australien. C'était un travail considérable et que le nombre des espèces dans chaque genre, leur variabilité suivant les lieux d'origine et l'indécision de leurs limites rendaient particulièrement difficile. Plus d'un botaniste en aurait été découragé, mais Fd Müller n'a pas reculé devant cette tâche; il l'a vaillamment entreprise et a réussi à la mener à bien en nous donnant quatre grandes monographies accompagnées de figures lithographiques de toutes les espèces décrites: les Salsolacées australiennes, au nombre de 90; les Myoporinées, plus nombreuses encore; les Acacias et genres voisins, en 13 fascicules contenant chacun la description et la figure de douze espèces; enfin, de 1879 à 1884, sous le titre d'*Eucalyptographia*, une oeuvre magistrale, comprenant la description très détaillée de cent espèces d'Eucalyptus.

Ce travail, très soigné, est peut-être le plus important de Fd. Müller, qui pressentait depuis long-temps le rôle que les Eucalyptus devaient un jour remplir dans la sylviculture de touts les pays à climat tempéré-chaud de la terre. 95.06.24

Je passe sous silence une multitude de notes détachées et d'observations de tout genre communiquées par notre auteur aux sociétés savantes, pour rappeler un des traits les plus caractéristiques de sa vie de botaniste explorateur, et qui le distingue de la plûpart de ceux qui l'ont précédé. Plus qu'aucun autre il a compris que la Science ne doit pas se borner à des théories plus ou moins philosophiques, et qu'elle n'est complète qu'à condition de fournir des applications pratiques qui contribuent à la richesse publique comme au bien-être des particuliers. A ce point de vue, le rôle essentiel de la botanique est la diffusion, dans le monde entier, des plantes qui peuvent avoir une utilité quelconque, suivant les lieux, les climats, les besoins et les goûts de la société civilisée, surtout en ce qui concerne l'agriculture et le jardinage. Sous l'empire de cette idée Fd. Müller s'est appliqué d'abord à introduire les plantes de l'agriculture européenne en Australie, puis celles de l'Australie en Europe, visant surtout celles qui devaient trouver d'utiles emploies chez nous, ou fournir aux botanistes des types nouveaux d'organisation. Depuis plus de 30 ans il nous inonde, peut-on dire, de plantes australiennes, les unes pour les serres du nord, les autres pour les climats plus doux du midi méditerranéen en France et dans le nord de l'Afrique. On n'oubliera jamais que c'est à lui que la Provence, l'Algérie et la Corse sont redevables de ces superbes Eucalyptus, arbres précieux pour l'assainessement des localités insalubres, non moins précieux comme arbres de reboisement, par la rapidité avec laquelle ils produisent des bois d'oeuvre et du combustible. C'est par ses soins que la Villa Thuret, pour n'en pas citer d'autres, possède aujourd'hui plus de Quatre-vingts espèces de ces arbres, qui, déjà en grand nombre, produisent des graines, qu'on expédie jusque dans nos colonies d'Amérique. Faut-il ajouter que nous devons encore à Fd Müller une quantité d'Acacias, les uns de simple ornament, les autres plus utiles comme arbrisseaux ou arbres tannifères par leurs écorces, ou fourragers par leur feuillage, sans compter que quelques uns produisent une gomme presque ou tout à fait identique à celle du

95.06.24 Sénégal? Même dans la famille des Salsolacées Fd Müller a trouvé des plantes d'une haute valeur (*Atriplex halimoides, Chenopodium nitrariaceum* &c) qui feront un jour la fond de l'agriculture des terrains salants, et qui commencent à être appréciées en Algérie.

Fd Müller, anobli par le Gouvernement anglais qui lui a conféré le titre de Baron, est membre de toutes les Académies scientifiques et sociétés savantes de l'Europe et de l'Amérique, un seul honneur lui manque, celui d'être associé à notre Académie des Sciences.

Fin!
Ch. Naudin

Finis coronat opus.

Note on the works of Baron Ferdinand Müller

Baron Ferdinand von Müller will count among the most industrious savants of this century. Having arrived in Australia young, he devoted all his efforts to studying the plants of this continent, where some botanists had preceded him but whom the difficulty of penetrating the interior and the danger of exploring had scarcely allowed to venture far from the coast. Thanks to the increasing immigration of Europeans, the retreat of the indigenous populations and the foundation of colonies the development of which has been rapid, thanks also to the benevolence and encouragement of the local authorities, the researches of the explorers have been gradually facilitated and Ferdinand Müller knew how to profit from these advantages in order to penetrate much further than his predecessors into the interior of the country. Without quite reaching its northern extremity,[2] he has scrutinized with meticulous care its entire length from East to West, from Queensland and New South Wales to the colony of Perth, in the south-west of the continent. One imagines without difficulty that in such a vast extent of country, so little explored until then, Ferdinand Müller must have made large collections of plants, the majority little or not at all known, that he has hastened to share with the museums of Europe, thereby furnishing inestimable resources for monographs. But, even as he collected plants for herbaria, our botanist was already preparing, through observations made on the spot, the works of systematic botany that he was to publish some years later.

2 Naudin was evidently unaware of M's travels on the north coast of Australia as a member of the North Australian Exploring Expedition, 1855-6.

Moreover, his personal situation did not take long to consolidate. The English Government, while making him its accredited botanist, at the same time charged him with directing the Botanic Garden of Melbourne, the great and flourishing capital of the Colony of Victoria.[3] It is in this relative peace and quiet that he has published the works that have brought him to rank among the most prominent botanists. They are too numerous and too varied to be able to mention them all; it will suffice to cite, between the years 1860 and 1862, the two volumes entitled *Plants of the Colony of Victoria*, accompanied by a third volume of illustrations;[4] in 1882 the *Systematic Census of Australian Plants*, a general inventory of all the plants of Australia known at that date, set out in dichotomous form with geographical and chronological annotations;[5] from 1885 to 1888 a real flora of the Colony of Victoria, in two volumes with a botanical map of the country, under the title *Key of the system of Victorian Plants*.[6] Shortly before this last publication, the author was appointed by the Government to go and inspect the new colony of Perth[7] and to make known its forestry resources and the state of the timber trade. The result of this exploration became the subject of a large folio volume[8] containing, in addition to a minute description of the principal trees of the country, a series of very well executed plates that represent them. Reading it is instructive from more than one connection. 95.06.24

The Australian flora is so specific, and so clearly different from that of other regions of the Earth through the predominance of certain types of organization, that the idea must have come naturally to Ferdinand Müller to undertake monographic studies of the most characteristic groups of plants of the Australian continent. This was a considerable labour that the number of species in each genus, their variability according to place of origin and the uncertainty of their limits rendered particularly difficult. More than one botanist would have been discouraged, but Ferdinand Müller did not shrink from the task; he undertook it valiantly and succeeded in bringing it off well in giving us four large monographs accompanied by lithographic figures of the species described: the Australian Salsolaceae, 90 of them;[9] the Myoporineae, more numerous

3 M was not appointed Director of the Melbourne Botanic Garden until several years after he was appointed Victoria's Government Botanist.

4 B62.03.03, B63.13.06, B65.13.04.

5 B82.13.16, which is not, however, set out in dichotomous form. Naudin has here evidently confused M's *Census* with his *Key to the sytem of Victorian plants* (B88.13.03), which is set out dichotomously.

6 B86.13.01, B88.13.03.

7 i.e. WA.

8 B79.13.10.

9 B89.13.04, B90.13.14, B91.13.24.

95.06.24 still;[10] the Acacia and neighbouring genera, in 13 fascicles each containing the description and illustration of a dozen species;[11] finally from 1879 to 1884, under the title *Eucalyptographia*, a magisterial work comprising the very detailed description of a hundred species of Eucalyptus.[12] This work, very carefully done, is perhaps the most important of Ferdinand Müller, who sensed long ago the role the Eucalypts might one day play in the sylviculture of every country in the world with a warm-temperate climate.

I pass silently over a multitude of separate notes and observations of all kinds communicated by our author to learned societies to recall one of the most characteristic features of his life as botanist-explorer, one that distinguishes him from the majority of those who have preceded him. More than any other he has understood that Science ought not to confine itself to more or less philosophical theories, and that it is complete only in so far as it furnishes practical applications that contribute to both public wealth and the well-being of individuals. From this point of view, botany's essential role is the diffusion, throughout the whole world, of plants that have any kind of utility, according to the places, the climates, the needs and tastes of civilized society – above all in relation to agriculture and gardening. Under the impress of this idea Ferdinand Müller applied himself first to introduce European agricultural plants to Australia, then those of Australia to Europe, aiming above all at those that might find useful employment with us or furnish botanists with new types of organization. For more that 30 years he has inundated us, may it be said, with Australian plants, some for the greenhouses of the north, others for the gentler climates of the Mediterranean south of France and north Africa. It will never be forgotten that it is to him that Provence, Algeria and Corsica owe those superb Eucalypts, precious trees for the cleaning up of unhealthy localities, no less precious as reafforesting trees on account of the speed with which they produce working and fuel timbers. It is thanks to him that the Villa Thuret, not to cite others, today possesses over eighty species of these trees, which already produce seeds in great number that are sent to our American colonies. Need one add that we likewise owe to Ferdinand Müller a great many Acacias, some simply ornamental, others more useful as shrubs or trees in tanning through their bark or as forage through their foliage, without taking into account that some produce a gum almost or quite identical to that of Senegal? Even in the family of Salsolaceae Ferdinand Müller has found plants of high value (*Atriplex halimoides, Chenopodium nitrariaceum* &c) that will one day provide a basis for agriculture of marshy terrains, and that are beginning to be appreciated by Algeria.

10 B86.13.21.

11 B87.13.04, B88.13.01.

12 B79.13.11, B80.13.14, B82.13.17, B83.13.07, B84.13.19.

Ferdinand Müller, ennobled by the English Government which has conferred the title of Baron on him,[13] is a member of all the scientific Academies and learned societies of Europe and America. One honour only is wanting, that of being associated with our Academy of Sciences.[14] 95.06.24

The end!
Ch. Naudin

Finis coronat opus.[15]

Grave Monument Appeal[1] 97.00.00

In Memoriam.

"*Blessed are the pure in heart: for they shall see God.*" – Matt. V. 8.

BARON SIR FERDINAND VON MUELLER,
K.C.M.G., M.D., F.R.S., &c., Government Botanist of Victoria.
Born at Rostock, Germany, 30th June, 1825. Died at South Yarra, Victoria, 10th October, 1896.

The death of Baron Sir Ferdinand von Mueller has bereft Victoria of its most illustrious citizen, Australia of its most distinguished geographer, and the Scientific world of one of the most erudite, industrious, open-handed, pure-hearted and lovable phytologists

13 In fact the title had been conferred on M by the King of Württemberg. M's knighthood (KCMG) was his highest award from the British government.

14 M was elected a Corresponding Member of the Académie des Sciences, Paris, on 1 July 1895.

15 Traditional Latin proverb: 'The end crowns the work'.

1 MS is an illustrated pamphlet, each page black edged, published by Alex M'Kinley & Co., Printers, Melbourne, n.d. The illustrations are: on the first page, a portrait of M by Troedel [see Figure 19]; on the second page, a photograph captioned 'Floral tributes on the grave.'; and on the fourth (and last) page a picture captioned 'Monument to be erected on the grave.' The monument erected (Figure 20) closely followed that depicted.

97.00.00 the present century has produced. The dominant wish expressed by the deceased *savant* of late years was, that he might be privileged to die in harness as Government Botanist, and the knowledge that his wish was gratified somewhat mitigates the sadness of the event. Although not in good health he continued to discharge his official duties until seized with brain paralysis, brought on by official worry and overwork, when, after a brief illness, he peacefully "fell asleep," at his residence, Arnold-street, South Yarra, at 20 minutes past one o'clock on the morning of Saturday, 10th October, 1896. The news of his unexpected death evoked expressions of deep sorrow from every section of the public Press, and his Executors (Rev. W. Potter, F.R.G.S., Alexander Büttner, M.D., F.R.C.S., etc., and Hermann Büttner, Esq.) received telegrams and floral tributes from the Governments and the Learned Societies of all the Australian Colonies, from the Consuls of several Foreign Powers, and from a large body of representative citizens – showing the esteem in which his many virtues and high scientific attainments were held.

On Sunday (11th October) reference was made in many of the Melbourne churches to the loss the community had sustained by Sir Ferdinand's demise. At the close of the evening service in the Australian Church, the Rev. Charles Strong, D.D., announced that, by request, "the organist would play the 'Dead March,' as a tribute to the memory of the late Baron von Mueller, a gentleman who, by his high attainments, had added largely to the scientific knowledge of the world. His life had been stainless, and he was now at rest. His death was a national loss." The spacious church was filled to over-flowing, and the whole congregation stood reverently during the performance.

Amongst the most exquisitely beautiful of the numerous floral devices received for the obsequies were those sent by His Excellency the Governor and Lady Brassey, the Governments of Victoria, South Australia, Western Australia, New South Wales and Queensland, the Consuls for Germany, Austria-Hungary, France and Denmark, the Council of the Victorian Branch of the Royal Geographical Society of Australasia, the Director of the Melbourne Botanical Gardens, the Director of the Sydney Botanical Gardens,

Figure 20. The Governor-General, Lord Hopetoun, unveiling the monument over Mueller's grave, 26 November 1901. Mueller was buried at St Kilda Cemetery, Melbourne, on 13 October 1896. Courtesy of the Library, Royal Botanic Gardens Melbourne.

97.00.00 the German Club, the Melbourne University Science Club, the Melbourne Liedertafel, the Melbourne,[2] Deutscher Turn Verein, the Field Naturalists Club, Sir Arthur and Lady Snowden, Sir William and Lady Clarke, the deceased nobleman's widowed sister and nephew (Mrs. Dr. Wehl and Mr. W. M. Wehl, South Australia), Mr. and Mrs. E. G. Fitz Gibbon, the Baron's executors, Dr. and Mrs. Rudall, Dr. and Mrs. Lewellyn, Reverends Hermann Herlitz and W. Fielder, Mrs. and Miss Ward Cole, Miss J. Law, Mrs. Varley, Mr. and Mrs. Troedel, Madame and Mademoiselle Dreyfus, and the Right Worshipful the Mayor and Lady Mayoress of Melbourne.

The funeral took place on Wednesday, 13th October, when the body was interred in the St. Kilda Cemetery, in a prominently situated plot of ground which had been presented to the Baron by the Trustees only a few days before the attack of his fatal illness. The cortége was solemnly imposing. Preceding the hearse were the band and members of the Deutscher Turn Verein, carrying the banner of the society, of which the deceased was a member. Both hearse and floral car were heavily laden with floral tributes. The coaches for the chief mourners came next, in which were seated the late Sir Ferdinand's nephew, his executors, members of the Council of the Geographical Society, and some of his oldest personal friends: then followed the coaches occupied by representatives sent by the Cabinet of this and of the other Australian colonies, and in addition to these a long line of private carriages and other vehicles. The band played the "Dead March" *en route* to the cemetery and also at the grave.

The following were the pall-bearers: – Professor Sir Frederick M'Coy, K.C.M.G., M.A., D.Sc., F.R.S. (representing the Royal Society), Professor William C. Kernot, M.A., C.E. (representing the University of Melbourne), Charles A. Topp, Esq., M.A. (representing the Government of Victoria), Joseph A. Panton, C.M.G., P.M. (representing Western Australian Government), Consul W. A. Brahe

2 *sic.*

(Imperial German), Consul Carl Pinschof (Austria-Hungary), Hon. 97.00.00
A. O. Sachse, M.L.C. (representing Victorian Branch of the Royal Geographical Society of Australasia), Capt. F. B. Gipps, C.E. (representing New South Wales Branch of the Royal Geographical Society of Australasia), J. M. Bruce, Esq. (President of the Melbourne Liedertafel), the Hon. Sir William Clarke, Bart., M.L.C., Alexander Büttner, M.D., F.R.C.S., etc., and Mr. W. Rudall, F.R.C.S., Eng., (representing oldest private friends).

The grave was bedded with roses, and around it was a raised platform for the officiating clergyman and the chief mourners. Standing near the bier, the members of the Melbourne Liedertafel gave a most touching rendering of the beautiful part song (Fr. Abt, "Evening"): –

Day again his race has run,
And the tranquil night comes on,
Soothingly her pinions waving,
Heated hill and valley laving,
Spreading from her starry chart
Holy peace to every dwelling,
Holy peace to every heart.

Followed by the "Hymn to Night" (Beethoven): –

Holy night, O soothe my breast
With thy solemn tranquil calm;
Give the weary wanderer rest,
O'er his sorrows shed thy balm.
Stars brightly o'er me gleaming,
From azure fields are dreaming;
They set my soul a-dreaming
Of rest above.

The Rev. Hermann Herlitz read the burial service, and delivered an impressive address, expatiating upon the nobleness of the deceased botanist's life, the absorption of mind and energy in his beloved science and its cognate sciences, his sympathy with the distressed, and his unbounded generosity. After which the requiem (Beethoven, arr. by Mr. J. King) was rendered with great sweetness by the Liedertafel –

97.00.00 Now the labourer's task is o'er,
Now the battle day is past;
Now upon the farther shore,
Lands the voyager at last.
Father, in Thy gracious keeping
Leave we now Thy servant sleeping.

There the tears of earth are dried,
There its hidden things are clear;
There the work of life is tried
By a juster Judge than here.
Father, in Thy gracious keeping
Leave we now Thy servant sleeping.

"Earth to earth and dust to dust:"
Calmly now the words we say,
Leaving him to sleep in trust
Till the resurrection day.
Father, in Thy gracious keeping
Leave we now Thy servant sleeping. – AMEN.

In his address at the Annual Commencement (1897) of the University of Melbourne, the Chancellor (Sir Anthony Brownless K.C.M.G., M.D.) said: "When speaking of science I cannot refrain from paying a tribute of respect to the memory and to the great talents of my late dear old friend, the Baron Sir Ferdinand von Mueller, whose fame as one of the greatest of Scientific Botanists of the present age was world-wide. Whilst the benefits which he conferred on the Australian colonies by the dissemination of the most practical information to agriculturists and others ought to cause him to be remembered with the deepest gratitude in these colonies, he will also be long and kindly remembered in many other countries for the benefits he conferred upon them by the introduction of the Eucalyptus as the destroyer of malaria and the conservator of human life."

Looking back upon the world-wide influence on botanical questions exerted by the late eminent phytologist while living who that enjoyed the favour of his personal friendship or appreciated his rare gifts of mind but must join in the aspiration so aptly expressed in the columns of the *Australasian*:[3] –

3 *Australasian*, 5 December 1896, p. 1115.

FERDINAND VON MUELLER. 97.00.00

'Twere sad if minds endowed with all the lore
Of science should at death forever die:
Fain would one hope that, far beyond the shore
Of Death's black river some fair land may lie
Where the fresh realms of nature may explore
With more than mortal power of brain and eye:
So may a flower transplanted bloom more fair
On richer soil, in a more genial air.

J. D. WOOD.

Scientists throughout the world, and all the Baron's personal friends, will be pleased to hear that his Executors (Rev. W. Potter, F.R.G.S., Alexander Büttner, M.D., F.R.C.S., etc., and Hermann Büttner, Esq.) are now making an effort to erect over the grave a monument worthy of the deceased *savant's* fame. The monument will be of gray granite, 23 feet in height, all highly polished, and surrounded by an ornamental iron railing. They will also be pleased to know that the Baron's supplemental volume of the *Flora Australiensis*, upon which he had worked for years and was preparing for the Press at the time of his death, together with two volumes on his administration as Director of the Botanical Gardens, embracing a biography and complete bibliograph of his writings, are to be published.[4] His executors will feel favoured by the loan of any of his letters, or the communication of incidents in the Baron's life which his friends deem to be worthy of notice in his biography.

W. POTTER.

ADDRESS – REV. W. POTTER,
"VONMUELLER," ARNOLD-ST., SOUTH YARRA.

4 None of these promised works were ever published.

97.11.00 *Baron von Mueller National Memorial Fund*[1]

E97/11925, unit 972, VPRS 3992/P inward registered correspondence, VA 475 Chief Secretary's Department, Public Record Office, Victoria.

BARON VON MUELLER NATIONAL MEMORIAL FUND.

At a public meeting held in the Town Hall of Melbourne on Monday, 15th November, 1897, at which the Mayor of Melbourne presided, a Committee[2] was appointed to secure the establishment of some permanent memorial to commemorate the great services rendered by the late Baron von Mueller.[3]

A movement has already been initiated by the Executors of the late Baron to secure funds with which to erect a Tombstone, and, whilst sympathising with this object, the Committee now appointed, to avoid confusion, desires it to be understood that the **Tombstone Fund**, with which the Executors are alone concerned, and the **National Memorial Fund**,[4] with which the present Committee is concerned, are quite distinct from one another.[5]

Whilst nominally the Government Botanist of Victoria, it is well known that the Baron von Mueller's assistance was sought by and always freely given not only to public bodies but to private individuals in all parts of Australia. Apart from his purely scientific work, upon the value of which it is unnecessary to dwell, the Baron von Mueller devoted himself to the development of the more practical side of various branches of work, such as those connected with Forestry, Agriculture, Horticulture, Pharmacy and, not least,

1 Printed circular sent to C. Topp, Under Secretary, as an enclosure with W. Baldwin Spencer to C. Topp, 26 November 1897.

2 The names of the members of the General Committee and those of the Executive Committee are listed on a printed sheet enclosed with this letter.

3 some permanent ... Mueller *is underlined presumably by C. Topp.*

4 Tombstone Fund *and* National Memorial Fund *are printed in bold type.*

5 *Most of this paragraph has been underlined, presumably by Topp.*

Geographical Exploration. His own explorations in early days, both in Northern Australia as botanist in the expedition under Mr. A. C. Gregory,[6] and when, subsequently, he traversed alone the then little known wilds of Gippsland,[7] were of considerable importance, and his deep interest in and the practical assistance which he rendered to the explorations of others are well known. 97.11.00

Not only did he spend his whole life in the furtherance of the work in which, from the nature of his position, he was most deeply interested, but he devoted practically the whole of his income to the assistance of those who were engaged in work the object of which was to increase our knowledge of the nature and products of Australasian lands.

The object of this Committee is to secure sufficient funds to allow of the establishment of some permanent Memorial which shall worthily perpetuate his name;[8] and whilst it is not possible as yet to state definitely the form which the Memorial will take, it is hoped that sufficient funds will be forthcoming to provide for (1) the erection of some form of statue, and (2) the endowment of a Medal, Prize or Scholarship, to be associated with Baron von Mueller's name and to be awarded from time to time in recognition of distinguished work in the special branches in which he was most deeply interested, and which shall be open to workers throughout the Australasian colonies.[9]

6 North Australian Exploring Expedition, 1855-6.

7 M's major exploring expeditions in Victoria were undertaken before he went with Gregory.

8 The object ... name *is underlined presumably by Topp.*

9 No statue was erected. At the meeting of the Council of the Australasian Association for the Advancement of Science held in Hobart on 8 January 1902, the proposal of the Committee of the Baron von Mueller National Memorial Fund was agreed to, that the money collected by the committee should be placed in the hands of the Association 'for the purpose of founding a medal ... to be awarded not more frequently than every second year to the author of the most important contribution, or series of contributions, to natural knowledge, published originally within His Majesty's dominions within a period of not more than five, nor less than one year, of the date of the award, preference being always given to work having special reference to Australasia'.

97.11.00 *Subscriptions to the Fund may be sent to the Hon. Treasurer, addressed to the College of Pharmacy, Swanston Street, Melbourne, or to the Hon. Secretaries, addressed to the University of Melbourne, and will be duly acknowledged.*

We have the honour to remain,
Your obedient Servants
W. WIESBADEN.
W. BALDWIN SPENCER. } Hon. Secs.

Appendix A: Biographical Register

This appendix provides biographical notes for Mueller correspondents and individuals mentioned in the present volume of letters. If, however, an entry was provided in Volumes 1 or 2, it is not repeated here. Entries are not provided, either, for individuals for whom no further information could be found, except in cases where there is evidence that the person was a correspondent of Mueller, where Mueller named a plant after the person, or where there is reference to the person in several items of correspondence.

The entries are arranged alphabetically by surname. The main entries for titled individuals are given under their first names, e.g. HAWAII, KING OF. See KALAKAUA, or under the names of their seats, e.g. LORNE, Marquis of. The main entries for women are usually given under their maiden names, e.g. RYAN, CLARA OAKLEY. See Clara Oakley GRAHAM.

Wherever possible all given names are included. Where the most commonly used name is not the first name, it is in italics, e.g. KÖRNER, KARL *THEODOR*. Variations in spellings are given in square brackets, e.g. HOLTZE MORITZ WILHELM [MAURICE WILLIAM].

Where an individual's gender is known but the given name is not, titles such as 'Mr' and 'Miss' are included, e.g. BAHSE, MR F.

Typical entries include an individual's years of birth and death and main occupations, and brief notes that focus on the nature and period of his or her contact with Mueller. Plant names introduced by Mueller to commemorate an individual are also listed, based on Appendix C, Volume 1. The words 'M correspondent' indicate that at least one letter to or from Mueller and the individual named is known to exist; 'M correspondent*' indicates that at least one letter is in Volume 3.

Sources are given in square brackets. Author (date) references are cited in full in the Bibliography. 'B' numbers refer to items in Appendix B, Volume 1. 'L' numbers refer to items (usually letters) in the main sequence.

The names of non-English universities, botanic gardens and herbaria have been translated and standardized. The names of other non-English institutions and societies are translated into English, usually with the non-English title given in parentheses, e.g. Natural History Museum, Paris (Muséum d'Histoire Naturelle).

Place-names are given in English, e.g. 'Florence' not 'Firenze'. The names of individual Australian states are preferred to the use of 'Australia'. Modern changes in place names are noted in parentheses, e.g. Buitenzorg (now Bogor). *Australia 1:250,000 map series gazetteer* (1975) has been used as a guide in determining modern Australian place-names, and *The Times index-gazetteer of the world* (1965) for other place-names.

Abbreviations

AAAS = Australasian Association for the Advancement of Science.
AB = BA.
ADB = *Australian dictionary of biography* (1966-90), 12 vols.
ALS = Associate of the Linnean Society, London.
AM = MA.
ANB = *American national biography* (1999), 10 vols.
b. = born.
BA = Bachelor of Arts.
BCE = Bachelor of Civil Engineering.
BD = Bachelor of Divinity.
BCL = Bachelor of Civil Law.
BS = BSc.
BSc = Bachelor of Science.
c. = circa.
Capt. = Captain.
CB = Companion of the Order of the Bath.
CBE = Commander of the Order of the British Empire.
ChB = Bachelor of Surgery.
ChD = Doctor of Surgery.
ChM = Master of Surgery.
CMG = Companion of the Order of St Michael and St George.
DAB = *Dictionary of American biography* (1955-).
DBE = *Deutsche biographische Enzyklopaedie* (1995-).
DBF = *Dictionnaire de biographie française* (1933-).
DBI = *Dizionario biografico degli Italiani* (1960-).
DCL = Doctor of Civil Law.
DD = Doctor of Divinity.
Desmond = Desmond (1994).
d. = died.
DNB = *Dictionary of national biography* (1917-90), 22 vols.
DrMed = Doctor of Medicine.
DrPhil = Doctor of Philosophy.
DrSciNat = Doctor of Natural Science.
DSB = *Dictionary of scientific biography* (1970-90), 17 vols.
DSc = Doctor of Science.
FHS (Vic.) = Fellow of the Horticultural Society of Victoria.
fl. = flourished.
FLS = Fellow of the Linnean Society, London.
FNCV = Field Naturalists Club of Victoria.

FPS = Fellow of the Pharmaceutical Society.
FRAS = Fellow of the Royal Astronomical Society, London.
FRCP = Fellow of the Royal College of Physicians, London.
FRCS = Fellow of the Royal College of Surgeons, London.
FRGS = Fellow of the Royal Geographical Society, London.
FRGSA = Fellow of the Royal Geographical Society of Australasia.
FRHS = Fellow of the Royal Horticultural Society, London.
FRS = Fellow of the Royal Society, London.
FZS = Fellow of the Zoological Society, London.
GCB = Knight Grand Cross of the Order of the Bath.
GCMG = Knight Grand Cross of the Order of St Michael and St George.
ISO = Companion of the Imperial Service Order.
JP = Justice of the Peace.
Jr = junior.
KBE = Knight Commander of the Order of the British Empire.
KCB = Knight Commander of the Order of the Bath.
KCIE = Knight Commander of the Order of the Indian Empire.
KCMG = Knight Commander of the Order of St Michael and St George.
Lic. Med. = Licentiate of medicine.
Lieut. = Lieutenant.
LLB = Bachelor of Laws.
LLD = Doctor of Laws.
LRCP = Licentiate of the Royal College of Physicians.
LSA = Licentiate of the Society of Apothecaries, London.
M = Mueller.
MA = Master of Arts.
MB = Bachelor of Medicine.
MCE = Master of Civil Engineering.
MD = Doctor of Medicine.
MEL = National Herbarium of Victoria, Melbourne.
MLA = Member of the Legislative Assembly.
MLC = Member of the Legislative Council.
MP = Member of Parliament.
MRCS = Member of the Royal College of Surgeons.
NBG = *Nouvelle biographie générale* (1853-66), 46 vols.
NDB = *Neue deutsche Biographie* (1952-).
NSW = New South Wales, Australia.
NT = Northern Territory, Australia.
NTDB = *Northern Territory dictionary of biography* (1990-6), 3 vols.
NZ = New Zealand.
NZDB = *Dictionary of New Zealand biography* (1990), 2 vols.
OM = Order of Merit.
PhD = Doctor of Philosophy.
PROV = Public Record Office, Victoria.
PRS = President of the Royal Society, London.
Qld = Queensland, Australia.
RBG, Melb. = Royal Botanic Gardens Melbourne.
RGS = Royal Geographical Society, London.
RGSA = Royal Geographical Society of Australasia.
SA = South Australia.
Tas. = Tasmania, Australia.
TL2 = Stafleu & Cowan (1976-88), 7 vols; Stafleu & Mennega (1992-2000), 6 vols.
VBOT = Library, Royal Botanic Gardens Melbourne.
Vic. = Victoria, Australia.
WA = Western Australia.
WW = *Who was who* (1916-).
WWA = *Who's who in America* (1900-).

ADANSON, MICHEL, 1727-1806, French naturalist. Educated at the Royal College (Collège Royal) and the Royal Garden (Jardin du Roi), Paris. Scientific expedition to Senegal, 1749-53. Elected to the Royal Academy of Sciences, Paris (Académie Royale des Sciences), in 1759. In his *Familles des plantes* (1763-4) he was an early advocate of a natural system of classification [DSB].

AGNEW, JAMES WILSON, 1815-1901, medical practitioner and politician. Born in Ireland. Studied medicine in London (MRCS, 1838), Paris and Glasgow (MD, 1839). Went to Australia in 1839. Assistant surgeon, later colonial surgeon, Tas., 1841-53. Medical practitioner, 1853-77. Secretary, Royal Society of Tas., 1861-81, 1884-94. MLC, 1877-81, 1884-7; Premier, 1886-7. KCMG, 1895 [ADB]. M correspondent*.

ALBERTIS, LUIGI MARIA D'. See Luigi Maria D'ALBERTIS.

ALDRIDGE, ROSA ANNIE. Married John Harry Grainger, 1880. Mother of Percy GRAINGER [ADB]. M correspondent*.

ALLIONI, CARLO, 1728-1804. Italian botanist. Studied medicine at Turin. Profesor of botany at Turin from 1760 [DBI].

ANDREWS, HENRY CHARLES, fl.1799-1830, English botanical painter and engraver [TL2].

ARMSTRONG, WILLIAM E., farmer, of Waranga, Vic. M correspondent*.

ARMYTAGE, ADA, 1859-1939, of Como, South Yarra, Vic. Daughter of Charles Henry Armytage, pastoralist [Selzer (2003)]. M correspondent*.

ASCHERSON, PAUL FRIEDRICH AUGUST, 1834-1913, German botanist. Born in Berlin. Studied at Berlin (DrMed, 1855). Assistant at the Berlin Botanic Garden, 1860-76; professor of botany and plant geography, Berlin, from 1863 [NDB]. M correspondent*.

ASKENASY, EUGEN, 1845-1903, German botanist. Studied agriculture at Hohenheim and then botany at Heidelberg (DrPhil, 1866). Professor of botany at Heidelberg from 1881. An expert on algae [NDB].

AUBLET, JEAN BAPTISTE CHRISTOPHE FUSÉE, 1723-78, French explorer and botanist. Studied pharmacy at Montpellier and botany at Paris. Pharmacist and botanist for the Company of the Indies (Compagnie des Indes) at Île-de-France (Mauritius), 1752-61, and in Guyana, 1762-5 [DBF].

AUCHER-ÉLOY, PIERRE MARTIN RÉMI, 1793-1838, French traveller and botanist. Studied pharmacy at Orléans and Paris. Proprietor of a bookshop and printery at Blois and then from 1826 in Paris. Thereafter travelled extensively in western Asia until his death [DBF]. Specimens are in the Melbourne Herbarium [L80.08.04].

AYRES, PHILLIP BURNARD, 1813-63, English medical officer and botanist. MD, London, 1841. Physician, Islington Dispensary, 1851. Superintendent of quarantine at Mauritius, 1856-63. An expert on cryptograms [Desmond]. Plant contributor to the Melbourne Herbarium [B61.02.01].

BABINGTON, CHURCHILL, 1821-89, British clergyman and botanist. Educated at Cambridge (MA, 1846; DD, 1879). Rector at Cockfield, Suffolk, from 1866. Expert on lichens [Desmond].

BÄUERLEN, WILHELM [WILLIAM], 1840-1917, botanical collector. Born in Germany. In Australia by 1883. Botanical collector on *Bonito* expedition to New Guinea, 1885. Collector in NSW and Vic. for Technological Museum, Sydney, 1890-1905 [Cohn (forthcoming), Desmond, Hall (1978)]. Plant contributor to the Melbourne Herbarium. *Correa bauerlenii, Dendrobium baeuerleni, Eucalyptus baeuerleni, Eugenia baeuerlenii, Eugenia bauerlenii, Halorgis baeuerlenii*. M correspondent*.

BAGE, EDWARD, engineer and surveyor. Born in Britain. In Sierra Leone in government service, 1850. Went to Melbourne, 1853. Government surveyor in western Vic. Surveyed suburbs in outer Melbourne from 1860 [Ryan (2001)]. M correspondent*.

BAHSE, MR F., Commissioner for Saxony at Centennial Exhibition, Melbourne, 1889.

BALE, WILLIAM MOUNTIER, 1851-1940, microscopist and inspector of customs. Specialized in the study of hydroids. Fellow, Microscopical Society, London, 1882. A founding member of the Field Naturalists Club, Vic. [Barnard, (1906)]. Named *Sertullaria muelleri* for M [Darragh (1996)]. Specimens are in the Melbourne Herbarium.

BALFOUR, ISAAC BAYLEY, 1853-1922, Scottish botanist. Son of J. H. BALFOUR (vol. 1). Educated at Edinburgh (BSc, 1873; DSc, 1875; MD, 1877). Professor of botany, Glasgow, 1879-84; Oxford, 1884-8; Edinburgh, 1888-1922. FRS, 1884; FLS, 1875. KBE, 1920. Authority on Asiatic plants [Desmond]. M correspondent*.

BALL, JOHN, 1818-89, Irish botanist. MA, Dublin, 1847. Trained as a barrister but never practised. Travelled in Morocco with Joseph HOOKER (vol. 1) in 1871, and in North and South America in the 1880s. FLS, 1856; FRS, 1868 [Desmond].

BANCROFT, JOSEPH, 1836-94, medical practitioner and experimental biologist. Born in England. Educated at Manchester Royal School of Medicine and Surgery (MRCS, LSA, 1859); MD, St Andrews, 1859. Joined Philosophical Society of Qld in 1864, president in 1882-3 when it became the Royal Society of Qld. President, Qld Medical Board, 1882-94. Discovered parasitic worm that causes filariasis and found *Duboisia myoporoides* to be a source of atropine [ADB]. Plant contributor to the Melbourrne Herbarium [B86.06.02, p.142]. *Isandra bancroftii.* M correspondent.

BARNARD, FRANCIS GEORGE ALLMAN, 1857-1932, pharmacist and naturalist. Born in Melbourne. A founder of the Field Naturalists' Club of Victoria and office-bearer for 40 years [Hall (1978)]. Specimens are in the Melbourne Herbarium. M correspondent*.

BASEDOW, MARTIN PETER *FRIEDRICH*, 1829-1902, teacher, newspaper proprietor and politician. Born in Germany. Arrived in SA, 1848. Opened school in Tanunda, SA; chairman, Tanunda district council, 1864-76. To Adelaide, 1874. Member, SA House of Assembly, 1876-90; MLC, 1894-1900. President and trustee of the SA German Club [ADB].

BAUDIN, NICOLAS THOMAS, 1754-1803, French surveyor and naturalist. Entered French navy, 1774. Collected plants in an expedition to the Indian Ocean, 1794. Led expedition to complete the French cartographic survey of the coast of Australia, 1800-3 [ADB].

BAUHIN, GASPARD [CASPAR], 1560-1624, Swiss anatomist and botanist. Educated at Basle, Padua, Bologna and Paris. Professor of Greek from 1582 and *consiliarius* in anatomy, Basle, 1584-1624; professor of the practice of medicine from 1615 [DSB].

BEHM, ERNST, 1830-84, German geographer, publisher and editor. Studied medicine at Jena, Berlin and Würzburg; graduated 1853. Became interested in geography and in 1856 took over the leadership of the *Geographischen Mitteilungen* of the Geographical Institute of Gotha. Founded the *Geographische Jahrbuch*, 1866 [DBE]. M correspondent*.

BENGOUGH, GEORGIANA, wife of FERDINAND VON HOCHSTETTER (vol. 1). M correspondent*.

BERNAYS, ALBERT JAMES, 1823-92, British chemist. DrPhil, Giessen. Analyst at Derby, 1852. Lecturer in chemistry, St Mary's Hospital, London, 1855-60. Lecturer/professor of chemistry, St Thomas' Hospital, London, 1860-92. Public analyst to various bodies. Brother of Lewis BERNAYS.

BERNAYS, LEWIS ADOLPHUS, 1831-1908, public servant. Born in England. Educated at King's College London. Went to NZ in 1850, then to Sydney in 1852. Clerk of Qld Legislative Assembly, 1860-1908. One of the founders of the Qld Acclimatisation Society. CMG, 1892. FLS. Brother of Albert BERNAYS [ADB]. *Microstylis bernaysii, Uncaria bernaysii.* M correspondent.

BERRY, ANDREW, 1825-1905, administrator. School's collector, Ballarat district, Vic., 1882. Acting Registrar, Ballarat School of Mines, 1882, then Registrar, 1882-95 [Perry (1984)]. M correspondent*.

BERTHOLD, GOTTFRIED DIETRICH WILHELM, 1854-1937, German botanist and algologist. Studied at Würzburg, 1874-7; Göttingen, 1877-8; DrPhil, 1878; assistant, Institute of Plant Physiology, 1878-9. Assistant, Zoological station, Naples, 1879-81, 1884-6. Extraordinary professor, Göttingen, 1885-7; regular professor and director of institute for plant physiology, 1887-1923 [TL2]. M correspondent.

BETCHE, ERNST, 1851-1913, gardener. Born at Potsdam, Germany. In Municipal Gardens, Berlin and in Ghent, 1874. Collected plants in Samoa, 1880-1, and for Sydney Botanic Garden, 1881-97. Botanical Assistant, Sydney, 1897 [Desmond]. Plant contributor to the Melbourne Herbarium [e.g. B85.11.02]. *Bulbophyllum betchei, Corysanthes betchei, Jasminum betchei, Oncinocalyx betchei.*

BEVAN, THEODORE FRANCIS, 1860-1907, botanist, explorer and writer. Born in England. Went to NZ, 1861, then to Australia. Made several voyages to New Guinea and explored many south-coast rivers. Collected plants in New Guinea, 1884-7 [Desmond]. New Guinea plants are in the Melbourne Herbarium. *Mussaenda bevani.* M correspondent.

BLACKETT, CUTHBERT ROBERT, 1831-1902, pharmacist and politician. Born in England. Apprenticed to a pharmacist in Norfolk. Went to Melbourne in 1853; established a pharmacy in Fitzroy. MLA, 1879-80, 1881-3. President of Pharmacy Board from 1883; Government analytical chemist from 1887 [Thomson & Serle].

BOCK, HIERONYMUS, 1498-1554, German physician, botanist and teacher. Prefect of the Zweibrücken Garden [TL2].

BOISSIER, PIERRE EDMOND, 1810-85, Swiss botanist and traveller [TL2].

BOIVIN, LOUIS-HYACINTHE, 1808-52, French botanist. Studied medicine for a time. Collected in Senegal, East Africa, Zanzibar, Île de Bourbon (Réunion) and Madagascar [DBF]. Specimens are in the Melbourne Herbarium [L80.08.04].

BOJER, WENCESLAS, 1797-1856, Czech naturalist. Settled on Mauritius; director of the Port Louis Botanic Garden [TL2].

BORCHGREVINK, CARSTEN EGEBERG, 1864-1934, surveyor, scientist and Antarctic explorer. Born in Norway. Educated at Gjertsen College and in Tharandt, Saxony. Went to Australia, 1888. With survey teams in Qld and NSW. Taught languages in NSW, 1892-4. Explored in Antarctica, 1895, 1899. FRGS [ADB].

BORNET, JEAN-BAPTISTE *ÉDOUARD*, 1828-1911, French botanist. Collaborator for many years with G. Thuret at the 'Villa Thuret' at Antibes. Later settled in Paris. An expert on cryptogams. Member of the Academy of Sciences (Académie des Sciences), Paris, 1886 [DBF]. M correspondent*.

BOURGEAU, EUGÈNE, 1813-77, French botanical collector. Collector for the Botanical Association of France (Association Botanique); collected in Spain on several occasions; in North America, 1857-9; around the Mediterranean basin, 1860-2; in Mexico, 1865-7; in the Balearic Islands, 1869 [DBF]. Specimens are in the Melbourne Herbarium [L80.08.04].

BOYD, WILLIAM ALEXANDER JENYS, 1842-1928, journalist, schoolmaster and soldier. Born in Paris. Educated in England, Germany, Switzerland and France. Went to Qld, 1860. Headmaster in Qld schools, 1874-5, 1883-9, 1891-3. Editor, *Queensland agricultural journal*, 1897-1921. Amateur soldier, 1885-97; commanded Brisbane Garrison Artillery from 1891. Member, RGSA, Qld branch [ADB].

BRANDIS, DIETRICH, 1824-1907, forester. Born and educated at Bonn (DrPhil, 1848). Superintendent of Forests, Pegu, India, 1856. Inspector-General of Indian Forests, 1864-83; founded the forestry school at Dehra Dun, 1878. FLS, 1860; FRS, 1875. KCIE, 1887 [Desmond]. M correspondent.

BRASSEY, THOMAS, 1836-1918, English colonial governor. Educated at Rugby and University College, Oxford; BA, 1859; MA, 1864; DCL, 1888. MP for Hastings, 1868-86. Civil Lord of the Admiralty, 1880-4, and its parliamentary secretary, 1884-5. KCB, 1881. Baron Brassey, 1886. Governor of Victoria, 1895-1900. GCB, 1906. Created 1st Earl Brassey, 1911 [ADB].

BRIDE, THOMAS FRANCIS, 1849-1927, librarian. Born in Ireland. Went to Melbourne as an infant with his parents. Educated at University of Melbourne (LLB, 1873; LLD, 1879). Assistant librarian there, 1873-81. Librarian, Melbourne Public Library, 1881-95 [ADB].

BRITTEN, JAMES, 1846-1924, English botanist. FLS, 1870. Assistant at Kew Herbarium, 1869-71, and in Botany Department, British Museum (Natural History), 1871-1909 [Desmond]. *Ternstroemia britteniana*.

BRODRIBB, William Adams, 1809-86, pastoralist and politician. Born in England. Went with his family to Tas. in 1818 to join his father, who had been transported in 1817. Went to NSW in 1835, managing and later acquiring various pastoral properties in the Riverina district. MLA, Vic., 1861-2. MLA, NSW, 1880-1; MLC, 1881-6 [ADB]. M correspondent*.

BROOKING, John Sherlock, 1840-1916, surveyor. Born in England. Arrived in WA from NZ, 1863. Surveyed in the Kimberley district and laid out the town of Broome. Became Surveyor-General of WA [Gibbney & Smith (1987)]. M correspondent*.

BROOKS, John Paul, 1848-1930, pastoralist and telegraph linesman. Born in Melbourne. Educated at Geelong Grammar School, Vic. At first a jackaroo and then a dairy farmer. Went to WA, 1874, and was pioneer pastoralist at Balbinia station. Linesman at the Israelite Bay telegraph station, 1877-84. Brother of Sarah Brooks [Hall (1984)]. M correspondent*.

BROOKS, Sarah Theresa, 1850-1928, artist and botanical collector. Born in England. In Vic. 1851-74, WA from 1874. Plant contributor to the Melbourne Herbarium. Sister of John Brooks [Archer & Maroske (1996), Cohn (forthcoming), Desmond]. *Hakea brookeana*, *Scaevola brookeana*. M correspondent*.

BROOME, Frederick Napier, 1842-96, colonial governor. Educated in Shropshire. Went to NZ, 1857. Returned to England, 1864, and to NZ 1865, farming in the South Island until 1868. Took up writing as a career. Appointed Colonial Secretary, Natal, 1875. CMG, 1877. Colonial Secretary, Mauritius, 1878; Lieutenant-governor, 1880. Governor, WA, 1883-9. Acting Governor, Barbados, 1890. Governor, Trinidad, 1891-96 [ADB]. M correspondent*.

BROTHERUS, Viktor Ferdinand, 1849-1929, Finnish bryologist. High school teacher in Helsinki [TL2].

BROWN, George, 1835-1917, English missionary. Went to NZ, 1855, to Samoa, 1860, and to New Britain, 1877. Published scientific papers on collections from Pacific islands. General Secretary of Missions, NSW, 1887-1908. [ADB]. M correspondent*.

BROWN, John Ednie, 1848-99, forest conservator and botanist. Born in Scotland and educated at Edinburgh. Managed estates in Yorkshire and Sussex. Visited Canada and USA, 1871-2, 1878. Went to SA, 1878. FLS, 1879; FRHS. Conservator of Forests, SA, 1879-90; Director-General of Forests, NSW, 1890-92; Conservator of Forests, WA, 1895-9 [ADB, Desmond, Hall (1978), TL2]. M correspondent.

BROWNLESS, Anthony Colling, 1817-97, medical practitioner. Born in England. MRCS, LSA, 1841; MD, St Andrews, 1847. Practised as a surgeon in London, 1842-52. Went to Melbourne in 1852. Physician at Melbourne Hospital from 1854. Vice-Chancellor, University of Melbourne, 1858-87; Chancellor, 1887-97 [ADB].

BRUCE, John Munro, 1840-1901, businessman. Born in Ireland. Went to Melbourne in 1858. Became a partner in a successful softgoods firm. President of the Young Men's Christian Association and the Melbourne Liedertafel [ADB].

BRUNET DE LA GRANGE, fl. 1840s, French author. Wrote extensively on silk culture [Website: ccfr.bnf.fr/ (April 2004)].

BUETTNER, Alexander, c.1837-1914, medical practitioner. FRCS, Edinburgh, 1872; MD, Berlin, 1873; FPS, Glasgow, 1874. Went to Melbourne by 1877. An executor of M's will [*Medical register*, 1882-92].

BUETTNER, Hermann, brother of Alexander Buettner. An executor of M's will.

BULL, Henryk J., Norwegian ship's master. Part-owner of *Antarctic* and a member of the party that in 1895 made the first landing on Antarctica [L95.10.17].

BURLEY, Mr J., senior messenger and caretaker at Melbourne Observatory [Vic. *Government gazette* (1884)].

BURN, Jemima France, 1822-1918, conchologist and artist [Willis et al. (1986)]. Married Mr Irvine. M correspondent*.

CALDWELL, William Hay, 1859-1941, Scottish zoologist. Educated at Cambridge. Became a demonstrator in comparative anatomy. Visited Australia in 1883-4. Established that Monotremes lay eggs. Paper manufacturer in Scotland from 1887 [Gibbney & Smith (1987)].

CAMBRIDGE, George William Frederick Charles, 2nd Duke of, 1819-1904, British aristocrat. Cousin of Queen Victoria (vol. 1). Succeeded to dukedom, 1850. Commander-in-Chief of British Army, 1856-95 [DNB].

CAMPBELL, Frederick Alexander, 1849-1930, engineer and educationist. Born in Scotland. Educated at Geelong College, Vic. and University of Melbourne (MCE, 1898). Collected in the New Hebrides and Loyalty Islands, 1871, and contributed plants to the Melbourne Herbarium [B93.13.01]. Assistant Engineer, NSW, 1879-86. Director, Working Men's College, Melbourne, 1887-1913 [ADB].

CAMPBELL, John Douglas Sutherland. See Marquis of Lorne.

CANDOLLE, Anne *Casimir* Pyramus De, 1836-1918, Swiss botanist. Son of Alphonse de Candolle (vol. 1). Educated at the Free Gymnasium, Geneva (Gymnasium libre); at the University of Paris, 1853-7. Spent some time in England with Miles Berkeley (vol. 1). Travelled in Algeria, 1859. Studied in Berlin, 1860. Thereafter in Geneva [*Proc. Linnean Soc.* (1919), pp. 51-2]. M correspondent*.

CANNSTADT, Baron Pavel Lvovich Schilling von, 1786-1837. Attaché to Russian embassy in Munich, 1812. Councillor of State [Mottelay (1922)].

CAVANILLES, Antonio José, 1745-1804, Spanish priest and botanist. At Paris, 1777-81, later at Madrid. Director of Madrid Botanic Garden from 1801 [TL2].

CHALMERS, James, 1841-1901, Scottish clergyman. Missionary in NZ, 1866. Collected plants in New Guinea and neighbouring islands, 1877-1901 [ADB, Cohn (forthcoming), Desmond]. Plant contributor to the Melbourne Herbarium. *Appendicula chalmersiana, Dendrobium chalmersii, Dichrotrichum chalmersii, Nauclea chalmersii, Vandopsis chalmersiana*. M correspondent.

CHEESEMAN, Thomas Frederick, 1846-1923, New Zealand botanist. Born in England. Went to NZ, 1854. Self-taught in botany; botanized extensively in NZ and Pacific islands. FLS, 1873. Curator, Auckland Museum from 1874 [NZDB]. Plant contributor to the Melbourne Herbarium. M correspondent*.

CHEVREUL, Michel-Eugène, 1786-1889, French chemist. Professor at the Natural History Museum (Muséum d'Histoire Naturelle), Paris. Member of the Academy of Sciences (Académie des Sciences), Paris, 1826 [DSB]. M correspondent.

CHICKERING, John White, 1831-1913, American educator. Educated at Bowdoin College, AB, 1852; AM, 1855; at Bangor Theological Seminary, 1860. Taught in various academies. Congregational minister at Springfield, Vermont, 1860; at Exeter, New Hampshire, 1865. From 1870 professor of natural science, Gaudelet College, Washington [WWA]. Plants received via Asa Gray are in the Melbourne Herbarium [B63.05.01].

CLARKE, Charles Baron, 1832-1906, English botanist. Studied at Cambridge (MA, 1859). Lecturer in mathematics there, 1857-65. Joined staff of Presidency College, Calcutta, 1865. FLS, 1867. Superintendent, Calcutta Botanic Garden, 1869-71. Inspector of Schools, East Bengal, 1883-7. At Kew, 1887, working on Indian botany. President, Linnean Society, 1894-6. FRS, 1882 [Desmond].

CLARKE, WILLIAM JOHN, 1831-97, pastoralist. Born in Tas. Acquired property at Sunbury, Vic, eventually becoming the largest landholder in the colony. MLC, 1878-97. Created baronet, 1882 [ADB].

CLUSIUS, CAROLUS [CHARLES DE L'ESCLUSE], 1526-1609, Dutch botanist [TL2].

COBBOLD, THOMAS SPENCER, 1828-1886, British helminthologist. MD, Edinburgh, 1851. Curator, Edinburgh Anatomical Museum, 1851-6. FLS, 1857; FRS, 1864. Professor of Botany, Royal Veterinary College, London, 1873 [Desmond].

COLMEIRO Y PENIDO, MIGUEL, 1816-1901, Spanish botanist. Director of the Madrid Botanic Garden [TL2]. *Colmeiroa*. M correspondent but none has been found [L96.01.23].

COMMERSON, PHILIBERT, 1727-73, French naturalist. Studied at Montpellier (MD, 1755). Naturalist on Bougainville's voyage of circumnavigation in *La Boudeuse* and *L'Étoile*, 1766-8. Left the expedition at Île-de-France (Mauritius) to investigate the botany of Madagascar. Died at Île-de-France, his collections being returned to France [DBF].

COOKE, JOSIAH PARSONS, 1827-94, American chemist, teacher and author. Studied at Harvard University, 1844-8; Erving professor of chemistry and mineralogy, 1850. Secretary of the American Academy of Arts and Sciences, 1873-92; president, 1892-4 [DAB]. M correspondent*.

COOKE, MORDECAI CUBITT, 1825-1914, British botanist. Worked at India Museum of India Office, London, 1861-80. ALS, 1877. In charge of lower cryptogams in Kew Herbarium, 1880-92. Expert on fungi. Began *Grevillea*, the first journal devoted to cryptogamic botany [Desmond, TL2]. Duplicate specimens were sent by Cooke to the Melbourne Herbarium. M correspondent*.

CORNU, MAXIME, 1843-1901, French agricultural scientist. Studied at the École Normale Supérieure, Paris (Lic. ès sciences, 1868; DSc, 1872). Assistant at the Natural History Museum (Muséum d'Histoire Naturelle), Paris, 1874; professor of cultivation, 1881-1901. Inspector-general of agriculture, 1881. An expert on fungi in relation to agriculture [DBF].

CORRENTI, CESARE, 1815-88, Italian patriot and politician. Educated at Pavia. Member of the Italian Parliament, 1860-86. Chancellor of the Order of St Maurizio from 1876 [DBI].

COSSON, ERNEST SAINT-CHARLES, 1819-89, French botanist. Studied medicine at Paris (MD, 1847). Botanist to the scientific commission for Algeria, 1852-8. Elected to the Academy of Sciences (Académie des Sciences), Paris, 1873 [DBF]. Plant contributor to the Melbourne Herbarium [B65.10.01, B69.07.03].

COULTER, John Merle, 1851-1928. American botanist. Professor of natural science, Hannover College, 1874-9. Professor of biology, Wabash College, 1879-91. PhD, Indiana University, 1882. President and professor of botany, Indiana University, 1891-3. President, Lake Forest University, Illinois, 1893-6. Foundation professor of botany, University of Chicago, 1896-1925 [TL2].

COX, Charles, c.1810-92, British civil servant. Appointed to Colonial Office in 1829, rising to Chief Clerk, 1872-9. Secretary and Registrar of the Order of St Michael and St George, 1872-7; Chancellor of the Order from 1877. KCMG, 1887 [*The Times* (London), 14 July 1892, p. 9].

CRAMER [KRAMER], Carl Eduard, 1831-1901, Swiss algologist. Professor of botany at Zürich [TL2].

CRUMMER, Henry Samuel Walker, 1839-1921, surveyor and anthropologist. Born in Sydney. An officer of the Survey Department, NSW, 1868-1914. Became an authority on early pastoral holdings. A founding member of the NSW Branch, RGSA; honorary secretary and librarian for 20 years. An authority on aboriginal and Pacific ethnology [Gibbney & Smith (1987)]. M correspondent.

CUMING, Hugh, 1791-1865, English natural history collector. Went to South America in 1819. Dealer in natural history objects, Valparaiso, Chile, 1822-30; collected in eastern Polynesia, 1827-8; along the west coast of South America, 1828-30. In England, 1831-5. FLS, 1832. Collected in the Philippines, 1836-40 [Cohn (forthcoming), Desmond]. Specimens are in the Melbourne Herbarium [L80.08.04].

CURR, Edward Micklethwaite, 1820-89, squatter, author and ethnologist. Born in Hobart. Educated in Lancashire and France. Returned to Tas., 1839, then to Vic. and managed sheep properties for his father, 1841-50. In pastoral activities in NZ, Qld and NSW until 1862; then inspector, later chief inspector of sheep. Chief inspector of stock, Vic., 1871 [ADB]. M correspondent.

CUTHBERTSON, Walter Robert, fl. 1877-98, plant collector and explorer. Collected in Papua. Climbed Mt Obree, Papua, 1877 [Hall 1984)]. *Cassia cuthbertsoni, Dendrobium cuthbertsoni*. M correspondent.

DAINTREY, Edwin, 1814-87, lawyer and botanist. Born in England. Went to Sydney in 1840s. A founder of the Linnean Society of NSW [Desmond; Gibbney & Smith (1987)]. M correspondent*.

D'ALBERTIS, Luigi Maria, 1841-1901, Italian zoologist and ethnographer. Made several expeditions in New Guinea 1871-78, the first in company with O. Beccari (vol. 2) [ADB, Cohn (forthcoming), Willis et al. (1986)]. Specimens from the 1875-6 exploration of the Fly River are in the Melbourne Herbarium [L80.08.04]. *Dendrobium undulatum var. albertisiana, Faredaya albertisii, Hibiscus d'albertisii, Mucuna albertisi, Quercus dalbertisii.*

D'ALTON, JOHANN SAMUEL EDUARD, 1803-54, German medical professor. Studied at Bonn (DrMed, 1824) and in Holland. Extraordinary professor of anatomy, Berlin, 1833-4; professor of anatomy and physiology and director of the anatomical institute at Halle, 1834-54 [Univ. Halle archives].

D'ALTON, ST ELOY, 1847-1930, engineer. Born in Ireland. Went to Australia c.1875. Official for several shires in Vic. At Dimboola until c. 1918. Studied the flora of the Wimmera, Grampians and Mallee regions of Vic. [Cohn (forthcoming), Desmond]. Plant contributor to the Melbourne Herbarium. *Trymalium daltoni.* M correspondent.

DATTARI, PAOLO, c. 1850-?, architect. Born in Italy. Went to Melbourne in 1877. Foundation member of the Field Naturalists' Club of Victoria, 1880. Returned to Italy in 1890s [Dwyer (1996a)]. M correspondent*.

DAVIDSON, ALEXANDER, fl. 1880s. Collected plants with W. A. SAYER in north Qld, 1886-7 [Desmond]. Plant contributor to the Melbourne Herbarium [B87.13.06, p.1106]. *Davidsonia, Spiraenthemum davidsonii.*

DAVIDSON, THOMAS, 1817-85. Scottish palaeontologist. Educated in Edinburgh, France, Italy and Switzerland. Published monograph of British fossil *Brachiopoda*, 1850-86. FRS, 1857. Awarded Wollaston Medal by the Geological Society of London, 1865; Royal Medal by the Royal Society, 1870 [DNB].

DAVIS, GEORGE S., American medical publisher. Co-founder of Parke Davis pharmaceutical company, 1868. Editor, *Therapeutic Gazette*, Detroit [Website: www23.brinkster.com/beneathdetroit/davisg.htm (April 2004)]

DAWSON, JOHN *WILLIAM*, 1820-99, Canadian palaeobotanist. Born in Nova Scotia. MA, Edinburgh, 1842; LLD, 1884. FRS, 1862. Professor of geology and Principal, McGill College, Montreal, 1855-93. Knighted 1884 [Desmond, TL2]. M correspondent.

DEAKIN, ALFRED, 1856-1919, barrister and politician. Born at Melbourne. Educated at the University of Melbourne; admitted to Victorian Bar, 1877. MLA, Vic., 1879, 1880-1901; Chief Secretary, Vic., 1885-90. Chairman of Royal Commission on Irrigation, 1884. A leading figure in the Australian federation movement. MP, 1901-13; Attorney-general, 1901-3; Prime Minister 1903-4, 1905-8, 1909-10 [ADB]. M correspondent*.

DEANE, HENRY, 1847-1924, engineer and scientist. Born in England. MA, Galway, 1882. Went to Sydney, 1880. Engineer-in-chief, NSW railway construction, 1891. Authority on Tertiary flora of Australia. FLS, 1886. President, Royal Society of NSW, 1897, 1907. President, Linnean Society of NSW, 1895-7 [ADB, Desmond, Hall (1978), TL2]. Plant contributor to the Melbourne Herbarium [B87.13.06, p. 1106]. *Melaleuca deanei*. M correspondent*.

DEJARDIN, LÉON, French consular official. French consul-general in Melbourne, 1888-1901. M correspondent*.

DE SALIS, LEOPOLD FABIUS DIETEGAN FANE, 1816-98, pastoralist and politician. Born in Italy. Educated at Eton, studied sheep farming in Scotland. Arrived in Sydney, 1840. Acquired property on Murrumbidgee River, NSW, and later other properties. A magistrate from 1844. MLA, NSW, 1864-9; MLC from 1874 [ADB].

DE TONI, GIOVANNI BATISTA, 1864-1924, Italian phycologist. Studied at Venice and Padua. Botanical assistant at Padua, 1885-92; at Parma, 1892-4. Librarian at Padua, 1894-9. Professor of botany at Modena from 1900 [DBI]. M correspondent but none has been found [L94.01.16 (Coll. Corr.)].

DIBBS, GEORGE RICHARD, 1834-1904, politician and businessman. Born in Sydney. Sailed to the Victoria River, NT, with Augustus GREGORY (vol. 1), 1855. MLA, NSW, 1874-7. Re-entered Parliament and became Colonial Treasurer, 1882. Held various parliamentary positions until 1895; Premier, 1885, 1889, 1891-4. Knighted, 1892 [ADB].

DICKSON, OSCAR, BARON, 1823-97, merchant and philanthropist. Born in Scotland. Settled in Gothenburg, Sweden. Established large trading establishment with his brother. Created Baron by King Oscar II [Website: electricscotland.com/history/sweden (March 2004)].

DIETRICH, AMALIE. See Amalie NELLE.

DISNEY, THOMAS ROBERT, 1842-1915, soldier. Joined British Army, 1861. Served in India and Abyssinia. In command of the Victorian Military Forces, 1883-8 [ADB].

DITTRICH, H., fl. 1886, pastoralist. Member of David Lindsay's expedition to northern Australia, 1886. Collected grasses, some of which are in the Melbourne Herbarium. *Calocephalus dittrichii*.

DOBSON, WILLIAM LAMBERT, 1833-98, politician and judge. Born in England. Went to Hobart, 1834. Educated at Hutchins School, Hobart. Served in Police Magistrate's Office. Went to England, 1853. Called to the bar, 1856. Returned to Hobart as a barrister, 1857. Member, House of Assembly, 1861-2, 1864-70; Attorney-general, 1861-3. Puisne judge in

Supreme Court, 1870; Chief Justice, 1886-98. KCMG, 1897 [ADB]. M correspondent*.

DOUGHTY, GEORGE FERDINAND ST HELEN, 1857-93, M's nephew, son of Bertha DOUGHTY, M's sister [Statton (1986)]. M correspondent*.

DOUGLAS, JOHN, 1828-1904, politician and administrator. Born in England. Went to NSW in 1851. Acquired sheep station on Darling Downs, Qld, in 1854. MLA, NSW, 1859. MLA, Qld, 1863-9, 1875-80; Treasurer, 1866-7; Colonial Secretary, 1877-9. Agent-General for Qld, London, 1869-70. Government resident, Thursday Island, Qld, 1885-1904; special commissioner of British New Guinea, 1885-8 [ADB].

DOW, JOHN LAMONT, 1837-1923, journalist and politician. Born in Scotland. Went to Vic in 1848. In Qld, 1862-8. Agricultural editor of the *Leader* (Melbourne). MLA, 1877-93; President of Board of Land & Works, 1886-90 [Thomson & Serle (1972)].

DUCHARTRE, PIERRE-ÉTIENNE-SIMON, 1811-94, French botanist. Studied at the Garden of Plants (Jardin des plantes), Toulouse, and at Montpellier (doctorat ès sciences, 1841). Professor of botany at the agricultural college at Versailles, 1849-52; at Faculté des Sciences, Paris, 1861-86. Elected to the Academy of Sciences (Académie des Sciences), Paris, 1861 [DBF].

DU PETIT-THOUARS, LOUIS-MARIE *AUBERT* AUBERT, 1758-1831, French traveller and botanist. Studied at the Collège de La Flèche. Captain in the military garrison at Lille, 1786-92. Botanized at Île-de-France (Mauritius), Île de Bourbon (Réunion) and Madagasacar, 1792-1802. Director of the Imperial Plant Nursery at Roule, 1807-27. Elected to the Academy of Sciences (Académie des Sciences), Paris, 1820 [DBF].

DURIEU DE MAISONNEUVE, MICHEL CHARLES, 1796-1878, French botanist. Infantry officer, 1815-48. Member of scientific commission to Algeria, 1840-2; revisited Algeria, 1844-8. Director of the Bordeaux Botanic Garden from 1853 [DBF]. Specimens are in the Melbourne Herbarium [B65.10.01, L80.08.04]. *Thryptomene maisonneuvii*. M correspondent.

DYER, WILLIAM TURNER THISELTON. See WILLIAM TURNER THISELTON-DYER.

EARL, GEORGE SAMUEL WINDSOR, c.1805-65. Born in England. Went to Swan River Colony, WA, 1829. Appointed Commissioner of Crown Lands, Port Essington, North Australia, 1838. Discovered Kimberley district, WA, 1857 [Gibbney & Smith (1987)]. *Earlia*.

EATON, Daniel Cady, 1834-95, American botanist. Studied at Yale and then at Harvard with Asa Gray (vol. 1). Professor of botany at Yale from 1864. An expert on ferns [ANB]. Plant contributor to the Melbourne Herbarium [B61.02.01, B63.05.01]. M correspondent but none has been found.

EDELFELDT, Erik Gustaf, d. 1895, government officer. Born in Sweden. Went to New Guinea in 1884, collected natural history specimens. Later in Qld. [ADB, Cohn (forthcoming)]. Plant contributor to the Melbourne Herbarium. *Coelogyne edelfeldtii.*

EICHLER, August Wilhelm, 1839-87, German botanist. Educated at University of Marburg, 1857-61. Assistant editor to C. Martius (vol. 1) of *Flora brasiliensis*, 1861-8; editor, 1868-71. Professor of botany at Graz, 1871-2; at Kiel, 1872-7; at Berlin from 1877 [DSB]. M correspondent.

EIMER, Gustav Heinrich *Theodor*, 1843-98, German zoologist. Educated at Tübingen, Freiburg, Heidelberg and Berlin. Graduated in medicine and natural history (DrMed, 1868; DrPhil, 1869). At Darmstadt Museum and Professor at the Polytechnikum, 1874. Professor of zoology & director of Zoological Institute at Tübingen from 1875 [NDB].

ELDER, Thomas, 1818-97, pastoralist, explorer and financier of scientific expeditions. Born in Scotland. Went to Adelaide, 1854. Imported the camels used by Ernest Giles (vol. 2) and others in their journeys of exploration. MLC, SA, 1863-9, 1871-8. KCMG, 1878; GCMG, 1887 [ADB]. *Banksia elderiana, Cloanthes elderi, Corchorus elderi, Eremophila elderi, Erodiophyllum elderi, Goodenia elderi*. M correspondent*.

ELLIOTT, Frederick John. Doctor and photographer on Elder expedition to north-west Australia, 1891-2 [Feeken, Feeken & Spate (1970)].

ENGELHART [ENGELHARDT], August, medical practitioner. Collector of algae in Lacepede and Guichen Bay area, SA [Willis et al. (1986)].

ENGLER, Adolf Gustaf Heinrich, 1844-1930, German botanist. Educated at Breslau. *Privatdozent* and curator of the Munich Herbarium, 1871-8. Professor of botany, Kiel, 1878-84; Breslau, 1884-9; Berlin, 1889-1930. Editor (with K. Prantl) of *Die natürlichen Pflanzenfamilien*, 1887-1930. Leader of Berlin school of taxonomy and plant geography [Cohn (forthcoming), NDB, TL2]. Duplicate specimens sent from Berlin are in the Melbourne Herbarium. M correspondent*.

ERNST, Adolf, 1832-99, German botanist. Settled in Venezuela, 1861 [TL2].

ETHERIDGE, Robert, 1819-1903, British palaeontologist. Assistant Palaeontologist (1857) then Palaeontologist (1863) to Geological Society of Great Britain. Assistant Keeper of Geology, British Museum

(Natural History), London, 1881-91. President, Geological Society, 1880-82. FRS, 1871 [Sackler Archive, Royal Society of London].

ETHERIDGE, ROBERT, Jr, 1846-1920, palaeontologist. Born in England. Son of Robert ETHERIDGE. Educated at Royal School of Mines. Went to Vic. in 1866. With Geological Survey of Vic., 1866-9. In Britain, 1871-87; assistant at British Museum, 1874-86. Palaeontologist to Geological Survey of NSW and Australian Museum, Sydney, 1887-95. Director, Australian Museum, 1895-1920 [ADB]. M correspondent*.

ETTINGSHAUSEN, CONSTANTIN, BARON, 1826-97, palaeobotanist and mineralogist at Graz, Austria [TL2]. *Gentiana ettingshauseni.* M correspondent.

EVERILL, HENRY CHARLES, c.1847-1900, shipmaster and explorer. Born in England. Commanded *Bonito* in expedition to Fly River, New Guinea, 1885 [Gibbney & Smith (1987)]. M correspondent but none has been found.

FAIRFAX, JAMES READING, 1834-1919, newspaper proprietor. Born in England. Went to Sydney with his family in 1838. Became a partner with his father and elder brother in John Fairfax & Sons, owner of the *Sydney morning herald*, in 1856; proprietor from 1877. Knighted, 1898 [ADB].

FALCONER, HUGH, 1808-65, Scottish botanist. Studied at Aberdeen (MA, 1826), and Edinburgh (MD, 1829). Assistant surgeon for the East India Company, 1830. Superintendent, Saharanpur Garden, 1832-41. FRS, 1845. Superintendent, Calcutta Botanic Garden, and professor of botany at Calcutta Medical College, 1848-55. [Desmond]. Specimens are in the Melbourne Herbarium [L80.08.04].

FALKENBERG, PAUL, 1848-1925, German botanist [Barnhart (1965)].

FARADAY, MICHAEL, 1791-1867, British natural philosopher, chemist and electrician. Assistant, Royal Institution, 1813; professor of chemistry from 1832. Made numerous discoveries in relation to electricity and magnetism [DSB]. *Faradaya.*

FAWCETT, CHARLES HUGH, 1812-90, public servant and police magistrate. Born in Ireland. Went to NSW, 1843. Acting government agent and police magistrate in Richmond River area, NSW, 1860s, later police magistrate at Bulledalah, NSW. Member, NSW parliament, 1880-82 [Desmond, Connolly (1983)]. Plant contributor to the Melbourne Herbarium. *Clematis fawcettii, Cylicodaphne fawcettiana, Fawcettia.* M correspondent*.

FAYENZ, Heinrich, 1836-1905, Austrian naval officer. Retired in 1886 [Website: kfunigraz.ac.at/ub/ausstellungen/kriegsmarine/html (March 2004)]. M correspondent*.

FAYRER, Sir Joseph, 1824-1907, English physician. Studied at Charing Cross Hospital, London. Medical officer, HMS *Victory*, 1847. Assistant surgeon, Bengal, 1850. Professor of surgery, Medical College, Calcutta, 1859-72. President, Medical Board of the India Office, 1874-95. FRS, 1877. Created baronet, 1896. Studied poisonous snakes of India and the physiological effects produced by their venom [DNB].

FEOKTISTOV, Alexandr Evgheniievich, author of works on cures for snakebite [VBOT].

FEUILLÉE, Louis Éconches, 1660-1732, French priest and botanist. Travelled in South America and the West Indies [TL2].

FICALHO, Francisco Manuel De Mello Breyner, Conde de, 1837-1903, Portuguese botanist and writer. Studied at the Polytechnic School (Escola Politécnica), Lisbon, 1855-60; professor of botany there from 1864 [Website: www.arqnet.pt/dicionatio/ (December 2004)].

FIELDER, Walter, 1858-1913, clergyman and naturalist. Born in England. Went to Qld in 1883 as a missionary, and later to Vic. Secretary of Vic. Field Naturalists' Club [Gibbney & Smith (1987)].

FINDLAY, James, 1821-1905, pastoralist and horse breeder. Born in Scotland. MA, Edinburgh. Arrived in Melbourne, 1843. Managed a property in north-east Vic. and later acquired a property on the upper Murray River, Vic. Accompanied M on plant collecting trip in Snowy Mountains, NSW, 1874. [Desmond]. Plant contributor to the Melbourne Herbarium. *Bertya findlayi*. M correspondent.

FISCHER, Johann Gustav, 1819-89?, German zoologist. Educated at Halle, Leipzig and Berlin (PhD, 1843). Teacher in Hamburg from 1847. Named several species of reptiles after M [Darragh (1996)].

FISCHER-BENZON, Rudolf von, 1839-1911, German teacher and librarian. Studied at Kiel (DrPhil, 1865). *Privatdozent* at Kiel, 1866-9. Schoolteacher in Schleswig-Holstein, 1869-93; at Husum, 1874-8; at Kiel, 1878-93. Provincial librarian at Kiel from 1895 [Vollbehr & Weyl (1956)]. M correspondent*.

FISHER, William Rogers, 1846-1910, forester. Born in Sydney. BA, Cambridge, 1867. Indian Forest Service, 1866. Director, Forest School, Dehra Dun. Assistant professor of forestry, Coopers Hill College, 1890. Professor of forestry, Oxford, from 1905 [Desmond]. M correspondent*.

FITZGERALD, ROBERT DAVID, 1830-92, surveyor, naturalist and botanical artist. Born in Ireland. Went to Sydney in 1856. Surveyor, Department of Lands, 1856-73; Deputy Surveyor-General, 1873-87. FLS, 1874. Member of Public Service Commission, NSW, 1888-92. Expert on orchids [ADB, Desmond, TL2]. Specimens are in the Melbourne Herbarium. *Dacrydium fitzgeraldi*, *Dracophyllum fitzgeraldii*, *Eugenia fitzgeraldi*, *Hibiscus fitzgeraldi*, *Sarcochilus fitzgeraldii*. M correspondent*.

FITZGERALD, WILLIAM VINCENT, 1867-1929, forest botanist and explorer. Born in Tas. On staff of Sir William MACGREGOR in New Guinea, 1895-6, conducting exploratory expeditions. Chairman of Forests Advisory Board of WA from 1904; undertook two expeditions to the Kimberley district, 1905-6 [VBOT].

FLETCHER, JOSEPH JAMES, 1850-1926, naturalist. Born in NZ. Moved with his family to Australia in 1860. Educated at Sydney University. Taught at Wesley College, Melbourne, then studied biology at London University (BSc). Returned to Australia, 1881. In 1885 became the Director and Librarian of the Linnean Society of NSW (the title later changed to Secretary) and served until 1919; president, 1920 [Hall (1978)]. *Isopogon fletcheri*. M correspondent.

FOELSCHE, PAUL HEINRICH MATTHIAS, 1831-1914, police inspector, magistrate, botanical collector and photographer. Born in Germany. Went to SA, 1856, then NSW and NT, 1870 [ADB, Hall (1978)]. Plant contributor to the Melbourne Herbarium. *Dendrobium foelschei*, *Eucalyptus foelscheana*. M correspondent*.

FORBES, HENRY OGG, 1851-1932, naturalist and collector. Born in Scotland. Educated at Aberdeen and Edinburgh. Travelled in Portugal, 1875-7; Indonesia, 1878; New Guinea, 1885-6, 1889; Socotra, 1898-9. Director of Canterbury Museum, NZ, 1890-3; Liverpool Museums, UK, 1894-1911 [ADB, Cohn (forthcoming), Desmond]. Plant contributor to the Melbourne Herbarium. *Agapetes forbesii*, *Dimorphanthera forbesii*, *Helicia forbesiana*, *Pterygota forbesii*, *Sloanea forbesii*. M correspondent*.

FORDE, HELENA. See Helena SCOTT.

FORSTER, JOHANN *GEORGE* ADAM, 1754-94, writer, traveller and revolutionary. Son of J. R. FORSTER. Born near Danzig. Accompanied his father on a journey of investigation in Russia, 1765. Went with his family to England, 1766. Assistant to his father on James COOK's (vol. 1) second voyage of exploration, 1772-5. Professor of natural science, Kassel, 1779-84; Vilna, 1784-7; university librarian, Mainz, 1788-92. Supported the French revolutionary army when it invaded Mainz in 1792. Died in Paris. FRS, 1777; member of Leopoldina, 1780; Academy of Sciences (Akademie der Wissenschaften), Berlin, 1786. [ADB, Desmond, NDB, TL2]. *Cordyline forsteri*.

FORSTER, Johann Reinhold, 1729-98, German polymath. Born in Prussia. Pastor at Nassenhüben, near Danzig, 1753-65. Undertook a journey of scientific investigation in Russia, 1765. Went to England in 1766. Lecturer at Dissenting Academy at Warrington, Lancashire, 1766-7. Naturalist on James Cook's second voyage of exploration, 1772-5. Professor of natural history, Halle, 1780-98. FRS, 1772 [ADB, Desmond, NDB, TL2].

FORTUNE, Robert, 1812-80, botanist. Born in Scotland. Gardener at the Edinburgh Botanic Garden and later at the Horticultural Society of London's garden at Chiswick, London. Collected plants on five journeys to East Asia between 1843 and 1861. Introduced tea plant from China to India. Curator, Chelsea Physic Garden, London, 1846-8 [Desmond, TL2]. Specimens are in the Melbourne Herbarium [L80.08.04].

FOWLER, James, 1829-1923, Canadian naturalist. Educated at Halifax. Presbyterian minister 1857-76. Lecturer in natural history, Queen's University, later professor, 1880-1907 [Website: biology.queens.ca/~fowler/history.htm (April 2004)].

FRANQUEVILLE, Albert, Comte De, 1814-91, French botanist. Mayor of Bizanos, south-west France, 1871-91 [TL2]. Specimens are in the Melbourne Herbarium [L80.08.04].

FRASER, Malcolm, 1834-1900, civil engineer, administrator and naturalist. Born in England. Surveyor in NZ, 1857-69. Surveyor-general, WA, 1870. MLC and commissioner of Crown Lands, 1872; Colonial Secretary, 1883-90. CMG, 1881, KCMG, 1887 [ADB]. *Eremophila fraseri*. M correspondent*.

FRENCH, Charles Hamilton, 1868-1950, naturalist. Son of Charles French (vol. 2). Born in Melbourne. Assistant at the Melbourne Herbarium, 1883-95, until transferred to the Department of Agriculture. Vic Government Entomologist, later Biologist, 1908-33 [VBOT].

FRIEDRICH Franz III, 1851-97, Grand Duke of Mecklenburg-Schwerin. Succeeded his father, Friedrich Franz II (vol. 1) in 1883 [*Brockhaus' Konversations-Lexikon*]. M correspondent*.

GABRIEL, Joseph, c.1847-1922, pharmacist. Active in Field Naturalists' Club of Vic., 1883-1922 [*Vic. Naturalist*, vol. 39, 1922, pp. 101-2].

GÄRTNER [GAERTNER], Joseph, 1732-91, German physician and botanist. Private tutor at Calw, Baden [TL2].

GEOFFROY SAINT-HILAIRE, Albert, 1835-?, French zoologist. Son of Isidore Geoffroy Saint-Hilaire. Director of the Acclimatisation Garden (Jardin d'acclimatation), Paris, 1865-93; President of the Acclimatisation Society (Société d'acclimatation), 1887-95 [Osborne, 1994].

GEOFFROY SAINT-HILAIRE, ÉTIENNE, 1772-1844, French zoologist. Curator of the zoo at the Garden of Plants (Jardin des Plantes), Paris; professor at the Natural History Museum (Muséum d'Histoire Naturelle), Paris, 1793-1844. Elected to the Academy of Sciences (Académie des Sciences), Paris, 1807; President, 1833 [DSB].

GEOFFROY SAINT-HILAIRE, ISIDORE, 1805-61, French zoologist. Son of Étienne GEOFFROY SAINT-HILAIRE. Professor at the Faculté des Sciences, Paris, from 1837 and at the Natural History Museum (Muséum d'Histoire Naturelle), Paris, from 1841. Elected to the Academy of Sciences (Académie des Sciences), Paris, 1833; President, 1856-7. Founding president of the Acclimatisation Society (Société d'acclimatation), Paris, 1854-61 [DSB].

GILBERT, JOSEPH HENRY, 1817-1901, English agronomist. Born in Yorkshire. DrPhil, Giessen, 1840. FRS, 1860; FLS, 1875. Knighted, 1893. Professor of Rural Economy, Oxford, 1884-90 [Desmond].

GILL, WALTER, 1851-1929, gardener and nurseryman. Born in England. Began work in a Dorset nursery. Went to SA, 1876. Sub-inspector of Crown Lands, SA; Chief Forester, Wirrabara, SA, 1886; Conservator of Forests, SA, 1890-1923 [Desmond, Hall (1978)]. M correspondent*.

GILL, WILLIAM WYATT, 1828-96, English missionary. Educated at Highbury College and New College, London (BA, 1850). Joined London Missionary Society in 1851. In Cook Islands, 1852-72; Raratonga, 1877-83. Retired to Sydney [ADB].

GINGINS, BARON FREDERIC CHARLES JEAN DE LA SARRAZ, 1790-1863, Swiss historian and botanist. Pupil of A. P. de CANDOLLE (vol. 1) [TL2].

GIPPS, FREDERICK BOWDLER, 1840-1903, engineer. Born in England. Officer in British Army; invalided after Indian uprising of 1857. Studied at University of London (BE). Went to Sydney in 1874. Engineer with NSW Water Supply Department. Went to Melbourne in 1896 [Gibbney & Smith (1987)].

GLADMAN, FREDERICK JOHN, 1839-84, educationist. Born and educated in England. Headmaster at Great Yarmouth, 1863. BA, University of London, 1871; BSc, 1875. Principal of Training Institution, Melbourne, 1877-84 [ADB]. M correspondent*.

GLAZIOU, AUGUSTE FRANÇOIS MARIE, 1818-1906, botanical traveller. Born in France. Collected plants in Brazil, 1858-97. Director of the Paseo publico and of the Acclimatisation Ground (Campo do Acclimaçao), Rio de Janeiro, 1873 [TL2]. Specimens are in the Melbourne Herbarium [L80.08.04].

GOEZE, EDMUND, 1838-1929, German botanist. Director of the Coimbra botanic garden, 1866-76; later in Greifswald [www.triplov.com/botanica/historia1.html (January 2005)]. Translator of German edition of M's *Select extra-tropical plants*. M correspondent but none has been found.

GOLDIE, ANDREW, 1840-91, naturalist and merchant. Born in Scotland. In NZ, 1862-74. Explored in New Guinea, 1877-9, and settled there [ADB, Cohn (forthcoming), Desmond]. Plant contributor to RBG, Melb. and the Melbourne Herbarium. *Combretum goldieanum.* M correspondent but none has been found [e.g. L78.06.04].

GOLDSWORTHY, ROGER TUCKFIELD, 1839-1900, soldier and public servant. Born in Calcutta. Joined colonial service and became Colonial Secretary, WA, 1877-80. Later Governor, British Honduras and then Falkland Is. KCMG, 1889 [Gibbney & Smith (1987)]. M correspondent.

GOODALE, GEORGE LINCOLN, 1839-1923, American botanist. Educated at Amherst College (BS, 1860) and Harvard Medical School (MD, 1863). Lecturer at Bowdoin College, Mass., 1868-72. Instructor in botany, Harvard University, from 1872; professor of botany, 1878-88; professor of natural history, 1888-1909 [DAB].

GOODENOUGH, JAMES GRAHAM, 1830-75, English naval officer. Entered Royal Navy, 1844. Served in various areas; Commodore of the Australian Station, 1873. Died after being wounded in the Santa Cruz Islands [ADB]. M correspondent but none has been found.

GOOLD, JAMES ALPIUS, 1812-86, Roman Catholic priest. Born in Ireland. Went to Sydney in 1838. Bishop of Melbourne, 1847; Archbishop, 1874-86 [ADB]. M correspondent.

GOSCHEN, GEORGE JOACHIM, 1831-1907, British statesman. Educated at Oxford (BA, 1853). With father's financial firm, Frühling & Göschen, 1853-66. Appointed director of Bank of England, 1858. MP, 1863-1900; President of Poor Law Board, 1868-71; First Lord of the Admiralty, 1871-4, 1895-1900; Chancellor of the Exchequer, 1886-92. Created Viscount Goschen, 1900 [DNB].

GRAHAM, CLARA OAKLEY, 1866-1945. Born in SA. Married Michael Ryan, 1889; postmaster at Onslow, WA, 1889-90; at Israelite Bay, WA, 1890-5; at Eucla, WA, 1895-1907 [Willis et al. (1986), VBOT biog file, *Bicentennial dictionary of Western Australians*]. M correspondent*.

GRAINGER, GEORGE PERCY, 1882-1961, pianist and composer. Born in Melbourne. Began musical studies at age 10, performed throughout the world [ADB].

GRAINGER, ROSA [ROSE] ANNIE. See Rosa Annie ALDRIDGE.

GRAY, JANE. See Jane LORING.

GREEN, TRAILL, 1813-97, American botanist [Barnhart (1965)]. Plant contributor to the Melbourne Herbarium [B63.05.01].

GREENE, EDWARD LEE, 1843-1915, American botanist and clergyman. Professor of botany at University of California, Berkeley, until 1895 [TL2]. M correspondent but none has been found [L94.07.13 (Coll. Corr.)].

GRIFFITH, WILLIAM, 1810-45, botanist. Born in England. Appointed assistant surgeon at Madras, India, 1832. Collected in Assam, 1835-6; Bhutan, 1837-8; Afghanistan, 1839-41; Malaya, 1841-2, 1845. FLS, 1840. Acting superintendent of the Calcutta Botanic Garden, 1842-4 [Desmond]. Specimens are in the Melbourne Herbarium [L80.08.04].

GRIFFITHS, GEORGE SAMUEL, stockbroker. Deputy chairman of Melbourne Stock Exchange until 1895. Declared insolvent, 1896. An enthusiastic amateur geologist and a leading advocate of Antarctic exploration for scientific purposes [Home et al., 1992].

GRUNOV, ALBERT, 1826-1914, German chemist. Worked in Vienna. An expert on diatoms [TL2].

GUNDERSEN, HANS. Consul for Sweden and Norway in Melbourne. M correspondent*.

HACKEL, EDUARD, 1850-1926, Austrian agrostologist. Born in Bohemia. High school teacher at St Pölten, 1871-1900; at Graz from 1904 [TL2].

HAGENAUER, FRIEDRICH AUGUST, 1829-1909, Moravian missionary. Born in Saxony. Missionary trainee at Herrnhut, Ebersdorf, 1851-6. Went to Vic. in 1856. Established Ebenezer mission at Wimmera River, 1859; Ramahyuck mission, Lake Wellington, Gippsland, 1863-1909 [ADB]. M correspondent*.

HAIG, WILLIAM, 1823-93, medical practitioner. Born in Britain. Went to Melbourne in 1853. Practised medicine at Emerald Hill (South Melbourne); surgeon at Alfred Hospital, Melbourne. President, Medical Society of Victoria, 1885 [Gibbney & Smith (1987)]. *Helipterum haigii*. M correspondent.

HAMILTON, Robert George Crookshank, 1836-95, Scottish civil servant and colonial governor. Educated at University of Aberdeen (MA, 1857). Accountant to Board of Education, 1861; to the Board of Trade, 1869. Accountant-general of the navy, 1878. Permanent secretary to the navy, 1882. CB, 1883. KCB, 1884. LLD, 1885. Governor of Tasmania, 1887-93 [ADB].

HARRIS, George, 1874-1922, married Bertha Wehl, 1874.

HARRIS, Ottilie *Bertha*. See Bertha Wehl.

HART, Thomas Stephen, 1871-1960, school teacher and naturalist. Born at Melbourne. Educated at University of Melbourne (BA 1890, MA 1892, BCE 1901). Lecturer, Ballarat School of Mines, 1896-1913; promoted to professor, 1908. Senior master, Creswick School of Forestry, 1913-6. Science teacher, Bairnsdale School of Mines, 1917-31 [ADB]. Plant contributor to the Melbourne Herbarium. M correspondent*.

HASWELL, William Aitcheson, 1854-1925, biologist. Born in Scotland. Educated at Edinburgh (MA, 1877; BSc, 1878; DSc, 1887). Went to Australia in 1878. Curator, Qld Museum, 1880. Professor of biology, University of Sydney, from 1890. President, Linnean Society of NSW, 1892-93. FRS, 1897. Awarded Clarke Medal of Royal Society of NSW, 1915 [ADB]. M correspondent.

HAUCK, Ferdinand, 1845-89, Austrian botanist. Born at Brno. Post office official at Trieste, 1866-89. An expert on algae [TL2]. Algae specimens are in the Melbourne Herbarium [L80.08.04].

HAWKESWORTH, John, c.1720-73, English professional writer. Editor of Pacific journals of James Cook (vol. 1) and Joseph Banks (vol. 1) [DNB].

HAYNALD, Lajos, 1816-91, Hungarian Catholic priest. Entered holy orders in 1839. DTheol, 1841. Bishop of Transylvania, 1852-64. Archbishop of Kalocsa-Bacs from 1867; appointed Cardinal, 1879. An enthusiastic amateur botanist [*Catholic encyclopedia*]. M correspondent*.

HAYTER, Henry Heylyn Ludwig, 1821-95, statistician. Born in England. Went to Melbourne in 1852. Assistant Registrar-General, Vic, 1859-74; Government Statist, 1874-95. CMG, 1882 [ADB].

HEISTER, Lorenz, 1683-1758, German botanist. Professor of medicine at Helmstädt [TL2].

HELDREICH, Theodor von, 1822-1902, botanist. Born in Germany. In Greece 1843-9, 1851-1902. Plant collector and director of the Athens Botanic Garden and Natural History Museum [TL2].

HELMS, RICHARD, 1842-1914, naturalist. Born in Germany. Went to Australia in 1858. On staff of Australian Museum, Sydney, 1888. Naturalist on Elder expedition to north-west Australia, 1891-2. Government fruit inspector, WA, 1896; bacteriologist, NSW Department of Agriculture, 1900 [Cohn (forthcoming), Desmond, Gibbney & Smith (1987), Hall (1978)]. Plant contributor to the Melbourne Herbarium. *Calycopeplus helmsii*, *Grevillea helmsiana*, *Plagianthus helmsii*, *Thryptomene helmsii*. M correspondent*.

HEMSLEY, WILLIAM BOTTING, 1843-1924, English botanist. Gardener at Kew from 1860. Assistant, Kew Herbarium, 1865-7, 1883-1908; Keeper from 1899 [Desmond].

HENRIQUES, JULIO AUGUSTO, 1838-1928, Portuguese botanist. Obtained degree in law, 1860, DrPhil, 1865. Assistant professor of botany, Coimbra, 1866; professor of botany and director of the Coimbra Botanic Garden, 1874 [TL2]. M correspondent.

HENRY, LEON. Assistant at the Melbourne Herbarium, 1881-4.

HEPP, JOHANN ADAM PHILIPP, 1797-1867, German physician, lichenologist and political activist. Lived in exile in Switzerland, 1851-67 [TL2]. Lichen specimens are in the Melbourne Herbarium [L80.08.04].

HERBERT, ROBERT GEORGE WYNDHAM, 1831-1905, politician and public servant. Born in England. Educated at Oxford (BA,1854; BCL, 1856). Colonial Secretary, Qld, 1859, then Premier until 1866, when he returned to England. Held positions in various UK government offices; permanent Under-Secretary in Colonial Office, 1871-92. KCB, 1882; GCB, 1892 [ADB]. M correspondent.

HERLITZ, HERMANN, 1834-1920, Lutheran pastor. Born in Saxony. Ordained, 1862, and went to Vic. Pastor at Germantown, near Geelong, Vic, 1862-8; Melbourne, 1868-1914 [ADB].

HEURCK, HENRI FERDINAND VAN. See Henri Ferdinand VAN HEURCK.

HEYDRICH, FRANZ, ?-1911, German algologist [TL2]. M correspondent.

HICKS-BEACH, MICHAEL EDWARD, 1837-1916, British statesman. Educated at Oxford (MA, 1861). MP, 1864-1906; Chief Secretary for Ireland, 1874-8; Secretary for the Colonies, 1878-80; Chancellor of the Exchequer, 1885-6, 1895-1902. Created Viscount St Aldwyn, 1906; promoted to earl, 1916 [DNB]. *Hicksbeachia*. M correspondent.

HILGARD, EUGENE N. WALDEMAR, 1833-1916, American agricultural chemist and geologist. Born in Germany. Went to America in 1836. Educated at Belleville, Illinois and at Heidelberg (DrPhil, 1853). Studied also at Zürich and Freiberg. State geologist, Missouri, and professor of chemistry, 1855-73. Professor of geology and natural history, University of Michigan, 1873-5. Professor of agriculture, University of California and director, Californian Agricultural Experimental Station, 1875-1904. Member, National Academy of Sciences from 1866. Liebig medal, Munich Academy of Science, 1894 [WWA]. M correspondent*.

HIMLY, AUGUST FRIEDRICH *KARL*, 1811-85, German chemist. Born and educated at Göttingen (DrPhil, 1835). *Privatdozent*, Göttingen, 1837-42; Professor of chemistry, Kiel, 1842-84 [Vollbehr & Weyl (1956)].

HOCHSTETTER, GEORGIANA. See Georgiana BENGOUGH.

HODGKINSON, MRS MARY H., botanical collector in Richmond River district, NSW.

HODGKINSON, WILLIAM OSWALD, 1835-1900, sailor, explorer, journalist, miner and politician. Born in England. Educated in Birmingham. Went to Vic, 1853. Returned to UK, 1854; clerk in War Office to 1859. Returned to Vic, 1860. Joined Burke & Wills expedition to cross Australia from south to north, 1860-1. Led expedition in Qld, 1875. Mining warden, 1878-83. MLA, Qld, 1874-5 and 1888-93 [ADB].

HOHENACKER, RUDOLF FRIEDRICH, 1798-1874, German botanical collector. Missionary, 1822-30. Collected plants in association with the Esslingen Travel Union, 1829-42. Dealer in plant collections at Esslingen, 1842-58; at Kirchheim unter Teck from 1858 [TL2]. Specimens are in the Melbourne Herbarium [L80.08.04].

HOLLAND, HENRY, 1788-1873, Scottish physician and scientist. Educated in London and at University of Edinburgh (MD, 1811). FRS, 1815; FRCP, 1828. Physician in ordinary to Prince ALBERT (vol. 1), 1840; to Queen VICTORIA (vol. 1), 1852. Created baronet, 1853 [DNB].

HOLLRUNG, UDO MAX, 1835-1937, German agriculturalist. Professor of plant diseases, Halle. Took part in German expedition to New Guinea, 1886-7 [Cohn (forthcoming)]. Plant contributor to the Melbourne Herbarium. M correspondent but none has been found.

HOLMES, EDWARD MORELL, 1843-1930, British pharmaceutical botanist and algologist. Curator of the Museum of the Pharmaceutical Society of Great Britain, 1872-1922 [TL2]. M correspondent.

HOLTZE, MORITZ WILHELM [MAURICE WILLIAM], 1840-1923, botanist. Born in Hanover. Studied botany and horticulture in the royal gardens at Hanover and St Petersburg. Went to Australia in 1872. Curator, Darwin Botanic Garden, 1878-91; Adelaide Botanic Garden, 1891-1917. FLS, 1888. Father of Nicholas and Waldemar HOLTZE [ADB, Cohn (forthcoming), Desmond]. Plant contributor to the Melbourne Herbarium. *Calochilus holtzei, Clerodendron holtzei, Eugenia holtzei, Eulophia holtzei, Habenaria holtzei, Hibbertia holtzei, Holtzea, Polyalthia holtzeana, Sida holtzei, Sterculia holtzei, Trichosanthes holtzei, Utricularia holtzei, Uvaria holtzei.* M correspondent*.

HOLTZE, NICOLAS [NICHOLAS], 1868-1913, gardener and botanical collector. Son of Moritz HOLTZE. Born in Russia. Went to Australia with his family in 1872. Curator, Darwin Botanic Garden, 1891-1913 [Cohn (forthcoming), Desmond]. Plant contributor to the Melboune Herbarium. *Aristolochia holtzei, Habenaria holtzei, Sida hottzei, Sterculia Holtzei, Utricularia holtzei.* M correspondent*.

HOOKER, BRIAN HARVEY HODGSON, 1860-1932, mine manager. Third son of Sir Joseph HOOKER (vol. 1) [Allan, 1967]. Met M at Melbourne in 1885 [L85.11.28]. Worked at mines in Vic, Qld, NZ and WA. M correspondent but none has been found [L86.06.22 (Coll. Corr.)].

HOOKER, REGINALD, 1867-1944, British statistician. Fourth son of Joseph HOOKER (vol. 1). Educated at Paris (B-ès-Sc) and Cambridge. Assistant in Intelligence Department of the Board of Agriculture; later head of its statistics branch [Allan (1967)].

HOOKER, SOPHIE. See Sophie WILLAN.

HOPPE, DAVID HEINRICH, 1760-1846, German botanist. Trained as a pharmacist at Celle, 1775-80. Joined a pharmacy at Regensburg, 1786. Founder of the botanical society at Regensburg. Teacher at the lyceum and examiner in pharmacy, 1803-25. One of the founders of the journal *Flora* [NDB]. Met M at Kiel in 1846 [L87.12.16].

HOUGH, FRANKLIN BENJAMIN, 1822-85, American forester. Educated at Union College, Schenectady (BA, 1843) and Western Reserve Medical College (MD, 1848). Physician at Somerville, New York, 1848-52. Directed State censuses, New York, 1855, 1865; District of Columbia, 1867; US census, 1870. Forestry agent, US Department of Agriculture from 1876 [ANB].

HUDDLESTON, [FRANK], New Zealand surveyor. Ranger in Mt Cook area, 1884-94 [Website rootsweb.com/~nzl/scant/artists.htm (March 2004)].

HUME, MARY BOZZAM, 1838-1915. Married Robert Kennedy, 1858. *Grevillea kennedyana.* M correspondent [VBOT].

HUSSEY, JESSIE LOUISA [LOUISE], 1862-99, botanical collector. Born in SA. Collected marine algae and land plants in SA for M and J. AGARDH (vol. 2) [Cohn (forthcoming), Desmond]. Plant contributor to the Melbourne Herbarium. M correspondent*.

INGHAM, William Bairstow, 1850-78, trader and government agent. Born in England. Went to Tas, 1873, and to north Qld, 1874. Government agent at Port Moresby, New Guinea, 1877-8 [ADB].

IRVINE, Jemima France. See Jemima Burn.

IVEY, William Edward, 1838-92, agricultural scientist. Born in Tas. Educated at Royal Agricultural College, Cirencester, England. Chemist, Vic. Department of Agriculture, and later superintendent of experimental farm reserves, 1874-8. Director, Lincoln College, NZ, 1878-92 [DNZB].

JACKSON, Benjamin Daydon, 1846-1927, English botanist. FLS, 1868. Botanical Secretary, Linnean Society, 1880-1902; General Secretary, 1902-26. Editor, *Index Kewensis* [Desmond].

JACKSON, Charles Loring, 1847-1935, American chemist. Educated at Harvard University (AM, 1870), Heidelberg and Berlin. Assistant, later professor of chemistry, Harvard University, 1871-1911 [WWA]. M correspondent*.

JACQUIN, Nikolaus Josef, 1727-1817, botanist. Born in the Austrian Netherlands. Studied medicine at Paris and Vienna. Natural history expedition to the West Indies and South America, 1754-9. Professor of chemistry at Schemnitz, Hungary, 1753-8; Vienna 1768-96. Created baron, 1806. Prolific author on chemical and botanical subjects [DSB].

JAMES, John Stanley, 1843-96, writer. Born in England. Became a journalist with limited success. Went to America in 1872 and changed his name to Julian Thomas. Went to Australia in 1875, travelling widely in eastern states. Achieved a successful career, using the alias 'The Vagabond' [ADB]. M correspondent*.

JEPHCOTT, Edwin, 1823-1911, pastoralist. Born in England. Went to Brisbane, 1861, working at the botanic garden. Settled at Ournie, NSW, 1864, where he established an arboretum [Willis et al. (1986)]. M correspondent*.

JOCHMUS VON CATIGNOLA, August von, 1808-81, soldier. Born in Germany. Studied military science in Paris, 1827. Served in Greece from 1827; in War Ministry, 1832. Went to England in 1835 and served in Spain; brigadier-general, 1837. Returned to England, 1838. Chief of the General Staff, Turkish-English-Austrian army, 1840. Briefly Minister of Marine and Foreign Affairs, Austria, 1849, but then retired to private life. Travelled extensively. Created Baron, 1866 [DBE].

JOHNSON, Sidney, fl. 1880s-90s, schoolteacher. At Meredith, Vic., 1883-4; at Portland, Vic., 1895. Collected extensively in the Brisbane Ranges, Vic [VBOT]. Plant contributor to the Melbourne Herbarium. M correspondent*.

JOHNSON, William, 1825-87, chemist. Went to Melbourne in the 1850s. Employed by Victorian government to analyse water of Yan Yean reservoir, 1858; appointed government analyst, 1886; member of Royal Commission on noxious trades, 1879. Foundation member of the Pharmaceutical Society of Victoria; member of the Board of Pharmacy [Gibbney & Smith (1987)]. *Thryptomene johnsonii.* M correspondent but none has been found [L82.05.02].

JOHNSTON, Robert Mackenzie, 1844-1918, public servant, geologist, palaeobotanist and statistician. Born in Scotland. Went to Vic. in 1870, later to Tas. Registrar-General and Government Statistician, Tas., 1881. Wrote on palaeontology, geology, zoology and botany. FLS 1879, FRGS. ISO, 1903 [ADB]. *Araucaria johnstonii.* M correspondent*.

JONES, Oliver Llewellyn, 1849-1933, policeman. Born at Guildford, WA. Member of WA Police Force 1872-84. Accompanied M on his travels in WA [Hall (1984)]. Plant contributor to the Melbourne Herbarium. M correspondent*.

JUDD, John Wesley, 1840-1916, English geologist. Educated at Royal School of Mines. With Geological Survey of England and Wales, 1867-70. Professor of Geology, Royal College of Science, London, 1876-1905. FRS, 1877. CB, 1895 [WW].

JUSSIEU, Antoine Laurent de, 1748-1836, French botanist. Nephew of B. De Jussieu, father of A. H. L. De Jussieu (vol. 1). Assistant at King's Garden (Jardin du Roi), Paris, 1770-93; professor at the Natural History Museum (Muséum d'Histoire Naturelle), Paris, 1793-1826. An early promoter of the 'natural system' of botanical classification [DSB].

JUSSIEU, Bernard de, 1699-1777, French botanist. Assistant at King's Garden (Jardin du Roi), Paris, 1722-77. Developed the Trianon garden, Versailles, as a demonstration of the 'natural system' of botanical classification [DSB].

JUST, Leopold J., 1841-91, German botanist. Studied plant physiology at Breslau (DrPhil, 1870). Worked at various mines and metallurgical works in Upper Silesia, 1870-2. Professor of botany, Technical High School (Technische Hochschule), Karlsruhe, from 1877. Founded the Botanical Institure, Karlsruhe. Editor of *Botanischer Jahresbericht* from 1873 [DBE].

KALAKAUA, King of Hawaii, reigned 1874-93 [*Everyman's Encyclopaedia* (1949-50)]. M correspondent*.

KARSTEN, Herman Rudolf Jakob, 1801-82, German church official. Born at Rostock. Deacon at St Mary's Church, Rostock, 1828-48;

church superintendent at Doberan, 1848-50; at Schwerin, 1850-6; pastor to the administration of nursing and the insane, Schwerin, while retaining the rank of Superintendent, 1857-80 [Rostock city archives]. M employed his great-nephew as a collector.

KAUP, ELIZABETH VON. See Elizabeth KRICHAUFF.

KELTIE, JOHN SCOTT, 1840-1927, Scottish geographer. Editor with Macmillan publishers from 1871. Librarian and later secretary, RGS, London, 1885-1915. Editor, *Statesman's yearbook*, 1884-1927; editor, *Geographical journal*, 1893-1917. Knighted, 1918 [DNB]. M correspondent*.

KELVIN, LORD. See William THOMSON.

KEMPE, FRIEDRICH ADOLF *HERMANN*, 1844-1929, Lutheran missionary. Born in Germany. Went to SA, 1875. A founder of the Hermannsburg Mission, NT, 1877 [Hall (1984)]. Plant contributor to the Melbourne Herbarium [Cohn (forthcoming)]. *Acacia kempeana*, *Calotis kempei*, *Commerconia kempeana*, *Helichrysum kempei*, *Justicia kempeana*, *Millotia kempei*, *Pentatropis kempeana*. M correspondent*.

KENNEDY, MARY BOZZAM. See Mary Bozzam HUME.

KERNER VON MARILAUN, ANTON JOSEPH, 1831-98, Austrian botanist. DrMed, Vienna, 1854. High school teacher Budapest 1855-8; at the Polytechnic, 1858-60. Director, Innsbruck Botanic Garden, 1860-78; Vienna Botanic Garden, 1878-98 [TL2]. M correspondent.

KERNOT, WILLIAM CHARLES, 1845-1909, engineer and educationist. Born in England. Went to Geelong, Vic., with his family in 1851. Educated at the University of Melbourne (BA, 1864; MA, 1866). In Vic. government service, 1865-75. Professor of engineering, Melbourne, 1883-1909. President, Royal Society of Vic., 1885-1900 [ADB]. *Aster kernotii*.

KING, GEORGE, 1840-1909, British botanist. MB, Aberdeen, 1865. Superintendent of Calcutta Botanic Garden and responsible for cinchona cultivation in Bengal, 1871-98. Director, Botanical Survey of India, 1891-8 [TL2].

KING, GEORGINA, 1845-1932, anthropologist and geologist. Born in WA. Went to Sydney in 1847. Wrote extensively on geological and anthropological subjects [Gibbney & Smith (1987)]. M correspondent*.

KING, HENRY SANDFORD, 1862-1931, public servant and surveyor. Born In Vic. Educated in Geelong and Melbourne. Joined WA Lands Department, 1884. Led surveying operations in WA, 1885. Surveyor-general, WA, 1918-23 [Gibbney & Smith (1987)]. Plant contributor to the Melbourne Herbarium. *Stemodia kingii*.

KING, Joseph, 1839-1923, Congregational minister. Born in England. Became a missionary of the London Missionary Society in 1860. Went to Melbourne in 1863, then to Samoa. Returned to England in 1872, but went again to Melbourne in 1874 as minister at Sandhurst (Bendigo), 1874-81, and at South Melbourne, 1881-9 [ADB].

KING, Mark Last, 1809-79, actor, auctioneer and politician. Born in England. Went to Melbourne in 1838. Associated with George Coppin. MLA, Vic., 1859-61, 1864-74, 1875-9 [Thomson & Serle (1972)].

KING, Philip Gidley, 1817-1904, pastoralist. Born at Parramatta, NSW. Educated in England. Midshipman in HMS *Beagle*, 1831-5, when he became a friend of Charles Darwin (vol. 1). Returned to Sydney, 1836. Engaged in pastoral activities in NSW, 1837-80. Commissioner for the Sydney International Exhibition, 1879. Appointed to NSW Legislative Council, 1880 [ADB].

KIRCHHOFF, Karl Reinhold *Alfred*, 1838-1907, geographer. Studied at Jena and Bonn (PhD, 1861). Teacher in Berlin, 1865-73. Professor of Geography at Halle, 1873-1904 [Univ. Halle archives]. M correspondent*.

KITTEL, Baldwin Martin, 1798-1885, German naturalist. Rector of Agricultural and Trade School (Landwirthschafts- und Gewerbsschule) at Aschaffenburg, 1834-69 [TL2].

KJELLMAN, Frans Reinhold, 1846-1907, Swedish botanist. Educated at Uppsala (DrPhil, 1872). Collected in Spitzbergen, 1872-3; Novaya Zemlya, 1875. Botanist with Nordenskiöld on the *Vega* expedition, 1878-80. Professor of Botany, Uppsala, 1883-1907 [TL2].

KNOBLAUCH, Karl Hermann, 1820-95, German physicist. Professor of physics at Halle [Poggendorff (1863-)].

KOENIG, Johann Gerhard, 1728-85, Baltic-German physicist and physician. In Denmark, 1759-67; from 1767 in India, Siam, Malacca and Ceylon as missionary and collector [TL2].

KÖLLIKER, Rudolph Albert von, 1817-1905, anatomist. Born in Switzerland. DrPhil, Zürich, 1841; MD, Heidelberg, 1842. Professor of anatomy, Heidelberg, 1844-7; Würzburg from 1847 [TL2].

KÖRNER, Karl *Theodor*, 1791-1813, German poet. Studied at mining school, Freiberg, 1808-10, and at Leipzig and Vienna. Killed while fighting with the Prussian army against the French [DBE].

KOTSCHY, Carl Georg Theodor, 1813-66, Austrian botanist. Botanical explorer in western Asia, 1835-43. Assistant at the herbarium, Vienna Natural History Museum, 1847-52; curator, 1852-66 [Cohn (forthcoming), TL2]. Specimens are in the Melbourne Herbarium [L80.08.04]. M correspondent.

KREFFT, ANNIE. See Annie MCPHAIL.

KRICHAUFF, ELIZABETH, sister of Friedrich KRICHAUFF (vol. 1). Married Reinhard von Kaup, burgomaster at Husum.

KRISTENSEN, LEONARD, Norwegian ship's captain. Commander, whaling ship *Antarctic*. Made first landing on the Antarctic continent, 1895 [*Trans. RGSA (Vic)*, v. 12/13 (1896), p. 73].

KUCKUCK, ERNST HERMANN PAUL, 1866-1918, German algologist. DrPhil, Kiel, 1891. With Reinke at Kiel, 1888-92; at the Heligoland biological station, 1892-1914; at the Botanical Museum, Dahlem (Berlin), 1914-8 [TL2].

KUNTZE, CARL ERNST *OTTO*, 1843-1907, German botanist. DrPhil, Freiburg, 1878. Travelled around the world, 1874-6, 1904. Polemical nomenclatural reformer [TL2].

LAMARCK, JEAN BAPTISTE PIERRE ANTOINE DE MONET DE, 1744-1829, French naturalist. Member of the Academy of Sciences (Académie des Sciences), Paris, from 1779. Botanist at the King's Garden (Jardin du Roi), Paris, 1788-93; professor of invertebrate zoology, Natural History Museum (Muséum d'Histoire Naturelle), Paris, 1793-1829. Developed theory of evolution through inheritance of acquired characteristics [DSB].

LA MESLÉE, EDMOND MARIE MARIN. See Edmond Marie MARIN LA MESLÉE.

LAMPERT, KURT, 1859-1918, German zoologist. Studied at Erlangen and Munich. Assistant, natural history museum, Stuttgart, 1884-90; curator, zoological and botanical department, 1890-4; director of the natural history collection, 1894-1918 [DBE].

LANGRIDGE, GEORGE DAVID, 1829-91, politician. Born in England. Went to NSW in 1853, then to Vic. Worked in Melbourne as a carpenter, later as an estate agent. Mayor of Collingwood, Melbourne, 1867 and 1872. MLA, 1874-91. Held various senior positions in government, being Chief Secretary, 1890, and Acting Premier at the time of his death [ADB]. M correspondent*.

LARNACH, JAMES MACDONALD, 1837-87. Born In Scotland. Went to NSW, then Vic. Manager with Bank of Victoria, 1864-87. Founding member of the RGSA and a member of the Society's Council [*Trans. & Proc. RGSA (Vic)*, v.4 (1886), p. 163].

LAUTERER, JOSEPH, 1848-1911, surgeon. Born in Germany. MD. Surgeon in Franco-Prussian war. Went to NSW in 1885 and then settled in Qld. President, Royal Society of Qld, 1896. Interested in flora of Moreton Bay [Desmond].

LAWES, JOHN BENNETT, 1814-1900, English agricultural chemist. Educated at Oxford. Founded Rothamsted Experimental Station on his estate, 1843. FRS, 1854; Royal Medal, 1867. Created baronet, 1882 [DNB].

LAWES, WILLIAM GEORGE, 1839-1907, English missionary. Joined London Missionary Society, 1858. In Niue 1861-72; New Guinea from 1874. Wrote first book in a Papuan language. Adviser to administrators of Papua. Retired to Sydney in 1906. FRGS [ADB]. *Dendrobium lawesii*, *Didymocarpus lawesii* . M correspondent*.

L'ECLUSE [L'ESCLUSE], CHARLES de. See Carolus CLUSIUS.

LE CREN, CHARLES, 1831-89, public servant. Born in England. Went to Melbourne, 1849, as a customs agent. Joined Vic. public service, 1858; Secretary, Department of Public Works, 1878-88 [Gibbney & Smith (1987)].

LEECH, FREDERICK WILLIAM, explorer. Second-in-command of Elder expedition to north-western Australia, 1891-2 [Feeken, Feeken & Spate (1970)].

LE MOUAT, JEAN EMMANUEL MAURICE, 1799-1877, French physician and botanist in Paris [TL2].

LENDENFELD, ROBERT IGNAZ LENDLMAYR VON, 1858-1913, Austrian phycologist and geologist. Educated at University of Graz; DrPhil in zoology. In Australia, 1881-5, studying invertebrate marine fauna. At University of Innsbruck, 1887-95; professor of zoology at German University, Prague, 1895-1913 [Walker & Tampke (1991)]. *Veronica lendenfeldii.*

LE SOUEF, ALBERT ALEXANDER COCHRANE, 1828-1902, zoo administrator. Born in England. Went to Melbourne in 1840. Overseer on various pastoral properties in northern Victoria. Usher of the Black Rod, Vic. Legislative Council, 1863-93. Director of the Melbourne Zoological Gardens, 1882-1902 [ADB]. M correspondent.

LÉVEILLÉ, AUGUSTIN ABEL HECTOR, 1863-1918, French botanist and clergyman. At Pondicherry, 1887-92, later at Le Mans. Founder and director of the *Monde des plantes* [TL2]. M correspondent.

LEWELLIN, AUGUSTUS JOHN RICHARD, medical practitioner. Son of J. H. H. and G. E. G. LEWELLIN. Educated at Melbourne (MB, 1873, ChB, 1879) [*Medical register* (1882-92)].

LEWELLIN, GRACE E. G. Wife of Dr J. LEWELLIN.

LEWELLIN, JOHN HENRY HILL, 1818-86, medical practitioner. Born in India. Trained as a surgeon (MRCS, 1842). Went to Australia in 1852; registered as a medical practitioner in Vic, 1854 [www.medicalpioneers.com/cgi-bin/index (January, 2005)].

LIEBMANN, FREDERIK MICHAEL, 1813-56, Danish botanist. Lecturer at Copenhagen veterinary college, 1837-40, 1845-53. Travelled in Cuba and Mexico, 1840-3. Extraordinary professor of botany, Copenhagen, 1845-52; professor, 1852-6 [TL2].

LINDERN, FRANZ BALTHASAR VON, 1682-1755, German botanist [Barnhart (1965)].

LINDHEIMER, FERDINAND JACOB, 1801-79, botanist. Born in Germany. Went to USA in 1834 as a political refugee. Settled in Texas and collected there [Cohn (forthcoming), TL2]. Plant contributor (via Asa GRAY, vol. 1) to the Melbourne Herbarium [L80.08.04].

LINDLEY-COWEN, LANCELOT, 1858-1902, journalist and public servant. Lieutenant, US Navy. Settled in Vic., 1880; agricultural writer for *Leader* (Melbourne). Went to WA, 1889; journalist in Albany and later in Perth. Appointed Secretary of WA Agriculture Department, 1894 [Gibbney & Smith (1987)]. M correspondent*.

LINDSAY, DAVID, 1856-1922, explorer and surveyor. Born at Goolwa, SA. Senior surveyor, SA Survey Department, 1875. Surveyor-General for NT, 1878-82. Led expedition into Arnhem Land, NT, 1883. Later explored territory between the Overland Telegraph Line and the Qld border, and in 1885-6 explored in the region of the MacDonnell Ranges, NT. Surveyed township of Alice Springs, 1888. Led Elder Exploring Expedition to north-west Australia, 1891-2 [ADB]. M correspondent*.

LINDT, JOHN WILLIAM, 1845-1926, photographer. Born in Germany. Went to Qld in 1862. Operated photography business in Grafton, NSW, from c.1864. Went to Melbourne c.1876 and opened a studio there. Visited New Guinea, 1885, and New Hebrides, 1890. In 1895 established a home and pleasure resort at Blacks Spur, Vic. [ADB].

LISTER, JOSEPH, 1827-1912, British surgeon. BA, London, 1847. FRCS, 1852. Professor of surgery, Glasgow, 1860-9; Edinburgh, 1869-77; King's College, London, 1877-92. The founder of antiseptic surgery. FRS, 1860; PRS 1895-1900. Created baron, 1897; OM, 1902 [DNB].

LITTLE, JOHN ARCHIBALD GRAHAM, 1843-1906, telegraph official. Born in SA. Joined SA telegraph department, 1857. Telegraph stationmaster at Darwin, 1871-1906, responsible for the section of the overland telegraph line between Darwin and Attack Creek [NTBD].

LIVERSIDGE, ARCHIBALD, 1846-1927, chemist. Born in Britain. Educated at Royal School of Mines, London, and at Cambridge. Went to Sydney, 1872, as Reader in geology at University of Sydney; professor of geology and mineralogy, 1874; professor of chemistry, 1881-1909. Instrumental in establishment of the AAAS, 1888; Secretary, 1888-1909; President 1898. FRS, 1882; FRGS [ADB]. *Liversidgea.* M correspondent*.

LOCH, HENRY BROUGHAM, 1827-1900, British public servant and colonial governor. Born in Scotland. CB, 1860; KCB, 1880. Governor of Vic., 1884-9. GCMG, 1887; GCB, 1892. Governor of Cape Colony; 1889-95. Created Baron Loch of Drylaw, 1895 [ADB]. M correspondent.

LÖFLING [LOEFLING], PEHR, 1729-56, Swedish botanist and explorer. Studied at Uppsala with Carl LINNAEUS (vol. 1). Travelled in South America, 1754-6 [TL2].

LOJKA, HUGO, 1845-87, Hungarian lichenologist. Studied in Vienna. High school teacher, Budapest [TL2].

LONGFELLOW, HENRY WADSWORTH, 1807-82, American poet. Born in Maine. Professor of modern languages and literature, Harvard University, from 1835 [DAB].

LORING, JANE LATHROP, c.1825-93, married Asa GRAY (vol. 1), 1848. M correspondent*.

LORNE, MARQUIS OF, 1845-1914, British aristocrat. Baptised John Douglas Sutherland Campbell. Private secretary to his father when the latter was Secretary of State for India, 1868-71. MP, 1868-78, 1895-1900. Married Princess Louise, 4th daughter of Queen VICTORIA (vol. 1), 1871. Governor-General of Canada, 1878-83. Constable of Windsor Castle, 1892-1914. Succeeded his father as 9th Duke of Argyll, 1900 [DNB].

LOUREIRO, JOAO DE, 1717-91, Portuguese missionary and naturalist. Collected in Mozambique, Goa and Cochin China [TL2].

LOW, HUGH, 1824-1905, British colonial administrator. Visited Borneo and became acquainted with the Rajah of Sarawak, Sir James Brooke, 1845. Inspector to colony of Labuan, 1848; British Resident at Perak, Straits Settlements (Malaya), 1877-87. GCMG, 1883 [WW].

LUBBOCK, JOHN, 1834-1913, English banker and man of science. Joined family bank, 1849. MP, 1879-1900. FRS, 1858. President, Linnean Society, 1881-6. Created 1st Baron Avebury, 1900 [DNB]. M correspondent.

LUCAS, ARTHUR HENRY SHAKESPEARE, 1853-1936, teacher and naturalist. Born in England. Educated at Oxford (MA, 1877) and London

(BSc, 1879). Went to Melbourne, 1883. Mathematics and science master at Wesley College, Melbourne. Editor, *Vic. Naturalist*, 1884-92; President, Field Naturalists' Club, Vic., 1887-9. Headmaster, Newington College, Sydney, 1892-8; science master, Sydney Grammar School, 1899-1920; headmaster, 1916-23. President, Linnean Society of NSW, 1907-9 [ADB]. M correspondent*.

LUERSSEN, Christian, 1843-1916, German botanist. DrPhil, Jena, 1866. Lecturer at Leipzig, 1869-84; at forestry school, Eberswalde, 1884-88. Professor of botany at Königsberg, 1888-1910 [TL2]. M correspondent but none has been found [e.g. L78.12.28 (Coll. Corr.)].

McALPINE, Daniel, 1849-1932, mycologist. Born in Scotland. Lecturer, Heriot Watt College, Edinburgh, 1877. Went to Melbourne in 1884. Lecturer in biology, University of Melbourne, 1884; College of Pharmacy, Melbourne, 1886. Plant pathologist, Department of Agriculture, Vic., 1890-1915 [ADB, Desmond, TL2]. M correspondent*.

MACARTNEY, Hussey Burgh, 1799-1894, Church of England clergyman. Born in Ireland. Educated at Trinity College, Dublin (BA, 1821; DD, 1847). Ordained 1823. Went to Vic. in 1847. Archdeacon of Geelong, 1848-52; dean of Melbourne, 1852-94 [ADB].

McCANN, Annie W. D. See Annie Bellew McDonald.

McCLINTOCK, Francis Leopold, 1819-1907. British naval officer. Born in Ireland. Entered Royal Navy, 1831. Served on several Arctic expeditions. Discovered remains of Sir John Franklin (vol. 2), 1859. Knighted, 1860. Eventually Vice-Admiral in command of the North America and West Indies station. Retired, 1884. KCB, 1891 [DNB].

McCREA, William, 1814-99, medical practitioner. Born in Ireland. MRCS England, 1833; LSA, London, 1834; MB, London, 1848; MB, Melbourne, 1868.Went to Castlemaine, Vic., 1852; later in Melbourne [*Medical Register* (1882-92)]. *Melodorum maccreai*. M correspondent*.

MACDONALD, Alexander Cameron, 1828-1917, accountant and geographer. Born in NSW. Overlanded stock to Vic., 1847-8. Surveyor and accountant at Geelong, Vic., 1850-76. Moved to Melbourne in 1876. Historian of Australian exploration. Secretary, Vic. Branch, RGSA, 1883-1906 [ADB]. *Coelogyne macdonaldi*, *Wormia macdonaldi*. M correspondent*.

McDONALD, Annie Bellew, 1838-1924, writer, flower painter and postmistress. Born in Ireland. Married T. McCann c.1856. Went to Vic., 1863 [Gibbney & Smith (1987)]. M correspondent*.

MACDONALD, DONALD, 1846-1927. Missionary to the New Hebrides. Author of an etymological dictionary of three languages of the New Hebrides. Representations about French military activity in the New Hebrides, 1886, led to an agreement for joint English-French government of the islands [*Argus*, 20 April 1927].

MACGREGOR, WILLIAM, 1846-1919, medical practitioner and colonial administrator. Born in Scotland. MB, Aberdeen; LRCP, Edinburgh. Went to Seychelles, Mauritius, Fiji. Administrator of British New Guinea, 1887-1898. KCMG, 1889. Governor, Newfoundland, 1904. GCMG, 1907. Governor, Qld, 1909-14. PC, 1914 [ADB, Cohn (forthcoming), Desmond]. Plant contributor to the Melbourne Herbarium. *Hypericum macgregorii, Quinetia macgregorii, Quintinia macgregorii, Rubus macgregorii, Santalum macgregori.* M correspondent.

MACKAY, JOHN, farmer, of Clover Creek, NSW. M correspondent*.

MCLEAN, ALLAN, 1840-1911, stock and station agent and politician. Born in Scotland. Went to Sydney with his parents, 1842, then to Gippsland, Vic. Established his business in Sale, Vic., 1873. MLA, 1880. Chief Secretary and Minister for Agriculture, 1891; Premier and Chief Secretary, 1899-1900 [ADB]. M correspondent*.

MACLEAY, WILLIAM JOHN, 1820-91, pastoralist, politician and entomologist. Born in Scotland. Went to Sydney, 1839. Pastoralist in NSW from 1840. Member of NSW Parliament, 1855-74; first president of Linnean Society of NSW, 1874. MLC, 1877. Knighted, 1889 [ADB]. M correspondent*.

MCLELLAN, WILLIAM, 1831-1906, politician. Born in Scotland. Went to Vic in 1850. Miner, 1851-9. MLA, 1859-77, 1883-97; Commissioner of Public Works, 1870; Minister of Mines, 1871-2, 1875-7 [ADB].

MCPHAIL, ANNIE. Married Johann Ludwig Gerard KREFFT (vol. 2) in 1869 [ADB]. M correspondent*.

MAGAREY, ALEXANDER THOMAS, 1849-1905, businessman. Born in Adelaide. Worked in the family milling business established by his father. Secretary, SA Branch, RGSA [Verco, (1935)]. M correspondent*.

MAIDEN, JOSEPH HENRY, 1859-1925, botanist. Born in England. Went to Australia, 1880. Curator, Technological Museum, Sydney, 1881-96. Director of Sydney Botanic Garden and NSW Government Botanist, 1896-1924. FLS, 1889; FRS, 1916 [Gilbert (2001)]. Plant contributor to the Melbourne Herbarium. *Acacia maidenii, Eucalyptus maideni, Medinilla maidenii.* M correspondent.

MAILLARD, PIERRE NÉHÉMIE, 1813-83, French botanist. Priest at La Mothe-St.-Héray (Deux-Sèvres), 1842-83 [TL2]. Specimens are in the Melbourne Herbarium [L80.08.04].

MAINGAY, ALEXANDER CARROLL, 1836-69, physician and botanical collector. Born in England. Educated at Edinburgh (MD, 1858). Joined the Indian Medical Service, 1859. Went to China, 1860. In charge of the jail at Malacca, Malaya, 1862-7 [Desmond].

MAISTRE, PAUL LOUIS PHILIBERT, c.1851-?, French consular official. Secretary in French Embassy, London, 1881-5. Vice-consul for France at Melbourne, 1889-98; at London, 1898-1901; Consul at Melbourne, 1901-8 [Thornton-Smith (1994)]. M correspondent*.

MANCINI, PASQUALE STANISLAO, 1817-88, Italian patriot and politician. Professor of international law, Turin, 1850-60; Rome 1872-. A leading spokesman of the Italian *Risorgimento*. A member of the democratic left in the Italian Parliament from 1860; Minister of Justice, 1876-8; Foreign Minister, 1881-5. Tutor of King UMBERTO I [*Enciclopedia Italiana*].

MANN, JOHN FREDERICK, 1819-1907, explorer and surveyor. Born in England. Went to Sydney, 1842. Took part in early expeditions led by LEICHHARDT (vol. 1). Licensed surveyor, 1848-79 [ADB]. M correspondent.

MARIN LA MESLÉE, EDMOND MARIE, 1852-93, public servant. Born in France. Went to Mauritius, 1870, then to Melbourne, 1876 and Sydney, 1878. Travelled extensively in western NSW, 1880. Founding member of the RGSA, 1883 [ADB]. M correspondent.

MARINUCCI, LUIGI, c.1836-79, Italian consul. Born in Naples. Member of Italian consular service for 20 years. Consul to Victoria, 1873; Consul-General to Australian colonies, 1877 [Gibbney & Smith (1987)]. M correspondent.

MARKHAM, CLEMENTS ROBERT, 1830-1916, British civil servant and geographer. Served in Royal Navy, 1844-51. Travelled in Peru, 1852-3. Joined British civil service, 1853; in India Office from 1858 and head of its geographical section, 1867-77. In South America, 1860; involved in the transfer of cinchona plants from Peru to India. Secretary, Royal Geographical Society, from 1863; President, 1893-1905 [DNB]. M correspondent.

MARSH, GEORGE PERKINS, 1801-82, American geographer and writer. Educated at Dartmouth College (AB, 1820). Practising lawyer, Vermont, 1825-42. Member, US Congress, 1843-9. Regent of Smithsonian Institution, 1847. US minister to Turkey, 1849-54; minister to Italy, 1861-82. Inspiration behind establishment of US Forest Commission [DAB].

MASSMANN, MAGNUS CHRISTIAN, 1835-1915, German official. Educated at Rostock (LLD, 1864) and Berlin. City councillor at Rostock from 1864; director of Rostock police office, 1872-80; mayor, 1889-1914 [Rostock city archives]. M correspondent*.

MATTHIESSEN, LOUISE THERESA. See Louise WEHL (vol. 1). M correspondent*.

MAXIMOWITZ [MAXIMOWICZ], CARL IVANOVICH, 1827-91, Russian botanist and explorer. Travelled in East Asia, 1853-7; China and Japan, 1859-64. Curator of the St Petersburg Herbarium [TL2].

MECKLENBURG, GRAND DUKE OF. See FRIEDRICH FRANZ III.

MEEHAN, THOMAS, 1826-1901, American nurseryman and botanical collector. Born in England. Gardener at Kew, 1846-8. Went to USA, 1848. Established nursery at Germantown, Pennsylvania c. 1853. State Botanist. Collected plants in New England, Colorado, Utah and Alaska [Desmond, TL2]. M correspondent.

MENZIES, ARCHIBALD, 1754-1842, British botanical collector. Born in Scotland. Assistant surgeon, Royal Navy, from 1782; surgeon-naturalist on Vancouver's exploring expedition, 1791-4, during which he collected extensively on the south-west coast of WA [Desmond].

MEZ, CARL CHRISTIAN, 1866-1944, German botanist. DrPhil, Berlin, 1888. At Munich, 1888-90; at Breslau, 1890-9. Professor of botany, Halle, 1899-1910. Director of the botanic garden, Königsberg, 1910-35 [TL2].

MIKLOUHO-MACLAY, NIKOLAI NIKOLAYEVITCH, 1846-88, Russian naturalist. Educated at St Petersburg, Heidelberg and Leipzig. Went to New Guinea, 1871, and over several years pursued anthropological and ethnological studies throughout the south-west Pacific area. Instrumental in the establishment of a marine laboratory in Sydney. Returned to Russia, 1886 [ADB].

MILL, JOHN STUART, 1806-73, British philosopher. Educated by his father. Clerk at India Office, London, from 1823. Known especially for his writings on logic and the philosophy of science. An enthusiastic amateur botanist [DNB]. Specimens are in the Melbourne Herbarium [L80.08.04].

MILNE-EDWARDS, ALPHONSE, 1835-1900, French palaeontologist. DrMed, Paris, 1860; DSc, Paris, 1861. Professor at the Paris School of Pharmacy, 1865-95. From 1876 in various functions at the Natural History Museum (Muséum d'Histoire Naturelle), Paris; director, 1891-1900 [TL2]. M correspondent*.

MINCHIN, JAMES HENRY, 1859-1919, clerk. Assistant, Melbourne Herbarium, 1879-97. Seconded to Department of Agriculture briefly in 1895 [Cohn (forthcoming)].

MOLINEUX, ALBERT, 1832-1909, promoter of agriculture. Born in England. Went with his family to SA in 1839. Founder-editor of *Garden and field*, 1875-97. Secretary, Central Bureau, SA Bureaux of Agriculture, 1888-1902 [ADB].

MÖLLER, ADOLPHO FREDERICO, 1842-1920, Portuguese forester. Trained in Germany. Worked in the Portuguese forestry commission from 1860. Curator of the Coimbra botanic garden, 1874-1914 [Website: wwwuser.gwdg.de/~sysbot/index_coll/Search_M.htm (April 2004)]. M correspondent*.

MOLONEY, CORNELIUS *ALFRED*, 1848-1913, British colonial administrator. In West Africa from 1867. Governor, Lagos, 1887-91. Governor in the West Indies, 1891-1904 [TL2]. M correspondent*.

MOORE, THOMAS BATHER, 1850-1919, prospector, explorer, geologist and botanist. Born in Tas. Educated in England. Returned to Tas., 1868. Explored west and south-west Tas. [Hall (1978)]. Plant contributor to the Melbourne Herbarium. M correspondent.

MOORS, HENRY, c.1827-1906, public servant. Born in England. Went to Melbourne, 1852, and joined Victorian Public Service. Chief clerk, Police Department, 1883; Chief Secretary's Department, 1883-91 [Gibbney & Smith (1987)]. M correspondent*.

MORAN, MR R.W. Policeman. M correspondent*.

MORESBY, JOHN, 1830-1922, British naval officer. Joined Royal Navy as a cadet, 1842. Commander on the China Station, 1861-4; promoted to post captain, 1865. Surveyed in Torres Strait and on the New Guinea coast, 1872-4. Rear admiral, 1881; Vice-admiral, 1888 [ADB].

MORITZI, ALEXANDER, 1806-50, Swiss botanist. Worked with A.P. de CANDOLLE (vol. 1) in Geneva. Later high school teacher at Solothurn [Cohn (forthcoming), TL2]. Plant contributor to the Melbourne Herbarium (via herbarium of Joachim STEETZ (vol. 1)) [L63.12.12].

MORONEY, BARTHOLOMEW T., c.1850-1886, singer. Born in Ireland. Went to Vic. in 1856. Held a minor appointment in Vic. public service. Much in demand at concerts as a bass singer. Went to London in 1882 to pursue a singing career but returned to Melbourne c.1883 on account of ill health [Liedertafel Collection, Grainger Museum, University of Melbourne].

MORRIS, DANIEL, 1844-1933, British botanist. Born in Wales. MA, Dublin, 1876. Assistant, Peradeniya garden, Ceylon, 1877-9. Director, public gardens, Jamaica, 1879-86. Assistant director, Kew Gardens, 1886-98. CMG, 1893. Imperial Commissioner for Agriculture, West Indies, 1898-1908 [Desmond]. M correspondent.

MUELLER, AUGUSTUS, 1828-98, medical practitioner. Born in Germany. MD and ChD, Giessen, 1854. Went to Vic., 1855. Practised at Yackandandah, Vic. for 40 years. Developed proposal to cure snakebite by strychnine [Gibbney & Smith (1987), *Medical Register* (1882-92)].

MÜLLER, FRIEDRICH *MAX*, 1823-1900, orientalist and philologist. Born in Germany. Educated at Leipzig (PhD, 1843). Became an expert in the study of Sanskrit. Went to England, 1846. Deputy Taylorian Professor of modern languages, Oxford, 1850-4; professor, 1854-68; professor of comparative philology, 1868-1900 [DNB].

MULLEN, SAMUEL, 1828-90, bookseller and publisher. Born in Ireland. Went to Melbourne in 1852. Assistant at George ROBERTSON's bookshop from 1853; opened his own business in 1859 [ADB].

MUNRO, WILLIAM, 1818-80, British botanist. In the Indian colonial army. FLS, 1840; CB, 1857. Collected in Barbados, 1870-5, Crimea and India [Desmond, TL2]. *Panicum munroi.* M correspondent*.

MURRAY, GEORGE ROBERT MILNE, 1858-1911, Scottish cryptogamist. Assistant, 1876-95, and keeper, 1895-1905, Department of Botany, British Museum (Natural History). Naturalist on the West Indian Eclipse Expedition of 1886. Scientific Director, National Antarctic Expedition in *Discovery*, 1901 [TL2]. M correspondent*.

MURRAY, JOHAN ANDREAS, 1740-91, botanist. Born in Sweden. Pupil of Carl LINNAEUS (vol. 1). Studied at Göttingen, 1760-3. Professor of medicine and botany, Göttingen, 1764-91 [TL2].

MUSSON, CHARLES TUCKER, 1856-1928, teacher and botanist. Born in England. Educated at Nottingham; lecturer in botany, Nottingham University College. Went to Australia, 1887. Science master, Hawkesbury Agricultural College, NSW, 1891-1919 [Desmond; Gibbney & Smith (1987)]. M correspondent*.

NACHTIGAL, GUSTAV, 1834-85, German physician and botanist. Military surgeon at Cologne from 1858. Between 1861 and 1875 travelled extensively in southern Mediterranean countries. Entered German consular service in 1882, serving in various parts of northern Africa [*Brockhaus' Konversations-Lexikon*].

NECKER, NOEL JOSEPH DE, 1730-93, French botanist and historiographer. Born in Lille. Court botanist at Mannheim [TL2].

NEGRI, CRISTOFORO, 1809-96, Italian geographer. Studied at Pavia. Bureaucrat in Piedmont from 1848 and later in the government of a united Italy, with special responsibilities for foreign affairs. Founder and president for life of the Italian Geographical Society, 1867. Created baron,

1882 [*Enciclopedia italiana*]. Father of Maria NEGRI. *Negria.* M correspondent.

NEGRI, MARIA LUISA, 1852-95. Daughter of Cristoforo NEGRI. M bequeathed to her his private correspondence, certificates and notebooks but she predeceased him. M correspondent.

NELLE, AMALIE, c. 1821-91, German naturalist. Married Wilhelm August Salomon Dietrich c.1846. In Qld, 1863-73, as natural history collector for Godeffroy Museum, Hamburg, then returned to Germany [ADB]. Plant specimens collected in Qld are in the Melbourne Herbarium. *Acacia dietrichiana.*

NEWBERY, JAMES COSMO, 1843-95, industrial chemist and public servant. Born in Italy of British parents. Went to US, then to England. Educated at Harvard University (BSc, 1864) and the Royal School of Mines, London. Went to Melbourne in 1865 to be analyst to Geological Survey of Vic. Scientific superintendent of Industrial and Technological Museum, Melbourne, 1870-95. CMG, 1881 [ADB]. M correspondent.

NORDENSKIOLD, NILS ADOLF ERIK, 1832-1901, Finnish-born Swedish explorer, geographer and geologist. Leader of the *Vega* expedition (1878-80) that discovered the northeastern passage from Europe around north Asia [TL2].

NORDSTEDT, CARL FREDRIK *OTTO*, 1838-1924, Swedish algologist. Amanuensis at Lund botanical institute, 1878-86; curator, 1886-1909; professor from 1903 [TL2]. M correspondent*.

NORMANBY, GEORGE AUGUSTUS CONSTANTINE PHIPPS, 2ND MARQUIS OF, 1819-90, British colonial governor. Officer in Scots Guards, 1838-46. MP, 1847-50, 1852-7. Lieut-governor of Nova Scotia, 1858-63; governor of Qld, 1871-4; NZ, 1874-8; Vic, 1879-84. GCMG, 1877; GCB, 1885 [ADB].

NORTON, JAMES, 1824-1906, solicitor. Born in Sydney. Admitted as a solicitor, 1848. Trustee of the Australian Museum, 1874-1906. A founder of the Linnean Society of NSW; president, 1899, 1900 [ADB].

NYLANDER, WILLIAM, 1822-99, botanist. Born in Finland. MD, Helsinki, 1847. Professor of botany, Helsinki, 1857-63. In Paris, 1863-99 [TL2].

NYULASY, MR C.W., surveyor and botanical collector. Draughtsman in Lands & Survey Dept., Vic., from 1861; later with Survey Dept., WA. Collected plants in Kimberley District, WA. Plant contributor to the Melbourne Herbarium [VBOT biog. files].

ÖRSTED, ANDERS SANDO, 1816-72, Danish botanist. Travelled in Central America, 1845-8. DrPhil, Copenhagen, 1854. Lecturer, later professor of botany at Copenhagen, 1851-62 [TL2].

O'HALLORAN, THOMAS JOSEPH SHULDHAM, 1835-1922, magistrate. Born in England. Son of T. S. O'HALLORAN. Went to SA with his family in 1838. Educated at St Peter's College, Adelaide. Treasury cadet, 1853. Employed in National Bank, 1865-72. Stipendiary magistrate, 1874, with country appointments [Gibbney & Smith (1987)].

O'HALLORAN, THOMAS SHULDHAM, 1797-1870, police commissioner. Born in India. Educated at Royal Military College, UK. British Army, 1813-38; in India, 1813-34. Retired as major. Went to SA in 1838. Commissioner of police, 1840-3. MLC, 1843-63. Father of T. J. S. O'HALLORAN [ADB].

O'LOGHLEN, BRYAN, 1828-1905, politician. Born in Dublin. Educated at Trinity College, Dublin (BA, 1856). Went to Melbourne, 1862. MLC, Vic., 1878. Premier, 1881-2. Baronet, 1877. Continued in Parliament until 1900 [ADB]. M correspondent*.

ORD, HARRY ST GEORGE, 1819-85, English soldier and colonial governor. Educated at the Royal Military Academy, Woolwich. Served in Ireland, the West Indies, West Africa and Ascension. Special Commissioner, Gold Coast; then Lieut-Governor, Dominica, 1857. Governor, Bermuda, 1861; Straits Settlements, 1867. Lieut-Governor, SA, 1877. KCMG, 1877. Governor, WA, 1878-80. GCMG, 1881 [ADB]. M correspondent*.

PAGET, JAMES, 1814-99, surgeon. Studied at St Bartholomew's Hospital, London (MRCS, 1836). Surgeon at St Bartolomew's Hospital from 1861. FRCS, 1843. Surgeon to Queen VICTORIA from 1858. FRS, 1851. Created Baronet, 1871 [DNB]. *Pagetia*.

PALISOT DE BEAUVOIS, AMBROSE MARIE FRANÇOIS JOSEPH, 1752-1820. French traveller and botanist. Educated at Paris. Government official, 1772-7. Travelled in West Africa, 1786-8; Saint Domingue (Haiti), 1788-91; USA, 1791-8. Elected to the Academy of Sciences (Académie des Sciences), Paris, 1806 [NBG].

PANCHER, JEAN ARMAND *ISODORE*, 1814-77, French botanist and collector. Gardener and collector in Tahiti 1849-57; in New Caledonia, 1857-69, 1874-7 [Cohn (forthcoming), TL2]. Plant contributor to the Melbourne Herbarium. *Pancheria*. M correspondent.

PANTON, JOSEPH ANDERSON, 1831-1913, public servant. Born in Scotland. Went to Sydney in 1851, then to Vic. Goldfields commissioner,

1852-8; later magistrate in various districts. Senior magistrate, Melbourne, 1874-1907. Leaseholder in Kimberley district, WA. CMG, 1895 [ADB].

PARKES, HENRY, 1815-96, journalist and politician. Born in England. Went to Sydney, 1839. Became a writer in 1840s; entered politics in 1849; MLC, NSW, 1854. Went to England, 1861-3. Re-entered NSW Parliament, 1864-70; Colonial Secretary, 1866; Premier, 1878-81, 1887-9. A leading figure in the Federation movement in Australia. KCMG, 1877; GCMG, 1888 [ADB]. M correspondent*.

PARRY, CHARLES CHRISTOPHER, 1823-90, American botanist and explorer. Born in England. Went to USA, 1832. MD, Columbia College, 1846. Botanist to Mexican Boundary Survey, 1849-61. Explored Rocky Mountains, 1861-7; botanist with Pacific Railway Survey, 1867-9. Botanist to Agricultural Department, Washington, 1869-72; to San Domingo Commission, 1871. Explored in Utah, Nevada, California and Mexico from 1872 [Desmond, TL2]. Plant contributor (via Asa GRAY [vol. 1]) to the Melbourne Herbarium [B63.05.01].

PASCO, CRAWFORD ATCHISON DENMAN, 1818-98, naval officer and police magistrate. Born in England. Joined the Royal Navy as a midshipman in 1830. Served with Owen Stanley in founding Port Essington, NT, 1838, and with J. L. STOKES (vol. 1) in HMS *Beagle* in surveying the north coast of Australia, 1839-42. Settled in Vic., 1852. Police magistrate at Williamstown, 1852-7, and at various country towns until 1878. A founding member of the RGSA (Vic.) and Chairman of its Antarctic Committee [ADB]. M correspondent.

PASSOW, EDUARD HANS FRIEDRICH, 1800-75, German official. City councillor at Rostock from 1840; mayor, 1871-5 [Rostock city archives].

PASTEUR, LOUIS, 1822-95, French chemist and microbiologist. Studied at École Normale Supérieure, Paris, 1843-6. Professor of chemistry, Strasbourg, 1849-54; Lille, 1854-7. Director of scientific studies, École Normale, 1857-67; director of physiological chemistry laboratory, 1867-88. Professor of chemistry, Sorbonne, 1867-74. Director of Institut Pasteur, Paris, 1888-95 [DSB].

PATTERSON, JAMES BROWN, 1833-95, butcher, auctioneer and politician. Born in England. Went to Vic., 1852. Farmed in Vic. before settling in Melbourne in 1870. MLA, Vic., 1870-95. Commissioner for Public Works, 1875, 1877-80; Commissioner for Railways, 1880-1. Premier and Chief Secretary, 1893-4. KCMG, 1894 [ADB]. M correspondent*.

PÉRON, FRANÇOIS, 1775-1810, French naturalist. Served in French army, 1792-4. Educated at Paris medical school (École de Medecine) and the Natural History Museum (Muséum d'Histoire Naturelle), Paris.

Naturalist and anthropologist on Nicolas BAUDIN's expedition to survey the coast of Australia, 1800-3 [ADB].

PERRIN, GEORGE SAMUEL, 1846-1900, forester. Born in England. Went to Vic., 1853; educated in Melbourne. Forester in SA, 1880; Chief Forester at Wirrabara, SA, 1885. FLS, 1885. Conservator of Forests, Tas., 1886-7; Vic., 1888-1900 [Desmond, Hall (1978)]. M correspondent*.

PERSIEH, W. ANTHONY, 1826-?, pharmacist and surveyor. Born in Belgium. Went to Vic., then to Qld by 1878. Laid out nursery of Cooktown Botanic Garden, Qld, 1885 [Cohn (forthcoming), Willis et al. (1986)]. Plant contributor to the Melbourne Herbarium. *Bauhinia persiehii*, *Hakea persiehana*. M correspondent*.

PERTHES, BERNHARD, 1858-1919, German publisher. Director of Justus Perthes Geographische Anstalt, Gotha, 1881-1919 [Painke (1985)]. M correspondent*.

PFAFF, CHRISTOPH HEINRICH, 1773-1852, German medical professor. Born at Stuttgart. Professor of medicine at Kiel, 1797-1845, where he established the chemical laboratory [DBE].

PHILIPPI, RUDOLPH AMANDUS [RODOLFO AMANDO], 1808-1904, botanist. Born in Germany. Studied in Berlin (DrPhil, 1830). In Italy 1830-2, 1838-40. Teacher at the Technical School, Kassel, 1835-8, 1840-51. Emigrated to Chile, 1851. Professor of botany and zoology, Santiago, 1853-74. Director, National Museum (Museo nacional) until 1897. Writer on Chilean flora [Cohn (forthcoming), TL2]. Plant contributor to the Melbourne Herbarium [B65.10.01]. M correspondent but none has been found [L58.04.01 (Coll. Corr.)].

PICHLER, KAROLINE, 1769-1843, Austrian novelist. Educated in languages, literature and music. A leading figure in Viennese culture. Author of ballads and historical romances [DBE].

PIERRE, JEAN BAPTISTE LOUIS, 1833-1905, French botanist. Born on Réunion. Studied in Paris and Strasbourg. At Calcutta botanic garden, 1861-5. Director of Saigon botanic garden, 1865-77. In France from 1877 [TL2].

PINSCHOF, CARL LUDWIG, 1855-1926, merchant. Born in Austria. Went to Melbourne in 1880; became a partner in an importing firm. Honorary consul in Vic. for Austria-Hungary from 1885 [ADB].

POIRET, JEAN LOUIS MARIE, 1755-1834, French priest and botanist. Travelled in North Africa, 1785-6 [TL2].

POLAK, JOSEPH, fl. 1880s. Gardener in WA. Recommended by M to be collector on proposed expedition to Kimberley district of north-west

Australia [L83.01.02]. Plant contributor to the Melbourne Herbarium. *Ptilotus polakii.*

POTTER, WILLIAM, 1836-1908, Baptist clergyman. Born in Ireland. Went to Australia, 1839. Educated in Hobart. Worked on Bendigo goldfields, Vic., 1852-3. Studied at University of Melbourne. Ordained in the Baptist Church. Founder of the Victorian Education League [Gibbney & Smith (1987)]. One of the executors of M's estate. M correspondent*.

PRAHL, PETER, 1843-1911, German physician and botanist. Born in Schleswig. Studied medicine at Wandsbeck. Military physician at Flensburg, 1868-70, 1876-9; at Hadersleben, 1871-6; at Kiel, 1879-88; at Stettin, 1890; at Wandsbeck, 1890-2; at Rostock, 1892-9. Retired to Lübeck, 1899. An ardent amateur botanist from his student days [TL2].

PRANTL, KARL ANTON EUGEN, 1849-93, German botanist. Dr Phil, Munich, 1870. Studied with von Naegeli and Ludwig RADLKOFER, and in 1871-6 at Würzburg with Sachs; professor of botany at the forestry college of Aschaffenburg, 1876-89 and at Breslau, 1889-93. Founder with Adolf ENGLER of *Die natürlichen Pflanzenfamilien* [TL2]. M correspondent but none has been found [L87.12.12 (Coll. Corr.)].

PRINGSHEIM, NATHANAEL, 1823-94, German botanist. DrPhil, Berlin, 1848. Lectured at Leipzig and Berlin, 1850-64. Professor of botany and director of the Jena Botanic Garden, 1864-8. Worked privately at Berlin from 1868. One of the founders of the German Botanical Society; President, 1882-94 [TL2]. M correspondent but none has been found [L92.02.04 (Coll. Corr.)].

RADLKOFER, LUDWIG ADOLPH TIMOTHEUS VON, 1829-1927, German botanist. DrMed, Munich, 1854. DrPhil, Jena, 1855. Professor of botany at Munich, 1863-1913; director of the Botanical Museum and State Herbarium, 1891-27 [TL2]. M correspondent.

RAMMELSBERG, KARL FRIEDRICH, 1813-99, German chemist. Studied chemistry at Berlin (DrPhil, 1837). Professor of chemistry, Berlin, 1846-99. Member of Berlin Academy of Sciences, 1855. Met M at Kiel in 1846 [L87.12.16, NDB].

RANFORD, HENRY SAMUEL, 1854-1934, surveyor. Born in WA. In government service in north-west Australia, 1884. Inspector of Lands & Surveys and Acting Surveyor-General, WA. Land agent at Katanning, WA, 1897. Education Officer, London, 1909 [*Bicentennial dictionary of Western Australians*]. M correspondent*.

RASMUSSEN, HANS PETER, 1861-1937, herbalist. Born in Denmark. Went to Qld, 1879, and to Sydney, 1884 [Gibbney & Smith (1987)].

REDDIN, FRANCIS, 1843-1926, public servant. Born in England. Went to Vic., 1859. Entered Vic. Public Service, 1875, employed in Education Department; later served in Premier's office. Transferred to Public Service Board, 1886; secretary, 1894. Secretary to Public Service Commissioner, Vic., c.1904-8 [Gibbney & Smith (1987)]. M correspondent*.

REICHENBACH, HEINRICH *GUSTAV*, 1824-89, German botanist. DrPhil, Leipzig, 1852. Extraordinary professor of botany, Leipzig, 1855-63; professor of botany and director of the botanic garden, Hamburg, 1863-89 [TL2].

REID, ROBERT, 1842-1904, merchant and politician. Born in Scotland. Went to Melbourne in 1855. Drapery importer from 1874, later with branches in Sydney, Adelaide and Brisbane. Represented Melbourne Chamber of Commerce at exhibitions in London, 1886, and Paris, 1889. MLA, Vic., 1892-1903 [ADB].

REINBOLD, THEODOR, d. 1918, German algologist. At Kiel University Botanic Garden [TL2]. M correspondent.

REINKE, JOHANNES, 1849-1931, German botanist. DrPhil, 1871. Assistant at Göttingen, 1871-2. Lecturer at Bonn, 1872-3. Professor of botany at Göttingen, 1873-84; at Kiel and director of Botanic Garden, 1887-1921. Member for Kiel University in Prussian parliament, 1894-1912 [TL2]. M correspondent.

RIDLEY, HENRY NICHOLAS, 1855-1956, British botanist. Assistant, Botany Department, British Museum (Natural History), 1880-8. FLS, 1881. Director, Botanic Gardens, Singapore, 1888-1911. FRS, 1907. Helped to introduce rubber into Malaya [Desmond, Hall (1984), TL2]. M correspondent*.

RIEDEL, LUDWIG, 1790-1861, botanist. Born in Germany. Served in the Prussian Army, 1813-5. Collected for St Petersburg in Brazil, 1820-36. At the Rio de Janeiro National Herbarium, 1836-8; director of the botanical section, National Museum (Museo nacional), 1842-61 [TL2]. Specimens are in the Melbourne Herbarium [L80.08.04].

RILEY, CHARLES VALENTINE, 1843-95, entomologist. Born in England. Went to USA, 1860. Agricultural writer, 1863-8. State entomologist, Missouri, 1868-77. Head, Division of Entomology, US Department of Agriculture, 1878-94 [ANB]. M correspondent*.

RITCHIE, WILLIAM, 1790-1837, British physicist. Professor of natural philosophy, University of London, 1833-7 [DNB].

RIVIÈRE, CHARLES MARIE, 1845-?, French agronomist. Director of the Experiment Garden (Jardin d'essai), Algiers [Osborne, 1994].

ROBERTSON, GEORGE, 1825-98, bookseller and publisher. Born in Scotland. Learned the bookseller's trade in Ireland. Went to Melbourne in 1852 and established a successful bookselling business [ADB].

ROBINSON, BENJAMIN LINCOLN, 1864-1935, American botanist. DrPhil, Strassburg, 1889. Assistant to Sereno Watson at the Gray Herbarium, Cambridge, Mass., 1890-2; professor of systematic botany, Harvard, 1899-1935 [TL2]. M correspondent*.

ROBINSON, WILLIAM CLEAVER FRANCIS, 1834-97, British colonial governor. Born in Ireland. Governor of the Falkland Islands, 1866-70; Prince Edward Island, 1870-3; Leeward Islands, 1874; WA, 1875-7; Straits Settlements, 1877-80; WA, 1880-3; SA, 1883-9; WA, 1890-5. CMG, 1873; KCMG, 1877; GCMG, 1887 [ADB].

ROCHEL, ANTON, 1770-1847, Austrian surgeon and botanist. Surgical assistant in the Austrian army, 1788-98. ChM, Vienna, 1792. Physician in Moravia, 1798-1820. Curator of the Pest Botanic Garden, 1820-40 [TL2]. Specimens are in the Melbourne Herbarium [L80.08.04].

ROTHSCHILD, LIONEL WALTER, 1868-1937, British banker and zoologist. Studied at Cambridge. Worked in family bank, N. M. Rothschild & Sons, 1889-1908. Established zoological museum on family estate at Tring, 1889. MP, 1899-1910. FRS, 1911. Succeeded as Baron Rothschild of Tring, 1915 [WW]. M correspondent*.

ROUGIER, ÉMILE, French microbiologist. Australian agent of Institut Pasteur, 1893-8 [Todd (1995)].

ROUY, GEORGES C., 1851-1924, French botanist [Barnhart (1965)].

ROWAN, MARIAN ELLIS. See Marian Ellis RYAN.

RUDALL, JAMES FERDINAND, 1865-1944, ophthalmologist. Born in Melbourne. Son of James Thomas RUDALL (vol. 2). Honorary medical officer, Victorian Eye and Ear Hospital [ADB].

RUSSELL, ARTHUR EDWARD JOHN, EARL, 1824-92, British politician. Son of Lord George William Russell. Educated and travelled widely in Europe. MP for Tavistock, 1857-85. FRGS, 1858; Foreign Secretary or member of the Council, RGS, 1875-91. FLS and FZS and on council of each [*Proceedings RGS* (1892), pp. 328-34]. M correspondent*.

RYAN, CLARA OAKLEY. See Clara Oakley GRAHAM.

RYAN, MARIAN *ELLIS*, 1848-1922, artist, naturalist and explorer. Born in Melbourne. Renowned painter of flowers, travelling to and exhibiting in many countries, winning many medals. Many of the plants she painted were named by M. Married Frederic Charles Rowan, 1873 [ADB]. M correspondent*.

SACCARDO, Pier Andrea, 1845-1920, Italian mycologist. Professor of natural sciences at Padua Technical Institute, 1869-79; Professor of botany and director of the Padua University Botanic Garden, 1879-1915 [TL2]. M correspondent*.

SACHSE, Arthur Otto, 1860-1920, engineer and politician. Born in Qld. Constructed sugar mills in Qld. Worked as an engineer in south-east Asia, 1879-85. Went to Melbourne in 1885 and established a practice as a consulting engineer. MLC, 1892-1920; minister of public instruction, 1903-8 [ADB].

SALISBURY, Richard Anthony, 1761-1829, British botanist. Studied at Edinburgh. Early supporter of the natural system of classification. Secretary, Horticultural Society of London, 1809-10 [DNB].

SALZMANN, Philipp, 1781-1851, German botanist. Studied medicine at Göttingen, Vienna, Halle and Paris. Hospital physician at Montpellier from 1806, where he devoted himself to botany and entomology. Collected in the south of France and North Africa; in Spain 1823-5; and in Brazil (mainly Bahia), 1827-30 [TL2]. Specimens are in the Melbourne Herbarium [L80.08.04].

SAMPSON, G. Theophilus, 1831-97, British colonial official. In British government service at Canton, China, 1858-89. Returned to England, 1889 [Desmond]. Specimens are in the Melbourne Herbarium [L80.08.04].

SARTWELL, Henry Parker, 1792-1867, American physician and botanist. Lic. med., Oneida County Medical Society, New York, 1811. Practised medicine at New Hartford, NY, 1811-2; Springfield, NY; Bethel, NY; and Penn Yang, NY, 1832-67 [TL2]. Specimens are in the Melbourne Herbarium [L80.08.04].

SAUERBECK, Friedrich Wilhelm, 1801-82, German botanist. At one time a law court official at Freiburg [TL2].

SAYER, William A., fl. 1886-8, botanical collector. Collected in north Qld, Vic. and King Island (Tas.) Naturalist on second RGSA expedition to New Guinea [Cohn (forthcoming), Hall (1978)]. Plant contributor to the Melbourne Herbarium. *Dracophyllum sayeri*, *Elaeocarpus sayeri*, *Gymnogramme sayeri*, *Helicia sayeriana*, *Saccolabium sayerianum*.

SCARR, Frank. Member, Council, Vic. Branch, RGSA. JP [*Proceedings, RGSA (Vic)*, v.4 (1886)].

SCHEFFER, Rudolph Herman Christian Carel, 1844-80, Dutch botanist. Studied at Utrecht with Friedrich Miquel (vol. 1). DrPhil, 1867. Director of the Buitenzorg (now Bogor) Botanic Garden, Java, 1869-80 [TL2]. Specimens are in the Melbourne Herbarium. *Solanum schefferi*. M correspondent.

SCHENK, JOSEPH AUGUST VON, 1815-91, botanist and palaeontologist. Born in Austria. Studied medicine at Munich (DrMed, 1837). DrPhil, Erlangen, 1840. Extraordinary professor of Botany, Würzburg, 1844; from 1850 regular professor and director of the Würzburg Botanic Garden. At Leipzig Botanic Garden, 1868-87 [TL2].

SCHIMPER, GEORG HEINRICH WILHELM, 1804-78, German botanist and plant collector. Collected in southern France, Algeria, Egypt, Arabia, Abyssinia. Governor of the Abyssinian district of Antitscho, 1840-55 [Cohn (forthcoming), TL2]. Plant contributor (from Abyssinia via Ferdinand HOCHSTETTER, vol.1) to the Melbourne Herbarium [B62.03.01, B63.05.01]. M correspondent but none has been found [L74.05.10 (Col. Corr.)].

SCHLICH, WILHELM PHILIPP DANIEL, 1840-1925, forester. Born in Germany. DrPhil, Giessen, 1867. In the Indian Forest Service, Burma, 1867-70; Sind, 1870-2; conservator of forests, Bengal, 1872-8; conservator of forests, Punjab, 1880-1; inspector-general of forests, 1883-5. At Royal Indian Engineering College, head of the Forestry Department, Coopers Hill, 1885-1906, and at Oxford, 1906-11. KCIE, 1909. Professor of Forestry at Oxford, 1911-9 [Desmond, TL2]. M correspondent but none has been found [L89.03.24].

SCHMIDT, ADOLF WILHELM FERDINAND, 1812-99, German pastor and botanist. Studied theology at Halle, 1832-7. Private teacher at Merseburg, subsequently curate at Halberstadt. Deacon at Aschersleben, 1846-8; archdeacon, 1848-95. An expert on diatoms [TL2].

SCHMITZ, CARL JOHANN *FRIEDRICH*, 1850-95, German algologist. Studied botany at Bonn (DrPhil, 1871). Assistant at Halle and Strassburg, 1872-4. Lecturer at Halle, 1874-8; at Naples zoological station, 1878; extraordinary professor of botany, Bonn, 1878-84; professor of botany, Greifswald, 1884-95 [TL2]. M correspondent*.

SCHREBER, JOHANN CHRISTIAN DANIEL VON, 1739-1810, German naturalist. Studied medicine, natural science and theology at Halle. MD, Uppsala, 1760. Practising physician at Bützow, Mecklenburg, 1761. Professor of medicine and botany at Erlangen, 1770. Director of the Erlangen Botanic Garden 1773; professor of natural history, 1776; of medicine, 1791. President of the Leopoldina, 1791-1810 [TL2].

SCHUSTER, ARTHUR, 1851-1934, British physicist. Born in Germany. Went with his family to England, 1870. Studied at Owens College, Manchester, and Heidelberg (DrPhil, 1873). Worked at Cavendish Laboratory, Cambridge, 1876-81. FRS, 1879. Professor of applied mathematics, Owens College, 1881-8; of physics, 1888-1907. Knighted, 1920. Secretary, International Research Council, 1919-28 [DNB].

SCHWEINFURTH, Georg August, 1836-1925, botanist, geographer and explorer. Born in Latvia. Studied at Heidelberg, Munich and Berlin; DrPhil, Heidelberg, 1862. Travelled in the Nile countries 1863-66; in Equatorial Africa, 1868-71. Founded the Cairo Geographical Society, 1875. Explored the area between the Nile and the Red Sea, 1876-86. Domiciled in Berlin from 1888 in association with the Botanical Museum, still making frequent trips abroad until 1914 [TL2]. M correspondent.

SCLATER, Philip Lutley, 1829-1916, barrister and naturalist. MA, DSc, DrPhil. FRS, 1861. Secretary, Zoological Society of London, 1859-1902 [WW]. M correspondent*.

SCOPOLI, Giovanni Antonio [Johannes Antonius], 1723-88, physician and botanist. Born in the Tyrol. Educated at Trento and Innsbruck (DrMed, 1743; state doctor's examination, Vienna, 1754). State physician, Idria, 1754-67. Professor of mineralogy and metallurgy, Schemnitz, 1767-76; professor of chemistry and botany, Pavia, 1776-88 [TL2].

SCORTECHINI, Benedetto, 1845-86, naturalist and Catholic priest. Born in Italy. Parish priest in Qld, 1871-84. FLS, 1881. Government botanist of Perak, Strait Settlements (Malaya), 1884-6. Collaborated with M and others [Desmond, TL2]. Plant contributor to the Melbourne Herbarium. *Agonis scortechiniana*, *Bossiaea scortechinii*, *Brachyloma scotechinii*, *Cryptendra scortechinii*, *Mezoneuron scortechinii.* M correspondent*.

SCOTT, Harriet, 1830-1907, artist and naturalist. Born in Sydney. Sister of Helena Scott. Collected for and corresponded with leading colonial scientists. For paintings of Australian insects was elected an honorary member of the Entomological Society. Married Dr C. W. Morgan, 1882 [ADB]. Plant contributor to the Melbourne Herbarium [B81.08.03, p.120]. *Dampiera scottiana*, *Elettaria scottiana*. M correspondent.

SCOTT, Helena, 1831?-1910, artist and naturalist. Born in Sydney. Sister of Harriet Scott. Collected plants and painted both fauna and flora. Married Edward Forde, 1864 [Scott & Scott (1988)]. Plant contributor to the Melbourne Herbarium [B63.05.01]. *Poa fordeana*.

SEDDON, Richard John, 1845-1906, politician. Born in England. Went to Melbourne in 1863, and to NZ in 1866. Member of NZ parliament from 1879; Minister for mines, defence and public works, 1891; Acting Premier, 1892-3; Premier, 1893-1906 [NZDB].

SETCHELL, William Albert, 1864-1943, American algologist and plant geographer. Assistant at Yale University, 1891-5, professor of botany at University of California, Berkeley, 1895-1934 [TL2]. M correspondent.

SHORT, CHARLES WILKINS, 1794-1863, American medical practitioner and botanist. Born in Kentucky. Educated at Transylvania University (BA, 1810) and the University of Pennsylvania (MD, 1815). Practised medicine in Kentucky, 1815-25. Collected and distributed large numbers of plant specimens. Professor of materia medica, Transylvania University, 1825-38; Louisville Medical Institute, 1838-46; University of Louisville, 1846-9 [DAB, Cohn (forthcoming)]. Specimens are in the Melbourne Herbarium [L80.08.04].

SHUTTLEWORTH, ROBERT JAMES, 1810-74, naturalist. Born in Britain. Educated in Geneva, where he developed a love of botany. Studied medicine at Edinburgh, 1830-2. At Bern, Switzerland, 1833-66; Hyères, France, 1866-74. Formed a large herbarium, partly from collectors whom he financed in other parts of the world. FLS, 1856 [DNB]. Specimens are in the Melbourne Herbarium [L80.08.04].

SISLEY, THOMAS A., b. 1850, speech and music teacher. Born in England. Educated in Paris and London. Civil servant and teacher until 1889. Went to Melbourne in 1889. Music teacher and lecturer in elocution [*Table Talk* (Melbourne), 21 November 1890]. M correspondent*.

SMITH, LOUIS LAWRENCE, 1830-1910, medical practitioner and politician. Born in England. Studied medicine in Paris and London (LSA, 1852). Went to Melbourne in 1852. MLA, Vic., 1859-65, 1871-4, 1877-83, 1886-94. Opened a Museum of Anatomy, 1862; closed, 1869. Member of the Phylloxera Board until 1893 [ADB]. M correspondent*.

SNOWDEN, ARTHUR, 1829-1918, lawyer and politician. Born in England. Went to Vic. in 1852. Solicitor's clerk in Melbourne, 1854-66; admitted as a lawyer, 1867, and developed a successful partnership. Melbourne City Councillor, 1891-1918. MLC, 1895-1904. Knighted, 1895 [ADB].

SOLMS-LAUBACH, HERMANN MAXIMILIAN CARL LUDWIG FRIEDRICH ZU, 1842-1915, German botanist. Studied in Berlin, DrPhil, 1865; in Freiburg, 1866-7; in Halle, 1867-72. Professor of botany and director of the Göttingen Botanic Garden, 1879-88; professor of botany, Strassburg, 1888-1907 [TL2]. *Pandanus solms-laubachii*. M correspondent but none has been found [L86.08.24 (Coll. Corr.)]

SPENCER, WALTER WILLIAM BALDWIN, 1860-1929, biologist and anthropologist. Born in England. Educated at Oxford. Professor of biology at Melbourne, 1887-1919. Member of Horn Expedition to central Australia, 1894. Collected plants in King Island, Tas. and central Australia. A notable contributor to Australian Aboriginal anthropology. FRS, 1900. CMG, 1904; KCMG, 1916 [ADB, Desmond, Hall (1978)]. Plant contributor to the Melbourne Herbarium. *Decatoca spencerii*. M correspondent*.

SPICER, William Webb, 1820-79, clergyman and botanist. Born in England. BA, Oxford, 1843; MA, 1848. Rector, Itchen Abbas, 1850-74. Travelled and collected plants in Tas., 1874-8 [Desmond, TL2]. *Helichrysum spiceri.* M correspondent.

SPRAGUE, Isaac 1811-95, American botanical and zoological artist, who worked for Audubon and Asa Gray (vol. 1) [TL2].

STAPF, Otto, 1857-1933, botanist. Born in Austria. Assistant in Vienna, 1882-9; in Persia, 1885. Assistant at Kew, 1890; principal assistant, 1899; keeper of the Herbarium, 1909-22; botanical secretary of the Linnean Society, 1908-16; editor of the *Botanical magazine*, 1922-33, and of the *Index londinensis* (1929-31) [TL2, DBE].

STEENSTRUP, Japetus Smith, 1813-97, Danish zoologist. An expert on Scandinavian fossils. Professor of zoology at Copenhagen, 1846-85. Met M at Kiel in 1846 [L85.07.11, DSB].

STEPHANI, Franz, 1842-1927, German botanist and businessman. Employee and later director of a woollen spinning mill, Dessau, 1859-69; in New York, 1864-6. Merchant in Waldkirchen, Saxony, 1869-80; in Leipzig, 1880-4. With Leipzig publishing house, 1884-1907. Expert on hepaticae [TL2]. M correspondent*.

STIRLING, Edward Charles, 1848-1919, surgeon and naturalist. Born in SA. Educated at Cambridge (MA, 1873; MB, 1874; MD, 1880). FRCS, 1874. Consulting surgeon at Adelaide Hospital from 1881; later also professor of physiology at the University of Adelaide. MHA, 1884-7. Director of SA Museum, 1884-1912. Member of Horn Expedition to Central Australia, 1894. FRS, 1893. CMG, 1893; KCMG, 1917 [ADB].

STIRLING, James, 1852-1909, geologist, botanist and surveyor. Born in Vic. District Surveyor and Lands Officer, Omeo, Vic, 1875. FLS, 1883. Retired as Government Geologist and Chief Inspector of Mines [Cohn (forthcoming), Gibbney & Smith (1987)]. Plant contributor to the Melbourne Herbarium. *Helichrysum stirlingi.* M correspondent*.

STIRTON, James, 1833-1917, Scottish physician and cryptogamist. MD, Edinburgh, 1858. Lecturer on gynaecology at the Royal Infirmary, Glasgow, 1879-89; professor of gynaecology, Glasgow, 1889-c.1904 [TL2].

STOBBS, John G., 1840-82, Presbyterian minister. Born in Scotland, educated at Glasgow (MA). Assistant minister, Greenock, 1868-74. Went to Melbourne in 1874, as minister of the Scot's Church; then minister at West Melbourne church, 1875-82 [*Age*, 11 August 1882].

STOKES, George Gabriel, 1819-1903, British mathematical physicist. Educated at Cambridge; senior wrangler and first Smith's prizeman, 1841. Lucasian professor of mathematics, Cambridge, 1849-1903. FRS,

1851; Secretary, 1854-85; President, 1885-90. MP, 1887-91. Knighted, 1889 [DSB]. M correspondent*.

STRAHAN, GEORGE CUMINE, 1838-87, colonial administrator. Born in Scotland. Educated at Royal Military College, Woolwich. Colonial Secretary, Bahamas, from 1868; acting Governor, 1872-3. Administrator, Lagos, 1873-4. Governor, Gold Coast, 1874-6; of Windward Islands, 1876-80. CMG, 1875; KCMG, 1880. Governor, Tasmania, 1881-6. GCMG, 1887 [ADB].

STRICKLAND, EDWARD, 1821-89, army officer. Born in Ireland. Served in Canada, 1838-9; in NSW, 1840-1; in Tas., 1842-4; in the Middle East, 1844-60; in Melbourne, 1863-4; in NZ, 1864-7; in Malta, 1874-6; in Cape Colony, 1877-9. CB, 1866; KCB, 1879. Retired 1881 and went to live in Sydney. FRGS, 1860. Vice-president, RGSA, and President of NSW Branch [ADB]. *Leptosiphonium stricklandi*. M correspondent*.

STRONG, CHARLES, 1844-1942, clergyman. Born in Scotland. Educated at Glasgow, 1859-64. Ordained in 1868; Presbyterian minister in Glasgow, 1871-5; at Scots Church, Melbourne, 1875-83. Minister of the Australian Church, Melbourne, 1885-1942 [ADB].

SWARTZ, OLOF PETER, 1760-1818, Swedish botanist. Travelled in Sweden, 1779-83; USA, 1783; West Indies and South America, 1784-6; worked in London, 1786-7. Studied at Uppsala (DrMed, 1785); Curator and professor at the Bergius horticultural school (Bergianska trädgårdsskolan), 1791-1806. Curator of the natural history cabinet of the Swedish Academy of Sciences, Stockholm, 1806; permanent secretary of the Academy, 1811-8. Professor of botany, Karolinska Institute, 1813-8 [TL2].

TATE, RALPH, 1840-1901, geologist and botanist. Born in England. Educated at Royal School of Mines, London. Assistant curator, museum of Geological Society of London, 1864-7. ALS, 1867. Worked as a geologist in Nicaragua and Venezuela, 1867-8, and later in England. Professor of natural science, Adelaide, 1874-1901. FLS, 1883. President, AAAS, 1893. Member of Horn Expedition to central Australia, 1894 [ADB, Desmond, Hall (1978), TL2]. Plant contributor to the Melbourne Herbarium [Cohn (forthcoming)]. *Achnophora tatei, Bassia tatei, Epaltes tatei, Phyllanthus tatei, Strobilanthes tatei, Tatea, Xanthorrhoea tateana*. M correspondent*.

TAYLOR, CAMPBELL, 1842-1900. Of Albany, WA. Brother of Kate TAYLOR [Battye Library].

TAYLOR, HELEN, 1831-1907, step-daughter of John Stuart MILL [DNB].

TAYLOR, KATE (i.e. CATHERINE LOUISA?), 1839-?. Plant collector at Albany, WA. Sister of Campbell TAYLOR [Battye Library]. M correspondent*.

TEIJSMANN [TEYSMANN], JOHANNES ELIAS, 1809-82, Dutch botanist. Went to Java as a gardener in 1830. Curator of the Buitenzorg (now Bogor) Botanic Garden, 1831-69 [Cohn (forthcoming), TL2]. Plant contributor to the Melbourne Herbarium. M correspondent but none has been found [L67.12.21a (Coll. Corr.)]

TEPPER, JOHANN GOTTLIEB OTTO, 1841-1923, entomologist and school teacher. Born in Prussia. Went to SA, 1847. School teacher at Ardrossan, SA, and later at Clarendon, SA. FLS, 1879. Collector, SA Museum, 1883, later entomologist at Museum [Cohn (forthcoming), Desmond, Gibbney & Smith (1987)]. Plant contributor to the Melbourne Herbarium. *Candollea tepperiana*, *Helichrysum tepperi*, *Lasiopetalum tepperi*, *Polygala tepperi*, *Schoenus tepperi*. M correspondent*.

THISELTON-DYER, WILLIAM TURNER, 1843-1928, British botanist. Educated at Oxford (BA, 1867) and London (BSc, 1870). Professor of natural history, Royal Agricultural College, Cirencester, 1868. Professor of botany, Royal College of Science, Dublin, 1870-2; Royal Horticultural Society, London, 1872-5. FLS, 1872. Assistant director, Kew Botanic Garden, London, 1875-85; director, 1885-1905. FRS, 1880. KCMG, 1899. Son-in-law of Joseph HOOKER (vol. 1) [Desmond, DSB, TL2]. *Encephalartos dyeri*. M correspondent*.

THOMAS, EBENEZER J., public servant. Joined Vic. public service, 1852. Private secretary to Premier and Secretary of Premier's Department at its formation in 1883 [Victorian public service lists]. M correspondent*.

THOMAS, JULIAN. See John Stanley JAMES.

THOMSON, JAMES PARK, 1854-1921, geographer and public servant. Born in Scotland. In USA and South America, 1872-4; NZ, 1876; NSW, 1877-9; Fiji, 1880-4. Joined Qld Department of Public Lands as draftsman. Founded Qld branch, RGSA; president, 1894-7. CBE, 1920 [ADB]. M correspondent*.

THOMSON, ROBERT, 1840-1908, British gardener. At Kew, 1862. At Castleton Botanic Garden, Jamaica, 1862-79 [Desmond].

THOMSON, WILLIAM, Baron KELVIN of Largs, 1824-1907, Scottish physicist. Educated at Glasgow and Cambridge, 1841-5. Professor of natural philosophy, Glasgow, 1846-99. FRS, 1851; president, 1890-4. Made important contributions to the study of electricity, magnetism, light and heat. Knighted in 1866 for his role in the laying of the trans-Atlantic telegraph cable. Raised to the peerage, 1892. OM, 1902 [DNB]. M correspondent*.

THÜMEN [THUEMEN], Felix Karl Albert Ernst Joachim von, 1839-92, German botanist. Lived on his estate in Silesia, devoting much of his time to botany. Adjunct at Klosterneuberg Experiment Station near Vienna from 1876. An expert on fungi [TL2]. Specimens of fungi are in the Melbourne Herbarium [L80.08.04].

THUNBERG, Carl Peter, 1743-1828, Swedish botanist. Studied natural sciences and medicine with Carl Linnaeus (vol. 1) at Uppsala (DrPhil, 1767; DrMed, 1772). Studied in Amsterdam and Paris, 1770-2. At the Cape of Good Hope, 1772-5; Java, 1775; Japan 1775-6; Java, 1776-7; Ceylon, 1777-8; Holland, 1778; Sweden, 1779. Botanical demonstrator, Uppsala, 1777; extraordinary professor, 1781; professor of botany and medicine, 1784-1828 [TL2].

THURSTON, John Bates, 1836-97, planter, politician and administrator. Born in Britain. In the merchant marine, 1850-5. Sheep farmer near Sydney, c.1859-62. Acting consul, Ovalau, Fiji, 1867-9; colonial secretary Fiji, 1874; lieutenant-governor, 1886; governor and commander-in-chief, Fiji and western Pacific, 1887-97. Collected plants in Fiji between 1863 and 1897. CMG, 1880; KCMG, 1887 [TL2, DNB]. *Pritchardia thurstoni*. M correspondent*.

TIETKENS, William Harry, 1844-1933, surveyor, prospector and explorer. Born in England. Went to Australia, 1859. Second-in-charge of Ernest Giles's (vol. 2) two exploring expeditions, 1873, 1875. Leader of exploring expedition westward from Alice Springs into the western desert, 1889. FRGS, 1889. Surveyor, NSW Department of Lands, 1891-1909 [ADB, Cohn (forthcoming)]. Plant contributor to the Melbourne Herbarium. *Aotus tietkensii*, *Eremophila tietkensii*, *Helipterum tietkensii.* M correspondent*.

TIMBRELL, Ann, of Collingwood, Melb. Experimented in silk culture. M correspondent*.

TODARO, Agostino, 1818-92, botanist. Born in Sicily. Studied at Palermo. Assistant at the Palermo Botanic Garden, 1848-56; acting director, 1856-60; director, 1860-92 [TL2]. Plant contributor to the Melbourne Herbarium [B69.07.03].

TODD, Charles, 1826-1910, astronomer, meteorologist and telegraph engineer. Born in England. Went to Adelaide in 1856. Director, Adelaide Observatory, 1856-1905. Postmaster-general, SA, 1870-1905. Supervised construction of trans-Australian telegraph lines, Adelaide-Darwin, 1870-2, and Adelaide-WA, 1877. CMG, 1872; KCMG, 1893. FRS, 1889 [ADB]. *Conospermum toddii*. M correspondent.

TORELLI, LUIGI, 1810-87, Italian politician. Was at times Minister for Agriculture, Industries and Commerce. Became a Senator of united Italy, 1860. Promoted study of *Phylloxera* infestation of vines [Hall (1978)]. *Eucalyptus torrelliana.* M correspondent*.

TOURNEFORT, JOSEPH PITTON DE, 1656-1708, French botanist. Educated at Montpellier. Assistant at the King's Garden (Jardin du Roi), Paris, 1683-1708. Elected to the Academy of Sciences (Académie des Sciences), Paris, 1691. Travelled in the Levant, 1700-2 [DSB].

TREUB, MELCHIOR, 1851-1910, Dutch botanist. Assistant at Leiden botanical institute, 1873-80. Director, Buitenzorg (now Bogor) Botanic Garden, 1880-1909 [TL2]. M correspondent.

TROEDEL, JOHANNES THEODOR *CHARLES*, 1836-1906, lithographer. Born in Hamburg. Trained under his father as a lithographer. Went to Melbourne in 1860; established his own lithographic printing firm in 1863 [ADB].

TYNDALL, JOHN, 1820-93, British physicist. Worked as surveyor in Ireland and England, 1839-47. Studied at Marburg (DrPhil, 1850). Professor of natural philosophy, Royal Institution, London, 1853-87. FRS, 1852 [DSB].

UMBERTO I, KING OF ITALY, 1844-1900. Reigned 1878-1900 [*Everyman's Encyclopaedia*]. M correspondent*.

URBAN, IGNATZ, 1848-1931, German botanist. Educated at Bonn and Berlin, 1866-9. On military service, 1869-71. DrPhil, Berlin, 1873. Teacher in Berlin, 1873-78. Assistant at the Berlin Botanic Garden, 1878-83; curator, 1883-89; second director and titular professor 1889-1913 [TL2]. M correspondent*.

VALE, RICHARD TAYLER, 1836-1916, bookseller and politician. Born in England. Went to Melbourne in 1853 and to Ballarat, 1869; bookseller there, 1872-1900. Represented Ballarat in Vic. Legislative Assembly, 1886-9, 1892-1902. Member of Council and Vice-president for some years of the Ballarat School of Mines [ADB].

VAN HEURCK, HENRI FERDINAND, 1838-1909, Belgian industrialist and botanist. In charge of his family's dye factory and chemical firm, 1869-96. Self-taught as a botanist. Director of the Antwerp Botanic Garden, 1877-1909. An expert on diatoms [TL2]. Specimens are in the Melbourne Herbarium [L80.08.04]. M correspondent but none has been found [L94.10.07 (Coll. Corr.)].

VENTENAT, ÉTIENNE PIERRE, 1757-1808, French priest and botanist. Librarian of his congregation; librarian of the Panthéon from 1796 [TL2, NBG].

VERCO, JOSEPH COOKE, 1851-1933, physician and conchologist. Born in Adelaide. Educated at University of London; MB, 1875; MD, 1876; BS, 1877. LRCP, 1875; FRCS, 1877. General practitioner, SA, from 1878. Co-founder of University of Adelaide medical school. Collector of shells, corals and other marine life [ADB].

VIEILLARD, EUGÈNE DEPLANCHE ÉMILE, 1819-96, French naval surgeon and botanist. Collected in Tahiti, 1855-7; in New Caledonia 1857-9, 1861-9. Director of the Caen Botanic Garden, 1871-95 [Cohn (forthcoming), TL2]. Specimens are in the Melbourne Herbarium [L80.08.04].

VIRCHOW, RUDOLPH CARL, 1821-1902, German medical scientist. Born in Pomerania. Studied in Berlin (MD, 1843). Lecturer in pathological anatomy, Berlin, 1847-9; professor at Würzburg, 1849-56; at Berlin from 1856. Co-founder of the German Anthropological Society, 1869. Member of the Reichstag, 1861-1902 [DSB]. *Ceratopetalum virchowii.* M correspondent*.

VOGAN, ARTHUR JAMES, 1859-1948, writer and traveller. Born in England. Went to NSW c.1885. Artist with New Guinea expedition, 1885. Travelled widely; adviser to British mining interests [Gibney & Smith (1987)]. M correspondent.

VOGRICH, MAXIMILIAN CARL WILHELM, 1852-1916, pianist and composer. Born in Transylvania. Went to Australia c.1881; conductor for Sydney Philharmonic Society. Married the Australian soprano Alice Rees. Went to America in 1885.

VOLGER, GEORG HEINRICH OTTO, 1822-97, German geologist and mineralogist. Met M at Kiel in 1846. Studied at Göttingen (PhD, 1845). Lecturer at Aargau and later at Zürich. At Senckenberg Museum, Frankfurt, from 1856. Founder of Freie Deutsche Hochstift, Frankfurt, 1859-81 [DBE]. M correspondent.

WALKER, ANNIE FRANCES, 1831-1913, artist. Born in NSW. Won gold medal for flower paintings, London Exhibition, 1873. Presented paintings to Queen VICTORIA for her jubilee; 1700 of her paintings are in the Mitchell Library, Sydney [Gibbney & Smith (1987), Willis et al. (1986)]. M correspondent*.

WALKER, JAMES BACKHOUSE, 1841-99, solicitor and historian. Born at Hobart. Solicitor in Hobart from 1876. Member of the Tas. Board of Education. Wrote extensively on the early history of Tas. [ADB]. Brother of Mary Augusta WALKER.

WALKER, MARY AUGUSTA, 1856-1952, governess and teacher. Sister of James Backhouse WALKER. Studied art in Melbourne, London and

Paris [Archives of the Religious Society of Friends, University of Tasmania Library].

WATSON, ROBERT, 1822-91, engineer. Born in England. Trained as a surveyor and engineer. Went to Melbourne in 1854. Joined Vic. Survey Department, 1854. Resident engineer of Vic. Railways; Engineer-in-chief, 1878-9, 1881-91. Led surveying expedition for Qld government, 1881 [Gibbney & Smith (1987)]. M correspondent*.

WATSON, SERENO, 1826-92, American botanist. Studied at Yale (BA, 1847). Practised as a physician, 1852-6. Botanist to King's expedition to survey the Fortieth Parallel, 1867. Added to the *Synoptical flora of North America*, begun by Asa GRAY (vol. 1). Assistant, Gray Herbarium, Harvard University, 1870-4; curator, 1874-92 [ANB].

WATSON & SCULL, M's London agents, superseding Blackith & Co. in 1880 [L80.02.16 (Coll. Corr.)]. M correspondent*.

WEBB, WILLIAM HEATON, 1834-97. Born in Britain. Transported to WA as a convict. Went to Albany, WA, 1862, employed as a shepherd. Later lived in Albany. Became an expert taxidermist [*Albany advertiser*, 19 November 1998]. Plant and seed contributor to the Melbourne Herbarium. *Bossiaea webbii*. M correspondent.

WEDDELL, HUGH ALGERNON, 1819-77, botanist and physician. Born in Britain. Grew up and worked in France. DrMed, Paris, 1841. Associated with Adrien JUSSIEU (vol. 1); went with François CASTELNAU (vol. 2) to S. America, 1843-8 (Brazil, Bolivia, Peru). *Aide-naturaliste* at the Natural History Museum (Muséum d'Histoire Naturelle), Paris, 1850-7. Visited Bolivia again in 1853 in search of Cinchona. FLS, 1859 [TL2].

WEHL, FERDINAND EDWARD, 1855-1908, M's nephew, first son of Clara WEHL, M's sister. Born in SA [VBOT]. M correspondent*.

WEHL, HENRIETTA JANE [ETTIE], 1868-1953, M's niece, sixth daughter of Clara WEHL, M's sister. Born in SA. Married Donald Sinclair, 1891 [VBOT]. M correspondent*.

WEHL, MARIA [MARIE] MAGDALENA, 1862-1960, M's niece, fourth daughter of Clara WEHL, M's sister. Born in SA. Botanical artist [Cohn (forthcoming), VBOT]. Plant and fungi contributor to the Melbourne Herbarium [May, Maroske & Sinkora (1995)].

WEHL, OTTILIE *BERTHA*, 1854-1922, M's niece, first daughter of Clara WEHL, M's sister. Born in SA. Married George HARRIS, 1874 [VBOT]. M correspondent.

WELD, FREDERICK ALOYSIUS, 1823-91, English colonial governor. Educated at Stonyhurst and the University of Fribourg. Went to NZ,

1843, and became a sheep-farmer. MLA, NZ, 1852; Minister for Native Affairs, 1860-1; Prime Minister, 1864-7. Governor, WA, 1869-74; Tas., 1875-80; Straits Settlements, 1880-7. CMG, 1875; KCMG, 1880; GCMG, 1885 [ADB]. *Eremophila weldii.* M correspondent.

WELWITSCH, FRIEDRICH MARTIN JOSEF, 1806-72, botanist and doctor. Born in Austria. MD, Vienna, 1836. Went to Portugal, 1839. Curator at Lisbon Polytechnic. Collected in Angola, 1853-61. ALS, 1858. In Portugal, 1861-3; in London, 1863-72. FLS, 1865 [Cohn (forthcoming), Desmond, TL2].

WHEATSTONE, CHARLES, 1802-75, English physicist. Maker of musical instruments, without formal scientific training. Experimented on acoustics. Professor of natural philosophy, King's College, London, 1834. FRS, 1836. Invented the system of electric telegraphs, 1837. Knighted, 1868 [DSB].

WHITELEGGE, THOMAS, 1850-1927, naturalist, museum assistant and bryologist. Born in England. Collected plants for his own herbarium. Went to Sydney, 1883. Senior scientific assistant, department of lower invertebrates, Australian Museum. Member of Council, Linnean Society, NSW, 1890-6. Transferred to National Herbarium, Sydney, 1908 [VBOT]. M correspondent*.

WILLAN, SOPHIE CATHERINE, 1863?-1943, married Brian HOOKER, 1888 [Allan (1967)]. M correspondent*.

WILLDENOW, CARL LUDWIG, 1765-1812, German botanist. Studied at Halle (MD, 1789). In charge of family pharmacy, Berlin, 1790-8. Professor of natural history, Medical and Surgical College (Collegium medico-chirurgicum), Berlin, 1798; botanist at the Berlin Academy of Sciences and director of the Berlin Botanic Garden, 1801; professor of botany, University of Berlin, 1810-2 [TL2].

WILLIAMS, DAVID JOHN, 1819-1902, medical practitioner. Born in Britain. MRCS, 1841; MD, St Andrews, 1842; LSA, London, 1845. Went to Melbourne in 1853. FRCS, 1861. MD, Melbourne, 1882. Coroner and district surgeon, Ballarat, Vic., 1854. At Queenscliff, Vic., from 1863 [*Medical Register, Victoria* (1882-92)].

WILLIAMSON, ALEXANDER WILLIAM, 1824-1904, English chemist. Fellow, Chemical Society of London, 1848; President, 1863-5, 1869-71. Professor of practical chemistry, University College London, 1855-87; of chemistry, 1876-1901. FRS, 1855; Council member, 1859-61, 1869-71, Foreign Secretary, 1873-89; Royal Medal, 1862 [DNB].

WILSON, JOHN BRACEBRIDGE, 1828-95, schoolmaster and naturalist. Born in England. Educated at Cambridge (BA, 1852). Went to

Melbourne in 1857. Teacher at Geelong Grammar School, Vic., from 1858; headmaster, 1863-95. An enthusiastic naturalist, especially of marine organisms. FLS, 1882 [ADB, Cohn (forthcoming), Desmond]. M correspondent.

WILSON, THOMAS RICHARD, 1834-95, public servant. Born in Ireland. Went to Melbourne in 1852. Joined Vic. public service in Lands Department; Under-secretary, Chief Secretary's Department, from 1881 [Gibbney & Smith (1987)]. M correspondent*.

WINNECKE, CHARLES GEORGE ALEXANDER, 1856-1902, explorer, geologist and botanist. Born in SA. Educated at St Peter's College, Adelaide. In charge of expedition to determine the border between SA and Qld, 1881. Collected plants in Central Australia, 1883, and near Stuart's Range, 1885. Leader of Horn Expedition, 1894 [Desmond, Gibbney & Smith (1987)]. Plant contributor to the Melbourne Herbarium. *Isotropis winneckei*, *Triumfetta winneckeana*. M correspondent.

WINTER, HEINRICH *GEORG*, 1848-87, German mycologist. Studied at Leipzig (DrPhil, 1873), Munich and Halle. Lecturer in botany, Zürich Polytechic, 1876-83. Editor of *Hedwigia*, 1879-87. In Leipzig from 1883 [TL2].

WINTLE, MR S. H., 1830-fl.1880s, naturalist. Born in Hobart. Fellow, Royal Society of Tas., 1864-1882 [Willis et al. (1986)]. M correspondent but none has been found [L83.12.24 (Coll. Corr.)].

WISE, MAY (i.e. MARY ISABEL?), 1882-1955. [Vic. register of births, deaths and marriages]. Plant collector at Sale, Vic. M correspondent*.

WITTMACK, MARX CARL LUDWIG, 1839-1929, German botanist. Studied at Jena and Berlin; DrPhil, Göttingen, 1867. Curator, Agricultural Museum, Berlin, 1871. Teacher, Berlin Agricultural Institute, 1874. Professor at University of Berlin, 1880-1913; at Agricultural University, 1881-1913 [TL2]. M correspondent.

WITTROCK, VEIT BRECHER, 1839-1914, Swedish botanist. Studied at Uppsala, 1857-65 (DrPhil 1866). High school teacher at a private school in Uppsala, 1865-78. Extraordinary professor of botany, Uppsala, 1878. Professor and curator of the botanical collections at the Stockholm Riksmuseum, 1879-1904; also 'Professor Bergianus' and director of the Bergius horticultural school (Bergianska trädgårdsskolan), 1879-1914, and botany teacher at the Institute of Advanced Studies (högskola), Stockholm, 1879-83 [TL2]. M correspondent*.

WOODWARD, BERNARD HENRY, 1832-1921, British palaeontologist. Educated at the Royal Agricultural College, Cirencester. Bank officer, Norwich, 1850-7. British Museum, 1858-1901; keeper of geology, 1880-1901. FRS, 1873. President of the Geological Society, London, 1894-6 [WW]. M correspondent.

WRIGHT, EDWARD PERCEVAL, 1834-1910, Irish naturalist and ophthalmic surgeon. MA, Dublin, 1857; MD, 1862. FLS, 1859. Lecturer in zoology, Trinity College, Dublin, 1858-68; professor of botany 1869-1904; Keeper of Herbarium, 1870-1910. Collected in the Seychelles, 1867, and in Sicily and Portugal, 1868 [Desmond, TL2].

YOUNG, ALLEN WILLIAM, 1827-1915, British polar explorer. In merchant navy, 1842-57. Took part in search for Sir John FRANKLIN (vol. 2), 1857-9. In China, 1862-4. Spent several years in the Arctic, 1875-82. Knighted, 1877 [DNB].

YOUNG, JESS [JESSE], 1851-1909, naturalist and explorer. Born in England. Trained as a naval officer. Went to Australia, 1874. Accompanied Ernest GILES (vol. 2) on his fifth expedition, from the overland telegraph line to Perth across the Great Victoria Desert, 1875. In the USA, 1877-98. Went to WA goldfields, 1898. Mining Registrar, Bangemall, WA, 1898. Acting warden of the Gascoyne goldfield, 1898. At Kanowna, WA, 1900-01. FRGS, FRAS [Hall (1978), Young (1978)]. Plant contributor to the Melbourne Herbarium. *Eucalyptus youngiana*. M correspondent.

ZEYHER, CARL LUDWIG PHILIPP, 1799-1858, German plant collector. Born in Hessen. Collected in Mauritius with Franz SIEBER (vol. 1), 1822; in South Africa, 1822-3, 1825-8, 1829-38 (with Christian ECKLON, vol. 2), 1842-3 (with J. Burke). In London and Hamburg, 1843-7. Botanist at the Cape of Good Hope Botanic Garden, 1849-51 [Cohn (forthcoming), Gunn & Codd (1981), TL2]. Specimens are in the Melbourne Herbarium (via herbarium of STEETZ (vol. 1)) [L63.12.12].

Appendix B: Orders, Offices, Affiliations and Sundry Honours of Ferdinand Mueller

This appendix is based on a manuscript list of the same name compiled in the 1940s by the late Jim Willis, Assistant Government Botanist of Victoria, but with corrections and additions. Willis's list, now in the library, Royal Botanic Gardens Melbourne, was in turn based chiefly on a printed list in Latin entitled 'Liber baro Ferdinandus de Mueller', compiled by Mueller in 1892, identified here as B92.06.01 in preparation for its inclusion in an updated list of Mueller's publications that we intend publishing with his Collected Correspondence. Annotations by Mueller on a copy of the printed list that is now at the Royal Society of London, apparently sent as an enclosure with Mueller's letter to Lord Kelvin of 23 April 1893, have provided additional information (indicated in the references by 'B92.06.01 annotation'). The entries are arranged alphabetically by country or colony as these were recognized in the year of Mueller's death, 1896 (e.g. New South Wales, Queensland etc. and not Australia), and within these countries by city. For each city, entries are arranged chronologically according to Mueller's earliest known links with the institutions listed. Names of institutions are as given at the time of Mueller's election, appointment or so on, with some later variations noted. Scudder (1879) has been used to verify names of scientific societies wherever possible. Institutions the names of which have not been verified independently of Mueller's own claims are given in square brackets, as are references; 'certificate' indicates that details have been drawn from a diploma now held in the library, Royal Botanic Gardens Melbourne. Institutions with no fixed address are listed at the end of the appendix under 'Other'; memberships that Mueller claimed of societies that he appears to have mis-named and that we have been unable to identify are listed under 'Doubtful'. Where there is doubt about the city in which an organization was based, it is entered under the major or capital city of the relevant state. The names of all foreign institutions are given first in English with the name in the original language following in brackets. The names of countries and cities are given in English, e.g. 'Florence' not 'Firenze'. The exhibition awards, conference attendances and committee memberships noted are unlikely to be an exhaustive list. References to Mueller's *Fragmenta phytographiae australiae* give the first volume only in which the information cited appears on the title page.

Argentina

Buenos Aires

Rural Society of Argentina (Sociedad Rural Argentina). Honorary member, 1 February 1873 [certificate].

Algeria

Algiers

Agricultural Society of Algeria (Société d'Agriculture d'Alger). Corresponding member, c. 1876 [*Bulletin* (1876) p. 258].

Algerian Society of Climatology (Société de Climatologie Algérienne). Honorary member, 30 December 1879 [certificate].

Austria

Vienna

Imperial-Royal Zoological and Botanical Society (Kaiserlich-Königliche Zoologisch Botanische Gesellschaft). Member, 9 April 1859 [certificate].

Imperial-Royal Geographical Society (Kaiserlich Königliche Geographische Gesellschaft). Member by 1861 [*Frag.* vol. 2].

Imperial-Royal Geological State Institute (Kaiserlich Königliche Geologische Reichsanstalt). Corresponding member by 1861 [*Frag.* vol. 2; B92.06.01].

Imperial-Royal Horticultural Society (Kaiserlich Königliche Gartenbau Gesellschaft). Foreign ordinary member, 13 March 1863 [[...] Beer to M, 13 March 1863 (Collected Correspondence)]; honorary member, 25 April 1881 [certificate].

Order of Franz Joseph (Franz-Josephs Orden). Knight's Cross, 16 July 1863 [Baron von Lichtenfels to M, 17 July 1863 (Collected Correspondence)].

General Austrian Apothecaries' Union (Allgemeiner Österreichischer Apotheker Verein). Honorary member, 15 September 1868 [certificate].

World Exhibition (Weltausstellung), 1873. Bronze medal 'for merit' (dem Verdienste) [medal at Melbourne Museum].

Belgium

Antwerp

Phytological and Micrographical Society (Société Phytologique et Micrographique). Corresponding member by 1874 [*Frag.* vol. 8; B92.06.01].

Brussels

Royal Horticultural Society of Belgium (Société Royale d'Horticulture de Belgique). Honorary member by 1892 [B92.06.01].

Ghent

Royal Society of Agriculture and Botany (Société Royale d'Agriculture et de Botanique). Honorary member by 1887 [M to R. von Fischer-Benzon, 16 December 1887; B92.06.01].

Liège

Royal Society of Sciences (Société Royale des Sciences). Corresponding member, 7 March 1871 [certificate].

Canada

Kingston

Botanical Society of Canada. Member by 1861 [*Frag.* vol. 2; B92.06.01].

Montreal

McGill University. Doctor of Laws (LLD honoris causa), 30 April 1892 [certificate].

Denmark

Copenhagen

Royal Society of Northern Antiquaries (Kongelige Nordiske Oldskrift Selskab). Foundation fellow, 23 October 1860 [certificate].

Order of Dannebrog (Dannebrogsorden). Knight, 3rd class, 17 April 1863 [A. Moltke to M, 23 April 1863 (Collected Correspondence)].

Copenhagen Botanical Society (Botaniske Forening i Kjöbenhavn). Corresponding member, 30 April 1866 [certificate].

Royal Danish Academy of Sciences (Kongelige Danske Videnskabernes Selskab). Foreign member 11 April 1890 [M to [...] Zeuthen, 3 June 1890 (Collected Correspondence)].

Egypt

Cairo

Egyptian Institute (Institut Égyptien). Honorary member, 5 January 1894 [certificate].

France

Bordeaux

Physical and Natural Sciences Society (Société des Sciences Physiques et Naturelles). Honorary member by 1887 [M to R. Fischer-Benzon, 16 December 1887].

Cherbourg

Natural Sciences Society (Société des Sciences Naturelles). Corresponding member by 1867 [*Frag.* vol. 6; B92.06.01].

Le Mans

International Academy of Botanical Geography (Académie Internationale de Géographie Botanique). Founding member, 1 July 1892 [certificate]; Director, 1893 but resigned before the end of the year [M to W. Thiselton-Dyer, 3 November 1893].

Montpellier

Society of Horticulture and Natural History (Société d'Horticulture et d'Histoire Naturelle). Corresponding member by 1892 [B92.06.01].

Nice

City of Nice International Exhibition, 1883-4 (Ville de Nice Exposition Internationale). Certificate and silver medal for Australian wood exhibited in groups 6, 7, 8, 9 and 10, 18 March 1884 [certificate].

Paris

Universal Exhibition, 1855 (Exposition Universelle). Commissioner for Victoria, 17 March 1854 [J. Foster to M, 17 March 1854].

Imperial Zoological Society of Acclimatisation (Société Impériale Zoologique d'Acclimatation). Honorary member, 14 February 1861 [certificate]. Later National Society of Acclimatisation of France (Société Nationale d'Acclimatation de France). Gold medal, 1885 [DL Sp.3/56, Mitchell Library, Sydney].

Museum of Natural History (Muséum d'Histoire Naturelle). Correspondent, 15 July 1862 [certificate].

Order of the Legion of Honour (Ordre de la Légion d'Honneur). Knight's cross, 1863 [certificate].

African Institute (Institut d'Afrique). Ordinary member, 10 December 1864 [certificate].

Universal Exhibition, 1867 (Exposition Universelle). Silver Medal, group 9, class 88, *Balantium antarcticum*, 31 October 1867 [certificate].

Geographical Society (Société de Géographie). Foreign corresponding member, 22 April 1887 [certificate].

Officer's Cross of Agricultural Merit, French Republic (Croix d'Officier du Mérite Agricole), 16 September 1889 [certificate].

Universal Exhibition, 1889 (Exposition Universelle). Gold Medal diploma in group 5, class 42, 29 September 1889 [certificate].

Academy of Sciences (Académie des Sciences). Corresponding member, 1 July 1895 [*Index biographique de l'Académie des Sciences* (1979), p. 389].

GERMANY

Altenburg

Naturalists Society of Osterland (Naturforschende Gesellschaft des Osterlandes). Honorary member, 1849 [*Mitteilungen aus dem Osterlande* (1894) p. 38].

Baden

Order of the Zähringen Lion (Ordens vom Zähringer Löwen). Knight's cross, 1st class with oak leaves, 10 October 1879 [certificate].

Berlin

Society of Friends of Nature Research (Gesellschaft Naturforschender Freunde). Honorary member, 19 July 1864 [certificate].

Geographical Society (Gesellschaft für Erdkunde). Foreign member, 4 November 1865 [certificate].

Order of the Crown (Kronen-Orden). Knight, 3rd class, 9 June 1866 [certificate].

Botanical Union of the Province of Brandenburg (Botanischer Verein für die Provinz Brandenburg). Honorary member, 1868 [*Verhandlungen* (1868 [1869]) pp. VIII, XVIII.

Berlin Society for Anthropology, Ethnology and Prehistory (Berliner Gesellschaft für Anthropologie, Ethnologie und Urgeschichte). Corresponding member, 12 October 1872 [certificate].

Order of the Red Eagle (Orden des Roten Adler). Knight, 3rd class, 8 March 1876 [certificate].

Union for the Promotion of Horticulture in the Royal Prussian State (Verein zur Beförderung des Gartenbaues in den Königlich Preussischen Staaten). Corresponding member, 29 November 1876 [certificate].

German Botanical Society (Deutsche Botanische Gesellschaft). Honorary member, 24 September 1891 [certificate].

Blankenburg

Natural Science Union (Naturwissenschaftlicher Verein). Honorary member, 18 August 1858 [certificate].

Braunschweig

[Society for Natural Science]. Honorary member by 1892 [B92.06.01].

Bremen

Natural Science Union (Naturwissenschaftlicher Verein). Corresponding member, 4 May 1868 [certificate].

Breslau

Silesian Society for National Culture (Schlesische Gesellschaft für Vaterländische Cultur). Corresponding member, 29 June 1860 [certificate]; honorary member, 17 December 1878 [certificate].

Cologne

Zoological Garden (Zoologischer Garten). Honorary member, 29 May 1869 [certificate].

Danzig

Naturalists' Society (Naturforschende Gesellschaft). Corresponding member, 20 October 1886 [certificate].

Darmstadt

Union for Geography and Related Sciences (Verein für Erdkunde und Verwandte Wissenschaften). Corresponding member, 11 January 1862 [certificate].

Order of Merit of Philipp the Magnanimous (Verdienstordens Philipps des Grossmüthigen). Commander's cross, 21 June 1887 [certificate].

Dresden

Isis Natural Science Society (Naturwissenschaftliche Gesellschaft Isis). Honorary member, 1849 [*Sitzungberichte und Abhandlungen* (July-December 1883) p. XI].

Geographical Union (Verein für Erdkunde). Corresponding member, 16 February 1866 [certificate].

Botanical and Horticultural Society of the Kingdom of Saxony (Gesellschaft für Botanik und Gartenbau im Königreiche Sachsen). Honorary member, 20 February 1891 [certificate].

Erlangen

Physico-Medical Society (Societas Physico-Medica). Corresponding member, 9 June 1860 [certificate].

Frankfurt am Main

Free German Foundation for Sciences, Arts and General Culture in Goethe's Parental Home (Freies Deutsches Hochstift für Wissenschaften, Künste und Allgemeine Bildung in Goethe's Vaterhause). Honorary member and master, 10 November 1863 [certificate].

Senckenberg Naturalists' Society (Senckenbergische Naturforschende Gesellschaft). Corresponding member, 24 April 1871 [certificate].

Freiburg

Naturalists' Society (Naturforschende Gesellschaft). Corresponding member, 2 June 1865 [certificate].

Görlitz

Naturalists' Society (Naturforschende Gesellschaft). Honorary member, 21 October 1887 [certificate].

Gotha

Order of the Saxe-Ernestine House (Sachsen-Ernestinischer-Haus-Ordern). Knight, 5 January 1865 [R. von Seebach to M, 5 January 1865 (Collected Correspondence)].

Göttingen

Royal Society of Sciences (Königliche Gesellschaft der Wissenschaften). Corresponding member, 1 December 1866 [certificate].

Halle

Naturalists' Society (Naturforschende Gesellschaft). Ordinary member, 19 December 1855 [certificate].

Geographical Union (Verein für Erdkunde). Corresponding member, 14 April 1880 [certificate].

Hamburg

Natural Science Union (Naturwissenschaftlicher Verein). Corresponding member, 30 June 1852 [*Verhandlungen* (1894)].

Botanical Society (Gesellschaft für Botanik). Honorary member, 19 January 1882 [certificate].

Hanover

Natural History Society (Naturhistorische Gesellschaft). Honorary member, 28 April 1867 [certificate].

Kiel

24th Congress of German Scientists and Medical Doctors [Versammlung Deutscher Naturforscher und Ärzte], September 1846. Registered participant [*Amtlicher Bericht*, p. 290].

State examination in pharmacy, 15 March 1847 (Staatsprüfung, Schleswig-Holstein) [Achelis (1952), p. 17].

Kiel University. Doctor of Philosophy (PhD), 2 August 1847 [certificate].

Natural Science Union of Schleswig-Holstein (Naturwissenschaftlicher Verein für Schleswig-Holstein). Member, 20 January 1896 [certificate].

Königsberg

East Prussian Physico-Economical Society (Ostpreussische Physikalisch-Ökonomische Gesellschaft). Honorary member, 22 February 1890 [certificate].

Landau

Pollichia, a Natural Science Union of the Rhineland Palatinate (Pollichia, ein Naturwissenschaftlicher Verein der Rheinpfalz). Honorary member, 7 November 1860 [certificate].

Leipzig

Museum of Ethnology (Museum für Völkerkunde). Member and patron by 1881 [*Frag.* vol. 11; B92.06.01].

Marburg

Society for the Promotion of all Natural Sciences (Gesellschaft zur Beförderung der Gesammten Naturwissenschaften). Ordinary member, 1 December 1876 [certificate].

Munich

Royal Bavarian Academy of Letters and Sciences (Academia Literarum et Scientiarum Regia Boica). Carl Friedrich Philipp von Martius commemorative medal, 30 March 1864 [medal at Melbourne Museum]. Corresponding member, 23 June 1866 [certificate]; Foreign member, 20 June 1881 [certificate].

Carl Friedrich Philipp von Martius commemorative medal, 30 March 1864 [medal at Melbourne Museum].[1]

Order of Merit of St Michael (Verdienst-Orden vom Heiligen Michael). Knight, 1st class, 14 January 1874 [certificate].

Oldenburg

House and Service Order of Duke Peter Friedrich Ludwig (Haus- und Verdienst-Orden des Herzogs Peter Friedrich Ludwig). Knight's cross, 1st class, 17 January 1887 [certificate].

Regensburg

Royal Botanical Society (Regia Societas Botanica Ratisbonensis/Königlich Botanische Gesellschaft). Corresponding member, 13 September 1853 [certificate]; honorary member, 10 December 1890 [certificate].

Rostock

Rostock University. Doctor of Medicine (MD), 7 November 1857 [certificate].

Mecklenburg Naturalists' Society (Mecklenburgische Naturforschende Gesellschaft). Honorary member [B92.06.01].

Schwerin

Order of the Wendish Crown (Orden der Wendischen Krone). Knight, 9 November 1865 [certificate].

Grand Duchy of Mecklenburg-Schwerin, Medal for Arts and Sciences, 1889 [M to Friedrich Franz III, Grand-Duke of Mecklenburg-Schwerin, 12 July 1889].

Stuttgart

Order of the Crown of Württemberg (Orden der Württembergischen Krone). Knight's cross, 1st class, 20 December 1867 [A. von Egloffstein to M, 20 December 1867 (Collected Correspondence)].

Baron (Freiherr), 6 July 1871 [Karl I, King of Württemberg, to M, 6 July 1871 (*Selected correspondence*, vol. 2, pp. 580-2)].

1 Two separate medals were issued to mark the fiftieth anniversary of Martius's doctorate, one by the Bavarian Academy and a rather grander one as the result of a public subscription (Wesche [1997]). M had both.

Union for National Natural History of Württemberg (Verein für Vaterländische Naturkunde in Württemberg). Honorary member, 2 June 1888 [certificate].

Weimar

Order of Vigilance or of the White Falcon (Hausorden der Wachsamkeit oder vom Weissen Falken). Knight, 17 December 1874 [certificate].

Botanical Union for all Thuringia (Botanischer Verein für Gesamt-Thüringen). Honorary member, 6 October 1889 [certificate].

Hawaii

Honolulu

Order of Kalakaua. Knight Companion, 13 February 1882 [certificate].

Royal Agricultural Society of Hawaii. Silver medal, 1884 [M to L. McCully, 9 September 1884 (Collected Correspondence)].

India

Calcutta

Agricultural and Horticultural Society of India. Honorary member, 24 July 1879 [*Journal* (1880) p. xx].

Calcutta International Exhibition, 1883-4. Certificate of merit of the first class and a gold medal for 'A collection of Australian flowers not yet introduced into horticulture, preserved in an Album', 18 July 1884 [certificate].

Ireland

Dublin

Royal Irish Academy. Honorary member, 16 March 1894 [certificate].

Italy

Order of St Maurice and St Lazarus (Ordine dei Santis Maurizio e Lazzaro). Knight, 29 July 1865 [certificate].

Order of the Crown of Italy (Ordine della Corona d'Italia). Knight, 20 September 1868 [certificate]; Officer, 22 July 1881 [certificate].

Brescia

Athenaeum of Brescia (Ateneo di Brescia). Honorary member, 6 March 1887 [F. Cazzago to M, 6 March 1887].

Florence

Italian Geographical Society (Società Geografica Italiana). Honorary member, 14 March 1870 [certificate].

Royal Museum of Physics and Natural History of Florence (Reale Museo di Fisica e Storia Naturale di Firenze). Medal, April 1870 [medal at Melbourne Museum].

International Exhibition of Horticulture (Esposizione Internazionale di Orticultura), 1874. Invited to serve as juror [F. Parlatore to M, 23 December 1873 (Collected Correspondence)]. The appointment was declined [M to J. Francis, 12 February 1874 (Collected Correspondence)].

Naples

International Maritime Industry Exhibition (Esposizione Internazionale delle Industrie Maritime), 1871. Silver medal, second class, for 'a collection of timbers' [certificate].

Palermo

Society of Acclimatization and Agriculture (Società di Acclimazione e di Agricoltura). Honorary member, 27 March 1863 [certificate].

Siena

Royal Academy of the Physiocritics (Reale Accademia dei Fisiocritici). Corresponding member, 17 June 1868 [certificate].

Venice

Royal Venetian Institute of Science, Letters and Arts (Reale Istituto Veneto di Scienze, Lettere ed Arti). Corresponding member, 25 March 1877 [certificate].

Luxembourg

Order of the Oak Crown (Ordre de la Couronne de Chêne). Knight by 1874 [*Frag.* vol.8; B92.06.01].

Mauritius

Port Louis

Royal Society of Arts and Sciences of Mauritius. Corresponding member, 5 September 1860 [certificate].

[Society for the Acclimatisation of Animals and Plants]. Corresponding member by 1877 [*Frag.* vol. 10; B92.06.01].

Mexico

Mexico City

Mexican Society of Geography and Statistics (Sociedad Méxicana de Geografia y Estadistica). Corresponding member, 28 June 1873 [certificate].

Mexican Society of Natural History (Sociedad Méxicana de Historia Natural). Corresponding member, 25 September 1873 [certificate].

Netherlands

Amsterdam

International Exhibition, 1883. Diploma of honour for an exhibit of 'botanical specimens &c &c' [certificate].

Royal Netherlands Academy of Arts and Sciences (Koninklijke Nederlandse Akademie van Wetenschappen). C. H. D. Buys Ballot commemorative medal, 1887 [medal at Melbourne Museum].

Netherlands Society for Horticulture and Botany (Nederlandche Maatschappij voor Tuinbouw eu Plantkunde). Honorary member, 28 August 1891 [certificate].

Rotterdam

Rotterdam Zoo (Rotterdamsche Diergaarde). Honorary member, 3 August 1863 [certificate].

Netherlands East Indies

Batavia

Royal Natural Science Society of the Netherlands Indies (Koninklijke Natuurkundige Vereeniging in Nederlandsch-Indië). Corresponding member, 21 December 1872 [certificate].

New South Wales

Sydney

Agricultural Society of New South Wales. Bronze medal for collection of plants, 1870 [medal at Melbourne Museum].

Royal Society of New South Wales. Honorary member, 4 August 1875 [A. Liversidge to M, 4 August 1875 (Collected Correspondence)]; Clarke Medal, 13 December 1882 [*Journal and proceedings* (1883) p. 212].

Linnean Society of New South Wales. Honorary member (one of the first two), 26 January 1876 [*Proceedings* (1877) p. 419].

University of Sydney. Examiner in 1884 and subsequently [University of Sydney Archives].

New Zealand

Wellington

New Zealand Institute. Honorary member, 6 January 1871 [certificate].

Portugal

Coimbra

Society of the Institute of Coimbra (Conimbricensis Instituti Societas). Corresponding fellow, 26 March 1872 [certificate].

Lisbon

Order of St James of the Sword (Ordem Militar de Sant' Iago da Espada). Commander, 29 November 1870 [Luis I, King of Portugal, to M, 29 November 1870 (Collected Correspondence); cross and badge in private hands].

Geographical Society (Sociedade de Geographia). Corresponding member, 12 August 1878 [certificate].

Order of Christ (Ordem Militar de Christo). Grand Cross, 28 April 1887 [certificate].

QUEENSLAND

Brisbane

Exhibition of Colonial Products, 1866. Bronze medal [medal at Melbourne Museum].

Royal Society of Queensland. Corresponding member, 1884 [*Proceedings* (1885) p. viii].

Charters Towers

Philosophical Society. Corresponding member by 1892 [B92.06.01].

RÉUNION

Acclimatization and Natural History Society of Réunion Island (Société d'Acclimatation et d'Histoire Naturelle de l'Ile de la Réunion). Delegate of the Committee of Foreigners, 1862 [*Bulletin* (1863) p. 15; M to A. Berg, 23 February 1863 (Collected Correspondence)]; member [*Bulletin* (1863) p. 50]; honorary member by 1892 [B92.06.01].

RUSSIA

Moscow

Committee of Acclimatization of Fauna and Flora of the Imperial Moscow Society of Agriculture (Komitet' Akklimatizatsii Zhivotnykh' i Rastenii Imperatorskago Moskovskago Obshchestva Sel'skago Khozyaistva). Member, 28 October 1862 [certificate].

Society of Lovers of Natural Knowledge (Obshchestvo Lybitelei Estestvoznaniya). Member, 14 May 1864 [certificate].

Imperial Society of Naturalists (Société Impériale des Naturalistes/Caesarea Naturae Curiosorum Mosquensis). Member, 1 March 1866 [certificate].

St Petersburg

Russian Imperial Society of Horticulture. Corresponding member by 1861 [*Frag.* vol. 2; B92.06.01].

[International Exhibition, 1869]. Gold medal [M to R. Barry, 5 September 1872 (Collected Correspondence)].

International Exhibition of Horticulture, Imperial Russian Society of Horticulture, 1884. Certificate and gold medal of the Ministry for Government Properties for *Todea barbara*, 30 May 1884 [certificate].

Ekaterinburg

Uralian Society of Natural Science (Uralskoye Obshtchestvo Estestvoznaniya). Member by 1895 [M to L. Dejardin, 17 May 1895 (Collected Correspondence)].

Salvador

San Salvador

[Geographical and Historical Society]. Honorary member by 1892 [B92.06.01].

South Australia

Adelaide

Naturalized British subject in the Colony of South Australia, 1849 [*South Australian government gazette*, 16 August 1849, p. 368 and 4 October 1849, p. 451].

Horticultural Society of Adelaide. Honorary member by 1868 [*Frag.* vol. 6].

Royal Society of South Australia. Honorary fellow, 1879 [*Transactions and proceedings* (1880) p. iv].

University of Adelaide. Honorary examiner [M to J. Hooker, 28 December 1885 (Collected Correspondence)].

Adelaide Jubilee International Exhibition, 1887. First order of merit for 'albums of dried specimens of ferns, grasses &c' [certificate], first order of merit for 'samples of Australian Woods in book form' [certificate], first order of merit for 'Photograph views of trees' [certificate], first order of merit for 'Vegetable products' [certificate], first order of merit for 'Educational collections of dried specimens' [certificate].

Penola

Penola Mechanics' Institute. Honorary member, 12 July 1864 [*Border watch*, 15 July 1864].

Spain

Madrid

Museum of Natural History (Museo de Historia Natural). Corresponding member, 30 June 1871 [certificate].

Order of Isabella the Catholic (Order de Isabel la Cátolica). Commander, 8 March 1872 [certificate; medal at Royal Botanic Gardens Melbourne].

Sweden

Gothenburg

Gothenburg Horticultural Society (Göteborgs Trädgårdsförening). Honorary member, 12 September 1881 [certificate].

Lund

Physiographical Society (Physiographiska Sällskapet/Physiografischen Gesellschaft). Corresponding member, 9 December 1863 [Lund University Library].

Stockholm

Order of the North Star (Nordstierne-Orden). Knight, 9 December 1871 [certificate].

Swedish Society of Medicine (Svenska Läkaresällskapet). Jacob Berzelius commemorative medal [medal at Melbourne Museum].

Royal Swedish Academy of Sciences (Regia Scientiarum Academia Suecica/ Kongliga Svenska Vetenskapsakademien). Honorary member, 9 November 1892 [certificate].

Uppsala

Royal Society of Sciences of Uppsala (Regia Scientiarum Societas Upsaliensis). Member, 14 March 1862 [certificate]; ordinary member, 25 November 1887 [certificate].

SWITZERLAND

Geneva

Genevan National Institute (Institut National Genevois). Corresponding member, 11 November 1893 [certificate].

Neuchatel

Neuchatel Geographical Society [Société Neuchâteloise de Géographie]. Honorary member by 1892 [B92.06.01].

St Gallen

[Swiss Society for Economic Geography]. Honorary member by 1881 [*Frag.* vol. 11; B92.06.01].

TASMANIA

Hobart

Royal Society of Tasmania. Member, 9 February 1858 [J. Milligan to M, 27 February 1858 (Collected Correspondence)]; honorary corresponding member, 11 May 1858 [J. Milligan to M, 18 May 1858 (*Selected correspondence*, vol. 1, p. 379)]; honorary member, 5 September 1882 [J. Barnard to M, 12 September 1882 (Collected Correspondence)].

[Royal Agricultural and Horticultural Society of Tasmania]. Honorary member by 1892 [B92.06.01].

Launceston

Launceston Horticultural Society. Member by 1864 [*Frag.* vol. 4].

Launceston Gardeners' and Amateurs' Horticultural Society. Honorary member, 4 January 1865 [S. Hannaford to M, 4 January 1865 (Collected Correspondence)].

Tasmanian Union Jack Field Club. Honorary member, 9 July 1884 [certificate].

Tasmanian Exhibition, 1892. Special first certificate of merit awarded in Group E, Class 37 for samples of wood [certificate].

United Kingdom

Bradford

Bradford Natural History and Microscopical Society. Honorary member, 7 May 1894 [certificate].

Edinburgh

Botanical Society. Foreign member by 1860 [*Transactions* (1860) p. 439], honorary member by 1892 [B92.06.01].

Edinburgh Geological Society. Foreign corresponding fellow, 15 May 1884 [certificate].

Scottish Geographical Society. Corresponding member by 1892 [B92.06.01].

London

Royal Geographical Society. Fellow, 25 January 1858 [Archives, Royal Geographical Society, London].

Linnean Society. Fellow, 20 January 1859 [Archives, Linnean Society of London].

Royal Society. Fellow, 6 June 1861 (*Record* p. 478); Royal Medal, 1888 'for his long services in Australian Exploration, and for his Investigations of the flora of the Australian Continent' [*Record* p. 351].

Zoological Society. Corresponding member, 17 October 1861 [certificate].

International Exhibition, 1862. Honorable mention, class 3, 11 July 1862 [certificate]. Honourable mention, Class 4 [medal at Melbourne Museum].

Anthropological Society. Local secretary (abroad), 20 January 1864 [certificate].

Order of St Michael and St George. Companion (CMG), 24 April 1869 [Royal warrant, 24 April 1869 (*Selected correspondence*, vol. 2, p. 497)]; Knight-Commander (KCMG), 24 May 1879 [Queen Victoria to M, 24 May 1879].

Royal Botanic Society. Corresponding member by 1871 [*Frag.* vol. 7, B92.06.01].

Royal Horticultural Society. Corresponding member, 17 December 1878 [certificate].

Victoria Institute or Philosophical Society of Great Britain. Associate, 1880 [*Journal* (1881) p. 396].

Society for the Encouragement of Arts, Manufactures and Commerce. Honorary member, 1883 [Royal Society of Arts archives].

Pharmaceutical Society of Great Britain. Corresponding member, 4 May 1892 [certificate].

Chemical Society. Member by 1892 [B92.06.01].

Geological Society. Member by 1892 [B92.06.01].

Royal Colonial Institute. Ordinary member by 1892 [B92.06.01].

[International Society of Letters, Science, and Arts]. Honorary member by 1892 [B92.06.01].

United States of America

Boston

Boston Society of Natural History. Corresponding member, 6 January 1886 [certificate].

American Academy of Arts and Sciences. Foreign honorary member in Class II Natural and Physiological Sciences, Section II Botany by 1893 [*Proceedings* (1893) p. 457].

New Orleans

New Orleans Academy of Sciences. Corresponding member, 1 July 1872 [certificate].

New York

American Chemical Society. Ordinary member by 1887 [M to R. von Fischer-Benzon, 16 December 1887; B92.06.01].

New York Academy of Sciences. Honorary member by 1892 [B92.06.01].

Philadelphia

Academy of Natural Sciences. Correspondent, 28 March 1876 [certificate].

Philadelphia College of Pharmacy. Honorary member, 27 March 1893 [certificate].

San Francisco

Californian Academy of Natural Sciences. Honorary member, 19 October 1863 [certificate].

[Californian Horticultural Society]. Honorary member by 1892 [B92.06.01].

Venezuela

Carácas

Society of Physical and Natural Sciences (Sociedad de Ciencias Físicas y Naturales). Corresponding member, 12 January 1880 [certificate].

Victoria

Ballarat

Agricultural Society. Honorary member by 1877 [*Frag.* vol. 10; B92.06.01].

School of Mines. Examiner in general botany, February 1881 [W. Barnard to M, 21 February 1881 (Collected Correspondence)].

[Science Society]. Honorary member by 1892 [B92.06.01].

Geelong

[Literary Society]. Honorary member by 1858 [*Frag.* vol. 1].

Horticultural Improvement Association. Corresponding member, 23 January 1861 [*Victorian agricultural and horticultural gazette*, 14 February 1861, p. 19].

Melbourne

Government Botanist of Victoria, 26 January 1853 to 10 October 1896 [W. Lonsdale to M, 26 January 1853 (*Selected correspondence*, vol. 1, p. 138)].

Philosophical Society of Victoria. Member, 17 June 1854; council member, 15 July 1854 [Minutes, State Library of Victoria]. [See also Philosophical Institute of Victoria.]

Victorian Institute for the Advancement of Science. Member and council member, 31 July 1854 [Minutes, State Library of Victoria]. [See also Philosophical Institute of Victoria.]

Philosophical Institute of Victoria. Member and council member, 28 June 1855; honorary member, 10 July 1855; president, 9 March 1859 to March 1860 [Minutes, State Library of Victoria]. [Institute formed in 1855 from a merger of the Philosophical Society of Victoria and Victorian Institute for the Advancement of Science; from 1860 Royal Society of Victoria (see below).]

German Club (Deutscher Verein). Honorary member, 26 January 1857 [certificate].

Melbourne Botanic Garden, Director 13 August 1857 to 30 June 1873 [T. Balmain to M, 13 August 1857 (*Selected correspondence*, vol. 1, p. 318); C. Hodgkinson to M, 31 May 1873 (*Selected correspondence*, vol. 2, pp. 671-2)]; Melbourne Botanic and Zoological Garden, Director August 1858 to August 1861 [J. Moore to M, 5 August 1858 (Collected Correspondence)].

Naturalized British subject in the Colony of Victoria, 14 September 1857 [certificate].

Horticultural Society of Victoria. Vice-president in 1858 [*Frag* vol. 1]; Vice-president, 7 January 1861 [W. Clarke to M, 11 January 1861 (Collected Correspondence)], Vice-president, 19 July 1865 [J. Toon to M, 29 July 1865 (Collected Correspondence)], honorary member February 1882 [M to J. Kirkland, February 1882 (Collected Correspondence)].

Board of Science, Government of Victoria. Member, 15 February 1858 [J. Moore to M, 15 February 1858 (Collected Correspondence)].

Pharmaceutical Society of Victoria. Honorary member, March 1858 [MS 9601, box 1/1 (a), La Trobe Australian Manuscripts, State Library of Victoria, Melbourne]. A certificate for honorary membership was sent on 19 August 1863 but it is not clear if it was for the 1858 election or a re-election [J. Bosisto to M, 19 August 1863 (Collected Correspondence)].

Microscopical Society of Victoria. Member, 20 April 1858 [*Victorian agricultural and horticultural gazette* (1858) 21 June, p. 48; society defunct by 1859].

Zoological Gardens, Committee of Management. Member, 20 July 1858 [J. Moore to M, 20 July 1858 (Collected Correspondence)].

Medical Society of Victoria. Honorary member, 22 November 1858 [Archives, Australian Medical Association, Melbourne].

Musical Society of Victoria. Honorary member, 1859 [A. Fisher to M, 8 January 1859 (Collected Correspondence)].

Board of Agriculture, Government of Victoria. Member, 5 July 1859 [*Government gazette* (1859) p. 1397].

Royal Society of Victoria. Honorary member, 1860; life member, 1867 [*Transactions and proceedings* (1868) p. 345]. [See also Philosophical Institute of Victoria.]

Acclimatisation Society of Victoria. Member and Vice-president, 1861 [*First annual report* (1862) pp. 3, 13]; honorary member 1890s [B92.06.01 annotation].

Victorian Exhibition, 1861. Commissioner [H. Barkly to M, 8 January 1861 (Collected Correspondence)]; committee member Class II 'Horticultural products and the manufactures and processes connected therewith'; and committee member Class III 'Indigenous vegetable products, and the manufactures and processes connected therewith' [*Catalogue* (1861) pp. 217, 225].

University of Melbourne. Honorary examiner in Botany, Materia Medica, Therapeutics and Prescriptions, November 1861 [J. James to M, 4 November 1861 (*Selected correspondence*, vol. 2, p. 119)].

German Health Society (Deutscher Kranken Verein). Honorary member, 1862 [*Germania*, 11 April 1862]; Honorary life member, May 1869 [Deutscher Kranken Verein to M, May 1869 (Collected Correspondence)].

Society for the Improvement of Horticulture. Failed to be elected a corresponding member in 1862 [Minute books, State Library of Victoria]; honorary member by 1892 [B92.06.01].

Royal Park, Trustees. Honorary secretary from 1862 [Letter and minute book, Mitchell Library, Sydney].

Gymnastic Club (Turn Verein). Honorary member, 1866 [*Germania*, 17 May 1866]; declined presidency in 1884 [M to J. Hooker, 11 March 1884 (Collected Correspondence)]; honorary president [B92.06.01].

Intercolonial Exhibition of Australasia, 1866-7. Commissioner and Colonial Representative for Western Australia [*Catalogue* (1867) p. iii]. Honorable mention, Class III, section 7A, for 'Sassafras hops & Medicinal Herbs'; Medal, class III, section 7 for 'Paper, and Materials used in making'; Medal, class III, section 9, for 'Chemical Products'; Honorable mention, class VI, section 27, for 'a collection of native war implements' [*Catalogue* (1867) pp. 208, 266, 321, 397; certificates; J. Knight to M, 4 February 1868 (Collected Correspondence].

Victorian Medical Benevolent Association. Vice-president, 15 July 1868 [J. Neild to M, 18 July 1868 (*Selected correspondence*, vol. 2, p. 474)].

West Melbourne Presbyterian Literary Association. Patron from the foundation of the Association in 1876 [Martin (1967) p. 26].

[Assistant Pharmacists' Society]. Honorary president and patron by 1877 [*Frag.* vol. 10; B92.06.01].

British Medical Association (Victorian Branch). [*Frag.* vol. 11].

Field Naturalists' Club of Victoria. Member (no. 36), June/July 1880 [Taylor (1996) p. 133]; declined presidency in 1884 [M to J. Hooker, 11 March 1884 (Collected Correspondence)]; patron, June 1886 [Taylor (1996) p. 135].

Melbourne International Exhibition, 1880-1. Commissioner, member of the Intercolonial Committee, Districts of Victoria Committee, Vegetable Products Committee [*Catalogue* (1880) edn 2, p. vi], Juror in the chemical branch and two other branches [M to T. Cheeseman, 16 February 1882 (Collected Correspondence)]; silver medal 'for services' [certificate], bronze medal 'for services' [certificate; bronze medal for services to 'Australian International Exhibitions Sydney 1879-80 Melbourne 1880-81' at Melbourne Museum].

Victorian Intercolonial Exhibition of Wine, Fruit, Grain &c, Melbourne, 1884. Certificate and gold medal, 1884 [certificate].

Geological Society of Australasia. Inaugural vice-president, 1886 [Litton (1886)].

[Young Christian Gentlemen's Association, Presbyterian Church]. Patron in 1887 [M to (P. Sclater), December 1887]. Probably the Young Men's Sabbath Morning Fellowship Association, West Melbourne Presbyterian Church, founded in 1885 [see Martin (1967)].

Centennial International Exhibition, 1888. First order of merit in Jury Section no. 24 for 'timber &c' [certificate]; first order of merit in Jury Section no. 29 for 'models of fruit' [certificate]; first order of merit in Jury Section no. 29 for 'Todea ferns' [certificate]; commissioner [M to (P. Sclater), December 1887].

Intercolonial Medical Congress, January 1889. President of therapeutic section [B89.13.16].

Choral Society (Liedertafel). Vice-president in 1889 [M to W. Woolls, 15 February 1889 (Collected Correspondence)]; President in 1890 [B92.06.01].

[Agricultural Students' Society]. Honorary president by 1892 [B92.06.01].

Concordia Club. Honorary member by 1892 [B92.06.01].[2]

[Melbourne Ladies' Club Salon]. Honorary member by 1892 [B92.06.01].

Veterinary College. Honorary examiner in 1893 [M to O. Tepper, 12 November 1893 (Collected Correspondence); B92.06.01 annotation].

Education Department, Government of Victoria. Examiner [M to O. Tepper, 12 November 1893 (Collected Correspondence)].

Veterinary Society of Victoria. Honorary member [B92.06.01].

[Victorian Viticultural Association]. Honorary member by 1892 [B92.06.01].

2 There appear to have been two Concordia Clubs in Melbourne in the early 1890s, one of them based in South Yarra, the district in which M lived. However, little is known about either club, and no membership records have been found. See Darragh (2000).

Stawell
[Mining Institute]. Honorary member by 1892 [B92.06.01].

Williamstown
Horticultural Spring Show, 1868. Honorary certificate for 'Collection of Rare Plants' [certificate].

WESTERN AUSTRALIA

Perth
West Australian Natural History Society. Honorary member, 1891 [M to B. Woodward, 21 March 1891 (Collected Correspondence)].

OTHER

Imperial Leopoldine-Carolinian Academy of Naturalists (Academia Caesarea Leopoldino-Carolina Naturae Curiosorum). Member, with cognomen Leschenault, 1 May 1857 [certificate]. Mueller was elected a member for a second time in 1865 (his earlier membership having apparently been overlooked) [Jahn & Schmidt (1996) p. 15].

One of five special votes of thanks awarded to men who were considered to have rendered exceptional services to the cause of geographical science, third International Geographical Congress, Venice, 1881 [*Proc. Roy. Geog. Soc.*, n.s., vol. 3, no. 12 (1881), p. 749].

Geographical Society of Australasia (later Royal Geographical Society of Australasia). Founder and life member, 15 December 1883 [certificate]; declined presidency in 1884 [M to J. Hooker, 11 March 1884 (Collected Correspondence)]; Vice-president and President of Victorian branch from 1883 [*Proc. Roy. Geog. Soc. Australasia*, vol. 1 (1883-4)].

Friedrich Traugott Kützing commemorative medal, 8 December 1887 [medal at Melbourne Museum].

Australasian Association for the Advancement of Science. Second meeting, 1890: President, member of Antarctic Exploration Committee, member of committee 'to investigate the Fertilisation of the Fig in the Australasian Colonies' [*Report* (1890) p. xxi]. Third meeting, Christchurch, 1891: member of Recommendation Committee [*Report* (1891) p. xv]. Fourth meeting, Hobart, 1892: presided over opening meeting of General Council in place of retiring President, Sir James Hector, who was unable to attend [*Report* (1892), p. xvii]. Fifth meeting, Adelaide, 1893: Vice-president but did not attend [*Report* (1894) p. xv; M to A. Macdonald, 25 September 1893 (Collected Correspondence)]. Sixth meeting, Brisbane, 1895: Vice-president and President of Section E (Geography) but did not attend [*Report* (1895) p. xiv; M to W. Thiselton-Dyer, 1 January 1895 (Collected Correspondence)].

International Commission for Botanical Nomenclature. Appointed member 9 September 1892 [Ascherson (1892) p. 37].

Doubtful

[German Medical Society]. Honorary fellow [*Frag.* vol. 1; not included in subsequent vols.]. No society has been identified to which this might refer.

[German Pharmaceutical Society]. Honorary fellow [*Frag.* vol. 1; not included in subsequent vols.]. The Deutsche Pharmaceutische Gesellschaft was not formed until 1890; no earlier body with such a name has been identified.

[Historical Society of Texas]. Honorary member by 1892 [B92.06.01]. The modern society of this name was not formed until some years after this date; no earlier body with such a name has been identified.

[Philosophical and Acclimatization Society of New Zealand]. Honorary member [*Frag.* vol. 4; B92.06.01]. There were societies of this name in several of the New Zealand provinces but not one for New Zealand as a whole.

Bibliography

The bibliography lists works, other than those by Mueller, cited in footnotes. Annotations are included if a work is known to have been part of M's library, either because it was listed as part of the donations he made to his departmental library in 1865 (B65.10.01) or 1869 (B69.07.03), and/or because a copy inscribed to or by M is now part of the Library, Royal Botanic Gardens Melbourne. Other annotations comment on, or provide sources for full discussions of, publication dates of works issued in parts.

Abbreviations

col. = column.
edn = edition.
G. P. = Government Property.
M = Mueller.
no. = number.
NSW = New South Wales.
NT = Northern Territory, Australia.
pt, pts = part, parts
TL2 = Stafleu & Cowan (1976-88), Stafleu & Mennega (1992-2000).
unpubl. = unpublished.
SA = South Australia.
Vic. = Victoria.
vol, vols = volume, volumes.
WA = Western Australia.

Achelis, T. O. (1952) *Prüflingen der Pharmazie in Schleswig-Holstein, 1804-1866*, Eutin.

Adams, N. M. (2002) 'John Buchanan F.L.S., botanist and artist (1819-1898)', *Tuhinga* vol. 13, pp. 71-115.

Adanson, M. (1763) [ie 1763-4] *Familles des plantes*, 2 vols, Paris.

Agardh, J. G. (1879) 'Florideernes morphologi', *Kongl. Svenska Vetenskaps Academiens handlingar ny fjöld* vol. 15, 1-199.

Agardh, J. G. (1889) 'Species sargassorum australiae descriptae et dispositae', *Kongliga Svenska vetenskaps-akademien handlingar* vol. 23, part 3, pp. 1-113.

Agardh, J. G. (1890) 'Til algernes systematik', *Acta universitatis lundensis* vol. 26, pp. 1-125.

Agardh, J. G. (1892-1903) *Analecta algologica: observationes de speciebus algarum minus cognitis earumque dispositione, continuatio I(-V)*, Lund.

Aiton, W. T. (1812) *Hortus kewensis ..., vol. IV*, edn 2, London.
Allan, M. (1967) *The Hookers of Kew, 1785-1911*, London.
Allioni, C. (1766) 'Stirpium aliquot descriptiones ...', *Mélanges de philosophie et mathématique de la société royale de Turin* vol. 3, pp. 176-84.
American national biography (1999), 13 vols, New York.
Archer, B. & Maroske, S. (1996) 'Sarah Theresa Brooks – plant collector for Ferdinand Mueller', *Victorian naturalist* vol. 113, pp. 188-94.
Ascherson, P. (1882) 'On the propagation of *Cymodocea antarctica*', *Transactions of the Royal Society of South Australia* vol. 5, pp. 37-9.
Ascherson, P. (1892) 'Rapport sur la question de la nomenclature', in Penzig, O. (ed.) *Atti del congresso botanico internazionale di Genova 1892*, Genoa, pp. 85-113.
Ascherson, P. et al. ([1892]) *Propositions to an amendement [sic] of the 'Lois de [la] nomenclature botanique'*, [Berlin].
Aublet, J. B. C. F. (1775) *Histoire des plantes de la Guiane françoise ...*, 4 vols, London & Paris.
Australian dictionary of biography (1966-90) vols 1-14, Melbourne.

Bailey, F. M. (1886) *A synopsis of the Queensland flora, containing the phænogamous and cryptogamous plants: first supplement*, Brisbane.
Bailey, F. M. (1892) *Botany bulletin no. V: contributions to the flora of Queensland*, Brisbane.
Baillon, H. E. (1867-95) *Histoire des plantes*, 13 vols, Paris.
Baillon, H. E. (1876-92) *Dictionnaire de botanique*, 4 vols, Paris.
Balfour, J. H. (1875) *Manual of botany*, 5th ed., Edinburgh.
Bancroft, J. (1877) *Pituri and Duboisia*, Brisbane.
Banks, J. (1896) *Journal of the Right Hon. Sir Joseph Banks, Bart., K.B., P.R.S. during Captain Cook's first voyage in H.M.S. Endeavour in 1768-71 to Tierra del Fuego, Otahite, New Zealand, Australia, the Dutch East Indies etc.*, edited by J. D. Hooker, London.
Barclay-Lloyd, J. (1994) 'SS Vincenzo e Anastasio alle Tre Fontane near Rome: the Australian connection', *Tjurunga* vol. 46, pp. 57-70.
Barnard, F. G. A. (1906) 'The first quarter of a century of the Field Naturalists' Club of Victoria', *Victorian naturalist* vol. 23, pp. 63-77.
Barnhart, J. H. (1965) *Biographical notes on botanists*, Boston, Mass.
Bentham, G. (1833) [Epitome of the characters of Californian Polemoniaceae], *Edwards's botanical register* vol. 19, t. 1622.
Bentham, G. (1845) 'Polemoniaceae', in Candolle, A. P. de *Prodromus systematis naturalis regni vegetabilis ...*, vol. 9, Paris, pp. 302-22.
Bentham, G. (1848) 'Labiatae' in Candolle, A. P. de *Prodromus systematis naturalis regni vegetabilis ...*, vol. 12, Paris, pp. 27-603.

Bentham, G. (1863-78) *Flora australiensis ...*, 7 vols, London. For publication dates of individual parts see below.
Bentham, G. (1863) *Flora australiensis ..., vol. 1 Ranunculaceae to Anacardiaceae*, London.
Bentham, G. (1864) *Flora australiensis ..., vol. 2 Leguminosae to Anacardiaceae*, London.
Bentham, G. (1866) [(i.e. [1867]) *Flora australiensis ..., vol. 3 Myrtaceae to Compositae*, London.
Bentham, G. (1870) *Flora australiensis ..., vol. 5 Myoporineae to Proteaceae*, London.
Bentham, G. (1878) *Flora australiensis ..., vol. 7 Roxburghiaceae to Filices*, London.
Bentham, G. (1997) *Autobiography 1800-1834*, edited by M. Filipiuk, Toronto.
Bentham, G. and Hooker, J. D. (1862-83) *Genera plantarum*, London.
Berg, O. (1857-9) 'Myrtaceae' in Martius, C. F. P. (1840-1906) *Flora brasiliensis* ..., vol. 14, part 1, Leipzig.
Bevan, T. F. (1890) *Toil, travel and discovery in British New Guinea*, London.
Bicentennial dictionary of Western Australians, pre-1829-1888 (1987-97), 10 vols, Nedlands, W.A.
Blasius, W (1883) 'Über die letzten Vorkommnisse des Reisen-Alks *Alca impennis* und die in Braunschweig und an anderen Orten befindlichen Exemplare dieser Art', *Jahresbericht der Vereins für Naturwissenschaften zu Braunschweig* vol. 3, pp. 89-115.
Blume, C. L. (1849-51) [i.e. [1849-57] *Museum botanicum lugduno-batavum* ..., 2 vols, Leiden.
Bojer, W. (1837) *Hortus mauritianus, ou énumération des plantes, exotiques et indigènes, qui croissent à l'île Maurice*, Mauritius.
Borchgrevink, C. E. (1901) *First on the Antarctic continent: being an account of the British Antarctic expedition, 1898-1900*, London.
Britten, J. (1886) 'On the nomenclature of some Proteaceæ', *Journal of botany* vol. 24, pp. 296-306.
Brockhaus, F. A. (1866) *Conversations-Lexikon, neunte Band, Konradin bis Mauer*, 11th ed., Leipzig.
Brockhaus' Konversations-Lexikon (1888-95), edn 14, 16 vols, Leipzig.
Brown, G. (1877) 'Notes on the Duke of York group, New Britain, and New Ireland', *Journal of the Royal Geographical Society* vol. 47, pp. 137-50.
Brown, J. C. (1877) *The schools of forestry in Europe: a plea for the creation of a school of forestry in connection with the arboretum at Edinburgh*, Edinburgh.

Brown, J. E. (1881) *A practical treatise on tree culture in South Australia*, Adelaide. Inscription: 'To. Baron Ferd. Von Mueller | K. C. M. G. | &. &. &. | Government Botanist. | Melbourne. | with Author's Compts: | Adelaide. | 17th May. 1881.'.

Brown, R. (1810) *Prodromus florae novae hollandiae* ..., London. B65.10.01. M inscription: 'Ferd. M. | E. bibliotheca Sprengelii | mecum communicavit | Guil.Ott. Sonder.'.

Brown, R. (1810a) 'On the natural order of plants called Proteaceae', *Transactions of the Linnean Society* vol. 10, pp. 15-226.

Brown, R. (1817) 'Some observations on the natural family of plants called Compositæ', *Transactions of the Linnean Society of London* vol. 12, pp. 76-142.

Brown, Robert (1874) *Manual of botany*, Edinburgh.

Caldwell, W. H. (1888) 'The embryology of Monotremata and Marsupialia, part 1', *Philosophical transactions of the Royal Society of London (B)* vol. 178, pp. 463-86.

Campbell, F. A. (1874) *A year in the New Hebrides, Loyalty Islands and New Caledonia*, Geelong.

Candolle, A. C. P. de (1869) 'Piperaceae', in Candolle, A. P. de *Prodromus systematis naturalis regni vegetabilis* ..., vol. 16, Paris, pp. 235-471.

Candolle, A. L. P. P. de (1844) 'Apocynaceae', in Candolle, A. P. de *Prodromus systematis naturalis regni vegetabilis* ..., vol. 8, Paris, pp. 317-489.

Candolle, A. L. P. P. de (1867) *Lois de la nomenclature botanique adoptées par le Congrès International de Botanique, tenu en août 1867*, Paris.

Candolle, A. L. P. P. de (1880) *La phytographie: ou l'art de décrire les végétaux considérés sous différents points de vue*, Paris.

Candolle, A. P. de (1813) *Théorie élémentaire de la botanique* ... , Paris.

Candolle, A. P. de (1816-21) *Regni vegetabilis systema naturale*, 2 vols, Paris. B65.10.01. M. inscription: vol. 1 : 'Ferd. Müller | Stud. botanicae | Kiliae 1846'; vol. 2: 'Stud. botanicae | Ferd. Müller | Kiliae | 1846'.

Candolle A. P. de (1819) *Théorie élémentaire de la botanique* ... , edn 2, Paris.

Candolle, A. P. de (1823-73) (i.e. [1824]-73) *Prodromus systematis naturalis regni vegetabilis* ..., 17 vols, Paris. B65.10.01 (vols 1-14?), B69.07.03 (vols 15 & 16). M annotations in vols 1-2. Printed labels in vols 2 & 3: 'FERD. MÜLLER'.

Caruel, T. (1881) *Pensieri sulla tassinomia botanica*, Florence.

Cavanilles, A. J. (1798) *Icones et descriptiones plantarum* ..., vol. 4, Madrid.

Chambers' encyclopaedia (1868), 10 vols, London.

Chapman, A. D. (1991) *Australian plant name index*, 4 vols, Canberra.

Charsley, F. A. (1867) *The wildflowers around Melbourne*, London.

Churchill, D. M., Muir, T. B. and Sinkora, D. M. (1978) 'The published works of Ferdinand J. H. Mueller (1825-1896)', *Muelleria* vol. 4, pp. 1-120.

Clarke, W. B. (1876) 'Effects of forest vegetation on climate', *Journal and proceedings of the Royal Society of New South Wales* vol. 10, pp. 179-235.

Clements, G. (1998) 'Colonial science: F. M. Bailey and the proposed supplement to *Flora australiensis*', *Historical records of Australian science* vol. 12, pp. 149-62.

Cohn, H. M. (2005) 'Watch dog over the Herbarium: Alfred Ewart, Victorian Government Botanist 1906-1921', *Historical records of Australian science* vol. 16.

Cohn, H. M. (forthcoming) *Novelty to rarity: a history of the National Herbarium of Victoria.*

Cohn, H. M. & Maroske, S. (1996) 'Relief from duties of minor importance: the removal of Baron von Mueller from the directorship of the Melbourne Botanic Gardens', *Victorian historical journal* vol. 67, pp. 103-127.

Collins, J. (1872) *Report on the caoutchouc of commerce* ..., London. M inscription: 'Private Property of | Baron Von Mueller'.

Connolly, C. N. (1983) *Biographical register of the New South Wales parliament, 1856-1901*, Canberra.

Cooke, M. C. (1872-92) *Grevillea: a monthly record of cryptogamic botany and its literature*, 20 vols, London.

Cooke, M. C. (1892) *Handbook of Australian fungi*, London.

Cooke, M. C. (1892a) 'Australian fungi: supplement to the *Handbook*', *Grevillea* vol. 21, pp. 35-9.

Crantz, H. J. N. von (1768) *De duabus draconis arboribus botanicorum* ... , Vienna.

Curr, E. M. (1886-7) *The Australian race: its origin, languages, customs, places of landing in Australia, and the routes by which it spread itself over that continent*, Melbourne.

Darragh, T. A. (1996) 'Mueller and personal names in zoology and palaeontology', *Victorian naturalist* vol. 113, pp. 195-7.

Darragh, T. A. (2000) 'The Deutsche Vereine of Victoria in the nineteenth century', in Mitchell, E. I. ed. *Baron von Mueller's German Melbourne*, Bundoora, Vic., pp. 66-86.

Darragh, T. A. (2001) 'Ferdinand Hochstetter's notes of a visit to Australia and a tour of the Victorian goldfields in 1859', *Historical records of Australian science* vol. 13, pp. 383-437.

Darragh, T. A. (2003) 'Bishop Goold and Ferdinand Mueller: a 30 year acquaintance', *Footprints: the journal of the Melbourne Diocesan Historical Commission* vol. 20, pp. 3-9.

Darwin, C. R. (1859) *On the origin of species by natural selection ...*, London.

Davidson, J. W. (1967) *Samoa mo Samoa: the emergence of the independent state of Western Samoa*, Melbourne.

Deane, H. and Maiden, J. H. (1901) 'Observations on the Eucalypts of New South Wales, part VIII', *Proceedings of the Linnean Society of New South Wales* vol. 26, pp. 122-44.

Desmond, R. (1994) *Dictionary of British and Irish botanists ...*, edn 2, London.

Desvaux, N. A. (1813-4) *Journal de botanique, appliquée à l'agriculture, à la pharmacie, à la médecine et aux arts*, 4 vols, Paris.

Deutsche biographische Enzyklopaedie (1995-), Munich.

Dictionnaire de biographie française (1933-), Paris.

Dictionary of American biography (1955-), London.

Dictionary of national biography (1917-90) 22 vols, London.

Dictionary of New Zealand biography (1990-) 5 vols, Wellington.

Dictionary of scientific biography (1970-90) 17 vols, New York.

Dizionario biografico degli Italiani (1960-), Rome.

Don, D. (1825) *Prodromus florae nepalensis ...*, London. B65.10.01.

Donn, J. (1800) *Hortus cantabrigiensis: or a catalog of plants indigenous and foreign, cultivated in the Walkerian botanic garden, Cambridge*, edn 2, Cambridge.

Ducker, S. C. (1988) *The contented botanist: letters of W. H. Harvey about Australia and the Pacific*, Melbourne.

Duff, U. G. ed. (1924) *The life-work of Lord Avebury (Sir John Lubbock) 1843-1913 ...*, London.

Du Petit-Thouars, L. M. A. A. (1822) *Histoire particulière des plantes orchidées ...* , Paris.

Dwyer, R. (1996) '"Baron von Mueller" at Adelaide', *Victorian naturalist* vol. 113, p. 139.

Dwyer, R. (1996a) 'To honour a noted botanist', *Victorian naturalist* vol. 113, pp. 208-9.

Eichler, A. W. (1875-8) *Blüthendiagramme*, 2 vols, Leipzig..

Enciclopedia italiana di scienze, lettere ed arti (1929-39), 36 vols, Rome.

Endlicher, S. L. (1833) (i.e. 1833[-1835]) *Atakta botanica ...*, Vienna. B65.10.01.

Endlicher, S. L. (1833a) *Prodromus florae norfolkicae*, Vienna.

Endlicher, S. L. (1836-40) *Genera plantarum ...*, Vienna. The first supplement was issued with this work. For publication dates of individual parts see TL2. B65.10.01.

Engler, H. G. A. (1884) 'Beiträge zur Kenntniss der Araceae V: über den Entwicklungsgang in der Familie der Araceen und über die Blütenmorphologie derselben', *Botanische Jahrbücher für Systematik, Pflanzengeschichte und Pflanzengeographie* vol. 5, pp. 287-336.

Engler, H. G. A. and Prantl, K. A. E. (1887-1915) *Die natürlichen Pflanzenfamilien ...* , 23 vol., Leipzig.

English, M. P. (1987) *Mordecai Cubitt Cooke: Victorian naturalist, mycologist, teacher and eccentric*, Bristol.

Ettingshausen, C. von (1883) *Beiträge zur Kenntniss der Tertiarflora australiens*, Vienna. Inscription: 'Herrn Baron Ferdinand Mueller in Melbourne | hochachtungsvollst der Verfasser'.

Everyman's encyclopaedia (1949-50), edn 3, 12 vols, London

Feeken, E. H. J., Feeken, G. E. E. and Spate, O. H. K. (1970) *The discovery and exploration of Australia*, Melbourne.

Feoktistov, A. E. (1889) 'Eine vorläufige Mittheilung über die Wirkung des Schlangengiftes auf den thierischen Organismus', *Mémoires de l'Académie Impériale des Sciences de St Pétersbourg* ser. 7, vol. 36, 22 p.

Feuillé, L. E. (1714-25) *Journal des observations physiques, mathématiques et botaniques*, 3 vols, Paris.

Fischer, J. G. (1885) 'Herpetologische Bemerkungen', *Jarhbuch der Hamburgischen wissenschaftlichen Anstalten* vol. 2, pp. 82-121.

Fitzgerald, R. D. (1882) [ie 1875-94] *Australian orchids*, 2 vols, Sydney.

Forbes, F. B. & Hemsley, W. B. (1886-1905) 'An enumeration of all the plants known from China proper, Formosa, Hainan, Corea, the Luchu Archipelago ... and Hongkong ...', *Journal of the Linnean Society: botany* vol. 23, pp. 1-521, vol. 26, pp. 1-529 and vol. 36, pp. 1-686. For publication dates of individual parts see TL2.

Forster, J. G. A. (1786) *Florulae insularum australium prodromus*, Göttingen. M inscription: 'Collections of Forsters plants are from Prof Hoffmanns Herb through Trinius in the Herb of Moscow now'.

Forster, J. R. & Forster, J. G. A. (1776) *Characteres generum plantarum quas in itinere ad insulas maris australis, collegerunt, descripserunt delinearunt ...*, London. B65.10.01.

Fraser, M. (1882) *General information respecting the present condition of the forests and timber trade of the southern parts of the colony ... together with a Report on the forest resources of the Colony by Baron Ferdinand von Mueller*, Perth.

Frost, W. & Harvey, S. (1997) 'Forest industries or dairy pastures? Ferdinand von Mueller and the 1885-1893 Royal Commission on Vegetable Products', *Historical records of Australian science* vol. 11, pp. 431-7.

Gaertner, C. F. von (1805-7) *Supplementum carpologicae seu continuati operis Josephi Gaertner de fructibus et seminibus plantarum ...*, Leipzig. Inscription: 'G. P.'. B65.10.01.

Gaertner, J. (1787-91) [i.e. 1787-92] *De fructibus et seminibus plantarum*, 2 vols, Tübingen. B65.10.01.

Gardner, C. A. (1930-1) *Enumeratio plantarum australiae occidentalis: a systematic census of the plants occurring in Western Australia*, 3 parts, Perth.

Gibbney, H. J. & Smith, A. G. (1987) *A biographical register 1788-1939: notes from the name index of the Australian dictionary of biography*, 2 vols, Canberra.

Gilbert, L. A. (1985) *William Woolls: "a most useful colonist"*, Cook, A. C. T.

Gilbert, L. A. (2001) *The little giant: the life and work of Joseph Henry Maiden, 1859-1925*, Armidale, N.S.W.

Gillbank, L. (1992) 'Alpine botanical expeditions of Ferdinand Mueller', *Muelleria* vol. 7, pp. 473-89.

Gillbank, L. (1993) 'Nineteenth-century perceptions of Victorian forests: ideas and concerns of Ferdinand Mueller', in Dargavel, J. and Feary, S. eds *Australia's ever-changing forests II*, Canberra, pp. 3-14.

Gillbank. L. (1996) 'A tale of two animals – alpaca and camel: zoological shaping of Mueller's Botanic Gardens', *Victorian historical magazine* vol. 67, pp. 83-102.

Gingins de la Sarraz, F. C. J. (1824) 'Violarieae', in Candolle, A. P. de *Prodromus systematis naturalis regni vegetabilis ...*, vol. 1, Paris, pp. 287-316.

Goodale, G. L. (1891) '[Address to the meeting]', *Report of the third meeting of the Australasian Association for the Advancement of Science* pp. xxxii-xxxiii.

Gray, A. (1879) '[Review of] F. von Mueller, Eucalyptographia, decades 1 and 2', *American journal of science* ser. 3, vol. 18, pp. 485-8.

Gray, A. (1880) 'Biographical memoir of Joseph Henry', *Annual report of the Board of Regents of the Smithsonian Institute ... for the year 1878* pp. 143-77.

Gray, A. (1880a) *Natural science and religion: two lectures delivered to the theological school of Yale College*, New York. Inscription: 'Sir Ferd. Müller | with kind regards of | the author'.

Gray, A. (1883) 'Some points in botanical nomenclature: "Nouvelles remarques sur la nomenclature botanique", par M. Alphonse de Candolle', *American journal of science and arts* 3rd ser, vol. 26, pp. 417-37.

Greenhalgh, P. (1988) *Ephemeral vistas: the expositions universelles, great exhibitions and world's fairs, 1851-1939*, Manchester.

Grey, J. G. (1879) [i.e. 1880] *His island home, and, Away in the far north: a narrative of travels in that part of the colony north of Auckland*, Wellington.

Griffiths, G. S. (1891) 'Address by the president (Section E: Geography)', *Report of the third meeting of the Australasian Association for the Advancement of Science* pp. 232-50.

Guilfoyle, W. R (1869) 'A botanical tour among the South Sea islands', *Journal of botany* vol. 7, pp. 117-36.

Guilfoyle, W. R. (1882) 'Some curious plants', *Southern science record* vol. 2, pp. 58-9.

Guilfoyle, W. R. (1883) *Catalogue of plants under cultivation in the Melbourne Botanic Gardens, alphabetically arranged*, Melbourne.

Guillemin, J. B. A. (1827) *Icones lithographicae plantarum australasiae rariorum ...*, Paris. B65.10.01

Gunn, M. & Codd, L. E. (1981) *Botanical exploration of southern Africa*, Cape Town.

Hall, N. (1978) *Botanists of the eucalypts: short biographies of people who have named Australian species ...*, Melbourne.

Hall, N. (1984) *Botanists of Australian acacias: short biographies of people who have named Australian species ...*, Melbourne.

Hampe, E. (1852) 'Lichenes', *Linnaea* vol. 25, pp. 709-15.

Harvey, W. H. (1858-63) *Phycologia australica ...*, 5 vols, London. For publication dates of individual pts see TL2. B65.10.01. M inscription: 'Private property of Baron von Mueller'.

Harvey, W. H. & Sonder, O. W. (1859-65) *Flora capensis: being a systematic description of the plants of the Cape colony ...*, 3 vols, Dublin.

Haswell, W. A. (1882) *Catalogue of the Australian stalk- and sessile-eyed Crustaceae*, Sydney.

Hawkesworth, J. (1773) *An account of the voyages undertaken ... for making discoveries in the southern hemisphere ...* , edn 2, 3 vols, London.

Heathcote, J. & Maroske, S. (1996) 'Drifting sand and Marram Grass on the south-west coast of Victoria in the last century', *Victorian naturalist* vol. 113, pp. 10-15.

Hemsley, W. B. (1883) 'The vegetation of Australia', *Gardeners' chronicle* ns, vol. 20, no. 508, pp. 390-1.

Henslow, G (1880) *Botany for children*, London.

Herbert, W. (1837) *Amaryllidaceae: preceded by an attempt to arrange the monocotyledonous orders, and followed by a treatise on cross-bred vegetables and supplement*, London.

Hilgard, E. W. (1906) *Soils: their formation, properties, composition, and relations to climate and plant growth in the humid and arid regions*, New York.

Home, R. W. et al. (1992) 'Why explore Antarctica?: Australian discussions in the 1880s', *Australian journal of politics and history* vol. 38, pp. 386-413.

Home, R. W. & Maroske, S. (1997) 'Ferdinand von Mueller and the French consuls', *Explorations* no. 18, pp. 3-50.
Hooker, J. D. (1844-7) *Flora Antarctica*, 2 vols, in *The botany of the Antarctic voyage of H. M. discovery ships* Erebus *and* Terror *in the years 1839-1843 ... part I*, London. For publication dates of individual pts see TL2. B65.10.01. M inscription vol. 1: 'Ferd. Mueller.'; vol. 2: 'Ferd. Mueller, M.D.'
Hooker, J. D. (1853-5) (i.e. [1852]-5) *Flora Novae-Zelandiae*, 2 vols, in *The botany of the Antarctic voyage of H. M. discovery ships* Erebus *and* Terror *in the years 1839-1843 ...part II*, London. For publication dates of individual pts see TL2. B65.10.01.
Hooker, J. D. (1855-60) *Flora Tasmaniae*, 2 vols, in *The botany of the Antarctic voyage of H. M. discovery ships* Erebus *and* Terror *in the years 1839-1843 ... part III*, London. For publication dates of individual pts see TL2. B65.10.01, B69.07.03.
Hooker, J. D. (1864-7) *Handbook of the New Zealand flora ...* , London.
Hooker, J. D. (1875-97) [i.e. 1872-97] *The flora of British India*, 7 vols, London. For dates of publication of individual pts see TL2.
Hooker, J. D. (1877) '*Xanthorrhoea minor*: native of south-western Australia and Tasmania', *Curtis's botanical magazine* vol. 103, t. 6297.
Hooker, J. D. (1878) *The student's flora of the British islands*, edn 2, London.
Hooker, J. D. (1879) '*Lasiopetalum baurei*: native of southern Australia', *Curtis's botanical magazine* vol. 105, t. 6445.
Hooker, J. D. (1882) 'On geographical distribution: address to the Geographical Section of the British Association', York, 1881, *Reports of the British Association for the Advancement of Science* vol. 50, pp. 727-38.
Hooker, J. D. (1884) *The student's flora of the British islands*, 3rd ed., London.
Hooker, J. D. (1897) *The flora of British India, vol. VII*, London.
Hooker, W. J. (1846-64) [ie 1844-64] *Synopsis filicum ...*, London.
Howitt, A. W. (1890) [i.e. 1891] 'The eucalypts of Gippsland', *Transactions of the Royal Society of Victoria* vol. 2, pp. 81-120.
Humboldt, F. H.A. von (1805-34) *Voyage aux régions équinoxiales du nouveau continent, fait en 1799-1804*, 24vols, Paris.

Intercolonial Exhibition of Australasia (1867) *Official record containing introduction, catalogues, reports and awards of the jurors and essays and statistics on the social and economic resources of the Australasian colonies*, Melbourne.

Jackson, B. D. (1881) *Guide to the literature of botany ...* , London.
Jackson, B. D. (1893-5) *Index kewensis plantarum phanerogamarum*, 2 vols, Oxford.
Jackson, B. D. (1906) *George Bentham*, London.
Jackson, B. D. (1924) 'The history of the compilation of the "Index kewensis"', *Journal of the Royal Horticultural Society* vol. 49, pp. 224-9.
Jacquin, N. J. (1760) *Enumeratio systematica plantarum, quas in insulis caribaeis vicinaque americes continente detexit novas ...*, Leiden.
Jaeger, A. & Sauerbeck, F. W. (1870-9) *Genera et species muscorum systematice disposita seu adumbratio florae muscorum totius orbis terrarum*, 2 vols, St Gallen. For publication dates of individual parts see TL2.
Jahn, I. & Schmidt, I. eds (1996) *Ferdinand Jacob Heinrich von Müller (1825-1896): ein Australienforscher aus Rostock und die Universität Rostock*, Rostock.
Johnston, R. (1884) 'Discovery of a cone, probably of a species of *Lepidostrobus*, in the sandstones of Campania', *Papers and proceedings of the Royal Society of Tasmania* 1884, p. 225.
Jussieu, A. L. de (1789) *Genera plantarum secundum ordines naturales disposita*, Paris.

Kernot, W. C. (1888) 'President's address for the year 1887', *Transactions and proceedings of the Royal Society of Victoria* vol. 24, pt 2, pp. ix-xxii.
Kirchhoff, K. R. A. (1886-93) [ie 1884-93] *Unser Wissen von der Erde: allgemeine Erdkunde und Länderkunde*, Prague.
Kirk, T. (1899) *The students' flora of New Zealand and the outlying islands*, Wellington.
Kittel, B. M. (1843) *Taschenbuch der Flora Deutschlands ...*, edn 2, 2 vols, Nuremberg. B65.10.01. M inscription vol. 1: 'Husum, December 1843 Ferd. Müller'; vol. 2: 'Ferd Müller.'
Knight, J. (1809) *On the cultivation of plants belonging to the natural order of Proteae*, London.
Kobert, R. (1911) *Pharmakobotanisches aus Rostocks Vergangenheit: ein im Rostocker Altertumsverein gehaltener Vortrag*, Stuttgart.
Koch, W. D. J. (1844) *Taschenbuch der deutschen und schweizer Flora ...* , Leipzig.
Kölliker, R. A. von (1872) 'Anatomisch-systematische Beschreibung der Alyconarien, I: die Pennatuliden [part 3]', *Abhandlungen von der senckenbergischen naturforschenden Gesellschaft* vol. 8, pp. 85-275.
Krempelhuber, A. von (1881) 'Ein neuer Beitrag zur Flechten-Flora Australiens', *Verhandlungen der kaiserlich-königlichen zoologisch-botanischen Gesellschaft in Wien* vol. 30, pp. 329-42.

Kristensen, L. (1896) 'Journal of the Right-Whaling cruise of the Norwegian steamship "Antarctic" in south polar seas under the command of Captain Leonard Kristensen during the years 1894-5', *Transactions of the Geographical Society of Australasia (Victorian Branch)* vol. 12/13, pp. 73-104.

Kützing, F. T. (1846-69) *Tabulæ phycologicae* ..., 19 vols, Nordhausen.

Kuntze, C. E. O. (1891) *Revisio generum plantarum* ..., Leipzig.

Labillardière, J. J. H. de (1800) *Relation du voyage à la recherche de La Pérouse* ..., 2 vols, Paris. M owned a copy of the 1800 English translation of this work. B65.10.01.

Labillardière, J. J. H. (1804-6) *Novae Hollandiae plantarum specimen* ..., 2 vols, Paris. B65.10.01. M inscription vol. 1: 'Ferd. von Mueller'; vol. 2: 'Ferd. Mueller'.

Labillardière, J. J. H. (1805) 'Sur un nouveau genre de plantes nommé *Candollea*', *Annales du muséum d'histoire naturelle* vol. 6, pp. 451-6. B65.10.01.

Lamarck, J. P. A. P. M. de (1783-1808) *Encyclopédie méthodique: botanique*, 8 vols, Paris. B65.10.01

Lamarck, J. B. A. P. M. de & Poiret, J. L. M. (1791-1823) *Tableau encyclopédique des trois règnes de la nature: botanique*, 3 vols, Paris.

L'Ecluse, Charles de (1601) *Rariorum plantarum historia* ..., Antwerp.

L'Ecluse, Charles de (1605) *Exoticorum libri decem* ..., Leiden.

Lehmann, J. C. G. (1844) *Novarum et minus cognitarum stirpium pugillus VIII*, Hamburg.

Lehmann, J. G. C. (1844-7) (i.e. 1844-[8]) *Plantae Preissianae* ..., 2 vols, Hamburg. B65.10.01. M inscription vol. 1: 'Ferdinandus Müller, philos. doct. actg. 1847 seq. Adelaide'. Vols 1 & 2 are interleaved copies with copious M annotations.

Le Maout, J. E. M. & Decaisne, J. (1868) *Traité général de botanique descriptive et analytique*, Paris. Mueller inscription: 'Private property of | Baron von Mueller, 1872'.

Lendenfeld, R. von (1887) *Descriptive catalogue of the Medusæ of the Australian seas*, 2 vols, Sydney.

Lendenfeld, R. von (1888) *Descriptive catalogue of sponges in the Australian Museum, Sydney,* London.

Lendenfeld, R. von (1889) *A monograph of the horny sponges*, London.

Lessing, C. F. (1831) 'De plantis in expeditione speculatoria romanzoffiana observatis disserere: Synanthereae Rich.', *Linnaea* vol. 6, pp. 209-60.

Lewis, D. (2006) 'The fate of Leichhardt', *Historical records of Australian science* vol. 17.

Liebmann, F. M. (1869) *Chênes de l'Amérique tropicale: iconographie des espèces nouvelles ou peu connues*, ed. by A. S. Örsted, Leipzig.

Lindley, J. (1846) *Vegetable kingdom: or, the structure, classification and uses of plants*, London.
Lindley, J. (1862) *School botany, descriptive botany and vegetable physiology ...*, 15th ed., London.
Lindsay, D. (1889) 'An expedition across Australia from north to south, between the telegraph line and the Queensland boundary, in 1885-6', *Proceedings of the Royal Geographical Society and monthly record of geography* vol. 11, pp. 650-71.
Lindsay, D. (1893) *Journal of the Elder Scientific Exploring Expedition 1891-2*, Adelaide.
Linnaeus. C. (1737) *Genera plantarum*, Leiden.
Linnaeus, C. (1745) *Öländska of Gothländska resa ...*, Stockholm.
Linnaeus, C. (1747) *Wästgöta-resa ...*, Stockholm.
Linnaeus, C. (1751) *Skånska resa ...*, Stockholm.
Linnaeus, C. (1753) *Species plantarum ...*, 2 vols, Stockholm. M inscription: vol. 1 'Private property | of Ferd. von Mueller | 1871'; vol. 2 'Private property of | Ferd. von Mueller | 1871'.
Linnaeus, C. (1767) *Systema naturae ..., vol. 2*, ed 12, Stockholm.
Linnaeus, C. (1771) *Mantissa plantarum altera generum editionis VI & specierum editionis II*, Stockholm. B65.10.01. Inscription: 'G.P.'.
Linnaeus, C. (1773) *Materia medica per regna tria naturæ ...* , edition altera, ed. J. C. D. Schreber, Vienna. B65.10.01.
Linnaeus, C. (1791) *Genera plantarum ... juxta thunbergii emendationes digesta, editio octava ...*, ed. T. Haenke. 2 vols, Vienna.
Linné, C. von. (1779) *Dissertatio botanico, illustrans nova graminium genera quam ... praeside ... Carolo à Linné ...*, Uppsala.
Linné, C. von (1781) *Supplementum plantarum*, Braunschweig.
Litton, R. T (1886) *List of members of the Geological Society of Australasia ...*, Melbourne.
Loefling, P. (1758) *Iter hispanicum, eller resa til spankska läderna uti Europa och America ...*, Stockholm.
Lorne, Marquis of (1886) 'Annual address on the progress of geography', *Proceedings of the Royal Geographical Society* ns, vol. 8, pp. 417-37.
Loureiro, J. de (1790) *Flora cochinchinensis: sistens plantas in regno cochinchina nascetes ...*, 2 vols, Lisbon. M owned the 2nd (Berlin) edition of 1793. Inscription vol. 1: G.P.; vol. 2: G.P. B65.10.01.
Lubbock, J. (1886) *Flowers, fruits and leaves*, London.
Lucas, A. M. (1988) 'Baron von Mueller: protégé turned patron', in Home, R. W. ed. *Australian science in the making*, Cambridge, pp. 133-52.

Lucas, A. M. (1995) 'Letters, shipwrecks and taxonomic confusion: establishing a reputation from Australia', *Historical records of Australian science* vol. 10, pp. 207-21.

Lucas, A. M. (2003) 'Assistance at a distance: George Bentham, Ferdinand von Mueller and the production of *Flora australiensis*', *Archives of natural history* vol. 30, pp. 255-81.

McAlpine, D. (1904) 'Native or Blackfellows' bread', *Journal of the Department of Agriculture of Victoria* vol. 2, pp. 1012-20.

McCoy, F. (1878-90) *Prodromus of the zoology of Victoria*, Melbourne.

Macleay, W. (1885) *Zoology of Australia*, Sydney.

Macleay, W. (1887) 'The insects of the Fly River, New Guinea: "Coleoptera"', *Proceedings of the Linnean Society of New South Wales* new ser., vol. 1, pp. 183-204.

MacLeod, R. (1974) 'The Ayrton incident: a commentary on the relations of science and government in England 1870-1873' in Thackray, A. and Mendelsohn, E. eds *Science and values: patterns of tradition and change*, New York, pp. 45-78.

Maiden, J. H. (1889) *The useful native plants of Australia (including Tasmania)*, London.

Maiden, J. H. (1903) 'The common Eucalyptus flora of Tasmania and New South Wales', *Report of the 9th meeting of the Australasian Association for the Advancement of Science* pp. 350-80.

Markham, C. (1895) 'The need for an Antarctic expedition', *Nineteenth century* vol. 38, no. 224 (October), pp. 706-12.

Maroske, S. (1992) 'The queen of aquatics: *Victoria amazonica*', *Australian garden history* vol. 3, no. 5, pp. 3-6.

Maroske, S. (1995) 'Mueller's educational collection of plants', *Botanic magazine* no. 6, p. 35.

Marsh, G. P. (1864) *Man and nature, or physical geography as modified by human action*, London.

Martin, J. S. (1967) *A tale of two churches: from West Melbourne to Box Hill*, Box Hill, Vic.

May, T. W., Maroske, S. & Sinkora, D. (1995) 'The mycologist, the Baron, the fungi hunters and the mystery artist', *Botanic magazine* no. 6, pp. 36-9.

Mez, C. C. (1896) 'Bromeliaceae', in Candolle, A. L. P. P. de ed., *Monographiae phanerogamarum*, vol. 9, Paris.

Miquel, F. A. W. (1843) [ie 1843-4] *Systema piperacearum*, Rotterdam. B65.10.01 M inscription: 'Ferd. Mueller.'.

Miquel, F. A. W. (1855-59) *Flora van Nederlandsch Indië*, 3 vols, Utrecht.

Mitten, W. (1883) 'Australian mosses', *Transactions and proceedings of the Royal Society of Victoria* vol. 19, pp. 49-96.
Mitten, W. (1883a) 'Record of new localities of Polynesian mosses, with descriptions of some hitherto undefined species', *Proceedings of the Linnean Society of New South Wales* vol. 7, pp. 98-104.
Montrouzier, X. (1860) 'Flore de l'île Art (prés de la Nouvelle-Calédonie)', *Mémoires de l'académie royale des sciences, belle-lettres at arts de Lyon, Section des sciences* ser. 2, vol. 10, pp. 173-254.
Moore, C. (1884) *A census of the plants of New South Wales*, Sydney.
Moore, C. (1893) *Handbook of the flora of New South Wales ...*, Sydney.
Moore, T. B. (1887) 'Notes on the discovery of a new Eucalyptus', *Papers and proceedings of the Royal Society of Tasmania* 1886, pp. 207-9.
Mottelay, P. F. (1922) *Bibliographical history of electricity and magnetism ...*, London.
Moulds, F. K. (1991) *The dynamic forest: a history of forestry and forest industries in Victoria*, Melbourne.
Moyal, A. (2001) *Platypus*, Crows Nest, N.S.W.
Mueller, A. (1888) 'On the action of snake-poison and the use of strychnine as an antidote', *Australian medical journal* ns, vol. 10, pp. 196-206.
Mueller, A. (1888a) '[Letter to Ferdinand Mueller]', *Archiv für pathologische Anatomie und Physiologie und für klinische Medizin*, vol. 113, pp. 393-4.
Mueller, A. (1893) *On snake-poison: its action and its antidote*, Sydney.
Müller, J. (1894) *Conspectus systematicus lichenum novae zeylandiae*, Geneva.
Muir, T. B. (1979) 'An index to the new taxa, new combinations and new names published by Ferdinand J. H. Mueller', *Muelleria* vol. 4, pp. 123-68. See Volume 1, Appendix C.
Murray, G. ed. (1892-5) *Phycological memoirs: being researches made in the Botanical Department of the British Museum*, 3 pts, London.

Necker, N. J. de (1790) *Elementa botanica ...*, 3 vols, Neuwied.
Nees von Esenbeck, C. G. D. (1841) 'Gramineae', *Nova acta leopoldina* vol. 19, supplement 1, pp. 135-208.
Neue deutsche Biographie (1952-), Berlin.
New South Wales Fisheries (2002) *Threatened species in New South Wales: Bennetts seaweed,* Vanvoorstia bennettiana, Sydney. (Fishnote, no. NSWF 1121.)
Nordstedt, C. F. O. (1891) *Australian Characeae described and figured, part 1*, Lund.
Nordenskiöld, A. E. (1890) 'Utkast till en svensk antarktisk expedition', *Ymer: tidskrift utgifven av Svenska Sällskapet för Antropologi och Geografi* vol. 9, pp. 122-8.

Northern Territory dictionary of biography, vol. 1: to 1945 (1990), Casuarina, N.T.

Nouvelle biographie générale (1852-68), 46 vols, Paris.

Nylander, W. (1858-69) *Synopsis methodica lichenum omnium hucusque cognitorum ...*, Paris.

Nylander, W. (1868) 'Synopsis lichenum novae caledonicae', *Bulletin de la société linnéenne de Normandie* ser. 2, vol. 2, pp. 39-140.

O'Neill, G. (1929) *Life of the Reverend Julian Edmund Tenison Woods (1832-1889)*, Sydney.

Osborne, M. A. (1994) *Nature, the exotic, and the science of French colonialism*, Bloomington.

Owen, R. (1887) 'Description of a newly-excluded young of the *Ornithorhynchus paradoxus*', *Proceedings of the Royal Society of London* vol. 42, p. 391.

Painke, W. (1985) 200 *Jahre Perthes geographische Verlagsanstalt Gotha-Darmstadt*, Darmstadt.

Palisot de Beauvois, A. M. F. J. (1806) *Flore d'Oware et de Benin, en Afrique*, vol. 1, part 6, Paris.

Palisot de Beauvois, A. M. F. J. (1812) *Essai d'une nouvelle agrostographie ...*, Paris. B65.10.011. Inscription: 'G. P'.

Palmer, V. (1940) *National portraits*, Melbourne.

Papenfuss, G. F. (1956) 'On the nomenclature of some Delesseriaceae', *Taxon* vol. 5, pp. 158-62.

Perry, W. (1984) *The School of Mines and Industries, Ballarat: a history of its one hundred and twelve years, 1870-1982*, Ballarat.

Persoon, C. H. (1805-7) *Synopsis plantarum ...*, Paris. B65.10.01.

Pierre, J. B. L. (1887) 'Sur le genre *Stixis* Lour.', *Bulletin mensuel de la société Linnéenne de Paris* vol. 1, pp. 652-6.

Poggendorff, J. C. (1863-) *J. C. Poggendorff's biographisch-literarisches Handwörterbuch zur Geschichte der exacten Wissenschaften ...*, Leipzig.

Poiret, J. L. M. (1810-17) *Encyclopédie méthodique: botanique. Supplément*, 5 vols. Paris.

Poiret, J. L. M. (1814) *Encyclopédie méthodique: botanique. Supplément 5*, Paris.

Powell, J. M. (1978) 'A letter from the Baron: von Mueller's Royal Medal, 1888', *Journal of the Royal Australian Historical Society* vol. 63, pp. 270-274.

Prahl, P. (1888-90). *Kritische Flora der Provinz Schleswig-Holstein des angrenzenden Gebiets der Hansestädte Hamburg und Lübeck und des Fürstenthums Lübeck*, 2 vols, Kiel.

Presl, K. B. (1844) [ie 1845] 'Botanische Bemerkungen', *Abhandlungen der königlichen Böhmischen Gesellschaft der Wissenschaften* ser. 5, vol. 3, pp. 431-584.
Press, M. M. (1994) *Julian Tenison Woods: father founder*, 2nd edn, North Blackburn, Vic.

Radlkofer, L. A. T. (1890) *Ueber die Gattungen der Familie der Sapindaceen* ..., Munich.
Reichenbach, H. G. (1882) 'Dendrobium macfarlanei, n. sp.', *Gardener's chronicle* n.s., vol. 18, p. 520.
Richard, L. C. M. (1826) *Commentatio botanica de conifereis et cycadeis* ..., Stuttgart.
Ridley, H. N. (1886) 'On the monocotyledonous plants of New Guinea collected by Mr H. O. Forbes', *Journal of botany* vol. 24, pp. 321-7, 353-60.
Rivière, C. (1885) 'Essais d'une végétation assainissante Australia Gabon', *Bulletin mensuel de la Société National d'acclimatation de France*, ser. 4, vol. 2, pp. 12-28.
Robertson, E. L. (1986) 'Botany', in Twidale, C. R., Tyler, M. J. and Davies, M. eds *Ideas and endeavours: the natural sciences in South Australia*, Adelaide, pp. 101-49.
Royal Botanic Gardens, Kew (1885) *Official guide to the Royal Botanic Gardens and Arboretum*, 29th ed., London.
Royal Botanic Gardens, Kew (1886) *Official guide to the Museums of Economic Botany, no. 1: dicotyledons and gymnosperms*, London.
Ryan (2001) *Edward Bage, Diamond Creek: surveyor, adventurer and gentleman*, Diamond Creek, Vic.

Saccardo, P. A. (1893) 'Mycetes aliquot australienses, series quarta', *Hedwigia* vol. 32, pp. 56-9.
Salisbury, R. A. (1805) [ie 1805-8] *The paradisus londinensis* ..., London.
Salisbury, R. A. (1807) 'The characters of several genera in the natural order of Coniferae: with remarks on their stigmata, and cotyledons', *Transactions of the Linnean Society* vol. 8, pp. 308-17.
Salisbury, R. A. (1866) *The genera of plants* ..., London.
Schimper, W. P. & Schenk, A. (1890) 'Palaeophytologie', in Zittel, K. A. ed. *Handbuch der Palaeontologie, II Abtheilung*, Munich and Leipzig.
Schmidt, A. W. F. (1874-1959) *Atlas der Diatomaceenkunde*, Leipzig.
Schmitz, F. & Hauptfleisch, P. (1897) 'Ceramiaceae', in Engler, H. G. A. and Prantl, K. A. E. *Die natürlichen Pflanzenfamilien*, Teil. 1, Abteilung 2, Leipzig, pp. 481-504.

Schreber, J. C. D. von (1789-91) *Genera plantarum ...*, 8th edn, 2 vols, Frankfurt am Main.

Schumann, K. & Hollrung, M. (1889) *Die Flora von Kaiser Wilhelms Land*, Berlin.

Scopoli, J. A. (1777) *Introductio ad historia naturalem ...*, Prague. M inscription: 'Baron v Müller'.

Scott, H. & Scott, H. (1988) *Historical drawings of native flowers, introduced and selected by M. Ord*, Roseville, N. S. W.

Scudder, S. H. (1879) *Catalogue of scientific serials of all countries*, Cambridge, Mass.

Seemann, B. C. (1865-73) *Flora vitiensis: a description of the plants of Viti or Fiji Islands*, 3 vols, London. M inscription: 'Private property of | Baron Ferd von Mueller'.

Selzer, A. (2003) *The Armytages of Como: pastoral pioneers*, Rushcutters Bay, N.S.W.

Smith, J. E. (1798) 'The characters of twenty new genera of plants ...', *Transactions of the Linnean Society of London* vol. 10, pp. 213-24. B65.10.01.

Smith, J. E. (1802) 'Botanical characters of four New-Holland plants, of the natural order Myrti', *Transactions of the Linnean Society of London* vol. 6, pp. 299-302.

Sonder, O. W. (1853) 'Plantae muelleriana: algae', *Linnaea* vol. 25, pp. 657-703.

Spencer, W. B. & French C. (1889) 'Trip to Croajingalong', *Victorian naturalist* vol. 6, pp. 1-38.

Spicer, W W. (1878) *A handbook of the plants of Tasmania*, Hobart Town.

Sprague, I. & Gray, A. (1848-9) *Genera florae americae boreali-orientalis illustrata*, Boston.

Sprengel, K. P. J. (1825-8) [ie 1824-8] *Systema vegetabilium ...* , Göttingen. B65.10.01. Inscription: 'G. P.'.

Stafleu, F. A. & Cowan, R. S. (1976-1988) *Taxonomic literature: a selective guide to botanical publications and collections with dates, commentaries and types*, 2nd edn, 7 vols, Utrecht.

Stafleu, F. A. & Mennenga, E. A. (1992-2000) *Taxonomic literature: a selective guide to botanical publications and collections with dates, commentaries and types: supplement 1(-6)*, 6 vols, Königstein.

Stapf, O. (1897-1900) 'Gramineae', in Thiselton-Dyer, W. T. ed. *Flora capensis ..., vol. VIII, Pontederiaceae to Gramineae*, London, pp. 310-750.

Stapf, O. (& Hubbard, C. E.) (1917-34) 'Gramineae', in Prain, D. ed. *Flora of tropical Africa, vol. IX*, London.

Statton, J. ed. (1986) *Biographical index to South Australians 1836-1885*, 4 vols, Marden, S. A.

Steudel, E. G. (1855) (i.e. [1853]-5) *Synopsis plantarum glumaceum ...*, 2 vols, Stuttgart. B65.10.01. M inscription vol. 1: 'Ferd. Mueller, M.D., Ph. D. 1857'; vol. 2: 'Ferd. Mueller M. & Ph. D.'

Stevens, P. F. (1991) 'George Bentham and the Kew rule', in Hawksworth, D. L. ed. *Improving the stability of names: needs and options*, Königstein, pp. 157-68.

Stevens, P. F. (1997) 'J. D. Hooker, George Bentham, Asa Gray and Ferdinand Mueller on species limits in theory and practice: a mid-nineteenth-century debate and its repercussions', *Historical records of Australian science* vol. 11, pp. 345-70.

Stirton, J. (1881) 'Additions to the flora of Queensland', *Transactions of the Royal Society of Victoria* vol. 17, pp. 66-78.

Sturt, C. (1849) *Narrative of an expedition into Central Australia ... during the years 1844, 5, and 6 ...*, London.

Swan, R. A. (1961) *Australia in the Antarctic: interest, activity and endeavour*, Melbourne.

Swartz, O. P. (1799) 'Dianome epidendri generis', *Nova acta regiae societatis scientiarum upsaliensis* [ser. 2], vol. 6, pp. 61-85.

Swartz, O. P. (1807) 'Stylidium, eine neue Pflanzengattung', *Gesellschaft Naturforschenden Freunde zu Berlin Magazin* vol. 1, pp. 47-53.

Tate, R (1880) 'A census of the indigenous flowering plants and ferns of extra-tropical South Australia', *Transactions of the Royal Society of South Australia* vol. 3, pp. 46-90.

Tate, R. (1883) 'A list of unrecorded plants and of new localities for rare plants in the south-east part of this colony', *Transactions of the Royal Society of South Australia* vol. 6, pp. 95-9.

Tate, R. (1883a) 'The botany of Kangaroo Island, prefaced by a historical sketch of its discovery and settlement, and by notes on its geology', *Transactions of the Royal Society of South Australia* vol. 6, pp. 116-71.

Tate, R. (1889) 'The gastropods of the older Tertiary of Australia (part II)', *Transactions of the Royal Society of South Australia* vol. 11, pp. 116-174.

Tate, R. (1896) *Botany of the Horn Expedition. 1, the Larapintine flora. 2, the central Eremian flora*, London & Melbourne, pp. 117-197.

Taylor, A. (1996) 'Baron von Mueller in the Field Naturalists' tradition', *Victorian naturalist* vol. 113, pp. 131-9.

Taylor, T. (1846) 'New Hepaticae', *London journal of botany* vol. 5, pp. 258-84, 365-417.

Tepper, J. G. O. (1882) 'Further observations on the propagation of Cymodocea antarctica', *Transactions of the Royal Society of South Australia* vol. 4, pp. 47-9.

Thiselton-Dyer, W. T. (1880) *The botanical enterprise of the Empire*, London.

Thomson, K. & Serle, G. (1972) *A biographical register of the Victorian parliament, 1859-1900*, Canberra.

Thornton-Smith, C. B. (1994) 'Paul Maistre, Vice-consul and later Consul for France in Victoria, 1886-1898, 1901-1908', *Explorations* no. 17, pp. 1-48.

Thunberg, C. P. (1781) *Nova genera plantarum, quorum partum premium* ..., Uppsala. B65.10.01.

Todd, J. (1995) *Colonial technology: science and the transfer of innovation to Australia*, Cambridge.

Tournefort, J. P. (1700) *Institutiones rei herbariae*, Paris.

Turczaninow, N. (1851) 'Synanthereae quaedam hucusque indescriptae', *Bulletin de la Société impériale des naturalistes de Moscou* vol. 24, no. 1, pp. 166-214; no. 2, pp. 59-95.

Ventenat, E. P. (1803-4) [ie 1803-5] *Jardin de la Malmaison* ..., 2 vols, Paris.

Verco, J. C. ([1935]) *Thomas and Elizabeth Magarey*, Adelaide.

Vilmorin, H. L. (1864) *Exposé historique et descriptive de l'école forestière des Barres*, Paris.

Vollbehr, F. and Weyl, R. (1956) *Professoren und Studenten der Christian-Albrechts-Universität zu Kiel 1665-1954*, edn 4, Kiel.

Voigt, J. H. (1996) *Die Erforschung Australiens: der Briefwechsel zwischen August Petermann und Ferdinand von Mueller 1861-1878*, Gotha.

Walker, A. F. (1887) *Flowers of New South Wales*, Sydney. Inscription: To | Baron Sir F. von Müeller K.C.M.G. | with grateful remembrance | Annie F. Walker | Rhodes. | Ryde. | New South Wales. | May the 18th 1893.

Walker, D. & Tampke, J. eds (1991) *From Berlin to the Burdekin: the German contribution to the development of Australian science, exploration and the arts*, Kensington, N.S.W.

Wallich, N. (1828) [ie 1818-49] *A numerical list of dried specimens of plants in the East India Company's museum, collected under the superintendance of Nathaniel Wallich*, London. Inscription: G.P. B65.10.01.

Wesche, M. (1997) *Die Bayerische Akademie der Wissenschaften und ihre Mitglieder im Spiegel von Medaillen und Plaketten*, Munich.

Who's who in America (1900-), Chicago.

Who was who (1916-), London.

Willdenow, C. L. (1805) *Species plantarum* ..., 4th [ie 5th] ed., vol. 4, Berlin. B65.10.01.

Willdenow, C. L. (1809) *Enumeratio plantarum horti regii botanici berolinensis*, Berlin.

Willis, J. H. et al. (1986) *Australian plants: collectors and illustrators 1780s-1980s*, [edn 2], Perth.

Willkomm, H. M. & Lange, J. (1861-80) *Prodromus florae hispanicae, seu methodica omnium plantarum in hispania sponte nascentium* ... , 3 vols, Stuttgart. B65.10.01 (vol. 1). Inscription vol. 1: 'Celeberrimo Dr. F. Müller | amice offert | Joh Lange.'.

Wilson, J. S. (1890) *The creation story and nebular theory*, Manchester.

Wittrock, V. B. (1891) *De horto botanico bergiano*, Stockholm.

Wittstein (1878) *The organic constituents of plants and vegetable substances and their chemical analysis*, translated by F. von Mueller, Melbourne.

Woods, J. E. T. (1865) *History of the discovery and exploration of Australia* ..., 2 vols, London.

Woolls, W. (1880) 'Eucalypts of the county of Cumberland: their classification, habitat and uses, part 1(-5)', *Proceedings of the Linnean Society of New South Wales* vol. 5, pp. 288-94, 448-58, 463-9, 488-93, 503-10.

Woolls, W. (1880a) *Plants indigenous in the neighbourhood of Sydney arranged according to the system of Baron F. von Mueller*, Sydney.

Young, J. (1978) *Recent journey of exploration across the continent of Australia: its deserts, native races, and natural history*, with an introduction by A. Potts, Melbourne.

Index of Botanical Names

Botanical names mentioned in the correspondence are indexed by item number. Some names used have not been validly published in the botanical literature; such cases are indicated in the footnotes to individual items. Spelling, including capitalization, is preserved, whether or not this is the currently accepted spelling. Contractions and abbreviations are expanded in the index. Plural forms, e.g., *Eucalypti*, have been converted to the singular. If the same name was used more than once in the same item, only one entry per item is given in the index. Where a generic name without a specific epithet, e.g., *Acacia*, is given in an item where a complete name in that genus is also used, e.g., *Acacia dimidiata,* an entry for the generic name used alone is not given for that item. Generic names used alone in an item follow in the index complete binomials in that genus.

Botanical names given in the footnotes have not been indexed.

General Index

An index of proper names and selected subjects, indexed by page number. The index covers only the Introduction, the Selected Correspondence, and the additional correspondence included in letters about M. It does not cover the Biographical Index or the Bibliography, neither does it include any botanical names, these being in the Index of Botanical Names. Letters marked 'to' and 'from' refer to M, unless marked otherwise.